Operations Research

Operations Research

A. M. Natarajan
Principal, Kongu Engineering College
Perundurai, Erode
Tamil Nadu, India

P. Balasubramani
Professor, Department of CSE – PG
Kongu Engineering College
Perundurai, Erode
Tamil Nadu, India

A. Tamilarasi
Professor, Department of CSE – PG
Kongu Engineering College
Perundurai, Erode
Tamil Nadu, India

PEARSON

ISBN 978-81-317-0000-6

First Impression, 2006
Eleventh Impression, 2012
Twelfth Impression

Published by Dorling Kindersley (India) Pvt. Ltd., licensees of Pearson Education in South Asia.

Head Office: 7th Floor, Knowledge Boulevard, A-8(A), Sector-62, Noida – 201309, U.P, India.
Registered Office: 11 Community Centre, Panchsheel Park, New Delhi 110 017, India.

Laser typeset by Anvi Composers, Delhi

Printed in India by Rahul Print O Pack.

Preface

There has been a long-felt need for a simple book on operations research covering the syllabuses of almost all the Indian universities. This book caters to the needs of the students of BE, MCA, MBA, M.Sc. and M.Com. courses. This book has grown out of the long-standing experience and interest of the authors in teaching operations research. The salient feature of the book is that each chapter includes many solved problems which are explained in a simple way to make the student understand and comprehend the subject. Owing to space constraints, we have accommodated many related concepts, solved problems, and exercises in the companion Web site for the benefit of the students as well as the teachers.

Operations research is the study of optimization techniques. It is applied to decision theory. The existence of optimization techniques can be traced at least to the days of Newton and Lagrange. Rapid development and invention of new techniques have occurred since the Second World War essentially because of the necessity to win the war with limited resources.

Chapter 1 defines operations research and gives a brief historical background of it. The chapter acquaints the reader with not only the meaning and purpose but also limitations of operational research. It also gives insight into the approaches and tools of operations research and describes the relationship between operational research specialist and manager.

Chapter 2 discusses the formulation of a linear programming problem (LPP) with different types of linear constraints. The graphical solution to an LPP is discussed with the aid of a number of examples. The cases of multiple, unbounded solution and infeasible problems are also illustrated graphically. The linear programming problems involving more than two decision variables can be solved with the help of simplex method. The slack variables are introduced to convert less than or equal to type constraints into equations. When some of the constraints are of greater than or equal to type, the surplus variables are introduced to convert inequalities to equations. Here the concept of artificial variables is introduced to initiate the simplex computations. LPPs in such cases are solved by two methods, namely, the Two-Phase method and Big M method. The cases of multiple solution, unbounded solution and infeasible problems are discussed with the help of appropriate examples. The chapter concludes by showing the applications of LP in business and industry.

The dual to an LPP finds a place in Chapter 3. The economic interpretation of dual variables which can be used by management for planning its resources are also explained in this chapter. If an LPP involving a large number of variables and constraints is to be solved by this method, it will require a large storage space and time on a computer. Some computational techniques have been developed which require much less computer storage and time than that required by the simplex method. The two important and efficient computational techniques are the revised simplex method or simplex method with multipliers and the bounded variables method. These techniques are dealt within this chapter. The sensitivity analysis of the problem to study the effect of change in various resource levels is also discussed. This chapter also describes some integer programming formulation techniques and highlights its applications in managerial problems where LP techniques fail. Two more methods, namely, cutting plane method and branch and bound method are also discussed.

In Chapter 4, the formulation of a transportation problem is discussed. Five methods, namely, the North West Corner Rule, Matrix Minima Method, Row Minima Method, Column Minima Method and Vogel's Approximation Method to determine an initial basic feasible solution are discussed. The Modified Distribution Method and Stepping Stone Method for obtaining the optimal solution of a transportation problem are explained. The chapter also explains unbalanced transportation problem, degenerate transportation problem and trans-shipment problem.

Chapter 5 is on assignment problem. The chapter explains the formulation of an assignment problem with the help of an example. The Hungarian Method for solving an assignment problem is discussed. The chapter also explains how to convert an unbalanced assignment problem into a balanced assignment problem and to handle problems with infeasible assignment. The modification of the assignment problem when the objective is to maximize the objective function is also discussed. The chapter concludes by showing the application of Hungarian method for solving crew assignment problem.

Chapter 6 is on dynamic programming. The methodology relating to dynamic programming along with various concepts has been explained. The chapter concludes by illustrating various dynamic programming applications.

The previous chapters dealt with formulation and solution of models under conditions of perfect information. This is usually referred to as decision-making under certainty. Many managerial decisions, however, are made with some uncertainty. Chapter 7 deals with decision under uncertainty, decision under certainty and decision under risk. Any decision-making process can be represented by means of a tree known as decision tree. This chapter also deals with game theory. The theory of games aids managerial decision-making in a comprehensive environment. After producing a conceptual framework on games, this chapter covers the salient elements of the theory of games, namely, saddle mixed strategies, dominance, $2 \times n$ games and $n \times 2$ games.

Consider the problem of a publisher who has to perform two operations, namely, printing and binding. Let us assume that the publisher has 10 books to be published (i.e., 10 jobs) and the time required for printing each book is known. Obviously there is an order in using the printing machine and the binding machine—print first and find next. Now the problem of the publisher is to decide the order in which the ten jobs should be done so as to minimize the total time taken for all jobs. Such a problem is called sequencing problem. Theoretically the number of ways in which this could be done is $(10!)^2$ since there are 10 jobs and two machines. If there are n jobs and m machines, then the sequencing problem is to determine the optimum sequence selected from among $(n!)^m$ theoretical alternatives that is technologically feasible. So easier methods of dealing with such problems are needed. Such methods are dealt with in Chapter 8.

Replacement problems are concerned with the situations that arise when some items such as men, machines, and electric bulbs need replacement due to their decreased efficiency, failure or breakdown. Such decreased efficiency or complete breakdown may either be gradual or sudden. Thus the problem of replacement is to decide the best policy to determine an age at which the replacement is most economical instead of continuing at increased cost. Chapter 9 deals with replacement of items with gradual deterioration, item deterioration with money value and item that failed completely and suddenly. This chapter concludes with staff replacement problem.

Chapter 10 discusses the various concepts pertaining to inventory. It explains the need, objectives and functions of inventory. Three-way classification of inventory and factors influencing inventory are also discussed. This chapter also describes inventory models with probabilistic demand. The chapter concludes by discussing Always Better Control (ABC) analysis.

Chapter 11 discusses the basic characteristics of a queuing problem and explains its occurrence in real-life situations. The queuing models under M/M/1, M/M/C and M/Ek/1 systems are explained with suitable examples.

A project is defined as a combination of interrelated activity which must be executed in a certain order before the entire task can be completed. The activities are interrelated in a logical sequence called network. Growing complexities of modern projects need more systematic and effective planning techniques with the objective of optimizing the efficiency of executing the project. Efficiency implies effecting the utmost reduction in the time required to complete the project. Project management has evolved as a new field with the development of two "analytic" techniques for planning, scheduling and controlling projects. These are the critical path method (CPM) and the project evaluation and review technique (PERT). These techniques are dealt in Chapter 12.

Chapter 13 deals with simulation. This technique is used in management problems where it is not possible to use any precise mathematical model. The applications of simulation model in practical business problems are illustrated in this chapter.

The last chapter of this book, Chapter 14, is on non-linear programming problem. The necessary and sufficient conditions to obtain optimal solution for NLP problems given by Kuhn-Tucker are discussed. Further the concepts of quadratic programming and separable programming are also discussed in this chapter along with their applications.

The corrections, suggestions and comments for the improvement of the book will be appreciated and duly acknowledged. We wish to thank our Secretary, Correspondent and all KVIT Trust members for their encouragement and moral support in bringing out this book successfully. We would like to thank Pearson Education for their untiring efforts in bringing out this book elegantly and on time.

A.M. NATARAJAN
P. BALASUBRAMANI
A. TAMILARASI

Contents

1

Basics of Operations Research

1.1 DEVELOPMENT OF OPERATIONS RESEARCH

Operations research (OR) has its beginning in World War II. The term, operations research, was coined by McClosky and Trefthen in 1940 in the UK. British scientists set up the first field installations of radars during the war and observed the air operations. Their analysis of these led to suggestions that greatly improved and increased the effectiveness of British fighters, and contributed to the success of British defence. Operations research was then extended to anti-submarine warfare and to all phases of military, naval, and air operations, both in Britain and in the United States, and was incorporated in the post-war military establishments of both the countries.

The effectiveness of operations research in military was instrumental in spreading interest in it to other governmental departments and industry. In the USA, the National Research Council formed a committee on operations research in 1951, and the first book on the subject, *Methods of Operations Research* by Morse and Kimball, was published. In 1952, the Operations Research Society of America came into existence. Success of OR in military attracted the attention of industrial managers who were seeking solutions to their complex problems.

Today, almost every large organisation or corporation in affluent nations has staff applying operations research, and in government the use of operations research has spread from military to widely varied departments at all levels. This general acceptance to OR has come as the managers have learned the advantage of the scientific approach on which OR is based. Availability of faster and flexible computing facilities and the number of qualified OR professionals has enhanced the acceptance and popularity of the subject. The growth of OR has not been limited to the USA and the UK. It has reached to many countries of the world. Indicative of this is that the International Federation of Operations Research Societies, which was founded in 1959, now comprises member societies from many countries of the world.

India was one of the first few countries who started using OR. In 1949, the first OR unit was established in the Regional Research Laboratory at Hyderabad. At about the same time, another group was set up in the Defence Science Laboratory to solve the problems of stores, purchase and planning. In 1953, an OR unit was established in Indian Statistical Institute, Calcutta, with the aim of using OR method in national planning and survey. The OR Society of India was formed in 1955. The society is one of the first members of the International Federation of OR Societies. The society started publishing

Opsearch, a learned journal on the subject in 1963. Today, OR is a popular subject in management institutes and schools of mathematics and is gaining currency in industrial establishments.

Towards the application of OR in India, Prof. Mahalonobis made the first important application. He formulated the Second Five-Year Plan with the help of OR techniques to forecast the trends of demand, availability of resources and for scheduling the complex schemes necessary for developing our country's economy. It was estimated that India could become self-sufficient in food merely by reducing the average food by 15%. Operations research techniques are being used to achieve this goal. Planning Commission made the use of operations research techniques for planning the optimum size of the Caravelle fleet of Indian Airlines.

In the industrial sector, in spite of the fact that opportunities of OR work at present are very much limited, organised industries in India are gradually becoming conscious of the role of operations research and a good number of them have well trained OR teams. Most popular practical application of OR in India has been mainly of linear programming. With the exception of the government and textile industries, applications of OR in other industries have been more or less equally distributed.

1.2 DEFINITION OF OPERATIONS RESEARCH

Operations research, rather simply defined, is the research of operations. An operation may be called a set of acts required for the achievement of a desired outcome. Such complex interrelated acts can be performed by four types of systems: man, machine, man–machine unit and any organisation of men, machines, and man–machine units. Operations research is concerned with the operations of the last type of system.

Many definitions of OR have been suggested from time to time. On the other hand, a number of arguments have been put forward as to why it cannot be defined. Perhaps, the subject is too young to be defined in an authoritative way. Some of the different definitions suggested are:

1. Operations research is a scientific method of providing executive departments with a quantitative basis for decisions regarding the operations under their control. —*Morse and Kimball*
2. Operations research in the most general sense, can be characterised as the application of scientific methods, tools and techniques to problems involving the operations of systems so as to provide those in control of the operations with optimum solutions to the problems. —*Churchman, Ackoff, Arnoff*
3. Operations research is applied decision theory. It uses any scientific, mathematical or logical means to attempt to cope with the problems that confront the executive when he tries to achieve a thoroughgoing rationality in dealing with his decision problems. —*Miller and Starr*
4. Operations research is a scientific approach to problem solving for executive management. —*H M Wagner*
5. Operations research is the art of giving bad answers to problems, to which, otherwise, worse answers are given. —*Thomas L Saaty*
6. Operations research is an aid for the executive in making his decisions by providing him with the needed quantitative information based on the scientific methods of analysis. —*C Kittes*
7. Operations research is the systematic, method-oriented study of the basic structure, characteristics, functions and relationships of an organisation to provide the executive with a sound, scientific and quantitative basis for decision-making. —*E L Arnoff and M J Netzorg*

8. Operations research is the application of scientific methods to problems arising from operations involving integrated systems of men, machines and materials. It normally utilises the knowledge and skill of an interdisciplinary research team to provide the managers of such systems with optimum operating solutions. —*Fabrycky and Torgersen*
9. Operations research is an experimental and applied science devoted to observing, understanding and predicting the behaviour of purposeful man-machine systems; and operations research workers are actively engaged in applying this knowledge to practical problems in business, government and society. —*Operations Research Society of America*
10. Operations research is the application of scientific method by interdisciplinary teams to problems involving the control of organised (man-machine) systems so as to provide solutions which best serve the purpose of the organisation as a whole. —*Ackoff and Sasieni*
11. Operations research utilises the planned approach (updated scientific method) and an interdisciplinary team in order to represent complex functional relationships as mathematical models for the purpose of providing a quantitative basis for decision-making and uncovering new problems for quantitative analysis. —*Thierauf and Klekamp*
12. Operations research is the application of modern methods of mathematical science to complex problems involving management of large systems of men, machines, materials and money in industry, business, government and defence. The distinctive approach is to develop a scientific model of the system incorporating measurement of factors such as chance and risk to predict and compare the outcomes of alternative decisions, strategies or controls. —*J O R Society, U.K.*

1.3 NECESSITY OF OPERATIONS RESEARCH IN INDUSTRY

After having studied as to what is operations research, we shall now try to answer as to why to study OR or what is its importance or why its need has been felt by the industry.

As already pointed out, science of OR came into existence in connection with war operations, to decide the strategy by which enemy could be harmed to the maximum possible extent with the help of the available warfare. War situation required reliable decision-making. The need of OR has been equally felt by the industry due to the following reasons:

(a) Complexity In a big industry, the number of factors influencing a decision have increased. Situation has become big and complex because these factors interact with each other in a complicated manner. There is, thus, great uncertainty about the outcome of the interaction of factors like technology, environment, competition, and so on. For instance, consider a factory production schedule which has to take into account:

(i) Customer demand
(ii) Requirements of raw materials
(iii) Equipment capacity and possibility of equipment failure, and
(iv) Restrictions on manufacturing process.

Evidently, it is not easy to prepare a schedule which is both economical and realistic. This needs mathematical models, which in addition to optimisation, help to analyse the complex situation. With such models, complex problems can be split up into smaller parts, each part can be analysed separately and then the results can be synthesised to give insights into the problem.

(b) Scattered responsibility and authority In a big industry, responsibility and authority of decision-making is scattered throughout the organisation and thus the organisation, if it is not conscious, *may be following* inconsistent goals. Mathematical quantification of OR overcomes this difficulty also to a great extent.

(c) Uncertainty There is a great uncertainty about economic and general environment. With economic growth, uncertainty is also growing. This makes each decision costlier and time consuming. Operations research is thus, quite essential from reliability point of view.

(d) Knowledge explosion Knowledge is increasing at a very fast rate. Majority of the industries are not up-to-date with the latest knowledge and are, therefore, at a disadvantage. Operations research teams collect the latest information for analysis purpose which is quite useful for the industries.

1.4 SCOPE/APPLICATIONS OF OPERATIONS RESEARCH

Although the complete list of OR techniques and their applications would fill volumes in itself, the following is an abbreviated set of applications to show how widely these techniques are used today:

1. **Accounting**	Cash flow planning Credit policy analysis Planning of delinquent account strategy
2. **Construction**	Allocation of resources to projects Determination of proper workforce Deployment of workforce Project scheduling, monitoring and control
3. **Facilities planning**	Factory size and location decision Hospital planning International logistics system design Estimation of number of facilities required Transportation loading and unloading Warehouse location decision
4. **Finance**	Dividend policy making Investment analysis Portfolio analysis
5. **Manufacturing**	Inventory control Projection marketing balance Production scheduling Production smoothing
6. **Marketing**	Advertising budget allocation Product introduction timing Selection of product mix
7. **Organisational behaviour**	Personal justification/planning Scheduling of training programmes Skills balancing Recruitment of employees

8. **Purchasing**	Material transfer Optimal buying Optimal reordering
9. **Research and Development**	Control of R&D projects Product introduction planning.

A similar list can be prepared for any major field of human endeavour. Military activities alone would cover an entire book.

1.5 OPERATIONS RESEARCH AND DECISION-MAKING

Operations research uses the method of science to understand and explain the phenomena of operating systems. It devices the theories (models) to explain these phenomena, uses these theories to describe what takes place under altered conditions, and checks these predictions against new observations. Thus, operations research is a tool employed to increase the effectiveness of managerial decisions as an objective supplement to the subjective feeling of the decision-maker.

For instance, in distribution or allocation areas, OR may suggest the best locations for agencies, warehouses as well as the most economical kind of transportation; in marketing areas, it may aid in indicating the most profitable type, use and size of advertising compaigns in regard to available financial limit. Operations research may suggest alternative courses of action when a problem is analysed and a solution is attempted. However, the study of complex problems by OR techniques becomes useful only when a choice between two or more courses of action is possible.

Operations research may be regarded as a tool that enables the decision-maker to be objective in creating alternatives and choosing an alternative which is best from among these.

Decision-making is not only the headache of management, rather all of us make decisions. We decide daily about many minor and major issues. The essential characteristics of all decisions are:

1. objective,
2. alternatives at the disposal, and
3. influencing factors.

Once these characteristics are known, one can think of improving the characteristics so as to improve upon the decision itself.

Let us consider a situation where a decision concerns spending summer vacations at a hill resort. The next problem may be to decide the mode of conveyance from amongst the alternatives: train, bus and a taxi.

At the first level of decision-making, bus is chosen as the mode of conveyance just by intuition (may be at random). At the second level of decision-making, the three conveyances are compared and it is decided qualitatively that the bus will be preferred since it is less time consuming than the train and cheaper than a taxi. At the third level of decision-making, the three alternatives are compared and it is suggested that the bus will be chosen, as it will be taking only half the time taken by train and shall be 40% less costlier than the taxi.

Although outcome of all these decisions is the same, one can easily judge the quality of each decision. We may brand the first decision as 'bad' since it is highly emotional, while we may call the second decision as 'good' since it is scientific, though qualitative. The third decision is undoubtedly the best as it is scientific as well as quantitative.

It is this scientific quantification used in OR that helps management to make better decisions.

Advantages of Operations Research Approach in Decision-Making

Following are the salient advantages of an operations research study approach in decision-making:

(i) **Better decisions** Operations research models frequently yield actions that do improve on intuitive decision-making. A situation may be complex so that the human mind can never hope to assimilate all the significant factors without the aid of OR guided computer analysis.

(ii) **Better Coordination** Sometimes operations research has been instrumental in bringing order out of chaos. For instance, an OR oriented planning model becomes a vehicle for coordinating marketing decisions within the limitations imposed on manufacturing capabilities.

(iii) **Better control** The managements of large organisations recognise that it is extremely costly to require continuous executive supervision over routine decisions. An OR approach thereby gained new freedom to the executives to devote their attention to more pressing matters. The most frequently adopted application in this category deals with production scheduling and inventory replenishment.

(iv) **Better system** Often, an OR study is initiated to analyse a particular decision problem, such as whether to open a new warehouse. Afterwards the approach is further developed into a system to be employed repeatedly. Thus, the cost of undertaking the first application may produce benefits.

1.6 OPERATIONS RESEARCH IN MODERN MANAGEMENT

The following are some of the roles of operations research in business and management:

1. **Marketing management**
   - (a) product selection
   - (b) competitive strategies
   - (c) advertising strategy
2. **Production management**
   - (a) production scheduling
   - (b) project scheduling
   - (c) allocation of resources
   - (d) location of factories and their sizes
   - (e) equipment replacement and maintenance
   - (f) inventory policy
3. **Finance management**
   - (a) cash flow analysis
   - (b) capital requirement
   - (c) credit policies
   - (d) credit risks
4. **Personal management**
   - (a) recruitment policies
   - (b) assignment of jobs
5. **Purchasing and procurement**
   - (a) rules of purchasing
   - (b) determining the quality
   - (c) determining the time of purchases

6. Distribution
   (a) location of warehouses
   (b) size of the warehouses
   (c) rental outlets
   (d) transportation strategies.

1.7 PHASES OF OPERATIONS RESEARCH

Operations research is a logical and systematic approach to provide a rational basis for decision-making. The phases and processes of OR study must also be quite logical and systematic. There are six important steps in OR study, but it is not necessary that in all the studies each and every step is invariably present. These steps are arranged in following logical order.

Step I Observe the Problem Environment

Step I in the process of OR study is observing the problem environment. The activities that constitute this step are visits, conferences, observations, research and so on. With the help of such activities, the OR scientist gets sufficient information and support to proceed and is better prepared to formulate the problem.

Step II Analyse and Define the Problem

Step II is analysing and defining the problem. In this step not only the problem is defined, but also uses, objectives and limitations of the study are stressed in the light of the problem. The end result of this step is a clear grasp of need for a solution and understanding its nature.

Step III Develop a Model

Step III is to construct a model. A model is representation of some real or abstract situation. Operations research models are basically mathematical models representing systems, processes or environment in the form of equations, relationships or formulae. The activities in this step include defining inter-relationships among variables, formulating equations, using known OR models or searching suitable alternate models. The proposed model may be field tested and modified in order to work under environmental constraints. The model may also be modified if the management is not satisfied with the answer that it gives.

Step IV Select an Appropriate Data Input

Garbage in and garbage out is a famous saying. No model will work appropriately if data input is not appropriate. Hence, tapping the right kind of data is a vital step in OR process. Important activities in this step are analysing internal–external data and facts, collecting opinions using computer data banks. The purpose of this step is to have a sufficient input to operate and test the model.

Step V Provide a Solution and Test Reasonableness

Step V in OR process is to get a solution with the help of a model and data input. Such a solution is not implemented immediately. First, the solution is used to test the model and to find limitations, if any. If the solution is not reasonable or if the model is not behaving properly, updating and modification of the model is considered at this stage. The end result of this step is a solution that is desirable and supports the current organisational objective.

Step VI Implement the Solution

Implementation of the solution obtained in previous step is the last step of OR process. In OR the decision-making is scientific and implementation of decision involves so many behavioural issues. Therefore, the implementing authority has to resolve the behavioural issues. He has to sell the idea of use of OR not only to the workers but also to the superiors. Distance between management and OR scientist may offer a lot of resistance. The gap between one who provides a solution and one who wishes to use it should be eliminated. To achieve this, OR scientist as well as management should play a positive role. A properly implemented solution obtained through OR techniques results in improved working and wins the management support.

1.8 MODELS IN OPERATIONS RESEARCH

1.8.1 Classification of Models

The first thing one has to do to use OR techniques after formulating a practical problem is to construct a suitable model to represent the practical problem. A model is a reasonably simplified representation of a real-world situation. It is an abstraction of reality. The models can broadly be classified as

- Iconic (physical) models
- Analogue models
- Mathematical models
- Static models
- Dynamic models
- Deterministic models
- Stochastic models
- Descriptive models
- Prescriptive models
- Predictive models
- Analytic models
- Simulation models.

Iconic model This is a physical, or pictorial representation of various aspects of a system.

Example: Toys, miniature model of a building, scaled up model of a cell in biology, etc.

Analogue or schematic model This uses one set of properties to represent another set of properties which a system under study has.

Example: A network of water pipes to represent the flow of current in an electrical network or graphs, organisational charts and so on.

Mathematical model or symbolic model This uses a set of mathematical symbols (letters, numbers, etc.) to represent the decision variables of a system under consideration. These variables are related by mathematical equations or inequations which describe the properties of the system.

Example: A linear programming model, a system of equations representing an electrical network or differential equations representing dynamic systems, etc.

Static model This is a model which does not take time into account. It assumes that the values of the variables do not change with time during a certain period of time horizon.

Example: A linear programming problem, an assignment problem, transportation problem.

Dynamic model This model considers time as one of the important variables.

Example: A dynamic programming problem, a replacement problem

Deterministic model Deterministic model is a model which does not take uncertainty into account.

Example: A linear programming problem, an assignment problem etc.

Stochastic model This is a model which considers uncertainty as an important aspect of the problem.

Example: Any stochastic programming problem, stochastic inventory models, etc.

Descriptive model Descriptive model is one which just describes a situation or system.

Example: An opinion poll, any survey.

Predictive model This is one which predicts something based on some data.

Example: Predicting election results before actually the counting is completed.

Prescriptive model Prescriptive model is one which prescribes or suggests a course of action for a problem.

Example: Any programming (linear, nonlinear, dynamic, geometric) problem.

Analytic model This is a model in which exact solution is obtained by mathematical methods in closed form.

Example: General linear programming model, specially structured transportation and assignment models.

Simulation model This is a representation of reality through the use of a model or device which will react in the same manner as reality under a given set of conditions. Once a simulation model is designed, it takes only a little time, in general, to run a simulation on a computer.

It is usually less mathematically and less time consuming and generally least expensive as well in many situations.

Example: Queueing problems, inventory problems.

1.8.2 Characteristics of a Good Model

1. It should be reasonably simple.
2. A good model should be capable of taking into account new changes in the situation affecting its frame significantly with ease; that, is updating the models should be as simple and easy as possible.
3. Assumptions made to simplify the model should be as small as possible.
4. Number of variables used should be small in number as possible.
5. The model should be open to parametric treatment.

1.8.3 Principles of Modelling

1. Do not build up a complicated model when a simple one would suffice.
2. Beware of moulding the problems to fit a technique.
3. Deductions must be made carefully.
4. Models should be validated prior to implementation.
5. A model should neither be pressed to do; nor criticised for failing to do that for which it was never intented.
6. Beware of overselling the model in cases where assumption made for the construction of the model can be challenged.
7. The solution of a model cannot be more accurate than the accuracy of the information that goes into the construction.
8. Models are only aids in decision-making.
9. Models should not be complicated. It should be as simple as possible.
10. Models should be as accurate as possible.

1.8.4 General Methods for Solving Operations Research Models

1. **Analytic procedure** Solving models by classical mathematical techniques like differential calculus, finite differences and so on, to obtain analytic solutions.
2. **Iterative procedure** Starts with a trial solution and a set of rules for improving it by repeating the procedure until further improvement is not possible.
3. **Monte-Carlo technique** Taking sample observations, computing probability distributions for the variable using random numbers and constructing some functions to determine values of the decision variables.

1.9 ROLE OF OPERATIONS RESEARCH IN ENGINEERING

The following are some of the applications of operations research in engineering:

1. Optimal design of water resources systems
2. Optimal design of structures
3. Production, planning, scheduling and control
4. Optimal design of electrical networks
5. Inventory control
6. Planning of maintenance and replacement of equipment
7. Allocation of resources of service to maximise the benefit
8. Design of material handling
9. Optimal design of machines
10. Optimal design of control systems
11. Optimal selection of sites for an industry.

1.10 LIMITATIONS OF OPERATIONS RESEARCH

Operations research has certain limitations. However, these limitations are mostly related to the problems of model building and the time and money factors involved in its application rather than its practical utility. Some of them are as follows:

1. **Magnitude of computation** Operations research tries to find out the optimal solution taking all the factors into account. In the modern society, these factors are numerous and expressing them in quantity and establishing relationship among these, requires huge calculations. All these calculations cannot be handled manually and require electronic computers which bear very heavy cost. Thus the use of OR is limited only to very large organisations.
2. **Non-quantifiable factors** Operations research provides solution only when all elements related to a problem can be quantified. All relevant variables do not lend themselves to quantification. Factors which cannot be quantified, find no place in OR models. Operations research does not take into account qualitative factors or emotional factors which may be quite important.
3. **Distance between a manager and operations researcher** Operations research being a specialist's job requires a mathematician or a statistician who might not be aware of the business problems. Similarly, a manager fails to understand the complex working of OR. Thus, there is a gap between the two. Management itself may offer a lot of resistance due to conventional thinking.
4. **Money and time costs** When the basic data is subjected to frequent changes, incorporating them into the OR models is a costly affair. Moreover, a fairly good solution at present may be more desirable than a perfect OR solution available after some time.
5. **Implementation** Implementation of decisions is a delicate task. It must take into account the complexities of human relations and behaviour. Sometimes, resistance is offered only due to phychological factors.

EXERCISES

1. What is operations research?
2. What are the characteristics of OR?
3. Discuss the limitations of operations research.
4. Enumerate with brief description some of the techniques of OR.
5. What is meant by a mathematical model of a real situation? Discuss the importance of models in the solution of operations research problems.
6. State the different types of models used in operations research.
7. Discuss the various classification schemes of models.
8. Explain briefly the general methods for solving OR models.
9. Describe the methodology of OR and enumerate the models used in production management.
10. What are the steps involved in operations research?
11. Describe briefly the different phases of operations research.
12. Discuss the importance of operations research in decision-making process.
13. Discuss scientific method in OR.
14. Discuss the significance and scope of operations research in modern management.
15. 'Operations research is a bunch of mathematical techniques.' Comment.
16. What are the essential characteristics of OR? Explain the role of computers in this field.
17. Write a note on application of various quantitative techniques in different fields of business decision-making.

18. Explain various types of OR models and indicate their application to production, inventory and distribution systems.
19. Explain how and why OR methods have been valuable in aiding executive decision.
20. Discuss the advantages and limitations of using results from mathematical model to make decisions about operations.
21. Discuss the significance and scope of operations research in modern management.
22. Is operations research a discipline, or a profession, or set of techniques, or a philosophy, or a new name for an old thing?
23. How can operations research models be classified? Which is the best classification in terms of learning and understanding the fundamentals of operations research?
24. List any three operations research techniques and state in what conditions they can be used.
25. State three properties and three advantages of an OR model.

2

Linear Programming Problem (LPP)

2.1 INTRODUCTION

Many business and economic activities are concerned with the problem of planning. If the supply of the resources are unlimited, the need for linear programming problems would not arise at all. In each case, limited resources are available, and to make use of this resource, programming and planning problems could be formulated to maximise/minimise a linear form of profit/cost function whose variables are restricted to values satisfying a system of linear constraints (a set of linear equations/or inequations). The term 'programming' refers to the process of determining a plan of action. Linear programming is a technique for determining an optimum schedule of independent activities in view of the available resources. Linear relationship between two or more variables is the one in which the variables are directly or precisely proportional.

The general LPP calls for optimising (maximising/minimising) a linear function of variables called the 'objective function' subject to a set of linear equations and/or inequalities called the constraints or restrictions.

The linear programming can be used for optimisation problems if the following conditions are satisfied:

1. There must be a well defined objective function which is to be either maximised (in the case of profit, production) or minimised (in the case of men, materials and so on) and has to be expressed as a linear function of the decision variables.
2. There must be constraints on the objective function and these constraints are expressed as linear equalities or inequalities in terms of variables.
3. There must be an alternative course of action. For example, a given product can be processed by two different machines and the problem may be as to how much of the product is to be allocated to each machine.
4. Another necessary requirement is that decision variables should be interrelated and non-negative. The non-negativity condition shows that linear programming deals with real life situations for which negative quantities are generally illogical.
5. The resources must be limited.

In most linear programming problems, the decision variables are permitted to take any non-negative values that satisfy the constraints. There are some problems in which the variables have integral values only. These problems are not LPP, but very often they can be solved by LPP techniques and the values are then approximated to nearest integers to satisfy the constraints.

Mathematically, the constraints that restrict the decision variables to be non-negative do not differ from other constraints. However, while solving the problems they are handled differently from the other constraints, they have been termed as non-negativity restrictions, and the term constraints has been used to represent constraints other than the non-negativity restriction.

In this chapter, mathematical formulation of LPP properties, graphical solution, simplex method artificial variable techniques are discussed.

2.2 MATHEMATICAL FORMULATION OF LINEAR PROGRAMMING PROBLEM

The procedure for mathematical formulation of LPP consists of the following steps:

Step 1 Identify the decision variables of the problem.

Step 2 Formulate the objective function to be optimised (maximised or minimised) as a linear function of the decision variables.

Step 3 Formulate the constraints of the problem such as resource limitations, market conditions, interrelation between variables and others as linear equation or inequations in terms of the decision variables.

Step 4 Add the non-negativity constraint so that negative values of the decisions variables do not have any valid physical interpretation.

The objective function, the set of constraint and the non-negative constraint together form a linear programming problem.

Example 1 Production Allocation Problem

A firm manufactures two types of products, *A* and *B* and sells them at a profit of Rs 2 on type *A* and Rs 3 on type *B*. Each product is processed on two machines *G* and *H*. Type *A* requires one minute of processing time on *G* and two minutes on *H*; type *B* requires one minute on *G* and one minute on *H*. The machine *G* is available for not more than 6 hours 40 minutes while machine *H* is available for 10 hours during any working day. Formulate the problem as a linear programming problem.

Formulation

Decision variables. Let x_1 be the number of products of type *A* to be produced and let x_2 be the number of units of type *B* to be produced.

The given information can be tabulated as follows:

table 2.1

Machine	*Time on products (minutes)*		*Available time (minutes)*
	Type A (x_1 units)	*Type B (x_2 units)*	
G	1	1	400
H	2	1	600

Objective function Since the profit on type A is 2 per product, $2x_1$ will be the profit on selling x_1 units of type A. Similarly, $3x_2$ will be profit on selling x_2 units of type B. Therefore, total profit z on selling x_1 units of A and x_2 units of B is given by

$$z = 2x_1 + 3x_2$$

Hence, the objective function is maximise $z = 2x_1 + 3x_2$.

Constraints Since machine G takes 1 minute time on type A and 1 minute time on type B, the total number of minutes required on machine G is given by $x_1 + x_2$. But machine G is available for not more than 6 hours 40 minutes (400 minutes). Therefore $x_1 + x_2 \le 400$.

Similarly, the total number of minutes required on machine H is given by $2x_1 + x_2$. But machine H is available for 10 hours on a working day. Therefore, $2x_1 + x_2 \le 600$.

Non-negativity restrictions Since it is not possible to produce negative quantities we have $x_1 \ge 0$, $x_2 \ge 0$. Hence the allocation problem can be represented as

$$\begin{aligned} \text{Maximise } z &= 2x_1 + 3x_2 \\ \text{subject to } x_1 + x_2 &\le 400 \\ 2x_1 + x_2 &\le 600 \\ \text{and } x_1, x_2 &\ge 0 \end{aligned}$$

Example 2 Blending Problem

A firm produces an alloy having the following specifications:

(i) specific gravity ≤ 0.98
(ii) chromium $\ge 8\%$
(iii) melting point $\ge 450°C$

Raw materials A, B, C having the properties shown in the Table 2.2 can be used to make the alloy.

table 2.2

Property	*Properties of raw materials*		
	A	B	C
Specific gravity	0.92	0.97	1.04
Chromium	7%	13%	16%
Melting point	440°C	490°C	480°C

Cost of the various raw materials per unit ton are: Rs 90 for A, Rs 280 for B and Rs 40 for C. Find the proportions in which A, B, C be used to obtain an alloy of desired properties while the cost of raw materials is minimum.

Formulation

Decision variables Let the percentage contents of A, B, C be x_1, x_2, x_3, respectively.

Objective function To minimise the cost

$$z = 90x_1 + 280x_2 + 40x_3 \tag{2.2.1}$$

Constraints Constraints are imposed by the specification required for the alloy. They are

$$\left.\begin{aligned} 0.92x_1 + 0.97x_2 + 1.04x_3 &\le 0.98 \\ 7x_1 + 13x_2 + 16x_3 &\ge 8 \\ 440x_1 + 490x_2 + 480x_3 &\ge 450 \\ x_1 + x_2 + x_3 &= 100 \end{aligned}\right\} \quad (2.2.2)$$

Non-negative restrictions

$$x_1, x_2, x_3 \ge 0 \quad (2.2.3)$$

The relations (2.2.1), (2.2.2) and (2.2.3) together represent the linear programming model for the given problem.

Example 3 Media Selection Problem

The owner of Metro sports wishes to determine how many advertisements to place in three selected monthly magazines *A*, *B*, *C*. His objective is to advertise in such a way that total exposure to principal buyers of expensive sports goods is maximised. Percentages of readers for each magazine are known. Exposure in any particular magazine is the number of advertisements placed multiplied by the number of principal buyers. The following data may be used:

table 2.3

Exposure category	*Magazines*		
	A	*B*	*C*
Readers	1 lakh	0.6 lakh	0.4 lakh
Principal buyers	10%	15%	7%
Cost per advertisement (Rs)	5000	4500	4250

The budgeted amount is Rs 1 lakh for the advertisements. The owner has already decided that magazine *A* should have no more than six advertisements and that *B* and *C* each have atleast two advertisements. Formulate a linear programming model for the problem.

Formulation

Decision variables Let x_1 be the number of advertisements placed in magazine *A* and let x_2 be the number of advertisements placed in magazine *B*; x_3 be the number of advertisements placed in magazine *C*.

Objective function The objective is to advertise in such a way that total exposure to principal buyers of expensive sports goods is maximised.

The number of readers who buy magazine *A* is 1,00,000 and so the readers who have seen the advertisements are $1,00,000x_1$. Similarly, the number of advertisements seen by the readers of magazine *B* is $60,000x_2$ and the number of advertisements seen by the readers of magazine *C* is $40,000x_3$. Hence, the objective functions is:

$$\begin{aligned} \text{Maximise } z &= 0.10 \times 1,00,000x_1 + 0.15 \times 60,000x_2 + 0.7 \times 40,000\, x_3 \\ &= 10,000x_1 + 9,000x_2 + 2,800x_3 \end{aligned} \quad (2.2.4)$$

Constraints Given the budgeting amount is at most Rs 1 lakh for the advertisements. The cost for all advertisements in all the three magazines is $500x_1 + 4500x_2 + 4250x_3$.

Hence, $5000x_1 + 4500x_2 + 4250x_3 \leq 1{,}00{,}00$ (2.2.5)
(budgeting)

Given that the owner have decided that magazine *A* should not have more than six advertisements and *B*, *C* have atleast two advertisements each.

Hence, $x_1 \leq 6,\ x_2 \geq 2,\ x_3 \geq 2$ (2.2.6)
(advertising)

Non-negative restrictions

$$x_1 \geq 0,\ x_2 \geq 0,\ x_3 \geq 0 \quad (2.2.7)$$

The relations from (2.2.8) through (2.2.11) gives the linear programming model for the given problem.

Example 4 Diet Problem

A person wants to decide the constituents of a diet which will fulfil his daily requirements of proteins, carbohydrates at the minimum cost. The choice is to be made from four different types of foods. The yield per unit of these foods are given in the Table 2.4.

table 2.4

Food type	*Yield per unit*			*Cost per units*
	Proteins	*Fats*	*Carbohydrates*	*(Rs)*
1	3	2	6	45
2	4	2	4	40
3	8	7	7	85
4	6	5	4	65
Minimum requirement	800	200	700	

Formulate the linear programming model for the problem.

Formulation

Decision variables Let x_1, x_2, x_3, x_4 be the units of food types 1, 2, 3, 4, respectively.

Objective function Since the cost of these food of type 1, 2, 3, 4 are Rs 45, Rs 40, Rs 85, Rs 65 per unit, the total cost is Rs $45x_1 + 40x_2 + 85x_3 + 65x_4$. As the objective function is to minimise the total cost, the objective function is

$$\text{Minimise } z = 45x_1 + 40x_2 + 85x_3 + 65x_4 \quad (2.2.8)$$

Constraints From these units of food types 1, 2, 3, 4, he requires

$3x_1 + 4x_2 + 8x_3 + 6x_4$ Proteins/day
$2x_1 + 2x_2 + 7x_3 + 5x_4$ Fats/day
$6x_1 + 4x_2 + 7x_3 + 4x_4$ Carbohydrates/day

Since the minimum requirement of these proteins, fats and carbohydrates are 800, 200 and 700, respectively, the constraints are

$$\left.\begin{aligned} 3x_1 + 4x_2 + 8x_3 + 6x_4 &\geq 800 \\ 2x_1 + 2x_2 + 7x_3 + 5x_4 &\geq 200 \\ 6x_1 + 4x_2 + 7x_3 + 4x_4 &\geq 700 \end{aligned}\right\} \quad (2.2.9)$$

Non-negativity restrictions

$$x_1, x_2, x_3, x_4 \geq 0 \quad (2.2.10)$$

Relations from (2.2.12) through (2.2.14) give the complete formulation of LPP.

Example 5 Warehouse Problem

A person running a warehouse purchases and sells identical items. The warehouse can accommodate 1,000 such items. Each month, the person can sell any quantity he has in stock. Each month, he can buy as much as he likes to have in stock for delivery at the end of the month, subject to a maximum of 1,000 items. The forecast of purchase and sale prices for the next six months is given below:

table 2.5

Month i	1	2	3	4	5	6
Purchase price C_i (Rs)	12	14	17	19	20	21
Sale price Si (Rs)	13	15	16	20	21	23

If at present he has a stock of 200 items, what should be his policy?

Formulation

Decision variables Let x_j and y_j ($j = 1, 2, \cdots, 6$) be the number of items purchased and sold for each of the six months.

Objective function The objective is to maximise the profit.

That is Maximise

$$z = 13y_1 + 15y_2 + 16y_3 + 20y_4 + 21y_5 + 23y_6 - (12x_1 + 14x_2 + 17x_3 + 19x_4 + 20x_5 + 21x_6) \quad (2.2.11)$$

Constraints The person cannot sell anything that is not in stock. Therefore, for each month $n = 1, 2, \cdots 6$.

$$200 + \sum_{j=1}^{n-1} (x_j - y_j) \geq y_n$$

$$\Rightarrow \quad 200 + \sum_{j=1}^{n-1} x_j \geq \sum_{j=1}^{n-1} y_j + y_n$$

$$\Rightarrow \quad 200 + \sum_{j=1}^{n-1} x_j \geq \sum_{j=1}^{n} y_j$$

$$\left.\begin{aligned} \Rightarrow \quad & \sum_{j=1}^{n} y_j - \sum_{j=1}^{n-1} x_j \le 200 \\ \Rightarrow \quad & \sum_{j=1}^{6} y_j - \sum_{j=1}^{5} x_j \le 200 \end{aligned}\right\} \quad (2.2.12)$$

Further, since we cannot over stock beyond 1,000 items.

$$200 + \sum_{j=1}^{n} (x_j - y_j) \le 1{,}000$$

$$\text{or,} \quad \sum_{j=1}^{n} x_j - \sum_{j=1}^{n} y_j \le 800$$

$$\Rightarrow \quad \sum_{j=1}^{6} x_j - \sum_{j=1}^{6} y_j \le 800 \quad (2.2.13)$$

Non-negativity restriction

$$x_j \ge 0,\ y_j \ge 0 \text{ for } j = 1, 2, \cdots, 6 \quad (2.2.14)$$

The equations (2.2.19) – (2.2.22) gives the complete LPP.

Example 6 Product Mix Problem

A firm manufactures three products A, B, C. Time to manufacture product A is twice that for B and thrice that for C and they are to be produced in the ratio 3 : 4 : 5. The relevant data is given in Table 2.6. If the whole raw material is engaged in manufacturing product A, 1600 units of this product can be produced. There is demand for atleast 300, 250, 200 units of products A, B and C and the profit earned per unit is Rs 50, Rs 40, Rs 70, respectively. Formulate the problem as a linear programming problem.

table 2.6

Raw material	Requirement per unit of product (kg)			Total availability (kg)
	A	B	C	
P	6	5	9	5,000
Q	4	7	8	6,000

Formulation

To determine the number of units of each of the products A, B, C to be manufactured.

Decision variables Let x_1, x_2, x_3 denote the number of units of products A, B, C to be manufactured.

Objective function To maximise the profit,

$$\text{Maximise } z = 50x_1 + 40x_2 + 70x_3 \quad (2.2.15)$$

Constraints

For raw material P, $6x_1 + 5x_2 + 9x_3 \le 5{,}000$

For raw material Q, $4x_1 + 7x_2 + 8x_3 \le 6{,}000$

As product B requires 1/2 and product C requires 1/3 of the time required for product A, the constraint on number of units manufactured may be expressed as

$$x_1 + \frac{1}{2}x_2 + \frac{1}{3}x_3 \le 1{,}600 \tag{2.2.16}$$

Market demand requires

$$\left.\begin{aligned} x_1 &\ge 300 \\ x_2 &\ge 250 \\ x_3 &\ge 200 \end{aligned}\right\} \tag{2.2.17}$$

Since products A, B, C are to be produced in the ratio 3 : 4 : 5,

$$x_1 : x_2 : x_3 = 3 : 4 : 5$$

or, $$\frac{x_1}{3} = \frac{x_2}{4} \text{ and } \frac{x_2}{4} = \frac{x_3}{5}$$

$\Rightarrow$ $$4x_1 - 3x_2 = 0 \text{ and } 5x_2 - 4x_3 = 0. \tag{2.2.18}$$

Non-negativity restrictions

$$x_1 \ge 0,\ x_2 \ge 0,\ x_3 \ge 0 \tag{2.2.19}$$

The relations from (2.2.15) through (2.2.19) gives the complete linear programming model.

Example 7 Product Mix Problem

A chemical company produces two products X and Y. Each unit of product X requires 3 hours on operation I and 4 hours on operation II, while each unit of product Y requires 4 hours on operation I and 5 hours on operation II. Total available time for operations I and II is 20 hours and 26 hours, respectively. The production of each unit of product Y also results in the production two units of a by-product Z at no extra cost. Product X sells a profit of Rs 10/unit, while Y sells at profit of Rs 20/unit. By-product Z brings a unit profit of Rs 6 if sold; in case it cannot be sold, the destruction cost is Rs 4/unit. Forecasts indicate that more than 5 units of Z can be sold. Determine the quantities of X and Y to be produced, keeping Z in mind, so that the profit earned is maximum.

Formulation

To determine the number of units of products X, Y, Z to be produced.

Decision variables Let the number of units of products X, Y, Z produced be x_1, x_2 and x_3, where x_3 is the number of units of Z to be produced.

That is x_3 = number of units of Z sold + number of units of Z destroyed

$= x_4 + x_5$ (say)

Objective function To maximise the profit earned. Objective function (profit function) for products X and Y is linear because their profits are constant irrespective of the number of units produced. A graph between the total profit and quantity produced will be a straight line. But a similar graph of Z shows that Z is nonlinear, since it has slope +6 for the first part and –4 for the second. However, it is piecewise linear, since it is linear in the region (0, 5) and (5, Y). So, splitting x_3 into two parts, namely, the number of units of Z sold (x_4) and number of units of Z destroyed (x_5) makes the objective function for product Z also linear. Thus the objective function is

$$\text{Maximise } Z = 10x_1 + 20x_2 + 6x_3 - 4x_4 \quad (2.2.20)$$

Constraints Since each unit of product X requires 3 hours on operation I and 4 hours on operation II each unit of product Y requires 4 hours on operation I and 5 hours on operation II, the total time available for operations I and II is 20 hours and 26 hours, respectively, we have the constraints

$$\left.\begin{aligned} 3x_1 + 4x_2 &\le 20 \text{ (for operation I)} \\ 4x_1 + 5x_2 &\le 26 \text{ (for operation II)} \end{aligned}\right\} \quad (2.2.21)$$

Since not more than 5 units of Z can be sold,

$$x_3 \le 5 \quad (2.2.22)$$

The production of each unit of product Y also results in two units of a byproduct Z at no extra cost. So,

$$x_2 = 2x_3 = x_4 + x_5$$

or

$$x_4 + x_5 - 2x_3 \quad (2.2.23)$$

Non-negativity restriction

Let

$$x_1, x_2, x_3, x_4, x_5 \ge 0 \quad (2.2.24)$$

The relations from (2.2.20) through (2.2.24) gives the complete formulation of linear programming model for this problem.

Example 8

A farmer has a two acre farm. He can sell all the tomatoes, drumstick or radishes he can raise. The price he can obtain is Re 1.00 per kg for tomatoes, Re 0.75 for a head of drumstick and Rs 2.00 per kg for radishes. The average yield peracre is 2,000 kg of tomatoes, 3,000 heads of drumstick, and 1,000 kgs of radishes. Fertilizer is available at Re 0.50 per kg and the amount required per acre is 100 kg each for tomatoes and drumstick, and 50 kg for radishes. Labour required for sowing cultivating and harvesting per acre is 5 man-days for tomatoes and radishes and 6 man-days for drumstick. A total of 400 man-days of labour are available at Rs 20.00 per man-day. Formulate this problem as a linear programming model to maximise the farmer's total profit.

Formulation

Decision variables Let x_1 denote the number of acres allotted for tomato cultivation, x_2 denote the number of acres alloted for drumstick cultivation and let x_3 denote the number of acres allotted for radish cultivation.

Objective function The objective is to maximise the farmer's profit. The farmer's total profit is = sale – expenditure.

The total sale of farmer will be

$$\text{Rs } [1 \times 2{,}000x_1 + 0.75 \times 3{,}000x_2 + 2 \times 1{,}000x_3]$$
$$= \text{Rs } 2{,}000x_1 + 2{,}250x_2 + 2{,}000x_3$$

The fertilizer expenditure will be

$$\text{Rs } 0.50\ (100x_1 + 100x_2 + 50x_3)$$
$$= \text{Rs } (50x_1 + 50x_2 + 25x_3)$$

Labour expenditure will be

$$\text{Rs } 20.00\ (5x_1 + 6x_2 + 5x_3)$$
$$= \text{Rs } (100x_1 + 120x_2 + 100x_3)$$

The total expenditure is:

$$\text{Rs } (150x_1 + 170x_2 + 125x_3)$$

The total profit is:

$$\text{Rs [sale – expenditure]}$$
$$= \text{Rs } [2{,}000x_1 + 2{,}250x_2 + 2{,}000x_3] - \text{Rs } [150x_1 + 170x_2 + 125x_3]$$
$$= \text{Rs } (1{,}850x_1 + 2{,}080x_2 + 1{,}875x_3)$$

Therefore, the objective function is

$$\text{Maximise } z = 1{,}850x_1 + 2{,}080x_2 + 1{,}875x_2 \qquad (2.2.25)$$

Constraints Since the total area is restricted to 100 acres and the total man-days labour is restricted to 400 man-days, the constraints are

$$\left.\begin{aligned} x_1 + x_2 + x_3 &\le 100 \\ 5x_1 + 6x_2 + 5x_3 &\le 400 \end{aligned}\right\} \qquad (2.2.26)$$

Non-negativity restrictions

$$x_1, x_2, x_3 \ge 0 \qquad (2.2.27)$$

The relations (2.2.25) through (2.2.26) gives a complete linear programming model for the problem.

2.3 STATEMENTS OF BASIC THEOREMS AND PROPERTIES

2.3.1 General Formulation of Linear Programming Problems

In order to find the values of n decision variables $x_1, x_2, \cdots, x_n \cdots$ to maximise or minimise the objective function

$$z = c_1\, x_1 + c_2\, x_2 + \cdots + c_n\, x_n \qquad (2.3.1)$$

and also satisfy m-constraints

$$\left.\begin{array}{l} a_{11}x_1 + \cdots + a_{ij}x_j + \cdots + a_{1n}x_n\ (\leq = \geq)\ b_1 \\ a_{21}x_1 + \cdots + a_{2j}x_j + \cdots + a_{2n}x_n\ (\leq = \geq)\ b_2 \\ \vdots \qquad\qquad \cdots \qquad\qquad \vdots \\ a_{i1}x_1 + \cdots + a_{ij}x_j + \cdots + a_{in}x_n\ (\leq = \geq)\ b_i \\ \vdots \qquad\qquad \cdots \qquad\qquad \vdots \\ a_{m_1}x_1 + \cdots + a_{mj}x_j + \cdots + a_{mn}x_n\ (\leq = \geq)\ b_m \end{array}\right\} \quad (2.3.2)$$

The above formulation may be put as

$$z = \sum_{j=1}^{n} c_j x_j \text{ subject to } \sum_{j=1}^{n} a_{ij} x_{ij}\ (\leq = \geq)\ b_i,\ i = 1, 2, \cdots, m$$

and, $x_j \geq 0,\ j = 1, 2, \cdots, n.$

where, constraints may be in the form of an inequality (≤ or ≥) or even in the form of an equation (=) and finally satisfying the non-negativity restrictions

$$x_1 \geq 0,\ x_2 \geq 0,\ \cdots,\ x_j \geq 0,\ \cdots\ x_n \geq 0 \quad (2.3.3)$$

However, by convention, the values of right hand side parameters b_i $(i = 1, 2, 3, \cdots, m)$ are restricted to non-negative values only. Note that any negative b_i can be changed to positive by multiplying – 1 on both sides (in this case direction of inequality will change).

2.3.2 Standard Form of Linear Programming Problem

The standard form of the linear programming problem is used to develop the general procedure for solving any linear programming problem. The characteristics of the standard form are explained in the following steps:

Step 1 *All the constraints should be converted to equations excepts for the non-negativity restrictions which remain as inequalities (≥ 0).*

Constraints of the inequality type can be changed to equations by augmenting (adding or subtracting) the left side of each such constraint by non-negative variables. These variables are called *slack variables* and are added if the constraints is of (≤) type or subtracted if the constraint is of (≥) type. Since in the case of ≥ type constraint, the subtracted variable represents the surplus of the left side over the right side, it is common to refer to it as *surplus variables*. A surplus is nothing but a negative slack.

For example, consider the constraints

$$3x_1 - 4x_2 \geq 7$$
$$x_1 + 2x_2 \leq 3$$

These constraints can be changed to equations by introducing a surplus variables s_1 and a slack variable s_2 as follows:

$$3x_1 - 4x_2 - s_1 = 7 \qquad s_1 \geq 0$$
$$5x_1 + 2x_2 + s_2 = 3 \qquad s_2 \geq 0$$

Step 2 *The right side element of each constraint should be made non-negative, if required.*

This is the standard constraint equation.

For example, the inequation

$$3x_1 - 4x_2 \geq -2$$

can be written in the form

$$3x_1 - 4x_2 - s_1 = -2$$

by introducing surplus variable $s_1 \geq 0$. Again, multiplying both sides by – 1, we get

$$-3x_1 + 4x_2 + s_1 = 2$$

which is in the standard form.

Step 3 *All variables must have non-negative values*

A variable which is unrestricted in sign (that is + ve, – ve, or 0) is equivalent to the difference between two non-negative variables. Thus, if n is unconstrained in sign, it can be replaced by $(x' - x'')$ where x' and x'' are both non-negative, that is $x' \geq 0$ and $x'' \geq 0$.

Step 4 *The objective function should be maximisation form*

The minimisation of a function $f(x)$ is equivalent to the maximisation of the negative expression of this function $f(x)$, that is

$$\text{Min } f(x) = -\text{ Max } \{-f(x)\}$$

For example, the linear objective function

$$\text{Minimise } z = c_1 x_1 + c_2 x_2 + \cdots + c_n x_n \text{ is equivalent to}$$

$$\text{Max } z' = \max(-z) = -c_1 x_1 - \cdots - c_n x_n$$

with $z = -z'$. So, the general form of LPP with constraints is:

$$\text{Maximise } z = c_1 x_1 + c_2 x_2 + \cdots + c_n x_n + 0x_{n+1} + \cdots + 0x_{n+m}$$

subject to

$$a_{11} x_1 + \cdots + a_{1n} x_n + x_{n+1} = b_1$$
$$a_{21} x_1 + \cdots + a_{2n} x_n + x_{n+2} = b_2$$
$$\vdots \quad \cdots$$
$$a_{m1} x_1 + \cdots + a_{mn} x_n + x_{n+m} = b_m$$
$$x_1 \geq 0, x_2 \geq 0, \cdots, x_{n+m} \geq 0.$$

Note that the coefficient of slack variables $x_{n+1}, x_{n+2}, \cdots, x_{n+m}$ in the objective function are assumed to be zero, so that the conversion of constraints to a system of simultaneous linear equations does not change the function to be optimised.

Definition If the constraints of a general LPP be

$$\sum_{j=1}^{n} a_{ij} x_j \leq b_i \quad i = 1, 2, \cdots, k \tag{2.3.4}$$

then non-negative variables s_i which are introduced to convert the inequalities (2.3.4) to the equalities

$$\sum_{j=1}^{n} a_{ij} x_j + s_i = b_i \qquad (i = 1, 2, \cdots, k)$$ are called *slack variables.*

Definition If the constraints of a general LPP be

$$\sum_{j=1}^{n} a_{ij} x_j \geq b_i \, (i = k, k+1, \cdots) \tag{2.3.5}$$

then the non-negative variables s_i which are introduced to convert the inequalities (2.3.5) to the equalities

$$\sum_{j=1}^{n} a_{ij}\, x_j - s_i = b_i \ (i = k, k+1, \cdots)$$ are called *surplus variables.*

Example 1

Express the following LPP in the standard form

Maximise $z = 2x_1 + 3x_2 + x_3$

subject to the constraints
$$4x_1 - 3x_2 + x_3 \le 6$$
$$x_1 + 5x_2 - 7x_3 \ge -4$$
$x_1, x_3 \ge 0$, x_2 is unrestricted

Solution: As x_2 is unrestricted $x_2 = x'_2 - x''_2$, where $x'_2, x''_2 \ge 0$.

The given LPP becomes

Maximise $z = 2x_1 + 3(x'_2 - x''_2) + x_3$

subject to
$$4x_1 - 3x'_2 + 3x''_2 + x_3 \le 6$$
$$x_1 + 5x'_2 - 5x''_2 - 7x_3 \ge -4$$
$$x_1, x'_2, x''_2, x_3 \ge 0$$

Convert the second constraint into $\le$ type by multiplying both sides by -1. Now, the LPP becomes

Maximise $z = 2x_1 + 3x'_2 - 3x''_2 + x_3$

subject to
$$4x_1 - 3x'_2 + 3x''_2 + x_3 \le 6$$
$$-x_1 - 5x'_2 + 5x''_2 + 7x_3 \le 4$$
$$x_1, x'_2, x''_2, x_3 \ge 0$$

which is in the standard form.

Example 2

Express the following LPP in the standard form

Maximise $z = 3x_1 + 2x_2$

subject to
$$2x_1 + x_2 \le 2$$
$$3x_1 + 4x_2 \ge 12$$
$$x_1, x_2 \ge 0$$

Solution: Introducing slack and surplus variables, the problem in the standard form can be expressed as

Maximise $z = 3x_1 + 2x_2$

subject to
$$2x_1 + x_2 + s_1 = 2$$
$$3x_1 + 4x_2 - s_2 = 12$$
$$x_1, x_2, s_1, s_2 \ge 0$$

Example 3

Express the following linear programming problem in the standard form.

Maximise $z = 3x_1 + 2x_2 + 5x_3$

subject to
$$2x_1 - 3x_2 \le 3$$
$$x_1 + 2x_2 + 3x_3 \le 5$$

$$3x_1 + 2x_3 \leq 2$$
$$x_1 \geq 0, x_2 \geq 0.$$

Solution: Here, x_1 and x_2 are restricted to be non-negative, while x_3 is unrestricted. Let us express x_3 as $x_3 = x'_3 - x''_3$, where $x'_3 \geq 0$, $x''_3 > 0$. Thus, the above constraints can be written as

$$2x_1 + 3x_2 \leq 3$$
$$x_1 + 2x_2 + 3x'_3 - 3x''_3 \geq 5$$
$$3x_1 + 2x'_3 - 2x''_3 \leq 2$$
$$x_1, x_2, x'_3, x''_3 \geq 0$$

Introducing the slack and surplus variables then the standard form is

$$\text{Maximise } z = 3x_1 + 2x_2 + 5x'_3 - 5x''_3$$

subject to
$$2x_1 - 3x_2 + s_1 = 3$$
$$x_1 + 2x_2 + 3x'_3 - 3x''_3 - s_2 = 5$$
$$3x_1 + 2x'_3 - 2x''_3 + s_3 = 2$$
where, $x_1, x_2, x'_3, x''_3, s_1, s_2, s_3 \geq 0$

Example 4

Reformulate the problem into the standard form

$$\text{Minimise } z = 2x_1 + 3x_2$$

subject to
$$2x_1 - 3x_2 - x_3 = -4$$
$$3x_1 + 4x_2 - x_4 = -6$$
$$2x_1 + 5x2 + x_5 = 10$$
$$4x_1 - 3x_2 + x_6 = 18$$
where, $x_3, x_4, x_5, x_6 \geq 0.$

Solution: Here, x_3, x_4, x_5, x_6 (which are all non-negative) are the slack variables. The decision variables x_1, x_2 are unrestricted in sign.

Putting $x_1 = y_1 - y_2$, $x_2 = y_3 - y_4$, $x_5 = y_5$, $x_4 = y_6$, $x_5 = y_7$, $x_6 = y_8$, the problem in standard form is

$$\text{Minimise } z = 2y_1 - 2y_2 + 3y_3 - 3y_4$$

subject to
$$-2y_1 + 2y_2 + 3y_3 - 3y_4 + y_5 = 4$$
$$-3y_1 + 3y_2 - 4y_3 + 4y_4 + y_6 = 6$$
$$2y_1 - 2y_2 + 5y_3 - 5y_4 + y_7 = 10$$
$$4y_1 - 4y_2 - 3y_3 + 3y_4 + y_8 = 18$$
where $y_1, y_2, y_3, y_4, y_5, y_6, y_6, y_8 \geq 0$

Example 5

Find all the basic solutions to the following LPP

$$\text{Maximise } z = x_1 + 3x_2 + 3x_3$$

subject to
$$x_1 + 2x_2 + 3x_3 = 4$$
$$2x_1 + 3x_2 + 5x_3 = 7$$

Also, find which of the basic solutions are

(i) basic feasible
(ii) non-degenerate basic feasible
(iii) optimal basic feasible

Since $m + n = 2$ and $m = 2$ in this problem, a basic solution can be obtained by setting any of the n variables equal to zero and then solving the constraint equations. The total number of basic solutions are

$$3C_2 = \frac{3!}{2!1!} = 3.$$

Out of these three solutions, solutions in which all basic variables (x_j) are ≥ 0 will be basic feasible; solutions in which all basic variables are > 0 will be non-degenerate basic feasible and the basic feasible solution that optimises the objective function will be the optimal basic feasible solution.

Table 2.7 gives a summary of the characteristics of the various basic solutions.

table 2.7

S.No.	Basic variables	Non-basic variables	Value of the basic variables given by the constraint equations	Value of the objective function	Is the solution feasible	Is the solution non-degenerate	Is the solution feasible and optimal
1.	x_1, x_2	$x_3 = 0$	$x_1 + 2x_2 = 4$ $2x_1 + 3x_2 = 7$ $\Rightarrow x_1 = 2, x_2 = 1$	5	Yes	Yes	Yes
2.	x_1, x_3	$x_2 = 0$	$x_1 + 3x_3 = 4$ $2x_1 + 5x_3 = 7$ $\Rightarrow x_1 = 1, x_3 = 1$	4	Yes	Yes	No
3.	x_2, x_3	$x_1 = 0$	$2x_2 + 3x_3 = 4$ $3x_2 + 5x_3 = 7$ $\Rightarrow x_2 = -1, x_3 = 2$	3	No	No	No

The first two solutions are basic feasible; they are also non-degenerate basic feasible solutions. The first solution is the optimal one. Thus, the optima solution is $x_1 = 2, x_2 = 1, x_3 = 0$, with maximum of $z = 5$.

Example 6

A firm manufactures four different machine parts M_1, M_2, M_3 and M_4 made of copper and zinc. The amounts of copper and zinc required for each machine part, their exact availability and the profit earned from one unit of each machine part are as follows:

table 2.8

	M_1 (kg)	M_2 (kg)	M_3 (kg)	M_4 (kg)	Availability (kg)
Copper	5	4	2	1	100
Zinc	4	3	8	1	75
Profit	12	8	14	10	

How many of each part has to be manufactured to maximise profit? For this problem find

(i) basic solutions
(ii) basic feasible solutions
(iii) non-degenerate basic feasible solutions and
(iv) optimal basic feasible solutions.

Solution: Let x_1, x_2, x_3, x_4 represent the quantities to be manufactured by machines M_1, M_2, M_3 and M_4, respectively. The formation of linear programming problem is

$$\text{Maximise } z = 12x_1 + 8x_2 + 14x_3 + 10x_4$$

subject to $5x_1 + 4x_2 + 2x_3 + x_4 = 100$

$2x_1 + 3x_2 + 8x_3 + x_4 = 75$

where, $x_1, x_2, x_3, x_4 \geq 0.$

Here, $m + n = 4$, $m = 2$. A basic solution can be obtained by setting any of the $n = 2$ non-basic variables equal to zero and then solving the constraint equations. The total number of basic solution is $\frac{4!}{2!2!} = 6$. The following Table gives the characteristics of the various solutions:

table 2.9

S.No. of the basic solution	*Basic variables*	*Non-basic variables*	*Values of the basic variables given by the constraint equation*	*Value of objective function*	*Is the solution feasible*	*Is the solution non-degenerate?*	*Is the solution feasible and optimal?*
1	x_1, x_2	x_3, x_4	$5x_1 + 4x_2 = 100$ $2x_1 + 3x_2 = 75$ $x_1 = 0, x_2 = 25$	200	Yes	No	No
2	x_1, x_3	x_2, x_4	$5x_1 + 2x_3 = 100$ $2x_1 + 8x_3 = 75$ $\Rightarrow x_1 = 325/18$ $x_3 = 175/36$	$\frac{5125}{18}$	Yes	Yes	No
3	x_1, x_4	x_2, x_3	$5x_1 + x_4 = 100$ $2x_1 + x_4 = 75$ $\Rightarrow x_1 = 25/3$ $x_4 = 175/3$	$\frac{2050}{3}$	Yes	Yes	Yes
4	x_2, x_3	x_1, x_4	$4x_2 + 2x_3 = 100$ $3x_2 + 8x_3 = 75$ $\Rightarrow x_2 = 25$ $x_3 = 0$	200	Yes	No	No

(Contd)

S.No. of the basic solution	Basic variables	Non-basic variables	Values of the basic variables given by the constraint equation	Value of objective function	Is the solution feasible	Is the solution non-degenerate?	Is the solution feasible and optimal?
5	x_2, x_4	x_1, x_3	$4x_2 + x_4 = 100$ $3x_2 + x_4 = 75$ $\Rightarrow x_2 = 25$ $x_4 = 0$	200	Yes	No	No
6	x_3, x_4	x_1, x_2	$2x_3 + x_4 = 100$ $8x_3 + x_4 = 75$ $\Rightarrow x_3 = -25/6$ $x_4 = 325/3$	1025	No	No	No

(i) All solutions are basic
(ii) Solutions 1, 2, 3, 4, 5 are basic feasible
(iii) Solutions 2, 3 are non-degenerate basic feasible solutions.

Hence, the optimal solution is $x_1 = 25/3, x_2 = x_3 = 0, x_4 = 175/3$ and maximum of $z = 2050/3$.

Example 7

Show, using matrix-vector notation the following system of linear equations has degenerate solutions

$$2x_1 + x_2 - x_3 = 2$$
$$3x_1 + 2x_2 + x_3 = 3$$

Solution: The given system of equations can be written as $AX = b$, where

$$A = \begin{pmatrix} 2 & 1 & -1 \\ 3 & 2 & 1 \end{pmatrix}; \quad X = \begin{pmatrix} x_1 \\ x_2 \\ x_3 \end{pmatrix} \quad b = \begin{pmatrix} 2 \\ 3 \end{pmatrix}$$

Since A is of order 2×3, there can be $3C_2 = \frac{3!}{2!1!} = 3$ submatrices of the order 2×2

They are

$$\begin{pmatrix} 2 & 1 \\ 3 & 2 \end{pmatrix}, \begin{pmatrix} 1 & -1 \\ 2 & 1 \end{pmatrix}, \begin{pmatrix} 2 & -1 \\ 3 & 1 \end{pmatrix}$$

Any of them can be taken as a basis matrix B. The variables not associated with the columns of these submatrices are respectively x_3, x_1 and x_2.

First, let $B = \begin{pmatrix} 2 & 1 \\ 3 & 2 \end{pmatrix}$, a basic solution to the given system is obtained by setting $x_3 = 0$ and solving the system

$$\begin{pmatrix} 2 & 1 \\ 3 & 2 \end{pmatrix} \begin{pmatrix} x_1 \\ x_2 \end{pmatrix} = \begin{pmatrix} 2 \\ 3 \end{pmatrix}$$

$$\Rightarrow \quad 2x_1 + x_2 = 2$$
$$3x_1 + 2x_2 = 3$$

$\Rightarrow x_1 = 1, x_2 = 0$ (basic); $x_3 = 0$ (non-basic)

Similarly, for other two sub-matrices we get the solutions:

$$x_2 = 5/3;\ x_3 = -1/3 \text{ (basic)};\ x_1 = 0 \text{ (non-basic)}$$

and $$x_1 = 1,\ x_3 = 0 \text{ (basic)};\ x_2 = 0 \text{ (non-basic)}$$

Since in two of these basic feasible solutions, one basic variable is zero, they are degenerate solutions. The second solution (0, 5/3, – 1/3) is not feasible. Thus, the remaining two degenerate basic feasible solutions are (1, 0, 0) and (1, 0, 0).

Example 8

Express the following LPP in the standard from

$$\text{Minimise } z = x_1 - 2x_2 + x_3$$

subject to
$$2x_1 + 3x_2 + 4x_3 \geq -4$$
$$3x_1 + 5x_2 + 2x_3 \geq 7$$

$x_1, x_2 \geq 0$ and x_3 is unrestricted.

Solution: Since x_3 is unrestricted, put $x_3 = x'_3 - x''_3$ where $x'_3, x''_3 \geq 0$.

Then, the given LPP becomes

$$\text{Maximise } z' = -x_1 + 2x_2 - (x'_3 - x''_3) \text{ where } z' = -z.$$

subject to
$$-2x_1 - 3x_2 - 4(x'_3 - x''_3) + s_1 = 4$$
$$3x_1 + 5x_2 + 2(x'_3 - x''_3) - s_2 = 7$$
$$x_1, x_2, x'_3, x''_3, s_1, s_2 \geq 0,$$

where s_1 is the slack variable and s_2 is the surplus variable.

2.3.3 Matrix Form of Linear Programming Problem

The linear programming problem in the standard form (2.3.1), (2.3.2), (2.3.3) can be expressed in the matrix form as follows:

$$\text{Maximise } z = CX \text{ (objective function)}$$

subject to
$$AX = b,\ b \geq 0 \text{ (constraint equation)}$$
$$X \geq 0 \text{ (non-negativity restriction)}$$

where,
$$X = (x_1, x_2, \cdots, x_n, x_{n+1}, \cdots, x_{n+m})$$
$$C = (c_1, c_2, \cdots, c_n, 0, 0, \cdots, 0)$$
$$b = (b_1, b_2, \cdots, b_m)$$

$$A = \begin{bmatrix} a_{11} & a_{12} & \cdots & a_{1n} & 1 & 0 & \cdots & 0 \\ a_{21} & a_{22} & \cdots & a_{2n} & 0 & 1 & \cdots & 0 \\ \vdots & & \cdots & & & & & \vdots \\ a_{m1} & a_{m2} & \cdots & a_{m2} & 0 & 0 & \cdots & 1 \end{bmatrix}$$

The vector X is assumed to include the decision variables, that is slack variables. For convenience, X is used to represent all types of variables. The vector C gives the corresponding coefficient of the objective function. For example, if the variable is slack, its corresponding element is zero.

Example

Express the following LPP in the matrix form

$$\text{Maximise } z = 2x_1 + 3x_2 + 4x_3$$

subject to

$$\begin{aligned} x_1 + x_2 + x_3 &\geq 5 \\ x_1 + 2x_2 &= 7 \\ 5x_1 - 2x_2 + 3x_3 &\leq 9 \\ x_1 \geq 0, x_2 \geq 0, x_3 &\geq 0. \end{aligned}$$

Solution: This problem can be written in standard form as

$$\text{Maximise } z = 2x_1 + 3x_2 + 4x_3 + 0x_4 + 0x_5$$

or,

$$\text{Maximise } z = (2, 3, 4, 0, 0) \begin{pmatrix} x_1 \\ x_2 \\ x_3 \\ x_4 \\ x_5 \end{pmatrix}$$

subject to

$$\begin{aligned} x_1 + x_2 + x_3 - x_4 &= 5 \\ x_1 + 2x_2 &= 7 \\ 5x_1 - 2x_2 + 3x_3 &= 9 \end{aligned}$$

or,

$$\begin{bmatrix} 1 & 1 & 1 & -1 & 0 \\ 1 & 2 & 0 & 0 & 0 \\ 5 & -2 & 3 & 0 & 1 \end{bmatrix} \begin{bmatrix} x_1 \\ x_2 \\ x_3 \\ x_4 \\ x_5 \end{bmatrix} = \begin{bmatrix} 5 \\ 7 \\ 9 \end{bmatrix}$$

Thus, $X = (x_1, x_2, x_3, x_4, x_5)$

$C = (2, 3, 4, 0, 0)$

$b = (5, 7, 9)$ and

$$A = \begin{bmatrix} 1 & 1 & 1 & -1 & 0 \\ 1 & 2 & 0 & 0 & 0 \\ 5 & -2 & 3 & 0 & 1 \end{bmatrix}$$

2.3.4 Definitions

1. **Solution to a LPP** A set $X = \{x_1, x_2, \cdots, x_{n+m}\}$ of variables is called a solution to a LPP, if it satisfies the set of constraints (2.3.2) only.
2. **Feasible solution** Any set $X = \{x_1, x_2, \cdots, x_{n+m}\}$ of variables is called a feasible solution of the LPP, if it satisfies the set of constraints (2.3.2) and non-negativity restrictions (2.3.3).

3. **Basic solution** A basic solution to (2.3.2) is a solution obtained by setting any n variables (among $m + n$ variables) equal to zero and solving for remaining m variables provided the determinant of the coefficients of these m variables is non-zero. Such m variables (any of them may be zero) are called *basic variables*, and the remaining variables which are set as zero are called *non-basic variables*.
   Thus, the number of basic solutions must be atmost $m + nC_m$.
4. **Basic feasible solution** A basic solution to a LPP is called as a basic feasible solution if it satisfies the non-negative restriction.
   There are two types of basic feasible solutions.
   (i) **Non-degenerate** All m basic variables are positive, and remaining n variables will be zero.
   (ii) **Degenerate** A basic feasible solution is degenerate, if one or more basic variables are zero.
5. **Optimum basic feasible solution** A basic feasible solution to a LPP is said to be its optimum solution if it optimises (maximises or minimises) the objective function.
6. **Unbounded solution** If the value of the objective function z can be increased or decreased indefinitely, such solutions are called unbounded solutions.

Note: An optimum solution to a LPP means that z has a finite maximum or finite minimum.

Example

Find all basic solution for the system of simultaneous equation

$$2x_1 + 3x_2 + 4x_3 = 5$$
$$3x_1 + 4x_2 + 5x_3 = 6$$

Solution: First decide the maximum possible number of basic solutions. The possible number of basic solutions will be $m + nCm$, the number of combination of things taken m at a time. Here, we have $m + n = 3$ variables and $m = 2$ constraints. Hence, the possible number of basic solutions are

$$3C_2 = \frac{3!}{2!(3-2)!} = 3.$$

Put $x_1 = 0$, solve for x_2 and x_3. The values obtained are $x_2 = -1$, $x_3 = 2$. If we put $x_2 = 0$, then we get $x_1 = -2$, $x_3 = 3/2$. Similarly, if we put $x_3 = 0$, then we get $x_1 = -2$, $x_2 = 3$.

Note that the three basic solutions are non-degenerate and feasible.

2.3.5 Basic Assumptions

The linear programming problems are formulated on the basis of the following assumptions.

1. *Proportionality*: The contribution of each variable in the objective function on its usage of the resources is directly proportional to the value of the variable. That is, if resource availability increases by some percentage, then the output shall also increase by the same percentage.
2. *Additivity*: Sum of the resources used by different activities must be equal to the total quantity of resources used by each activity for all the resources individually or collectively.
3. *Divisibility*: The variables are not restricted to integer values.
4. *Certainity or Deterministic*: Coefficients in the objective function and the constraints are completely known and do not change during the period under study in all the problems that are considered.
5. *Finiteness*: Variables and constraints are finite in number.

6. *Optimality*: In a linear programming problem we determine the decision variables so as to optimise the objective function of the LPP.
7. The problem involves only one objective function namely profit maximisation or cost minimisation.

Convex set A set C in an n-dimensional space is said to be convex if for any two points x_1 and x_2 in the set C, the line joining x_1 and x_2 is also in the set C.

Examples The following diagrams represent convex/non-convex sets.

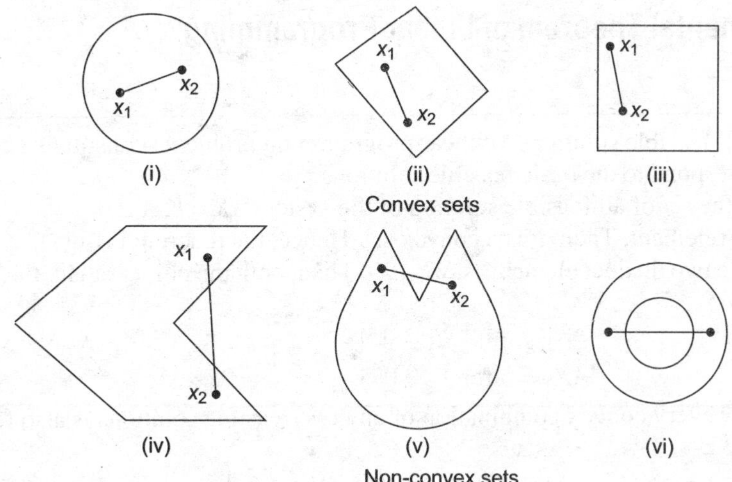

figure 2.1 **Convex and non-convex sets**

Note that the set C containing only single point is convex. Also, the expression $\lambda x_2 + (1 - \lambda)x_1$ is called the convex combination of x_1, x_2 for a given λ.

Note that the intersection of two convex sets is also convex.

Definition A point x in a convex set C is called an *extreme point* if x cannot be expressed as a convex combination of any two distinct points x_1 and x_2 in C. Mathematically, a point x is an extreme point of a convex set if there do not exist other points x_1, x_2 with $x_1 \neq x_2$ such that $x = \lambda x_2 + (1 - \lambda)x_1$, $0 < \lambda < 1$.

Example Consider a polygon, which is a convex sets having the extreme points as their vertices.

Definition A convex combination of a finite number of points $x_1, \cdots, x_n$ is defined as a point x such that $x = \lambda_1 x_1 + \cdots + \lambda_n x_n$ with $\lambda_1, \cdots, \lambda_n \geq 0$ and $\lambda_1 + \cdots + \lambda_n = 1$.

Definition The *convex hull* $C(X)$ of any given set of points X is the set of all convex combinations of sets of points from X. Precisely, convex hull is the smallest convex set containing a point X.

Example If X is the eight vertices of a cube, then the convex hull $C(X)$ is the whole cube.

Definition If the set X consists of a finite number of points, the convex hull of X is called a *convex polyhedran.*

Example Convex hull of the eight vertices of a cube is a convex polyhedran.

Definition A convex cone is a cone which is convex. A *simplex* is an n-dimensional convex polyhedran having exactly $(n + 1)$ vertices.

Example In two dimensional space the simplex is a triangle.

2.3.6 Fundamental Theorem of Linear Programming

Theorem 2.1

The collection of all feasible solutions to linear programming problems constitute a convex set whose extreme points correspond to the basic feasible solutions.

Proof: Let F be the set of all feasible solution of the system $AX = b, X \geq 0$.

Case (i): F has only element. Then, F is a convex set. Hence, the theorem is true.

Case (ii): Let F have two distinct elements say X_1, X_2. Then, both the points satisfy the equation $AX = b$. That is

$$\left.\begin{aligned} AX_1 &= b \text{ for } X_1 \geq 0 \\ AX_2 &= b \text{ for } X_2 \geq 0 \end{aligned}\right\} \tag{2.3.6}$$

We have to show every convex combination of any two feasible solutions is also feasible. Define a point X_0 as

$$X_0 = \lambda X_1 + (1 - \lambda)X_2,\ 0 \leq \lambda \leq 1$$

We show that X_0 satisfies the equation $AX = b, X \geq 0$

Now,
$$\begin{aligned} AX_0 &= A[\lambda X_1 + (1 - \lambda)\, X_2] \\ &= \lambda AX_1 + (1 - \lambda)\, AX_2 \\ &= \lambda b + 1\,(1 - \lambda)b \\ &= b \text{ (by (2.3.6))} \end{aligned}$$

Also, since $0 \leq \lambda \leq 1$ and $X_1, X_2 \geq 0$, then $X_0 \geq 0$. Hence $X_0 \in F$ implying F is convex.

Next, we show that the extreme points corresponds to the basic feasible solutions. Suppose that $X = [X_B, 0]$ is a basic feasible solution, where X_B is a $m \times 1$ vector, such that for a non-singular submatrix B of A,

$$BX_B = b$$

Let us suppose X be a point of F such that for any two points X_1 and $X_2 \in F$,

$$X = \lambda X_1 + (1 - 1\,\lambda)X_2,\ 0 < \lambda < 1$$

Let $X_1 = [u_1, v_1]$, $X_2 = [u_2, v_2]$ where u_1, u_2 are $m \times 1$ vectors and v_1, v_2 are $(n - m) \times 1$ vectors. Then,

$$[X_B, 0] = \lambda\,[u_1, v_1] + (1 - \lambda)\,[u_2, v_2]$$

$$\Rightarrow \quad X_B = \lambda u_1 + (1 - \lambda)\, u_2 \text{ and}$$

$$0 = \lambda v_1 + (1 - \lambda)\, v_2,\ 0 < \lambda < 1$$

Since X_1, X_2 are feasible solutions, therefore $u_1, u_2, v_1, v_2 \geq 0$.

Now, $0 < \lambda < 1$ and $0 = \lambda v_1 + (1 - \lambda)v_2$. Therefore, we must have $v_1 = v_2 = 0$. Thus, $X_1 = [u_1, 0]$, $X_2 = [u_2, 0]$. Since X_1 and X_2 satisfy $AX = b$, we have $Bu_1 = b$, $Bu_2 = b$. Also, since $BX_B = b$ and since expressions of b as a linear combination of basic vectors must be unique, therefore $u_1 = u_2 = X_B$.

Hence, $X = X_1 = X_2$ which is a contradiction to the hypothesis that $X_1 \neq X_2$. Here, X is an extreme point of F. Hence the theorem.

Remark 2.2 There is only one extreme point for a given feasible solution and one basic feasible solution corresponds to any one extreme point. Hence, the number of extreme points of the feasible region is finite and that it cannot exceed the number of its basic solutions. So, the maximum number of extreme points is given by

$$(m + n)C_m = \frac{(m+n)!}{n!\,m!}.$$

Next, we quote the statement of a theorem without proof.

Theorem 2.3

If the convex set of the feasible solutions of $AX = b$, $X \geq 0$ is a convex polyhedran, then at least one of the extreme points gives an optimal solution. If the optimal solution occurs at more than one extreme point, the value of the objective function will be the same for all convex combinations of these extreme points.

2.3.7 Fundamental Properties of Solutions

Theorem 2.4 Reduction of a Feasible Solution to a Basic Feasible Solution

If an LPP has a feasible solution, then it also has a basic feasible solution.

Proof: Consider the LPP to determine X so as to maximise $z = CX$, where $C, X^T \in R^n$ subject to the constraints: $AX = b$, $X \geq 0$, where A is an $m \times n$ real-matrix and b, C are $m \times 1$, $1 \times n$ real-matrices, respectively. Let the rank of the matrix A, denoted by $\rho(A)$ be m. As a feasible solution exist, we must have

$$\rho(A, b) = \rho(A) \text{ and } m < n.$$

Let $X = (x_1, x_2, \cdots, x_n)$ be a feasible solution so that $x_j \geq 0$ for all j. That is, suppose X has p positive components and the remaining $n - p$ components be zero. Relabel the positive components as the first p components and assume that the columns of A have been relabelled accordingly. Then, $a_1 x_2 + a_2 x_2 + \cdots + a_p x_p = b$ where $a_1, \cdots, a_p$ are the first p columns of A. The vectors $a_1, \cdots a_p$ are either linearly independent or linearly dependent.

Case (i): Let $a_1, \cdots, a_p$ be linearly independent vectors. When $p \leq m$.

If $p = m$, the given solution is a non-degenerate basic feasible solution with $x_1, x_2, \cdots, x_p$ as the basic variables.

If $p < m$, then the set $\{a_1, \cdots, a_p\}$ can be extended to $\{a_1, \cdots, a_p, a_{p+1}, \cdots, a_m\}$ to form a basis for the columns of A. Then, we have

$$a_1 x_1 + \cdots + a_m x_m = b$$

where $x_j = 0$ for $j = p + 1, \cdots, m$.

In this case, we have a degenerate basic feasible solution with $m - p$ of the basic variables as zero.

Case (ii): Let $a_1, \cdots, a_p$ be linearly dependent vectors. Obviously, $p > m$. Let $\{\alpha_1, \cdots, \alpha_p\}$ be a set of constants (not all zero) such that

$$\alpha_1 a_1 + \cdots + \alpha_p a_p = 0.$$

Suppose that for any index r, $\alpha_r \neq 0$. Then,

$$\alpha_1 a_1 + \alpha_2 a_2 + \cdots + \alpha_r a_r + \alpha_{r+1} a_{r+1} \cdots \alpha_p a_p = 0$$

$$\Rightarrow \quad \alpha_r a_r = -[\alpha_1 a_1 + \alpha_2 a_2 + \cdots + \alpha_{r+1} a_{r+1} + \cdots + \alpha_p a_p]$$

$$\Rightarrow \quad a_r = -\frac{1}{\alpha_r}[\alpha_1 a_1 + \cdots \alpha_{r-1} a_{r-1} + \alpha_{r+1} a_{r+1} + \cdots + \alpha_p a_p]$$

That is, $a_r = -\sum_{j \neq r}^{p} \frac{\alpha_j a_j}{\alpha_r}$

So, $\quad a_1 x_1 + a_2 x_2 + \cdots + a_p x_p = b$

$$\Rightarrow \quad \sum_{j \neq r}^{p} a_j x_j + a_r x_r = b$$

$$\Rightarrow \quad \sum_{j \neq r}^{p} a_j x_j + \left[-\sum_{j \neq r}^{p} \frac{\alpha_j a_j}{\alpha_r}\right] x_r = b$$

$$\Rightarrow \quad \sum_{j \neq r}^{p} \left(x_j - \frac{\alpha_j}{\alpha_r} x_r\right) a_j = b$$

Thus, we have a solution with not more than, $p-1$ non-zero components. In order that these should be non-zero, choose a_r such that

$$x_j - \frac{\alpha_j}{\alpha_r} x_r \geq 0, \text{ for } j \neq r.$$

This requires that either $\alpha_j = 0$ or

$$\frac{x_j}{\alpha_j} \geq \frac{x_r}{\alpha_r}, \text{ if } \alpha_j > 0$$

$$\frac{x_j}{\alpha_j} \leq \frac{x_r}{\alpha_r} \text{ if } \alpha_j < 0$$

Thus, if we select a_r such that

$$\frac{x_r}{\alpha_r} = \min_j \left\{\frac{x_j}{\alpha_j}, \alpha_j > 0\right\}$$

then, for each of the $p-1$ variables $x_j - x_r \dfrac{\alpha_j}{\alpha_r}$ is non-negative, and so we have a feasible solution with not more than $p-1$ non-zero components.

Consider this feasible solution with atmost $p-1$ non-zero components. If the corresponding set of $p-1$ columns of A is linearly independent *case (i)* applies and we get a basic feasible solution. If case (ii) holds, repeat the process to arrive at a feasible solution with not more than $p-2$ non-zero components. Keep on repeating the process until we arrive at a feasible solution with an associated set of column vectors of A which is linearly independent. The discussion of *case (i)* applies and we get a basic feasible solution.

Note: There exists only finite number of basic feasible solutions to a LPP.

Example

Let $x_1 = 2, x_2 = 4, x_3 = 1$ be a feasible solution to the system of equations

$$2x_1 - x_2 + 2x_3 = 2$$
$$x_1 + 4x_2 = 18$$

Reduce the given feasible solution to a basic feasible solution.

Solution: The given system of equations can be written as $AX = b$, where

$$A = \begin{pmatrix} 2 & -1 & 2 \\ 1 & 4 & 0 \end{pmatrix}, b = \begin{pmatrix} 2 \\ 8 \end{pmatrix}$$

The rank $(A) = 2$, there are only two linearly independent columns of A. Thus, if the column vectors of A be denoted by a_1, a_2, a_3 respectively then

$$\alpha_1 a_1 + \alpha_2 a_2 + \alpha_3 a_3 = 0$$

where α_i's are scalars not all zero. If we let $\alpha_3 = -1$, then $a_3 = \alpha_1 a_1 + \alpha_2 a_2$.

Now, $\quad 2x_1 - x_3 + 2x_3 = a_1 x_1 + a_2 x_2 + a_3 x_3$

$\Rightarrow \quad a_1 = 2, a_2 = -1, a_3 = 2$

So, $\quad 2 = a_3 = \alpha_1 a_1 + \alpha_2 a_2 = 2\alpha_1 - \alpha_2 \qquad (2.3.7)$

Also, $\quad x_1 + 4x_2 = a_1 x_1 + a_2 x_2 + a_3 x_3$

$\Rightarrow \quad a_1 = 1, a_2 = 4, a_3 = 0$

In this case

$$0 = a_3 = \alpha_1 a_1 + \alpha_2 a_2 = \alpha_1 + 4\alpha_2 \qquad (2.3.8)$$

Solving (2.3.7) and (2.3.8) we get

$$\alpha_1 = 8/9, \alpha_2 = -2/9$$

Now, $\quad \dfrac{x_r}{\alpha_r} = \min_j \left\{ \dfrac{x_j}{\alpha_j}, \alpha_j > 0 \right\}$

$=\min \left\{ \dfrac{x_1}{\alpha_1}, \text{ where } \alpha_1 > 0 \right\}$, since α_2, α_3 are negative

$$= \min \left\{ \frac{2}{8/9} \right\}$$

$$= 9/4.$$

This indicates that vector a_1 can be removed. Hence, the new solution with not more than two non-zero variables is given by

$$\hat{x}_1 = x_1 - \frac{x_r}{\alpha_r}\alpha_1 = 2 - \frac{9}{4} \times \frac{8}{9} = 0$$

$$\hat{x}_2 = x_2 - \frac{x_r}{\alpha_r}\alpha_2 = 4 - \frac{9}{4} \times \left(\frac{-2}{9} \right) = \frac{9}{2}$$

$$\hat{x}_3 = x_3 - \frac{x_r}{\alpha_r}\alpha_3 = 1 - \frac{9}{4} \times (-1) = \frac{13}{4}.$$

Thus, a basic feasible solution to the given LPP is $x_2 = 9/2$, $x_3 = 13/4$, with $x_1 = 0$ (non-basic).

If we let $\alpha_3 = 1$, another basic feasible solution is obtained as $x_1 = 26/9, x_2 = 34/9, x_3 = 0$ (non-basic).

Theorem 2.5 Replacement of a Basis Vector

Let an LPP have a basic feasible solution. If we drop one of the basis vectors and introduce a non-basis vector in the basis set, then the new solution obtained is also a basic feasible solution.

Proof: Let us consider the LPP

$$\text{Maximise } z = CX$$

subject to $$AX = b,\ X \geq 0$$

where $C, X^T \in R^n$, A, b are $m \times n$ and $m \times 1$ real matrices, respectively. Let the rank of A be m. Suppose there exist a basic feasible solution X_B so that

$$BX_B = b,\ X_B \geq 0 \tag{2.3.9}$$

where, B forms a basis set for the column vectors of A. Clearly, for any column vector $a_j \in A$. We have $a_j = y_{1j}\, b_1 + y_{2j}\, b_2 + \cdots + y_{mj}\, b_m = By_j$ where $b_i \in B$, y_{ij} are suitable scalars.

We know that if the basis vector b_r for which the coefficient y_{rj} is non-zero, is replaced by $a_j \in A$, then the new set of vectors also forms a basis. Now, for $y_{rj} \neq 0$,

$$a_j = y_{1j}\, b_1 + y_{2j}\, b_2 + \cdots + y_{rj}\, b_r + \cdots + y_{mj}\, b_m$$

$$\Rightarrow \quad y_{rj}\, b_r = a_j - [y_{1j}\, b_1 + y_{2j}\, b_2 + \cdots + y_{r-1j}\, b_{r-1} + y_{r+1j}\, b_{r+1} + \cdots + y_{mj}\, b_{mj}]$$

$$\Rightarrow \quad b_r = \frac{a_j}{y_{rj}} - \frac{1}{y_{rj}} [y_{1j}\, b_1 + y_{2j}\, b_2 + \cdots + y_{mj}\, b_m]$$

That is, $$b_r = \frac{a_j}{y_{rj}} - \sum_{\substack{i=1 \\ i \neq r}}^{m} \left(\frac{y_{ij}}{y_{rj}}\right) b_i$$

Therefore, (2.3.9) gives

$$b = BX_B = b_1\, X_{B_1} + b_2\, X_{B2} + \cdots + b_m\, X_{B_m}$$

$$= b_1\, X_{B_1} + b_2\, X_{B_2} + \cdots + b_{r-1}\, X_{B_{r-1}} + \left[\frac{a_j}{y_{rj}} - \sum_{\substack{i=1 \\ i \neq r}}^{m} \left(\frac{y_{ij}}{y_{rj}}\right) b_i\right] X_{B_r}$$

$$+ b_{r+1}\, X_{B_{r+1}} + \cdots + b_m\, X_{B_m}$$

$$\Rightarrow \quad b = \sum_{\substack{i=1 \\ i \neq r}}^{m} X_{B_i}\, b_i + X_{B_r} \left[\frac{a_j}{y_{rj}} - \sum_{\substack{i=1 \\ i \neq r}}^{m} \left(\frac{y_{ij}}{y_{rj}}\right) b_i\right]$$

$$= \sum_{\substack{i=1 \\ i \neq r}}^{m} \left[X_{B_i} - X_{B_r} \frac{y_{ij}}{y_{rj}}\right] b_i + \frac{X_{B_r}}{y_{rj}}\, a_j$$

Thus, the new basic solution is $\hat{X}_B$ having its components as the variables

$$\hat{X}_{B_i} = X_{B_i} - X_{B_r}\frac{y_{ij}}{y_{rj}} \quad (i = 1, \cdots, m, i \neq r)$$

and, $$\hat{X}_{B_r} = \frac{x_{B_r}}{y_{rj}}$$

Next, we show $\hat{X}_B$ is set of feasible basic solution.

Case (i): $X_{B_r} = 0$. In this case the new set of basic variables is obviously non-negative, since we have assumed the existence of a basic feasible solution X_B.

Case (ii): $X_{B_r} \neq 0$. In this case we must have $y_{rj} > 0$. This requires that for the remaining y_{ij} $(i \neq r)$ either

$$y_{ri} = 0 \text{ for } i \neq r \text{ or, } \frac{X_{B_i}}{y_{ij}} \geq \frac{X_{B_r}}{y_{rj}} \text{ for } y_{ij} > 0, i \neq r$$

and, $$\frac{X_{B_i}}{y_{ij}} \leq \frac{X_{B_r}}{y_{rj}} \text{ for } y_{ij} < 0, i \neq r.$$

So, if we select the index r $(y_{rj} \neq 0)$ in such a way that

$$\frac{X_{B_r}}{y_{rj}} = \min_i \left\{\frac{X_{B_i}}{y_{ij}}, y_{ij} > 0, i \neq r\right\}$$

Then, the new set of basic variables is non-negative and hence the basic solution $\hat{X}_B$ is feasible.

Note: After replacement of basis vector, the new basis matrix is

$$\hat{B} = \left(\hat{b}_1, \hat{b}_2, \cdots \hat{b}_m\right)$$

where, $\hat{b}_i = b_i$ for $i \neq r$, $\hat{b}_r = a_j$; then the new basic feasible solution is

$$\hat{X}_B = \hat{B}^{-1} b$$

where, $$\hat{X}_{B_i} = X_{B_i} - X_{B_r}\left(\frac{y_{ir}}{y_{rj}}\right), i \neq r,$$

$$\hat{X}_B = \frac{X_{B_r}}{y_{rj}} \text{ are the basic variables.}$$

Definition: Let $X_B = B^{-1} b$ be a basic feasible solution to the LPP. Maximise $z = CX$ where $AX = b$, and $X \geq 0$, let C_B be the cost vector corresponding to X_B. For each column vector a_j in A, which is not a column vector of B, let $a_j = \sum_{j=1}^{m} a_{ij} b_i$.

Then, the number $z_j = \sum_{i=1}^{m} C_{B_i} a_{ij}$ is called the evaluation corresponding to a_j and the number $(z_j - c_j)$ is called the net evaluation corresponding to a_j where c_j is the jth component of C.

Definition: *(Improved basic feasible solution).* Let X_B and $\hat{X}_B$ be two basic feasible solutions to the standard LPP. Then, $\hat{X}_B$ is said to be an improved basic feasible solution as compared to X_B, if

$$\hat{C}_B \hat{X}_B \geq C_B \hat{X}_B$$

where, $\hat{C}_B$ is the cost component corresponding to $\hat{X}_B$.

Theorem 2.6 Initial Basic Feasible Solution IBFS

Let X_B be a basic feasible solution to a LPP

$$\text{Maximise } z = CX$$

subject to $$AX = b, X \geq 0$$

Let $\hat{X}_B$ be another basic feasible solution obtained by admitting a non-basis column vector a_j in the basis, for which the net evaluation $z_j - c_j$ is negative. Then, $\hat{X}_B$ is an IBFS.

Proof: Let the LPP be

$$\text{Maximise } z = CX,\ C, X^T \in R^n$$

subject to $$AX = b, X \geq 0, b^T \in R^m$$

where, A is an $m \times n$ real matrix. Given X_B is a basic feasible solution. Let $z_0 = C_B X_B$. Let $\hat{a}_j$ be the column vector introduced in the basis, such that $z_j - c_j < 0$. Let b_r be the vector removed from the basis and let $\hat{X}_B$ be the new basic feasible solution then

$$\hat{X}_{B_i} = X_{B_i} - X_{B_r} \frac{y_{ij}}{y_{rj}}$$

$$\hat{X}_{B_r} = \frac{X_{B_r}}{y_{rj}}$$

Hence, the new value of the objective function is

$$\hat{z} = \sum_{i=1}^{m} \hat{C}_{B_i} \hat{X}_{B_i}$$

$$= \sum_{i=1}^{m} C_{B_i} \left(X_{B_i} - X_{B_r} \frac{y_{ij}}{y_{rj}} \right) + \hat{C}_{B_r} \frac{X_{B_r}}{y_{rj}}$$

$$= \sum_{i=1}^{m} C_{B_i} \left(X_{B_i} - X_{B_r} \frac{y_{ij}}{y_{rj}} \right) + c_j \frac{X_{B_r}}{y_{rj}} \quad \text{since } \hat{C}_{B_r} = c_j$$

$$= z_0 - (z_j - c_j) \cdot \frac{X_{B_r}}{y_{rj}}$$

$$> z_0 \qquad \left[\because \frac{X_{B_r}}{y_{rj}} > 0\right]$$

Hence, the new basic feasible solution $\hat{X}_B$ gives an improved value of the objective function.

Corollory 2.7

If $z_j - c_j = 0$ for at least one j, for which $y_{ij} > 0 \;\; i = 1, 2, \cdots, m$, then another basic feasible solution is obtained which gives an unchanged value of the objective function.

Proof: From Theorem 2.6 we have,

$$\hat{z} = z_0 - (z_j - c_j)\frac{X_{B_r}}{y_{rj}}$$

$$= z_0 \qquad \text{since } z_j - c_j = 0$$

Theorem 2.8 Unbounded Solution

Let there exist a basic feasible solution to a given LPP. If for at least one j, for which $y_{ij} < 0$ $(i = 1, 2, \cdots, m)$ and $z_j - c_j$ is negative, then there does not exist any optimum solution to this LPP.

Proof: Let a given LPP be to determine X so as to maximise $z = CX$, $C, X^T \in R^n$, subject to the constraints $AX = b$, $X \geq 0$, $b^T \in R^m$ where, A is an $m \times n$ real matrix.

Let a basic feasible solution to this problem be X_B, so that $BX_B = b$ and $X_B \geq 0$

with, $$z_0 = C_B X_B = \sum_{i=1}^{m} C_{B_i} X_{B_i}$$

Write

$$b = BX_B + \xi a_j - \xi a_j,\; a_j \in A,\; \xi \text{ is a scalar.}$$

$$= \sum_{i=1}^{m} X_{B_i} b_i + \xi a_j - \xi \sum_{i=1}^{m} y_{ij} b_i$$

$$= \sum_{i=1}^{m} (X_{B_i} - \xi y_{ij}) b_i + \xi a_j$$

If $\xi > 0$, then $(X_{B_i} - \xi y_{ij}) > 0$, since $y_{ij} < 0$. That is, there exists a feasible solution whose $(m + 1)$ components may be strictly positive. But in general, it may not be a basic solution. The value of the objective function for the $(m + 1)$ variable is given by

$$\hat{z} = \sum_{i=1}^{m} C_{B_i}(X_{B_i} - \xi y_{ij}) + \xi c_j$$

$$= \sum_{i=1}^{m} C_{B_i} X_{B_i} - \xi\left(\sum_{i=1}^{m} C_{B_i} y_{ij} - c_j\right)$$

$$= z_0 - \xi(z_j - c_j)$$

But $z_j - c_j < 0$ and $\xi > 0$

$$\therefore \qquad \hat{z} \to \infty \text{ as } \xi \to \infty.$$

Hence, there is no limit to the optimum value of z and so there exists an unbounded solution.

Remark: If for some a_j for which $z_j - c_j > 0$ and $y_{ij} \le 0$, $i = 1, 2, \cdots, m$ then there exists a feasible solution to the LPP such that the $(m + 1)$ components of the solution may be strictly positive and

$$\hat{z} \to -\infty \text{ as } \xi \to +\infty$$

If $y_{ij} \le 0$ for at least one j for which $z_j - c_j = 0$, then $\hat{z} = z_0$.

Theorem 2.9 Conditions of Optimality

A sufficient condition for a basic feasible solution to an LPP to be an optimum (maximum) is that $z_j - c_j \ge 0$ for all j for which the column vector $a_j \in A$ is not in the basis B.

Proof: Let the LPP be to determine X so as to

$$\text{Maximise } z = CX,\ C, X^T \in R^n$$

subject to $$AX = b,\ X \ge 0$$

A, b are $m \times n$ and $m \times 1$ real matrices, respectively. Let the rank of A be m, so that we can choose $m \times m$ submatrix B of A as a basis matrix.

Let us assume that there exists a basic feasible solution X_B to this LPP. Let C_B be the cost vector corresponding to X_B.

$\therefore$ $$BX_B = b,\ X_B \ge 0,\ Z_0 = C_B X_B.$$

For all those j, for which $a_j \notin B$, we are given that $z_j - c_j \ge 0$. Let $a_j = b_j$ for all such j for which $a_j \in B$. Then, $y_j = B^{-1} b_j$, since $y_j = B^{-1} a_j = e_j$, the unit vector.

And, $$z_j - c_j = C_B y_j - c_j = C_B e_j - c_j = C_{B\,j} - c_j = 0 \qquad (\because C_{B_j} = c_j)$$

Thus, $z_j - c_j \ge 0$ for all j for which $a_j \in A$. Now, let X be a feasible solution. Then,

$$\sum_{j=1}^{n} (z_j - c_j)\, X_J \ge 0, \text{ since } X_j \ge 0$$

$$\Rightarrow \quad \sum_{j=1}^{n} z_j X_j \ge \sum_{j=1}^{n} c_j X_j$$

$$\Rightarrow \quad \sum_{j=1}^{n} C_B Y_j X_j \ge \sum_{j=1}^{n} C_j X_j, \text{ since } z_j = C_B y_j$$

$$\Rightarrow \quad \sum_{i=1}^{m} C_{B_i} \sum_{j=1}^{n} y_{ij} X_j \ge \sum_{j=1}^{n} c_j X_j, \text{ since } C_B y_j = \sum_{i=1}^{n} C_{B_i}. \tag{2.3.10}$$

for all j for which $a_j \notin B$. Now, since

$$X_B = B^{-1} (AX) = (B^{-1} A)\, X = YX$$

or, $$X_{B_i} = \sum_{j=1}^{n} y_{ij} X_j,\ i = 1, 2, \cdots, m$$

Equation (2.3.10) can be written as

$$\sum_{i=1}^{m} Y_{B_i} C_{B_i} \ge \sum_{j=1}^{n} c_j X_j$$

or $$C_B X_B \ge CX,\ X z_0 \ge z^*$$

where z^* is the value of the objective function for the feasible solution. Here, z_0 is an optimum for that basic feasible solutions for which $z_j - c_j \geq 0$ for all j such that $a_j \notin B$.

Combining Theorems 2.6, 2.8 and 2.9, we get:

Theorem 2.10

A necessary and sufficient condition for a basic feasible solution to a LPP to be an optimum is that $z_j - c_j \geq 0$ for all $j \ni$ such that $a_j \notin B$.

2.4 GRAPHICAL SOLUTIONS TO A LINEAR PROGRAMMING PROBLEM

A linear programming problem involving two decision variables can be conveniently solved graphically. For models with three or more variables the graphical method is impractical. This method is easy to understand and useful to explain the basic concepts of LPP.

The redundant constraints are automatically eliminated from the system. Multiple solutions, unbounded solutions and infeasible solutions are clearly defined using graphical method. Sensitivity analysis can be represented by mean of graphs also.

Consider the LPP

$$\text{Maximise } z = 3x_1 + 2x_2$$

subject to

$$x_1 + 2x_2 \leq 6$$
$$2x_1 + x_2 \leq 8$$
$$-x_1 + x_2 \leq 1$$
$$x_2 \leq 2$$

and,

$$x_1, x_2 \geq 0$$

The first step in the graphical method is to plot the feasible solution space that satisfies all the constraints simultaneously.

The non-negativity restrictions $x_1 \geq 0$, $x_2 \geq 0$ confine all the feasible values to the first quadrant (which is defined by the $X_1\, X_2$ axis). The space enclosed by the remaining constraints is determined by first replacing $\leq$ (or $\geq$) by $=$ for each constraint thus yielding a straight line equation. Each straight line is plotted on the ($X_1\, X_2$) plane, and the region in which each constraint holds when the inequality is activated is directed by the arrow on the associated line. An easy way to determine the direction of the arrow is to use the origin as a reference point. If (0, 0) satisfies the inequality, the feasible direction should include the origin. Otherwise, it should be on the opposite side. If the constraint passes through the origin, in that case a reference point does not lie on the associated straight line.

A linear equation will always represent a straight line and a straight line can be completely described by any two points that lie on the line. How to get those two points? For example, consider the equation $x_1 + 2x_2 = 6$. Put $x_1 = 0$, then $x_2 = 3$. If $x_2 = 0$, then $x_1 = 6$, So, the required points for this equation are (0, 3) and (6, 0). Similarly, we can obtain for all other constraints also. Figure 2.2. depicts the required solution space.

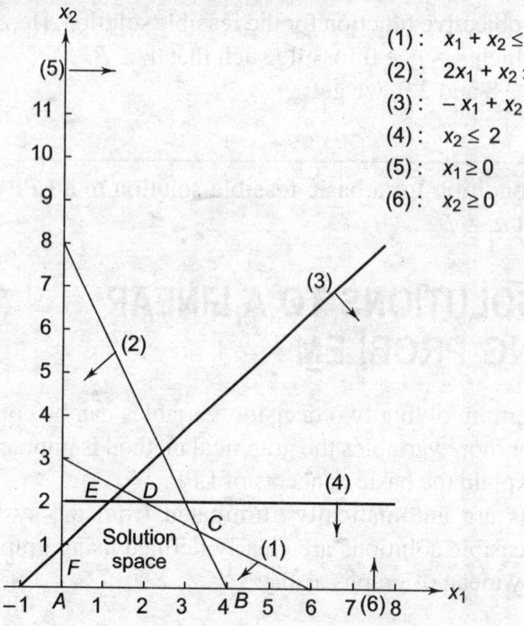

figure 2.2

The solution space is *ABCDEF* as shown in Fig. 2.2.

All the points in the solution space will be a feasible for the given problem. That is, each point within or on the boundary of the solution space satisfies all the constraints. Although there is infinity of feasible point in the solution space, the optimum solution can be determined by observing the direction in which the objective function increases.

Next, we give the algorithm to find the optimum solution of a LPP by graphical method.

Working procedure for graphical method:

Step 1: Formulate the appropriate LPP.

Step 2: Draw the graph of the LPP.

Step 3: Obtain a feasible region (a region which is common to all the constraints of the LPP which is a convex region).

Step 4: Obtain the solution points (the corner points of the feasible region)

Step 5: Calculate the values of objective function at the solution points.

Step 6: For maximisation problem, the optimum solution is the solution point which gives the maximum value of the objective function and for minimisation problems the optimum solution is the solution point that gives the minimum value of the objective function.

Example 1

Apply graphical method to solve the LPP

$$\text{Maximise } z = x_1 - 2x_2$$
$$\text{subject to } -x_1 + x_2 \le 1$$
$$6x_1 + 4x_2 \ge 24$$
$$0 \le x_1 \le 5,\ 2 \le x_2 \le 4$$

Solution: Since $x_1, x_2 \geq 0$ consider the $X_1 \, OX_2$ plane. Consider the inequalities as equalities.

$$-x_1 + x_2 = 1 \quad \Rightarrow \quad (0, 1), (-1, 0)$$
$$6x_1 + 4x_2 = 24 \quad \Rightarrow \quad (4, 0), (0, 6)$$
$$x_1 = 5$$
$$x_1 = 0 \text{ which is } x_2\text{-axis}$$
$$x_2 = 0 \text{ which is } x_1\text{-axis}$$
$$x_2 = 2$$
$$x_2 = 4$$

The graph representing the above LPP is given in Fig. 2.3.

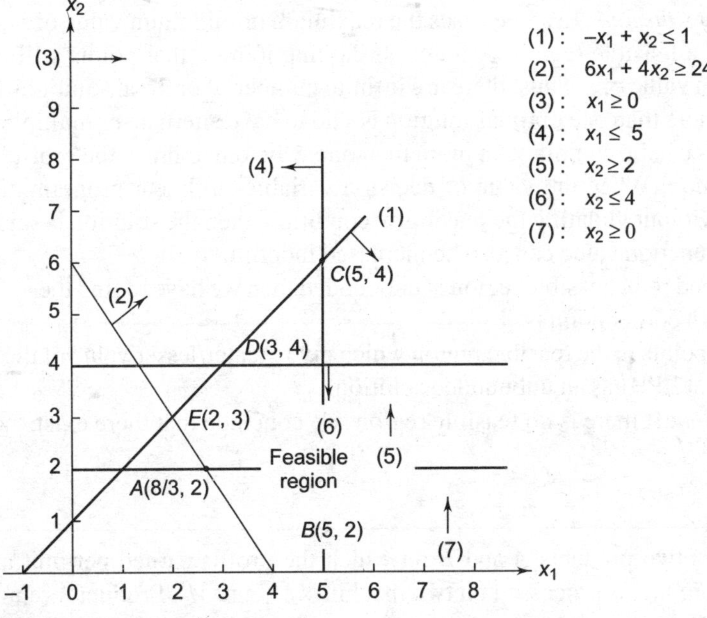

figure 2.3

The solution space (feasible region) is *ABCDE*. The co-ordinates of the corner points of the feasible region are obtained and is given in Fig. 2.3. The values of the objective function at the corner points are

$$z = x_1 - 2x_2$$

table 2.10

Vertex	*Value of z*
A(8/3, 2)	− 4/3
B(5, 2)	1
C(5, 4)	− 3
D(3, 4)	− 5
E(2, 3)	− 4

Since the problem is of maximisation type the optimal solution is

$$x_1 = 5, x_2 = 2 \text{ with maximum of } z = 1$$

Redundant constraint: In a given LPP if any constraint does not affect the feasible region, then the constraint is said to be a redundant constraint.

2.4.1 Some Exceptional Cases

So far, we have seen linear programming problem which may be rightly called as "well behaved" problems for which the solutions was obtained. But there may be some problems for which no solution exists or for which the only solution obtained is an unbounded one.

1. *Infinite number of solutions*: In some cases the maximum or minimum value of z occurs at more than one corner point of a feasible region. A point on the line joining that points will also give the same maximum/minimum value of z. Thus, there are infinite number of optimal solutions for this LPP.

An LPP having more than one optimal solution is said to have alternate or multiple optimal solutions. That is the resources can be combined in more than one way to maximise the profit.

2. *Unbounded solution*: When the value of decision variables in linear programming is permitted to increase infinitely without violating the feasibility condition, then the solution is said to be unbounded. Here the objective function value can also be increased indefinitely.

In graphical method if the feasible region is unbounded then we have to find the value of the objective function at the known corner points.

If there are some points in the feasible region which give greater/lesser value of the objective function then we conclude that LPP has an unbounded solution.

3. *No feasible solution*: If there is no feasible region, we conclude that there exists no feasible solution to the given LPP.

Example 2

A firm manufactures two products A and B on which the profits earned per unit are Rs 3 and Rs 4, respectively. Each product is processed on two machines M_1 and M_2. Product A requires one minute of processing time on M_1 and two minutes of processing on M_2 while processing of product B requires one minute on M_1 and one minute on M_2. Machine M_1 is available for not more than 7 hours 30 minutes while machine M_2 is available for 10 hours during any working day. Find the number of units of products A and B need to be manufactured to get maximum profit. Formulate the above as a LPP and solve by graphical method.

Solution: Let the firm decide to manufacture x_1 units of product A and x_2 units of product B. To produce these units of products A and B, it requires $x_1 + x_2$ hours of processing time on M_1, $2x_1 + x_2$ hours of processing time on M_2. But the availability of these two machines M_1 and M_2 are 450 minutes and 600 minutes, respectively, therefore the constraints are

$$x_1 + x_2 \leq 450$$

$$2x_1 + x_2 \leq 600$$

$$x_1, x_2 \geq 0$$

Since the profit from product A is Rs 3 per unit and from product B is Rs 4 per unit the total profit is Rs $3x_1 + 4x_2$. The objective is to maximise the profit Rs $3x_1 + 4x_2$. Hence, the LPP is

$$\text{Maximise } z = 3x_1 + 4x_2$$
$$\text{subject to } x_1 + x_2 \leq 450$$
$$2x_1 + x_2 \leq 600$$
$$x_1, x_2 \geq 0.$$

Graphical solution: The graph of the LPP is given in Fig. 2.4.

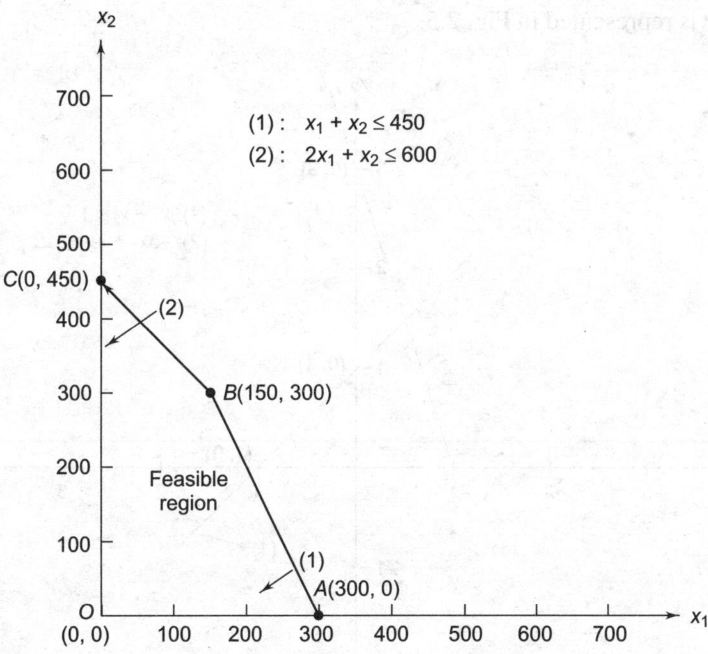

figure 2.4

The feasible region is *OABC*. The vertices of solution space are *O*(0, 0), *A*(300, 0), *B*(150, 300) and *C*(0, 450).

The values of the objective function at the solution points is given by Table 2.11.

table 2.11

Vertex	*Value of z*
O(0, 0)	0
A(300, 0)	900
B(150, 300)	1,650
C(0, 450)	1,800

Since the problem is of maximisation and the maximum value of z is attained at a single vertex, this problem has a unique optimal solution. The optimal solution is

$$x_1 = 0, x_2 = 450 \text{ and maximum of } z = 1800$$

Example 3

Solve the following LPP

$$\text{Maximise } z = x_1 + x_2$$
$$\text{subject to } x_1 + x_2 \leq 1$$
$$-3x_1 + x_2 \geq 3$$
and,
$$x_1, x_2 \geq 0.$$

Solution: The LPP is represented in Fig. 2.5.

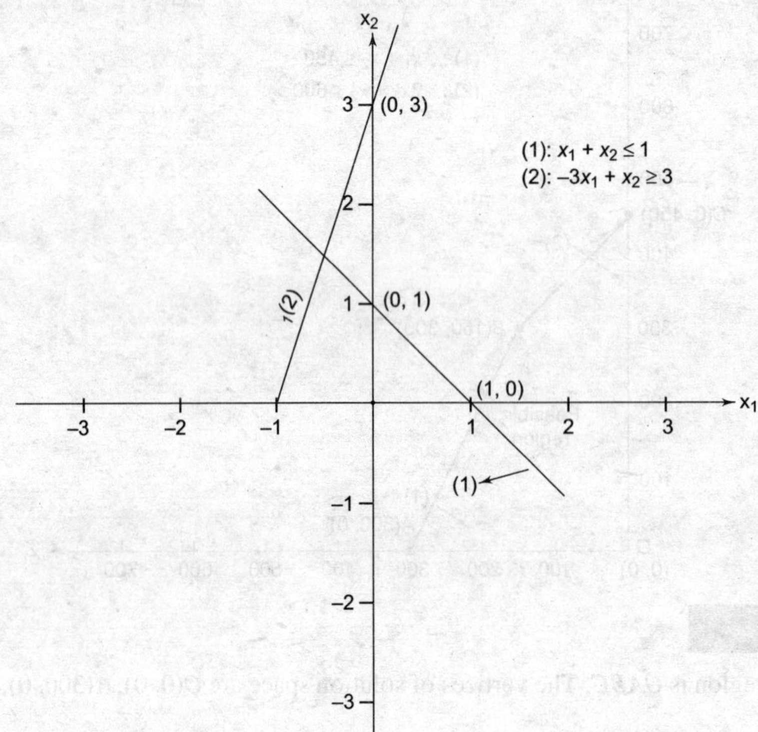

figure 2.5

Here, there exists no feasible region and so there exists no feasible solution to the given problem.

Example 4

Using graphical method, solve the following LPP

$$\text{Maximise } z = 2x_1 + 3x_2$$
$$\text{subject to } x_1 - x_2 \leq 2$$
$$x_1 + x_2 \leq 4$$
$$x_1, x_2 \geq 0.$$

Solution: By using graphical method the feasible region is the shaded portion in the Fig. 2.6 .

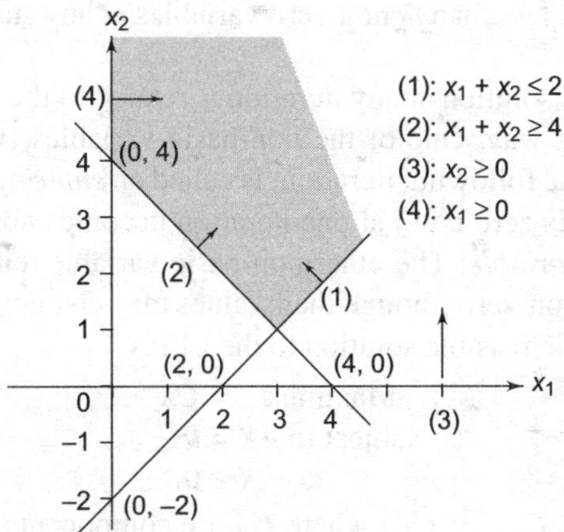

figure 2.6

Here, the feasible region is unbounded. The vertices of the feasible region (in the finite plane) are $A(3, 1)$ and $B(0, 4)$. And, the value of the objective function $z = 2x_1 + 3x_2$ at these vertices ae $z(A) = 9$, $z(B) = 12$. But there are infinite number of points in the feasible region (for example (3, 5), (3, 10), (2, 7), and so on (whose z value is greater than $z(A)$ and $z(B)$. In fact, the maximum value of z occurs at infinity. Hence, this problem has an unbounded solution.

2.5 SIMPLEX METHOD

It is not possible to obtain the graphical solution to the linear programming problem of more than two variables. The analytic solution is also not possible because the tools of analysis are not well studied to handle inequalities. The most commonly used method for locating the optimal vertex is the simplex method which was developed by G. Dantzig in 1947.

The simplex method provides an algorithm (a rule of iterative procedure) which is based on the fundamental theorem of linear programming.

The simplex method provides a systematic algorithm which consists of moving from one basic feasible solution (one vertex) to another in such a way that the value of the objective function at the succeeding vertex is more than at the preceeding vertex. The procedure of jumping from vertex to vertex is repeated. If the objective function is improved at each jump, then no basis can even repeat and no need to go back to vertex already covered. Since the number of vertices is finite, the method leads to an optimal vertex in a finite number of steps or indicates the existence of an unbounded solution.

The simplex method consists of:

(i) having a trial basic feasible solution to the constraint equations.
(ii) testing whether it is an optimal solution or not.
(iii) improving, if required, the first trial solution by a set of rules and repeating the process till an optimal solution is obtained.

The computation procedure requires atmost m (the number of equations) non-zero variables in the solution at any step. In case of less than m non-zero variables at any stage of computations the degeneracy arises in LPP.

Further, note that a feasible solution at any iteration is related to the feasible solution at the successive iteration in the following way. One of the non-basic variables (which is zero) at one iteration becomes basic (non-zero) at the following iteration, is called an *entering variable*. To compensate, one of the basic variables (which is zero now) at one iteration becomes non-basic (zero) at the following iteration, is called a *leaving variable*. The other non-basic variable remains zero, and the other basic variables, in general, remain non-zero (though their values may change).

Definition: Let X_B be a basic feasible solution to the LPP.

$$\text{Maximise } z = CX$$
$$\text{subject to } AX = b$$

and,
$$X \geq 0.$$

Then, the vector $C_B = (C_{B_1}, C_{B_2}, \ldots, C_{B_n})$ where C_{B_i} are components of C associated with the basic variables, is called the *cost vector* associated with the basic feasible solution X_B.

Remark:

1. If a LPP has a feasible solution, then it has a basic feasible solution.
2. There exists only finite number of basic feasible solutions to a LPP.
3. Let a LPP have a feasible solution. If we drop one of the basic variables and introduce another variable in the basis set, then the new solution obtained is also a basic feasible solution.
4. If the net evaluation $z_j - c_j = 0$ for at least one j for which $a_{ij} > 0, i = 1, 2, \ldots, m$ then another basic feasible solution is obtained which gives an unchanged value of the objective function. If for at least one j, for which $a_{ij} \leq 0, i = 1, 2, \ldots, m$ and $z_j - c_j$ is negative, then there does not exist any optimum solution.

The computational procedure: The two fundamental conditions on which the simplex method is based are:

(i) *Condition of feasibility*

It ensures that if the initial (starting) solution is basic feasible than during computation only basic feasible solutions will be obtained.

(ii) *Condition of optimality*

It guarantees that only better solutions will be encountered.

Starting simplex table:

$$c_j: (c_1\ c_2 \ldots c_n\ 0, 0 \ldots 0)$$

table 2.12

C_B	Y_B	X_B	$y_1 = a_1$	$y_2 = a_2$	...	$y_n = a_n$	s_1	s_2 ...	s_n	*Min Ratio*
			Body matrix				**Basic matrix**			
C_{B_1}	Y_{B_1}	X_{B_1}	$y_{11} = a_{11}$	$y_{12} = a_{12}$	...	$y_{1n} = a_{1n}$	1	0	...	0
C_{B_2}	Y_{B_2}	X_{B_2}	$y_{21} = a_{21}$	$y_{22} = a_{22}$	...	$y_{2n} = a_{2n}$	0	1	...	0
⋮	⋮	⋮	⋮	⋮	⋮	⋮	⋮	⋮	⋮	⋮
C_{B_m}	Y_{B_m}	X_{B_m}	$y_{m_1} = a_{m_1}$	$y_{m_2} = a_{m_2}$	...	$y_{mn} = a_{mn}$	0	0	0	1
$z_j - c_j$	z_0		$z_1 - c_1$	$z_2 - c_2$	...	$z_n - c_n$	0	0	...	0

where c_j – row denotes the coefficients of the variables in the objective function.
C_B – column denotes the coefficients of the basic variables in the objective function.
Y_B – column denotes the basic variables.
Body matrix – The coefficients of the non-basic variables in the constraints.
Basis matrix – Coefficients of the basic variables
$z_j - c_j$ – net evaluations.

Flow Diagram of a Simplex Algorithm for Maximisation LPP

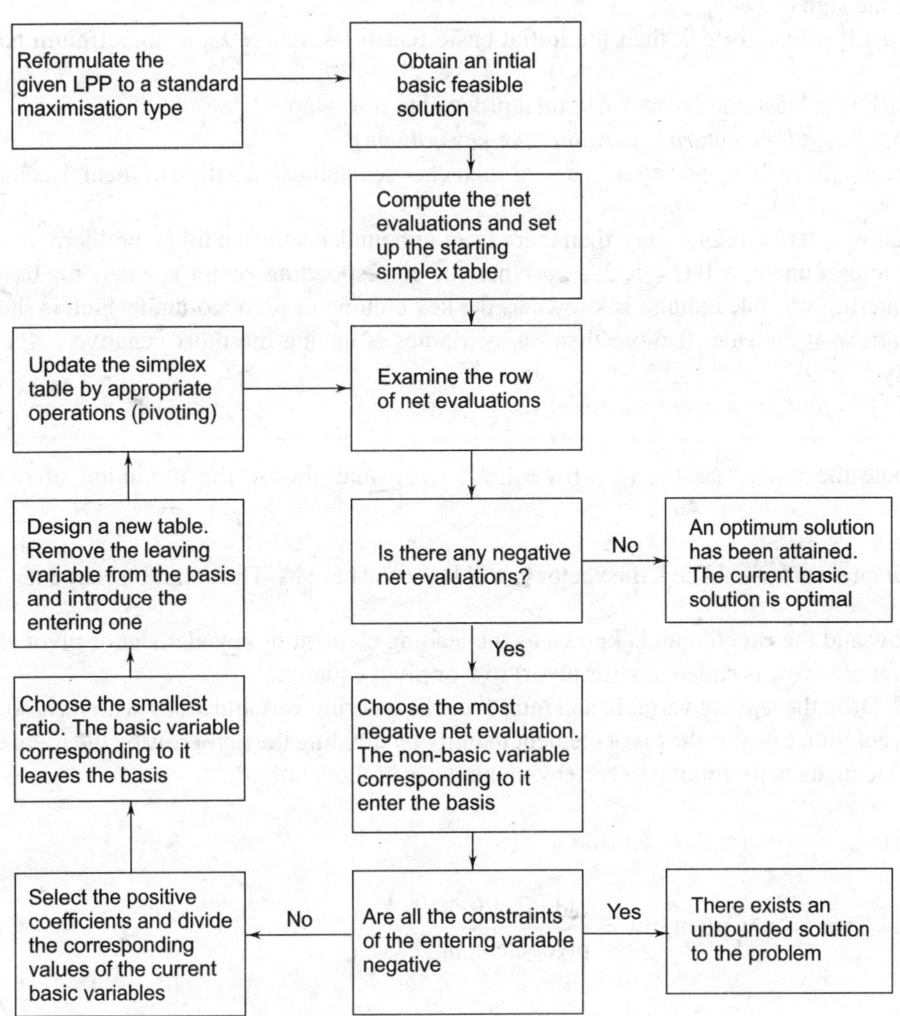

The Simplex Algorithm:

For the solution of any LPP by simplex method, the existence of an initial basic feasible solution is assumed. The steps for computing the optimal solution is as follows:

Step 1: Check whether the objective function of the given LPP is to be maximised or minimised. If it is to be minimised then convert it into a problem of maximisation by

$$\text{Minimum of } z = -\text{ Maximum } \{-z\}$$

Step 2: Check whether all $b_i (i = 1, 2, \ldots, m)$ are non-negative. If any one of b_i is negative, then multiply the corresponding inequation by – 1, in order to get all b_i's $(i = 1, 2, \ldots, m)$ as non-negative.

Step 3: Convert all the inequations of the constraints into equations by introducing slack and include the symbol surplus variables in the constraint. Put the cost of these variables equal to zero.

Step 4: Obtain an initial basic feasible solution to the problem in the form $X_B = B^{-1}\, b$ and put it in the first column of the simplex table.

Step 5: Compute the net evaluations $z_j - c_j, j = 1, 2, \ldots, n$ by using the relation $z_j - c_j = c_B\, y_j - c_j$. Examine the sign of $z_j - c_j$.

Case (i): If all $z_j - c_j \geq 0$, then the initial basic feasible solution X_B is an optimum basic feasible solution.

case (ii): If at least one $(z_j - c_j) < 0$, then proceed to next step.

Step 6: *(To find the entering variable, the key column)*

If there are more than one negative $z_j - c_j$, then choose the most negative of them. Let it be $z_r - c_r$ for some $j = r$.

(i) If all $y_{ir} > 0$ $(i = 1, 2, \ldots, m)$, then there is an unbounded solution to the problem.

(ii) If at least one $y_{ir} > 0$ $(i = 1, 2, \ldots m)$ then the corresponding vector y_r enters the basis y_B.

The entering variable column is known as the key column or pivot column which is shown marked with an arrow at the side. If more than one variables is having the most negative $z_j - c_j$ select one arbitrarily.

Step 7: *(To find the leaving variable)*

Compute the ratios $\left\{\dfrac{X_{Bi}}{y_{ir}}, y_{ir} > 0, i = 1, 2, \ldots, m\right\}$ and choose the minimum of them. Let the minimum ratio be $\dfrac{X_{Bk}}{y_{kr}}$. Then, the vector y_k will leave the base Y_B. The common element y_{kr} which is in the kth row and the rth column is known as the leading element or key element or pivot element. The leaving variable row is called key (or pivot) row or pivot equation.

Step 8: Drop the leaving variable and introduce the entering variable along with its associated value under C_B column. Convert the pivot element to unity by dividing the pivot row by the pivot element and all other elements in its column to zero by making use of the relation.

(i) $\hat{y}_{kj} = \dfrac{yk_j}{y_{kr}}, j = 0, 1, 2, \ldots, n$

That is, new pivot row $= \dfrac{\text{old pivot row}}{\text{pivot element}}$

(ii) $\hat{y}_{ij} = y_{ij} - \dfrac{y_{kj}}{y_{kr}}\, y_{ir}$

That is, new equation (all other rows including $z_j - c_j$ row)

$$= \text{old equation} - \begin{pmatrix}\text{Corresponding}\\ \text{Column}\\ \text{Coefficients}\end{pmatrix} \times \text{new pivot equation}$$

Step 9: Go to step 5 and repeat the procedure until either an optimum solution is obtained or there is an indication of an unbounded solution.

Remark:

1. For maximisation (minimisation problems if all $(z_j - c_j) \geq 0$ (≤ 0), then the current basic feasible solution is optimal.
2. For maximisation (minimisation) problem if at least one of $(z_j - c_j) < 0$ (> 0), then the current basic feasible solution is not optimal.
3. For maximisation (minimisation) problems the entering variable is the non-basic variable corresponding to the most negative (most positive) value of $z_j - c_j$.
4. For both maximisation and minimisation problems, the leaving variable is the basic variable corresponding to the minimum ratio θ.

Example 1

Use simplex method to solve the LPP

$$\begin{aligned} \text{Maximise } z &= 4x_1 + 10x_2 \\ \text{subject to } 2x_1 + x_2 &\leq 50 \\ 2x_1 + 5x_2 &\leq 100 \\ 2x_1 + 3x_2 &\leq 90 \\ x_1, x_2 &\geq 0 \end{aligned}$$

Solution:

Step 1: Introducing the slack variables $s_1, s_2, s_3 \geq 0$ the LPP becomes

$$\text{Maximise } z = 4x_1 + 10x_2 + 0s_1 + 0s_2 + 0s_3$$

$$\begin{aligned} \text{subject to } 2x_1 + x_2 + s_1 + 0s_2 + 0s_3 &= 50 \\ 2x_1 + 5x_2 + 0s_1 + s_2 + 0s_3 &= 100 \\ 2x_1 + 3x_2 + 0s_1 + 0s_2 + s_3 &= 90 \\ x_1, x_2, s_1, s_2, s_3 &\geq 0 \end{aligned}$$

Step 2: An obvious initial basic feasible solution is given by $X_B = B^{-1}\, b$ where,

$$B = I_3 = \begin{bmatrix} 1 & 0 & 0 \\ 0 & 1 & 0 \\ 0 & 0 & 1 \end{bmatrix} \text{ and } b = \begin{pmatrix} 50 \\ 100 \\ 90 \end{pmatrix} \text{ and } X_B^T = [s_1, s_2, s_3]$$

Step 3: We compute x_j and $z_j - c_j$ corresponding to the non-basic variables x_1, x_2 and the basic variable, s_1, s_2 and s_3.

$$\begin{aligned} y_1 &= B^{-1}\, a_1 = I_3\, [2\ 2\ 2]^T = [2\ 2\ 2]^T \\ y_2 &= B^{-1}\, a_2 = I_3\, [1\ 5\ 3]^T = [1\ 5\ 3]^T \\ s_1 &= B^{-1}\, e_1 = I_3\, [1\ 0\ 0]^T = [1\ 0\ 0]^T \\ s_2 &= B^{-1}\, e_2 = I_3\, [0\ 1\ 0]^T = [0\ 1\ 0]^T \\ s_3 &= B^{-1}\, e_3 = I_3\, [0\ 0\ 1]^T = [0\ 0\ 1]^T \\ z_1 - c_1 &= C_B\, x_I - c_1 = (0\ 0\ 0)\ (2\ 2\ 2)^T - 4 = -4 \\ z_2 - c_2 &= C_B\, x_2 - c_2 = (0\ 0\ 0)\ (1\ 5\ 3)^T - 10 = -10 \\ z_3 - c_3 &= C_B\, s_1 - c_3 = (0\ 0\ 0)\ (1\ 0\ 0)T - 0 = 0 \end{aligned}$$

$$z_4 - c_4 = C_B\, s_2 - c_4 = (0\ 0\ 0)\ (0\ 1\ 0)^T - 0 = 0$$
$$z_5 - c_5 = C_B\, s_3 - c_5 = (0\ 0\ 0)\ (0\ 0\ 1)^T - 0 = 0$$

The starting simplex table is

table 2.13

		c_j	(4	10	0	0	0)	
C_B	Y_B	X_B	x_1	$x_2 \downarrow$	s_1	s_2	s_3	Min Ratio
0	s_1	50	2	1	1	0	0	$\frac{50}{1} = 50$
0	$\overleftarrow{s_2}$	100	2	[5]	0	1	0	$\frac{100}{5} = 20$
0	s_3	90	2	3	0	0	1	$\frac{90}{3} = 30$
			– 4	– 10	0	0	0	

Since there are some $(z_j - c_j) < 0$, the current solution is not optimal.

Step 5: Since $z_2 - c_2 = -10$, is the most negative, the corresponding non-basic variable x enters the basis. The column corresponding to x_2 is called the key column.

Step 6: Find the ratio

$$\theta = \min\left\{\frac{X_{Bi}}{X_{i2}}, X_{i2} > 0\right\}$$

$$= \min\left\{\frac{50}{1}, \frac{100}{5}, \frac{90}{3}\right\} = 20, \text{ which corresponds to } s_2.$$

Hence, s_2 leaves the basis and $x_{22} = 5$ is the pivot element.

Step 8: Convert the pivot element 5 to unity and all other elements in its column to zero by the following transformations.

$$\text{New pivot equation} = \frac{\text{Old pivot equation}}{\text{Pivot element}}$$

The new elements corresponding to the other elements of the body matrix, basis matrix and X_B column of Table 2.22 can be obtained by using

$$\text{New element} = \text{old element} - \begin{pmatrix}\text{corresponding}\\ \text{new element}\\ \text{in the pivotal}\\ \text{row}\end{pmatrix} \times \begin{pmatrix}\text{corresponding}\\ \text{old element}\\ \text{in the pivotal}\\ \text{column}\end{pmatrix}$$

The improved basic feasible solution is given in Table 2.14

First iteration:

table 2.14

		c_j	(4	10	0	0	0)	
C_B	Y_B	X_B	x_1	x_2	s_1	s_2	s_3	Min Ratio
0	s_1	30	8/5	0	1	- 1/5	0	
10	x_2	20	2/5	1	0	1/5	0	
0	s_3	30	4/5	0	0	- 3/5	1	
	$z_j - c_j$	200	0	0	0	2	0	

Step 9: Since all $z_j - c_j \geq 0$, the current solution is optimal. The optimal solution is $x_1 = 0, x_2 = 20$ with maximum of $z = 200$.

Example 2

Solve the LPP

$$\text{Maximise } z = 107x_1 + x_2 + 2x_3$$
$$\text{subject to } 14x_1 + x_2 - 6x_3 + 3x_4 = 7$$
$$16x_1 + \frac{x_2}{2} - 6x_3 \leq 5$$
$$3x_1 - x_2 - x_3 \leq 0$$
$$x_1, x_2, x_3, x_4 \geq 0.$$

Solution: By introducing slack variables $s_1, s_2 \geq 0$ in the given constraints the LPP becomes

$$\text{Maximise } z = 107x_1 + x_2 + 2x_3 + 0s_1 + 0s_2$$
$$\text{subject to } \frac{14}{3}x_1 + \frac{1}{3}x_2 - 2x_3 + x_4 = 7$$
$$16x_1 + \frac{1}{2}x_2 - 6x_3 + s_1 = 5$$
$$3x_1 - x_2 - x_3 + s_2 = 0$$
$$x_1, x_2, x_3, x_4, s_1, s_2 \geq 0.$$

Note that in the first constraint the variable x_4 can be taken as a basic variable. An obvious initial basic feasible solution is $x_4 = 7$, $s_1 = 5$ and $s_2 = 0$.

Starting table:

table 2.15

		c_j	(107	1	2	0	0	0)	
C_B	Y_B	X_B	$x_1 \downarrow$	x_2	x_3	x_4	s_1	s_2	Min Ratio
0	x_4	7/3	14/3	1/3	- 2	1	0	0	$\frac{7/3}{14/3} = 1/2$
0	s_1	5	16	1/2	- 6	0	1	0	5/16
0	$\overleftarrow{s_2}$	0	3	- 1	- 1	0	0	1	0/3 = 0
	$z_j - c_j$		- 107	- 1	- 2	0	0	0	

Here the most negative net evaluation is 107 corresponding to x_1 and so x_1 enters into the basis. The minimum ratio column of Table 2.16 shows that the variable s_2 leaves the basis.

First iteration:

table 2.16

C_B	Y_B	X_B	x_1	x_2	$x_3\downarrow$	x_4	s_1	s_2	Min Ratio
0	x_4	7/3	0	17/9	- 4/9	1	0	- 14/9	
0	s_1	5	0	35/6	- 2/3	0	1	- 16/3	
107	x_1	0	1	- 1/3	- 1/3	0	0	1/3	
	$z_j - c_j$		0	- 110/3	- 113/3	0	0	- 107/3	

Since the most negative net evaluation is – 113/3 corresponding to the variable x_3. Since there is no positive element in the x_3 (pivotal) column there exists an unbounded solution to the given LPP.

2.6 ARTIFICIAL VARIABLE TECHNIQUES

The last section dealt with LPPs with less than or equal (≤) type constraints. This property together with the fact that the right-hand side of all the constraints is non-negative, provides us with a ready starting basic feasible solution that comprises of all slack variables. There are however, many linear programming problems for which mere slack variables cannot provide such a solution. For those, we accomplish by adding artificial variables which play the role of slack variables. However, since such artificial variables have no physical meaning in the original model (hence the name artificial), provisions must be made to make them zero level at the optimum iteration. In other words, we use them to start the solution, and abandon them once their work has been over. A logical way to achieve this objective is to penalise the artificial variables in the objective function. Two (closely related) networks based on the use of penalties are available to solve this type of LPP. They are

1. Big 'M' method or method of penalty due to A. Charnes
2. The Two-phase simplex method due to Dantzig, Orden and Wolfe.

2.6.1 The Big M Method

Step 1: Convert the LPP into maximisation form and then express it in the standard form by introducing slack and/or surplus variables as the case may be.

Step 2: Introduce non-negative variables to the left-hand side of all the constraints of (≥ or =) type. These variables are called artificial variables. The purpose of introducing artificial variables is just to obtain an initial basic feasible solution. In order to get rid of the artificial variables in the final optimum iteration, we assign a very large penalty '$-M$' to these artificial variables in the objective function for maximisation objective function.

Step 3: Solve the modified linear programming problem by simplex method. While making iterations by this method, one of the following three cases may arise.

(a) If no artificial variable remains in the basis and the optimality condition is satisfied, then the current solution is an optimal basic feasible solution.

(b) If at least one artificial variable appears in the basis at zero level (with zero value in the X_B column) and the optimality condition is satisfied, then the current solution is an optimal basic feasible (though degenerated) solution.

(c) If at least one artificial variable appears in the basis at non-zero level (with positive value in X_B column) and the optimality condition is satisfied, then the original problem has no feasible solution. The solution satisfies the constraints but does not optimise the objective function since it contains a very large penalty M and is called pseudo optimal solution.

While applying simplex method, whenever an artificial variable happens to leave the basis, we drop that artificial variable and omit all the entries corresponding to its column from the simplex table.

Step 4: Application of simplex method is continued until either an optimum basic feasible solution is obtained or there is an indication of the existence of an unbounded solution to the given LPP.

Example 1

Use penalty (or Big M) method to

$$\text{Minimise } z = 4x_1 + 3x_2$$
$$\text{subject to } 2x_1 + x_2 \geq 10$$
$$-3x_1 + 2x_2 \leq 6$$
$$x_1 + x_2 \geq 6$$
$$x_1, x_2 \geq 0$$

Solution: Convert the problem into maximisation type then objective function becomes

$$\text{Maximise } (-z) = -4x_1 - 3x_2$$

Introducing surplus variables $s_1, s_3 \geq 0$ and a slack variable $s_2 \geq 0$ in the given constraints the LPP becomes

$$\text{Maximise } (-z) = -4x_1 - 3x_2 + 0s_1 + 0s_2 + 0s_3$$
$$\text{subject to } 2x_1 + x_2 - s_1 = 10$$
$$-3x_1 + 2x_2 + s_2 = 6$$
$$x_1 + x_2 - s_3 = 6$$
$$x_1, x_2, s_1, s_2, s_3, \geq 0$$

Clearly, we do not have a basic feasible solution, since the surplus variables carry negative coefficients (– 1). Introduce two new artificial variables $A_1 \geq 0$ and $A_2 \geq 0$ in the first and third equation, respectively. It provides a starting basic feasible solution. We assign a very high penalty cost (say $-M$, $M \geq 0$) to these variables in the objective function so that they may be driven to zero at the optimality step. Then, the LPP becomes

$$\text{Maximise } (-z) = -4x_1 - 3x_2 + 0s_1 + 0s_2 + 0s_3 - MA_1 - MA_2$$
$$\text{subject to } 2x_1 + x_2 - s_1 + 0s_2 + 0s_3 + A_1 = 10$$
$$-3x_1 + 2x_2 + 0s_1 + 0s_2 + 0s_3 = 6$$
$$x_1 + x_2 + 0s_1 + 0s_2 - s_3 + A_2 = 6$$
$$x_1, x_2, s_1, s_2, s_3, A_1, A_2 \geq 0$$

If we take the basic variables as A_1, s_2, A_2 then a obvious initial basic feasible solution is $A_1 = 10$, $s_2 = 6$, $A_2 = 6$

Starting Table:

table 2.17

		c_j	(– 4	– 3	0	0	0	– M	– M)	
C_B	Y_B	X_B	$x_1 \downarrow$	x_2	s_1	s_2	s_3	A_1	A_2	Min Ratio
– M	$\overleftarrow{A_1}$	10	2	1	– 1	0	0	1	0	$\frac{10}{2} = 5$
0	s_2	6	– 3	2	0	1	0	0	0	
– M	A_1	6	1	1	0	0	– 1	0	1	$\frac{6}{1} = 6$
$z_j - c_j$			$-3M + 4$	$-2M + 3$	M	0		0	0	

From the Table we observe that the most negative net evaluation is $-3M + 4$ (since M is very large), the non-basic variable x_1 enters into the basis and the basic variable A_1 (variable with minimum ratio) enters into the basis.

First iteration:

table 2.18

		c_j	(– 4	– 3	0	0	0	– M)	
C_B	Y_B	X_B	x_1	$x_2 \downarrow$	s_1	s_2	s_3	A_2	Min Ratio
– 4	x_1	5	1	1/2	– 1/2	0	0	0	$\frac{5}{1/2} = 10$
0	s_2	21	0	7/2	– 3/2	1	0	0	$\frac{21}{7/2} = 42/7$
– M	$\overleftarrow{A_2}$	0	0	1/2	1/2	0	– 1	1	$\frac{0}{1/2} = 0$
$z_j - c_j$		$-M - 20$	0	$-\frac{M}{2} + 1$	$-\frac{M}{2} + 2$	0		0	

Note that in Table 2.42 we have omitted all entries of column vector A_1, because the artificial variable A_1 has left the basis and so the re-entering of the variable in any subsequent iterations can be avoided.

Since the most negative net evaluation is $-\frac{M}{2} + 1$, the non-basic vector x_2 enters the basis. Minimum ratio column implies that the variable A_2 leaves the basis.

Second iteration:

table 2.19

C_B	Y_B	X_B	x_1	x_2	s_1	s_2	s_3	Min. Ratio
– 4	x_1	4	1	0	– 1	0	1	
0	s_2	14	0	0	– 5	1	7	
– 3	x_2	2	0	1	1	0	– 2	
	$z_j - c_j$	– 22	0	0	1	0	2	

Since all $z_j - c_j \geq 0$ for all j, the optimum basic feasible solution is given by

$$x_1 = 4, x_2 = 2 \text{ with minimum of } z = 22$$

Example 2

Maximise $z = 3x_1 + 2x_2$

subject to
$$2x_1 + x_2 \leq 2$$
$$3x_1 + 4x_2 \geq 12$$
$$x_1, x_2 \geq 0.$$

Solution: Introducing slack variable $s_1 \geq 0$, surplus variable $s_2 \geq 0$ and an artificial variable $A_1 \geq 0$, the LPP can be written as

$$\text{Maximise } z = 3x_1 + 2x_2 + 0s_1 + 0s_2 - MA_1$$

subject to
$$2x_1 + x_2 + s_1 = 2$$
$$3x_1 + 4x_2 - s_2 + A_1 = 12$$
$$x_1, x_2, s_1, s_2 \geq 0 \text{ and } A_1 \geq 0.$$

An obvious initial basis feasible solution is $s_1 = 2, A_1 = 12$.

Starting Table:

table 2.20

		c_j	(3	2	0	0	− M)	
C_B	Y_B	X_B	x_1	$x_2 \downarrow$	s_1	s_2	A_1	Min. Ratio
0	$\overleftarrow{s_1}$	2	2	1	1	0	0	$\frac{2}{1} = 2$
− M	A_1	12	3	4	0	− 1	1	$\frac{12}{4} = 3$
$z_j - c_j$		− 12M − 2	− 3M − 3	− 4M	0	M	0	

From the Table 2.20 we observe that the non-basic variable x_2 enters into the basis and the basic variable s_1 leaves the basis.

First iteration:

table 2.21

		c_j	(3	2	0	0	− M)	
C_B	Y_B	X_B	x_1	x_2	s_1	s_2	A_1	Min. Ratio
2	x_2	2	2	1	1	0	0	
− M	A_1	4	− 5	0	− 4	− 1	1	
$z_j - c_j$		4M + 4	5M + 1	0	4M + 2	M	0	

Here, the coefficient of M in each $z_j - c_j \geq 0$, and an artificial variable A_1 appears in the basis at non-zero level. Thus, the given LPP does not possess any feasible solution. We can say the LPP possess a pseudo-optimal solution.

Example 3

Use penalty method to

$$\text{Maximise } z = 2x_1 + x_2 + x_3$$

subject to
$$4x_1 + 6x_2 + 3x_3 \le 8$$
$$3x_1 - 6x_2 - 4x_3 \le 1$$
$$2x_1 + 3x_2 - 5x_3 \ge 4$$
and,
$$x_1, x_2, x_3 \le 0.$$

Solution: By introducing slack variables $s_1, s_2 \ge 0$, surplus variable $s_3 \ge 0$ and artificial variable $A_1 \ge 0$, an obvious IBFS $s_1 = 8, s_2 = 1, A_1 = 4$.

Starting table:

table 2.22

	c_j		(2	1	1	0	0	0	− M)	
C_B	Y_B	X_B	x_1	$x_2\downarrow$	x_3	s_1	s_2	s_3	A_1	*Min. Ratio*
0	s_1	8	4	6	3	1	0	0	0	8/6 = 1.33
0	s_2	1	3	− 6	− 4	0	1	0	0	–
− M	$\overleftarrow{A_1}$	4	2	3	− 5	0	0	− 1	1	4/3 = 1.33
		− 4M	− 2M − 2	− 3M − 1	5M − 1	0	0	M	0	

First iteration: Introduce x_2 and drop A_1.

table 2.23

	c_j		(2	1	1	0	0	0)	
C_B	Y_B	X_B	x_1	x_2	$x_3\downarrow$	s_1	s_2	s_3	*Min. Ratio*
0	$\overleftarrow{s_1}$	0	0	0	$\boxed{13}$	1	0	2	0
0	s_2	9	7	0	− 14	0	1	− 2	—
1	x_2	4/3	2/3	1	− 5/3	0	0	− 1/3	—
			− 4/3	0	− 8/3	0	0	−1/3	

Second iteration: Introduce x_3 and drop s_1.

table 2.24

	c_j		(2	1	1	0	0	0)	
C_B	Y_B	X_B	$x_1\downarrow$	x_2	x_3	s_1	s_2	s_3	*Min. Ratio*
1	x_3	0	0	0	1	1/13	0	2/13	—
0	$\overleftarrow{s_2}$	9	$\boxed{7}$	0	0	4/13	1	2/13	9/7
1	x_2	4/3	2/3	1	0	5/39	0	− 1/13	2
$z_j - c_j$		4/3	− 4/3	0	0	8/39	0	1/13	

Third iteration: Introduce x_1 and drop s_2.

table 2.25

		c_j	(2	1	1	0	0	0)	
C_B	Y_B	X_B	x_1	x_2	x_3	s_1	s_2	s_3	Min. Ratio
1	x_3	0	0	0	1	1/13	0	2/13	
2	x_1	9/7	1	0	0	2/13	1/7	2/91	
1	x_2	10/21	0	1	0	1/39	- 2/21	-25/273	
		64/21	0	0	0	16/39	4/21	29/273	

Since all $z_j - c_j \geq 0$, the current solution is optimal. The optimal solution is

$$x_1 = 9/7, x_2 = 10/21, x_3 = 0 \text{ with maximum of } z = 64/21$$

2.6.2 The Two-phase Simplex Method

The two-phase simplex method is another method to solve an LPP in which some artificial variables are involved. The solution can be obtained in two phases.

Phase I It consists of the following steps:

Step 1: Convert the LPP into maximisation form and ensure that all b_i (constant terms) are non-negative. If some of them are negative, make them non-negative by multiplying both sides of those inequations/ equations by –1.

Step 2: Add artificial variables A_i (≥ 0) to the LHS of the constraints of (= and $\geq$) to complete the identity matrix I.

Step 3: Express the given LPP in standard form.

Step 4: Obtain an initial basic feasible solution.

Step 5: Assign a cost '–1' to each artificial variable and a cost 0 to all other variables (in place of their original cost) in the objective function. The new objective function is

$$\text{Max. } z^* = -A_1 - A_2 - \cdots - A_n \text{ where } A_i\text{'s are the artificial variables.}$$

Step 6: Write down the auxiliary LPP in which the new objective function is to be maximised subject to the given set of constraints.

Step 7: Solve the auxiliary LPP by simplex method until either of the following three possibilities arise.

(i) Max $z^* < 0$ and at least one artificial variable appears in the optimum basis at a + ve level. In this case no feasible solution exist, stop the procedure.

(ii) Max $z^* = 0$ and at least one artificial variable appears in the optimum basis at zero level.

(iii) Max $z^* = 0$ and no artificial variable appears in the optimum basis.

If case (ii) and (iii) arise proceed to phase II.

Phase II Use the optimum basic feasible solution of phase I as a starting solution for the original LPP. Assign the actual costs to the variables in the objective function and a cost 0 to every artificial variable in the basis at zero level. Delete the artificial variables column from the table which is eliminated from the basis in phase I. Apply simplex method to the modified simplex table obtained at the end of phase I till an optimum basic feasible solution is obtained or till there is an indication of unbounded solution.

Remark:

1. In phase-I, the iterations are stopped as soon as the value of the new (artificial) objective function becomes zero because this is its minimum value. There is no need to continue till the optimality is reached if this value becomes zero earlier than that.
2. Note that the new objective function is always of maximisation type whether the given original problem is maximisation or minimisation type.
3. Before starting phase II, remove all artificial variables from the table which were non-basic at the end of phase I.

Example 1

Use two-phase simplex method to solve

$$\text{Maximise } z = 5x_1 + 8x_2$$

subject to the constraints

$$3x_1 + 2x_2 \geq 3$$
$$x_1 + 4x_2 \geq 4$$
$$x_1 + x_2 \leq 5$$
$$x_1, x_2 \geq 0.$$

Solution: By introducing the non-negative slack, surplus and artificial variables, the standard form of the LPP becomes

$$\text{Maximise } z = 5x_1 + 8x_2 + 0s_1 + 0s_2 + 0s_3$$

subject to
$$3x_1 + 2x_2 - s_1 + 0s_2 + 0s_3 + A_1 = 3$$
$$x_1 + 4x_2 + 0s_1 - s_2 + 0s_3 + A_2 = 4$$
$$x_1 + x_2 + 0s_1 + 0s_2 + s_3 = 5$$
$$x_1, x_2, s_1, s_2, s_3, A_1, A_2 \geq 0.$$

An obvious initial basic feasible solution is $A_1 = 3, A_2 = 4, s_3 = 3$.

Phase I Construct the auxiliary LPP

$$\text{Maximise } z^* = -A_1 - A_2$$

subject to the given set of constraints.

Starting table:

table 2.26

		c_j	(0	0	0	0	0	-1	-1)	
C_B	Y_B	X_B	x_1	$x_2 \downarrow$	s_1	s_2	s_3	A_1	A_2	Min. Ratio
-1	A_1	3	3	2	-1	0	0	1	0	3/2
-1	$\overleftarrow{A_2}$	4	1	[4]	0	-1	0	0	1	4/4
0	s_3	5	1	1	0	0	1	0	0	5/1
$z_j - c_j$		–	-4	-6	1	1	0	0	0	

Since there are some $(z^*_j - c_j) < 0$, the current basic feasible solution is not optimal.

First iteration: Introduce x_2 and drop A_2.

table 2.27

C_B	Y_B	X_B	$x_1\downarrow$	x_2	s_1	s_2	s_3	A_1	A_2	Min. Ratio
-1	$\overleftarrow{A_1}$	1	[5/2]	0	-1	1/2	0	1	-1/2	2/5
0	x_2	1	1/4	1	0	-1/4	0	0	1/4	4
0	s_3	4	3/4	0	0	1/4	1	0	-1/4	16/3
$z_j - c_j$		-1	-5/2	0	1	-1/2	0	0	3/2	

Since there are $z^*_j - c_j < 0$, the current basic feasible solution is not optimal.
Second iteration: Introduce x_1 and drop A_1.

table 2.28

		c_j	(0	0	0	0	0	-1	-1)	
C_B	Y_B	X_B	x_1	x_2	s_1	s_2	s_3	A_1	A_2	Min. Ratio
0	x_1	2/5	1	0	-2/5	1/5	0	2/5	-1/5	
0	x_2	9/10	0	1	1/10	-3/10	0	-1/10	3/10	
0	s_3	37/10	0	0	3/10	1/10	1	-3/10	-1/10	
		0	0	0	0	0	0	1	1	

Since all $z^*_j - c_j \geq 0$, the current basic solution is optimum. Further, no artificial variable appears in the optimum basis so we proceed to phase II.

Phase II Consider the actual costs associated with the original variables. The new objective function is $z = 5x_1 + 8x_2 + 0s_1 + 0s_2 + 0s_3$. The initial basic feasible for this phase is the one obtained at the end of phase I.

The iterative simplex table for this phase is

table 2.29

		c_j	(5	8	0	0	0)	
C_B	Y_B	X_B	x_1	x_2	s_1	$s_2\downarrow$	s_3	Min. Ratio
5	$\overleftarrow{x_1}$	2/5	1	0	-2/5	1/5	0	$\frac{2/5}{1/5} = 2$
8	x_2	9/10	0	1	1/10	-3/10	0	
0	s_3	37/10	0	0	3/10	1/10	1	$\frac{37/10}{1/10} = 3$
$z_j - c_j$		46/5	0	0	-6/5	-7/5	0	

From Table 2.29, we observe that s_2 enters into the basis and x_1 leaves the basis.

First iteration: Introduce s_3 and drop x_1

table 2.30

		c_j	(5	8	0	0	0)	
C_B	Y_B	X_B	x_1	x_2	$s_1\downarrow$	s_2	s_3	*Min. Ratio*
0	s_2	2	5	0	-2	1	0	—
8	x_2	3/2	3/2	1	-1/2	0	0	—
0	$\overleftarrow{s_3}$	7/2	-1/2	0	[1/2]	0	1	7
	$z_j - c_j$	12	7	0	-4	0	0	

Since there are some $z_j - c_j < 0$, current basic feasible solution is not optimal. From Table 2.30, we observe that s_1 enters into the basis and s_3 leaves the basis.

Second iteration: Introduce s_1 and drop s_3.

table 2.31

C_B	Y_B	X_B	x_1	x_2	s_1	s_2	s_3	*Min. Ratio*
0	s_2	16	3	0	0	1	4	
8	x_2	5	1	1	0	0	1	
0	s_1	7	-1	0	1	0	2	
	$z_j - c_j$	40	3	0	0	0	8	

Since all $z_j - c_j \geq 0$, the current basic feasible solution is optimal. The optimal solution is

$$x_1 = 0, x_2 = 5 \text{ with maximum of } z = 40$$

Example 2

Use two phase simplex method to

$$\text{Maximise } z = 5x_1 + 2x_2$$

subject to
$$2x_1 + x_2 \leq 1$$
$$x_1 + 4x_2 \geq 6$$
$$x_1, x_2 \geq 0$$

Solution: Since the given LPP is having one constraint with $\geq$ type constraint we can solve it by using two phase simplex method. Convert the LPP into standard form by introducing slack variable $s_1 \geq 0$ and surplus variable $s_2 \geq 0$ and artificial variable $A_1 \geq 0$. An obvious initial basic feasible solution is $s_1 = 1$, $A_1 = 6$.

Phase I The auxiliary LPP is

$$\text{Maximise } z^* = 0x_1 + 0x_2 + 0s_1 + 0s_2 - A_1$$

subject to
$$2x_1 + x_2 + s_1 = 1$$
$$x_1 + 4x_2 - s_2 + A_1 = 6$$
$$x_1, x_2, s_1, s_2, A_1 \geq 0$$

The iterative simplex tables for the auxiliary LPP are:
Initial iteration:

table 2.32

		c_j	(0	0	0	0	– 1)	
C_B	Y_B	X_b	x_1	$x_2 \downarrow$	s_1	s_2	A_1	*Min. Ratio*
0	$\overleftarrow{s_1}$	1	2	[1]	1	0	0	1/1 = 1
– 1	A_1	6	1	4	0	– 1	1	6/4 = 1.5
		– 6	– 1	– 4	0	1	0	

From Table 2.32, we observe that x_2 enters into the basis and s_1 leaves the basis.
First iteration: Introduce x_2 and drop s_1.

table 2.33

		c_j	(0	0	0	0	– 1)	
C_B	Y_B	X_B	x_1	x_2	s_1	s_2	A_1	*Min. Ratio*
0	x_2	1	2	1	1	0	0	
– 1	A_1	2	– 7	0	– 4	– 1	1	
$z_j - c_j$		– 2	7	0	4	1	0	

Since all $(z_j - c_j) \geq 0$, an optimum basic feasible to the auxiliary LPP is obtained. But maximum of $z^* < 0$ and an artificial vector A_1 appears at the positive level. Therefore, the original LPP does not possess any feasible solution.

Example 3

Use two phase simplex method to

$$\text{Minimise } z = -2x_1 - x_2$$

subject to

$$x_1 + x_2 \geq 2$$
$$x_1 + x_2 \leq 4$$
$$x_1, x_2 \geq 0$$

Solution: Convert the LPP into maximisation by using Min $z = -$ Max $(-z)$, the LPP becomes

$$\text{Maximise } (-z) = 2x_1 + x_2$$

subject to

$$x_1 + x_2 \geq 2$$
$$x_1 + x_2 \leq 4$$
$$x_1, x_2 \geq 0$$

By introducing slack variable s_2, surplus variable s_1, artificial variable A_1, the standard form of LPP is

$$\text{Maximise } (-z) = 2x_1 + x_2$$

Subject to $x_1 + x_2 - s_1 + 0s_2 + A_1 = 2$

$$x_1 + x_2 + 0s_1 + s_2 = 4$$

$x_1, x_2, s_1, s_2, A_1 \geq 0$

An obvious IBFS is $s_2 = 4, A_1 = 2$.

Phase I The objective function of the auxiliary LPP is Maximise $(-z) = -A_1$.

Starting Table:

table 2.34

		c_j	(0	0	0	0	− 1)	
C_B	Y_B	X_B	x_1 ↓	x_2	s_1	s_2	A_1	Min. Rat
− 1	$\overleftarrow{A_1}$	2	[1]	1	− 1	0	1	2/1 = 2
0	s_2	4	1	1	0	1	0	4/1 = 4
			− 1	− 1	1	0	0	

From Table 2.34 we observe that the non-basic variable x_1 enters into the basis and the basic vari A_1 leaves the basis.

First iteration: Introduce x_1 and drop A_1.

table 2.35

		c_j	(0	0	0	0)	
C_B	Y_B	X_B	x_1	x_2	s_1	s_2	Min. Ratio
0	x_1	2	1	1	− 1	0	
0	s_2	2	0	0	1	1	
		0	0	0	0	0	

Since all $z^*_j - c_j \geq 0$ and no artificial variable appears in the optimum basis, the current basic feas solution is optimal to the auxiliary LPP and we proceed to phase II.

Phase II Assign the actual costs to the original variables. The new objective function then beco

$$\text{Maximise } (-z) = 2x_1 + x_2 + 0s_1 + 0s_2$$

The initial basic feasible solution for this phase is the one obtained at the end of phase I. The itera simplex tables for this phase is:

Initial iteration:

table 2.36

C_B	Y_B	X_B	x_1	x_2	s_1 ↓	s_2	Min. Ratio
2	x_1	2	1	1	− 1	0	–
0	$\overleftarrow{s_2}$	2	0	0	[1]	1	2/1 = 2
	$z^*_j - c_j$	4	0	1	− 2	0	

From Table 2.36, we observe that s_1 enters into the basis and s_2 leaves the basis.

irst iteration: Introduce s_1 and drop s_2.

ble 2.37

C_B	Y_B	X_B	x_1	x_2	s_1	s_2	Min. Ratio
2	x_1	4	1	1	0	1	
0	s_1	2	0	0	1	1	
	$z_j^* - c_j$	8	0	1	0	2	

ince all $z^*_j - c_j \geq 0$, the current basic feasible solution is optimal.
he optimal solution is

$$x_1 = 4, x_2 = 0 \text{ with Maximum of } (-z) = 8$$

hat is, $x_1 = 4, x_2 = 0$ with minimum of $z = -8$

advantages of Big-M Method over Two-phase Method

. Even though Big-M method can always be used to check the existence of a feasible solution it may be computationally inconvenient because of the manipulation of the constant M. On the other hand, two-phase method eliminates the constant M from the calculations.

. Another difficulty arises especially when the problem is to be solved on a digital computer. To use a digital computer, M must be assigned some numerical value which is much larger than the values $c_1, c_2, \cdots$, in the objective function. But a computer has only a fixed number of digits.

VARIANTS OF SIMPLEX METHOD

is section we discuss some complications and variations which are very often encountered during lex procedure. The following are some of them:

. Degeneracy and cycling
. Unbounded solution
. Multiple solution or alternative optimum solution
. No feasible solution
. Unrestricted variables.

Degeneracy (Tie for minimum ratio)

he concept of obtaining a degenerate basic feasible solution in a LPP is known as *degeneracy*.

t the stage of improving the solution during simplex procedure, minimum ratio $\frac{X_B}{y_{ik}}$, $y_{ik} > 0$ is mined in the last column of simplex table to find the key row. But sometimes this ratio may not be ue (i.e. more than one variable is eligible to leave the basis) or at the very first iteration, the value of or more basic variables in X_B column becomes zero, this causes the problem of degeneracy.

If the minimum ratio is zero, for two or more basic variables, degeneracy may result and the simplex routine to cycle indefinitely. That is, the solution which we have obtained in one iteration may repeat again after few iterations and therefore, no optimum solution may be obtained. This concept is known as *cycling* or *circling*.

Method to resolve degeneracy (Tie)

Step 1: First find out the rows for which the minimum non-negative ratio is the same (tie), for example, suppose there is a tie between first and second row.

Step 2: Now, rearrange the columns of the usual simplex table so that the columns forming the original unit matrix come first.

Step 3: Find the minimum of the ratio

$$\left[\frac{\text{elements of first column of unit matrix}}{\text{corresponding elements of key column}}\right]$$

Only for the rows for which the minimum ratio was not unique (i.e. for tied rows, in our example for first and second rows).

(i) If this minimum is attained for second row (say) then this row will determine the key element by intersecting with the key column.

(ii) If this minimum is also not unique, then go to the next step.

Step 4: Compute the minimum of the ratio

$$\left[\frac{\text{elements of second column of unit matrix}}{\text{corresponding elements of key column}}\right]$$

only for the rows for which minimum ratio is not unique in step 3.
If the minimum is also not unique, then go to next step.

Step 5: Next compute the minimum of the ratio

$$\left[\frac{\text{elements of third column of unit matrix}}{\text{corresponding elements of key column}}\right]$$

If this minimum is still not unique, then go on repeating the above procedure till the unique minimum ratio is obtained to resolve the degeneracy. After the resolution of this tie, simplex method is applied to obtain the optimum solution.

Note: If we have a tie for an artificial variable and some other variable we can choose the artificial variable to leave the basis.

Example 1

Solve the LPP

$$\text{Maximise } z = 3x_1 + 9x_2$$

subject to

$$x_1 + 4x_2 \le 8$$
$$x_1 + 2x_2 \le 4$$
$$x_1, x_2 \ge 0$$

Solution: Introduce the slack variables $s_1, s_2 \ge 0$, the LPP becomes

$$\text{Maximise } z = 3x_1 + 9x_2 + 0s_1 + 0s_2$$

subject to

$$x_1 + 4x_2 + s_1 = 8$$
$$x_1 + 2x_2 + s_2 = 4$$
$$x_1, x_2, s_1, s_2 \ge 0$$

An obvious initial basic feasible solution is $s_1 = 8, s_2 = 4$.

table 2.38

		c_j	(3	9	0	0)	
C_B	Y_B	X_B	x_1	$x_2 \downarrow$	s_1	s_2	*Min. Ratio*
0	s_1	8	1	4	1	0	8/4 = 2
0	s_2	4	1	2	0	1	4/2 = 2
			- 3	- 9	0	0	

From Table 2.37, we observe that the non-basic variable x_2 enters into the basis. Since the minimum ratio is 2 for both the slack variables s_1 and s_2, there is a tie for the variable to leave the basis. This is an indication for the existence of degeneracy in the given LPP. Rearrange the columns of the simplex table so that the initial identity matrix appears first.

table 2.39

		c_j	(0	0	3	9)	
C_B	Y_B	X_B	s_1	s_2	x_1	$x_2 \downarrow$	*Min. Ratio*
0	s_1	8	1	0	1	4	1/4
0	s_2	4	0	1	1	[2]	0/2 = 0
$z_j - c_j$		0	0	0	- 3	- 9	

Now, using the step 3 of the procedure for resolving degeneracy, we find

$$\text{Min}\left[\frac{\text{elements of first column}}{\text{corresponding elements of key column}}\right]$$

$$= \text{Min}\left[\frac{1}{4}, \frac{0}{2}\right] = 0.$$

which occurs for the second row. Hence, s_2 must leave the basis and the pivot element is 2.

First iteration:

table 2.40

C_B	Y_B	X_B	s_1	s_2	y_1	x_2	*Min Ratio*
0	s_1	0	1	- 2	- 1	0	
9	x_2	2	0	1/2	1/2	1	
$z_j - c_j$		18	0	9/2	3/2	0	

Since all $z_j - c_j \geq 0$, an optimum solution has been reached. Hence, the optimum basic feasible solution is

$$\boxed{x_1 = 0, x_2 = 2 \text{ with maximum of } z = 18}$$

xample 2

lve the following LPP

$$\text{Maximise } z = 5x_1 - 2x_2 + 3x_3$$

subject to the constraints

$$2x_1 + 2x_2 - x_3 \geq 2$$
$$3x_1 - 4x_2 \leq 3$$
$$x_2 + 3x_3 \leq 5$$
$$x_1, x_2, x_3 \geq 0$$

lution: Introducing surplus variable $s_1 \geq 0$, slack variables $s_2, s_3 \geq 0$ and an artificial variable $A_1 \geq 0$ the given constraints the LPP becomes

$$\text{Maximise } z = 5x_1 - 2x_2 + 3x_3 + 0s_1 + 0s_2 + 0s_3 - MA_1$$

subject to

$$2x_1 + 2x_2 - x_3 - s_1 + A_1 = 2$$
$$3x_1 - 4x_2 + s_2 = 3$$
$$x_2 + 3x_3 + s_3 = 5$$
$$x_1, x_2, x_2, s_1, s_2, s_3, A_1 \geq 0$$

An obvious initial basic feasible solution is $A_1 = 2$, $s_2 = 3$, $s_3 = 5$.

Starting table:

able 2.41

C_B	Y_B	X_B	x_1	x_2	x_3	s_1	s_2	s_3	A_1	Min Ratio
$-M$	A_1	2	2	2	-1	-1	0	0	1	$2/2 = 1$
0	s_2	3	3	-4	0	0	1	0	0	$3/3 = 1$
0	s_3	5	0	1	3	0	0	1	0	
		$-2M$	$-2M - 5$	$-2M + 2$	$M - 3$	M	0	0	0	

From the Table 2.41, we observe that the variable x_1 enters into the basis. Since the minimum ratio 1, for both the first and the second rows, therefore either A_1 or s_2 tends to leave the basis. This is an dication of the existence of degeneracy. But A_1, being an artificial variable will be preferred to leave e basis. Continuing the routine simplex procedure we get the optimal solution as

$$x_1 = 23/3, x_2 = 5, x_3 = 0 \text{ with maximum of } z = 85$$

2. Unbounded solutions: The case of unbounded solutions occur when the feasible region is nbounded such that the value of the objective function can be increased indefinitely. It is not necessary, owever, that an unbounded feasible region will give rise to an unbounded value of the objective funcon. The following example will illustrate these points.

xample 3

Unbounded solution space but bounded optimal solution).

Solve the following LPP

$$\text{Maximise } z = 3x_1 - x_2$$

subject to

$$x_1 - x_2 \leq 10$$

$$x_1 \le 20$$
$$x_1, x_2 \ge 0$$

lution: By introducing the slack variables $s_1, s_2 \ge 0$, the LPP becomes

$$\text{Maximise } z = 3x_1 - x_2 + 0s_1 + 0s_2$$

subject to
$$x_1 - x_2 + s_1 = 10$$
$$x_1 + s_2 = 20$$
$$x_1, x_2, s_1, s_2 \ge 0$$

An obvious initial basic feasible solution is $s_1 = 10, s_2 = 20$. The successive simplex iteration proce-re will give the optimum solution as

$$\boxed{x_1 = 20, x_2 = 10 \text{ with maximum of } z = 50}$$

The graphical solution of the LPP is given in Fig. 2.7.

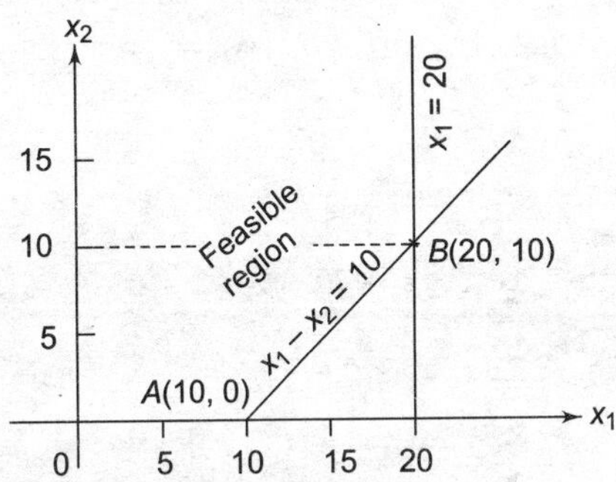

igure 2.7

We observe that the solution space is unbounded. But the optimal solution occurs at the vertex 0, 10).

xample 4 Unbounded Solution Space and Unbounded Optimal Solution

olve the following LPP

$$\text{Maximise } z = 2x_1 + x_2$$

subject to
$$x_1 - x_2 \le 10$$
$$2x_1 - x_2 \le 40$$
$$x_1 \ge 0, x_2 \ge 0$$

olution: Introducing the slack variables $s_1, s_2 \ge 0$ the LPP in standard form is

$$\text{Maximise } z = 2x_1 + x_2 + 0s_1 + 0s_2$$

subject to
$$x_1 - x_2 + s_1 = 10$$
$$2x_1 - x_2 + s_2 = 40$$
$$x_1, x_2, s_1, s_2 \ge 0$$

An obvious IBFS is $s_1 = 10, s_2 = 40$.

Starting Table:

table 2.42

C_B	Y_B	X_B	x_1 ↓	x_2	s_1	s_2	Min Ratio
		c_j	(2	1	0	0)	
0	$\overleftarrow{s_1}$	10	[1]	−1	1	0	$\frac{10}{1} = 10$
0	s_2	40	2	−1	0	1	$\frac{40}{2} = 20$
$z_j - c_j$		10	−2	−1	0	0	

First iteration: Introduce x_1 and drop s_1

table 2.43

C_B	Y_B	X_B	x_1	x_2 ↓	s_1	s_2	Min Ratio
		c_j	(2	1	0	0)	
2	x_1	10	1	−1	1	0	—
0	$\overleftarrow{s_2}$	20	0	[1]	−2	1	$\frac{20}{1} = 20$
		20	2	−2	2	0	

Second iteration: Introduce x_2 and drop s_2.

table 2.44

C_B	Y_B	X_B	x_1	x_2	s_1 ↓	s_2	Min Ratio
		c_j	(2	1	0	0)	
2	x_1	30	1	0	−1	1	
1	x_2	20	0	1	−2	1	
$z_j - c_j$		80	0	0	−4	3	

From Table 2.44 we see that s_1 column is the pivotal column, but there is no positive element in that column. Hence, there exists an unbounded solution to the given LPP. If we solve the LPP by graphical method we can see that the feasible region is unbounded [exercise to the reader].

3. Multiple Solutions or Alternate Optimal Solutions

While solving the LPP by simplex method, in the optimum simplex table, if the net evaluation $z_j - c_j \neq 0$ for all non-basic variables, then the problem is said to have a unique optimal solution. On the other hand, if the net evaluation $z_j - c_j = 0$ for at least one non-basic variable, then the problem is said to have an alternative or infinite number of solutions.

In a graphical method if the optimal solution occurs at a vertex of the solution space, then the problem is said to have a unique optimal solution. If the optimum solution occurs on an edge of the solution space, then the problem is said to have an alternative or infinite number of solutions.

Example 5

Consider the LPP represented in Example 2 of Section 2.5. From Table 2.36, we observe that the net evaluation for the non-basic variable s_1 is zero. By taking s_1 as the entering variable we can continue the iterative process.

Third iteration:

table 2.45

		c_j	(1	2	0	0	0)	
C_B	Y_B	X_B	x_1	x_2	s_1	s_2	s_3	*Min Ratio*
2	x_2	3	0	1	0	1/3	- 1/3	
1	x_1	6	1	0	0	1/3	2/3	
0	s_1	8	0	0	1	- 1/3	4/3	
$z_j - c_j$		12	0	0	0	1	0	

Since all $z_j - c_j \geq 0$, the current solution is optimal. Hence, an alternative optimal solution is

$$x_1 = 6,\ x_2 = 3 \text{ and maximum of } z = 12$$

We know that if two basic (optimum) feasible solutions are known, an infinite number of non-basic (optimum) feasible solutions can be obtained by taking any weighted average of the two solutions.

Thus if, $X_1 = [2, 5, 0, 0, 6]$, $X_2 = [6, 3, 8, 0\ 0]$ are two BFS, then a NBFS is

$$X^* = \lambda X_1 + (1-\lambda)X_2 = \begin{bmatrix} 2\lambda + 6(1-\lambda) \\ 5\lambda + 3(1-\lambda) \\ 8(1-\lambda) \\ 0 \\ 6\lambda \end{bmatrix} = \begin{bmatrix} 6-4\lambda \\ 3+2\lambda \\ 8(1-\lambda) \\ 0 \\ 6\lambda \end{bmatrix}, 0 \leq \lambda \leq 1.$$

We can verify that the solution X will always give the same value of z for all $0 \leq \lambda \leq 1$.

4. No feasible *solution or Non-existing feasible solution*

In an LPP, where there is no point in the solution space satisfying all the constraints, then the problem is said to have no feasible solution. In simplex method, if there exists at least one artificial variable in the basis at positive level, and even though optimality conditions are satisfied which is the indication of non-feasible solution.

Example 6

Consider the LPP given in Example 2 of Section 2.6. From Table 2.45 we have observed that even though all $z_j - c_j \geq 0$ (optimality condition is satisfied) there exists no feasible solution to the problem since the artificial variable A_1 appears at + ve level (since the value of $A_1 = 4$).

Unrestricted Variable Unconstrained Variables

an LPP, if any variable is unrestricted (it can have positive value or negative value or zero value) it n be expressed as the difference between two non-negative variables. The problem can be converted o an equivalent one involving only non-negative variables.

ample 7

lve the LPP

$$\text{Maximise } z = 2x_1 + 3x_2$$

subject to

$$-x_1 + 2x_2 \le 4$$
$$x_1 + x_2 \le 6$$
$$x_1 + 3x_2 \le 9$$

and x_1, x_2 are unrestricted.

lution: Since x_1, x_2 are unrestricted, we put $x_1 = x'_1 - x''_1$ and $x_2 = x'_2 - x''_2$ so that $x'_1, x''_1, x'_2,$ ≥ 0.

The given LPP becomes

$$\text{Maximise } z = 2x'_1 - 2x''_1 + 3x'_2 - 3x''_2$$

subject to

$$-x'_1 + x''_1 + 2x'_2 - 2x''_2 \le 4$$
$$x'_1 - x''_1 + x'_2 - x''_2 \le 6$$
$$x'_1 - x''_1 + 3x'_2 - 3x''_2 \le 9$$
$$x'_1, x''_1 x'_2, x''_2 \ge 0$$

If we solve the LPP by ordinary simplex method (an exercise to the reader) the optimum solution is

$$x_1 = x'_1 - x''_1 = 9/2 - 0 = 9/2$$
$$x_2 = x'_2 - x''_2 = 3/2 - 0 = 3/2$$
with maximum of $z = 27/2$

8 SOLUTION OF SIMULTANEOUS EQUATIONS BY SIMPLEX METHOD

the solution of n simultaneous linear equations in n variables a dummy objective function is intro- ed as

$$\text{Maximise } z = 0X - 1A_i$$

where A_i are artificial variables, and $x_r = x'_r - x''_r$ such that $x'_r > 0, x''_r > 0$.

The reformulated linear programming problem is then solved by simplex method.

ample 1

simplex method to solve the following system of linear equations

$$x_1 - x_3 + 4x_4 = 3$$
$$2x_1 - x_2 = 3$$
$$3x_1 - 2x_2 - x_4 = 1$$
$$x_1, x_2, x_3, x_4 \ge 0$$

tion: Since the objective function for the given constraint equation is not prescribed, so a dummy ctive function is introduced as:

$$\text{Maximise } z = 0x_1 + 0x_2 + 0x_3 + 0x_4 - 1A_1 - 1A_2 - 1A_3$$

where, $A_1 \geq 0, A_2 \geq 0, A_3 \geq 0$ are the artificial variables. Given

$$x_1 - x_3 + 4x_4 + A_1 = 3$$
$$2x_1 - x_2 + A_2 = 3$$
$$3x_1 - 2x_2 + A_3 = 1$$

If we apply simplex method to solve the reformulated LPP, we can obtain the optimum solution a

$$\boxed{x_1 = 5, x_2 = 7, x_3 = 2 \text{ and } x_4 = 0}$$

2.9 INVERSE OF A MATRIX BY SIMPLEX METHOD

Let A be an $n \times n$ matrix. Let b be any dummy $n \times 1$ real matrix. Consider the equation

$$AX = b, X \geq 0$$

Introduce a dummy objective function

$$z = 0X - 1A_1$$

where, A_1 is an artificial variable. Then, the inverse of the optimal basis B consists of those columns the last iterations of simplex method which were present in the initial basis. In addition, if in the l. iteration the columns of A become the columns of I, then $B^{-1} = A^{-1}$.

Example 1

Use simplex method to find the inverse of the matrix

$$A = \begin{pmatrix} 3 & 2 \\ 4 & -1 \end{pmatrix}$$

Solution: Consider

$$\begin{pmatrix} 3 & 2 \\ 4 & -1 \end{pmatrix}\begin{pmatrix} x_1 \\ x_2 \end{pmatrix} = \begin{pmatrix} 4 \\ 6 \end{pmatrix}, x_1, x_2 \geq 0$$

where, $\begin{pmatrix} 4 \\ 6 \end{pmatrix}$ is a dummy vector. Apply the simplex method to maximise the dummy objective funct

$z = 0x_1 + 0x_2 - 1A_1 - 1A_2$ subject to the above constraints on x_1, x_2 where $A_1 \geq 0$ and $A_2 \geq 0$ are artific variables.

Starting Table:

table 2.46

		c_j	(0	0	−1	−1)	
C_B	Y_B	X_B	$x_1 \downarrow$	x_2	A_1	A_2	*Min Ratio*
−1	$\overleftarrow{A_1}$	4	[3]	2	1	0	4/3
−1	A_2	6	4	−1	0	1	6/4
			−7	−1	0	0	

From Table 2.46, we observe that the non-basic variable x_1 enters into the basis and the basic variable A_1 leaves the basis.

First iteration: Introduce x_1 and drop A_1.

table 2.47

		c_j	(0	0	− 1	− 1)	
C_B	Y_B	X_B	x_1	x_2	A_1	A_2	*Min Ratio*
0	x_1	4/3	1	2/3	1/3	0	
− 1	A_2	2/3	0	11/3	− 4/3	1	
$z_j - c_j$		− 2/3	0	11/3	7/3	0	

From the Table 2.47, we observe that even though the optimality condition is satisfied the last columns of A does not become the columns of I. So introduce x_2 and drop A_2.

Second iteration:

table 2.48

		c_j	(0	0	− 1	− 1)	
C_B	Y_B	X_B	x_1	x_2	A_1	A_2	*Min Ratio*
0	x_1	16/11	1	0	1/11	2/11	
0	x_2	− 2/11	0	1	4/11	− 3/11	
		0	0	0	1	1	

Now, an optimum but infeasible solution has been obtained. Since the initial basis consists of the column vectors A_1 and A_2, the inverse of the matrix A is

$$A^{-1} = \begin{bmatrix} 1/11 & 2/11 \\ 4/11 & -3/11 \end{bmatrix}$$

EXERCISES

Section 2.2

1. A company produces two types of hats. Each hat of the first type requires twice as much labour time as the second type. If all hats are of the second type only, the company can produce a total of 500 hats a day. The market limits daily sales of the first and second type of 150 and 250 hats. Assuming that the priority per hat are Rs 8 for type A and Rs 5 for type B, formulate the problem as a linear programming model in order to determine the number of hats to be produced of each type so as to maximise the profit.
[*Hint*: x_1 = number of hats of type A, x_2 = the number of hats of type B. Objective function is $z = 8x_1 + 5x_2$, since hats of type A require twice as much time as that of type B, $2x_1 + x_2 \leq 500$. Further constraints are $x_1 \leq 150$, $x_2 \leq 250$, and non-negativity restriction is $x_1, x_2 \geq 0$]

2. A firm manufactures 3 products *A*, *B* and *C*. The profits are Rs 3, Rs 2 and Rs 4, respectively. The firm has two machines and below is the processing time required in minutes on each machine for each product.

		Product		
		A	*B*	*C*
Machine	*C*	4	3	6
	D	3	2	4

Machine *C* and *D* have 2,000 and 2,500 machine minutes, respectively. The firm must manufacture 100 A's, 200 B's and 500 C's, but not more than 150 *A*'s. Set up an linear programming problem to maximise profit.

[*Hint*: Let x_1, x_2, x_3 be the number of products of *A*, *B*, *C*, respectively. Objective function is to maximise $z = 3x_1 + 2x_2 + 4x_3$, since the total number of minutes required in producing these three products at machine *C* is given by $4x_1 + 3x_2 + 5x_3$ and it has 2,000 minutes, so we have $4x_1 + 3x_2 + 5x_3 \leq 2000$. Similarly, for *D*, $3x_1 + 2x_2 + 4x_3 \leq 2{,}500$. Also, since the firm manufactures 100 *A*'s, 200 *B*'s and 50 C's, but not more than 150 *A*'s and so $100 \leq x_1 \leq 150$, $200 \leq x_2$, $50 \leq x_3$.]

3. A complete unit of a certain product consists of four units of component *A* and three units of component *B*. The two components (*A* and *B*) are manufactured from two different raw materials of which 100 units and 200 units, respectively are available. Three departments are engaged in the production process with each department using a different method for manufacturing the components. The following Table gives the raw material requirements per production run and the resulting units of each component. The objective is to determine the number of production runs for each department which will maximise the total number of component units of the final product.

Department	Input per run		Output per run	
	Raw material I	Raw material II	Component A	Component B
1	7	5	6	4
2	4	8	5	8
3	2	7	7	3

[*Hint*: x_1, x_2, x_3 be the number of production runs for the departments 1, 2, 3, respectively. Then, the required LPP model is

$$\text{Maximise } z = \min\left\{\frac{1}{4}(6x_1 + 5x_2 + 7x_3),\ \frac{1}{3}(4x_1 + 8x_2 + 3x_3)\right.$$

subject to

$$6x_1 + 5x_2 + 7x_3 - 4z \geq 0$$
$$4x_1 + 8x_2 + 3x_3 - 4z \geq 0]$$

4. A dairy feed company may purchase and mix one more of the three types of grains containing different amounts of nutritional elements. The data is given in the Table below. The production manager specifies that any feed mix for his livestock must meet atleast minimal nutritional requirements, and seeks the least costly among all such mixes.

	Item	Grain1	One unit weight of Grain2	Grain3	Minimial requirement
Nutritional	A	2	3	7	1,250
ingredients	B	1	1	0	250
	C	5	3	0	900
Cost/unit	D	6	25	1	232.5
weight (Rs)		41	35	96	

Analyse the situation to recognise the key decision objective, alternatives and restrictions. F mulate a linear programming model for the problem.

[*Answer*: Minimise $z = 41x_1 + 35x_2 + 96x_3$

subject to $2x_1 + 3x_2 + 7x_3 \geq 1{,}250$

$$x_1 + x_2 \geq 250$$
$$5x_1 + 3x_2 \geq 900$$
$$6x_1 + 25x_2 + x_3 \geq 232.5$$
$$x_1, x_2, x_3, \geq 0]$$

5 A truck company requires the following number of drivers for its trucks during 24 hours.

Time	Required
00–04 hr	5
04–08 hr	10
08–12 hr	20
12–16 hr	12
16–20 hr	22
20–24 hr	8

According to the shift schedule a driver may join for duty at midnight 04, 08, 12, 16, 20 hours work continuously for 8 hrs. Formulate the problem as linear programming problem for opti shift plan.

[*Hint*: Let x_1, x_2, x_3, x_4, x_5 and x_6 denote the number of drivers joining duty at 00, 04, 08, 12, 20 hours, respectively. The objective function is to minimise the number of drivers required. Tha

Minimise $$z = x_1 + x_2 + x_3 + x_4 + x_5 + x_6$$

Drivers who join duty at 00 hrs and 04 hours shall be available between 04 and 08 hours. As number of drivers required during this interval is 10, we have the constraint

$$x_1 + x_2 \geq 10$$
$$x_2 + x_3 \geq 20$$
$$x_3 + x_4 \geq 12$$
$$x_4 + x_5 \geq 22$$
$$x_5 + x_6 \geq 8$$
$$x_6 + x_1 \geq 5]$$

6. *Investment Problem.* Mr Ramesh, a retired government employee has received his retirement benefits, namely, provident fund, gratuity and others. He is contemplating as to how much funds he should invest in various alternatives open to him so as to maximise return on investment. The investment alternatives are: government securities, fixed deposits of a public limited company, equity shares, time deposits in banks, national saving certificates and real estate. He has made a subjective estimate of the risk involved on a fine point scale. The data on the return on investments, the number of years for which the funds will be blocked to earn this return on investment and the subjective risk involved are as follows:

Investment Alternatives	*Return*	*Number of Years*	*Risk*
Govt. securities	6%	15	1
Company deposits	15%	3	3
Equity shares	20%	6	7
Time deposits	10%	3	2
NSC	12%	6	1
Real estate	25%	10	2

He was wondering what percentage of funds he should invest in each alternative so as to maximise the return on investment. He decided that average risk should not be more than 4, and funds should not be locked up for more than 15 years. He would necessarily invest at least 30% in real estate. Formulate on linear programming model.

[*Hint*: Let x_1, x_2, x_3, x_4, x_5 be the percentages of the total funds that should be invested in government securities, company deposits, equity shares, time deposits, national saving certificate and real estate, respectively. The linear programming model is

$$\text{Maximise } z = 0.06x_1 + 0.15x_2 + 0.20x_3 + 0.10x_4 + 0.12x_5 + 0.25x_6$$

subject to the constraints

$$15x_1 + 3x_2 + 6x_3 + 3x_4 + 6x_5 + 10x_6 \leq 15$$
$$x_1 + 3x_2 + 7x_3 + x_4 + x_5 + 2x_6 \leq 4$$
$$x_6 \geq 0.30$$
$$x_1, x_2, x_3, x_4, x_5, x_6 \geq 0]$$

Section 2.3

7. Find all basic feasible solutions of the equations

$$2x_1 + 6x_2 + 2x_3 + x_4 = 3$$
$$6x_1 + 4x_2 + 4x_3 + 6x_4 = 2$$

[*Answer*: Since A is of order 2×4, there can be $4C_2 = 6$, 2×2 submatrices as our basic matrix.

They are $\begin{bmatrix} 2 & 6 \\ 2 & 4 \end{bmatrix}, \begin{bmatrix} 2 & 2 \\ 6 & 4 \end{bmatrix}, \begin{bmatrix} 2 & 1 \\ 6 & 6 \end{bmatrix}, \begin{bmatrix} 6 & 2 \\ 4 & 4 \end{bmatrix}, \begin{bmatrix} 6 & 1 \\ 4 & 6 \end{bmatrix}$ and $\begin{bmatrix} 2 & 1 \\ 4 & 6 \end{bmatrix}$.

The basic solution to the given system of equations are

(i) $x_1 = 0, x_2 - 1/2$ (basic); $x_3 = x_4 = 0$ (non-basic)

(ii) $x_1 = -2, x_3 = 7/2$ (basic); $x_2 = x_4 = 0$ (non-basic)

(iii) $x_1 = 8/3,\ x_4 = -1/3$ (basic); $x_2 = x_3 = 0$ (non-basic)
(iv) $x_2 = 1/2,\ x_3 = 0$ (basic); $x_1 = x_4 = 0$ (non-basic)
(v) $x_2 = 1/2,\ x_4 = 0$ (basic); $x_1 = x_3 = 0$ (non-basic)
(vi) $x_3 = 2,\ x_4 = -1$ (basic); $x_1 = x_2 = 0$ (non-basic)

The solutions (–2, 0, 7/2, 0), (8/3, 0, 0, –7/3), (0, 0, 2, – 1) are not feasible. The degenerate basic feasible solutions of the system are (0, 1/2, 0, 0), (0, 1/2, 0, 0), (0, 1/2, 0, 0).

8. Is the following solution

$$x_1 = 1, \quad x_2 = 1/2, \quad x_3 = x_4 = x_5 = 0$$

a basic solution of the equations

$$x_1 + 2x_2 + x_3 + x_4 = 2$$
$$x_1 + 2x_2 + \frac{1}{2}x_3 + x_5 = 2$$

[*Hint*: Let the equation be $AX = b$, where

$$A = \begin{pmatrix} 1 & 2 & 1 & 1 & 0 \\ 1 & 2 & 1/2 & 0 & 1 \end{pmatrix}; \; X = \begin{pmatrix} x_1 \\ x_2 \\ x_3 \\ x_4 \\ x_5 \end{pmatrix}; \; b = \begin{pmatrix} 2 \\ 2 \end{pmatrix}$$

Since A is of order 2×5, we can take any of the following $5C_2 = 10$, 2×2 submatrices as the basis matrix B.

$$\begin{bmatrix} 1 & 2 \\ 1 & 2 \end{bmatrix}, \begin{bmatrix} 1 & 1 \\ 1 & 1/2 \end{bmatrix}, \begin{bmatrix} 1 & 1 \\ 1 & 0 \end{bmatrix}, \begin{bmatrix} 1 & 0 \\ 1 & 1 \end{bmatrix}, \begin{bmatrix} 2 & 1 \\ 2 & 1/2 \end{bmatrix}, \begin{bmatrix} 2 & 1 \\ 2 & 0 \end{bmatrix}, \begin{bmatrix} 2 & 0 \\ 2 & 1 \end{bmatrix}, \begin{bmatrix} 1 & 1 \\ 1/2 & 0 \end{bmatrix}, \begin{bmatrix} 1 & 0 \\ 1/2 & 1 \end{bmatrix}, \begin{bmatrix} 1 & 0 \\ 0 & 1 \end{bmatrix}$$

Dropping first matrix, using remaining matrix, we get the solution as

(i) $x_1 = 2,\ x_3 = 0$ (basic); $x_2 = x_4 = x_5 = 0$ (non-basic)
(ii) $x_1 = 2,\ x_4 = 0$ (basic); $x_2 = x_3 = x_5 = 0$ (non-basic)
(iii) $x_1 = 2,\ x_5 = 0$ (basic); $x_2 = x_3 = x_4 = 0$ (non-basic)
(iv) $x_2 = 1,\ x_3 = 0$ (basic); $x_1 = x_4 = x_5 = 0$ (non-basic)
(v) $x_2 = 1,\ x_4 = 0$ (basic); $x_1 = x_3 = x_5 = 0$ (non-basic)
(vi) $x_2 = 1,\ x_5 = 0$ (basic); $x_1 = x_3 = x_4 = 0$ (non-basic)
(vii) $x_3 = 4,\ x_4 = -2$ (basic); $x_1 = x_2 = x_5 = 0$ (non-basic)
(viii) $x_3 = 2,\ x_5 = 1$ (basic); $x_1 = x_2 = x_4 = 0$ (non-basic)
(ix) $x_4 = 2,\ x_5 = 2$ (basic); $x_1 = x_2 = x_3 = 0$ (non-basic)

Clearly, none of these corresponds to the given solution, namely, $x_1 = 1, x_2 = 1/2, x_3 = x_4 = x_5 = 0$. Hence, the given solution is not basic.]

9. Express the following problem in the standard form.

$$\text{Maximise } z = 4x_1 + x_2 - 3x_3$$

$$x_1 + 5x_2 - 3x_3 \leq 20$$
$$2x_1 + 7x_2 + 2x_3 \leq 10$$
$$x_1 - 5x_2 - 3x_2 \geq 3$$
$$x_1 \geq 0$$

[*Answer*: Maximise $z = 4y_1 + y_2 - y_3 - 3y_4 + 3y_5$

subject to $y_1 + 5y_2 - 5y_3 - 3y_4 + 3y_5 + y_6 = 20$

$$2y_1 + 7y_2 - 7y_3 + 2y_4 - 2y_5 + y_7 = 10$$
$$y_1 - 5y_2 + 5y_3 - 3y_4 + 3y_5 - y_8 = 3$$
$$y_1, y_2, \ldots, y_8 \geq 0]$$

10. Express the following linear programming problem in the matrix form

Minimise $z = 4x_1 + 5x_2 + 6x_3$

subject to $2x_1 + 3x_3 \geq 6$

$$2x_1 + 3x_2 + 2x_3 = 8$$
$$5x_1 - 2x_2 + 5x_3 \leq 10,$$
$$x_1, x_2, x_3 \geq 0$$

[*Answer*: Minimise $z = (4, 5, 6, 0, 0)\ (x_1, x_2, x_3, x_4, x_5)$

subject to

$$\begin{pmatrix} 2 & 0 & 3 & -1 & 0 \\ 2 & 3 & 2 & 0 & 0 \\ 5 & -2 & 5 & 0 & 1 \end{pmatrix} \begin{pmatrix} x_1 \\ x_2 \\ x_3 \\ x_4 \\ x_5 \end{pmatrix} = \begin{pmatrix} 6 \\ 8 \\ 10 \end{pmatrix}$$

$x_1, x_2, \ldots, x_5 \geq 0$

Section 2.4

11. Solve graphically the following L.P.P.

Maximise $z = 3x_1 + 2x_2$

subject to the constraints

$$-2x_1 + x_2 \leq 1$$
$$x_1 \leq 2$$
$$x_1 + x_2 \leq 3$$
$$x_1, x_2 \leq 0$$

[*Hint*: The graph is

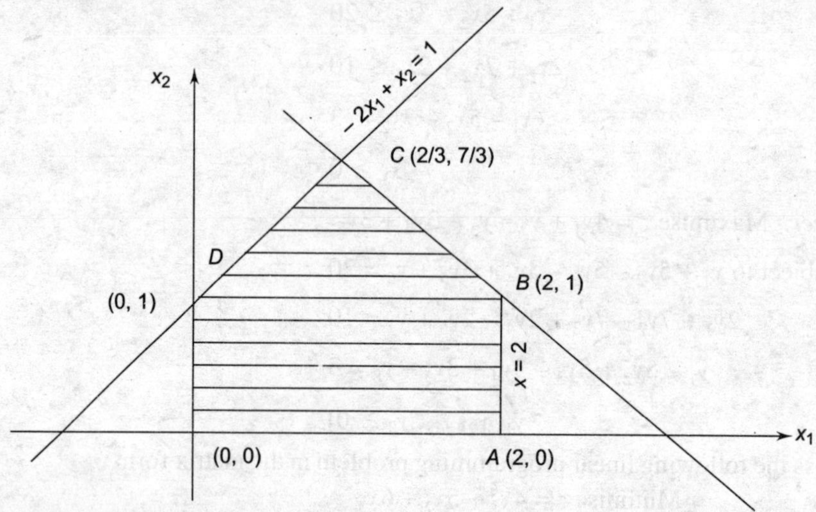

The maximum value of z is 8 and it occurs at the vector $B = (2, 1)$.]

12. Solve the following by graphical method

$$\text{Maximise } z = x_1 - 3x_2$$
$$\text{subject to } x_1 + x_2 \leq 300$$
$$x_1 - 2x_2 \leq 200$$
$$2x_1 + x_2 \geq 100$$
$$x_2 \leq 200$$
$$x_1, x_2 \geq 0$$

[*Answer*: Max $z = 200$, $x_1 = 200$, $x_2 = 0$]

13. Minimise $z = x - 3y$
subject to the constraints

$$x + y \leq 300$$
$$x - 2y \leq 200$$
$$2x + y \geq 100$$
$$x, y \geq 0$$

by graphical method.
[*Answer*: Min $z = -600$, $x = 0$, $y = 200$]

14. Solve graphically the following L.P.P.
Max $z = 3x + 2y$
subject to

$$-2x + 3y \leq 9$$
$$x - 5y \geq -20$$
$$x, y \geq 0$$

[*Answer*: Unbounded solution]

15. Solve graphically the following LPP

$$\text{Maximise } z = 3x_1 - 2x_2$$

subject to

$$x_1 + x_2 \leq 1$$

$$2x_1 + 2x_2 \geq 4, \quad x_1, x_2 \geq 0$$

[*Answer*: No solution]

16. A manufacturer of furniture makes two products, chairs and tables. Processing of these products is done on two machines *A* and *B*. A chair requires 2 hours on machine *A* and 6 hours on machine *B*. A table requires 5 hours on machine *A* and no time on machine *B*. There are 16 hours of time per day available on machine *A* and 30 hours on machine *B*. Profit gained by the manufacturer from a chair and a table is Rs 2 and Rs 10, respectively, what should be the daily production of each of the products.

[*Answer*: Maximise $z = 2x_1 + 10x_2$

subject to $2x_1 + 5x_2 \leq 6$

$6x_1 \leq 30$

$x_1, x_2 \geq 0$

Solving graphically, we get

$$\text{Maximum } z = 32, x_1 = 0, x_2 = 3.2.$$

17. Maximise $z = 3x + 2y$

subject to $-2x_2 + 3y \leq 9$

$3x - 2y \leq -20$

$x, y \geq 0.$

Hint:

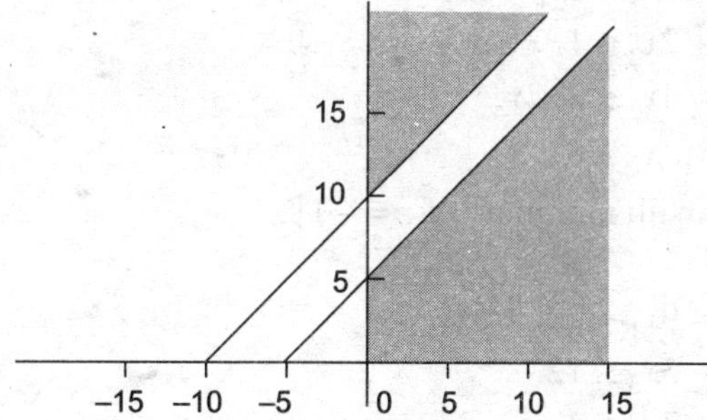

[*Answer*: No solution]

18. Maximise $z = 3x_1 + 4x_2$

subject to $x_1 - x_2 \geq 0$

$2.5x_1 - x_2 \leq -3$

$x_1, x_2 \geq 0.$

Hint:

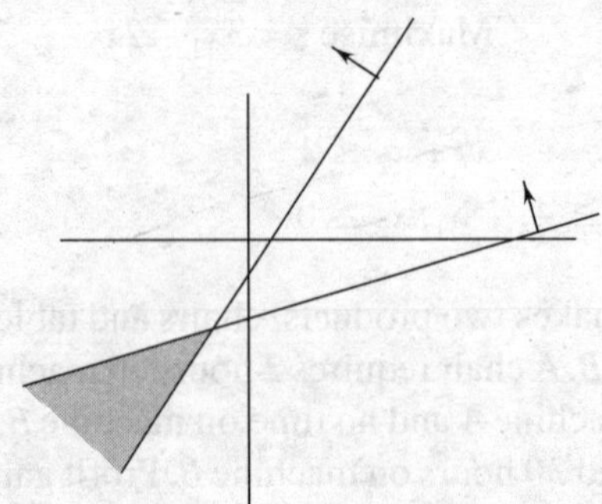

Any point within this region satisfies the constraints but not the non-negativity restrictions. Although the constraints are consistent, the problem does not possess a feasible solution.

Section 2.5

19. Find the non-negative values of x_1, x_2 an x_3 which maximise

$$z = 3x_1 + 2x_2 + 5x_3$$

subject to

$$x_1 + 4x_2 \leq 420$$
$$3x_1 + 2x_3 \leq 460$$
$$x_1 + 2x_2 + x_3 \leq 430$$
$$x_1, x_2, x_3 \geq 0$$

[*Answer*: $x_1 = 0$, $x_2 = 100$, $x_3 = 230$ with maximum of $z = 1{,}350$]

20. Solve the following LPP by simplex method

$$\text{Minimise } z = 8x_1 - 2x_2$$
$$\text{subject to } -4x_1 + 2x_2 \leq 1$$
$$5x_1 - 4x_2 \leq 3$$
$$x_1, x_2 \geq 0$$

[*Answer*: $x_1 = 0$, $x_2 = 1/2$ with minimum of $z = -1$]

21. Solve by simplex method

$$\text{Maximise } z = 9x_1 + 2x_2 + 5x_3$$
$$\text{subject to } 2x_1 + 3x_2 - 5x_3 \leq 12$$
$$2x_1 - x_2 + 3x_3 \leq 3$$
$$3x_1 + x_2 - 2x_3 \leq 2$$

[*Answer*: Maximum $z = 637/13$, $x_1 = 0$, $x_2 = 12$, $x_3 = 65/13$]

22. Use simplex method to solve the following LPPs

(i) Maximise $z = 3x_1 + 2x_2$

subject to $x_1 + x_2 \leq 4,$

$x_1 - x_2 \leq 2,$

$x_1 \geq 0$ and $x_2 \geq 0$

(ii) Maximise $z = 5x_1 + 3x_2$

subject to

$$x_1 + x_2 \le 2$$
$$5x_1 + 2x_2 \le 10$$
$$3x_1 + 8x_2 \le 12$$

and, $x_1, x_2 \ge 0$

(iii) Maximise $z = x_1 - x_2 + 3x_3$

subject to

$$x_1 + x_2 + x_3 \le 10$$
$$2x_1 - x_3 \le 2$$
$$2x_1 - 2x_2 + 3x_3 \le 0$$

and, $x_1, x_2, x_3 \ge 0$

[*Answer*: (i) $x_1 = 3$ and $x_2 = 1$ with maximum $z = 4$
(ii) $x_1 = 2$ and $x_2 = 0$ with maximum of $z = 10$.
(iii) $x_1 = 0, x_2 = 6$ and $x_3 = 4$ with maximum of $z = 6$]

23. Show that the following LPP

$$\text{Maximise } z = 4x_1 + x_2 + 3x_3 + 5x_4$$

subject to the constraints

$$4x_1 - 6x_2 - 5x_3 - x_4 \ge -20$$
$$3x_1 - 2x_2 + 4x_3 + x_4 \le 10$$
$$8x_1 - 3x_2 + 3x_3 + 2x_4 \le 20,$$
$$x_1, x_2, x_3, x_4, \ge 0$$

has an unbounded solution.

24. Show that the following LPP

$$\text{Maximise } z = 15x_1 + 6x_2 + 9x_3 + 2x_4$$

subject to the constraints

$$2x_1 + x_2 + 5x_3 + 0.6x_4 \le 10,$$
$$3x_1 + x_2 + 3x_3 + 0.25x_4 \le 12$$
$$7x_1 + x_4 \le 35,$$
$$x_j \ge 0 \text{ for } j = 1, 2, 3, 4$$

has an unbounded solution.

Section 2.6

25. Maximise using Big-*M* method

$$z = 3x_1 - x_2$$

subject to

$$2x_1 + x_2 \ge 2$$
$$x_1 + 3x_2 \le 3$$
$$x_2 \le 4$$

$$x_1, x_2 \geq 0.$$

[*Answer*: $z = 9$ at $x_1 = 3$, $x_2 = 0$]

26. Maximise $z = x_1 + 2x_2 + 3x_3 - x_4$

subject to $x_1 + 2x_2 + 3x_3 = 15$

$$2x_1 + x_2 + 5x_3 = 20$$

$$x_1 + 2x_2 + x_3 + x_4 = 10$$

$$x_1, x_2, x_3, x_4 \geq 0.$$

[*Answer*: $z = 15$, $x_1 = 5/2$, $x_2 = 5/2$, $x_3 = 5/2$, $x_4 = 0$]

27. Use M-technique to

Minimise $z = 4x_1 + x_2$

subject to

$$3x_1 + x_2 = 3$$

$$4x_1 + 3x_2 \geq 6$$

$$x_1 + 2x_2 \leq 3$$

$$x_1, x_2 \geq 0$$

[*Answer*: $z = 18/5$ at $x_1 = 3/5$, $x_2 = 6/5$]

28. Use Charnes penalty method to solve the problem

Maximise $z = 4x_1 + 5x_2 + 2x_3$

subject to $2x_1 + x_2 + x_3 \leq 10$

$$x_1 + 3x_2 + x_3 \leq 12$$

$$x_1 + x_2 + x_3 = 6$$

$$x_1, x_2, x_3 \geq 0$$

29. Solve the following problem by using x_4, x_5, x_6 for the starting basic feasible solution:

Maximise $z = 3x_1 + x_2 + 2x_3$

subject to $12x_1 + 3x_2 + 6x_3 + 3x_4 = 9$

$$8x_1 + x_2 - 4x_3 + 2x_5 = 10$$

$$3x_1 - x_6 = 0$$

$$x_1, x_2, x_3, x_4, x_5, x_6 \geq 0.$$

[*Answer*: $x_1 = x_2 = x_4 = x_6 = 0$, $x_3 = 3/2$, $x_5 = 8$, Maximum $z = 3$]

30. Minimise $z = 2x_1 + 9x_2 + x_3$,
subject to the constaints

$$x_1 + 4x_2 + 2x_3 \geq 5$$

$$3x_1 + x_2 + 2x_3 \geq 4$$

$$x_1, x_2, x_3 \geq 0$$

[*Answer*: $x_1 = 0$, $x_2 = 0$, $x_3 = 5/2$, $z = 5/2$]

31. Solve the LPP using two-phase method

Minimise $z = 2x_1 + x_2$

subject to $5x_1 + 10x_2 - x_3 = 8$

$x_1 + x_2 + x_4 = 1$

$x_1, x_2, x_3, x_4 \geq 0$

$$\left[Answer: x_1 = 0, x_2 = \frac{4}{5}, x_3 = 0, x_4 = \frac{1}{5}, z = \frac{4}{5}\right]$$

32. Minimise $z = \left(\frac{15}{2}\right) x_1 - 3x_2$

subject to $3x_1 - x_2 - x_3 \geq 3$

$x_1 - x_2 + x_3 \geq 2$

$x_1, x_2, x_3 \geq 0$

Using two-phase method

$$\left[Answer: x_1 = 5/4, x_2 = 0, x_3 = 3/4, z = \frac{75}{8}\right]$$

33. Minimise $z = 2x_1 + 3x_2 + 2x_3 - x_4 + x_5$

subject to $3x_1 - 3x_2 + 4x_3 + 2x_4 - x_5 = 0$

$x_1 + x_2 + x_3 + 3x_4 + x_5 = 2$

$x_j \geq 0, j = 1 \text{ to } 5$

$$\left[Answer: x_1 = x_2 = x_3 = 0, x_4 = \frac{2}{5}, x_5 = \frac{4}{5}, z = \frac{2}{5}\right]$$

34. Show that there does not exist any feasible solution to the LPP

Maximise $z = 2x_1 + 3x_2 + 5x_3$

subject to $3x_1 + 10x_2 + 5x_3 \leq 15$

$33x_1 - 10x_2 + 9x_3 \leq 33$

$x_1 + 2x_2 + x_3 \geq 4,$

$x_1, x_2, x_3 \geq 0$

3

Advanced Topics in Linear Programming

The dual of a linear programming problem finds a place in this chapter. The economic interpretation of dual variables that can be used by management for planning its resources is also explained in this chapter. If a LPP involving a large number of variables and constraints is solved by this method, it will require a large storage space and time on a computer. Some computational techniques have been developed which require much less computer storage and time than that required by the simplex method. The two important and efficient computational techniques are the revised simplex method or simplex method with multipliers and the bounded variables method. These techniques are dealt with in this chapter. The sensitivity analysis of the problem to study the effect of change in various resource levels is also discussed. This chapter also describes some integer programming formulation techniques and highlights its application in managerial problems where linear programming techniques fail. Further, two methods, namely, *cutting plane method* and *branch and bound* method are discussed to solve integer programming problems. Some advantages and limitation of LPP are also discussed.

3.1 DUALITY THEORY

Every linear programming problem is associated with another unique linear programming problem. The original (given) problem is then called the *primal problem* while the other is called its *dual problem*. The relationship between the solutions of a primal problem and its dual problem prove to be extremely useful in a variety of ways. The solution of one of the problem readily provides the solution of the other one. This fact is important because situations may arise where it will be easier to solve the dual rather than to solve the primal.

3.1.1 The Dual Problem

There are two important forms of primal-dual pairs, namely, symmetric form and unsymmetric form.

Symmetric form:

Let the primal problem be

Maximise $z = c_1 x_1 + c_2 x_2 + \cdots + c_n x_n$

subject to

$$\left.\begin{aligned} a_{11} x_1 + a_{12} x_2 + \cdots + a_{1n} x_n &\le b_1 \\ a_{21} x_1 + a_{22} x_2 + \cdots + a_{2n} x_n &\le b_2 \\ \vdots \qquad \cdots \qquad &\vdots \\ a_{m_1} x_1 + a_{m_2} x_2 + \cdots + a_{m_n} x_n &\le b_n \\ \text{and,} \qquad x_1, x_2, \cdots, x_n &\ge 0 \end{aligned}\right\} \tag{3.1.1}$$

In matrix notation the equation (3.1.1) can be written as

Maximise $z = CX$

subject to $AX \le b$

and, $X \ge 0,\ b^T \in R^m$

where, $A = \begin{bmatrix} a_{11} & \cdots & a_{1n} \\ a_{21} & \cdots & a_{2n} \\ \vdots & & \vdots \\ a_{m1} & \cdots & a_{m_n} \end{bmatrix}$; $X = \begin{bmatrix} x_1 \\ \vdots \\ x_n \end{bmatrix}$; $b = \begin{bmatrix} b_1 \\ \vdots \\ b_m \end{bmatrix}$ and $C = [c_1, \ldots, c_n]$

Remark: It is not necessary that the primal is to be maximised. It can be minimised too.

Steps to be followed to convert a primal to a dual.

Step 1: The maximisation problem in the primal becomes the minimisation problem in the dual and vice versa.

Step 2: The maximisation problem has less than or equal to (≤) type of constraints while the minimisation problem has greater than or equal to (≥) type of constraints.

Step 3: If the primal contains m constraints and n variables, then the dual will contain n constraints and m variables. That is, the transpose of the body matrix of the primal problem gives the body matrix of the dual and vice versa.

Step 4: The constants $c_1, \cdots, c_n$ in the objective function of the primal appear in the right-hand sides of the constraints of the dual.

Step 5: The constants $b_1, \cdots, b_m$ in the constraints of the primal appear in the objective function of the dual.

Step 6: Both the primal and the dual variables are non-negative.

The relationship between the primal and dual problems can be best described by

$$\begin{array}{cc} & \text{Primal variables} \\ & \left[\begin{array}{cccc} x_1 & x_2 & \cdots & x_n \\ a_{11} & a_{12} & \cdots & a_{1n} \\ a_{21} & a_{22} & \cdots & a_{2n} \\ \vdots & \cdots & & \vdots \\ a_{m1} & a_{m2} & \cdots & a_{mn} \\ \ge & \ge & \cdots & \ge \\ c_1 & c_2 & \cdots & c_n \end{array}\right] \begin{array}{l} \\ \le b_1 \ \text{R.H.S. of primal} \\ \le b_2 \ \text{constraints} \\ \vdots \\ \le b_m \\ \\ \\ \end{array} \\ \text{Dual variables} \begin{array}{c} y_1 \\ y_2 \\ \vdots \\ y_m \end{array} & \text{R.H.S. of dual constraints} \end{array}$$

The dual of equation (3.1.1) is

$$\left.\begin{aligned}
&\text{Minimise } w = b_1 y_1 + b_2 y_2 + \cdots + b_m y_m \\
&\text{subject to} \\
&\qquad a_{11} y_1 + a_{21} y_2 + \cdots + a_{m1} y_n \geq c_1 \\
&\qquad a_{12} y_1 + a_{22} y_2 + \cdots + a_{m2} y_m \geq c_2 \\
&\qquad \cdots \\
&\qquad \cdots \\
&\qquad a_{1n} y_1 + a_{2n} y_2 + \cdots + a_{mn} y_m \geq c_n \\
&\text{and,} \qquad y_1, \cdots, y_m \geq 0
\end{aligned}\right\} \quad (3.1.2)$$

In matrix form equation (3.1.2) can be represented in matrix forms as

Minimise $w = b^T Y$

subject to $A^T Y \geq C^T$ and $Y \geq 0$

Equations (3.1.1) and (3.1.2) are called symmetric primal-dual form.

Unsymmetric duals:

For primal problems in standard matrix form duals may be defined as follows:

Primal	Dual
Minimise $z = C^T X$	Maximise $w = b^T Y$
subject to $AX = b$	subject to $A^T Y \leq C$
and $X \geq 0$	and $Y \geq 0$
Maximise $z = C^T X$	Minimise $w = b^T Y$
subject to $AX = b$	subject to $A^T Y \leq C$
and $X \geq 0$	and $Y \geq 0$

Example 1

Determine the dual of the problem

Minimise $z = 5x_1 + 2x_2 + x_3$

subject to

$$\begin{aligned}
2x_1 + 3x_2 + x_3 &\geq 20 \\
6x_1 + 8x_2 + 5x_3 &\geq 30 \\
7x_1 + x_2 + 3x_3 &\geq 40 \\
x_1 + 2x_2 + 4x_3 &\geq 50 \\
\text{and } x_1, x_2, x_3 &\geq 0.
\end{aligned}$$

Solution: The given (primal) problem is in symmetric form. Since there are 4 constraints in the primal there will be 4 variables in the dual. Let it be y_1, y_2, y_3, y_4.

The dual problem is

Maximise $w = 20y_1 + 30y_2 + 40y_3 + 50y_4$

subject to

$$\begin{aligned}
2y_1 + 6y_2 + 7y_3 + y_4 &\leq 5 \\
3y_1 + 8y_2 + y_3 + 2y_4 &\leq 2 \\
y_1 + 5y_2 + 3y_3 + 4y_4 &\leq 1 \\
y_1, y_2, y_3, y_4 &\geq 0.
\end{aligned}$$

Note: We observe that the dual problem has three constraints since the primal problem has three variables.

Example 2

Write the dual of the following LPP

Maximise $z = 2x_1 + 5x_2 + 3x_3$

subject to

$$2x_1 + 4x_2 - 3x_3 \leq 8$$
$$-2x_1 - 2x_2 + 3x_3 \geq -7$$
$$x_1 + 3x_2 - 5x_3 \geq -2$$
$$4x_1 + x_2 + 3x_3 \leq 4$$
$$x_1, x_2, x_3 \geq 0.$$

Solution: First, we write the primal (given) problem in symmetric form.

Maximise $z = 2x_1 + 5x_2 + 3x_3$

subject to

$$2x_1 + 4x_2 - x_3 \leq 8$$
$$2x_1 + 2x_2 - 3x_3 \leq 7 \quad \text{(multiplying both sides by } -1)$$
$$-x_1 - 3x_2 + 5x_3 \leq 2 \quad \text{(multiplying both sides by } -1)$$
$$4x_1 + x_2 + 3x_3 \leq 4$$
$$x_1, x_2, x_3 \geq 0.$$

Then, the dual problem is

Minimise $w = 8y_1 + 7y_2 + 2y_3 + 4y_4$

subject to

$$2y_1 + 2y_2 - y_3 + 4y_4 \geq 2$$
$$4y_1 + 2y_2 - 3y_3 + y_4 \geq 5$$
$$-y_1 - 3y_2 + 5y_3 + 3y_4 \geq 3$$
$$y_1, y_2, y_3, y_4 \geq 0.$$

Unrestricted variables:

We recall that unrestricted variable is a variable that can take positive or negative or zero value.

Example 3

Write the dual of the following LPP

Maximise $z = 4x_1 + 2x_2$

subject to

$$x_1 - 2x_2 \geq 2$$
$$x_1 + 2x_2 = 8$$
$$x_1 - x_2 \leq 10$$
$$x_1 \geq 0, x_2 \text{ unrestricted in sign.}$$

Solution: The primal in symmetric form is

Maximise $z = 4x_1 + 2x_2$

subject to

$$\left.\begin{aligned} -x_1 + 2x_2 &\leq -2 \\ x_1 + 2x_2 &\leq 8 \\ -x_1 - 2x_2 &\leq -8 \\ x_1 - x_2 &\leq 10 \\ x_1, x_2 &\geq 0 \end{aligned}\right\} \quad (3.1.3)$$

Since x_2 is unrestricted put $x_2 = x'_2 - x''_2$ where $x'_2, x''_2 \geq 0$. Then, (3.1.3) becomes

Maximise $z = 4x_1 + 2x'_2 - 2x''_2$

subject to
$$\begin{aligned} -x_1 + 2x'_2 - 2x''_2 &\le -2 \\ x_1 + 2x'_2 - 2x''_2 &\le 8 \\ -x_1 - 2x'_2 + 2x''_2 &\le -8 \\ x_1 - x'_2 + x''_2 &\le 10 \\ x_1, x'_2, x''_2 &\ge 0. \end{aligned}$$

Then, the dual problem is

Minimise $w = -2y_1 + 8y_2 - 8y_3 + 10y_4$

subject to
$$\begin{aligned} -y_1 + y_2 - y_3 + y_4 &\ge 4 \\ 2y_1 + 2y_2 - 2y_3 - y_4 &\ge 2 \\ -2y_1 - 2y_2 + 2y_3 + y_4 &\ge -2 \\ y_1, y_2, y_3, y_4 &\ge 0. \end{aligned}$$

Put $y_2 - y_3 = y$. Then, the LPP becomes

Minimise $w = -2y_1 + 8y + 10y_4$

subject to
$$\begin{aligned} -y_1 + y + y_4 &\ge 4 \\ 2y_1 + 2y - 2y_4 &= 2 \end{aligned}$$

$y_1, y_4 \ge 0$ and y is unrestricted which is the dual of the given problem.

Remark 1: If the kth constraint of the primal problem is an equality, then the corresponding dual variable y_k is unrestricted in sign and vice versa.

Remark 2: If any variable of the primal problem is unrestricted in sign, the corresponding constraint in the dual problem will be an equality and vice versa.

3.1.2 Duality Theorems

Theorem 3.1

The dual of the dual is the primal.

Proof: Let the primal problem be

$$\left.\begin{aligned} &\text{Maximise } z = CX,\ C \in R^n \\ &\text{subject to } AX \le b,\ b^T \in R^m \end{aligned}\right\} \qquad (3.1.4)$$

and $X \ge 0$

Then, the dual of (3.1.4) is

$$\left.\begin{aligned} &\text{Minimise } w = b^T Y,\ b^T \in R^m \\ &\text{subject to } A^T Y \ge C^T,\ C \in R^n \end{aligned}\right\} \qquad (3.1.5)$$

and $Y \ge 0$

The dual of (3.1.5) is

$$\left.\begin{aligned} &\text{Minimise } s = (C^T)^T U,\ C \in R^n \\ &\text{subject to } (A^T)^T \le (b^T)^T,\ b^T \in R^m \end{aligned}\right\} \qquad (3.1.6)$$

and $U \ge 0$

By observing $(C^T)^T = C$ and $(A^T)^T = A$, (3.1.6) becomes

$$\left.\begin{aligned} &\text{Maximise } s = CU,\ C \in R^n \\ &\text{subject to } AU \le b,\ b^T \in R^n \end{aligned}\right\} \qquad (3.1.7)$$

and $U \ge 0$

(3.1.4) and (3.1.7) represents the same problem. Hence, the dual of the dual is the primal.

Theorem 3.2

Let X_0 be a feasible solution to the primal problem.

Maximise $z = CX$ subject to $AX \leq b$, $X \geq 0$, where X^T and $C \in R^n$, $b^T \in R^m$ and A is an $m \times n$ real matrix. If Y_0 be a feasible solution to the dual of the primal, namely Minimise $w = b^T Y$ subject to $A^T Y \geq C^T$, $Y \geq 0$, $Y^T \varepsilon R^m$ then $CX_0 \leq b^T Y_0$.

Proof: Since X_0 and Y_0 are the feasible solutions to the primal and its dual, respectively, we have

$$AX_0 \leq b,\ X_0 \geq 0 \text{ and } A^T Y \geq C^T,\ Y_0 \geq 0.$$

Hence, $\quad C \leq Y_0^T A$ or $CX_0 \leq Y_0^T AX_0 \leq Y_0^T b$

or, $\quad CX_0 \leq b^T Y_0 \qquad [\because Y_0^T b = b^T Y_0]$

Theorem 3.3

Let X_0 be a feasible solution to the primal problem. Maximise $Z = CX$ subject to $AX \leq b$, $X \geq 0$ and Y_0 be a feasible solution to its dual: Maximise $w = b^T Y$ subject to $A^T Y \geq C^T$, $Y \geq 0$ where $C, X^T \in R^n$ and Y^T, $b^T \in R^m$ and A is a $m \times n$ real matrix. If $CX_0 = b^T W_0$ then both X_0 and W_0 are optimum solutions to the primal and dual respectively.

Proof: Let X'_0 be any other feasible solution to the primal problem. Then, theorem 3.2 gives,

$$CX'_0 \leq b^T Y_0$$

Thus, $\quad CX'_0 \leq CX_0 \qquad [\because CX_0 = b^T Y_0]$

and hence X_0 is an optimum solution to the primal problem, because primal is of maximisation type.

Similarly, if Y'_0 is any other feasible solution to the dual problem, then $b^T Y_0 \leq b^T Y'_0$ and thus Y_0 is an optimum solution to the dual problem, because dual is a minimisation problem.

Theorem 3.4 Basic Duality Theorem

Let a primal problem be Maximise $z = CX$ subject to $AX \leq b$, $X \geq 0$, $C, X^T \in R^n$ and the associated dual be minimise $w = b^T Y$ subject to $A^T Y \geq C^T$, $Y \geq 0$, $Y^T, b^T \in R^m$. If X_0 (Y_0) is an optimum solution to the primal (dual) then there exists a feasible solution Y_0 (X_0) to the dual (primal) such that $CX_0 = b^T Y_0$.

Proof: The standard primal can be written as Maximise $z = CX$ subject to $AX + IX_s = b$, where $X_s^T \in R^m$ is the slack vector and I is the associated $m \times m$ identity matrix.

Let $X_0 = [X_B, 0]$ be an optimum solution to the primal, where $X_B^T \in R^m$ is optimum basic feasible solution given by $X_B = B^{-1} b$, B being the optimal basis of A. Then, the optimal primal objective function is

$z = CX_0 = C_B X_B$ where C_B is the cost vector associated with X_B.

Now, the net evaluations in the optimal simplex table are given by

$$z_j - c_j = C_B P_j - c_j = \begin{cases} C_B B^{-1} a_j - c_j \text{ for all } a_j \in A \\ C_B B^{-1} e_j - 0 \text{ for all } e_j \in I \end{cases}$$

Since X_B is optimal, we must have $z_j - c_j \geq 0$ for all j.

This gives $C_B B^{-1} a_j \geq c_j$ and $C_B B^{-1} e_j \geq 0$ for all j.

or, $\quad C_B B^{-1} A \geq C$ and $C_B B^{-1} \geq 0$ in matrix form.

or, $\quad A^T B^{-1} C_B^T \geq C^T$ and $B^{-1} C_B^T \geq 0$.

Now, if we let $B^{-1} C_B^T = Y_0$, the above become

$A^T W_0 \geq C^T$ and $W_0 \geq 0$, $W_0{}^T \in R^m$. This means that W_0 is a feasible solution to the dual problem. Moreover, the corresponding dual objective function value is $b^T Y_0 = Y_0^T b = C_B B^{-1} b = C_B X_B = CX_0$. Thus, given an optimal solution X_0 to the primal, there exists a feasible solution Y_0 to the dual such that $CX_0 = b^T Y_0$.

Similarly, starting with Y_0, the existence of X_0 can be proved.

Corollary 3.5

If X_0 is an optimal solution to the primal, an optimal solution to the dual is given by $W_0 = B^{-1} C_B$, where B is the primal optimal basis.

Theorem 3.6 Fundamental Theorem of Duality

If the primal or the dual has a finite optimum solution, then the other problem also pocesses a finite optimum solution and the optimum values of the objective functions of the two problems are equal.

Proof: Consider the primal-dual pair:

Primal: Maximise $z = CX$, subject to $AX \leq b$ and $X \geq 0$.

Dual: Minimise $w = b^T Y$, subject to $A^T Y \geq C^T$ and $W \geq 0$.

Necessary condition: Let a feasible solution X_0 (Y_0) be an optimum solution to the primal (dual) problem. It then follows from Theorem 3.4 that there exists a feasible solution Y_0 (X_0) to the dual (primal) problem, such that $CX_0 = b^T Y_0$.

It now follows from Theorem 3.3 that Y_0 must be optimal. This proves the necessity of the condition. Sufficiency follows from Theorem 3.3.

Theorem 3.7 Existence Theorem

If either the primal or the dual problem has an unbounded objective function value, then the other problem has no feasible solution.

Proof: Let the given primal problem have unbounded solution. Then, for any value of the objective function, say $+\infty$, there exists a feasible solution, say X yielding this solution that is, $CX \to \infty$. If possible, let the dual problem have a feasible solution. Then from Theorom 3.2, for each feasible solution Y_0 of the dual, there exists a feasible solution. X_0 to the primal such that $CX_0 \leq b^T Y_0$. That is $b^T Y \to \infty$. Now, as b is a constant and W_0 has to satisfy the constraint $A^T Y_0 \geq C^T$, therefore the dual objective function $b^T Y_0$ must be finite. This contradicts the result $b^T Y_0 \to \infty$. Hence, the dual problem has no feasible solution.

By a similar argument it can be shown that when the dual has an unbounded solution, the primal has no solution.

Remark: The above result shows the following relation between the solutions of the primal and dual.

Dual problem	*Primal problem*	
	Feasible solution	No feasible solution
Feasible solution	Optimum	Dual unbounded
No feasible solution	Primal unbounded	Unbounded or No feasible solution

Theorem 3.8 Complementary Slackness Theorem

Let X_0 and Y_0 be the feasible solutions to the primal Maximise $z = C^T X$, $AX \leq b$, $X \geq 0$ and its dual Minimise $w = b^T Y$, $A^T Y \geq C^T$, $Y \geq 0$, respectively. Then, a necessary and sufficient condition for X_0 and Y_0 to be optimal to their respective problem is that

$$Y_0^T (b - AX_0) = 0 \text{ and } X_0^T (A^T Y_0 - C^T) = 0.$$

Proof: *Necessity*: Let $\alpha = Y_0^T (b - AX_0)$ and $\beta = X_0^T (A^T Y_0 - C^T)$. Since X_0, Y_0 are solutions to the primal and dual respectively, we have, $\alpha \geq 0$, $\beta \geq 0$ and $\alpha + \beta = Y_0^T b - X_0^T C^T$.

Now, if X_0, Y_0 are optimal, then $CX_0 = b^T Y_0$ so that $\alpha + \beta = 0$. But since $\alpha \geq 0$ and $\beta \geq 0$ this gives $\alpha = 0$ and $\beta = 0$.

Thus, the conditions are necessary.

Sufficiency: Let the given conditions hold for the feasible solutions X_0 and W_0. That is, $\alpha = 0$ and $\beta = 0$.

Then, $0 = \alpha + \beta = Y_0^T b - X_0^T C^T$

$\Rightarrow \quad CX_0 = b^T Y_0$

$\Rightarrow \quad X_0$ and W_0 are optimal

Thus, the conditions are sufficient.

3.1.3 Duality and Simplex Method

Since any LPP can be solved by using simplex method, the method is applicable to both the primal as well as its dual. The fundamental theorem of duality suggests that an optimum solution to the associated dual can be obtained from that of its primal and vice versa.

Properties of Primal and Dual Optimal Solutions

1. If the primal (dual) variables correspond to the slack variable in the dual (primal) problem its optimum value is directly read off from the dual (primal) simplex table as the net evaluation corresponding to the slack variable.
2. If the primal (dual) variables correspond to the artificial variable in the dual (primal) problem its optimum value is directly read off from the optimum dual (primal) simplex table as the net evaluation corresponding to that artificial variable after deleting the constant M.

Example 1

Use duality to solve the following LPP

$$\text{Maximise } z = 4x_1 + 2x_2$$

$$\text{subject to the constraints} \quad -x_1 - x_2 \leq -3$$
$$-x_1 + x_2 \geq -2$$
$$x_1, x_2 \geq 0$$

Solution: The given problem can be rewritten as

$$\left.\begin{aligned} &\text{Maximise } z = 4x_1 + 2x_2 \\ \text{subject to} \quad &-x_1 - x_2 \leq -3 \\ &x_1 - x_2 \leq 2 \\ &x_1, x_2 \geq 0 \end{aligned}\right\} \quad (3.1.8)$$

The dual of (3.1.8) is

$$\left.\begin{aligned} &\text{Minimise } w = -3y_1 + 2y_2 \\ \text{subject to} \quad &-y_1 + y_2 \geq 4 \\ &-y_1 - y_2 \geq 2 \\ &y_1, y_2 \geq 0 \end{aligned}\right\} \quad (3.1.9)$$

Solution of (3.1.9) by Big M method:

Introduce the surplus variables s_1, $s_2 \geq 0$ and the artificial variables A_1, $A_2 \geq 0$ in the above constraints (3.1.9) can be written as

Minimise $w = -3y_1 + 2y_2 + 0s_1 + 0s_2 - MA_1 - MA_2$

subject to $-y_1 + y_2 - s_1 + A_1 = 4$

$-y_1 - y_1 - s_2 + A_2 = 2$

$y_1, y_2, s_1, s_2, A_1, A_2 \geq 0.$

An obvious IBFS is $A_1 = 4, A_2 = 2$

table 3.1

C_B	Y_B	X_B	y_1↓	y_2	s_1	s_2	A_1	A_2	Min. Ratio
- M	A_1	4	- 1	1	- 1	0	1	0	
- M	A_2	2	- 1	- 1	0	- 1	0	1	
$z_j - c_j$		- 6 M	- 2M + 3	- 2	M	M	0	0	

We observe that the column corresponding to y_1 is the pivotal column. Since there is no positive element in the pivotal column the dual problem has an unbounded solution. Hence, there exists no finite optimum solution to the primal problem.

Example 2

Use duality to solve the following LPP.

Maximise $z = 2x_1 + x_2$

subject to the constraints

$$x_1 + 2x_2 \leq 10$$
$$x_1 + x_2 \leq 6$$
$$x_1 - x_2 \leq 2$$
$$x_1 - 2x_2 \leq 1$$
$$x_1, x_2 \geq 0.$$

Solution: The dual of the given problem is

Minimise $w = 10y_1 + 6y_2 + 2y_3 + y_4$

subject to the constraints

$$y_1 + y_2 + y_3 + y_4 \geq 2$$
$$2y_1 + y_2 - y_3 - y_4 \geq 1$$
$$y_1, y_2, y_3, y_4 \geq 0.$$

Introducing surplus variables $s_1, s_2 \geq 0$ and artificial variables $A_1, A_2 \geq 0$ then the optimal simplex table is

table 3.2

		c_j	- 10	- 6	- 2	- 1	0	0	- M	- M
C_B	Y_B	X_B	y_1	y_2	y_3	y_4	s_1	s_2	A_1	A_2
- 2	y_3	1/2	- 1/2	0	1	3/2	- 1/2	1/2	1/2	- 1/2
- 6	y_2	3/2	3/2	1	0	- 1/2	- 1/2	- 1/2	1/2	1/2
$z_j - c_j$		- 10	2	0	0	1	4	2	M - 4	M - 2

Since all $z_j - c_j \geq 0$, the current solution is optimal. The optimum solution of the dual is

$$y_1 = 0, y_2 = 3/2, y_3 = 1/2, y_4 = 0 \text{ and Minimum of } w = -10.$$

To find the optimum solution of the primal:

Corresponding to the variable x_1 we have the dual constraint $y_1 + y_2 + y_3 + y_4 \geq 2$ and the basic variable associated with the constraint is A_1. Hence, the optimal value of x_1

$= M -$ (net evaluation value of A_1)

$= M - (M - 4) = 4$

Similarly, the optimal value of x_2 is

$= M - (M - 2) = 2$

$\therefore$ The optimal solution of the primal is

$$x_1 = 4, x_2 = 2 \text{ with Maximum of } z = -10$$

Example 3

A company makes three products X, Y, Z out of three raw materials A, B and C. The number of units of raw materials required to produce one unit of products x,y, z is given in the following Table:

table 3.3

	X	Y	Z
A	1	2	1
B	2	1	4
C	2	5	1

The profit per unit on the products X, Y and Z are Rs 40, 25 and 50, respectively. The number of units of raw materials available are 36, 60 and 45, respectively.

(i) Determine the product mix that will maximise the total profit.

(ii) Through the final simplex table, write the solution to the dual problem and give the economic interpretation.

Solution: *Formulation of the LPP*

Let the company produce x, y and z units of the products X, Y and Z, respectively. Then, according to the given conditions, the objective function is Maximise $z = 40x + 25y + 50z$.

Also, from the Table it is clear that the material constraints are

$$\left.\begin{aligned} x + 2y + z &\leq 36 \\ 2x + y + 4z &\leq 60 \\ 2x + 5y + z &\leq 45 \\ \text{and } x, y, z &\geq 0 \end{aligned}\right\} \quad (3.1.10)$$

Hence, the complete LPP is

Maximise $z = 40x + 25y + 50z$

Subject to (3.1.10).

Solution by simplex method:

Introduce slack variables $s_1, s_2, s_3 \geq 0$ in the given constraints. The LPP becomes

Maximise $z = 40x + 20y + 50z + 0s_1 + 0s_2 + 0s_3$

subject to
$$x + 2y + z + s_1 = 36$$
$$2x + y + 4z + s_2 = 60$$
$$2x + 5y + z + s_3 = 45$$
and $x, y, z, s_1, s_2, s_3 \geq 0$.

If we solve the problem by regular simplex method the optimum simplex table is

table 3.4

		cj	40	20	50	0	0	0	
C_B	Y_B	X_B	x	y	z	s_1	s_2	s_3	Min. Ratio
0	s_1	11	0	1/6	0	1	- 1/6	- 1/3	
50	z	5	0	- 4/3	1	0	1/3	- 1/3	
40	x	20	1	19/6	0	0	- 1/6	2/3	
	$z_j - c_j$	1050	0	35	0	0	10	10	

Since all $z_j - c_j \geq 0$ the current solution is optimum to the primal.
The optimum solution is

$$\boxed{x = 20,\ y = 0,\ z = 5 \text{ and maximum of } z = 1050}$$

The dual problem:
The dual of the problem is

Minimise $w = 36y_1 + 60y_2 + 45y_3$

subject to
$$\left.\begin{aligned} y_1 + 2y_2 + 2y_3 &\geq 40 \\ 2y_1 + y_2 + 5y_3 &\geq 25 \\ y_1 + 4y_2 + y_3 &\geq 50 \\ y_1, y_2, y_3 &\geq 0 \end{aligned}\right\} \qquad (3.1.11)$$

Solution of the dual from the optimum simplex table of the primal:
The dual variable y_1 corresponds to the constraints $x + 2y + z \leq 36$ and the initial slack variable corresponding to this constraint is s_1. Hence, the optimal solution of $y_1 = z_j - c_j$ value of $s_1 = 0$.
Similarly, the optimum value of $y_2 = z_j - c_j$ value of $s_2 = 10$
Optimum value of $y_3 = z_j - c_j$ value of $s_3 = 10$. Hence, the optimum solution of the (3.1.11) is

$$\boxed{y_1 = 0,\ y_2 = 10,\ y_3 = 10 \text{ and Minimum of } w = 1050}$$

Economic intrepretation: Suppose the manager of the company wishes to sell the three raw materials *A*, *B* and *C* instead of using them for production of the products *X*, *Y* and *Z*. Suppose the selling prices were Rs y_1, Rs y_2 and Rs y_3 per unit of raw material *A*, *B* and *C*, respectively. The cost to the purchaser of all three raw materials would be Rs $36y_1 + 60y_2 + 45y_3$. Of course, the purchaser would like to set the selling prices in such a way so as to minimise the total cost paid. Hence, the objective function becomes

Minimise $w = 36y_1 + 60y_2 + 45y_3$

The final table of the primal problem indicates that the marginal value of raw material *A* is Rs zero, for *B* is Rs 10 per unit; and for *C* is Rs 10 per unit. Thus, if the manager sells the raw materials *A*, *B* and *C* at price Rs 0, Rs 10 and Rs 10 per unit, respectively, he will get the same contribution of Rs 1050 which he going to fetch in case he utilises these resources for production of the three products *X*, *Y* and *Z*.

3.2 DUAL SIMPLEX METHOD

3.2.1 Introduction

In simplex method we have already seen that every basic solution with all $z_j - c_j \geq 0$ will not be feasible (since $z_j - c_j = C_B B^{-1} a_j - c_j$ is independent of right hand side vector b for all j), but any basic feasible solution with all $z_j - c_j \geq 0$ will certainly be an optimal solution. Such typical problems, for which it is possible to find infeasible but better than optimal (with all $z_j - c_j \geq 0$) starting basic solutions, can be solved more easily by dual simplex method. Such a situation is recognised by first expressing the constraints in the form (≤) and the objective function in the maximisation form. After adding the slack variables and putting the problem in the Table form, if any of the right hand side elements is negative and if the optimality condition is satisfied, then the problem can be solved by the dual simplex method. It is important to note that by this arrangement, negative element on the right hand side signifies that the corresponding slack variable is negative.

3.2.2 Difference between Regular Simplex Method and Dual Simplex Method

The regular simplex method starts with a basic feasible but a non-optimal solution and works towards optimality, whereas the dual simplex method starts with a basic infeasible but an optimal solution and works towards feasibility. Also, in regular simplex method we first determine the entering variable and then the leaving variable while in the case of dual simplex method we first determine the leaving variable and then the entering variable.

3.2.3 Dual Simplex Algorithm

Step 1: Convert the problem into a maximisation type if it is in the minimisation type.

Step 2: Convert ≥ type inequalities (if any) into ≤ type inequalities by multiplying both sides by – 1.

Step 3: Express the LPP in standard form by introducing slack variables. Obtain the initial basic solution (not feasible) and express these information in the simplex table.

Step 4: (Optimality condition). Test the nature of $z_j - c_j$.

Case (i): If all $(z_j - c_j) \geq 0$ and all $X_{B_i} \geq 0$, then the current solution is an optimum basic feasible solution.

Case (ii): If all $(z_j - c_j) \geq 0$ and atleast one $X_{B_i} < 0$ then the current solution is not an optimum basic feasible solution. Go to the next step.

Case (iii): If any $(z_j - c_j) < 0$ then this method fails.

Step 5: (Feasibility conditions).

(i) *(Leaving variable)*: The leaving variable is the basic variable corresponding to the most negative value of X_{B_i}. Let x_k be the leaving variable.

(ii) *(Entering variable)*: Compute the ratio between $(z_j - c_j)$ row and the key row, that is,

$$\text{Compute } \theta = \max\left\{\frac{z_j - c_j}{a_{ik}}, a_{ik} < 0\right\}$$

(Consider the ratios with negative denominators alone). The entering variable is the non-basic variable corresponding to the maximum ratio θ. If there is no such ratio with negative denominator, then the problem has no feasible solution.

Step 6: Carry out the row operations as in the regular simplex method and repeat the procedure until either an optimum feasible solution is obtained or there is an indication of non-existence of a feasible solution.

Example 1

Use dual simplex method to solve the LPP

Maximise $z = 2x_1 + 2x_2$

subject to
$$2x_1 - x_2 - x_3 \geq 3$$
$$x_1 - x_2 + x_3 \geq 2$$
$$x_1, x_2, x_3 \geq 0.$$

Solution: First convert all the inequalities into $\leq$ type.

Maximise $z = 2x_1 + 2x_2$

subject to
$$-2x_1 + x_2 + x_3 \leq -3$$
$$-x_1 + x_2 - x_3 \leq -2$$
$$x_1, x_2, x_3 \geq 0.$$

Introduce slack variables $s_1, s_2 \geq 0$ an obvious initial basic non-feasible solution is $s_1 = -3, s_2 = -2$.

Starting Table:

table 3.5

		c_j	2	1	0	0	0
C_B	Y_B	X_B	x_1	x_2	x_3	s_1	s_2
0	s_1	-3	-2	1	1	1	0
0	s_2	-2	-1	1	-1	0	1
$z_j - c_j$		0	-2	-1	0	0	0

Since there are some $(z_j - c_j) < 0$ this method fails. That is, we cannot solve this problem by dual simplex method.

Remark: From the above example we observe that if a LPP is of maximisation type with positive objective function coefficient, then the LPP cannot be solved by dual simplex method.

Example 2

Use dual simplex method to solve

Maximise $z = -3x_1 - 2x_2$

subject to
$$x_1 + x_2 \geq 1$$
$$x_1 + x_2 \geq 7$$
$$x_1 + 2x_2 \geq 10$$

$$x_2 \le 3$$
$$x_1, x_2 \ge 0.$$

Solution: First, we convert all the inequalities into ≤ type. The given LPP becomes

Maximise $z = -3x_1 - 2x_2$

subject to
$$-x_1 - x_2 \le -1$$
$$x_1 + x_2 \le 7$$
$$-x_1 - 2x_2 \le -10$$
$$x_2 \le 3.$$

Introducing slack variables $s_1, s_2, s_3, s_4 \ge 0$ in the given constraints the LPP becomes

Maximise $z = -3x_1 - 2x_2 + 0s_1 + 0s_2 + 0s_3 + 0s_4$

subject to
$$-x_1 - x_2 + s_1 = -1$$
$$x_1 + x_2 + s_2 = 7$$
$$-x_1 - 2x_2 + s_3 = -10$$
$$x_2 + s_4 = 3$$
$$x_1, x_2, s_1, s_2, s_3, s_4 \ge 0.$$

An obvious initial basic non-feasible solution is

$$s_1 = -1, s_2 = 7, s_3 = -10, s_4 = 3.$$

table 3.6

		c_j	−3	−2	0	0	0	0
C_B	Y_B	X_B	x_1	x_2↓	s_1	s_2	s_3	s_4
0	s_1	−1	−1	−1	1	0	0	0
0	s_2	−7	1	1	0	1	0	0
0	s_3 ←	−10	−1	[−2]	0	0	1	0
0	s_4	3	0	1	0	0	0	1
		0	3	2	0	0	0	0

Since all $z_j - c_j \ge 0$ the current solution is optimal and since some X_{B_i} are negative the solution is non-feasible. From the Table we see that s_3 is the leaving variable.

To find the entering vector:

Since the maximum of $\left\{\frac{3}{-1}, \frac{2}{-2}\right\} = -1$

corresponding to the variable x_2. Hence, x_2 enters into the basis.

First iteration: Introduce x_2 and drop s_3.

table 3.7

		c_j	-3	-2	0	0	0	0
C_B	Y_B	X_B	$x_1 \downarrow$	x_2	s_1	s_2	s_3	s_4
0	s_1	4	$-1/2$	0	1	0	$-1/2$	0
0	s_2	2	$1/2$	0	0	1	$1/2$	0
-2	x_2	5	$1/2$	1	0	0	$-1/2$	0
0	$\bar{s}_4$	-2	[$-1/2$]	0	0	0	$1/2$	1
			2	0	0	0	1	0

From Table 3.7, it is clear that s_4 is the leaving variable and x_1 is the entering variable.

Second iteration: Introduce x_1 and drop s_4.

table 3.8

		c_j	-3	-2	0	0	0	0
C_B	Y_B	X_B	x_1	x_2	s_1	s_2	s_3	s_4
0	s_1	6	0	0	1	0	-1	-1
0	s_2	0	0	0	0	1	1	1
-2	x_2	3	0	1	0	0	0	1
-3	x_1	4	1	0	0	0	-1	-2
	$z_j - c_j$	-18	0	0	0	0	3	4

Since all $z_j - c_j \geq 0$, the optimality condition is satisfied. All $X_{B_i} \geq 0$ implies that the feasibility condition is also satisfied. Hence, the optimum solution is

$$\boxed{x_1 = 4, x_2 = 3 \text{ with maximum of } z = -18}$$

Example 3

Using dual simplex method solve the following LPP

Minimise $z = x_1 + x_2$

subject to
$$2x_1 + x_2 \geq 2$$
$$-x_1 - x_2 \geq 1$$

and,
$$x_1, x_2 \geq 0$$

Solution: Convert the minimisation LPP into a maximisation LPP and convert all the inequalities into $\leq$ type.

Maximise $z^* = -x_1 - x_2$ where $z^* = -z$

subject to
$$-2x_1 - x_2 \leq -2$$
$$x_1 + x_2 \leq -1$$

and,
$$x_1, x_2 \geq 0$$

Introduce slack variables $s_1, s_2 \geq 0$ in the given constraints. The LPP becomes

Maximise $z^* = -x_1 - x_2 + 0s_1 + 0s_2$

subject to
$$-2x_1 - x_2 + s_1 = -2$$
$$x_1 + x_2 + s_2 = -1$$

and, $x_1, x_2, s_1, s_2 \geq 0$

An obvious initial basic (non-feasible) solution is $s_1 = -2, s_2 = -1$.

Starting Table:

table 3.9

		c_j	-1	-1	0	0
C_B	Y_B	X_B	$x_1 \downarrow$	x_2	s_1	s_2
0	$\bar{s}_1$	-2	[-2]	-1	1	0
0	s_2	-1	1	1	0	1
			1	1	0	0

s_1 is the entering variable since s_2 is the variable with most negative X_{B_i}. Also maximum of $\left\{\frac{1}{-2}, \frac{1}{-1}\right\}$ $= \frac{-1}{2}$ corresponding to x_1. Hence, x_1 is the entering variable.

First iteration: Introduce x_1 and drop s_1.

table 3.10

		c_j	-1	-1	0	0
C_B	Y_B	X_B	x_1	x_2	s_1	s_2
-1	x_1	1	1	1/2	-1/2	0
0	$\bar{s}_2$	-2	0	1/2	1/2	1
$z_j - c_j$		1	0	1/2	1/2	0

Since all $z_j - c_j \geq 0$ the optimality condition is satisfied. Also, s_2 is the only variable with negative X_{B_i} and so s_2 is the leaving variable. But there is no negative entries in the key row. Hence, there exists no feasible solution to the given problem.

3.3 REVISED SIMPLEX METHOD

3.3.1 Introduction

The original simplex method is a straight forward algebraic procedure. However, this way of executing the algorithm is not the most efficient computational procedure for electronic digital computers. It computes and stores many numbers that are needed at the current iteration and which may not become relevant for decision-making at subsequent iterations. The only piece of information relevant at each iteration are the coefficients of the non-basic variables, the coefficient of the entering basic variable in the other equations, and the right hand side of the equation. It would be very useful to have a procedure that could obtain this information efficiently without computing and storing all the other coefficients.

These considerations motivated the development of the revised simplex method. This method was designed to accomplish exactly the same thing as the original method, but in a way that is more efficient for execution on a digital computer. Thus, it is a streamlined version of the original procedure. It computes the essential data in a more compact form.

3.3.2 Revised Simplex Algorithm

Step 1: Introduce slack or surplus variables and the artificial variables whenever necessary. Express the given LPP in standard form after converting it into maximisation type.

Step 2: Obtain an initial basic feasible solution with an initial basis $B = I_m$ (identity matrix of order m) and find B^{-1} and then calculate $\bar{b} = B^{-1}\, b$ where b is the right hand side constant vector.

Step 3: Calculate the simplex multiplier $M = (M_1, M_2, \cdots, M_m)$ (where m denotes the number of constraints) using $M = C_B\, B^{-1}$ where C_B is the cost vector associated with the basis B.

Step 4: Calculate the net evaluations of the non-basic variables using $\bar{z}_i - \bar{c}_i = B^{-1}\, P_i - c_i$ where P_i is the column vector of the variable x_i and c_i is the cost of the variable x_i in the objective function.

Case (i): If there is no negative $\bar{z}_i - \bar{c}_i$ then the current solution is optimum.

Case (ii): If there is some negative $\bar{z}_i - \bar{c}_i$ then the non-basic variable x_k with $\bar{z}_k - \bar{c}_k$ as most negative enters into the basis.Proceed to Step 5.

Step 5: Calculate the pivotal column vector.

$$\bar{P}_k = B^{-1}\, P_k$$

Case (i): If all $x_{ik} \le 0$, then there exists an unbounded solution to the given LPP.

Case (ii): If at least one $x_{ik} < 0$, consider the current X_B and determine the leaving variable. Proceed to Step 6.

Step 6: Write down the results obtained from step (2) to step (5) in the revised simplex table.

Step 7: Convert the leading element to unity and all other elements of the entering column to zero and update the current basic feasible solution.

Step 8: Go to step (4) and repeat the procedure until an optimum basic feasible solution is obtained or till there is an indication of an unbounded solution.

Example 1

Use revised simplex method to solve the LPP.

Maximise $z = 6x_1 - 2x_2 + 3x_3$

subject to
$$2x_1 - x_2 + 2x_3 \le 2$$
$$x_1 + 4x_3 \le 4$$
$$x_1, x_2, x_3 \ge 0$$

Solution: Convert the LPP into standard form by introducing slack variables x_4 and x_5.

Maximise $z = 6x_1 - 2x_2 + 3x_3 + 0s_1 + 0s_2$

subject to
$$2x_1 - x_2 + 2x_3 + s_1 = 2$$
$$x_1 + 0x_2 + 4x_3 + s_2 = 4$$

and,
$$x_1, x_2, x_3, s_1, s_2 \ge 0$$

Let $P_1, \ldots, P_5$ denote the column vectors corresponding to the variables, $x_1, \cdots, x_5$ and let b denote the right hand constants, respectively. Then,

$$P_1 = \begin{bmatrix} 2 \\ 1 \end{bmatrix}, P_2 = \begin{bmatrix} -1 \\ 0 \end{bmatrix}, P_3 = \begin{bmatrix} 2 \\ 4 \end{bmatrix}, P_4 = \begin{bmatrix} 1 \\ 0 \end{bmatrix}, P_5 = \begin{bmatrix} 0 \\ 1 \end{bmatrix} \text{ and } b = \begin{bmatrix} 2 \\ 4 \end{bmatrix}$$

Now, (x_4, x_5) forms the initial basis $B = [P_4, P_5] = \begin{bmatrix} 1 & 0 \\ 0 & 1 \end{bmatrix}$

Therefore, $B^{-1} = B$

Now, $\bar{b} = B^{-1}\, b = \begin{bmatrix} 1 & 0 \\ 0 & 1 \end{bmatrix}\begin{bmatrix} 2 \\ 4 \end{bmatrix} = \begin{bmatrix} 2 \\ 4 \end{bmatrix} = b$

Starting Table: (The last three columns are added later)

table 3.11

Basis	B^{-1}		Constants	Variable to enter	Pivot column	Ratio
$\bar{x}_4$	1	0	2		$\boxed{2}$	$\frac{2}{2} = 1$
				x_1		
x_5	0	1	4		1	$\frac{4}{1} = 4$

The simplex multipliers are

$$M = (M_1, M_2) = C_B\, B^{-1} = (0\ 0)\begin{pmatrix} 1 & 0 \\ 0 & 1 \end{pmatrix} = (0\ 0)$$

Now, $\bar{z}_1 - \bar{c}_1 = MP_1 - C_1 = (0\ 0)\begin{pmatrix} 2 \\ 1 \end{pmatrix} - 6 = -6$

$$\bar{z}_2 - \bar{c}_2 = MP_2 - C_2 = (0\ 0)\begin{pmatrix} -1 \\ 0 \end{pmatrix} - (-2) = 2$$

$$\bar{z}_3 - \bar{c}_3 = MP_3 - C_3 = (0\ 0)\begin{pmatrix} 2 \\ 4 \end{pmatrix} - 3 = -3$$

Since $\bar{z}_1 - \bar{c}_1$ is most negative the variable to enter is x_1. Hence, the pivotal column is

$$\bar{P}_1 = B^{-1}\, P_1 = \begin{bmatrix} 1 & 0 \\ 0 & 1 \end{bmatrix}\begin{bmatrix} 2 \\ 1 \end{bmatrix} = \begin{bmatrix} 2 \\ 1 \end{bmatrix}$$

After calculating the minimum ratio we see that x_4 leaves the basis. Now, the pivotal column $\begin{bmatrix} 2 \\ 1 \end{bmatrix}$ must be reduced to $\begin{bmatrix} 1 \\ 0 \end{bmatrix}$. This is obtained by dividing row 1 by 2 and then subtracting row 2 from row 1.

First iteration:

table 3.12

Basis	B^{-1}		Constants	Variable to enter	Pivot column	Ratio
x_1	1/2	0	1	x_2	- 1/2	
x_5	- 1/2	1	3		1/2	

The simplex multipliers corresponding to the first iteration are $M = (M_1, M_2) = C_B\ B^{-1}$ $= (6\ 0)\begin{pmatrix} 1/2 & 0 \\ -1/2 & 1 \end{pmatrix} = (3\ 0)$

$$\text{Now,} \qquad \bar{z}_2 - \bar{c}_2 = (3\ \ 0)\begin{pmatrix} -1 \\ 0 \end{pmatrix} + 2 = -1$$

$$\bar{z}_3 - \bar{c}_3 = (3\ \ 0)\begin{pmatrix} 2 \\ 4 \end{pmatrix} - 3 = 3$$

$$\bar{z}_4 - \bar{c}_4 = (3\ \ 0)\begin{pmatrix} 1 \\ 0 \end{pmatrix} - 0 = 3$$

$\therefore$ x_2 enters into the basis
Now, the pivotal column

$$\bar{P}_2 = B^{-1}\ P_2 = \begin{bmatrix} 1/2 & 0 \\ -1/2 & 1 \end{bmatrix}\begin{bmatrix} -1 \\ 0 \end{bmatrix} = \begin{bmatrix} -1/2 \\ 1/2 \end{bmatrix}$$

Hence, x_2 enters and x_5 leaves the basis. Now, the pivotal column is $\begin{bmatrix} -1/2 \\ 1/2 \end{bmatrix}$ must be reduced to $\begin{bmatrix} 0 \\ 1 \end{bmatrix}$.

Second iteration:

table 3.13

Basis	B^{-1}		Constants	Variable to enter	Pivot column	Ratio
x_1	0	1	4			
x_2	- 1	2	6			

The simplex multipliers corresponding to the second iteration are

$$M = (M_1\ M_2) = C_B\ B^{-1} = (6 - 2)\begin{pmatrix} 0 & 1 \\ -1 & 2 \end{pmatrix} = (2\ \ 2)$$

$$\bar{z}_3 - \bar{c}_3 = (2\ \ 2)\begin{pmatrix} 2 \\ 4 \end{pmatrix} - 3 = 9$$

$$\bar{z}_4 - \bar{c}_4 = (2 \ 2)\begin{pmatrix}1\\0\end{pmatrix} - 0 = 2$$

$$\bar{z}_5 - \bar{c}_5 = (2 \ 2)\begin{pmatrix}0\\1\end{pmatrix} - 0 = 2$$

Since all $\bar{z}_i - \bar{c}_i$ is non-negative the current solution is optimum.

The optimum solution is

$$\boxed{x_1 = 4, x_2 = 6, x_3 = 0 \text{ with maximum of } z = 6 \times 4 - 2 \times 6 + 3 \times 0 = 12}$$

Example 2

Use revised simplex method to solve the following LPP.

Minimise $z = x_1 + 2x_2$

subject to

$$2x_1 + 5x_2 \geq 6$$
$$x_1 + x_2 \geq 2$$
$$x_1, x_2 \geq 0$$

Solution: Convert the minimisation LPP into maximisation type and then introduce surplus variables x_3, x_4 and the artifical variables x_5, x_6. The LPP becomes

Maximise. $z^* = -x_1 - 2x_2 + 0x_3 + 0x_4 - Mx_5 - Mx_6$

subject to

$$2x_1 + 5x_2 - x_3 + x_5 = 6$$
$$x_1 + x_2 - x_4 + x_6 = 2$$
$$x_1, x_2, x_3, x_4, x_5, x_6 \geq 0$$

Let $P_1, P_2, \cdots, P_6$ denote the column vectors corresponding to the variables $x_1, \cdots, x_6$ and let b denote the right hand constants, respectively.

Then, $P_1 = \begin{bmatrix}2\\1\end{bmatrix}, P_2 = \begin{bmatrix}5\\1\end{bmatrix}, P_3 = \begin{bmatrix}-1\\0\end{bmatrix} P_4 = \begin{bmatrix}0\\-1\end{bmatrix}, P_5 = \begin{bmatrix}1\\0\end{bmatrix}, P_6 = \begin{bmatrix}0\\1\end{bmatrix}$ and $b = \begin{bmatrix}6\\2\end{bmatrix}$

Now, $(x_5 \, x_6)$ forms the basis

$$B = [P_5, P_6] = \begin{bmatrix}1 & 0\\0 & 1\end{bmatrix} = B^{-1}$$

$$\bar{b} = B^{-1} b = \begin{bmatrix}1 & 0\\0 & 1\end{bmatrix}\begin{bmatrix}6\\2\end{bmatrix} = \begin{bmatrix}6\\2\end{bmatrix} = b$$

Initial Table:

table 3.14

Basis	B^{-1}		Constants	Variable to enter	Pivot column	Ratio
x_5	1	0	6		[5]	6/5
x_6	0	1	2	x_2	1	2/1 = 2

The simplex multipliers for Table 3.14 are

$$M = (M_1\ M_2) = C_B\, B^{-1} = (-M\ -M)\begin{pmatrix}1 & 0\\ 0 & 1\end{pmatrix} = (-M\ -M)$$

$$\bar{z}_1 - \bar{c}_1 = MP_1 - c_1 = (-M\ -M)\begin{pmatrix}2\\1\end{pmatrix} - (-1) = -3M + 1$$

$$\bar{z}_2 - \bar{c}_2 = (-M\ -M)\begin{pmatrix}5\\1\end{pmatrix} + 2 = -6M + 2$$

$$\bar{z}_3 - \bar{c}_3 = (-M\ -M)\begin{pmatrix}-1\\0\end{pmatrix} - 0 = M$$

$$\bar{z}_4 - \bar{c}_4 = (-M\ -M)\begin{pmatrix}0\\-1\end{pmatrix} - 0 = M$$

Hence, x_2 enters in the basis and hence the pivotal column is

$$\bar{P}_2 = B^{-1}\, P_2 = \begin{bmatrix}1 & 0\\ 0 & 1\end{bmatrix}\begin{bmatrix}5\\1\end{bmatrix} = \begin{bmatrix}5\\1\end{bmatrix}$$

After calculating the ratio we see that x_5 leaves the basis. Now, the pivotal column is $\begin{bmatrix}5\\1\end{bmatrix}$. This must be reduced to $\begin{bmatrix}1\\0\end{bmatrix}$.

First iteration:

table 3.15

Basis	B^{-1}		Constants	Variable to enter	Pivot column	Ratio
x_2	1/5	0	6/5	x_1	2/5	3
x_5	- 1/5	1	4/5		3/5	4/3

The simplex multipliers for Table 3.15 are

$$M = (M_1\ M_2) = (-2\ -M)\begin{pmatrix}1/5 & 0\\ -1/5 & 1\end{pmatrix} = \begin{pmatrix}\dfrac{M-2}{5} & -M\end{pmatrix}$$

Hence,
$$\bar{z}_1 - \bar{c}_1 = \begin{pmatrix}\dfrac{M-2}{5} & -M\end{pmatrix}\begin{pmatrix}2\\1\end{pmatrix} - (-1) = \frac{-3M+1}{5}$$

$$\bar{z}_3 - \bar{c}_3 = \begin{pmatrix}\dfrac{M-2}{5} & -M\end{pmatrix}\begin{pmatrix}-1\\0\end{pmatrix} - 0 = \frac{-M+2}{5}$$

$$\bar{z}_4 - \bar{c}_4 = \left(\frac{M-2}{5} - M\right)\begin{pmatrix}0\\-1\end{pmatrix} - 0 = M - 0 = M$$

$$\bar{z}_5 - \bar{c}_5 = \left(\frac{M-2}{5} - M\right)\begin{pmatrix}1\\0\end{pmatrix} - 0 = \frac{M-2}{5}$$

Hence, x_1 enters. The pivotal column is

$$B^{-1} P_1 = \begin{bmatrix}1/5 & 0\\-1/5 & 1\end{bmatrix}\begin{bmatrix}2\\1\end{bmatrix} = \begin{bmatrix}2/5\\3/5\end{bmatrix}$$

and the ratio implies that x_6 leaves the basis. We must reduce the pivotal column to $\begin{bmatrix}0\\1\end{bmatrix}$.

Second iteration:

table 3.16

Basis	B^{-1}		Constants	Variable to enter	Pivot column	Ratio
x_2	1/3	- 2/3	2/3			
x_1	- 1/3	5/3	4/3			

The simplex multipliers are

$$M = (M_1\ M_2) = (-2\ -1)\begin{pmatrix}1/3 & -2/3\\-1/3 & 5/3\end{pmatrix} = (-1/3 \quad -1/3)$$

$$\bar{z}_3 - \bar{c}_3 = (-1/3 \quad -1/3)\begin{pmatrix}-1\\0\end{pmatrix} - 0 = 1/3$$

$$\bar{z}_4 - \bar{c}_4 = (-1/3 \quad -1/3)\begin{pmatrix}0\\-1\end{pmatrix} - 0 = 1/3$$

$$\bar{z}_5 - \bar{c}_5 = (-1/3 \quad -1/3)\begin{pmatrix}1\\0\end{pmatrix} + M = M - 1/3$$

$$\bar{z}_6 - \bar{c}_6 = (-1/3 \quad -1/3)\begin{pmatrix}0\\1\end{pmatrix} + M = M - 1/3$$

Since all $\bar{z}_i - \bar{c}_i$ are non-negative the current solution is optimum. The optimum solution is

$$x_1 = 4/3,\ x_2 = 2/3,\ x_3 = 0 \text{ and minimum of } z = 4/3 + 2(2/3) = 8/3$$

3.3.3 Advantages of the Revised Simplex Method Over the Regular Simplex Method

1. For LPP involving large number of variables the number of constraints and the amount of

comptations is very much reduced because the revised simplex method works with a reduced Table whose size is determined by the number of constraints.

2. The revised simplex method works with a reduced Table as it stores only the basic variables, the basis inverse and the constants. Hence, less amount of new information needs to be stored in the memory of the computer from one iteration to the other.
3. Data can be stored in lesser space. The original data is usually given in fixed decimals of three or four digits. The storage of data is more accurate and compact since the revised simplex method works only with the original data.
4. There is less accumulation of round off errors since no calculations are done on a column unless it is ready to enter the basis.
5. The theory of the revised simplex method, especially the importance of the basis inverse and the simple multipliers is quite helpful in understanding the advanced topics in linear programming such as duality theory and sensitivity analysis.

3.4 BOUNDED VARIABLE TECHNIQUE

3.4.1 Introduction

A linear programming problem may have, in addition to the regular constraints, some or all variables with lower and upper limits. That is, constraints of the type $l_j \leq x_j \leq u_j$ where l_j and u_j are the lower and upper bounds of the variable x_j, respectively. The standard form of such a problem may be expressed as

Maximise $z = CX$

subject to $AX = b$

$$l_j \leq x_j \leq u_j \qquad (j = 1, 2, \cdots, n)$$

where $C = (c_1, \cdots, c_n)$

$$A = \begin{pmatrix} a_{11} & a_{12} & \cdots & a_{1n} \\ a_{21} & a_{22} & \cdots & a_{2n} \\ & \cdots & & \\ & \cdots & & \\ a_{m_1} & a_{m_2} & \cdots & a_{mn} \end{pmatrix} \quad X = \begin{pmatrix} x_1 \\ \vdots \\ x_n \end{pmatrix} \quad b = \begin{pmatrix} b_1 \\ \vdots \\ b_m \end{pmatrix}$$

The equality constants $L \leq X \leq U$ can be converted into equality constraints by introducing slack and/or surplus variables X' and X'' as shown below.

$$X + X' = U$$
$$X - X'' = L$$
$$X, X' X'' \geq 0$$

The given problem, thus contains $3n$ variables and $m + 2n$ constraint equations. However, this problem can be reduced in size by using a special technique which ultimately reduces the constraints to the form $AX = b$.

First, considering the lower bound constraints, the equality constraint $X - X'' = L$ could be written as $X = L + X''$ and thus X can be eliminated from all the remaining constraints. The problem will now have X' and X'' as variables and since both are non-negative, X will always be non-negative.

The real problem arises when dealing with upperbound variable, because the substitution $X = U - X'$ does not guarantee that X will remain non-negative. This difficulty is overcome by using a special technique called the bounded variable simplex method.

The upperbound problem can be written as

Maximise $z = CX$

subject to $AX = b$

$X + X' = U$

$X, X' \geq 0$

However, instead of considering the constraints $X + X' = U$ in the simplex table, their effect can be accounted for by modifying the feasibility condition of the simplex method, the optimality condition remaining the same. The upperbound condition requires a basic variable to become non-basic at its upperbound. Moreover, when a non-basic variable enters the solution, its value should not exceed its upperbound. Thus, while developing the new feasibility condition, two things are to be taken care of:

1. The non-negativity and upperbound constraints of the entering variable.
2. The non-negativity and upperbound constraints of all those basic variables that are affected by the entering variable.

3.4.2 Bounded Variable Algorithm (or) Upperbound Algorithm (or) Lower and Upperbound Technique

Step 1: If the LPP is of minimisation type convert it into maximisation type.

Step 2: If RHS of any constraint is negative, make it positive by multiplying the constraint by – 1.

Step 3: Convert the LPP in standard form by adding slack and/or surplus variables and obtain an initial basic feasible solution.

Step 4: If any of the variables is at a positive lowerbound, substitute it out at its lowerbound by

$$l_j = x_j - x'_j \text{ (or) } x_j = l_j + x'_j \text{ or } x'_j = x_j - l_j$$

where $x'_j \geq 0$.

Step 5: Compute the net evaluation $(z_j - c_j)$ and examine their sign.

(i) If all $(z_j - c_j) \geq 0$, then the current basic feasible solution is optimum.

(ii) If at least one $(z_j - c_j) < 0$, then the current basic feasible solution is not optimal. Go to the next step.

Step 6: Choose the most negative $z_j - c_j$. Let it be $z_r - c_r$ for some $j = r$. The corresponding non-basic variable x_r enters into the basis.

Step 7: Compute the quantities

$$\theta_1 = \min_i \left\{ \frac{X_{B_i}}{a_{ir}}, a_{ir} > 0 \right\}$$

$$\theta_2 = \min_i \left\{ \frac{U_i - X_{B_i}}{-a_{ir}}, a_{ir} < 0 \right\}$$

u_r – the upperbound of the entering variable x_r and $\theta = \min(\theta_1, \theta_2, u_r)$. It is to be noted that if all $a_{ir} \geq 0$, then $\theta_2 = \infty$.

Case (i): If $\theta = \theta_1$ which corresponds to X_{B_k} then x_k leaves the basis. Use the regular operations of the simplex method.

Case (ii): If $\theta = \theta_2$ which corresponds to X_{B_k}, then x_k leaves the basis, x_k being non-basic at its upperbound must be substituted by using

$$(X_{B_k})_r = (X_{B_k})_r^1 - a_{k_r} u_r \text{ where } 0 \leq (X_{B_k})_r^1 \leq u_r$$

and $(X_{B_k})_r$ denotes that the variable x_r was x_k. The non-basic variable x_r at upperbound is put at zero level by using $x_r = u_r - x'_r$, $0 \leq x'_r \leq u_r$.

Case (iii): If $\theta = u_r$, x_r is substituted at its upperbound while remaining non-basic. That is, it can be put at zero level by using $x_r = u_r - x'_r$, $0 \leq x'_r \leq u_r$.

Step 8: Go to step (5) and repeat the procedure until an optimum basic feasible solution is obtained or there is an indication of unbounded solution.

Note: If any variable is without its upperbound, then assume the upperbound as ∞.

Example 1

Solve the following LPP by bounded variable simplex method.

Maximise $z = 3x_1 + 5x_2 + 3x_3$

subject to
$$x_1 + 2x_2 + 2x_3 \leq 14$$
$$2x_1 + 4x_2 + 3x_3 \leq 23$$
$$0 \leq x_1 \leq 4,\ 0 \leq x_2 \leq 5 \text{ and}$$
$$0 \leq x_3 \leq 3$$

Solution: Introducing slack variables $x_4, x_5 \geq 0$ in the given constraints the LPP becomes

Maximise $z = 3x_1 + 5x_2 + 3x_3 + 0x_4 + 0x_5$

subject to
$$x_1 + 2x_2 + 2x_3 + x_4 = 14$$
$$2x_1 + 4x_2 + 3x_3 + x_5 = 23$$
$$0 \leq x_1 \leq 4,\ 0 \leq x_2 \leq 5,\ 0 \leq x_2 \leq 3,\ x_4 \geq 0,\ x_5 \geq 0$$

An IBFS is $x_4 = 14$, $x_5 = 23$.

Since no upperbounds are specified for these basic variables, take their upperbound as ∞.

Initial Table:

table 3.17

		c_j	3	5	3	0	0
C_B	Y_B	X_B	x_1	x_2	x_3	x_4	x_5
0	x_4	14	1	2	2	1	0
0	x_5	23	2	4	3	0	1
$z_j - c_j$			-3	-5	-3	0	0

Now, $\theta_1 = \min(14/2, 23/4) = 23/4$

$\theta_2 = \infty$, since all elements in x_2 column are non-negative and $u_2 = 5$.

$$\therefore \quad \theta = \min(\theta_1, \theta_2, u_2)$$
$$= \min(23/4, \infty, 5) = 5$$

Since $\theta = u_2$, x_2 is substituted at its upperbound difference i.e $x_2 = 5 - x'_2$ but it remains non-basic. Then, the new Table is

table 3.18

		c_j	3	- 5	3	0	0
C_B	Y_B	X_B	x_1	x'_2	x_3	x_4	x_5
0	x_4	4	1	- 2	2	1	0
0	x_5	3	[2]	- 4	3	0	1
	$x_j - c_j$	0	- 3	5	- 3	0	0

Now, let x_1 be the incoming variable.

Then, $\theta_1 = \min \left\{\frac{4}{1}, \frac{3}{2}\right\} = 3/2$

$\theta_2 = \infty$ Since all in x_1 column are non-negative.

$u_1 = 4$

$\therefore$ $\theta = \min \{3/2, \infty, 4\} = 3/2 = \theta_1$

Since $\theta = \theta_1$, introduce x_1 and drop x_5. This yields

table 3.19

		c_j	3	- 5	3	0	0
C_B	Y_B	X_B	x_1	x'_2	x_3	x_4	x_5
0	x_4	5/2	0	0	1/2	1	- 1/2
3	x_1	5/2	1	- 2	3/2	0	1/2
	$z_j - c_j$	—	0	- 1	3/2	0	3/2

x'_2 is the entering variable.

Now, $\theta_1 = \infty$ [$\because$ there is no positive element in x'_2 column]

$$\theta_2 = \frac{4 - 3/2}{-(-2)} = \frac{5}{4}, \text{ corresponding to } x_1$$

$u_2 = 5$

$\therefore$ θ = minimum $\{\infty, 5/4, 5\} = 5/4$. Since $\theta = \theta_2$, introduce x'_2 into the basis and drop x_1 and then substitute it out at its upperbound $4 - x'_1$. Thus, by removing x_1 and introducing x'_2 the Table becomes

table 3.20

C_B	Y_B	X_B	x_1	x'_2	x_3	x_4	x_5
0	x_4	5/2	0	0	1/2	1	- 1/2
- 5	x'_2	- 3/4	- 1/2	1	- 3/4	0	1/4

Now, substituting $x_1 = 4 - x'_1$, the final Table becomes

table 3.21

C_B	Y_B	X_B	x'_1	x'_2	x_3	x_4	x_5
0	x_4	5/2	0	0	1/2	1	- 1/2
- 5	x'_2	5/4	1/2	1	- 3/4	0	- 1/4
$z_j - c_j$			1/2	0	3/4	0	5/4

Therefore, the optimal solution is $x'_1 = 0, x'_2 = 5/4, x_3 = 0$. In terms of original variables, the solution is $x_1 = 4 - 0 = 4$, $x_2 = 5 - 5/4 = 15/4$, $x_3 = 0$ and maximum $z = 3 \times 4 + 5 \times 15/4 + 3 \times 0 = 123/4$.

3.5 SENSITIVITY ANALYSIS

3.5.1 Introduction

After a linear programming problem has been solved, it is useful to study the effect of changes in the parameters of the problem on the current optimal solution. The LP solution must provide dynamic information. A static optimum solution will become obsolete as soon as the conditions under which the model is constructed change. Sensitivity analysis is concerned with studying possible changes in the available optimal solution as a result of making changes in the original model.

The change in parameters of the problem may be discrete or continuous. The study of the effect of discrete changes in parameters on the optimal solution is called sensitivity analysis or post-optimality analysis, while that of continuous changes in parameters is called parametric programming. One way to determine the effects of parameter changes is to solve the new problem anew, which may be computationally inefficient. Alternatively, the current optimal solution may be investigated, making use of the properties of the simplex criteria. The second method reduces additional computations considerably and hence forms the subject of present discussion.

The changes in the parameters of a LPP include

1. changes in the right hand side of the constraint b_i
2. changes in the cost coefficient c_j
3. addition of new variables
4. changes in the coefficients of constraints a_{ij}
5. addition of new constraints
6. deletion of variables
7. deletion of constraints.

Generally, these parameter changes result in one of the following three cases:

1. The optimal solution does not change.
2. The basic variables remains the same but their values change.
3. The basic variables as well as their values are changed.

While dealing with these changes, one important objective is to find the maximum extent to which a parameter or a set of parameters can change so that the current optimal solution remains optimal. In other words, the objective is to determine how sensitive is the optimal solution to the changes in those parameters. Such an analysis is called sensitivity analysis.

1. Variation in the right side of constraints Since the optimality condition $z_j - c_j$ does not involve any of b_i, if any component b_i of the requirement vector $b = (b_1, ..., b_m)$ is changed, then this change will not effect the conditions of optimality. Hence, if any b_i is changed to $b_i + \Delta b_i$ then the new solution thus obtained will remain optimal.

But $X_B = B^{-1}b$ depends on b. Therefore, any change in b may affect the feasibility of the current solution.

Example 1

Consider the LPP

Maximise $z = 5x_1 + 12x_2 + 4x_3$

subject to

$$x_1 + 2x_2 + x_3 \le 5$$
$$2x_1 - x_2 + 3x_3 = 2$$
$$x_1, x_2, x_3 \ge 0$$

(i) Discuss the effect of changing the requirement vector from $\begin{bmatrix}5\\2\end{bmatrix}$ to $\begin{bmatrix}7\\2\end{bmatrix}$ on the optimum solution.

(ii) Discuss the effect of changing the requirement vector from $\begin{bmatrix}5\\2\end{bmatrix}$ to $\begin{bmatrix}3\\9\end{bmatrix}$ on the optimum solution.

(iii) Which resource should be increased and by how much to achieve the best marginal increase in the value of the objective function?

(iv) Which resource should be decreased and by how much to achieve the best marginal increase in the value of the objective function?

Solution: If we solve the given problem by Big '*M*' method the final simplex table is

table 3.22

		c_j	5	12	4	0	0
C_B	Y_B	X_B	x_1	x_2	x_3	s_1	A_1
12	x_2	8/5	0	1	− 1/5	2/5	− 1/5
5	x_1	9/5	1	0	7/5	1/5	2/5
$z_j - c_j$			0	0	3/5	29/5	M − 2/5

and the optimal solution is

$$x_1 = 9/5,\ x_2 = 8/5,\ x_3 = 0 \text{ and maximum } z = 141/5$$

(i) If the requirement vector changes to $\begin{bmatrix}7\\2\end{bmatrix}$ then the new values of the current basic variables are given by

$$\begin{bmatrix}x_2\\x_1\end{bmatrix} = B^{-1}\,b = \begin{bmatrix}2/5 & -1/5\\1/5 & 2/5\end{bmatrix}\begin{bmatrix}7\\2\end{bmatrix} = \begin{bmatrix}12/5\\11/5\end{bmatrix}$$

Hence, the new solution is optimal and feasible and the new optimum solution is

$x_1 = 11/5, x_2 = 12/5$ with maximum of $z = 5 \times 11/5 + 12 \times 12/5 + 4 \times 0 = 199/5$

(ii) If the requirement vector changes to $\begin{bmatrix} 3 \\ 9 \end{bmatrix}$ then the new values of the basic variables are given by

$$\begin{bmatrix} x_2 \\ x_1 \end{bmatrix} = B^{-1}\, b = \begin{bmatrix} 2/5 & -1/5 \\ 1/5 & 2/5 \end{bmatrix} \begin{bmatrix} 3 \\ 9 \end{bmatrix} = \begin{bmatrix} -3/5 \\ 21/5 \end{bmatrix}$$

Hence, the current optimal solution becomes infeasible. Hence, we have to apply the dual simplex method.

table 3.23

			5	12	4	0	0
C_B	Y_B	X_B	x_1	x_2	x_3↓	s_1	A_1
12	x_2	- 3/5	0	1	[−1/5]	2/5	- 1/5
5	x_1	21/5	1	0	7/5	1/5	2/5
			0	0	3/5	29/5	M - 2/5

Introduce x_3 into the basis and drop x_2 from the basis then the next iteration is

table 3.24

		c_j	5	12	4	0	0
C_B	Y_B	X_B	x_1	x_2	x_3	s_1	A_1
4	x_3	3	0	- 5	1	- 2	1
5	x_1	0	1	7	0	3	- 1
	$x_j - c_j$		0	3	0	7	M - 1

Since all $z_j - c_j \geq 0$ and all $X_{B_i} \geq 0$ the current solution is optimum. The optimal solution is

$x_1 = 0, x_2 = 0, x_3 = 3$ with maximum of $z = 5(0) + 12(0) + 4 \times 3 = 12$

(iii) In order to find the resource that should be increased (or decreased) we shall write the final objective function, which is $G = 5y_1 + 2y_2$, where $y_1 = 29/5$ and $y_2 = -2/5$ are the optimal dual variables. Thus, the first resource increases the objective function by 29/5. Next, we are to find how much the first resource should be increased so that each additional unit continues to increase the objective function by 29/5. This requirement will be met so long as the primal problem remains feasible. If Δ be the increase in the first resource, it can be determined from the condition

$$\begin{bmatrix} x_2 \\ x_1 \end{bmatrix} = B^{-1}b = \begin{bmatrix} 2/5 & -1/5 \\ 1/5 & 2/5 \end{bmatrix} \begin{bmatrix} 5+\Delta \\ 2 \end{bmatrix} = \begin{bmatrix} \dfrac{8+2\Delta}{5} \\ \dfrac{9+\Delta}{5} \end{bmatrix} \geq \begin{bmatrix} 0 \\ 0 \end{bmatrix}$$

As x_1 and x_2 remain feasible (≥ 0) for all values of $\Delta > 0$, the first resource can be increased indefinitely while maintaining the condition that each additional unit will increase the objective function by 29/5.

(iv) The second resource should be decreased as each additional unit of the second resource decreases the objective function by 2/5. Let Δ be the decrease in the second resource. To find its extent, we make use of the condition that the current solution remains feasible so long as

$$\begin{bmatrix} x_2 \\ x_1 \end{bmatrix} = B^{-1}b = \begin{bmatrix} 2/5 & -1/5 \\ 1/5 & 2/5 \end{bmatrix}\begin{bmatrix} 5 \\ 2-\Delta \end{bmatrix} = \begin{bmatrix} \dfrac{8+\Delta}{5} \\ \dfrac{9-2\Delta}{5} \end{bmatrix} \geq \begin{bmatrix} 0 \\ 0 \end{bmatrix}$$

Evidently x_1 remains positive only so long as $\dfrac{9-2\Delta}{5} \geq c$ or $\Delta \leq 9/2$. If $\Delta > 9/2$, x_1 becomes negative and must leave the solution.

2. Addition of a new constraint

Addition of a new constraint may or may not affect the feasibility of the current optimal solution. The addition of a new constraint can result in one of the two conditions.

(i) If the new constraint is satisfied by the current optimal solution, then this new constraint is said to be redundant and its addition will not change the solution. That is, the current solution remains feasible as well as optimal.

(ii) If the new constraint is not satisfied by the current optimal solution, the solution becomes infeasible. In this case the new solution is obtained by using the dual simplex method.

However, in general, whenever a new constraint is added to a linear programming problem, the old optimal value will always be better or at least equal to the new optimal value. In other words, the addition of a new constraint cannot improve the optimal value of any LPP.

Example 2

Consider the LPP

Maximise $z = 4x_1 + 6x_2 + 2x_3$,

subject to

$$\left.\begin{aligned} x_1 + x_2 + x_3 &\leq 3 \\ x_1 + 4x_2 + 7x_3 &\leq 9 \\ x_1, x_2, x_3 &\leq 0 \end{aligned}\right\} \quad (3.1.12)$$

If we solve the problem by simplex method then the optimal simplex table will be

table 3.25

		c_j	4	6	2	0	0
C_B	Y_B	X_B	x_1	x_2	x_3	s_1	s_2
4	x_1	1	1	0	- 1	4/3	- 1/3
6	x_2	2	0	1	2	1/3	1/3
			0	0	6	10/3	2/3

and the optimal solution will be

$$x_1 = 1, x_2 = 2, x_3 = 0 \text{ with maximum of } z = 4 \times 1 + 6 \times 2 + 2 \times 0 = 16 \quad (3.1.13)$$

If we add the additional constraint $2x_1 + 3x_2 + 2x_3 \leq 4$ then the optimal solution given in (3.1.13) does not satisfy the additional constraint. The current optimal solution is not optimal for the modified problem. Let s_3 be the slack variable for this constraint then the modified Table will be

table 3.26

		c_j	4	6	2	0	0	0
C_B	Y_B	X_B	x_1	x_2	x_3	s_1	s_2	s_3
4	x_1	1	1	0	- 1	4/3	- 1/3	0
6	x_2	2	0	1	2	- 1/3	1/3	0
0	x_3	4	2	3	2	0	0	1
	$z_j - c_j$		0	0	6	10/3	2/3	0

Since x_1 and x_2 are basic variables the corresponding entries in the third column should be zero. To achieve this multiply the first row by – 2 and second row by – 3 and add them to the third row, the above Table becomes

table 3.27

		c_j	4	6	2	0	0	0
C_B	Y_B	X_B	x_1	x_2	x_3	s_1	s_2	s_3
4	x_1	1	1	0	- 1	4/3	- 1/3	0
6	x_2	2	0	1	2	- 1/3	1/3	0
0	s_3	- 4	0	0	- 6	- 5/3	- 1/3	1
	$z_j - c_j$		0	0	6	10/3	2/3	0

Since all $z_j - c_j \geq 0$, the optimality condition is satisfied but X_B value for the third row is negative. Now, we apply the dual simplex method then the optimal solution becomes

$$x_1 = 5/3, x_2 = 5/3 \text{ and } x_3 = 2/3 \text{ with maximum of } z = 12$$

3.5.2 Changes Affecting Optimality

1. Changes in the objective function (or) changes in the cost coefficient c_j:

Changes in the coefficients of the objective function may take place due to a change in cost or profit of either basic variables or non-basic variables.

If X_B is the optimal basic feasible solution, then we have $X_B = B^{-1} b$. It is clear that X_B is independent of C and therefore, any change in some component c_j of C will not change X_B. X_B will always remain basic feasible solution.

But however, the optimal condition $(z_j - c_j) \geq 0$ for all j which is satisfied for optimal solution c_j changes. So, any change in some component c_j of C may affect the optimality of the current solution and will not affect the feasibility of the current solution.

The variation in the price vector C may be made the following two ways:

(i) Variation $c_j \notin C_B$ (or) variation in the coefficient c_j of the non-basic variable x_j in the objective function.

If $c_j \notin C_B$ changes to $(c_j + \Delta c_j)$ such that $\Delta c_j \leq z_j - c_j$, then the value of the objective function and the optimal solution of the problem remain changed. The range over $c_j \notin C_B$ can vary maintaining the optimality of the solution given by $-\infty \leq c_j \leq c_j + \Delta c_j$. Note that there is a lowerbound to Δc_j.

(ii) Variation in $c_k \in C_B$ (or) variation in the coefficient c_k of the basic variable x_k in the objective function.

If $c_k \in C_B$ changes to $(c_k + \Delta c_k)$, then the range over $c_k \in C_B$ can vary so that the solution remains optimal is given by

$$\max_{a_{kj} > 0} \left\{ \frac{-(z_j - c_j)}{a_{kj}} \right\} \leq \Delta c_k \leq \max_{a_{kj} < 0} \left\{ \frac{-(z_j - c_j)}{a_{kj}} \right\}$$

If there is no $a_{kj} > 0$, there is a lowerbound to Δc_k and if no $a_{kj} < 0$, there is no upperbound to Δc_k. The value of the objective function will be improved further improved by an amount $X_{B_k} \cdot \Delta C_k$.

i.e. $z^* = z + X_{B_k} \cdot \Delta C_{B_k}$

Remark 1: Any change $c_j \notin C_B$ will affect only its net evaluation coefficients and not others.

2. Any change $c_j \in C_B$ will affect its net evaluation coefficients and the value of the objective function.

Example

Consider the LPP

Maximise $z = 4x_1 + 6x_2 + 2x_3$

subject to

$$x_1 + x_2 + x_3 \leq 3$$
$$x_1 + 4x_2 + 7x_3 \leq 9$$
$$x_1, x_2, x_3 \geq 0$$

(a) Solve the LPP

(b) (i) Find the range on the value of non-basic variable coefficient c_3 such that the current solution will still remains optimal.

(ii) What happens if c_3 in increased to 12? What is the new optimal solution?

(c) (i) Find the range on basic variable coefficient c_1 such that the current optimal solution remains optimal.

(ii) Find the effect of $c_1 = 8$ on the optimal solution.

(d) Find the effect of changing the objective function to $z = 2x_1 + 8x_2 + 4x_3$ on the current optimal solution.

Solution: (a) If we solve the LPP by simplex method (left as an exercise to the reader) the optimal simplex table is

table 3.28

	c_j		4	6	2	0	0
C_B	y_B	X_B	x_1	x_2	x_3	s_1	s_2
4	x_1	1	1	0	- 1	4/3	- 1/3
6	x_2	2	0	1	2	- 1/3	1/3
	$z_j - c_j$		0	0	6	10/3	2/3

(Here, s_1 and s_2 are slack variables) and the optimal solution is

$$x_1 = 1, x_2 = 2, x_3 = 0 \text{ and maximum of } z = 4 \times 1 + 6 \times 2 + 2 \times 0 = 16$$

(b) (i) The coefficient c_3 corresponds to the non-basic variable x_3 and its value is zero in the current optimal solution since the associated cost for x_3 is 2 (very low). Clearly if c_3 further decreases, it will have no effect on the current optimal solution. However, if c_3 is increased beyond a certain value it may become profitable for the third decisions variable.

As a rule, the sensitivity of the current optimal solution is determined by studying how the current optimal solution changes as a result of changes in the input data. When a value of c_3 changes, then the net evaluation value of x_3 also changes. The current solution remains optimal as long as $z_3 - c_3$ remains non-negative.

From the optimal simplex table

$$(4 \quad 6)\begin{bmatrix} 1 \\ -2 \end{bmatrix} - c_3 \geq 0$$

or, $$8 - c_3 \geq 0$$

or, $$c_3 \leq 8$$

Hence, as long as the coefficient of x_3 is less than or equal to 8, it is not profitable to produce it.

(ii) if $c_3 = 12$, $z_3 - c_3 = (4 \quad 6)\begin{bmatrix} -1 \\ 2 \end{bmatrix} - 12 = -4$

Since $z_3 - c_3$ is negative the current solution is not optimal. Since there is a negative net evaluation we continue the iteration process [left as an exercise to the reader]. The new optimal solution is

$$\boxed{x_1 = 2, x_2 = 0, x_3 = 1 \text{ with maximum of } z = 4 \times 2 + 6 \times 0 + 12 \times 1 = 20}$$

(c) (i) When c_1 decreases below a certain level, it may no longer remain profitable with respect to the decision variable x_1. On the other hand, if c_1 increases beyond a certain value, it may become so profitable that it is most paying to produce only the product corresponding to the decision variable x_1. In either case the optimal solution will change and hence there is lower as well as upper limit on c_1 within which the optimal product mix will not be affected.

It is clear from Table that any variation in c_1 (and/or in c_2 also) will not change $z_1 - c_1$ and $z_2 - c_2$ ($\because$ they remain zero), while $z_3 - c_3$, $z_4 - c_4$, $z_5 - c_5$ will change. However, as long as $z_j - c_j$ ($j = 3, 4, 5$) remain non-negative, Table will remain optimal.

Now, $$z_3 - c_3 = (c_1 \; 6)\begin{bmatrix} -1 \\ 2 \end{bmatrix} - 2 = -c_1 + 10$$

$$z_4 - c_4 = (c_1 \; 6)\begin{bmatrix} 4/3 \\ -1/3 \end{bmatrix} - 0 = 4/3\, c_1 - 2$$

$$z_5 - c_5 = (c_1 \; 6)\begin{bmatrix} -1/3 \\ 1/3 \end{bmatrix} - 0 = -1/3\, c_1 + 2$$

Hence, $$-c_1 + 10 \geq 0 \quad \text{i.e.} \quad c_1 \leq 10$$

$$\frac{4}{3} c_1 - 2 \geq 0 \quad \text{i.e.} \quad c_1 \geq 3/2$$

$$1/3\, c_1 + 2 \geq 0 \quad \text{i.e.} \quad c_1 \leq 6$$

$\therefore$ Range of c_1 for the optimal solution to remain optimal is $3/2 \leq c_1 \leq 6$.

Thus, so long as c_1 lies within these limits, the optimal solution in Table remains optimal. However, within this range, as the value of c_1 is changed, maximum of z undergoes a change. For example, when $c_1 = 4$, maximum of $z = 4 \times 1 + 6 \times 2 = 16$.

(ii) When $c_1 = 8$,

$$z_3 - c_3 = 10 - 8 = 2$$
$$z_4 - c_4 = -2 + 4/3\ c_1 = 26/3$$
$$z_5 - c_5 = -1/3\ c_1 + 2 = -2/3$$
$$z_1 - c_1 = 0$$
$$z_2 - c_2 = 0$$

As the net evaluation corresponding to the slack variable x_2 is $-2/3$, the current solution is not optimum. We can continue the simplex iteration process by taking s_2 as the entering variable. The new optimal solution is $x_1 = 3, x_2 = 0, x_3 = 0$ with maximum of $z = 24$ [left as an exercise to the reader].

(d) The effect on the optimal solution can be determined by checking whether the net evaluation row remains non-negative.

$$z_1 - c_1 = 0$$
$$z_2 - c_2 = 0$$
$$z_3 - c_3 = (2\ \ 8)\begin{bmatrix} -1 \\ 2 \end{bmatrix} - 4 = 10 > 0$$
$$z_4 - c_4 = (2\ \ 8)\begin{bmatrix} 4/3 \\ -1/3 \end{bmatrix} - 0 = 0$$
$$z_5 - c_5 = (2\ \ 8)\begin{bmatrix} -1/3 \\ 1/3 \end{bmatrix} - 0 = 2 > 0$$

Hence, the optimal solution does not change. The optimal solution remains as $x_1 = 1, x_2 = 2, x_3 = 0$ and maximum of $z = 1 \times 2 + 2 \times 8 + 0 \times 4 = 18$.

Note that there is also an indication of alternate optimum solution since $z_4 - c_4 = 0$.

2. *Variation in the coefficients a_{ij} of the constraints (or) variation in the component a_{ij} of the coefficient matrix A*:

Case (i): If the coefficients a_{ij} of a non-basic variable x_j ($a_{ij} \notin B$) gets changed in the current optimal solution, then this change will not affect the feasibility of the current solution but may affect the optimality of the current solution.

Case (ii): If the coefficient a_{ij} of a basic variable x_j ($a_{ij} \in B$) gets changed in the current optimal solution, then this may effect both feasibility and optimality of the current solution. As the basis matrix is affected, it may affect all the quantities given in the current optimal Table. Under this situation, it may be better to solve the problem again.

Example

Consider the LPP

Maximise $z = 45x_1 + 100x_2 + 30x_3 + 50x_4$

subject to

$$7x_1 + 10x_2 + 4x_3 + 9x_4 \leq 1200$$
$$3x_1 + 40x_2 + x_3 + x_4 \leq 800$$
$$x_1, x_2, x_3, x_4 \geq 0$$

Find the effect of the following on the optimal solution:

(a) If the x_1 column in the problem changes from $\begin{bmatrix} 7 \\ 3 \end{bmatrix}$ to $\begin{bmatrix} 7 \\ 5 \end{bmatrix}$

(b) If the x_1 column changes from $\begin{bmatrix} 7 \\ 3 \end{bmatrix}$ to $\begin{bmatrix} 5 \\ 8 \end{bmatrix}$.

Solution: (a) Now, x_1 is a non-basic variable and $z_1 - c_1 = C_B B^{-1} P_1 - c_1$

$= MP_1 - c_1$ [where M is the simplex multiplier]

Now, $$M = C_B B^{-1} = (30 \;\; 100) \begin{bmatrix} 4/15 & -1/15 \\ -1/150 & 2/75 \end{bmatrix}$$

as s_1, s_2 forms the initial basis.

$= (22/3 \;\; 2/3)$

$$\therefore \quad z_1 - c_1 = (22/3 \;\; 2/3) \begin{pmatrix} 7 \\ 5 \end{pmatrix} - 45 = 29/3$$

As this value is non-negative, the original optimum solution remains optimum for the new solution also.

(b) $z_1 - c_1 = C_B B^{-1} P_1 - c_1$

$= MP_1 - c_1$

$$= (22/3 \;\; 2/3) \begin{pmatrix} 7 \\ 8 \end{pmatrix} - 45 = -3$$

Hence, the solution can be improved by taking the variable x_1 as the entering variable. By applying the simplex iteration process the new optimum solution is

$x_1 = 2000/7$, $x_3 = 5600/27$, $x_2 = x_4 = 0$ and maximum of $z = 86000/7$

3. Addition of a new activity (or variable)

If a new variable is introduced in a LPP whose optimal solution has been obtained, then the solution of the problem will remain feasible. Addition of an extra variable x_{n+1} to the problem will introduce an extra column, say a_{n+1} to the coefficient matrix A and an extra cost c_{n+1} will be introduced in the cost vector c_k. Thus, the addition of this extra variable may effect the optimality of the problem. If $z_{n+1} - c_{n+1} \geq 0$ then the solution of the original problem will remain optimal for the new problem also. Otherwise we use simplex method to obtain the optimal solution to the new problem.

Further, it can be noted that the addition of a new variable can never worsen the value of the objective function.

Example

Consider the LPP given in the above example. Then, the optimum simplex table is given by Table. If a new variable x_5 is added to this problem with a column $\begin{bmatrix} 10 \\ 10 \end{bmatrix}$ and $c_5 = 120$, find the change in the optimal solution.

Solution: $z_5 - c_5 = C_B B^{-1} P_5 - c_5$

$$= (30 \;\; 100) \begin{bmatrix} 4/15 & -1/15 \\ -1/50 & 2/75 \end{bmatrix} \begin{bmatrix} 10 \\ 10 \end{bmatrix} - 120$$

using table.

Since $z_5 - c_5 = -40$ which is negative, the existing optimum solution can be improved. By applying the simplex iteration process (left as an exercise to the reader), the new optimum solution is

$$x_3 = 400/3,\ x_4 = 200/3 \text{ with maximum of } z = 30 \times 400/3 + 120 \times 200/3 = 12000$$

4. Deletion of a variable

Deletion of a non-basic variable is a totally superfluous operation and does not affect the feasibility and/or optimality of the current optimal solution. But the deletion of a basic variable may affect the optimality of the current solution. To find the new optimum solution assign a large penalty $-M$ to the variables under consideration and then use the regular simplex method to the modified simplex table.

Example

Consider the optimal Table of a maximisation problem.

table 3.29

		c_j	3	5	0	0	7
C_B	Y_B	X_B	x_1	x_2	x_3	x_4	x_5
7	x_5	4	1	0	1	0	1
5	x_2	5	1/2	1	-1	1/2	0
	$z_j - c_j$	53	13/2	0	2	5/2	0

Find the change in the optimal solution, when the basic variable x_2 is deleted.

Solution: Since this problem is of maximisation type we assign a cost $-M$ to the variable x_2. The modified simplex table becomes

table 3.30

		c_j	3	$-M$	0	0	7
C_B	Y_B	X_B	x_1	x_2	x_3	x_4 ↓	x_5
7	x_5	4	1	0	1	0	1
$-M$	$\bar{x}_2$	5	1/2	1	-1	[1/2]	0
			$\dfrac{-M+8}{2}$	0	$M+7$	$-M/2$	0

First iteration: Introduce x_4 and drop x_2.

table 3.31

C_B	Y_B	X_B	x_1	x_2	x_3	x_4	x_5
7	x_5	4	1	0	1	0	1
0	x_4	10	1	2	-2	1	0
		28	4	M	7	0	0

Since all $z_j - c_j \geq 0$, the current solution is optimum. The optimal solution to the new problem is

$$x_1 = 0,\ x_2 = 0 \text{ and } x_5 = 4 \text{ with maximum of } z = 28.$$

5. Deletion of a constraint

The constraint to be deleted may be either binding or unbinding (redundant) on the optimal solution. The deletion of an unbinding constraint can only enlarge the feasible region but will not affect the optimal solution. Moreover, if the constraint under consideration has a slack or surplus variable of zero value in the basis matrix, it cannot be binding and hence will not affect the optimal solution.

The deletion of a binding constraint will, however, cause post-optimality problem. The simplest way to proceed in this case is via the addition of one or two new variables.

3.6 PARAMETRIC PROGRAMMING

3.6.1 Introduction

Sensitivity analysis discussed the effect of discrete charges in the input coefficients of the linear programming problem on its optimal solution. However, if there is continuous change in the values of these coefficients, none of the results derived in that section are applicable. Parametric linear programming investigates the effect of predetermined continuous variations of these coefficients on the optimal solution. It is simply an extension of sensitivity analysis and aims at finding the various basic solutions that become optimal one after the other as the coefficients of the problem change continuously. The coefficients change as a linear function of a single parameter, hence the name parametric linear programming for this computational technique. As in sensitivity analysis, the purpose of this technique is to reduce the additional computations required to obtain the changes in the optimal solution. Here, we consider only two types of parametric problems.

1. Parametric cost problem in which cost vector c varies linearly as a function of parameter μ.
2. Parametric right hand side problem in which the requirement vector b varies linearly as a function of parameter μ.

3.6.2 Parametric Cost Problem

Consider the LPP

Minimise $z = CX$

subject to $AX = b$,

$X \geq 0$,

where C is the given cost vector.

Let this cost vector change to $c + \mu\bar{c}$. Then the parametric cost problem is Minimise $z = (c + \mu\bar{c})X$,

subject to $AX = b$,

$X \geq 0$

where $\bar{c}$ is the given predetermined cost variation vector and μ is an unknown (positive or negative) parameter. As μ changes, the cost coefficients of all variables also change. We want to find the family of optimal solutions as μ changes from $-\infty$ to $+\infty$.

The problem can be converted into maximisation type and can be solved by using simplex method and sensitivity analysis. When $\mu = 0$, the parametric cost problem reduces to the original linear programming problem. Let B and X_B represent the optimal basis matrix and the optimum basic feasible solution, respectively for $\mu = 0$. The net evaluations or relative cost coefficients are all non-negative and are given by $z_j - c_j = C_B B^{-1} \overline{P}_j - c_j$ where C_B is the cost vector of the basic variables and $\overline{P}_j$ is the jth column (corresponding to the variable x_j) in the optimal Table.

As μ changes from zero to a positive or negative value, the feasible region and values of the basic variables X_B remain unaltered, but the relative cost coefficients change. For any variable x_j, the relative cost coefficients is given by $(z_j - c_j)(\mu) = (C_B + \mu\overline{C}_B) B^{-1} P_j - (c_j + \mu\overline{c}_j)$

$$= (C_B B^{-1} P_j - c_j) + \mu\left(\overline{C}_B B^{-1} P_j - \overline{c}_j\right)$$

$$= (z_j - c_j) + \mu\left(\overline{z}_j - \overline{c}_j\right)$$

Since the vectors C and C' are known, $(z_j - c_j)$ and $\left(\overline{z}_j - \overline{c}_j\right)$ can be determined. For the solution to be optimum $(z_j - c_j)(\mu) \geq 0$ or $(z_j - c_j) + \mu\left(\overline{z}_j - \overline{c}_j\right) \geq 0$. In other words, for a given solution, we can determine the range for μ within which the solution remains optimal.

Example

Consider the linear programming problem

Maximise $z = 4x_1 + 6x_2 + 2x_3$

subject to $x_1 + x_2 + x_3 \leq 3,$

$x_1 + 4x_2 + 7x_3 \leq 9$

$x_1, x_2, x_3 \geq 0$

If we solve this problem by regular simplex method then the optimal simplex table is

table 3.32

		c_j	4	6	2	0	0
C_B	y_B	X_B	x_1	x_2	x_3	s_1	s_2
4	x_1	1	1	0	- 1	4/3	- 1/3
6	x_2	2	0	1	2	- 1/3	1/3
$z_j - c_j$			0	0	6	10/3	2/3

(where s_1 and s_2 are slack variables).

Solve this problem if the variation cost vector $C' = (2 \quad -2 \quad 2 \quad 0 \quad 0)$. Identify all critical values of the parameter μ.

Solution: The given parametric cost problem is

Maximise $z = (4 + 2\mu)\, x_1 + (6 - 2\mu)\, x_2 + (2 + 2\mu)\, x_3 + 0x_4 + 0x_5$

subject to $x_1 + x_2 + x_3 + s_4 = 3,$

$x_1 + 4x_2 + 7x_3 + s_2 = 9,$

$x_1, x_2, x_3, s_1, s_2 \geq 0$

when $\mu = 0$, the problem reduces to the linear programming problem, whose optimal solution is given in Table 3.32. For values of μ other than zero, the relative profit coefficients become linear functions of μ. To compute them we first add a new relative profit row called $\overline{z}_j - \overline{c}_j$ to Table 3.32. This is shown in the following Table.

table 3.33

		$\bar{c}_j$		2	-2	2	0	0
		c_j		4	6	2	0	0
$\bar{C}_B$	C_B	Y_B	X_B	x_1	x_2	x_3	s_1	s_2
2	4	x_1	1	1	0	-1	4/3	-1/3
-2	6	x_2	2	0	1	2	-1/3	1/3
	$z_j - c_j$		16	0	0	6	10/3	2/3
	$\bar{z}_j - \bar{c}_j$		-2	0	0	-8	10/3	-4/3

Note that $\bar{z}_j - \bar{c}_j$ is calculated similar to $z_j - c_j$ except that c is replaced by $\bar{c}$.

The above Table represents a basic feasible solution for the given parametric cost problem. It is given by $x_1 = 1, x_2 = 2, x_3 = s_1 = s_2 = 0$ and the value of the objective function $z(\mu) = z + \mu z' = 16 - 2\mu$.

The relative profit coefficients, which are linear functions of μ, are given by

$$\left(\bar{z}_j - \bar{c}_j\right)(\mu) = (z_j - c_j) + \mu\left(\bar{z}_j - \bar{c}_j\right)$$

$$j = 1, 2, 3, 4, 5.$$

The above Table will be optimal if $\left(\bar{z}_j - \bar{c}_j\right)(\mu) \geq 0$ for $j = 3, 4, 5$. Thus, we can determine the range of μ for which the above Table remains optimal as follows:

$$\left(\bar{z}_3 - \bar{c}_3\right)(\mu) = 6 - 8\,\mu \geq 0 \text{ or } \mu \leq 3/4$$

$$\left(\bar{z}_4 - \bar{c}_4\right)(\mu) = 10/3 + 10/3\,\mu \geq 0 \text{ or } \mu \geq -1$$

$$\left(\bar{z}_5 - \bar{c}_5\right)(\mu) = 2/3 - 4/3\,\mu \geq 0 \text{ or } \mu \leq 1/2$$

Thus, $x_1 = 1, x_2 = 2, x_3 = x_4 = x_5 = 0$ is an optimal solution for the given parametric problem for all values of μ between -1 and $1/2$ and maximum of $z = 16 - 2\mu$.

For $\mu > 1/2$, the relative profit coefficient of the non-basic variable s_2, namely, $\bar{z}_5 - \bar{c}_5$ becomes negative and Table 3.33 no longer remains optimal. Regular simplex method is used to iterate towards optimality. The entering variable is s_2 and the leaving variable is x_2.

table 3.34

		$\bar{c}_j$		2	-2	2	0	0
		c_j		4	6	2	0	0
$\bar{C}_B$	C_B	y_B	X_B	x_1	x_2	x_3	s_1	s_2
2	4	x_1	3	1	1	1	1	0
0	0	s_2	6	0	3	6	-1	1
	$z_j - c_j$			0	-2	2	4	0
	$\bar{z}_j - \bar{c}_j$			0	4	0	2	0

The above Table will be optimum if

$(\bar{z}_j - \bar{c}_j)\ \mu \geq 0$ for $j = 2, 3, 4$.

Now, $(\bar{z}_2 - \bar{c}_2)\ (\mu) = -2 + 4\mu \geq 0 \quad \therefore \mu \geq 1/2$

$(\bar{z}_3 - \bar{c}_3)\ (\mu) = 2 + \mu 0 \geq 0$ which is true

$(\bar{z}_4 - \bar{c}_4)\ (\mu) = 4 + 2\mu \geq 0 \quad \therefore \mu \geq -1/2$

$\therefore$ For all $\mu \geq 1/2$, the optimal solution is given by

$$x_1 = 3, x_2 = x_3 = s_1 = 0, s_2 = 6 \text{ and Maximum of } z = 12 + 6\mu.$$

For $\mu < -1$, the relatively profit coefficient of the non-basic variable s_1 namely, $(\bar{z}_4 - \bar{c}_4)\ (\mu)$ becomes negative. Then, the above Table is not optimal.

s_1 enters into the basis and x_1 leaves the basis.

table 3.35

$\bar{C}_B$	C_B	y_B	X_B	x_1	x_2	x_3	s_1	s_2
0	0	s_1	3/9	3/4	0	- 3/4	1	- 1/4
- 2	6	x_2	9/4	1/4	1	7/4	0	1/4
	$z_j - c_j$			- 5/2	0	17/2	0	3/2
	$\bar{z}_j - \bar{c}_j$			- 5/2	0	- 11/2	0	- 1/2

The above Table will be optimal if

$$(z_j - c_j)\ (\mu) \geq 0 \text{ for } j = 1, 3, 5.$$

Now, $(\bar{z}_1 - \bar{c}_1)\ (\mu) = -5/2 - 5/2\ \mu \geq 0 \quad \therefore \mu \leq -1$

$(\bar{z}_3 - \bar{c}_3)\ (\mu) = 17/2 + 11/2\ \mu \geq 0 \quad \therefore \mu \leq 17/11$

$(\bar{z}_5 - \bar{c}_5)\ \mu = \dfrac{3}{2} - 1/2\ \mu \geq 0 \quad \therefore \mu \leq 3.$

$\therefore$ For all $\mu \leq -1$, the optimal solution is given by

$$x_1 = 0, x_2 = 9/4, x_3 = 0, s_1 = 3/4, s_2 = 0 \text{ and maximum of } z = 27/2 - 9/2\ \mu.$$

The Tables 3.32, 3.33, 3.34 provide the families of optimal solution for $-1 \leq \mu \leq 1/2$, $\mu \geq 1/2$ and $\mu \leq -1$, respectively.

3.6.3 Parametric Right Hand Side Problem

The right hand side constraints in a linear programming problem represents the limits in the resources and the outputs. In some practical problems all the resources are not independent of one another. A shortage of one resource may cause shortage of the other resources at varying levels. Same is true for outputs also. For example, consider a firm manufacturing electrical appliances. A shortage in electric power will decrease the demand of all the electric items produced in varying degrees depending upon the electric energy consumed by them. In all such problems, we are to consider simultaneous changes in the right hand side constants, which are functions of one parameter and study how the optimal solution is affected by these changes.

Let the linear programming problem before parameterisation be

Maximise $z = CX$,

Subject to $AX = b$,

$X \geq 0$

where b is the known requirement (right hand side) vector. Let this requirement vector be changed to $b + \mu b'$ so that parameteric right hand side problem becomes

Maximise $z = CX$

subject to $AX = b + \mu b'$, $X \geq 0$

where b' is the given and predetermined variation vector and μ is an unknown parameter. As μ changes, the right hand constants also change. We wish to determine the family of optimal solutions as μ changes from $-\infty$ to $+\infty$. When $\mu = 0$, the parametric problem reduces to the original linear programming problem. Simplex method is used to find its optimal solution.

Let B and X_B represent the optimal basis matrix and the optimal feasible solution, respectively for $\mu = 0$. Then, $X_B = B^{-1}\, b$. As μ changes from zero to a positive or negative value, the value of the basic variables change and the new values are given by

$$X_B = B^{-1}\,(b + \mu b') = B^{-1}\, b + \mu B^{-1}\, b' = \bar{b} + \mu \bar{b_1}$$

A change in μ has no effect on the values of relative profit coefficients $\bar{c}_j$, that is, $\bar{c}_j$ values remain non-negative. For a given matrix B, values of $\bar{b}$ and $\bar{b}'$ can be calculated. The solution $X_B = \bar{b} + \lambda \bar{b}'$ is feasible and optimal as long as $\bar{b} + \mu \bar{b}' \geq 0$. In other words, for a given solution we can determine the range for μ within which the solution remains optimal.

Example

Consider the linear programming problem

Maximise $z = 4x_1 + 6x_2 + 2x_3$

subject to

$$x_1 + x_2 + x_3 \leq 3$$
$$x_1 + 4x_2 + 7x_3 \leq 9$$
$$x_1, x_2, x_3 \geq 0$$

If we solve this problem by simplex method by taking s_1 and s_2 as the slack variables then the optimal simplex table is

table 3.36

C_B	y_B	X_B	x_1	x_2	x_3	s_1	s_2
4	x_1	1	1	0	−1	4/3	−1/3
6	x_2	2	0	1	2	−1/3	1/3
$z_j - c_j$			0	0	6	10/3	2/3

Solve the problem if the variation right hand side vector $b' = \begin{bmatrix} 3 \\ -3 \end{bmatrix}$. Perform the complete parametric analysis and identify all critical values of parameter μ.

Solution: The given parametric right hand side problem is

Maximise $z = 4x_1 + 6x_2 + 2x_2 + 0s_1 + 0s_2$

subject to
$$x_1 + x_2 + x_3 + s_1 = 3 + 3\mu$$
$$x_1 + 4x_2 + 7x_3 + s_2 = 9 - 3\mu$$
$$x_1, x_2, x_3, s_1, s_2 \geq 0$$

When $\mu = 0$, the problem reduces to the linear programming problem whose optimal solution is given in Table 3.36. For values of μ other than zero, the values of right hand side constants change because of the variation vector b'. The vector X_B and X'_B are computed as follows:

$$X_B = B^{-1}\, b = \begin{bmatrix} 4/3 & 1/3 \\ -1/3 & 1/3 \end{bmatrix} \begin{bmatrix} 3 \\ 9 \end{bmatrix} = \begin{bmatrix} 1 \\ 2 \end{bmatrix}$$

$$X'_B = B^{-1}\, b' = \begin{bmatrix} 4/3 & -1/3 \\ -1/3 & 1/3 \end{bmatrix} \begin{bmatrix} 3 \\ -3 \end{bmatrix} = \begin{bmatrix} 5 \\ -2 \end{bmatrix}$$

Hence, the new Table is

table 3.37

		c_j		4	6	2	0	0
C_B	y_B	X_B	X'_B	x_1	x_2	x_3	s_1	s_2
4	x_1	1	5	1	0	- 1	4/3	- 1/3
6	$\bar{x}_2$	2	- 2	0	1	2	- 1/3	1/3
				0	0	6	10/3	2/3

For a fixed μ, the value of the basic variables in the above Table are given by

$$x_1 = X_B + \mu\, X'_B = 1 + 5\mu$$
$$x_2 = X_B + \mu\, X'_B = 2 - 2\mu$$

$\bar{z}_j - \bar{c}_j$ values are not affected as long as the basis consists of variables x_1 and x_2. As μ changes values of basic variables x_1 and x_2 change and the above Table remains optimal as long as basis (x_1, x_2) remains feasible. In other words, the above Table remains optimal as long as

$$x_1 = 1 + 5\mu \geq 0 \text{ or } \mu \geq -1/5$$
$$x_2 = 2 - 2\mu \geq 0 \text{ or } \mu \leq 1.$$

Thus, the above Table remains optimal as μ varies from – 1/5 to 1. Thus, for all $-1/5 \leq \mu \leq 1$, the optimal solution is given by

$$x_1 = 1 + 5\mu,\ x_2 = 2 - 2\mu,\ x_3 = s_1 = s_2 = 0, \text{ and maximum of } z = 16 + 8\,\mu.$$

For $\mu > 1$, the basic variable x_2 becomes negative. Dual simplex method can, therefore, be applied to find the new optimal solution for $\mu > 1$. Evidently, x_2 is the variable that leaves the basis. The ratio's of the non-basic variables are – 3, 10, – 2. Thus, variable s_1 is the entering variable. The next iteration is

table 3.38

	c_j			4	6	2	0	0
C_B	y_B	X_B	X'_B	x_1	x_2	x_3	s_1	s_2
4	x_1	9	- 3	1	4	7	0	1
0	s_1	- 6	6	0	- 3	- 6	1	- 1
	$z_j - c_j$	36	- 12	0	10	20	0	4

The basic solution given by the above Table is

$$x_1 = 9 - 3\,\mu,\ x_2 = 0,\ x_3 = 0,\ s_1 = -6 + 6\,\mu,\ s_2 = 0,\ \text{maximum of } z = 36 - 12\,\mu.$$

The solution is optimal as long as the basic variables x_1 and s_1 remain non-negative. That is, as long as

$$x_1 = 9 - 3\mu \geq 0 \text{ or } \mu \leq 3,$$
$$s_1 = -6 + 6\mu \geq 0 \text{ or } \mu \geq 1$$

The above solution is optimal for all $1 \leq \mu \leq 3$.

For $\mu > 3$, the basic variable x_1 becomes negative. As there is no negative coefficient in the first row the solution is infeasible. Hence, there is no optimal solution to the problem for all $\lambda > 3$.

For $\mu \leq -1/5$, the basic variable x_1 in Table 3.37 becomes negative. Dual simplex method can therefore be applied to find the new optimum solution for $\mu \leq -1/5$. Evidently, x_1 is the variable that leaves the basis and s_2 is the variable that enters into the basis.

table 3.39

	c_j			4	6	2	0	0
C_B	y_B	X_B	X'_B	x_1	x_2	x_3	s_1	s_2
0	s_2	- 3	- 15	- 3	0	3	- 4	1
6	x_2	3	3	1	1	1	1	0
	$z_j - c_j$	18	18	2	0	4	6	0

The basic solution given by the above Table is $x_1 = 0, x_2 = 3 + 3\mu, x_3 = 0, s_1 = 0, s_2 = -3 - 15\,\mu$ and maximum of $z = 18 + 18\,\mu$.

The solution is optimal as long as

$$x_2 = 3 + 3\,\mu \geq 0 \text{ or } \mu \geq -1$$
$$s_2 = -3 - 15\,\mu \geq 0 \text{ or } \mu \leq -1/5.$$

The above solution is optimal for all $-1 \leq \mu \leq -1/5$. For $\mu < -1$, the basic variable x_2 in the above Table becomes negative. As there is no negative coefficient in the second row, the solution is infeasible. Hence, there exists no optimal solution to the problem for all $\lambda < -1$.

3.7 GOAL PROGRAMMING

3.7.1 Introduction

A difficulty with a linear programming problem is that the objective function is measured in only one dimension such as profit or loss or production capacity and so on. It is impossible in linear programming to have multiple goals unless they can be measured in the same units.

The term goal programming was introduced by A. Charnes and W.W. Copper in 1961. The concept emerged as an issue for unsolvable linear programming problems. Organisational objectives vary according to the characteristics, types, philosophy of management, and particular environmental conditions of the organisation. Profit maximisation, which is regarded as the sole objective of the business firm in classical economic theory, is one of the most widely accepted goals of management. However, in the present dynamic business environment, profit maximisation, is not the only objective of management. In fact, business firms do on quite a few occasions place higher priorities on goals other than profit maximisation. We have seen for example, firms place great emphasis on social responsibilities, social contribution, public relations, industrial and labour relations and so on. Such goals are sought because of outside pressure or voluntary management decisions.

Thus, we see management has multiple conflicting objectives to achieve in the present day business scenario. This implies that the decision criteria should be multi-dimensional and the decision involves multiple goals. In this context goal programming assumes greater importance as a powerful quantitative technique capable of handling multiple decision criteria.

3.7.2 Concepts of Goal Programming

Linear programming technique that has been applied quite extensively to various decision problems has a limited value for problems involving multiple goals. The difficulty in using linear programming lies in the uni-dimensionality of the objective function, which requires maximisation or minimisation of one objective function—an over simplification of reality. However, in real life situations multiple goals and subgoals must be taken into account. Various goals may be expressed in different units of measurement such as rupees, hours, tons and others. Often, multiple goals of management are in conflict or are achievable only at the expense of other goals. Furthermore, these goals are incommensurable. Thus, the solution of the problem requires an establishment of a hierarchy of importance among these incompatible goals so that the lower-order goals are considered only after the higher-order goals are satisfied or have reached the point beyond which no further improvement are desired. If management can provide an ordinal ranking of goals in terms of their contributions or importance to the organisation and all goal constraints are in linear relationships, the problem can be solved by goal programming.

How is an objective function to be determined and expressed in algebraic form when there exists multiple incommensurable, and conflicting goals? Goal programming provides an answer in this case. In goal programming, instead of trying to maximise or minimise the objective criterion directly as in linear programming, the deviations between goals and what can be achieved within the given set of constraint are to be minimised.

In goal programming, the objective function contains primarily the deviational variables that represent each goal or subgoal. The deviational variable is represented in two dimensions in the objective function, a positive and a negative deviation from each subgoal and/or constraint. Then, the objective function becomes the minimisation of these deviations, based on the relative importance or priority assigned to them.

The solution of linear programming is limited by quantification. Unless the management can accurately quantify the relationship of the variables in cardinal numbers, the solution is only as good as inputs. The distinguishing characteristic of goal programming is that it allows for ordinal solution. Stated differently, management may be unable to specify the cost or utility of goal or a subgoal, but often upper or lower limits may be stated for each subgoal. The manager is usually called upon to specify the priority of the desired attainment of each goal or subgoal and rank them in ordinal sequence for decision analysis. The true value of goal programming is, therefore, the solution of problems involving multiple, conflicting goals according to the manager's priority structure.

3.7.3 Formulation of Goal Programming Problem

It is important to consider formulation of the model before getting into the details of goal programming solution. In fact, the most difficult problem in the application of management science to decision analysis is the formulation of the model for the practical problem in question. Model formulation is the process of transforming a real-word decision problem into a management science model. While the advent of micro-computers with powerful software have now made the solution an easy task, the problem formulation requires a great deal of effort in terms of conceptualisation. One key to successful application of goal programming is the ability to recognise when a problem can be solved by goal programming and to formulate the corresponding model.

Steps of goal programming formulation

Step 1: *Define variables and constants*:

Define the choice variables and the right hand side constants. The right hand side constants may be either available resources or specified goal levels. It requires a careful analysis of the problem in order to identify all relevant variables that have some effect on the set of goals stated by the decision-maker.

Step 2: *Formulation of constraints*:

Through an analysis of the relationships among choice variables and their relationships to the goals, a set of constraint should be formulated. A constraint may be either a system constraint representing a relationship among the variables or a goal constraints that defines the relationship between the choice variables and the goals. It should be remembered that if there is no deviational variable to minimise in order to achieve a certain goal, a new cosntraint must be created. Also, if further refinement of goals and priorities are required, it may be facilitated by decomposing certain deviational variables.

Step 3: *Developing of objective function*:

Through the analysis of the decision-maker's goal structure, the objective function must be developed. First, the pre-emptive priority factors should be assigned to certain deviational variables that are relevant to goal attainment. Second, if necessary differential weights must be assigned to deviational variables at the same priority level. It is imperative that goals at the same priority level be commensurable.

Example 1

The manufacturing plant of an electronics firm produces two types of television sets, both colour and black and white. According to the past experience, production of either a colour or a black and white set requires an average of one hour in the plant. The plant has a normal production capacity of 40 hours a week. The marketing department reports that, because of limited sales opportunity, the maximum number of colour and black and white sets that can be sold are 24 and 30, respectively for the week. The gross margin from the sale of a colour set is Rs 80, whereas it is Rs 40 from a black and white set.

The chairman of the company has set the following goals as arranged in the order of their importance to the organisation:

(i) First, he wants to avoid any underutilisation of normal production capacity (no layoffs of production workers).

(ii) Second, he wants to sell as many television sets as possible. Since the gross margin from the sale of a colour television set is twice the amount from a black and white set, he has twice as much desire to achieve sales for colour television sets as for black and white sets.

(iii) Third, the chairman wants to minimise the overtime operation of the plant as much as possible.

Formulate this as a goal programming problem.

Solution: First, we note that the gross margin shows that the colour set priority is two times that of the black and white set. Let x_1 denote the number of colour television sets and x_2 denote the black and white sets for production. Then,

(i) The production capacity of both types of sets is given as $x_1 + x_2 + d_1^- - d_1^+ = 40$ where d_1^-, d_1^+ are the deviational variables, representing underutilisation of normal operation and overtime operation of the plant, respectively.

(ii) The sales capacity for two type of sales are

$$x_1 + d_2^- - d_2^+ = 24$$

$$x_2 + d_3^- - d_3^+ = 30$$

where d_2^-, d_3^- are the deviational variables representing under achievement of sales towards goals and d_2^+, d_3^+ stand for deviational variables to represent over achievement of sales towards goals.

(iii) If $p_1, p_2,$ and p_3 designate the priority of the goals, the complete formulation of goal programming is as follows:

Minimise $z = p_1\, d_1^- + 2p_2\, d_2^- + p_2\, d_3^- + p_3\, d_1^+$

subject to the constraints

$$x_1 + x_2 + d_1^- - d_1^+ = 40$$

$$x_1 + d_2^- - d_2^+ = 24$$

$$x_2 + d_3^- - d_3^+ = 30$$

$$x_1, x_2, d_1^-, d_2^-, d_3^-, d_1^+; d_2^+, d_3^+ \geq 0$$

Example 2

A textile company produces two types of materials A and B. The material A is produced according to direct orders from furniture manufacturers. The material B, is distributed to retail stores. The average rate of production for material A and B are identical at 1000 metres/hour. By running two shifts the operational capacity of the plant is 80 hours per week.

The marketing department reports that the maximum estimated sales for the following week is 70,000 metres of material A and 45,000 metres of material B. According to the accounting department the profit from a metre of material A is Rs 2.50 and from a metre of material B is Rs 1.50.

The management of the company decides that a stable employment level is a primary goal for the firm. Therefore, whenever there is a demand exceeding normal production capacity, the management simply expands production capacity by providing overtime. However, management feels that overtime operation of the plant of more than 10 hours per week should be avoided because of the accelerating costs. The management has the following goals in the order of importance:

1. The first goal is to avoid any underutilisation of production capacity (i.e. to maintain stable employment at normal capacity).
2. The second goal is to limit the overtime operation of the plant to 10 hours.
3. The third goal is to achieve the sales goals of 70,000 metres of material *A* and 45,000 metres of material *B*.
4. The last goal is to minimise the overtime operation of the plant as much as possible.

Formulate this is a goal programming problem to help the management for the best solution.

Solution:

Constraint for production capacity:

The production capacity is limited to 80 hours at present by running two shifts. However, since overtime of the plant is allowed to a certain extent, the constraint may be written as

$$x_1 + x_2 + d_1^- - d_1^+ = 80$$

where, x_1 — Number of hours used for producing material *A*.
x_2 — Number of hours used for producing material *B*.
d_1^- — Underutilisation of production capacity as set at 80 hours of operation.
d_1^+ — overutilisation of normal production capacity beyond 80 hours.

Constraints for Sales:

The maximum sales for material *A* and material *B* are 70,000 and 45,000 metres, respectively. Hence, it is assumed that overachievement of sales beyond the maximum limits are impossible. Then, the sales constraints will be (as before x_1 and x_2 are expressed in thousands)

$$x_1 + d_2^- = 70$$
$$x_2 + d_3^- = 45$$

where, d_2^- = underachievement of sales goal of material *A*.
d_3^- = underachievement of sales goal of material *B*.

Constraint for overtime operation:

From the case itself, only production and sale constraints can be formulated. However, the analysis of goals indicates that overtime operation of the plant is to be minimised to 10 hours or less. To solve the problem by goal programming, we need a deviational variable that represents the overtime operation of the plant beyond 10 hours. By minimising this deviational variable to zero we can achieve the goal. Since there is no such deviational variable in the three constraints presented above, we must create a new constraint.

The overtime operation of the plant d_1^+ should be limited to 10 hours or less. However, it may not be possible to limit the overtime operation to 10 hours or less in order to meet higher order goals. Therefore, d_1^+ can be smaller than, equal to 10. Constraint regarding overtime can be expressed as

$$d_1^+ + d_{12}^- - d_{12}^+ = 10$$

where, d_{12}^- — negative deviation of overtime operation from 10 hours.
d_{12}^+ — overtime operation beyond 10 hours.

One metre of material *A* gives a profit of Rs 2.50 and one metre of *B* gives a profit of Rs 1.50. Since the production rate is same for both *A* and *B*, namely, 1000 metres per hour, the hourly profits of *A* and

B are in the ratio of 2.50 : 1.50 or 5 : 3. Hence, it is appropriate to assign these as differential weights in goal 3rd. The differential weights imply that management is relatively more concerned with the achievement of the sales goal for material A than that for material B.

Hence, the problem is

$$\text{Minimise } z = P_1 d_1^- + P_2 d_4^+ + 5P_3 d_3^- + P_4 d_1^+$$

subject to

$$x_1 + x_2 + d_1^- - d_1^+ = 80$$
$$x_1 + d_2^- = 70$$
$$x_2 + d_3^- = 45$$
$$d_1^+ + d_{12}^- - d_{12}^+ = 10$$
$$x_1, x_2, d_1^-, d_2^-, d_3^-, d_1^+, d_{12}^+, d_{12}^- \geq 0.$$

The simplex method of goal programming Simplex method can be used to solve a goal programming problem.

Example 1

A company manufactures two products, radios and transistors, which must be processed through assembly and finishing departments. Assembly department has 90 hours available, finishing department can handle up to 72 hours of work. Manufacturing one radio requires 6 hours in assembly and 3 hours in finishing department. Each transistor requires 3 hours in assembly and 6 hours in finishing. If profit is Rs 120 per radio and Rs 90 per transistor, determine the best combination of radios and transistors to realise a maximum profit of Rs 2,100.

Solution: This is a single goal (profit) problem. Let

d^- = amount by which profit goal is underachieved

d^+ = amount by which profit goal is overachieved.

The given problem can be expressed as the following goal problem:

Minimise $z = d^-$ (under achievement of the profit target)

subject to $120x_1 + 90x_2 + d^- - d^+ = 2100$

($\because$ profit earned + underachievement – overachievement = target)

Adding slack variables, the assembly and finishing constraints can be expressed as

$$6x_1 + 3x_2 + s_1 = 90$$
$$3x_1 + 6x_2 + s_2 = 72,$$
$$x_1, x_2, s_1, s_2 \geq 0$$

If we convert the problem as maximisation type, the objective function becomes maximise $z^* = -d^-$.

Solution by simplex method:

Initial Table:

table 3.40

		c_j	0	0	0	0	−1	0	
C_B	y_B	X_B	$x_1\downarrow$	x_2	s_1	s_2	d^-	d^+	Min Ratio
−1	d^-	2100	120	90	0	0	1	−1	35/2
0	$\overleftarrow{s_1}$	90	[6]	3	1	0	0	0	15
0	s_2	72	3	6	0	1	0	0	24
	$z_j - c_j$		−120	−90	0	0	0	1	

First iteration: Introduce x_1 and drop s_1.

table 3.41

		c_j	0	0	0	0	-1	0	
C_B	Y_B	X_B	x_1	x_2 ↓	s_1	s_2	d^-	d^+	Min Ratio
-1	d^-	300	0	30	-20	0	1	-1	10
0	x_1	15	1	1/2	1/6	0	0	0	30
0	s_2 ←	27	0	[9/2]	-1/2	1	0	0	6
			0	-30	20	0	0	1	

Second iteration: Introduce x_2 and drop s_2.

table 3.42

C_B	y_B	X_B	x_1	x_2	s_1	s_2	d^-	d^+
-1	d^-	120	0	0	-50/3	-20/3	1	-1
0	x_1	12	1	0	2/9	-1/9	0	0
0	x_2	6	0	1	-1/9	2/9	0	0
			0	0	50/3	20/3	0	1

Since all $z_j - c_j \geq 0$, the current solution is optimum and the optimal solution is $x_1 = 12, x_2 = 6$ and minimum of $z = 120$.

Minimum of z = Rs 120 means that the target profit of Rs 2,100 is underachieved by Rs 120, that is, profit actually earned (by manufacturing 12 radios and 6 transistors) is Rs 2,100 – 120 = Rs 1,980.

Example 2

If in example 1, the company sets two equally ranked goals, one to reach a profit goal of Rs 1,500 and the other to meet a radio goal of 10 sets. Find the optimal solution.

Solution: Since the two goals are equally ranked, a Re 1 deviation from the profit target is important (in the goal programming model) as a deviation of 1 radio set. Let

d_p^- = amount by which the profit goal is under achieved.
d_p^+ = amount by which the profit goal is over achieved.
d_r^- = amount by which the radio goal is under achieved.
d_r^+ = amount by which the radio goal is over achieved.

Then, the goal programming model becomes

Minimise $z = d_p^- + d_r^-$

subject to

$$120x_1 + 90x_2 + d_p^- - d_p^+ = 1{,}500$$
$$x_1 + d_r^- - d_r^+ = 10$$
$$6x_1 + 3x_2 \leq 90$$
$$3x_1 + 6x_2 \leq 72$$
$$x_1, x_2, d_p^-, d_r^-, d_r^+ \geq 0.$$

Convert the problem to maximisation, the objective function becomes maximise $z^* = -\bar{d}_p - \bar{d}_r$.

Adding slack variables, the assembly and finishing constraints can be written as

$$6x_1 + 3x_2 + s_1 = 90$$
$$3x_1 + 6x_2 + s_2 = 72$$

Note that d_p^- and d_r^- are the basic variables corresponding to the first and second constraint, respectively.

Initial table:

table 3.43

		c_j	0	0	0	0	−1	0	−1	0	
C_B	y_B	X_B	$x_1\downarrow$	x_2	s_1	s_2	d_p^-	d_p^+	d_r^-	d_r^+	Min Ratio
−1	d_p^-	1,500	120	90	0	0	1	−1	0	0	25/2
−1	$\overleftarrow{d_r^-}$	10	$\boxed{1}$	0	0	0	0	0	1	−1	10
0	s_1	90	6	3	1	0	0	0	0	0	15
0	s_2	72	3	6	0	1	0	0	0	0	24
			−121	−90	0	0	0	1	0	1	

First iteration: Introduce x_1 and drop s_2.

table 3.44

		c_j	0	0	0	0	−1	0	−1	0	
C_B	y_B	X_B	x_1	x_2	s_1	s_2	d_p^-	d_p^+	d_r^-	d_r^+	Min Ratio
−1	$\overleftarrow{d_p^-}$	300	0	90	0	0	1	−1	−120	$\boxed{120}$	5/2
0	x_1	10	1	0	0	0	0	0	1	−1	
0	s_1	30	0	3	1	0	0	0	−6	6	5
0	s_2	42	0	6	0	1	0	0	−3	3	14
		0	−90	0	0	0	1	120	−120		

Second iteration: Introduce d_r^+ and drop d_p^-.

table 3.45

		c_j	0	0	0	0	−1	0	−1	0
C_B	y_B	X_B	x_1	x_2	s_1	s_2	d_p^-	d_p^+	d_r^-	d_r^+
0	d_r^+	5/2	0	3/4	0	0	1/120	−1/120	−1	1
0	x_1	25/2	1	3/4	0	0	1/120	−1/120	0	0
0	s_1	15	0	−3/2	1	0	−1/20	1/20	0	0
0	s_2	69/12	0	15/4	0	1	−1/40	1/40	0	0
$z_j - c_j$			0	0	0	0	1	0	1	0

It may be noted from the above Table that the goal of 10 radios is achieved and in fact improved $(25/2 - 10 = 5/2)$, 5/2 appears in this Table as d_r^+ (overachievement of radios). Profit goal of Rs 1500 is also reached since both d_p^- and d_p^+ are zero (because they are not in the final solution) and therefore, profit is exactly Rs 1500.

3.7.4 Graphical Method of Goal Programming

The graphical method of solving goal programming problem is quite similar to the graphical method of linear programming. In linear programming, the method is used to maximise or minimise an objective function with one goal, whereas in goal programming, it is used to minimise the total deviation from a set of multiple goals. The deviation from the goal with the highest priority is minimised to the fullest extend possible before deviation from the next goal is minimised.

Example 1

A manufacturing firm produces two types of product A and B. According to the past experience, production of either product A or product B requires an average of one hour in the plant. The plant has a normal production capacity of 400 hours a month. The marketing department of the firm reports that because of limited market, the maximum number of product A and product B that can be sold in a month are 240 and 300, respectively. The net profit from the sale of product A and product B are Rs 800 and Rs 400, respectively. The manager of the firm has set the following goals arranged in the order of importance.

p_1: He wants to avoid any underutilisation of normal production capacity.

p_2: He wants to sell maximum possible units of product A and product B. Since the net profit from the sale of product A is twice the amount from that of product B, the manager has twice as much desire to achieve sales for product A as for product B.

p_3: He wants to minimise the overtime operation of the plant as much as possible.

Solution: Formulation:

Let x_1 and x_2 be the numbers of units of product A and product B to be produced, respectively. Since overtime operation is allowed, the plant capacity can be expressed as

$$x_1 + x_2 + d_1^- - d_1^+ = 400$$

where d_1^- is the underutilisation of production capacity and d_1^+ is the overtime operation of the normal production capacity. Since the sales goals given are the maximum possible sales volume, positive deviations will not appear in the sales constraints. The sales constraints can be expressed as:

$$x_1 + d_2^- = 240$$
$$x_2 + d_3^- = 300$$

where, d_2^- = underachievement of sales goal for product A

d_3^- = Underachievement of sales goal for product B

Hence, the complete model is

$$\text{Minimise } z = p_1\, d_1^- + 2p_2\, d_2^- + p_2\, d_3^- + p_3\, d_1^+$$

subject to $x_1 + x_2 + d_1^- - d_1+ = 400$

$$x_1 + d_2^- = 240$$

$$x_2 + d_3^- = 300$$

$$x_1,\, x_2,\, d_1^-,\, d_2^-,\, d_3^-,\, d_1^+ \geq 0$$

First plot all the goal constraints on a graph as shown in Fig. 3.1. Since the underutilisation and overutilisation of the plant capacity is permissible, the deviational variables d_1^- and d_1^+ are indicated by arrows in Fig. 3.1. Similarly, d_2^-, d_3^+ are also denoted by arrows in the diagram. The feasible region is *OABC*.

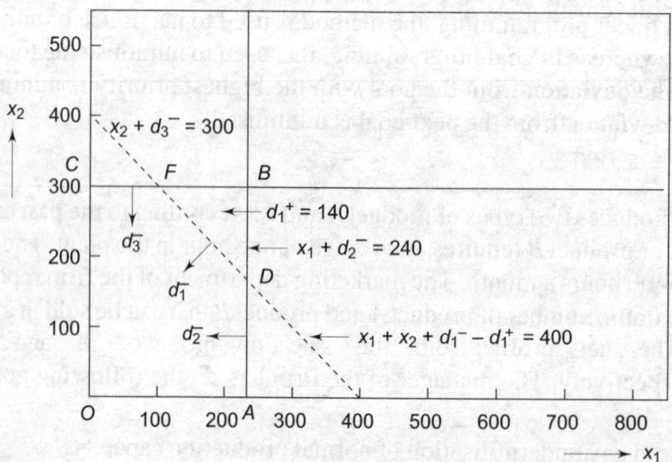

figure 3.1

The first goal is completely achieved at the line *EF* in the reduced feasible region *EBF*. Now, the manager would like to move to priority level 2, where the differential weight for product *A* is two times that of *B* (assigned in the ratio of the profits). The sales goal of product *A* is completely achieved at line *EB* in the feasible region *EBF*. So, we see that the first two goals are completely achieved at point *B* in the feasible region *EBF*. The third goal on overtime operation cannot be achieved without dispensing with the first two goals in the priority order. The optimal solution to the problem is

$$x_1 = 240,\ x_2 = 300,\ d_1^- = 0,\ d_2^- = 0,\ d_3^- = 0,\ d_1^+ = 140$$

Example 2

A small furnishing company manufactures tables and chairs. Each chair requires 4 man-hours of labour while each table requires 5 man-hours of labour. If only 40 man-hours are available each week and the owner of the company would neither hire additional labour nor utilise overtime, formulate the linear goal programming problem and solve it. Both the table and the chair fetch a profit of Rs 2000 per week. Also, he would like to supply 10 chairs, if possible, per week to a sister concern.

Solution: The goals of the company and their assigned priorities are

Priority	*Goal*
p_1	To avoid hiring extra labour or utilise overtime.
p_2	To reach a profit goal of Rs 2000 a week.
p_3	To supply 10 chairs a week to the sister concern.

Let x_1 and x_2 denote the number of chairs and tables to be produced per week. d_i^- and d_i^+ represent the amount by which the ith goal is underachieved and overachieved, respectively, then the goals are

g_1: $4x_1 + 5x_2 + d_1^- - d_1^+ = 40$ (man-hour goal)

g_2: $100x_1 + 100x_2 + d_2^- - d_2^+ =$ Rs 2,000 (profit goal)`

g_3: $x_2 + d_3^- - d_3^+ = 10$ (Supply of chair goal)

To achieve the first goal completely, the overtime d_1^+ must be minimised. Similarly, the second goal will be fulfilled when the underachievement in profit, d_2^- is minimised and the supply of chair goal will be attained when underachievement d_3^- is minimised. Therefore, the objective function of the problem may be written as

$$\text{Minimise } z = \{d_1^+, d_2^-, d_3^-\}$$

subject to $4x_1 + 5x_2 + d_1^- - d_1^+ = 40$

$100x_1 + 100x_2 + d_2^- - d_2^+ = 2{,}000$

$x_2 + d_3^- - d_3^+ = 10,$

$x_1, x_2, d_1^-, d_2^-, d_3^-, d_1^+, d_2^+, d_3^+ \geq 0$

The three goal constraints are plotted as straight lines by assuming each $d_i^- = d_i^+ = 0$. Arrows (← or →) are then associated with each line to represent underachievement or over achievement of the goal.

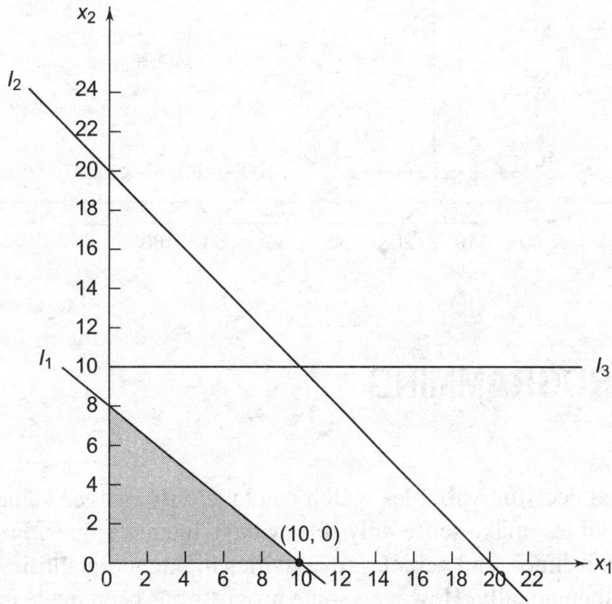

figure 3.2

Man-hour goal is the most important and is the first to be considered. Since this goal is achieved when d_1^- is minimised, we may set $d_1^+ = 0$ as its minimum value. The region satisfying $x_1, x_2 \geq 0$ and $d_1^+ = 0$ is shown shaded in the Fig. 3.2.

Any point within this region satisfies the man-hour goal and minimises d_1^+ at the priority level one.

The next important goal is the profit goal (p_2) and to attain it we must minimise d_2^+. Clearly, d_2^+ cannot be decreased drawn to zero since this would degrade our previous solution—a higher priority goal. The minimum value of d_2^+ without affecting the previously attained higher goal is the point ($x_1 = 4$, $x_2 = 0$) and is marked in Fig. 3.2.

The last important goal (in the order of priority) is g_3. To satisfy this we must minimise d_3^-. However, any minimisation of d_3^- would require a movement from the solution previously determined and would degrade at least one higher priority goal already attained.

Therefore, the optimal solution to the problem is $x_1 = 10$, $x_2 = 0$ and minimum of $z = (0, 1000, 10)$

The optimum values of z simply indicate that goal is completely attained, while g_2 and g_3 are only partially attained.

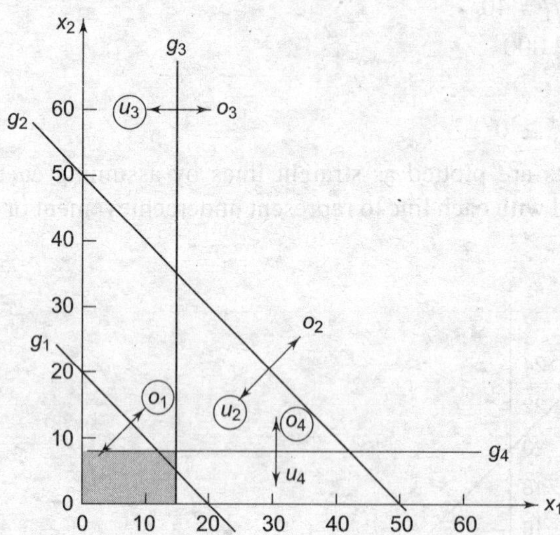

figure 3.3

3.8 INTEGER PROGRAMMING

3.8.1 Introduction

So far, we have considered decision variables which can take only integer values. In many practical problems the decision variables make sense only if they have integer values. For example, it is often necessary to assign men, machines, and vehicles to activities in integer quantities. This restriction is a difficult one to handle mathematically. However, some progress has been made in developing solution procedures for the case of linear programming problems subjected to this additional restriction that the decision variables must have integer values.

A linear programming problem in which all or some of the decision variables are restricted to assume non-negative integer values is called an integer programming problem [or IPP or integer linear programming].

Importance of Integer Programming Integer programming problem is of particular importance in business and industry where, quite often, the fractional solutions are unrealistic because the units are not divisible. For example, it is quite possible to use 2.34 kg of raw material, or 4.56 machine hours and so on. But it is meaningless to produce 5.67 chairs or 7.89 tables, or to open 4.23 shops or to run 1.23 and so on. Hence, a new procedure has been developed in this direction for the case of LPP subjected to the additional restriction that the decision variables must have integer values.

Applications of integer programming Integer programming problems occur quite frequently in business and industry. Transportation problems, assignment problems are integer programming problems, since the decision variables are either zero or one.

Many other decision problems can necessitate integer programming models. One category of such problems deals with the sequencing, scheduling and routing decisions. An example is the travelling salesman problem which aims at finding a least distance route for a salesman who must visit each of n cities, starting and ending his journey at home city. Larger expenditures of capital and resources are required in capital budgeting decisions. This is the main reason why integer programming is so important for marginal decisions.

An optimal solution to a capital budgeting problem may yield considerably more profit to a firm than an approximate or guessed solution. For example, fertilizer manufacturing firm with 15 plants may be able to substantially increase profits by cutting down to 10 plants or less-provided this reduction is planned optimally.

Rounding off Techniques and its Draw Backs IPP can be solved by the usual simplex method by ignoring the integer restriction and then rounding off the non-integer values to integers in the optimal solution obtained by the simplex method. But there is no guarantee that the integer valued solution thus obtained will satisfy one or more constraints and as such the new solution may not be feasible.

Definition of Integer Programming *Integer programming problem (IPP)* The linear programming problem, maximise $z = CX$, subject to $AX \leq b$, $X \geq 0$ and some $x_j \in X$ are integers where A is an $m \times n$ real matrix is called an integer programming problem.

All integer programming problem (All I.P.P. or pure I.P.P.) An integer programming problem is said to be an all integer programming problem if all $x_j \in X$ are integers.

Mixed integer programming problem (Mixed I.P.P.) An integer programming problem is said to be mixed integer programming problem if not all $x_j \in X$ are integers.

Zero-one programming problem or standard discrete problem: If all the variables in the optimum solution are allowed to take values either 0 or 1 as in 'do' or 'not to do' type decisions, then the problem is called the zero-one programming problem.

3.8.2 Formulation of Integer Programming Problem

Integer programming problem formulation do not pose problem when the variables inherently are discrete in nature. Problems arise in situations where direct LP or IP formulations are not possible. First, we see a direct IP formulation of a problem situation.

Example 1

Suppose we are landed on a treasure island full of three types of valuable stones, emerald (E), ruby (R) and topaz (T). Each piece of E, R and T weighs 3, 2, 2 kg and is known to have a value of 4, 3, 1, crore respectively. We have a bag that can carry a maximum of 11 kg. The stones cannot be made into smaller piecs. Formulate this situation as an IPP so as to maximise the total value carried.

Solution:

Decision variables

We have to decide how many pieces of each type of stone has to be carried so as to be within the capacity of the bag.

Let x_1 denote the number of emeralds to be carried and let x_2 be the number of rubies to be carried and let x_3 be the number of topaz to be carried.

Objective function

The total value carried is $4x_1 + 3x_2 + x_3$ and so the objective function is maximise $z = 4x_1 + 3x_2 + x_3$.

Constraints: The total weight to be carried is $3x_1 + 2x_2 + 2x_3$. Since the capacity of the bag is limited to 11 kg we have

$3x_1 + 2x_2 + 2x_3 \le 11$

Also, x_1, x_2, x_3 denote the number of stones, that cannot be negative and so $x_1, x_2, x_3 \ge 0$.

Hence, the complete IPP is

Maximise $z = 4x_1 + 3x_2 + x_3$

subject to $3x_1 + 2x_2 + 2x_3 \le 11$

$x_1, x_2, x_3 \ge 0$

Example 2 Capital Budgeting Problem

Consider a company which would like to take up three projects. However, because of budget limitations not all the projects can be selected. It is known that project j has a present value of c_j, and would require an investment of a_{jt} in period t. The capital available in period t is b_t. The problem is to choose projects so as to maximise present value, without violating the budget constraints. Formulate the problem as an IP.

Solution: For choice of situations of "yes-no", "go-no-go" type, where the objective is to determine whether or not a particular activity is to be undertaken, integer binary variables that can take a value of 0 or 1 can be used to represent the decision variables. Here, we find that for each project, we want to find out whether it should be taken up or not. Define three decision variables x_j ($j = 1, 2, 3$) corresponding to each project, and

$x_j = 1$, if project j is selected
$= 0$, otherwise

Then, the objective function and the constraints can be expressed in terms of the decision variables, to give the required formulation.

Maximise $z = \sum c_j x_j$

subject to $\Sigma a_{jt} x_j \le b_t$, for all period t

$x_j = 1$ or 0, for all projects.

Solution Methodologies of Integer Programming

Integer programming methods can be categorised as

1. Cutting plane methods
2. Search methods

Cutting methods: A systematic procedure for solving pure integer programming problem was first developed by RE Gomory in 1958. Later on, he extended the procedure to solve mixed IPP named as cutting plane algorithm; the method consists in first solving the IPP as ordinary LPP by ignoring the integerality restriction and then introducing additional constraints one after the other to cut (eliminate) certain part of the solution space until an integral solution is obtained.

Search method: It is an enumeration method in which all feasible integer points are enumerated. The widely used search method is the branch and bound technique. It also starts with the continuous optimum, but systematically partitions the solution space into subproblems that eliminate parts that contain no feasible integer solution. It was originally developed by AH Land and AG Doig.

3.8.3 Gomory's all IPP Method (or) Cutting Plane Method or Gomory's Fractional Cut Method

An optimal solution to an IPP is first obtained by simplex method ignoring the restriction of integral values. In the optimum solution if all the variables have integer values, the current solution will be the desired optimum integer solution. Otherwise, we add a new constraint to the problem such that the new set of feasible solutions includes all the original feasible integer solutions but does not include the optimum non-integer solution initially found. We then solve the revised problem using the simplex method and see if we can get an integer solution. If not, we add another fractional cut and repeat the process until an integer solution is found. Since we never eliminate any feasible integer solution from consideration when we add fractional cuts, the integer solution ultimately found must be optimum.

***Algorithm*:**

Step 1: Convert the problem into maximisation if it is in the minimisation form. All the coefficients and constants should be integers.

Step 2: Ignore the integerality condition and find the optimum solution of the resulting LPP by simplex method.

Step 3: *Test the integerality of the optimum solution*:

(i) If all $X_{B_i} \geq 0$ and are integers, an optimum integer solution is obtained.

(ii) If all $X_{B_i} \geq 0$ and at least one X_{B_i} is not an integer, then go to the next step.

Step 4: Rewrite each X_{B_i} as $X_{B_i} = [X_{B_i}] + f_i$, where $[X_{B_i}]$ is the integral part of X_{B_i} and f_i is the positive fractional part of X_{B_i}, $0 \leq f_i \leq 1$. Choose the largest fraction of X_{B_i}'s that is choose Max $\{f_i\}$. In case of a tie, select arbitrarily. Let Max $\{f_i\} = f_k$ corresponding to X_{B_k} (then k^{th} row is called as a source row).

Step 5: Express each of the negative fractions if any, in the kth row (source row) of the optimum simplex Table as the sum of a negative integer and a non-negative fraction.

Step 6: Find the fractional cut constraint (Gomorian constraint or secondary constraint).

From the source row $$\sum_{j=1}^{n} a_{k_j} x_j = X_{B_k}$$

that is, $$\sum_{j=1}^{n} ([a_{k_j}] + f_{k_j})\, x_j = [X_{B_k}] + f_k$$

in the form $$\sum_{j=1}^{n} f_{k\,j} x_j \geq f_k$$

or, $$-\sum_{j=1}^{n} f_{k_j}\, x_j \leq -f_k \quad \text{or,} \quad -\sum_{j=1}^{n} f_{k_j} + s_1 = -f_k$$

where s_1 is the Gomorian slack.

Step 7: Add the fractional cut constraint obtained in step 6 at the bottom of the optimum simplex table obtained in step 2. Find the new feasible optimum solution using dual simplex method.

Step 8: Go to step 3 and repeat the procedure until an optimum integer solution is obtained.

Remark: In cutting plane method, the fractional cut constraints cut the useless area of the feasible region in the graphical solution of the problem. That is, cut area which has no integer-valued feasible solution. Thus, these Gomorian constraints eliminate all the non-integral solutions without loosing any integer valued solution.

Example 1

Solve the following problem

Maximise $z = 6x_1 + 2x_2$

subject to $3x_1 + 2x_2 = 5$

$x_1, x_2 \geq 0$ and are integers.

Solution: (*By Gomory's cutting plane method*)

The given LPP can be written as

Maximise $z = 6x_1 + 2x_2$

subject to $x_1 + 2/3\ x_2 = 5/3$

$x_1, x_2 \geq 0$ and are integers.

By taking x_1 as the basic variable the initial basic feasible solution is $x_1 = 5/3$.

Initial table:

table 3.46

		c_j	6	2	
C_B	Y_B	X_B	X_1	X_2	Min ratio
6	x_1	5/3	1	2/3	
			0	2	

Since all $z_j - c_j \geq 0$ the current solution is optimum. The optimum solution is

$x_1 = 5/3, x_2 = 0$ and maximum of $z = 10$.

But the solution obtained is not an integer solution.

Construction of Gomory's Constraint:

Now, the constraint equation (x_1 – equation) can be written as $1 + 2/3 = (1 + 0)x_1 + (0 + 2/3)x_2$ and so the Gomory's constraint is

$0x_1 + 2/3x_2 \geq 2/3$

or $-2/3x_2 \leq -2/3$

or $-2/3x_2 + s_1 = -2/3$ where s_1 is a Gomory's slack variable.

We obtain $s_1 = -2/3$. Since the solution is infeasible apply dual simplex method. Then, the next Table is

table 3.47

		c_j	6	2	0
C_B	y_B	X_B	x_1	x_2 ↓	s_1
6	x_1	5/3	1	2/3	0
0	$\bar{s}_1$	- 2/3	0	$\boxed{\frac{-2}{3}}$	1
			0	2	0

Drop s_1 from the basis and introduce x_2. Then, the next iteration is

table 3.48

C_B	y_B	X_B	x_1	x_2	s_1
6	x_1	1	1	0	1
2	x_2	1	0	1	- 3/2
			0	0	3

Since all $X_{B_i} \geq 0$ and all $z_j - c_j \geq 0$, the current solution is optimal. The solution is also an integer solution.

The optimum integer solution is

$$\boxed{x_1 = 1, x_2 = 1 \text{ with maximum of } z = 8}$$

Example 2

Solve the integer programming problem:

Maximise $z = 7x_1 + 9x_2$

subject to

$$-x_1 + 3x_2 \leq 6$$
$$7x_1 + x_2 \leq 35$$
$$x_1, x_2 \leq 0 \text{ and are integers.}$$

Solution: Ignore the integer restriction and solve the problem by regular simplex method. Then, the optimal simplex table is

table 3.49

C_B	y_B	X_B	x_1	x_2	x_1	x_2	Min Ratio
9	x_2	7/2	0	1	7/22	1/22	
7	x_1	9/2	1	0	- 1/22	3/22	
$z_j - c_j$			0	0	28/11	15/11	

Now, $7/2 = 3 + 1/2 \quad (f_1 = 1/2)$

$9/2 = 4 + 1/2 \quad (f_2 = 1/2)$

and maximum of {1/2, 1/2} = 1/2. Hence, arbitrarily choose first row (x_2 row) for constructing the Gomory's constraint.

Now, x_2 equation is

$$3 + 1/2 = (0 + 0)\,x_1 + (1 + 0)\,x_1 + (0 + 7/22)\,s_1 + (0 + 7/22)\,s_2.$$

Hence, the Gomory's constraint is

$$7/22s_1 + 1/22s_2 \geq 1/2$$

or, $$-7/22s_1 - 1/22s_2 \leq -1/2$$

that is, $$-7/22s_1 - 1/22\,s_2 + s_3 = -1/2$$

where, s_3 is the Gomory's slack variable. Hence, $s_3 = -1/2$. As the solution is infeasible apply dual simplex method after including the Gomory's constraint equation in the above Table.

table 3.50

		c_j	7	9	0	0	0
C_B	y_B	X_B	x_1	x_2	$s_1\downarrow$	s_2	s_3
9	x_2	7/2	0	1	7/22	1/22	0
7	x_1	9/2	1	0	- 1/22	3/22	0
0	$\bar{s}_3$	- 1/2	0	0	$\boxed{\frac{-7}{22}}$	- 1/22	1
			0	0	28/11	15/11	0

Since maximum of $\left\{\frac{28/11}{-7/22}, \frac{15/11}{-1/22}\right\} = \frac{28/11}{-7/22}$ corresponding to s_1

Introduce s_1 and drop s_3.

table 3.51

		c_j	7	9	0	0	0
C_B	Y_B	X_B	x_1	x_2	s_1	s_2	s_3
9	x_2	3	0	1	0	0	1
7	x_1	32/7	1	0	0	1/7	6/7
0	s_1	11/7	0	0	1	1/7	- 22/7
			0	0	0	1	8

Since all $X_{B_i} \geq 0$ and all $z_j - c_j \geq 0$ the current solution is optimum and the optimum non-integer solution is

$$x_1 = 32/7,\ x_2 = 3 \text{ with maximum of } z = 59.$$

Since $32/7 = 4 + 4/7$

and, $11/7 = 1 + 4/7$

and maximum $\{4/7, 4/7\} = 4/7$.

Arbitrarily choose x_1 equation for constructing the next Gomory's constraint.

$$4 + 4/7 = (1 + 0)x_1 + (0 + 0)x_2 + (0 + 0)s_1 + (0 + 1/7)s_2 + (0 + 6/7)\ s_3$$

$$1/7s_2 + 6/7s_2 \geq 4/7$$

i.e. $$- 1/7s_2 - 6/7s_2 \leq - 4/7$$

i.e. $- 1/7s_2 - 6/7s_2 + s_4 = - 4/7$ where s_4 is a slack variable. The next iteration is

table 3.52

		c_j	7	9	0	0	0	0
C_B	y_B	X_B	x_1	x_2	s_1	$s_2\downarrow$	s_3	s_4
9	x_2	3	0	1	0	0	1	0
7	x_1	$4\frac{4}{7}$	1	0	0	1/7	1/7	0
0	s_1	$1\frac{4}{7}$	0	0	1	1/7	22/7	0
0	s_4	-4/7	0	0	0	$\boxed{\frac{-1}{7}}$	- 6/7	1
			0	0	0	1	8	0

Since maximum $\{1/-1/7, 8/-6/7\} = -7$ corresponding to s_2. Drop s_4 and introduce s_2.

table 3.53

		c_j	7	9	0	0	0	0
C_B	Y_B	X_B	x_1	x_2	s_1	s_2	s_3	s_4
9	x_2	3	0	1	0	0	1	0
7	x_1	4	1	0	0	0	- 1	1
0	s_1	1	0	0	1	0	- 4	1
0	s_2	1	0	0	0	1	6	- 7
	z_j-c_j	55	0	0	0	o	2	7

Since all $z_j - c_j \geq 0$, the current solution is optimum and the optimum integer solution is

$$\boxed{x_1 = 4, x_2 = 3 \text{ with maximum of } z = 55}$$

Gomory's Mixed Integer Method

In mixed integer programming problem only some of the variables are integer constrained, while the other variables may take integer or other real values. Like the pure integer problem, the mixed integer problem should be of the maximisation type and all the coefficients and constants should be integers.

The problem is first solved as a continuous LPP by ignoring the integerality condition. If the values of the integer constrained variables are integers, then the current solution is an optimum solution to the given mixed IPP. Otherwise, select the source row which corresponds to the largest fractional part f_k among those basic variables which are constrained to be integers. Then, construct the Gomorian constraint (secondary constraint) from the source row.

From the source row $$\sum_{j=1}^{n} a_{k_j} x_j = X_{B_k}$$

i.e. $$\sum_{j=1}^{n} ([a_{k_j}] + f_{k_j})\, x_j = [X_{B_k}] + f_k$$

in the form of $$\sum_{j=1}^{n} f_{k_j} x_j \geq f_k$$

i.e. $$\sum_{j \in J^+} f_{k_j} x_J + \left(\frac{f_k}{f_{k-1}}\right) \sum_{j \in J} f_k x_j \geq f_k$$

i.e. $$-\sum_{j \in J^+} f_{k_j} x_j - \left(\frac{f_k}{f_{k-1}}\right) \sum_{j \in J} f_k x_j \leq -f_k$$

i.e. $$-\sum_{j \in J^+} f_{k_j} x_j - \left(\frac{f_k}{f_{k-1}}\right) f k_j x_j + s_k = -f_k$$

where, s_k = Gomorian slack

$J^+ = \{J/f_{k_j} \geq 0\}$

$J^- = \{J/f_{k_j} < 0\}$

Add this secondary constraint at the bottom of the optimum simplex table and use dual simplex method to obtain the new feasible optimal solution. Repeat the procedure until the values of the integer restricted variables are integers in the optimum solution obtained.

Example 1

Solve the following mixed integer programming problem:

Maximise $z = x_1 + x_2$

Subject to the constraints

$2x_1 + 5x_2 \leq 16$

$6x_1 + 5x_2 \leq 30$

$x_2 \geq 0$, x_1 non-negative integer.

Solution: Ignoring the integer restriction, solve the problem by simplex method then the optimum simplex table is

table 3.54

		c_j	1	1	0	0
C_B	y_B	X_B	x_1	x_2	s_1	s_2
1	x_2	18/10	0	1	3/10	- 1/10
1	x_1	7/2	1	0	- 1/4	1/4
	$z_j - c_j$		0	0	1/20	3/20

But $x_1 = 7/2$ which is not an integer. Take the x_1 equation and construct the Gomory's constraint

i.e. $3 + 1/2 = x_1 + 0x_2 - 1/4x_3 + 1/4x_4$

The Gomorian constraint is

$$\left(\frac{1/2}{1/2-1}\right)(-1/4)\ s_1 + 1/4s_2 \geq 1/2$$

i.e. $$1/4s_1 + 1/4s_2 \geq 1/2$$

i.e. $$-1/4s_1 - 1/4s_2 \leq -1/2$$

i.e. $$-1/4s_1 - 1/4s_2 + s_3 = -1/2$$

where, s_3 is the Gomorian slack. Add this constraint at the bottom of the above optimum simple table. Since the solution is infeasible apply dual simplex method.

table 3.55

		c_j	1	1	0	0	0
C_B	y_B	X_B	x_1	x_2	s_1↓	s_2	s_3
1	x_2	9/5	0	1	3/10	- 1/10	0
1	x_1	7/2	1	0	- 1/4	1/4	0
0	s_3	- 1/2	0	0	[−1/4]	- 1/4	1
			0	0	1/20	3/20	0

Since maximum $\left\{\frac{1/20}{-1/4}, \frac{3/20}{-1/4}\right\} = -1/5$ which corresponds to the variable s_1. Hence, s_1 enters into the basis. The next iteration is

table 3.56

		c_j	1	1	0	0	0
C_B	y_B	X_B	x_1	x_2	s_1	s_2	s_3
1	x_2	6/5	0	1	0	- 2/5	6/5
1	x_1	4	1	0	0	1/2	- 1
0	x_3	2	0	0	1	1	- 4
		26/5	0	0	0	1/10	1/5

Since all $z_j - c_j \geq 0$ and all $X_{Bi} \geq 0$, the current solution is optimal and feasible. Since $x_1 = 4$ which is an integer the condition is satisfied. The required optimum solution is

$$x_1 = 4,\ x_2 = 6/5 \text{ and maximum of } z = 26/5$$

3.8.4 Branch and Bound Technique

This method is applicable to both pure (all) as well as mixed integer programming problems and involves the continuous version of the problem.

Branch and bound is a generic method for solving a class of optimisation problems. The optimal solution is selected by successive partitioning of the set of all possible solutions into two mutually exclusive and exhaustive subsets, and establishing limit for each set such that all solutions in the set are worse than the limit. The first part of the procedure involving partitioning is called *branching* while the second part of establishing limit is referred to as *bounding*.

The branch and bound method first divides the feasible region into smaller subsets and then examines each of them successively until a feasible solution that gives optimal value of objective function is obtained. The iterative procedure is as follows:

Algorithm:

Step 1: Obtain an optimum solution to the given LPP ignoring the restriction of integers.

Step 2: Test the integerality of the optimum solution obtained in step 1. There are two cases:

(i) If the solution is in integers, the current solution is optimum to the given integer programming problem.

(ii) If the solution is not in integers, go to next step.

Step 3: Considering the value of the objective function as upperbound, obtain the lowerbound by rounding off to integral values of the decision variables.

Step 4: Let the optimum value x_j^* of the variable x_j is not an integer. Then, subdivide the given LPP into two problems:

Sub-problem 1: Given LPP with an additional constraint $x^*_j \leq [x_j^*]$.

Sub-problem 2: given LPP with an additional constraint $x^*_j \geq [x_j^*] + 1$

where $[x_j^*]$ is the integer part of x_j^*.

Step 5: Solve the two sub-problems obtained in step 4.

There may arise three cases:

(i) If the optimum solution of the two sub-problems are integral, then the required solution is one that gives larger value of z.

(ii) If the optimum solution of one sub-problem is integral and other sub-problem has no feasible optimum solution, the required solution is same as that of the sub-problem having integer valued solution.

(iii) If the optimum solution of one sub-problem is integral while that of the other is not integer valued, then record the integer valued solution and repeat step 3 and 4 for the non-integer valued sub-problem.

Step 6: Repeat steps 3 to 5, until all integer valued solutions are recorded.

Step 7: Choose the solution amongst the recorded integer valued solutions that yields an optimum value of z.

Remark: The above method can be represented by an enumeration tree. Each node in the tree represents a sub-problem to be evaluated. Each branch of tree creates a new constraint which is added to the original problem.

Example 1

Use branch and bound technique to solve the following integer programming problem:

Maximise $z = 7x_1 + 9x_2$

subject to

$$-x_1 + 3x_2 \leq 6$$
$$7x_1 + x_2 \leq 35$$
$$x_1 \geq 0,\ x_2 \leq 7$$

x_1, x_2 are integers.

Solution: First, we ignore the integer restriction and solve the problem by graphical method.

Let P_0: Maximise $z = 7x_1 + 9x_2$

subject to $-x_1 + 3x_2 \leq 6$

$$7x_1 + x_2 \leq 35$$
$$x_1 \geq 0,\ x_2 \leq 7$$

x_1, x_2 are integers

$$-x_1 + 3x_2 = 6 \rightarrow (0, 2)\ (-6, 0)$$
$$7x_1 + x_2 = 35 \rightarrow (5, 0)\ (0, 35)$$

The feasible region for P_0 is *OABC*

$$z = 7x_1 + 9x_2$$

Now,
$$Z \text{ at } O(0, 0) = 7(0) + 9(0) = 0$$
$$Z \text{ at } A(5, 0) = 7(5) + 9(0) = 35$$
$$\boxed{Z \text{ at } B(9/2, 7/2) = 7(9/2) + 9(7/2) = 63}$$
$$Z \text{ at } C(0, 2) = 7(0) + 9(2) = 18$$

Hence, the optimum solution of P_0 is

$$x_1 = 9/2,\ x_2 = 7/2 \text{ with maximum of } z = 63$$

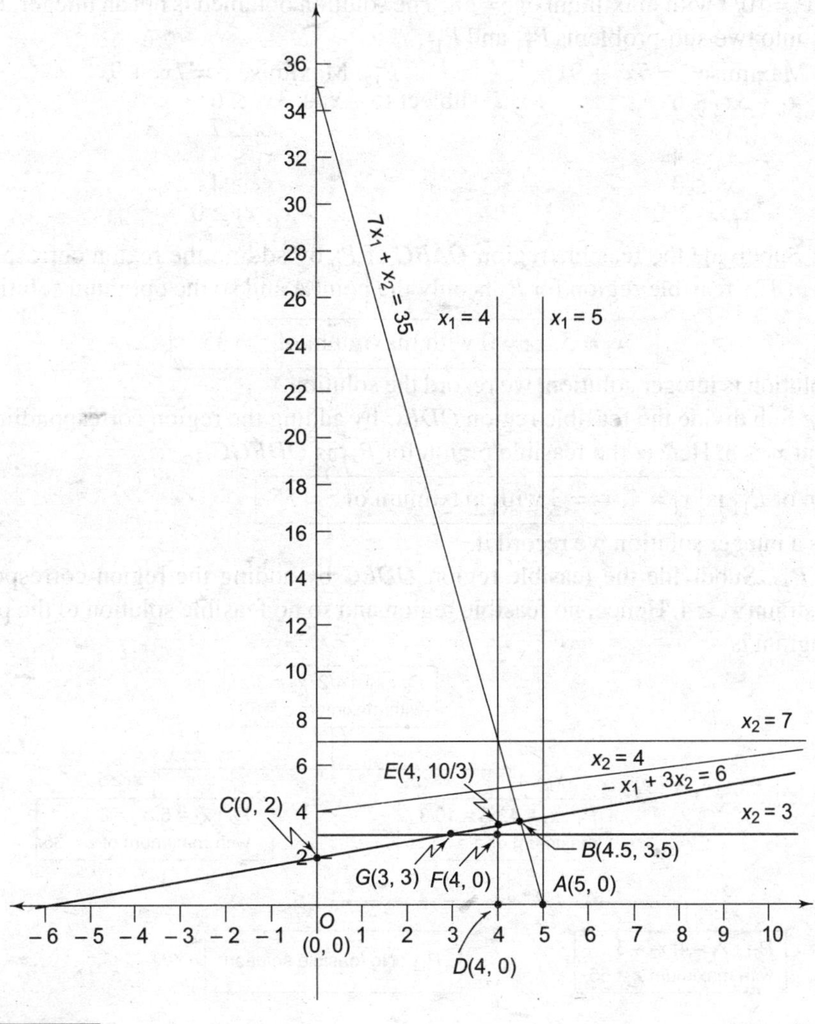

figure 3.4

The solution obtained is not an integer solution.

Now, $x_1 = 4 + 1/2$, $x_2 = 3 + 1/2$ and the maximum of $\{1/2, 1/2\} = 1/2$. Arbitrarily, choose x_1 for branching. Now, divide the problem P_0 into sub-problems:

P_1: Maximise $z = 7x_1 + 9x_2$

subject to

$$-x_1 + 3x_2 \le 6$$
$$7x_1 + x_2 \le 35$$
$$x_1 \ge 0, x_2 \le 7$$
$$\boxed{x_1 \le 4}$$

x_1, x_2 are integers

P_2: Maximise $z = 7x_1 + 9x_2$

subject to

$$-x_1 + 3x_2 \le 6$$
$$7x_1 + x_2 \le 35$$
$$x_1 \ge 0, x_2 \le 7$$
$$\boxed{x_1 \ge 5}$$

x_1, x_2 are integers

Solution of P_1: Sub divide the feasible region *OABC* of P_0 by adding the region corresponding to the new constraint $x_1 \le 4$. The feasible region is *ODEC* and the point *E* is (4, 10/3). The optimum solution of P_1 is $x_1 = 4, x_2 = 10/3$ with maximum of $z = 58$. The solution obtained is not an integer. Hence, divide the problem P_1 into two sub-problems P_{11} and P_{12}.

P_{11}: Maximise $z = 7x_1 + 9x_2$

subject to $-x_1 + 3x_2 \le 6$

$$x_2 \le 7$$
$$x_1 \le 4$$
$$x_2 \le 3$$
$$x_1, x_2 \ge 0$$

P_{12}: Maximise $z = 7x_1 + 9x_2$

subject to $-x_1 + 3x_2 \le 6$

$$x_2 \le 7$$
$$x_1 \le 7$$
$$x_2 \ge 4$$
$$x_1, x_2 \ge 0$$

Solution of P_2: Subdivide the feasible region *OABC* of P_0 by adding the region corresponding to the constraint $x_1 \ge 5$. The feasible region for P_2 is only the point *A* and so the optimum solution is

$$\boxed{x_1 = 5, x_2 = 0 \text{ with maximum of } z = 35}$$

Since the solution is integer solution, we record the solution.

Solution of P_{11}: Sub divide the feasible region *ODEC* by adding the region corresponding to the additional constraint $x_2 \le 3$. Hence, the feasible region for P_{11} is *ODFGC*.

The solution of P_{11} is $\boxed{x_1 = 4, x_2 = 3 \text{ with maximum of } z = 55.}$

Since this is a integer solution we record it.

Solution of P_{12}: Subdivide the feasible region *ODEC* by adding the region corresponding to the additional constraint $x_2 \ge 4$. Hence, no feasible region and so no feasible solution to the problem.

The tree diagram is

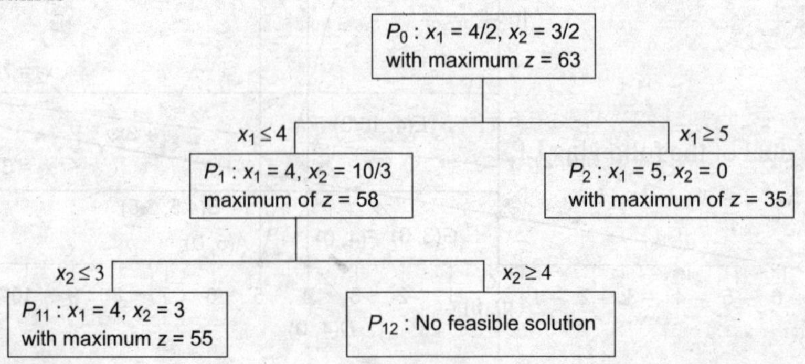

figure 3.5

If we examine the recorded solutions we observe that the optimum solution is

$$x_1 = 4, x_2 = 3 \text{ with maximum of } z = 55$$

3.9 LIMITATIONS OF LINEAR PROGRAMMING PROBLEM

In this section we will see some advantages and some limitations of linear programming problems.

3.9.1 Advantages of Linear Programming Problem

1. It provides an insight and perspective into the problem environment. This generally results in clear picture of the true problem.
2. It makes a scientific and mathematical analysis of the problem situations.
3. It gives an opportunity to the decision-maker to formulate his strategies consistent with the constraints and the objectives.
4. It deals with changing situations. Once a plan is arrived through the linear programming it can also be evaluated for changing conditions.
5. By using linear programming the decision-maker makes sure that he is considering the best solution.

3.9.2 Limitations of Linear Programming

1. The major limitation of linear programming is that it treats all relationship as linear. But it is not true in many real life situations.
2. The decisions variables in some LPP would be meaningful only if they are integer values. But sometimes we get fractional values to the optimal solution, where only integer values are meaningful.
3. All the parameters in the linear programming model are assumed to be known constants. But in real life they may not be known completely or they may be probabilistic and they may be liable for changes from time to time.
4. The problems are complex if the number of variables and constraints are quite large.
5. Linear programming deals with only a single objective problem whereas in real life situations, there may be more than one objective.

EXERCISES

Section 3.1

1. Write the dual of the following LPP

   Maximise $z = x_1 + 2x_2 + x_3$

   subject to
   $$2x_1 + x_2 - x_3 \le 2$$
   $$-2x_1 + x_2 - 5x_3 \ge -6$$
   $$4x_1 + x_2 + x_3 \le 6$$
   $$x_1, x_2, x_3 \ge 0$$

   [*Answer*: The dual is minimise $w = 2y_1 + 6y_2 + 6y_3$, subject to $2y_1 + y_2 + 4y_3 \ge 1$, $y_1 - y_2 + y_3 \ge 1$, $-y_1 + 5y_2 + y_3 \ge 1$ and $y_1, y_2, y_3 \ge 0$.]

2. Write the dual of the LPP

   Maximise $z = 3x + 10y + 2z$

   subject to
$$2x + 3y + 2z \leq 7$$
$$3x - 2y + 4z = 3$$
$$\text{and } x, y, z \geq 0$$

   [*Answer*: The dual problem is minimise $w = 7r + 3s$ subject to $2r + 3s \geq 3$, $3r - 2s \geq 10$, $2r + 4s \geq 2$ and $r \geq 0$, s is unrestricted.]

3. Obtain the dual of the following LPP

   Maximise $z = 3x_1 + x_2 + 2x_3 - x_4$

   subject to
$$2x_1 - x_2 + 3x_3 + x_4 = 1$$
$$x_1 + x_2 - x_3 + x_4 = 3$$
$$x_1, x_2, x_3 \geq 0 \text{ and } x_4 \text{ is unrestricted.}$$

   [*Answer*: The dual is minimise $w = y_1 + y_2$, subject to $2y_1 + y_2 \geq 3$, $-y_1 + y_2 \geq 1$, $3y_1 - y_2 \geq 2$, $y_1 + y_2 = -1$; y_1 and y_2 are unrestricted.]

4. Write the dual of the LPP

   Minimise $z = 7x_1 + 3x_2 + 8x_3$

   subject to
$$8x_1 + 2x_2 + x_3 \geq 3$$
$$3x_1 + 6x_2 + 4x_3 \geq 4$$
$$4x_1 + x_2 + 5x_3 \geq 1$$
$$x_1 + 5x_2 + 2x_3 \geq 7$$
$$x_1, x_2, x_3 \geq 0$$

   [*Answer:* The dual LPP is Maximise $w = 3y_1 + 4y_2 + y_3 + 7y_4$, subject to $8y_1 + 3y_2 + 4y_3 + y_4 \leq 7$, $2y_1 + 6y_2 + y_3 + 5y_4 \leq 3$, $y_1 + 4y_2 + 5y_3 + 2y_4 \leq 8$, $y_1, y_2, y_3, y_4 \geq 0$.]

5. Write the dual of the primal

   Maximise $z = 6x_1 + 6x_2 + x_3 + 7x_4 + 5x_5$

   subject to
$$3x_1 + 7x_2 + 8x_3 + 5x_4 + x_5 = 2$$
$$2x_1 + x_2 + 3x_4 + 9x_5 = 6$$
$$x_1, x_2, x_3, x_4 \geq 0$$
$$\text{and } x_5 \text{ unrestricted.}$$

   [*Answer*: The dual is minimise $w = 2y_1 + 6y_2$ subject to $3y_1 + 2y_2 \geq 6$, $7y_1 + y_2 \geq 7$, $8y_1 + 3y_2 \geq 1$, $5y_1 + 2y_2 \geq 7$, $y_1 + 9y_2 = 5$; y_1 and y_2 are unrestricted.]

6. Apply simplex method to solve the following LPP

   Maximise $z = 30x_1 + 23x_2 + 9x_3$, subject to
$$6x_1 + 5x_2 + 3x_3 \leq 26,\ 4x_1 + 2x_2 + 5x_3 \leq 7,\ \text{every } x_j \geq 0$$

   Also, read the solution to the dual from the final Table of primal.

   [*Answer*: Solution of the primal is $x_1 = 0$, $x_2 = 7/2$, $x_3 = 0$, maximum of $z = 161/2$. The solution of the dual problem is $y_1 = 0$, $y_2 = 23/2$ and minimum of $w = 161/2$.]

7. Applying the concept of duality, solve the LPP

   Maximise $z = 3x_1 + 4x_2$

subject to
$$x_1 - x_2 \le 1$$
$$x_1 + x_2 \le 4$$
$$x_1 - 3x_2 \le 3$$
$$x_1, x_2 \ge 0$$

[*Answer*: Since the dual problem does not possess any optimum basic feasible solution hence there exists an unbounded solution to the primal problem.]

8. Solve the following LPP by formulating its dual

Minimise $z = 50x_1 - 80x_2 - 140x_3$

subject to
$$x_1 - x_2 - 3x_3 \ge 4$$
$$x_1 - 2x_2 - 2x_3 \ge 3$$
$$\text{and } x_1, x_2, x_3 \ge 0$$

[*Answer*: Dual has no feasible solution and hence primal has no feasible solution.]

9. Use duality to solve the LPP

Minimise $z = 8x_1 - 2x_2 - 4x_3$

subject to
$$x_1 - 4x_2 - 2x_3 \ge 2$$
$$x_1 + x_2 - 3x_3 \ge -1$$
$$-3x_1 - x_2 + x_3 \ge 1$$
$$x_1, x_2, x_3 \ge 0$$

[*Answer*: Dual has an unbounded solution and hence primal has either an unbounded or no solution.]

10. Use duality to solve the LPP

Minimise $z = 4x_1 + 2x_2 + 3x_3$

subject to
$$2x_1 + 4x_3 \ge 5$$
$$2x_1 + 3x_2 + x_3 \ge 4$$
$$x_1, x_2, x_3 \ge 0$$

[*Answer*: The solution of the primal is $x_1 = 0, x_2 = 11/12, x_3 = 5/4$ with minimum of $z = 67/12$. The solution of the dual is $y_1 = 7/2$, $y_2 = 2/3$ and maximum of $w = 67/12$.]

Section 3.2

11. Solve the following linear programming problems using dual simplex method.

(i) Minimise $z = 2x_1 + x_2$

subject to
$$3x_1 + x_2 \ge 3$$
$$4x_1 + 3x_2 \ge 6$$
$$x_1 + 2x_2 \ge 3$$
$$x_1, x_2 \ge 0$$

[*Answer*: $x_1 = 3/5$, $x_2 = 6/5$ with minimum of $z = 12/5$.]

(ii) Minimise $z = 3x_1 + x_2$

subject to
$$x_1 + x_2 \ge 1$$
$$2x_1 + 3x_2 \ge 2$$
$$x_1, x_2 \ge 0$$

[*Answer*: $x_1 = 0$, $x_2 = 1$ with minimum of $z = 1$.]

(iii) Minimise $z = x_1 + 2x_2$
subject to $2x_1 + x_2 \geq 4$
$x_1 + 2x_2 \geq 7$
$x_1, x_2 \geq 0$

[*Answer*: $x_1 = 0, x_2 = 2$ with minimum of $z = 4$.]

(iv) Minimise $z = 3x_1 + 2x_2 + x_3 + 4x_4$
subject to $2x_1 + 4x_2 + 5x_3 + x_4 \geq 10$
$3x_1 - x_2 + 7x_3 - 2x_4 \geq 2$
$5x_1 + 2x_2 + x_3 + 6x_4 \geq 15$
$x_1, x_2, x_3, x_4 \geq 0$

[*Answer*: $x_1 = 65/23, x_2 = 0, x_3 = 20/23, x_4 = 0$ with minimum of $z = 215/213$.]

(v) Minimise $z = 2x_1 + 9x_2 + 24x_3 + 8x_4 + 5x_5$
subject to $x_1 + x_2 + 2x_3 - x_5 - x_6 = 1,$
$-2x_1 + x_3 + x_4 + x_5 - x_7 = 2,$
$x_j \geq 0, j = 1, \cdots, 7$

[*Answer*: $x_1 = x_2 = x_5 = x_6 = x_7 = 0, x_3 = 1/2, x_4 = 3/2$ with minimum of $z = 24$.]

(vi) Minimise $z = x_1 + 2x_2 + 3x_3$
subject to $x_1 - x_2 + x_3 \geq 4$
$x_1 + x_2 + 2x_3 \leq 8$
$x_2 - x_3 \geq 2$
$x_1, x_2, x_3 \geq 0$

[*Answer*: $x_1 = 6, x_2 = 2, x_3 = 0$ with minimum of $z = 10$.]

Section 3.3

12. Use revised simplex method to solve the following linear programming problems:

(i) Maximise $z = 3x_1 + 2x_2 + 5x_3$
subject to $x_1 + 2x_2 + x_3 \geq 430$
$3x_1 + 2x_3 \geq 460$
$x_1 + 4x_2 \leq 420$
$x_1, x_2, x_3 \geq 0$

[*Answer*: $x_1 = 0, x_2 = 100, x_3 = 230$ with maximum of $z = 1350$.]

(ii) Minimise $z = x_1 + x_2$
subject to $x_1 + 2x_2 \geq 7$
$4x_1 + x_2 \geq 6$
$x_1, x_2 \geq 0$

[*Answer*: $x_1 = 5/7, x_2 = 22/7$ with minimum of $z = 27/7$.]

(iii) Maximise $z = 2x_1 + 3x_2$
subject to $-x_1 + x_2 \geq 0$
$x_1 \leq 4$
and $x_1, x_2 \geq 0$

[*Answer*: Unbounded solution.]

(iv) Maximise $z = x_1 + x_2$
subject to
$$2x_1 + 5x_2 \geq 6,$$
$$x_1 + x_2 \geq 2,$$
$$x_1, x_2 \geq 0$$

[*Answer*: Unbounded solution.]

(v) Maximise $z = x_1 + x_2$
subject to
$$3x_1 + 2x_2 \leq 6$$
$$x_1 + 4x_2 \leq 4$$
$$x_1, x_2 \geq 0$$

[*Answer*: $x_1 = 8/5, x_2 = 3/5$ with maximum of $z = 11/5$.]

(vi) Maximise $z = x_1 + 2x_2$
subject to
$$3x_1 + 2x_2 \geq 6$$
$$x_1 + 6x_2 \leq 3$$
$$x_1, x_2 \geq 0$$

[*Answer*: Unbounded solution.]

Section 3.4

13. Use bounded variable technique to solve the following LPP.

(i) Maximise $z = x_2 + 3x_3$
subject to
$$x_1 + x_2 + x_3 \leq 10$$
$$x_1 - 2x_3 \geq 0$$
$$2x_1 - x_3 \leq 10$$
$$0 \leq x_1 \leq 8,\ 0 \leq x_2 \leq 4,\ x_3 \geq 0$$

[*Answer*: $x_1 = 20/3, x_2 = 0, x_3 = 10/3$ with maximum of $z = 10$.]

(ii) Maximise $z = 3x_1 + 5x_2 + 2x_3$
subject to
$$x_1 + 2x_2 + 2x_3 \leq 14$$
$$2x_1 + 4x_2 + 3x_3 \leq 23$$
$$0 \leq x_1 \leq 4,\ 2 \leq x_2 \leq 5,\ 0 \leq x_3 \leq 3$$

[*Answer*: $x_1 = 4, x_2 = 15/4, x_3 = 0$ with maximum of $z = 123/4$.]

(iii) Maximise $z = 6x_1 + 2x_2 + 8x_3 + 4x_4 + 2x_5 + 10x_6$
subject to the constraints
$$8x_1 + x_2 + 8x_3 + 2x_4 + 2x_5 + 4x_6 \leq 13,$$
$$0 \leq x_j \leq 1 \quad (j = 1, 2, 3, 4, 5, 6)$$

[*Answer*: $x_1 = x_5 = 0, x_2 = x_4 = x_6 = 1, x_3 = 3/4$ with maximum of $z = 22$.]

(iv) Maximise $z = 3x_1 + 2x_2$
subject to
$$x_1 - 3x_2 \leq 3$$
$$x_1 - 2x_2 \leq 4$$
$$2x_1 + x_2 \leq 20$$
$$x_1 + 3x_2 \leq 30$$
$$-x_1 + x_2 \leq 6$$
$$0 \leq x_1 \leq 8,\ 0 \leq x_2 \leq 6$$

[*Answer*: $x_1 = 7, x_2 = 6$ with maximum of $z = 33$.]

Section 3.5

14. Consider the LPP

Minimise $z = 2x_1 + x_2 + 4x_3 - x_4$

subject to
$$x_2 + x_2 + x_3 - 3x_4 \le 8$$
$$-x_1 + x_3 + 2x_4 \le 0$$
$$2x_1 + 7x_2 - 5x_3 - 10x_4 \le 21$$
$$\text{and } x_1, x_2, x_3, x_4 \ge 0$$

(a) Solve the LPP.
(b) Discuss the effect of change of b_2 to 11.
(c) Discuss the effect of change of b to [3, – 2, 4].

[*Answer*: The optimal solution is
(a) $x_1 = 8, x_2 = x_3 = x_4 = 0$, maximum of $z = 16$
(b) $x_1 = 49/2, x_2 = 0 = x_3, x_4 = 11/2$, maximum of $z = 87/2$
(c) No feasible solution.]

15. Consider the LPP

Maximise $z = 3x_1 + 5x_2 + 4x_3$

subject to
$$2x_1 + 3x_2 \le 8$$
$$2x_2 + 5x_3 \le 10$$
$$3x_1 + 2x_2 + 4x_3 \le 15$$
$$x_1, x_2, x_3 \ge 0$$

(a) Solve the LPP.
(b) Find the range over which b_2 can be changed maintening the feasibility of the solution.

[*Answer*: (a) $x_1 = 89/41, x_2 = 50/41, x_3 = 62/41$ with maximum of $z = 765/41$.
(b) $-25/4 \le \Delta b_2 \le 89/12$.]

16. Consider the LPP

Maximise $z = -x_1 + 2x_2 - x_3$

subject to
$$3x_1 + x_2 - x_3 \le 10$$
$$-x_1 + 4x_2 + x_3 \ge 6$$
$$x_2 + x_3 \le 4$$
$$x_1, x_2, x_3 \ge 0$$

(a) Determine an optimum solution to the problem.
(b) Determine the ranges for discrete changes in the components b_2 and b_3 of the required vector so as to maintain the optimality of the current optimum solution.

[*Answer*: (a) The optimal solution is $x_1 = 0, x_2 = 4, x_3 = 0$ with maximum of $z = 8$.
(b) $\Delta b_2 \le 10, -5/2 \le \Delta b_3 \le 6$].

17. Given the LPP

Minimise $z = 3x_1 + 6x_2 + x_3$

subject to the constraints
$$x_1 + x_2 + x_3 \ge 6$$

$$x_1 + 5x_2 - x_3 \geq 4$$
$$x_1 + 5x_2 + x_3 \geq 24$$
$$x_1, x_2, x_3 \geq 0$$

(a) Solve the LPP

(b) Discuss the effect of changing the required vector from [6, 4, 24] to [6, 2, 12] on the optimum solution.

[*Answer*: (a) $x_1 = 14, x_2 = 0, x_3 = 0$, minimum of $z = 52$
(b) $x_1 = 7, x_2 = 0, x_3 = 5$; minimum of $z = 26$.]

18. Solve the following linear programming problem
Maximise $3x_1 + 5x_2$
subject to
$$x_1 + x_2 \leq 1$$
$$2x_1 + x_2 \leq 1$$
$$x_1, x_2 \geq 0$$
Determine the range of variation of the cost coefficients c_j ($j = 1, 2$) corresponding to the optimal solution.
[*Answer*: $x_1 = 0, x_2 = 1$, maximum $z = 5, -\infty < c_1 < 5$ and $3 \leq c_2 \leq \infty$]

19. Consider the LPP
Minimise $z = 5x_1 + 3x_2$
subject to
$$3x_1 + 5x_2 \leq 15$$
$$5x_1 + 6x_2 \leq 10$$
and,
$$x_1, x_2 \geq 0$$
(a) Solve the LPP.
(b) Find how far the component c_1 of C can be increased without affecting the optimality of the solution.
[*Answer*: (a) $x_1 = 20/19, x_2 = 45/19$ with minimum of $z = 235/19$, (b) $5 \leq c_1 \leq 15/2$]

20. Consider the LPP.
Maximise $z = 3x_1 + 5x_2$
subject to
$$x_1 \leq 4$$
$$3x_1 + 2x_2 \leq 18$$
and,
$$x_1, x_2 \geq 0$$
(a) Solve this LPP.
(b) If a new variable x_5 is added to this problem with a column $\begin{pmatrix} 1 \\ 2 \end{pmatrix}$ and $c_5 = 7$, find the change in the optimal solution.
[*Answer*: (a) $x_1 = 0, x_2 = 9$ with maximum of $Z = 45$.
(b) $x_1 = 0, x_2 = 5, x_5 = 5$ with maximum of $Z = 53$.]

21. Consider the LPP.
Maximise $z = 5x_1 + 3x_2 + 7x_3$
subject to
$$x_1 + x_2 + 2x_3 \leq 22$$
$$3x_1 + 2x_2 + x_3 \leq 26$$
$$x_1 + x_2 + x_3 \leq 18$$
and,
$$x_1, x_2, x_3 \geq 0.$$

(a) Solve the LPP.

(b) What will be the solution if the first constraint changes to $x_1 + x_2 + 2x_3 \leq 26$.

[*Answer*: (a) $x_1 = 6, x_2 = 0, x_3 = 52/2$ and maximum of $z = 26$.

(b) $x_1 = 26/5, x_2 = 0, x_3 = 52/2$ and maximum of $z = 494/5$.]

22. (a) Solve the following LPP

Maximise $z = x_1 + 5x_2 + 3x_3$

subject to
$$x_1 + 2x_2 + x_3 = 3$$
$$2x_1 - x_2 = 4$$
and,
$$x_1, x_2, x_3 \geq 0$$

(b) If the objective function is changed to maximise $z = 2x_1 + 5x_2 + 2x_3$, find the new optimal solution.

[*Answer*: (a) $x_1 = 2, x_2 = 0, x_3 = 1$

(b) $x_1 = 11/5, x_2 = 2/5, x_3 = 0$ with maximum of $z = 32/5$.]

23. Given the LPP

Maximise $z = 10x_1 + 3x_2 + 6x_3 + 5x_4$

subject to the constraints
$$x_1 + 2x_2 + x_4 \leq 6$$
$$3x_1 + 2x_3 \leq 5$$
$$x_2 + 4x_3 + 5x_4 \leq 3$$
$$x_1, x_2, x_3 \geq 0$$

Compute the limits for a_{11} and a_{23} so that the new solution remains optimum feasible.

[*Answer*: $-\infty \leq a_{11} \leq 51/10, 39/40 \leq a_{23} < \infty$]

Section 3.6

24. Consider the parametric linear programming problem

Maximise $z = (\theta - 1)x_1 + x_2$

subject to
$$x_1 + 2x_2 \leq 10$$
$$2x_1 + x_2 \leq 11$$
$$x_1 - 2x_2 \leq 3$$
$$x_1, x_2 \geq 0$$

Perform a complete parametric analysis. Identify all critical values of the parameter θ and the optimal basic solutions.

[*Answer*: $x_1 = 0, x_2 = 5$ for $0 \leq \theta \leq 3/2$

$x_1 = 4, x_2 = 3$ for $3/2 \leq \theta \leq 3$

$x_1 = 5, x_2 = 1$ for $3 \leq \theta$]

25. For the following linear programming problem:

Minimise $z = \lambda x_1 - \lambda x_2 - x_3 + x_4$

subject to
$$3x_1 - 3x_2 - x_3 + x_4 \geq 5$$
$$2x_1 - 2x_2 + x_3 - x_4 \leq 3$$
$$x_1, x_2, x_3, x_4 \geq 0.$$

[*Answer*: For $-2 \leq \lambda < 3$, $x_1 = 8/5$, $x_2 = 0$, $x_3 = 0$, $x_4 = 1/5$ with minimum of $z = 1/5 + 8/5\,\lambda$. For $\lambda > 3$, no solution exists.]

26. Solve the problem
Maximise $z = 3x_1 + 2x_2 + 5x_3$
subject to
$$x_1 + 2x_2 + x_3 \leq 430 + 500\lambda$$
$$3x_1 + 2x_3 \leq 460 + 100\lambda$$
$$x_1 + 4x_2 \leq 420 - 200\lambda$$
$$x_1, x_2, x_3 \geq 0.$$
[*Answer*: For $0 \leq \lambda \leq 1/55$, $(x_1, x_2, x_3) = (0, 100 + 225\lambda, 230 + 5\lambda)$ and maximum of $z = 1{,}350 + 700\lambda$
for $1/55 \leq \lambda \leq 2.1$, $(x_1, x_2, x_3) = (0, 105 - 50\lambda, 230 + 50\lambda)$ and maximum of $z = 1{,}360 + 150\lambda$; for $\lambda > 2.1$, no feasible solution exists.]

Section 3.7

27. Solve the following linear goal programming problem
Minimise $z = \{o_3 + o_4, o_1, u_2, u_3 + 8/5\, u_4\}$
subject to
$$g_1\!: x_1 + x_2 + u_1 - o_1 = 20$$
$$g_2\!: x_1 + x_2 + u_2 - o_2 = 50$$
$$g_3\!: x_1 + u_3 - o_3 = 15$$
$$g_4\!: x_2 + u_4 - o_4 = 8,$$
$x_i, u_i, o_i \geq 0$ where the goals are written in order of priority.
[*Answer*: $x_1 = 12$, $x_2 = 8$ and maximum of $z = (0, 0, 30, 3)$.]

28. The manager of the only record shop in a town has a decision problem that involves multiple goals. The record shop employs five full-time and four part-time salesmen. The normal working hours per month for a full-time salesman and a part-time salesman are 160 hours and 80 hours, respectively. According to the performance records of salesmen, the average sales has been five records per hour for full-time salesmen and two records per hour for part-time salesmen. The average hourly wage rates are Rs 3 for full-time salesmen and Rs 2 for part-time salesmen. Average profit from the sales of a record is Rs 1.50.

In view of the past sales record and increased enrolment at the local college, the manager feels that the sales goal for the next month should be 5500 records. Since the shop is open six days a week, overtime is often required of salesmen. The manager believes that a good employer-employee relationship is an essential factor of business success.

Therefore, he feels that a stable employment level with occasional overtime requirement is a better practice than an unstable employment level with no overtime. However, he feels that overtime of more than 100 hours among the full-time salesmen should be avoided because of the declining sales effectiveness caused by fatigue.

The manager has set the following goals:

1. The first goal is to achieve a sales goal of 5500 records for the next month.
2. The second goal is to limit the overtime of full-time salesmen to 100 hours.
3. The third goal is to provide job security to salesmen. The manager feels that full utilisation of

employees' regular working hours (no layoffs) is an important factor for a good employer-employee relationship. However, he is twice as concerned with the full utilisation of full-time salesmen as with the full utilisation of part-time salesmen.

4. The last goal is to minimise the sum of overtime for both full-time and part-time salesmen. The manager desires to assign differential weights to the minimisation of overtime according to the net marginal profit ratio between the full-time and part-time salesmen.

Formulate the situation as a goal programming problem.

[*Answer*: The LPP is

Minimise $z = P_1 d_1^- + P_2 d_{21}^+ + 2P_3 d_2^- + P_3 d_3^- + 3P_4 d_3^+ + P_4 d_2^+$

subject to
$$5x_1 + 2x_2 + d_1^- - d_1^+ = 5500$$
$$x_1 + d_2^- - d_2^+ = 800$$
$$x_2 + d_3^- - d_3^+ = 320$$
$$d_2^+ + d_{21}^- - d_{21}^+ = 100$$
$$x_1, x_2, d_1^-, d_2^-, d_3^-, d_2^-, d_1^+, d_2^+, d_3^+, d_{21}^+ \geq 0.]$$

Section 3.8

29. Find the optimum integer solution to the following linear programming problems using Gomory's cutting plane method.

(i) Maximise $z = x_1 + x_2$

subject to
$$3x_1 + 2x_2 \leq 5$$
$$x_2 \leq 2$$

and $x_1 > 0$, $x_2 \geq 0$ and are integers.

[*Answer*: $x_1 = 0, x_2 = 2$ with maximum of $z = 2$.
(or) $x_1 = 1, x_2 = 1$ with maximum of $z = 2$.]

(ii) Maximise $z = 2x_1 + 2x_2$

subject to
$$5x_1 + 3x_2 \leq 8$$
$$2x_1 + 4x_2 \leq 8$$

and $x_1, x_2 \geq 0$ and are all integers.

[*Answer*: $x_1 = 1, x_2 = 1$ with maximum of $z = 4$ (or)
$x_1 = 0, x_2 = 2$ with maximum of $z = 4$.]

(iii) Maximise $z = 3x_1 + 4x_2$

subject to
$$3x_1 + 2x_2 \leq 8$$
$$x_1 + 4x_2 \leq 10$$

and $x_1, x_2 \geq 0$ and are integers.

[*Answer*: $x_1 = 0, x_2 = 3$ with maximum of $z = 12$.]

(iv) Maximise $z = x_1 - 2x_2$

subject to
$$4x_1 + 2x_2 \leq 15$$

$x_1, x_2 \geq 0$ and are integers.

[*Answer*: $x_1 = 3, x_2 = 0$ with maximum of $z = 3$.]

(v) Maximise $z = 7x_1 + 9x_2$

subject to
$$-x_1 + 3x_2 \leq 6$$
$$7x_1 + x_2 \leq 35$$

$x_1, x_2 \geq 0$ and are integers.

[*Answer*: $x_1 = 4$, $x_2 = 3$ with maximum of $z = 55$.]

(vi) Minimise $z = 9x_1 + 10x_2$

subject to
$$x_1 \le 9$$
$$x_2 \le 8$$
$$4x_1 + 3x_2 \ge 40$$
$$x_1, x_2 \ge 0 \text{ and are integers.}$$

[*Answer*: $x_1 = 9$, $x_2 = 2$ with minimum of $z = 101$.]

(vii) Maximise $z = 2x_1 + 20x_2 - 10x_3$

subject to
$$5x_1 + 20x_2 + 4x_3 \le 15$$
$$6x_1 + 20x_2 + 4x_3 = 20$$
$$x_1, x_2, x_3 \ge 0 \text{ and are integers.}$$

[*Answer*: $x_1 = 2$, $x_2 = 0$, $x_3 = 2$ with maximum of $Z = -16$.]

(viii) Maximise $z = 3x_2$

Subject to
$$3x_1 + 2x_2 \le 7$$
$$x_1 - x_2 \ge -2$$
$$x_1, x_2 \ge 0 \text{ and are integers.}$$

[*Answer*: Simplex method gives the integer solution: $x_1 = 0$, $x_2 = 2$ with maximum of $z = 6$.]

(ix) Maximise $z = x_1 + 5x_2$

subject to
$$x_1 + 10x_2 \le 20$$
$$x_1 \le 2$$
$$x_1, x_2 \ge 0 \text{ and are integers.}$$

[*Answer*: $x_1 = 2$, $x_2 = 1$ with maximum of $z = 7$.]

(x) Maximise $z = 2x_1 + 2x_2$

subject to
$$5x_1 + 3x_2 \le 8$$
$$x_1 + 2x_2 \le 4$$
$$x_1, x_2 \ge 0 \text{ and are integers.}$$

[*Answer*: $x_1 = 1$, $x_2 = 1$ and maximum of $z = 4$.]

30. Solve the following mixed integer programming problems:

(i) Maximise $z = x_1 + x_2$

subject to
$$3x_1 + 2x_2 \le 5$$
$$x_2 \le 2$$
$$\text{and } x_1, x_2 \ge 0 \text{ and } x_1 \text{ is an integer.}$$

[*Answer*: $x_1 = 0$, $x_2 = 2$ with maximum of $z = 2$.]

(ii) Minimise $z = 10x_1 + 9x_2$

subject to
$$x_1 \le 8$$
$$x_2 \le 10$$
$$5x_1 + 3x_2 \ge 45$$
$$x_1, x_2 \ge 0, x_1 \text{ is an integer.}$$

[*Answer*: $x_1 = 8$, $x_2 = 5/3$ with minimum of $z = -95$.]

(iii) Maximise $z = 4x_1 + 6x_2 + 2x_3$

subject to
$$4x_1 - 4x_2 \le 5$$

$$-x_1 + 6x_2 \leq 5$$
$$-x_1 + x_2 + x_3 \leq 5$$
$$x_1, x_2, x_3 \geq 0 \text{ and } x_1, x_3 \text{ are integers.}$$

[*Answer*: $x_1 = 2, x_2 = 1, x_3 = 6$ with maximum of $z = 26$.]

(iv) Maximise $z = 3x_1 + x_2 + 3x_3$

subject to
$$-x_1 + 2x_2 + x_3 \leq 4$$
$$4x_2 - 3x_3 \leq 2$$
$$x_1 - 3x_2 + 2x_3 \leq 3$$
$$x_2 \geq 0 \text{ and } x_1, x_3 \text{ are non-negative integers.}$$

[*Answer*: $x_1 = 5, x_3 = 3, x_2 = 2.75$ with maximum of $z = 26.75$.]

31. Use branch and bound technique to solve the following integer programming problem.

(i) Maximise $z = x_1 + x_2$

subject to
$$4x_1 - x_2 \leq 10$$
$$2x_1 + 5x_2 \leq 10$$
$$2x_1 - 3x_2 \leq 6$$
$$x_1, x_2 \geq 0 \text{ and are integers.}$$

[*Answer*: $x_1 = 2, x_2 = 1$ with maximum of $z = 3$.]

(ii) Maximise $z = 3x_1 + 4x_2$

subject to
$$3x_1 - x_2 + x_3 = 12$$
$$3x_1 + 11x_2 + x_4 = 66$$
$$x_j \geq 0 \text{ and are integers for } j = 1, 2, 3, 4.$$

[*Answer*: $x_1 = 5, x_2 = 4$ with maximum of $z = 31$.]

(iii) Maximise $z = x_1 + 4x_2$

subject to
$$2x_1 + 4x_2 \leq 7$$
$$5x_1 + 3x_2 \leq 15$$
$$x_1, x_2 \geq 0 \text{ and are integers.}$$

[*Answer*: $x_1 = 1, x_2 = 1$ with maximum of $z = 5$.]

(iv) Maximise $z = 3x_1 + 4x_2$

subject to
$$7x_1 + 16x_2 \leq 52$$
$$3x_1 - 2x_2 \leq 18$$
$$x_1, x_2 \geq 0 \text{ and are integers.}$$

[*Answer*: $x_1 = 5, x_2 = 1$, with maximum of $z = 19$.]

4

The Transportation Problem

4.1 INTRODUCTION

In the previous chapter we have used simplex method to solve any type of linear programming problems. However, there is another method, *Allocation method* which is applied to a lot of very practical problems generally called *Transportation problems.* The objective of the problem is to transport a commodity (single product) from more than one centre, called *origins* (or sources, or supply or capacity centres) to more than one places called *destinations* (sinks or demand or requirement centres) and the costs of transportation from each of the origins to each of the destinations being different and known. It is further assumed that

(i) The availability as well as requirements of the various centres are finite and contains the limited resources

(ii) The cost of transportation is linear.

The problem is to transport the goods from various origins to different destinations in such a manner that the cost of transportation is minimum.

The distinct feature of transportation problems is that sources and jobs must be expressed in terms of only one kind of unit.

4.2 MATHEMATICAL FORMULATION

Suppose that there are m sources and n destinations. Let a_i be the number of supply units available at source i ($i = 1, 2, \cdots, m$) and let b_j be the number of demand units required at destination j ($j = 1, 2, \cdots, n$). Let c_{ij} represent the cost of transporting one unit of commodity from source i to destination j. If x_{ij} ($x_{ij} \geq 0$) is the number of units transported from source i to destination j, then the equivalent linear programming model will be:

Determine non-negative values of x_{ij} satisfying both the availability constraints

$$\sum_{j=1}^{n} x_{ij} = a_i, \text{ for } i = 1, 2, \cdots, m \qquad (4.2.1)$$

as well as the requirement constraint:

$$\sum_{i=1}^{m} x_{ij} = b_j \text{ for } j = 1, 2, \cdots, n. \tag{4.2.2}$$

and minimising the total cost of transportation

$$z = \sum_{i=1}^{m} \sum_{j=1}^{n} x_{ij}\, c_{ij} \tag{4.2.3}$$

It is also assumed that total availabilities Σa_i satisfy the total requirement Σb_j. That is

$$\sum_{i=1}^{m} a_i = \sum_{j=1}^{n} b_j \tag{4.2.4}$$

Since the constraint equations (4.2.1), (4.2.2) and the objective function (4.2.3) are all linear in x_{ij}, so it may be viewed as a linear programming problem.

The above transportation problem can be represented by a table as:

table 4.1

		Destination						
		D_1	D_2	D_3			D_n	Supply
Source	S_1	x_{11} c_{11}	x_{12} c_{12}	x_{13} c_{13}	…	…	x_{1n} c_{1n}	a_1
	S_2	x_{21} c_{21}	x_{22} c_{22}	x_{23} c_{23}	…	…	x_{2n} c_{2n}	a_2
		⋮	⋮	⋮	⋮	⋮	⋮	⋮
	S_m	x_{m1} c_{m1}	x_{m2} c_{m2}	x_{m3} c_{m3}	…	…	x_{mn} c_{mn}	a_m
Demand		b_1	b_2	b_3	…		b_n	

The matrix consists of squares called 'cells' which stacks terms, columns vertically and rows horizontally. Unit costs are placed in each cell. The unit transportation cost c_{ij} from the ith source to the jth destination is displayed in the lower right hand corner of $(i, j)^{\text{th}}$ cell. The quantity x_{ij} transported from i^{th} source to j^{th} destination is displayed in the upper left side of the $(i, j)^{\text{th}}$ cell. Right and bottom sides of the transportation table point out the amounts of supplies a_i available at source i and the amount demanded b_j at destination j. The various a's and b's are called *rim requirements.*

4.2.1 Definitions

A *feasible solution* to a transportation problem is a set of non-negative allocations, x_{ij} that satisfies the rim (row and column) restrictions. A feasible solution is called a *basic feasible solution* if it contains no more than $m + n - 1$ non-negative allocations where m is the number of rows and n is the number of columns of the transportation problem. A feasible solution (not necessarily basic) is said to be *optimal* if it minimises (maximises) the transportation cost (profit). A basic feasible solution that contains less than $m + n - 1$ non-negative allocations is called *degenerate basic feasible*. A basic feasible solution to a ($m \times n$) transportation problem is said to be a *non-degenerate basic feasible* solution if it contains exactly $m + n - 1$ non-negative allocations in independent positions. The allocations are said to be in independent positions, if it is not possible to form a closed path. Closed path (loop) means by allowing horizontal and vertical lines and all the corner cells are occupied. Mathematically, an ordered set of at least four cells in a transportation table is said to form a loop provided:

(i) any two adjacent cells of the ordered set lie either in the same row or in the same column.

(ii) no three or more adjacent cells in the ordered set lie in the same row or column.

For example,

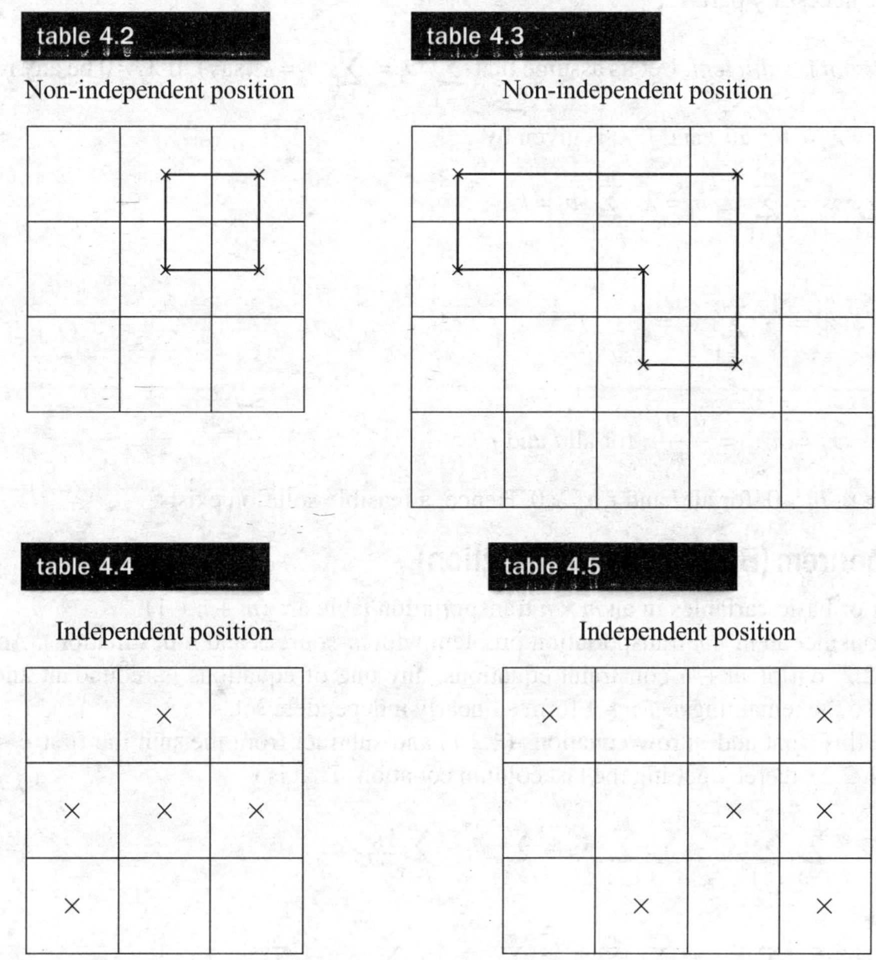

4.2.2 Theorem (Existence of Feasible Solution)

A necessary and sufficient condition for the existence of a feasible solution to a transportation problem is that

$$\sum_{i=1}^{m} a_i = \sum_{j=1}^{n} b_j$$

Proof: The condition is necessary. Suppose there exist a feasible solution to the transportation problem. Then,

$$\sum_{i=1}^{m}\sum_{j=1}^{n} x_{ij} = \sum_{i=1}^{m} a_i \quad \text{and} \quad \sum_{j=1}^{n}\sum_{i=1}^{m} x_{ij} = \sum_{j=1}^{n} b_j$$

$$\Leftrightarrow \quad \sum_{i=1}^{m} a_i = \sum_{j=1}^{n} b_j$$

Hence the necessary part.

The condition is sufficient. Let us assume that $\sum_{i=1}^{m} a_i = \sum_{j=1}^{n} b_j = k$ (say). If $\lambda_i \neq 0$ be any real number such that $x_{ij} = \lambda_i b_j$ for all i and j, λ_i is given by

$$\sum_{j=1}^{n} x_{ij} = \sum_{j=1}^{n} \lambda_i b_j = \lambda_i \sum_{j=1}^{n} b_j = k\lambda_i$$

or,

$$\lambda_i = \frac{1}{k}\sum_{j=1}^{n} x_{ij} = \frac{a_i}{k}$$

Thus,

$$x_{ij} = \lambda_i b_j = \frac{a_i b_j}{k}, \text{ for all } i \text{ and } j$$

Since $a_i > 0$, $b_j > 0$, for all i and j, $x_{ij} \geq 0$. Hence, a feasible solution exists.

4.2.3 Theorem (Basic Feasible Solution)

The number of basic variables in an $m \times n$ transportation table are $(m + n - 1)$.

Proof: Consider an $m \times n$ transportation problem with m-sources and n-destinations. According to Theorem 4.2.2, out of $m + n$ constraint equations, any one of equations is redundant and it can be eliminated, so the remaining $m + n - 1$ form a linearly independent set.

To prove this, first add m row equations (4.2.1) and subtract from the sum the first $n - 1$ column equations (4. 2.2), thereby getting the last column equation. That is,

$$\sum_{i=1}^{m}\sum_{j=1}^{n} x_{ij} - \sum_{j=1}^{n-1}\sum_{i=1}^{m} x_{ij} = \sum_{i=1}^{m} a_i - \sum_{j=1}^{n-1} b_j$$

$$\Rightarrow \quad \sum_{i=1}^{m}\sum_{j=1}^{n} x_{ij} - \left(\sum_{j=1}^{n}\sum_{i=1}^{m} x_{ij} - \sum_{i=1}^{m} x_{in}\right) = \sum_{i=1}^{m} a_i - \left(\sum_{j=1}^{n} b_j - b_n\right)$$

$$\Rightarrow \quad \sum_{i=1}^{m} x_{in} = b_n \qquad \text{since } \sum_{i=1}^{m} a_i = \sum_{j=1}^{n} b_j.$$

Thus, we have proved that out of $m + n$ constraint equations, only $m + n - 1$ equations are linearly independent. Hence, a basic feasible solution will consist of at most $m + n - 1$ positive variables, others being zero. In the degenerate case, some of the basic variables will be zero. By fundamental theorem of linear programming, one of the basic feasible solutions will be the optimal solution.

Remark: (i) When the total capacity equals total requirement, the problem is called *balanced.*

(ii) The allocated cells in the transportation table are called *occupied cells* and empty cells are called *non-occupied cells.*

4.2.4 Definition: Triangular Basis

The transportation problem has a triangular basis if the system of equations $AX = b$, is represented in terms of basic variables only, non-basic variables are set to zero in the system. In other words, the system has the property that there is at least one equation having exactly one basic variable; in a second equation not more than two basic variables occur, in a third equation not more than three basic variables occur, and so on.

4.2.5 Theorem

All the basis for a transportation problem are triangular.

Proof: Let the set of constraints of a transportation problem be

$$\sum_{j=1}^{n} x_{ij} = a_i,\ i = 1, 2, \cdots m \text{ and } \sum_{i=1}^{m} x_{ij} = b_j,\ j = 1, 2, \cdots n.$$

Suppose that these constraints have been displayed in the transportation table having m rows and n columns. We prove,

(i) each row and column of the transportation table contains at least one basic variable, and

(ii) it is not possible that all the rows and columns have two or more basic variables.

(i) follows from the fact that if any row or column does not contain a basic variable, then the corresponding rim requirement cannot be satisfied (as a_i's and b_j's are all non-zero).

We prove (ii) by contradiction. Suppose every equation has at least two basic variables. Then, there will be at least two basic variables in each row, and at least two basic variables in each column. So, the total number of basic variables will be at least $2(m + n)$ since there are $m + n$ equations. Therefore, if N denotes the total number of basic variables, we must have $N \geq 2m$ and $N \geq 2n$. Now, three cases may arise:

Case 1: If $m > n$, then $N \geq 2m \Rightarrow N > m + n$

Case 2: If $m < n$, then $N \geq 2n \Rightarrow N > n + m$

Case 3: If $m = n$, then $N \geq 2m \Rightarrow N = m + n$

We observe that in all cases $N \geq m + n$. But we know $N = m + n - 1$, which is a contradiction.

Thus, assumption of existence of at least two basic variables in each row and each column is not possible. Therefore, combining (i) and (ii) we conclude that there is at least one constraint equation that contains exactly one basic variable.

Let x_{rc} be the only variable in the rth row and c^{th} column. Then $x_{rc} = a_r$. The equation can be eliminated from the system by deleting the rth row and substituting $x_{rc} = a_r$ in the cth column equation. So r^{th} row gets cancelled, and b_c is replaced $b'_c = b_c - a_r$. The resulting system now consists of $m - 1$ row equations and n column equations of which $m + n - 2$ are linearly independent. Thus, there are $m + n - 2$

basic variables in the system. Repeating the process it is concluded that there is an equation in the reduced system which has only one basic variable. If this equation is suppose say c^{th} column equation, it contains two basic variables. Thus, the original system has an equation which has atmost two basic variables. Continuing the process, it can be shown that there is an equation which has atmost three basic variables, and so on.

4.3 METHODS FOR FINDING INITIAL BASIC FEASIBLE SOLUTION

Methods of finding an optimum solution to the transportation problem will consists of two steps:

(i) Find an Initial Basic Feasible Solution (IBFS).

(ii) Find an optimal solution by making successive improvements to IBFS.

The triangularity of all basis in a transportation problem says that there is an IBFS to a transportation problem in such a manner that all the rim requirements are satisfied with exactly $(m + n - 1)$ allocations in independent positions. We explain the different methods to find initial basic feasible solution of a transportation problem. In all the methods it is assumed at the beginning that the transportation table for x_{ij} is blank. That is, initially all $x_{ij} = 0$.

4.3.1 First Method—North-West Corner Rule

Step 1: Starting with the cell at the upper left (north-west) corner of the transportation matrix, we allocate as much as possible so that either the capacity of the first row is exhausted or the destination requirement of the first column is satisfied. That is, $x_{11} = \min(a_1, b_1)$.

Step 2: (i) If $\min\{a_1, b_1\} = a_1$, then put $x_{11} = a_1$, decrease b_1 by a_1 and move down vertically to the second row and make the second allocation of magnitude $x_{21} = \min(a_2, b_1 - x_{11})$ in the cell (2, 1). Cross out the first row.

(ii) If $\min\{a_1, b_1\} = b_1$, then put $x_{11} = b_1$, and decrease a_1 by b_1 and move horizontally to the second column and make the second allocation of magnitude $x_{12} = \min(a_1 - x_{11}, b_2)$ in the cell (1, 2). Cross out the first column.

(iii) If $a_1 = b_1$, then put $x_{11} = a_1 = b_1$, and proceed diagonally. That is, make the second allocation of magnitude

$$x_{12} = \min(a_1 - a_1, b_1) = 0 \text{ in the cell } (1, 2)$$

or,

$$x_{21} = \min(a_2, b_1 - b_1) = 0 \text{ in the cell } (2, 1)$$

cross out the first row and the first column.

Step 3: Repeat steps 1 and 2 moving down towards the lower right corner of the transportation table until all the rim requirements are satisfied.

4.3.2 Second Method—Least Cost or Matrix Minima Method

Step 1: Determine the smallest cost in the cost matrix of the transportation table and allocate

$$x_{ij} = \min(a_i, b_j) \text{ in the cell } (i, j)$$

Step 2: If $x_{ij} = a_i$, cross off the ith row and decrease b_j by a_i. Go to step 3.
If $x_{ij} = b_j$, cross off the jth column and decrease a_i by b_j. Go to step 3.
If $x_{ij} = a_i = b_j$, cross off either the ith row or the jth column, but not both.

Step 3: Repeat steps 1 and 2 for the resulting reduced transportation table until all the rim requirements are satisfied. Whenever, the minimum cost is not unique, make an arbitrary choice among the minima.

4.3.3 Third Method—Vogel's Approximation Method (VAM) or Unit Cost Penalty Method

Step 1: Calculate the penalties by taking the difference between the smallest and next smallest costs in each row (column) and write them in brackets against the corresponding row (column).

Step 2: Identify the largest row difference or column difference. In the case of a tie, choose either of them.

Step 3: Allocate as much as possible in the lowest cost cell of the row (or column) which are identified by step 2.

Step 4: In case the allocation is made fully to a row (or column), cross out the row (or column).

Step 5: Again, compute the column and row penalties for the reduced transportation table and then go to step 2. Continue the procedure until all rows and columns have been crossed out.

4.3.4 Fourth Method—Row Minima Method

Step 1: Allocate as much as possible in the lowest cost cell of the first row until the capacity of the source 1 (first row) is exhausted or the requirement at j^{th} distribution centre is satisfied or both. Three cases arise:

(i) If the capacity of the source 1 is completely exhausted, cross off the first row and proceed to second row.

(ii) If the requirement at *j*th destination is satisfied, cross off the jth column and reconsider the first row with the remaining capacity.

(iii) If the capacity of the source 1 as well as the requirement at j^{th} destination are completely satisfied, make a zero allocation in the second lowest cost cell of the first row. Cross off the row as well as the jth column and move down the second row.

Step 2: Continue the process for the resulting reduced transportation table until the rim conditions are satisfied.

4.3.5 Fifth Method—Column Minima Method

Step 1: Allocate as much as possible in the lowest cost cell of the first column so that the demand of the first distribution centre is satisfied or the capacity of the ith source is exhausted or both. Three cases arise:

(i) If the requirement of the first distribution centre is satisfied, cross out the first column and move right to the second column.

(ii) If the capacity of ith source is satisfied, cross out i^{th} row and reconsider the first column with the remaining requirement.

(iii) If the requirement of the first destination as well as the capacity of the *i*th source are completely satisfied, make a zero allocation in the second lowest cost cell of the first column. Cross out the column as well as the ith row and move right to the second column.

Step 2: Continue the process for the resulting reduced transportation table until all the rim conditions are satisfied.

Example 1

Determine the basic feasible solution to the following transportation problem:

table 4.6

Sources	D_1	D_2	D_3	D_4	Supply
	Distribution Centres				
S_1	2	3	11	7	6
S_2	1	0	6	1	1
S_3	5	8	15	9	10
Requirements	7	5	3	2	

Solution: Since total requirements = 7 + 5 + 3 + 2 = 17 = 6 + 1 + 10 = total supply, the given problem is balanced.

We find the initial basic feasible solution (IBFS) by all the five methods.

(a) North-West Corner Rule

Since $\Sigma a_i = \Sigma b_j = 17$, there exists a feasible solution to the problem. We find IBFS as follows.

Step 1: First, allocation is made in the cell (1, 1), with the magnitude of x_{11} = min (a_1, b_1) = min (6, 7) = 6. Cross out the first row, since this meet the supply position of S_1. No allocations are to be made to cells (1, 2), (1, 3), (1, 4). See Table 4.7.

table 4.7

6 2	3	11	7	
1	0	6	1	1
5	8	15	9	10
1	5	3	2	

Step 2: Proceed to cell (2, 1). The magnitude of the allocation is given by

$$\begin{aligned} x_{21} &= \min(a_2, b_1 - x_{11}) \\ &= \min(1, 7 - 6) \\ &= 1. \end{aligned}$$

Cross out first column and second row, since it meet the supply at S_2 and requirements at D_1. See Table 4.8.

table 4.8

[6] 2	3	11	7	
[1] 1	0	6	1	
5	8	15	9	10
	5	3	2	

Step 3: Proceed to cell (3, 2). The magnitude of third allocation in the cell (3, 2) is given by

$$x_{32} = \min(10, 5) = 5$$

The supply from S_3 is 10 units while the demand for D_2 is 5 units.See Table 4.9. So, set $x_{32} = 5$. Since the demand for D_2 is met, cross off the second column and proceed to the cell (3, 3). Allocate $x_{33} = \min$ (5, 3) = 3. See Table 4.10.

table 4.9

[6] 2	3	11	7	
[1] 1	0	6	1	
5	[5] 8	15	9	5
		3	2	

table 4.10

[6] 2	3	11	7	
[1] 1	0	6	1	
5	[5] 8	[3] 15	9	2
			2	

Step 4: Proceed to cell (3, 4) and allocate $x_{34} = 2$ there. See Table 4.11.

table 4.11

[6] 2	3	11	7
[1] 1	0	6	1
5	[5] 8	[3] 15	[2] 9

It can be seen that the proposed solution is a feasible solution, as all the supply and requirement constraints are fully satisfied. In this method allocations have been made to various cells without considering the cost of the transportation associated with them. Hence, the solution obtained may not be best, but it may be economical.

Note that any cell in which no allocation is made, the corresponding x_{ij} is equal to zero. A cell which gets allocation is called a basic cell. The transportation cost is

$$z = \text{Rs } 2 \times 6 + 1 \times 1 + 8 \times 5 + 15 \times 3 + 9 \times 2$$
$$= \text{Rs } 116.$$

(b) Least cost method

This method consists of maximum possible allocation in the lowest cost cell and then the next allocation is done in the cell with second lowest cost, and so on.

Consider Table 4.6. Here, the lowest cost cell is (2, 2) and maximum possible allocation (meeting supply and requirement positions) is made here.

So, maximum feasible allocation in cell (2, 2) is 1. That is, set $x_{22} = 1$. This meets the supply at S_2 and therefore, we cross off the second row from the Table [see Table 4.12], indicating no further allocations to the cells (2, 1), (2,3), (2, 4).

table 4.12

2	3	11	7	6
1	[1] 0	6	1	0
5	8	15	9	10
7	4	3	2	

table 4.13

[6] 2	3	11	7	
1	[1] 0	6	1	
5	8	15	9	10
	4	3	2	

table 4.14

[6] 2	3	11	7	
1	[1] 0	6	1	
[1] 5	8	15	9	9
	4	3	2	

The next lowest cost cell (excluding the cells in row 2) is (1, 1). Maximum possible allocation for (1, 1) is 6, and row 1 is crossed out. Next lowest cost cell in row 3 is (3, 1) and allocation of (1) is made here.

Likewise, allocations are made in the cell (3, 2), (3, 4) and (3, 3) with $x_{32} = 4$, $x_{34} = 2$, $x_{33} = 3$, respectively.

All the rim requirements have been satisfied and hence an initial basic feasible solution has been determined. The solution is displayed in Table 4.15.

Since the solution does not form a loop, the solution is basic. The transportation cost is

$$z = \text{Rs } [2 \times 6 + 1 \times 0 + 1 \times 5 + 4 \times 8 + 3 \times 15 + 2 \times 9]$$
$$= \text{Rs } 112.$$

table 4.15

[6] 2	3	11	7
1	[1] 0	6	1
[1] 5	[4] 8	[3] 15	[2] 9

(c) Vogel's Approximation Method

The difference between the smallest and next to the smallest costs in each row and each column are first computed and displayed inside the parenthesis against the respective rows and columns. For example, in column 1, the two lowest elements are 1 and 2 and their difference is 1 which is entered as (1) below column 1. Similarly, the two smallest elements in row 2 are 0 and 1, their difference is 1, which is entered as (1) to the right of row 2. See Table 4.16.

table 4.16

2	3	11	7	6(1)
1	0	6	1	1(1)
5	8	15	9	10(3)
7	5	3	2	
(1)	(3)	(5)	(6)	

table 4.17

2	3	11	7	6
1	0	6	1 [1]	
5	8	15	9	10
7	5	3	1	

Select the row or column with the greatest difference and allocate as much as possible within the restrictions of the rim conditions to the lowest cost cell in the row or column selected.

Since (6) is the greatest number in the parenthesis we choose column 4 and allocate as much as possible to the cell (2, 4) as it has lowest cost 1 in column 4. Since supply is 1, and requirement is 2, maximum possible allocation is 1. See Table 4.17. For the assignment made at (2, 4), supply of S_2 is completely satisfied. So, row 2 is crossed out. The row and column differences are now computed for the reduced transportation Table (4.18), the largest of these is (5) which is associated with the second column. So, we allocate 5 units to the cell (1, 2), since it has the smallest transportation cost 3 in column 2. Since requirements of column 2 are completely satisfied, this column is crossed out and the reduced transportation table is Table 4.19.

table 4.18

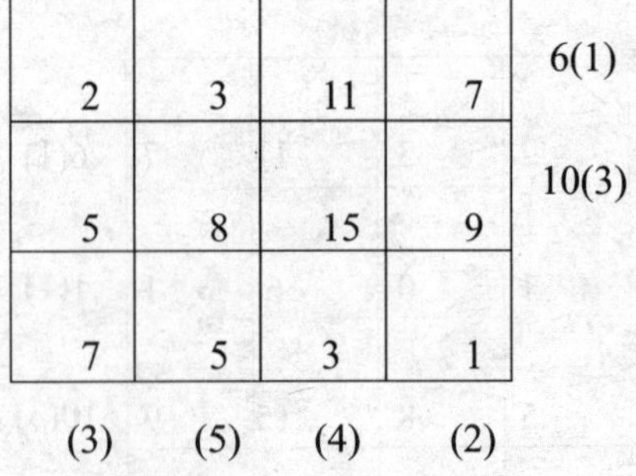

2	3	11	7	6(1)
5	8	15	9	10(3)
7	5	3	1	
(3)	(5)	(4)	(2)	

table 4.19

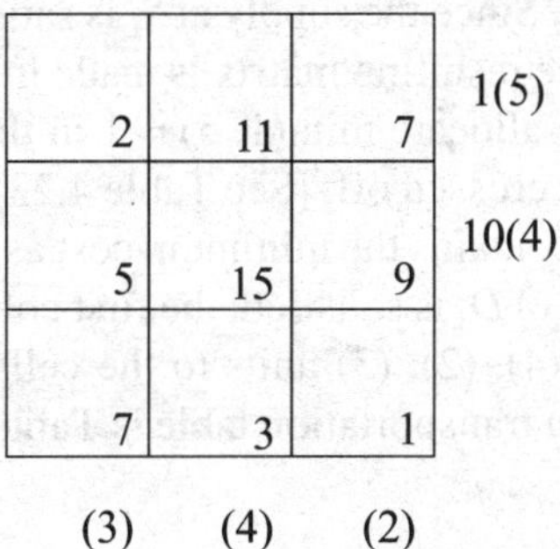

2	11	7	1(5)
5	15	9	10(4)
7	3	1	
(3)	(4)	(2)	

Differences are calculated again. The maximum difference is (5). Therefore we allocate (1) to the cell (1, 1) since it has the lowest cost in row 1. Since requirements of row 1 are fully satisfied, it is crossed out and the reduced table is Table 4.20. As cell (3, 1) has the lowest cost 5, maximum allocation of (6) is made here. Likewise, for (3, 4), the allocation is (1) and for (3, 3) the allocation is (3). Eventually, the basic feasible solution in Table 4.21 is obtained.

table 4.20

(6) 5	(3) 15	(1) 9	10|4|3|0
6/0	3/0	1/0	

table 4.21

(1) 2	(5) 3	11	7
1	0	6	(1) 1
(6) 5	8	(3) 15	(1) 9

The transportation cost is

$$z = \text{Rs } [2 \times 1 + 3 \times 5 + 1 \times 1 + 6 \times 5 + 3 \times 15 + 1 \times 9]$$

$$= \text{Rs } 102.$$

Note that the least of all the values of transportation cost found by three methods is VAM, which is having most economical initial feasible solution.

(d) Row Minima Method

Consider Table 4.6. We first allocate to cell (1, 1) in the first row as it contains the minimum cost 2. We allocate min (6, 7) = 6 in this cell. Since the supply at S_1 is satisfied, the first row is crossed off. See Table 4.22. The next allocation, in the resulting matrix is made in cell (2, 2) of row 2 as it contains the minimum cost 0 in the that row. We allocate min (1, 5) = 1 in this cell. The supply capacity of S_2 is satisfied and hence the second row is crossed off. [See Table 4.23]. The next allocation, in the reduced table is made in the cell (3, 1) which contains the minimum cost as 5. We allocate min (1, 10) = 1 to this cell. Since the requirement condition of D_1 is satisfied, the first column is crossed off. [See Table 4.24]. Proceeding in this way we allocate (4), (2), (3) units to the cells (3, 2), (3, 4), (3, 3) till all the rim conditions are satisfied. The resulting transportation table is Table 4.25.

table 4.22

	D_1	D_2	D_3	D_4	
S_1	[6] 2	3	11	7	0
S_2	1	0	6	1	1
S_3	5	8	15	9	10
	1	5	3	2	

table 4.23

1	[1] 0	6	1	1/0
5	8	15	9	10
1	5/4	3	2	

table 4.24

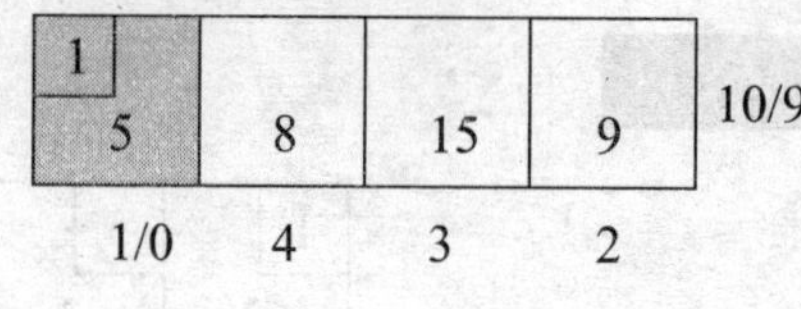

[1] 5	8	15	9	10/9
1/0	4	3	2	

table 4.25

[6] 2	3	11	7
1	[1] 0	6	1
[1] 5	[4] 8	[3] 15	[2] 9

The transportation cost is

$$z = \text{Rs } [2 \times 6 + 1 \times 0 + 1 \times 5 + 4 \times 8 + 3 \times 15 + 2 \times 9]$$
$$= \text{Rs } 112.$$

(e) Column Minima Method

Consider Table 4.6. Allocate first to cell (2, 1) in the first column as it contains the minimum cost 1. We allocate min (1, 7) = 1 in this cell. Since the supply at S_2 is satisfied the second row is crossed off. [See Table 4.26]

table 4.26

2	3	11	7	6
(1) 1	0	6	1	1/0
5	8	15	9	10
7/6	5	3	2	

The next allocation in the resulting Table is made in cell (1, 1) of column 1 as it contains the second lowest cost 2 in that column. We allocate min (6, 6) = 6 in that cell. Since the supply at S_1 well as the requirement at D_1 are satisfied, we allocate zero in cell (3, 1) of first column and cross off first row and first column and move on to second column [See Table 4.27].

table 4.27

(6) 2	3	11	7	6/0
(0) 5	8	15	9	10
7/6/0	5	3	2	

table 4.28

(6) 2	3	11	7
(1) 1	0	6	1
(0) 5	(5) 8	(3) 15	(2) 9

Proceeding in this way we allocate (5), (3), (2) to cells (3, 2), (3, 3) and (3, 4) till all the rim conditions are met with. The resulting Table is shown in Table 4.28. The transportation cost associated with this solution is

$$z = \text{Rs } [2 \times 6 + 1 \times 1 + 5 \times 8 + 3 \times 15 + 2 \times 9]$$
$$= \text{Rs } 116.$$

Note: Suppose the problem is unbalanced, a dummy origin or destination (as the case may be) is created to balance the supply and requirements.

4.4 OPTIMUM SOLUTION OF A TRANSPORTATION PROBLEM

Following the determination of an initial basic feasible solution to a transportation problem, we next obtain the optimum solution. An optimality test can be performed only on the feasible solution in which
(a) The number of allocations is $m + n - 1$, where m is the number of rows and n is the number of columns.
(b) These $(m + n - 1)$ allocations should be in independent positions.

The technique is similar to that of simplex method. That is, determine the net evaluations for the non-basic variables (here, empty cells), then determine the entering variable, then leaving variable, compute a better basic feasible solution, and repeat the steps until an optimum solution has been reached.

The two methods are commonly used. They are,

(i) The Stepping-stone Method
(ii) The Modified Distribution (MODI) Method or *u-v* Method.

First, we determine the net evaluations by making use of the properties of the primal and dual problems.

Let $u_1, u_2, \cdots u_m$ and $v_1, v_2, \cdots v_n$ be the dual variables associated with the origin and destination constraints, respectively.

The net evaluations for a transportation problem can be determined from the following theorem.

Theorem 2

If there is a feasible solution consisting of $m + n - 1$ independent allocations, and if u_i and v_j satisfies $c_{rs} = u_r + v_s$ for each occupied cell (r, s), then the evaluation d_{ij} corresponding to each empty cell (i, j) is given by

$$d_{ij} = c_{ij} - (u_i + v_j)$$

Proof: The transportation problem is to find $x_{ij} \geq 0$ in order to minimise

$$z = \sum_{i=1}^{m} \sum_{j=1}^{n} x_{ij}\, c_{ij} \tag{4.4.1}$$

subject to the restrictions

$$\sum_{j=1}^{n} x_{ij} = a_i, \quad i = 1, \quad 2, \cdots m \tag{4.4.2}$$

and,

$$\sum_{i=1}^{m} x_{ij} = b_j, \quad j = 1, 2, \cdots n \tag{4.4.3}$$

The restrictions (4.4.2) and (4.4.3) has been written as

$$0 = a_i - \sum_{j=1}^{n} x_{ij}, \quad i = 1, 2, \cdots m \tag{4.4.4}$$

and,

$$0 = b_j - \sum_{i=1}^{m} x_{ij}, \quad j = 1, 2, \cdots, n \tag{4.4.5}$$

Any multiple of each of these restrictions (4.4.4) and (4.4.5) can be added to the objective function (4.4.1) to eliminate the basic variable. Now, multiply u_i ($i = 1, 2, \cdots m$) and v_j ($j = 1, 2, \cdots n$), respectively so that

$$z = \sum_{i=1}^{m} \sum_{j=1}^{n} x_{ij} (c_{ij}) + \sum_{i=1}^{m} u_i \left[a_i - \sum_{j=1}^{n} x_{ij} \right] + \sum_{j=1}^{n} v_j \left[b_j - \sum_{i=1}^{m} x_{ij} \right]$$

or,

$$z = \sum_{i=1}^{m} \sum_{j=1}^{n} x_{ij} (c_{ij} - (u_i + v_j)) + \sum_{i=1}^{m} u_i a_i + \sum_{j=1}^{n} v_j b_j$$

For each basic cell (r, s), given $c_{rs} = u_r + v_s$. That is, the net evaluations must vanish for the basic variables. Since there are $m + n - 1$ dual equations in $m + n$ dual unknowns, if we assign arbitrary value to one of these unknowns u_i or v_j, the rest of the $m + n - 1$ variables can be solved uniquely. After an arbitrary assignment, say $u_1 = 0$ the rest of the values are obtained by simple addition and subtraction.

To prove $d_{ij} = c_{ij} - (u_i + v_j)$, consider that the empty cell (i, j) which is connected to occupied cells by a closed loop. See Table 4.29.

table 4.29

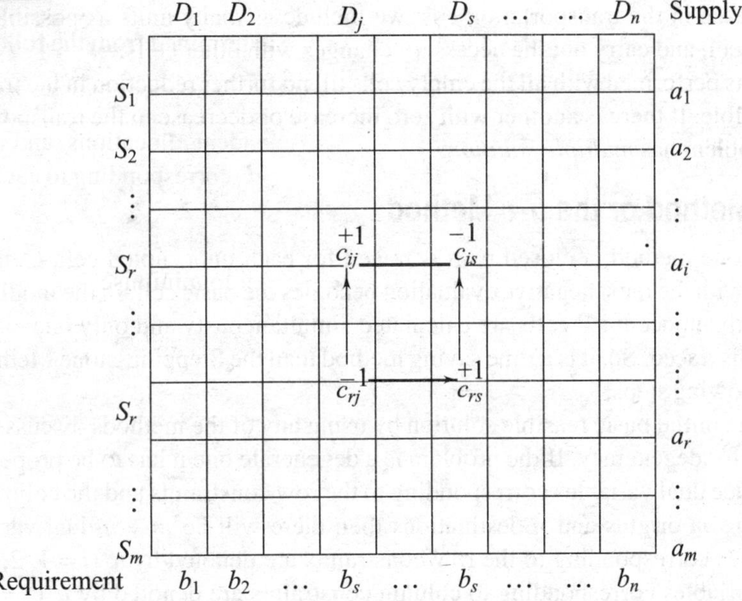

First, allocate + 1 unit to the empty cell (i, j) and in order to balance the total requirement of D_j, add − 1 unit to occupied cell (r, j). Consequently, the total amount available from source S_r will be balanced by adding + 1 unit to occupied cell (r, s), which cause the column D_s to become unbalanced. Balance the column D_s by adding − 1 unit to the occupied cell (i, s).

This process will give the cost difference d_{ij} [called the empty cell evaluation of (i, j)] between the new solution and the original solution. Thus,

$$d_{ij} = c_{ij} - c_{rj} + c_{rs} - c_{is}$$

Since $c_{rs} = u_r + v_s$ for all occupied cells such as (r, j), (r, s) and (i, s)

$$d_{ij} = c_{ij} - (u_r + v_j) + (u_r + v_s) - (u_i + v_s)$$
$$= c_{ij} - (u_i + v_j)$$

Similarly, we generalise the result for all empty cell (i, j) to occupied cells. This completes the theorem.

Note that if the cost difference $d_{ij} \geq 0$, then the basic feasible solution under test must be optimal. If $d_{ij} < 0$ (because negative difference implies decrease in cost) for one or more empty cells, then it is better to reduce the cost more by allocating as much as possible to the cell with the largest negative value of d_{ij}.

4.4.1 The Stepping-stone Method

Step 1: Find the initial basic feasible solution by any of the methods discussed above.

Step 2: Check the number of occupied cells. If there are less than $m + n - 1$, there exists degeneracy and introduce a very small positive assignment of $\varepsilon (= 0)$ in suitable independent positions, so that the number of occupied cells is exactly equal to $m + n - 1$.

Step 3: Each empty (non-allocated) cell is now examined for a possible decrease in the transportation cost. One unit is allocated to an empty cell. A number of adjacent cells are balanced so that the row and the column constraints are not violated. If the net result of these changes is a decrease in the transportation cost we include as many units as possible in the selected empty cell and carry out the necessary changes with other cells.

Step 4: Step 3 is performed with all the empty cells till no further reduction in the transportation cost is possible. If there is another with zero increase or decrease in the transportation cost, then the problem has *multiple solutions*.

4.4.2 MODI Method or the *u*-*v* Method

In the Stepping-stone method, a closed path is traced for each unoccupied cell. Cell evaluations are found and the cell with the most negative evaluation becomes the basic cell. In the modified method, cell evaluations of all the unoccupied cells are calculated simultaneously and only one closed path for the most negative cell is traced. So, it is a time saving method than the Stepping-stone Method. This method consists of the following steps:

Step 1: Find the initial basic feasible solution by using any of the methods discussed above.

Step 2: Check for degeneracy. If the problem is a degenerate one it has to be properly modified.

Step 3: Introduce dual variables corresponding to the row constraints and the column constraints. If there are m origins and n destinations then there will be $m + n$ dual variables. The dual variables corresponding to the row constraints are denoted by u_i $(i = 1, 2, \cdots m)$ while the dual variables corresponding to column constraints are denoted by v_j $(j = 1, 2, \cdots n)$. The values of the dual variables should be determined from the following equation:

$$u_i + v_j = c_{ij} \quad \text{if } x_{ij} > 0$$

starting initially with same $u_i = 0$ or $v_j = 0$ and entering successively the values of u_i and v_j in the transportation table margins.

Step 4: Compute the cell evaluations $d_{ij} = c_{ij} - (u_i + v_j)$ for each unoccupied cell (i, j) and enter them in the upper right corners of the corresponding cell (i, j).

Step 5: Examine the sign of each d_{ij}. If all $d_{ij} > 0$, then the solution under the test is optimal and unique. If all $d_{ij} > 0$ with at least one $d_{ij} = 0$, then the solution under the test is optimal and an alternative optimal solution exists. If at least one $d_{ij} < 0$, then the solution is not optimal. Go to the next step.

Step 6: Form a new basis feasible solution by giving maximum allocation to the cell for which d_{ij} is most negative by making an occupied cell empty. For that draw a closed path consisting of horizontal and vertical lines beginning and ending at the cell for which d_{ij} is most negative and having its other corners at some allocated cells. Along this closed loop indicate $+\theta$ and $-\theta$ alternatively at the corners. Choose minimum of θ from the cells having $-\theta$. Add this minimum of θ to the cells with $+\theta$ and subtract this minimum of θ from the allocation to the cells with $-\theta$.

Step 7: Repeat steps (2) to (6) to test the optimality of this new basic feasible solution.

Step 8: Continue the above procedure till an optimum solution is obtained.

Example 3

Using stepping-stone method find the optimum solution of the transportation problem.

table 4.30

	Distribution Centres				
Sources	D_1	D_2	D_3	D_4	Supply
S_1	2	3	11	7	6
S_2	1	0	6	1	1
S_3	5	8	15	9	10
Requirements	7	5	3	2	

Solution: We compute an initial basic feasible solution of the problem by Vogel's approximation method [See Example 4.3.6] which is given in Table 4.31.

table 4.31

[1] 2	[5] 3	11	7
1	0	6	[1] 1
[6] 5	8	[3] 15	[1] 9

The solution is non-degenerate since the number of non-zero basic variables is $m + n - 1 = 3 + 4 - 1 = 6$.

Let us start with any arbitrary empty cell (a cell without allocation), say (2, 3) and allocate + 1 to this cell. In order to keep up the row 3 restriction, + 1 must be allocated to cell (3, 4) and allocate – 1 to the cell (3, 3), – 1 to the cell (2, 4). This can be shown in Table 4.32.

table 4.32

1 2	5 3	11	7
1	0	1 6	0 1
6 5	8	2 15	2 9

The increase in the transportation cost per unit quantity of reallocation is 6 + 9 – 1 – 15 = – 1.

The total number of empty cells will be $mn - (m + n - 1) = (m - 1)(n - 1)$ such number of cell evaluations must be calculated. In this problem (3 – 1) (4 – 1) = 6 cell evaluations must be calculated. Table 4.33 shows the transportation table after reallocation for the cell (2, 1):

table 4.33

1 2	5 3	11	7
+1 1	0	0 6	1
5 5	8	3 15	2 9

The increase in the transportation cost after reallocation is 1 × 1 – 6 × 1 – 5 × 1 + 1 × 15 = 5. This cannot decrease the transportation cost. So, we choose the empty cell (2, 2), the transportation table after reallocation is

table 4.34

2 2	4 3	11	7
1	1 0	6	1
5 5	8	3 15	2 9

The increase in the transportation cost per unit of reallocation is $1 \times 0 + 1 \times 2 - 1 \times 1 - 3 \times 1 = -2$. This procedure was repeated with the remaining empty cells (1, 3), (1, 4), (3, 2). The reallocation tables are as follows:

table 4.35

2	[5] 3	[1] 11	7
1	0	6	[1] 1
[7] 5	8	[2] 15	[1] 9

table 4.36

2	[5] 3	11	[1] 7
[1] 1	0	6	1
[6] 5	8	[3] 15	[1] 9

table 4.37

[2] 2	[4] 3	11	7
1	0	6	[1] 1
[5] 5	[1] 8	[3] 15	[1] 9

The increase in cost per unit of reallocation for the cells (1, 3), (1, 4), (3, 2) are $1 \times 11 + 1 \times 5 - 1 \times 15 - 1 \times 2 = -1$, $1 \times 7 + 1 \times 5 - 1 \times 9 - 1 \times 2 = 1$, and $1 \times 8 + 1 \times 2 - 1 \times 3 - 1 \times 5 = 2$, respectively. From all these reallocation to unoccupied cells, we can see for the cell (2, 2), the transportation cost per unit of reallocation is – 2. Thus, the new route would be beneficial to the company. Hence, the minimum transportation cost is

$$z = 2 \times 2 + 4 \times 3 + 1 \times 0 + 5 \times 5 + 3 \times 15 + 2 \times 9$$

$$= \text{Rs } 104.$$

Example 4

Solve the transportation problem

table 4.38

	D_1	D_2	D_3	D_4	Supply
S_1	21	16	25	13	11
S_2	17	18	14	23	13
S_3	32	27	18	41	19
Demand	6	10	12	15	

Solution: Since $\Sigma a_i = \Sigma b_j = 43$, the given transportation problem is balanced. Hence, there exists a basic feasible solution to this problem.

By VAM, the initial basic feasible solution, with all allocations made during the procedure in a single Table is

table 4.39

S_1	21	16	25	11 13	11 (3)
S_2	6 17	3 18	14	4 23	13 (3) (3) (3) (4)
S_3	32	7 27	12 18	41	19 (9) (9) (9) (9)
	6	10	12	15	
	(4)	(2)	(4)	(10)	
	(15)	(9)	(4)	(18)	
	(15)	(9)	(4)	—	
	—	(9)	(4)		

Initial transportation cost is Rs 796.

To find optimum solution by MODI method

Since the number of basic cells = 6 = 3 + 4 – 1, the solution is non-degenerate.

We calculate u_i and v_j using $u_i + v_j = c_{ij}$ for basic cells. Arbitrarily, start with some u_i or v_j as zero. Here, start with $u_2 = 0$ as the second row has maximum number of allocation.

$$C_{21} = u_2 + v_1 \Rightarrow 17 = 0 + v_1 \Rightarrow v_1 = 17$$
$$C_{22} = u_2 + v_2 \Rightarrow 18 = 0 + v_2 \Rightarrow v_2 = 18$$
$$C_{24} = u_2 + v_4 \Rightarrow 23 = 0 + v_4 \Rightarrow v_4 = 23$$
$$C_{14} = u_1 + v_4 \Rightarrow 13 = u_1 + 23 \Rightarrow u_1 = -10$$
$$C_{32} = u_3 + v_2 \Rightarrow 27 = u_3 + 18 \Rightarrow u_3 = 9$$
$$C_{33} = u_3 + v_3 \Rightarrow 18 = 9 + v_3 \Rightarrow v_3 = 9$$

We find the net evaluations $d_{ij} = c_{ij} - (u_i + v_j)$ for each non-basic cells and enter at the upper right corner of that cell. Complete Table is given as:

table 4.40

(14) 21	(8) 16	(26) 25	11 13	$u_1 = -10$
6 17	3 18	(5) 14	4 23	$u_2 = 0$
(6) 32	7 27	12 18	(9) 41	$u_3 = 9$
$v_1 = 17$	$v_2 = 18$	$v_3 = 9$	$v_4 = 23$	

Since all $d_{ij} > 0$, the solution under the test is optimal and unique.
Therefore, the optimum allocation schedule is given by
$x_{14} = 11, x_{21} = 6, x_{22} = 3, x_{24} = 4, x_{32} = 7, x_{33} = 12$, and minimum transportation cost = Rs 796.

Example 5

The following Table shows all the necessary information on the available supply to each warehouse, the requirement of each market and the unit transportation cost from each warehouse to each market.

table 4.41

Warehouse	Market I	II	III	IV	Supply
A	5	2	4	3	22
B	4	8	1	6	15
C	4	6	7	5	8
Requirement	7	12	17	9	

The shipping clerk has worked out the following schedule from experience: 12 units from *A* to II, 1 unit from *A* to III, 9 units from *A* to IV, 15 units from *B* to III, 7 units from *C* to I, and 1 unit from C to III.

(a) Check that the clerk has the optimal schedule.
(b) Find the optimal schedule and minimum total shipping cost.
(c) If the clerk is approached by a carrier of route *C* to II who offers to reduce his rate in the hope of getting some business, by how much must the rate be reduced before the clerk should give him an order.

Solution: The basic feasible solution is given as follows.

table 4.42

	I	II	III	IV	
A	5	(12) 2	(1) 4	(9) 3	22
B	4	8	(15) 1	6	15
C	(7) 4	6	(1) 7	5	8
	7	12	17	9	

(a) We calculate optimal schedule by MODI's method. Starting iteration table is:

table 4.43

(4) 5	12 2	1 4	9 3	$u_1 = 0$
(6) 4	(9) 8	15 1	(6) 6	$u_2 = -3$
7 4	(1) 6	1 7	(–1) 5	$u_3 = 3$
$v_1 = 1$	$v_2 = 2$	$v_3 = 4$	$v_4 = 3$	

Since $d_{ij} < 0$ for the cell (3, 4), the clerk does not have the optimal schedule.
Trace the path and assign $+\theta$, $-\theta$ alternatively to each occupied cell at the corners of the path.
Here, $\theta = \min(9, 1) = 1$. Table 4.45 gives the result of second iteration.

table 4.44

5	12 2	1 $+\theta$ 4	9 $-\theta$ 3	0
4	8	15 1	6	–3
7 4	6	1 $-\theta$ 7	(–1) $+\theta$ 5	3
1	2	3	4	

table 4.45

(3) 5	12 2	2 4	8 3	$u_1 = 0$
(5) 4	(9) 8	15 1	(6) 6	$u_2 = -3$
7 4	(2) 6	(1) 7	1 5	$u_3 = 2$
$v_1 = 2$	$v_2 = 2$	$v_3 = 4$	$v_4 = 3$	

Since all $d_{ij} > 0$, the current solution is optimal. Hence, the optimal schedule is

$$x_{12} = 12,\ x_{13} = 2,\ x_{14} = 8,\ x_{23} = 15,\ x_{31} = 7,\ x_{34} = 1.$$

Minimum total shipping cost is

$$z = 12 \times 2 + 2 \times 4 + 8 \times 3 + 15 \times 1 + 7 \times 4 + 1 \times 5 = \text{Rs } 104.$$

(c) If the clerk decides to transport at the most 8 units from *C* to II (instead of 7 to I, and 1 to IV), then II may reduce his cost from $C_{32} = 6$ to at least 4 in order to have the improved cost. So, according to the given proposal, the total minimum transport cost will become Rs 103.

Example 6

Given the following data

table 4.46

		Destinations 1	2	3	Capacity
Sources	1	2	2	3	10
	2	4	1	2	15
	3	1	3	X	40
	Demand	20	15	30	

The cost of shipment from third source to the third destination is not known. How many units should be transported from the sources to the destinations so that the total cost of transporting all the units to their destinations is a minimum.

Solution: Since the cost c_{33} is unknown, assign a large cost, say M to this cell. Now, using VAM the initial basic feasible solution is obtained in Table 4.47.

By MODI's method, we obtain Table 4.48

table 4.47

2	2	(10) 3
4	1	(15) 2
(20) 1	(15) 3	(5) M

table 4.48

$2-(4-M)$ +ve; 2	$2-(6-M)$ +ve; 2	(10) 3	$u_1 = 3 - M$
$4-(3-M)$ +ve; 4	$1-(5-M)$ +ve; 1	(15) 2	$u_2 = 2 - M$
(20) 1	(15) 3	(5) M	$u_3 = 0$
$v_1 = 1$	$v_2 = 3$	$v_3 = M$	

Since all $d_{ij} > 0$, the current solution is optimum. But, it is to be noticed that cell (3, 3) also appears in the solution for which the cost of shipment is not known. Hence, there exists a pseudo optimum basic feasible solution:

$$x_{13} = 10, x_{23} = 15, x_{31} = 20, x_{32} = 15, x_{33} = 5.$$

4.5 DEGENERACY IN TRANSPORTATION PROBLEM

We have seen that for an $m \times n$ transportation table, the number of basic (or occupied) cells must be $m + n - 1$. The basic solution will degenerate whenever number of basic cells is less than $m + n - 1$. Degeneracy occur in two ways,

1. Degeneracy in the initial solution.
2. Degeneracy at any intermediate stage.

4.5.1 Resolution of Degeneracy in the Initial Stage

To resolve degeneracy, allocate a very small quantity ε (close to zero) to one or more of the unoccupied cells so that a number of occupied cells becomes $m + n - 1$. In a minimisation transportation problem it is better to allocate ε to unoccupied cells that have lowest transportation costs. The quantity ε is considered to be so small that if it is transferred to an occupied cell it does not change the quantity of allocation. That is, $x_{ij} + \varepsilon = x_{ij} - \varepsilon = x_{ij}$, but $\varepsilon - \varepsilon = 0$. Also, ε does not affect the total transportation cost of the allocation. The quantity ε is used to evaluate unoccupied cells and if we get the optimum solution, it will be removed.

4.5.2 Resolution of Degeneracy during Solution Stages

To resolve degeneracy which occur during optimality test, allocate ε to one or more of recently vacated cells so that the number of occupied cells is $m + n - 1$ in the new solution. It may be removed, once we obtain the optimal solution.

Example 1 Degeneracy Initial Stage

Find the non-degenerate basic feasible solution for the following transportation problem using (i) North west corner rule (ii) Vogel's Approximation Method.

table 4.49

		Destinations				
		1	2	3	4	Supply
	1	10	20	5	7	10
	2	13	9	12	8	20
Sources	3	4	5	7	9	30
	4	14	7	1	0	40
	5	3	12	5	19	50
Demand		60	60	20	10	

Solution: Since $\Sigma a_i = \Sigma b_j = 150$, the given transportation problem is balanced. Hence, there exists a basic feasible solution to this problem.

(i) *North-West Corner Rule*

table 4.50

(10) 10	20	5	7
(20) 13	9	12	8
(30) 4	5	7	9
14	(40) 7	1	0
3	(20) 12	(20) 5	(10) 19

Since the number of non-negative allocations is 7 which is less than $(m + n - 1) = 5 + 4 - 1 = 8$, the solution is degenerate.

Therefore, there is a need for making one infinitesimal allocation. Out of the unoccupied cells, cell (4, 4) has the least cost of Rs 0. The infinitesimal allocation should be made in this cell. However, allocating ε to this cell forms a closed loop among the cells (4, 2), (5, 2), (5, 4) and (4, 4), such that allocations in these cells do no remain in independent positions. Therefore, no allocation is made in cell (4, 4). Next higher cost cell is (4, 3) with cost 1. Allocation in this cell also form a loop. So, we choose next highest cost, namely, 3 in the cell (5, 1). Let us make the allocation in cell (5, 1).

table 4.51

(10) 10	20	5	7
(20) 13	9	12	8
(30) 4	5	7	9
14	(40) 7	1	0
(8) 3	(20) 12	(20) 5	(10) 19

The initial transportation cost

$= \text{Rs } [10 \times 10 + 20 \times 13 + 30 \times 4 + 40 \times 7 + \varepsilon \times 3 + 20 \times 12 + 20 \times 5 + 10 \times 19]$

$= \text{Rs } (1290 + 3\varepsilon)$

$= \text{Rs } 1290 \text{ as } \varepsilon \to 0.$

(ii) *Vogel's Approximation Method*: The starting solution in the method is given as:

table 4.52

(10) 10	20	5	7
13	(20) 9	12	8
4	(30) 5	7	9
14	(10) 7	(20) 1	(10) 0
(50) 3	12	5	19

Since the number of non-negative allocations is 7 which is less than $m + n - 1 = 8$, this basic solution is a degenerate one. To resolve this degeneracy, allocate a very small quantity ε to the unoccupied cell (3, 1) so that the number of occupied cells becomes $(m + n - 1)$.

Hence, the non-degenerate basic feasible solution is as shown in the following Table:

table 4.53

(10) 10	20	5	7
13	(20) 9	12	8
(ε) 4	(30) 5	7	9
14	(10) 7	(20) 1	(10) 0
(50) 3	12	5	19

The initial transportation cost

$= \text{Rs } [10 \times 10 + 20 \times 9 + 30 \times 5 + 10 \times 7 + 20 \times 1 + 10 \times 0 + 50 \times 3 + 4\varepsilon]$

$= \text{Rs } 670 + 4\varepsilon$

$= \text{Rs } 670 \quad \text{as } \varepsilon \to 0.$

Note: It is not possible to add ε to any unoccupied cells. ε must be added to those empty cells which makes it possible to determine a unique set of u_i and v_j. Applying MODI's method to find the optimal solution, the dual variables u_i and v_j are obtained from the c_{ij} values of the basic variables. If we do not choose the empty cell for the allocation of ε properly, it would not be possible to locate one or more c_{ij} value which should be equated to the corresponding $u_i + v_j$.

Example 2 Degeneracy at Intermediate Stage

Find the basic feasible solution of the following transportation problem by north-west corner rule. Also find the optimal transportation plan.

table 4.54

	1	2	3	4	5	Available
A	4	3	1	2	6	80
B	5	2	3	4	5	60
C	3	5	6	3	2	40
D	2	4	4	5	3	20
Demand	60	60	30	40	10	

Solution: Since $\Sigma a_i = \Sigma b_j = 200$, the problem is balanced. By following the north west corner rule (explained in Example 4.3.6), we get the non-degenerate initial basic feasible solution shown in Table 4.55.

table 4.55

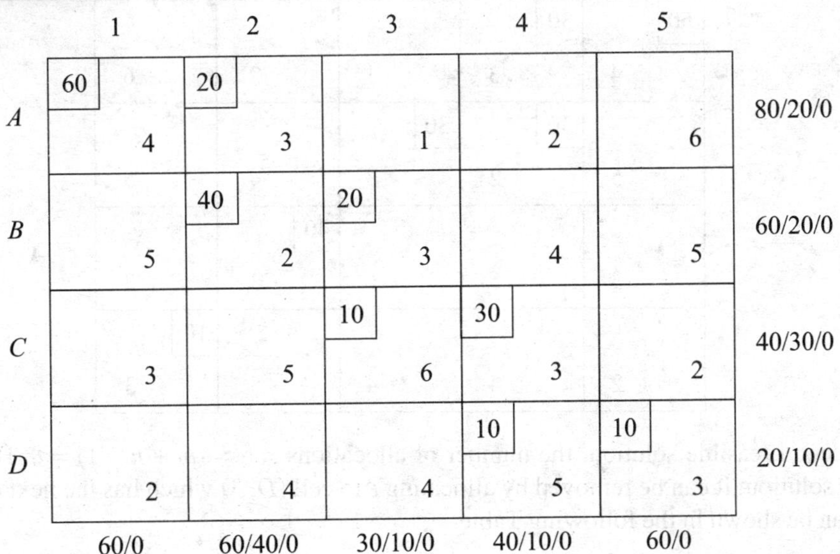

	1	2	3	4	5	
A	(60) 4	(20) 3	1	2	6	80/20/0
B	5	(40) 2	(20) 3	4	5	60/20/0
C	3	5	(10) 6	(30) 3	2	40/30/0
D	2	4	4	(10) 5	(10) 3	20/10/0
	60/0	60/40/0	30/10/0	40/10/0	60/0	

Required number of allocations = $m + n - 1 = 4 + 5 - 1 = 8$. Actual number of allocations = 8. Hence, the problem is non-degenerate.

To obtain optimal solution, apply MODI's method. The results of the iterations are displayed.

First iteration:

table 4.56

60 $-\theta$ 4	20 $+\theta$ 3	(–3) 1	(1) 2	(7) 6	$u_1 = 4$
(2) 5	40 $-\theta$ 2	20 $+\theta$ 3	(4) 4	(7) 5	$u_2 = 3$
(–3) 3	(0) 5	10 $-\theta$ 6	30 $+\theta$ 3	(1) 2	$u_3 = 6$
(–6) $-\theta$ 2	(–3) 4	(–4) 4	10 $-\theta$ 5	10 3	$u_4 = 8$
$v_1 = 0$	$v_2 = -1$	$v_3 = 0$	$v_4 = -3$	$v_5 = -5$	

Second iteration: $\theta = \min(40, 10, 10) = 10$

table 4.57

50 4	30 3	1	2	6
5	30 2	30 3	4	5
3	5	6	40 3	2
2	4	4	5	10 3

In the second feasible solution, the number of allocations = $6 \neq (m + n - 1) = 8$. Hence, this is a degenerate solution. It can be removed by allocating ε to cell $(D, 3)$ which has the next cell evaluation – 4. This can be shown in the following Table.

table 4.58

50 4	30 3	1	2	6
5	30 2	30 3	4	5
3	5	6	40 3	2
10 2	4	ε 4	5	10 3

The final iteration is

table 4.59

10 4	3	30 1	40 2	6
5	60 2	3	4	5
30 3	5	6	3	10 2
20 2	4	4	5	3

4.6 UNBALANCED TRANSPORTATION PROBLEMS

We solved the various transportation problems with the assumption that the total supply at the origins is equal to the total requirement at the destinations. If they are unequal the problem is known as an unbalanced transportation problem. If the total supply is more than the total demand we introduce an additional column with zero cost. The excess supply is entered as a rim requirement for the dummy destination. Likewise, if the total demand is more than the total supply an additional row is introduced with zero cost, the excess demand is entered as a rim requirement for this dummy source. Hence, the unbalanced transportation problem is converted into a balanced transportation problem.

Example 1

A company has three plants at locations A, B, C which supply to warehouses located at D, E, F, G, and H. Monthly plant capacities are 800, 500 and 900 units, respectively. Monthly warehouse requirements are 400, 400, 500, 400 and 800 units, respectively. Unit transportation costs (in Rs) are given below. Determine an optimum distribution for the company in order to minimise the total transportation cost.

table 4.60

		To				
		D	E	F	G	H
From	A	5	8	6	6	3
	B	4	7	7	6	5
	C	8	4	6	6	4

Solution: In this problem, $\Sigma a_i = 2200$ units, whereas $\Sigma b_j = 2500$. So, the problem is an unbalanced one.

table 4.61

	D	E	F	G	H	Plant capacity
A	5	8	6	6	3	800
B	4	7	7	6	5	500
C	8	4	6	6	4	900
P	0	0	0	0	0	300
Requirements	400	400	500	400	800	

So, introduce a dummy plant P having all transportation costs equal to zero and having the plant availability equal to (2500 – 2200) = 300 units. The modified table is shown in Table 4.62.

Using VAM the following IBFS is obtained.

table 4.62

	D	E	F	G	H	
A	5	8	[500] 6	6	[300] 3	800
B	[400] 4	7	7	[100] 6	[ε] 5	$500 + \varepsilon = 500$
C	8	[400] 4	6	6	[500] 4	900
P	0	0	0	[300] 0	0	300
	400	400	500	400	$800 + \varepsilon = 800$	

Since the number of occupied cells = 7 which is less than $(m + n - 1) = 8$, the solution is degenerate.

To make the number of allocations equal to 8 introduce a infinitesimal quantity ε in the independent cell (2, 5). Now, test the current solution for optimality, by MODI's method.

Starting iteration:

table 4.63

(3) 5	(5) 8	500 $-\theta$ 6	(2) 6	300 $+\theta$ 3	$u_1 = -2$
400 4	(2) 7	(–1) 7	100 $+\theta$ 6	ε $-\theta$ 5	$u_2 = 0$
(5) 8	400 4	(–1) 6	(1) 6	500 4	$u_3 = -1$
(2) 0	(1) 0	(–2) $+\theta$ 0	300 $-\theta$ 0	(1) 0	$u_4 = -6$
$v_1 = 4$	$v_2 = 5$	$v_3 = 8$	$v_4 = 6$	$v_5 = 5$	

Here, $\theta = \min(500, \varepsilon, 300) = \varepsilon$. So, enter the non-basic cell (4, *c*) and leave the basic cell (2, 5).

First iteration:

table 4.64

(1) 5	(5) 8	500–ε $-\theta$ 6	(0) 6	300+ε $+\theta$ 3	$u_1 = 0$
400 4	(4) 7	(1) 7	100+ε 6	(2) 5	$u_2 = 0$
(3) 8	400 4	(–1) 6	(–1) $+\theta$ 6	500 $-\theta$ 4	$u_3 = 1$
(2) 0	(3) 0	ε $+\theta$ 0	300–ε $-\theta$ 0	(3) 0	$u_4 = -6$
$v_1 = 4$	$v_2 = 3$	$v_3 = 6$	$v_4 = 6$	$v_5 = 3$	

$\theta = \min(500 - \varepsilon, 500 - \varepsilon, 300 - \varepsilon) = 300 - \varepsilon.$

Second iteration:

table 4.65

					u_i
(2) 5	(5) 8	200 $-\theta$ 6	(1) 6	600+2ε $+\theta$ 3	$u_1 = -1$
400 4	(3) 7	(0) 7	100+ε 6	(1) 5	$u_2 = 0$
(4) 8	400 4	(−1) $+\theta$ 6	300−ε 6	200−ε $-\theta$ 4	$u_3 = 0$
(3) 0	(3) 0	300 0	(1) 0	(3) 0	$u_4 = -7$
$v_1 = 4$	$v_2 = 4$	$v_3 = 7$	$v_4 = 6$	$v_5 = 4$	

Here, $\theta = \min(200, 200 - \varepsilon) = 200 - \varepsilon$
The third iteration is:

table 4.66

					u_i
(1) 5	(4) 8	ε 6	(0) 6	800+ε 3	$u_1 = 0$
400 4	(3) 7	(1) 7	100+ε 6	(2) 5	$u_2 = 0$
(4) 8	400 4	200−ε 6	300−ε 6	(1) 4	$u_3 = 0$
(2) 0	(2) 0	300 0	(0) 0	(3) 0	$u_4 = -6$
$v_1 = 4$	$v_2 = 4$	$v_3 = 6$	$v_4 = 6$	$v_5 = 3$	

Since all the net evaluations are non-negative, the optimum solution is

$x_{13} = 0$, $x_{15} = 800$, $x_{21} = 400$, $x_{24} = 100$, $x_{32} = 400$, $x_{33} = 200$, $x_{34} = 300$, $x_{43} = 300$.

The optimum transportation cost is given by

$$z = \text{Rs } 0 \times 6 + 800 \times 3 + 400 \times 4 + 100 \times 6 + 400 \times 4$$
$$+ 200 \times 6 + 300 \times 6 + 300 \times 0 = \text{Rs } 9200.$$

Example 2

Three warehouses supply five stores. The Table indicates the cost of shipment per unit between warehouses and stores, warehouse capacities and requirements of the stores. However, a major bridge has been damaged preventing deliveries from warehouse *A* to store 5, from warehouse *B* to store 2, and from warehouse *C* to store 4. Within these limitations determine the optimum delivery scheme to minimise cost.

table 4.67

		Warehouse A	Warehouse B	Warehouse C	Requirement
Store	1	2	4	6	75
	2	3	8	7	345
	3	4	3	8	180
	4	4	6	3	90
	5	2	6	5	210
Capacity		850	300	450	

Solution: Here, total supply = 1600 units, total requirement = 900 units

Since demand is less than the supply, a dummy warehouse *D* is created to absorb the additional supply of 700 units. The associated cost element will be zero. Further, to avoid allocation in cells (*A*, 5), (*B*, 2), (*C*, 4), a very heavy penalty, + *M* is allocated to these cells. The modified transportation matrix is shown in Table 4.68.

table 4.68

		1	2	3	4	5	6	
Warehouse	A	2	3	4	4	*M*	0	850
	B	4	*M*	3	6	6	0	300
	C	6	7	8	*M*	5	0	450
requirement		75	345	180	90	210	700	

Find the initial basic feasible solution by VAM: This is shown in Table 4.69.

table 4.69

	1	2	3	4	5	6	Supply
A	2	[345] 3	[90] 4	[90] 4	M	[250] 0	850/345/250/75 (2) (2) (2) (2) (0) (0)
B	4	[90] M	3	6	6	[210] 0	300/210/0 (3) (1) (1) (1) (1) (3) (6) 9 (0) (0) (4)
C	6	7	8	M	5	[450] 0	450/0 (5)
Demand	75	345	180/90	90/0	210/0	700/450/0	
	(2)	(4)	(1)	(2)	(1)	(0)	
	(2)	(M)	(1)	(2)	(M)	(0)	
	(2)		(1)	(2)	(M)	(0)	
	(2)		(1)	(2)		(0)	
	(2)		(1)	(2)			
			(1)	(2)			
			(4)	(2)			

Initial transportation cost is

$$z = \text{Rs } [75 \times 2 + 345 \times 3 + 90 \times 4 + 90 \times 4 + 250 \times 0 + 90 \times 3 + 210 \times 6 + 450 \times 0]$$
$$= \text{Rs } [150 + 1035 + 360 + 360 + 270 + 1260]$$
$$= \text{Rs } 3435$$

To find optimal solution, we use MODI's method.

Starting iteration:

table 4.70

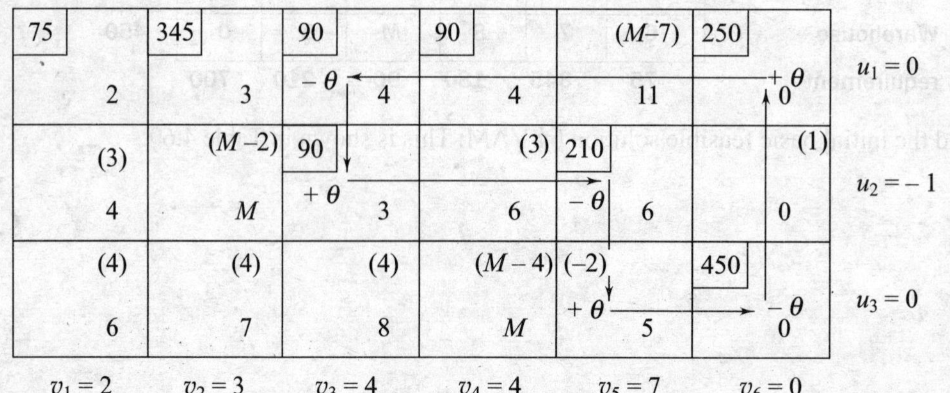

Here, $\theta = \min(450, 90, 210) = 90$.

First iteration:

table 4.71

75 2	345 3	4	90 4	(M–5) M	340 0	$u_1 = 0$
(1) 4	(M–4) M	180 3	(1) 6	120 $-\theta$ 6	(1) $+\theta$ 0	$u_2 = -1$
(4) 6	(4) 7	(6) 8	(M–4) M	90 $+\theta$ 5	360 $-\theta$ 0	$u_3 = 0$
$v_1 = 2$	$v_2 = 3$	$v_3 = 4$	$v_4 = 4$	$v_5 = 7$	$v_6 = 0$	

$\theta = \min(360, 120) = 120$

Second iteration:

table 4.72

75 2	345 3	4	90 4	(M–5) +ve M	340 0	$u_1 = 0$
(2) 4	(M–3) +ve M	180 3	(2) 6	(1) 6	120 0	$u_2 = 0$
(4) 6	(4) 7	(5) 8	(M–4) –ve M	210 5	240 0	$u_3 = 0$
$v_1 = 2$	$v_2 = 3$	$v_3 = 3$	$v_4 = 4$	$v_5 = 5$	$v_6 = 0$	

Since $d_{ij} > 0$ for all non-basic cells, the current solution is optimum. Hence,
$x_{11} = 75$, $x_{12} = 345$, $x_{14} = 90$, $x_{16} = 340$, $x_{23} = 180$, $x_{26} = 120$, $x_{35} = 210$, $x_{36} = 240$.
The minimum transportation cost is

$$z = \text{Rs } [75 \times 2 + 345 \times 3 + 90 \times 4 + 340 \times 0 + 180 \times 3 + 120 \times 0 + 210 \times 5 + 240 \times 0]$$
$$= \text{Rs } [150 + 1035 + 360 + 540 + 1050]$$
$$= \text{Rs } 3135.$$

Example 3

The unbalanced transportation problem is given in Table 4.73.

table 4.73

		To			
		1	2	3	Supply
From	1	5	1	7	10
	2	6	4	6	80
	3	3	2	5	15
Demand		75	20	50	

Since there is not enough supply, some of the demand at the destinations may not be satisfied. Suppose there are penalty costs for every unsatisfied demand unit which are given by 5, 3, 2 for destinations 1, 2, 3, respectively. Find the optimal solution.

Solution: In this problem, the demand exceeds the supply. The problem is unbalanced. Introduce a 'dummy source' whose transportation costs are given as 5, 3, 2, respectively, and supply is given as (145 – 105) = 40. The modified transportation table is given by Table 4.74.

table 4.74

		To			Supply
From	1	5	1	7	10
	2	6	4	6	80
	3	3	2	5	15
	4	5	3	2	40
Demand		75	20	50	

table 4.75

5	(10) 1	7	10
(60) 6	(10) 4	(10) 6	80
(15) 3	2	5	15
5	3	(40) 2	40
75	20	50	

Using VAM an initial basic feasible solution having the transportation cost Rs 595 is obtained in Table 4.100. The optimum table is given by Table 4.101. The optimum transportation cost is

$$z = 10 \times 1 + 60 \times 6 + 10 \times 4 + 10 \times 6 + 15 \times 3 + 40 \times 2$$

$$= \text{Rs } 595.$$

table 4.76

(2) 5	10 1	(4) 7	$u_1 = -3$
60 6	10 4	10 6	$u_2 = 0$
15 3	(1) 2	(2) 5	$u_3 = -3$
(3) 5	(3) 3	40 2	$u_4 = -4$
$v_1 = 6$	$v_2 = 4$	$v_3 = 6$	

Example 4

A product is produced by four factories F_1, F_2, F_3 and F_4. Their unit production cost are Rs 2, 3, 1 and 5, respectively. Production capacity of the factories are 50, 70, 30 and 50 units, respectively. The product is supplied to four stores S_1, S_2, S_3 and S_4 the requirements of which are 25, 35, 105 and 20, respectively. Unit cost of transportation are given in Table 4.102. Find the transportation plan such that the total production and transportation cost is minimum.

table 4.77

	S_1	S_2	S_3	S_4
F_1	2	4	6	11
F_2	10	8	7	5
F_3	13	3	9	12
F_4	4	6	8	3

table 4.78

	S_1	S_2	S_3	S_4	Capacity
F_1	4	6	8	13	50
F_2	13	11	10	8	70
F_3	14	4	10	13	30
F_4	9	11	13	8	50
Demand	25	35	105	20	

Solution: Form the transportation table which consists of both production and transportation costs (see Table 4.78). The total capacity = 200 and the total demand = 185. The problem is an unbalanced one which is converted into a balanced one by adding a dummy store S_5 with cost 0 and supply in excess is given to this store.

table 4.79

25 / 4	5 / 6	20 / 8	13	0	50
13	11	50 / 10	20 / 8	0	70
14	30 / 4	10	13	0	30
9	11	35 / 13	8	15 / 0	50
25	35	105	20	15	

The initial basic feasible solution is obtained by least cost method. The solution is non-degenerate. The total transportation cost is

$$z = \text{Rs } [4 \times 25 + 6 \times 5 + 8 \times 20 + 10 \times 50 + 8 \times 20 + 4 \times 30 + 13 \times 55 + 0 \times 15] = \text{Rs } 1525.$$

Optimum solution by MODI's method is obtained. The final iteration now becomes:

table 4.80

25 / 4	5 / 6	20 / 8	(10) 13	(5) 0	$u_1 = 0$
(7) 13	(3) 11	70 / 10	5 / 8	(3) 0	$u_2 = 2$
(12) 14	30 / 4	(4) 10	(12) 13	(7) 0	$u_3 = -2$
(0) 9	(1) 11	15 / 13	(0) 8	15 / 0	$u_4 = 5$
$v_1 = 4$	$v_2 = 6$	$v_3 = 8$	$v_3 = 3$	$v_5 = -5$	

Since all $d_{ij} \geq 0$, the solution is optimum but an alternative solution exists with transportation cost as Rs 1465.

4.7 MAXIMISATION IN TRANSPORTATION PROBLEMS

There are certain types of transportation problems where the objective function is to be maximised instead of being minimised. These problems can be solved by converting the maximisation problem into a minimisation problem by multiplying all the entries by – 1 or by subtracting all the entries from the highest entry in the given transportation table. The modified minimisation problem can be solved in the usual manner.

Example 1

A company has four factories F_1, F_2, F_3 and F_4 manufacturing the same product. The production and raw material costs differ from factory to factory and are given in the Table 4.81. In the first two rows the transportation costs from the factories to the sales depots S_1, S_2, S_3 are also given. The last two columns in the Table give the sales price and the total requirement at each depot. The production capacity of each factory is given in the last row.

table 4.81

	F_1	F_2	F_3	F_4	Sales price per unit	Requirement
Production cost/unit	15	18	14	13		
Raw material cost/unit	10	9	12	9		
Transportation cost/unit						
S_1	3	9	5	4	34	80
S_2	1	7	4	5	32	120
S_3	5	8	3	6	31	150
Supply	10	150	50	100		

Determine the most profitable production and distribution schedule and the corresponding profit. The deficit production should be taken to yield zero profit.

Solution: First, we construct a profit table in which

Profit = [Sales price/unit] – [Production cost/unit] – [raw material cost/unit] – [transportation cost/unit]

Profit from F_1 to $S_1 = 34 - 15 - 10 - 3 = 6$

Profit from F_2 to $S_1 = 34 - 18 - 9 - 9 = -2$

Profit from F_3 to $S_1 = 34 - 14 - 12 - 5 = 3$

Profit from F_4 to $S_1 = 34 - 13 - 9 - 4 = 8$

Profit from F_1 to $S_2 = 32 - 15 - 10 - 1 = 6$

Similarly, we can calculate the other profits and get the profit matrix as:

table 4.82

	Sales depots			
	S_1	S_2	S_3	Supply
F_1	6	6	1	10
F_2	– 2	– 2	– 4	150
F_3	3	2	2	50
F_4	8	5	3	100
Requirement	80	120	150	

Note that the total supply = 310 units and total requirement = 350 units. The given problem is of maximisation of profit. Convert this into minimisation type by subtracting all the profit values from the highest profit of 8. Now, in order to make the problem balanced, a dummy factory *d* is introduced to meet the extra requirement with zero profit coefficients. This is depicted in Table 4.83.

table 4.83

	S_1	S_2	S_3	Supply
F_1	2	2	7	10
F_2	10	10	12	150
F_3	5	6	6	50
F_4	0	3	5	100
d	0	0	0	40
Requirement	80	120	150	350 (Total)

The problem is balanced. So, we get the IBFS.
We can obtain the initial basic feasible solution by VAM.

table 4.84

2	[10] 2	7	10/0	(0) (0) (5)
10	[90] 10	[60] 12	150/60/0	(0) (0) (2) (2) (2)
5	6	[50] 6	50	(0) (0) (0) (0) (0)
[80] 0	[20] 3	5	100	(3) (3) (2) (2)
0	0	[40] 0	40	(0)
80	120	150		
(0)	(2)	(5)		
(2)	(1)	(1)		
	(1)	(1)		
	(3)	(1)		
	(4)	(6)		

Since the number of allocations = $m + n - 1 = 5 + 3 - 1 = 7$ the optimality test can be performed. By MODI's method we can perform the optimality test which is shown in Table 4.85. Then, Table 4.86 gives the optimal solution.

table 4.85

(3) 2	[10] 2	(3) 7	$u_1 = -1$
(3) 10	[90] 10	[60] 12	$u_2 = 7$
(4) 5	(2) 6	[50] 6	$u_3 = 1$
[80] 0	[20] 3	(0) 5	$u_4 = 0$
(5) 0	(2) 0	[40] 0	$u_5 = -5$
$v_1 = 0$	$v_2 = 3$	$v_3 = 5$	

table 4.86

6	[10] 6	1
–2	[90] –2	[60] –4
3	2	[50] 2
[80] 8	[20] 5	3
0	0	[40] 0

Total profit $= \text{Rs } [6 \times 10 - 2 \times 90 - 4 \times 60 + 2 \times 50 + 8 \times 80 + 5 \times 20 + 0 \times 40]$
$= \text{Rs } 480.$

Example 2

There are three factories *A*, *B*, and *C* which supply goods to four dealers D_1, D_2, D_3, D_4. The production capacities of these factories are 1000, 700, 900 units per month, respectively. The requirements from the dealers are 900, 800, 500 and 400 units per month, respectively. The per unit return (excluding transportation costs) are Rs 8, Rs 7, Rs 9 at the three factories. Table 4.111 gives the unit transportation costs from the factories to the dealers. Determine the optimum solution to maximise the profit.

table 4.87

	D_1	D_2	D_3	D_4
A	2	2	2	4
B	3	5	3	2
C	4	3	2	1

table 4.88

					Capacity
	6	6	6	4	1000
	4	2	4	5	700
	5	6	7	8	900
Demand	900	800	500	400	2600

Solution: Profit = Return – Transportation cost. Hence,

Profit from *A* to $D_1 = 8 - 2 = 6$

Profit from A to $D_2 = 8 - 2 = 6$
Profit from A to $D_3 = 8 - 2 = 6$
Profit from A to $D_4 = 8 - 4 = 4$
Profit from B to $D_1 = 7 - 3 = 4$
Profit from B to $D_2 = 7 - 5 = 2$ and so on.

Table 4.4.88 gives the profit matrix. The problem is a balanced one. The problem is of maximisation type. Convert this into minimisation by subtracting all the cost entries in Table 4.88 by the highest cost entry 8. Hence, the new table is given in Table 4.89.

table 4.89

	D_1	D_2	D_3	D_4
A	2	2	2	4
B	4	6	4	3
C	3	2	1	0

table 4.90

[200] 2	[800] 2	2	4	1000
[700] 4	6	4	3	700
3	2	[500] 1	[400] 0	900
900	800	500	400	2600

We can find the initial basic feasible solution by VAM, which is shown in Table 4.90. Since the number of allocated cells = 5 < $m + n - 1 = 6$, the solution is a degenerated one. To resolved the degeneracy, add ε to the empty cell (1, 3) which does not form a path, and which is the least cost cell. Now, optimality test can be performed using MODI's method.

table 4.91

	D_1	D_2	D_3	D_4	u_i
A	[200] 2	[800] 2	[ε] 2	(3) 4	0
B	[700] 4	(2) 6	(0) 4	(0) 3	2
C	(2) 3	(1) 2	[500] 1	[400] 0	–1
v_j	2	2	2	1	

table 4.92

[200] 6	[800] 6	[ε] 6	4
[700] 4	2	4	5
5	6	[500] 7	[400] 8

Since $d_{ij} \geq 0$, the initial solution is optimum. The optimum distribution is shown in Table 4.92.
Total profit or the maximum return

$$= \text{Rs } [6 \times 200 + 6 \times 800 + 4 \times 700 + 7 \times 500 + 8 \times 400]$$
$$= \text{Rs } 15{,}500.$$

4.8 THE TRANS-SHIPMENT PROBLEM

The transportation problem assumes that direct routes exist from each source to each destination. There are situations in which units may be shipped from one source to another or to other destinations before reaching their final destination. This is called a trans-shipment problem and is solved using a similar algorithm with a little modification. The points to be noted are:

1. The number of sources and destinations are say m and n, respectively in a transportation problem. In trans-shipment study, we have $m + n$ sources and $m + n$ destinations.
2. If S_i denotes the ith source, D_j the jth destination, then commodity can move along the routes:

$$S_i \to D_i \to D_j$$
$$S_i \to S_j \to D_i \to D_j$$
$$S_i \to D_i \to S_j \to D_j$$

   and so on. The transportation cost from S_i to S_i is zero and the transportation costs from S_i to S_j is different from S_j to S_i. Similarly, $D_i \to D_j \neq D_j \to D_i$.
3. In solving the trans-shipment problem, we first find the optimum solution to the transportation problem.
4. The basic feasible solution contains $2m + 2n - 1$ basic variables, if we omit the variables appearing in the $(m + n)$ diagonal cells, we have only $m + n - 1$ basic cells.

The formulation and solution of a trans-shipment can be illustrated in Example 4.8.1.

Example 1

Consider a transportation problem where the sources are plants and destinations are depots. The unit transportation cost, capacity at the plants and the requirement at the depots are given below.

table 4.93

	X	Y	Z	Capacity
A	1	3	15	150
B	3	5	25	300
Requirement	150	150	150	

If each plant is also considered as a destination and each depot is also considered as a source, then there are five sources and five destinations. Some additional cost data are given in Table 4.94, Table 4.95 and Table 4.96.

table 4.94

	A	B
A	0	65
B	1	0

table 4.95

	X	Y	Z
X	0	23	1
Y	1	0	3
Z	65	3	0

table 4.96

	A	B
X	3	15
Y	25	3
Z	45	55

table 4.97

	A	B	X	Y	Z	Capacity
A	0	65	1	3	15	150 + 450 = 600
B	1	0	3	5	25	300 + 450 = 750
X	3	15	0	23	1	450
Y	25	3	1	0	3	450
Y	45	55	65	3	0	450
	450	450	150 + 450 = 600	150 + 450 = 600	150 450 = 600	2700

Table 4.97 is the formulation of trans-shipment problem. A buffer stock of 450 which is the total capacity and total requirement in the original transportation problem is added to each row and column of the trans-shipment problem. The resulting transportation problem has $m + n = 5$ sources and 5 destinations. Solving the transportation problem in Table 4.125 as usual we obtain,

$x_{11} = 150$, $x_{13} = 300$, $x_{14} = 150$, $x_{21} = 300$, $x_{22} = 450$, $x_{33} = 300$, $x_{35} = 150$, $x_{44} = 450$, $x_{55} = 450$.

The description of the trans-shipment problem is

1. Transport $x_{21} = 300$ units from plant B to A. This increases the availability at plant A to 450 units including the 150 units originally available from A.

2. From plant A to depot X, transport $x_{13} = 300$
   From plant A to depot Y, transport $x_{14} = 150$
   From depot X to depot Z, transport $x_{35} = 150$

The total shipment cost is

$$1 \times 300 + 3 \times 150 + 1 \times 300 + 1 \times 150 = 1200.$$

If, however, the commodities are transported from plants A, B to depots X, Y, Z, according to Table 4.125, the minimum transportation cost is $x_{13} = 150, x_{21} = 150, x_{22} = 150$ with a minimum cost of 3450. Thus, the trans-shipment reduces the cost of cargo movement.

4.9 LEAST-TIME TRANSPORTATION PROBLEMS

There are some problems where the objective is to minimise time rather than transportation cost. In these problems, the transportation table contains the time t_{ij} required to transport the commodity from the ith origin to j^{th} destination. This time of shipment is, independent of the number of units shipped. These problems are solved by the following procedure. The initial basic feasible solution can be obtained by methods like North-West Corner Rule or VAM. If T_k is the largest transportation time corresponding to the kth feasible plan, then the objective is to find out the plan for which T_k is minimum for all values of k.

Find T_k for the initial feasible plan and cross off all non-basic cells for which $t_{ij} \geq T_k$.

Draw a closed path for the basic cell associated with T_k such that when the values at the corner elements are shifted, this basic cell reduces to zero and no other cell value becomes negative. This procedure ends if no such closed path can be traced out.

Example 1

A military equipment is to be transported from the sources to destinations as shown in Table 4.98. The units to be shipped as obtained by North-West Corner Rule are given in brackets. Find a plan so that the transportation time is minimum.

table 4.98

Sources \ Destinations	1	2	3	4	Supply
1	(12) 10	(3) 0	20	11	15
2	1	(5) 7	(15) 9	(5) 20	25
3	12	14	16	(5) 18	5
Demand	12	8	15	10	45 Total

In the Table, time for shipping are $t_{11} = 10$, $t_{12} = 0$, $t_{22} = 7$, $t_{23} = 9$, $t_{24} = 20$, $t_{34} = 18$ and the initial basic feasible solution is $x_{11} = 12$, $x_{12} = 3$, $x_{22} = 5$, $x_{23} = 15$, $x_{24} = 5$, $x_{34} = 5$ and the other variables are zero.

Now, $T_1 = \max\{t_{11}, t_{12}, t_{22}, t_{23}, t_{24}, t_{34}\}$

$= \max\{10, 0, 7, 9, 20, 18\}$

$= 20$ time units.

Therefore, corresponding to the rules we can cross off all non-basic cells for which $t_{ij} \geq T_1$. So, cross off the non-basic cell (1, 3) with $t_{13} = 20$. The closed path for $T_k = t_{24}$ is shown in Table 4.99.

table 4.99

	1	2	3	4
1	[12] 10	[3] $-\theta$ 0	20 (crossed)	$+\theta$ 11
2	1	[5] $+\theta$ 7	[15] 9	[5] $-\theta$ 20
3	12	14	16	[5] 18

$\theta = \min(3,5) = 3$.

It is clear from the closed path that x_{24} can be decreased by only 3 units, otherwise x_{12} will become negative. The new table is shown in Table 4.100. From the Table we observe that T_2 is still equal to 20. The next closed path is shown in Table 4.101, where x_{24} is decreased by 2 units, there by reducing it to zero, which is shown in Table 4.102.

table 4.100

	1	2	3	4
1	[12] 10	0	20 (crossed)	[3] 11
2	1	[8] 7	[15] 9	[2] 20
3	12	14	16	[5] 18

table 4.101

	1	2	3	4
1	[12] $-\theta$ 10	0	20	[3] $+\theta$ 11
2	$+\theta$ 1	[8] 7	[15] 9	[2] $-\theta$ 20
3	12	14	16	[5] 18

table 4.102

	1	2	3	4
1	[10] 10	0	20 (crossed)	[5] 11
2	[2] 1	[8] 7	[15] 9	20 (crossed)
3	12	14	16	[5] 18

Observe that $T_3 = t_{34} = 18$, corresponding to x_{34}. The closed path for x_{34} is shown in Table 4.103.

table 4.103

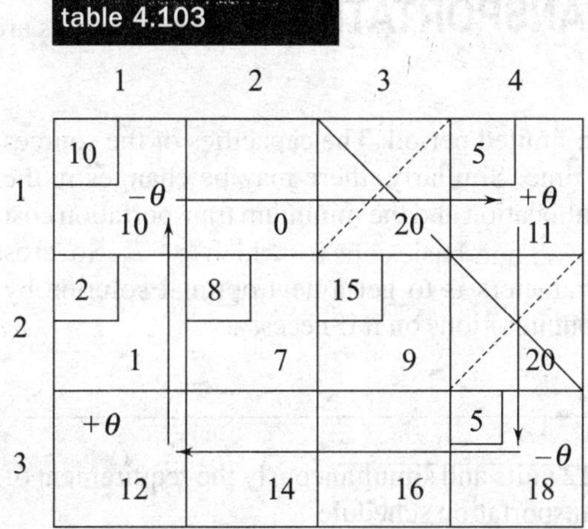

	1	2	3	4
1	(10) 10 $-\theta$	0	× 20	(5) 11 $+\theta$
2	(2) 1	(8) 7	(15) 9	× 20
3	$+\theta$ 12	14	16	(5) 18 $-\theta$

table 4.104

	1	2	3	4
1	(5) 10	0	× 20	(10) 11
2	(2) 1	(8) 7	(15) 9	× 20
3	(5) 12	14	16	× 18

table 4.105

	1	2	3	4
1	(5) 10	0	× 20	(10) 11
2	(2) 1	(8) 7	(15) 9	× 20
3	(5) 12	× 14	× 16	× 18

Since $\theta = \min(5, 10) = 5$, x_{34} can be reduced by 5 units and thus the allotment for the cell (3,4) is $5 - 5 = 0$. So, cross out and the resulting solution is shown in Table 4.104.

Now, $T_4 = t_{31} = 12$, hence the cells (3, 2), (3, 3) are crossed off since t_{32}, t_{33} are both $> t_{31}$, which is given in Table 4.105.

Since no closed path can be traced for x_{31}, the iteration ends. Thus, the above plan is optimal with total shipment time as $T_4 = 12$ units.

$x_{11} = 5$ with $t_{11} = 10$

$x_{13} = 10$ with $t_{13} = 11$

$x_{21} = 2$ with $t_{21} = 1$

$x_{22} = 8$ with $t_{22} = 7$

$x_{23} = 15$ with $t_{23} = 9$

$x_{31} = 5$ with $t_{31} = 12$

4.10 SENSITIVITY ANALYSIS IN TRANSPORTATION PROBLEM

So far we have seen problems which are valid only for a limited period. The capacities of the sources and/or requirements at the destinations may vary with time. Similarly, there may be changes in the transportation cost. Such changes may affect the optimal allocation and the minimum transportation cost associated with it.

One way to study the effect of changes in these parameters is to get a new optimal solution by modifying the current optimal solution and by carrying out iterations on it if necessary.

Example 1

Consider Example 4.4.7 and solve the following:

(d) Suppose the supply from warehouse *B* reduces to 12 units and simultaneously the requirement of market III reduces to 14 units. Find the optimal transportation schedule.

(e) Further, suppose the supply from warehouse *A* also reduces to 18 units and simultaneously the requirement of market III reduces further to 10 units. Will the optimal solution of part (d) change?

(f) Suppose the supply of warehouse *B* is increased [with respect to optimal solution of part (b)] by 1 unit and simultaneously the demand of market III is also increased by 1 unit. Will the optimal solution change?

(g) Suppose the supply of warehouse *B* increases by 3 units [instead of 1 unit in part (f)] and demand of market IV also increases by 3 units. Will there be change in the optimal solution?

Solution: (d) The modified transportation table is shown below.

table 4.106

	I	II	III	IV	Supply
A	5	(12) 2	(2) 4	(8) 3	22
B	4	8	(12) 1	6	12
C	(7) 4	6	7	(1) 5	8
	7	12	14	9	

As there is no change in the allocated cells, the above modification does not change the optimal schedule.

(e) As the allocation in the cell (*A*, III) is 2 units, the adjustment is not possible by considering the cell (*A*, III) alone. Even if this allocation is cancelled and the allocation in cell (*B*, III) is also reduced to 10 units to match the requirement of market III, this allocation can cause imbalance in the allocated cells of *B* and *C*. So, it is necessary to solve the problem anew.

(f) Making necessary changes in Table 4.54, we get the modified table as [See Table 4.107].

table 4.107

	I	II	III	IV	Supply
A	5	[12] 2	[2] 4	[8] 3	22
B	4	8	[12] 1	6	16
C	[7] 4	6	7	[1] 5	8
Requirement	7	12	18	9	

We can see that the modified solution does not change the optimal schedule.

(g) Table 4.54 shows that it will not be possible to confine the changes to allocated cells only. The optimal solution, therefore, changes and the problem has to be solved a new.

EXERCISES

Section 4.2

1. Describe the matrix form of the transportation problem. Illustrate with 2 origins and 3 destinations.
2. Show that all bases for transportation problem are triangular.
3. What do you mean by an unbalanced transportation problem and explain how to convert the unbalanced transportation problem into a balanced one.
4. Explain degeneracy in a transportation problem and how to resolve it.

Section 4.3

1. Solve the following transportation problem:

	1	2	3	Capacity
1	2	2	3	10
2	4	1	2	15
3	1	3	1	40
Demand	20	15	30	

(*Answer*: $x_{13} = 10$, $x_{23} = 15$, $x_{31} = 20$, $x_{32} = 15$, $x_{33} = 5$. The transportation cost is Rs 130.

2. Determine an IFBS using (a) North-West Corner Rule, (b) VAM.

	D_1	D_2	D_3	D_4	D_5	Supply
A	2	11	10	3	7	4
B	1	4	7	2	1	8
C	3	9	4	8	12	9
Demand	3	3	4	5	6	21

[*Answer*: (a) North-West Corner Rule: $x_{11} = 3, x_{12} = 1, x_{22} = 2, x_{23} = 4, x_{24} = 2, x_{34} = 3, x_{35} = 6$, with cost = Rs 153.]
(b) Vogel's Method: $x_{14} = 4, x_{22} = 2, x_{25} = 6, x_{24} = 3, x_{32} = 1, x_{33} = 4, x_{34} = 1$, cost = Rs 68.]

3. Goods have to be transported from factories F_1, F_2, F_3 to warehouses W_1, W_2, W_3 and W_4. The transportation cost per unit, capacities and requirement of the warehouses are given in the Table.

	W_1	W_2	W_3	W_4	Capacity
F_1	95	105	80	15	12
F_2	115	180	40	30	7
F_3	195	180	95	70	5
Requirements	5	4	4	11	

Find the distribution with minimum cost. Obtain an IFBS to the following transportation problem.

	D_1	D_2	D_3	D_4	Supply
O_1	1	2	3	4	6
O_2	4	3	2	0	8
O_3	0	2	2	1	10
Demand	4	6	8	6	

[*Answer*: By least cost method IFBS is Table 4.112]

1	6 2	3	4
4	3	3 2	6 0
4 0	2	6 2	1

Section 4.4

1. Solve the transportation problem

	1	2	3	4	
1	1	2	1	4	30
2	3	3	2	1	50
3	4	2	5	9	20
	20	40	30	10	100

[*Answer*: By least cost method, the IFBS is Table 4.114]

20 / 1	2	10 / 1	4
3	20 / 3	20 / 2	10 / 1
4	20 / 2	5	9

20 / 1	2	10 / 1	4
3	20 / 3	20 / 2	10 / 1
4	20 / 2	5	9

Optimal solution is given in Table 4.115.

2. The following Table gives the cost for transporting material from supply points *A*, *B*, *C*, *D* to demand points *E*, *F*, *G*, *H*, *J*.

	E	*F*	*G*	*H*	*J*
A	8	10	12	17	15
B	15	13	18	11	9
C	14	20	6	10	13
D	13	19	7	5	12

The present allocation is as follows.

A to *E* is 90; *A* to *F* is 10; *B* to *F* is 150; *C* to *F* is 10; *C* to *G* is 50, *C* to *J* is 120, *D* to *H* is 210; *D* to *J* is 70.

(a) Check if this allocation is optimum. If not, find an optimum schedule.

(b) If in the above problem the transportation cost from *A* to *G* is reduced to 10, what will be the new optimum schedule?

[*Answer*: (a) No. The optimum schedule is: $x_{12} = 100$, $x_{22} = 70$, $x_{25} = 80$, $x_{31} = 70$, $x_{33} = 50$, $x_{31} = 40$, $x_{44} = 210$, $x_{45} = 70$, with minimum cost = Rs 6810.]

(b) Same as part (a).

3. A company has three plants *A*, *B*, *C*, three warehouses *X*, *Y*, *Z*. The number of units available at the plants is 60, 70, 80 and the demand at *X*, *Y*, *Z* are 50, 80, 80, respectively. The unit cost of the transportation is given in the following Table.

	X	*Y*	*Z*
A	8	7	3
B	3	8	9
C	11	3	5

Find the allocation so that the total transportation cost is minimum.
Answer: By least cost method, the IBFS is

$x_{13} = 50; x_{21} = 50, x_{23} = 20, x_{32} = 80$ with cost Rs 750

Optimal solution is same as IBFS.

4. Solve the following transportation problem

	A	B	C	D	Supply
1	11	20	7	8	50
2	21	16	20	12	40
3	8	12	8	9	70
Demand	30	25	35	40	

Section 4.5

1. A company has four warehouses and six stores, the cost of shipping one unit from warehouse *i* to store *j* is c_{ij}.

		Store					
		1	2	3	4	5	6
Warehouse	1	7	10	7	4	7	8
	2	5	1	5	5	3	3
	3	4	3	7	9	1	9
	4	4	6	9	0	0	8

The requirements of six stores are 4, 4, 6, 2, 4, 2 and quantities at warehouses are 5, 6, 2, 9. Find the minimum cost solution.
[*Answer*: $x_{13} = 5, x_{22} = 4, x_{26} = 2, x_{31} = 1, x_{32} = \varepsilon, x_{33} = 1, x_{41} = 3, x_{44} = 2, x_{45} = 4$ and the optimal transportation cost = Rs $(68 + 3\varepsilon)$ = Rs 68.]

2. Solve the following transportation problem,

		To			Availability
From	I	50	30	220	1
	II	90	45	170	3
	III	250	200	50	4
Requirement		4	2	2	

[*Answer*: $x_{11} = 1, x_{21} = 3, x_{31} = \varepsilon, x_{32} = 2$, and $x_{33} = 2$ and minimum transportation cost = Rs 743.]

3. Solve the following transportation table using Vogel's approximation method.

	A	B	C	D	E	F	Available
1	9	12	9	6	9	10	5
2	7	3	7	7	5	5	6
3	6	5	9	11	3	11	2
4	6	8	11	2	2	10	9
Requirement	4	4	6	2	4	2	

[*Answer*: $x_{13} = 5$, $x_{22} = 4$, $x_{26} = 2$, $x_{31} = 1$, $x_{32} = \varepsilon$, $x_{33} = 1$, $x_{41} = 3$, $x_{44} = 2$, $x_{45} = 4$ and the minimum transportation cost is Rs 112.]

Section 4.6

1. Solve the transportation problem

	D_1	D_2	D_3	D_4	Supply
O_1	6	1	9	3	70
O_2	11	5	2	8	55
O_3	10	12	41	7	70
Demand	85	35	50	45	

[*Answer*: $x_{11} = 65, x_{12} = 5, x_{22} = 30, x_{23} = 25, x_{33} = 25, x_{34} = 45, x_{41} = 20$, minimum cost = Rs 1010.]

2. Solve the following transportation problem

	A	B	C	D	Supply
1	11	20	7	8	50
2	21	16	20	12	40
3	8	12	8	9	70
Demand	30	25	35	40	

[*Answer*: $x_{13} = 35, x_{14} = 15, x_{24} = 10, x_{25} = 30, x_{31} = 30, x_{32} = 25, x_{34} = 15$, minimum cost = Rs 1130.]

3. Solve the transportation problem

		Destination				
		1	2	3	4	Availability
	1	10	15	12	12	200
Factory	2	8	10	11	9	150
	3	11	12	13	10	120
Requirement		140	120	80	220	

[*Answer*: $x_{11} = 140, x_{13} = 60, x_{22} = 50, x_{24} = 100, x_{34} = 120, x_{42} = 70, x_{43} = 20$, the minimum cost is Rs 2940.]

Section 4.7

1. Solve the following transportation problem to maximise profit.

		Destination				
		A	B	C	D	Supply
Source	1	40	25	22	33	100
	2	44	35	30	30	30
	3	38	38	28	30	70
Demand		40	20	60	30	

[*Answer*: $x_{11} = 20$, $x_{14} = 30$, $x_{15} = 50$, $x_{21} = 20$, $x_{23} = 10$, $x_{32} = 20$, $x_{33} = 50$, optimum profit = Rs 5130.]

2. Consider four bases of operations B_i and three targets T_j. The tons of bombs per aircraft from any base that can be delivered to any target are given in the following Table.

		Target (T_j)		
		T_1	T_2	T_3
Base (B_i)	B_1	8	6	5
	B_2	6	6	6
	B_3	10	8	4
	B_4	8	6	4

The daily sortie capability of each of the four bases is 150 sorties per day. The daily requirement in sorties over each individual target is 200. Find the allocation of sorties from each base to each target which maximises the total tonnage over all the three targets explaining each step.
[*Answer*: $x_{11} = 50$, $x_{12} = 50$, $x_{23} = 150$, $x_{31} = 150$, $x_{42} = 150$, maximum tonnage = 4250.]

Section 4.8

1. Consider the following trans-shipment problem with two sources and two destinations, the costs for shipment in rupees are given below. Determine the optimum shipping schedule.

		Source		Destination		Supply
		S_1	S_2	D_1	D_2	
Source	S_1	0	1	3	4	5
	S_2	1	0	2	4	25
Destination	D_1	3	2	0	1	–
	D_2	4	4	1	0	–
Demand	–	–	–	20	10	

2. Consider the following trans-shipment problem with two sources and three destinations. The costs per unit for shipment in rupees are also given. Find the optimum shipping schedule.

	S_1	S_2	D_1	D_2	D_3	Supply
S_1	0	80	10	20	30	100
S_2	10	0	20	50	40	200
D_1	20	30	0	4	10	
D_2	40	20	10	0	20	
D_3	60	70	80	20	0	
Demand			100	100	100	

3. Given the following data, find the optimum transportation routes.

		Destinations				
		S_1	S_2	D_1	D_2	Available
Sources	S_1	0	2	3	4	5
	S_2	2	0	2	4	25
	D_1	3	2	0	1	
	D_2	4	4	1	0	
Demand				20	10	

Sections 4.9 and 4.10

1. The following Table gives the optimal cost of manufacturing and transporting from factories *A*, *B*, *C* and *D* to destinations *E*, *F*, *G*, *H* and *J* as Rs 6530.
   (a) Suppose some new equipments are installed that reduce the variable operating cost by Rs 2 in factory *B*, will the optimal shipping schedule change?
   (b) Is there an alternate optimal solution to part (a)? If so, find it.

	E	*F*	*G*	*H*	*J*	Supply
A	8	[100] 10	12	17	15	100
B	15	[70] 13	18	11	[80] 9	150
C	[70] 14	20	[50] 6	10	[40] 13	160
D	13	19	7	[210] 6	[70] 12	280
Demand	70	170	50	210	190	

(c) Suppose freight charges from *A* to *E* are reduced by Rs 2 unit, will it change the shipping schedule? If so what is the new shipping schedule?

(d) If the demand at *G* increases from 50 units to 60 units, what will be the shipping schedule if there is no increase in supply from any factory?

(e) If demand at *G* increases to 60 units as well as the supply at factory *D* is increased by 10 units to meet this increased demand, find the optimal shipping schedule.

[*Answer*: (a) The solution is still optimal.

(b) The alternate optimal solution is:

$$x_{12} = 100,\ x_{22} = 70,\ x_{25} = 80,\ x_{33} = 50,\ x_{35} = 110,\ x_{41} = 70,\ x_{44} = 210]$$

(c) Yes. The new optimal schedule is

$$x_{11} = 70,\ x_{12} = 30,\ x_{22} = 140,\ x_{25} = 10,\ x_{33} = 50,\ x_{35} = 110,\ x_{44} = 210,\ x_{45} = 70.$$

(d) Check whether the following Table is optimal.

8	[100] 10	12	17	15	100
15	[60] 13	18	11	[90] 9	150
[70] 14	20	[60] 6	10	[30] 13	160
13	19	7	[210] 6	[70] 12	280
70	160	60	210	190	

(e) Check whether the following Table is optimal.

8	(100) 10	12	17	15	100
15	(70) 13	18	11	(80) 9	150
(70) 14	20	(60) 6	10	(30) 13	160
13	19	7	(210) 6	(80) 12	290
70	170	60	210	190	

2. The following Table gives the cost of transporting material from supply points *A*, *B*, *C*, *D* to demand *E*, *F*, *G*, *H*, *J*.

	E	*F*	*G*	*H*	*J*
A	8	10	12	17	15
B	15	13	18	11	9
C	14	20	6	10	13
D	13	19	7	5	12

The present allocation is as follows:

A to *E* is 90, *A* to *F*: 10, *B* to *F* : 150, *C* to *F* : 10, *C* to *G*: 50, *C* to *J* : 120, *D* to *H* : 210, *D* to *J* : 70

(a) Check if this allocation is optimum. If not, find an optimum schedule.

(b) If in the above problem the transporting cost from *A* to *G* is reduced to 10, what will be the new optimum schedule?

[*Answer*: (a) No. *A* to *F* : 100, *B* to *F* : 70, *B* to *J* : 80, *C* to *E* : 90, *C* to *G* : 50, *C* to *J* : 40, *D* to *H* : 210, *D* to *J* : 70, minimum cost = 6.600. (b) same as part (a).]

3. A company has factories at four different places which supply to warehouses *A*, *B*, *C*, *D* and *E*. Monthly factory capacities are 200, 175, 150 and 325, respectively. Monthly warehouse requirements are 110, 90, 120, 230, and 160, respectively. The transport cost are given in the Table.

	A	B	C	D	E
1	13	-	31	8	20
2	14	9	17	6	10
3	25	11	12	17	15
4	10	21	13	-	17

Shipment from 1 to *B* and from 4 to *D* is not possible. Find the optimum solution.
[*Answer*: $x_{1D} = 200$, $x_{2D} = 30$, $x_{2E} = 145$, $x_{3B} = 90$, $x_{3C} = 45$, $x_{3E} = 15$, $x_{4A} = 110$, $x_{4C} = 75$ and minimum cost = 140.]

5

Assignment Model

5.1 INTRODUCTION AND FORMULATION

The assignment problem is a particular case of transportation problem in which the number of jobs (or origins or sources) are equal to the number of facilities (or destinations or machines or persons and so on). The objective is to maximise total profit of allocation or to minimise the total cost. An assignment problem is a completely degenerate form of a transportation problem. The units available at each origin and the units demanded at each destination are all equal to one.

Mathematical Formulation of an Assignment Problem

Given n jobs (or activities) and n persons (or facilities) and effectiveness (in terms of cost, profit, time and others) of each person for each job, the problem lies in assigning each person to one and only one job so that the given measure of effectiveness is optimised.

Let c_{ij} be the cost of assigning ith person to the jth job. The assignment problem can be stated in the form of $n \times n$ *cost matrix* or *effectiveness matrix* (c_{ij}) as shown in the Table.

		Jobs							
		1	2	3	...	j	...	n	Supply
	1	c_{11}	c_{12}	c_{13}	...	c_{1j}	...	c_{1n}	1
	2	c_{21}	c_{22}	c_{23}	...	c_{2j}	...	c_{2n}	1
Persons	3	c_{31}	c_{32}	c_{33}	...	c_{3j}	...	c_{3n}	1
	⋮	⋮							
	i	c_{i1}	c_{i2}	c_{i3}	...	c_{ij}	...	c_{in}	1
	⋮	⋮							
	n	c_{n1}	c_{n2}	c_{n3}	...	c_{nj}	...	c_{nn}	1
Demand		1	1	1	...	1	...	1	n

Let x_{ij} denote the assignment of person i to job j such that

$$x_{ij} = \begin{cases} 1 & \text{if person } i \text{ is assigned to job } j \\ 0 & \text{otherwise} \end{cases}$$

The assignment problem is simply the following LPP :

$$\text{Minimise } z = \sum_{i=1}^{n} \sum_{j=1}^{n} c_{ij} x_{ij}$$

subject to the constraints

$$\sum_{j=1}^{n} x_{ij} = 1, \text{ for } i = 1, 2, \cdots n$$

$$\text{and, } \sum_{i=1}^{n} x_{ij} = 1 \text{ for } j = 1, 2, \cdots n$$

with $x_{ij} = 0$ or 1, for all i and j.

Note that an assignment problem is an $n \times n$ transportation problem with each $a_i = b_j = 1$.

Assignment problem is completely a degenerate form of a transportation problem. The units available at each origin and units demanded at each destination are all equal to one. This means exactly one occupied cell in each row and each column of the transportation table, that is, only n occupied cells in place of the required $n + n - 1 = 2n - 1$ occupied cells.

Because of degeneracy, the problem cannot be solved by either simplex method or transportation method. For example, in a simplex method, for a problem involving five persons/jobs, there will be $5 \times 5 = 25$ decision variables and $2 \times 5 = 10$ inequalities. It is difficult to solve this problem manually.

In a transportation method, in order to remove degeneracy, $(n - 1)$ number of dummy allocations will be required to proceed with the transportation model. The problem of degeneracy at each solution makes the computation by the transportation method inefficient.

So, we go for a method called the **Hungarian method** developed by the Hungarian mathematician D. Konig to find the optimal solution without having to make a direct comparison of every solution.

5.2 HUNGARIAN ASSIGNMENT ALGORITHM

First, we prove the following theorems which form the basis of the assignment algorithm.

5.2.1 Theorem

The optimum assignment schedule remains unaltered if we add or subtract a constant to/from all the elements of the row or column of the assignment cost matrix.

Proof: Let we add or subtract a constant say u_i, v_j to/from all the elements of the ith row and j^{th} column of the cost matrix. The new cost matrix is $c^*_{ij} = c_{ij} \pm u_i \pm v_j$. The objective function is

$$z^* = \sum_{i=1}^{n} \sum_{j=1}^{n} (c_{ij} \pm u_i \pm v_j)\, x_{ij}$$

$$= \sum_{i=1}^{n} \sum_{j=1}^{n} c_{ij} x_{ij} \pm \sum_{i=1}^{n} \sum_{j=1}^{n} u_i x_{ij} \pm \sum_{i=1}^{n} \sum_{j=1}^{n} v_j x_{ij}$$

$$= z \pm U \pm V$$

where, $$U = \sum_{i=1}^{n} u_i \left[\sum_{j=1}^{n} x_{ij} \right] = \sum_{i=1}^{n} u_i \text{, since } \sum_{j=1}^{n} x_{ij} = 1$$

and, $$v = \sum_{j=1}^{n} v_j \left[\sum_{i=1}^{n} x_{ij} \right] = \sum_{j=1}^{n} v_j \text{, since } \sum_{i=1}^{n} x_{ij} = 1$$

Note that U and V are constants, hence an assignment (x_{ij}) which minimises z will also minimise z^*.

5.5.2 Theorem

If for an assignment problem all $c_{ij} \geq 0$ then an assignment schedule (x_{ij}) which satisfies $\Sigma\Sigma\, x_{ij}\, c_{ij} = 0$, must be optimal.

These theorems can be used in two different ways to solve the assignment problem. If in an assignment problem some cost elements are negative we convert the problem into an equivalent assignment problem with all cost elements as non-negative by adding a suitable large constant to the elements of the relevant row or column. Next, we look for a feasible solution which has *zero assignment cost* after adding suitable constants to the elements of various rows and columns. Since it has been assumed that all the cost elements are non-negative, this assignment must be optimum.

5.2.3 Hungarian Assignment Algorithm

Various steps of the computational procedure for obtaining an optimal solution may be summarised as follows.

Step 1: If the number of rows are not equal to the number of columns and vice versa, then a dummy row or dummy column must be added with zero cost elements.

Step 2: Find the smallest cost in each row of the cost matrix. Subtract this smallest cost element from each element in that row. Therefore, there will be at least one zero in each row of this new matrix which is called the first reduced cost matrix.

Step 3: In the reduced cost matrix, find the smallest element in each column. Subtract the smallest cost element from each element in that column. As a result, there would be at least one zero in each row and column of the second reduced cost matrix.

Step 4: Determine an optimum assignment as follows:

(i) Examine the rows successively until a row with exactly one zeros is found. Box around the zero element as an assigned cell and cross out all other zero in its column. Proceed in this manner until all the rows have been examined. If there are more than one zero in any row, then do not consider that row and pass on to the next row.

(ii) Repeat the procedure for the columns of the reduced cost matrix. If there is no single zero in any row or column of the reduced matrix, then arbitrarily choose a row or column having the minimum number of zeros. Arbitrarily, choose zero in the row or column and cross the remaining zeros in that row or column.

Repeat steps (i) and (ii) until all zeros are either assigned or crossed out.

Step 5: An optimal assignment is found, if the number of assigned cells equals the number of rows (and columns). If a zero cell is arbitrarily chosen, there may be an alternate optimum. If no optimum solution is found (some rows or columns without an assignment), then go to next step.

Step 6: Draw the minimum number of horizontal and/or vertical lines through all the zeros as follows:

(i) Mark (✓) to those rows where no assignment has been made.
(ii) Mark (✓) to those columns which have zeros in the marked rows.
(iii) Mark (✓) rows (not already marked) which have assignments in marked columns.
(iv) The process may be repeated until no more rows or columns can be checked.
(v) Draw straight lines through all unmarked rows and marked columns.

Step 7: If the minimum number of lines passing through all the zeros is equal to the number of rows or columns, the optimum solution is attained by an arbitrary allocation in the positions of the zeros not crossed in step 3. Otherwise go to the next step.

Step 8: Revise the cost matrix as follows:

(i) Find the elements that are covered by a line. Choose the smallest of these elements and subtract this element from all the uncrossed elements and add the same at the point of intersection of the two lines.
(ii) Other elements crossed by the lines remain unchanged.

Step 9: Go to step 4 and repeat the procedure till an optimum solution is attained.

Example 1

The assignment cost of assigning any one operator to any one machine is given in the following Table.

		Operators			
		I	II	III	IV
Machine	*A*	10	5	13	15
	B	3	9	18	3
	C	10	7	3	2
	D	5	11	9	7

Find the optimal assignment.

Solution:

Step 1: Subtracting the smallest element of each row from every element of the corresponding row, we get the reduced matrix as:

$$\begin{bmatrix} 5 & 0 & 8 & 10 \\ 0 & 6 & 15 & 0 \\ 8 & 5 & 1 & 0 \\ 0 & 6 & 4 & 2 \end{bmatrix}$$

Step 2: Subtracting the smallest element of each column of the reduced matrix from every element of the corresponding column, we get the following reduced matrix:

$$\begin{bmatrix} 5 & 0 & 7 & 10 \\ 0 & 6 & 14 & 0 \\ 8 & 5 & 0 & 0 \\ 0 & 6 & 3 & 2 \end{bmatrix}$$

Step 3: Starting with row 1, we box a single zero (i.e. make assignment), and cross (×) all other zeros in its column. Thus, we get

$$\begin{bmatrix} 5 & \boxed{0} & 7 & 10 \\ \not{0} & 6 & 14 & \boxed{0} \\ 8 & 5 & \boxed{0} & \not{0} \\ \boxed{0} & 6 & 3 & 2 \end{bmatrix}$$

Since each row and each column contain exactly one assignment, the current assignment is optimal. The optimum schedule is

$$A \to \text{II}, \quad B \to \text{IV}, \quad C \to \text{III}, \quad D \to \text{I}$$

and the optimum assignment cost is

$$\text{Rs } [5 + 3 + 3 + 5] = \text{Rs } 16.$$

Example 2

A department has five employees with five jobs to be performed. The time (in hours) each man will take to perform each job is given in the cost matrix.

		Employees				
		I	II	III	IV	V
Jobs	*A*	10	5	13	15	16
	B	3	9	18	13	6
	C	10	7	2	2	2
	D	7	11	9	7	12
	E	7	9	10	4	12

How should the jobs be allocated, one per employee, so as to minimise the total man-hours?

Solution: Applying step 1 and 2 of the algorithm, the reduced time matrix is shown below:

$$\begin{bmatrix} 5 & 0 & 8 & 10 & 11 \\ 0 & 6 & 15 & 10 & 3 \\ 8 & 5 & 0 & 0 & 0 \\ 0 & 4 & 2 & 0 & 5 \\ 3 & 5 & 6 & 0 & 8 \end{bmatrix}$$

Step 3: Starting with row 1, box a single zero, if any, and cross all other zeros in its column. Thus, we get

5	[0]	8	10	11
[0]	6	15	10	3
8	5	[0]	~~0~~	~~0~~
~~0~~	4	2	~~0~~	5
3	5	6	[0]	8

The solution is not optimal since only four assignments are made.

Step 4: Since row *D* does not have any assignment we tick this row (✓)

	I	II	III	IV	V	
A	5	[0]	8	10	11	
B	[0]	6	15	10	3	
C	8	5	[0]	0	~~0~~	
D	~~0~~	4	2	~~0~~	5	✓
E	3	5	6	[0]	8	

(i) Now, there is a zero in the Ist and IVth column of the ticked row. So, we tick I and IVth column.

(ii) Mark ✓ in the rows *B* and *E* since columns I and IV have an assignment in rows *B* and *E*, respectively.

(iii) Draw straight lines through all unmarked rows *A* and *C* and marked columns, I and IV as shown below:

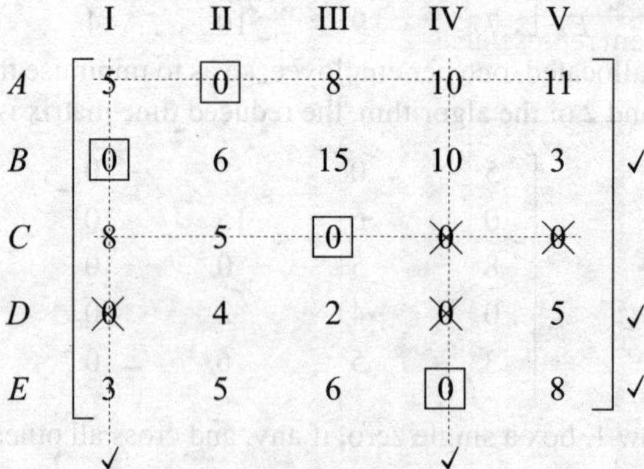

	I	II	III	IV	V	
A	5	[0]	8	10	11	
B	[0]	6	15	10	3	✓
C	8	5	[0]	~~0~~	~~0~~	
D	~~0~~	4	2	~~0~~	5	✓
E	3	5	6	[0]	8	✓
	✓			✓		

Step 5: In step 4, the minimum number of lines drawn is 4, which is less than the order of the cost matrix (order is 5), indicating that the current assignment is not optimum.

To increase the minimum number of lines, we generate new zeros in the modified matrix.

Step 6: Develop the new Table by selecting the smallest element among all uncovered elements by the lines. Here, the element is 2. Subtract this element from all the uncovered elements and add the same to all the elements lying at the intersection of the lines. We obtain the following new reduced cost matrix.

$$\begin{bmatrix} 7 & 0 & 8 & 12 & 11 \\ 0 & 4 & 13 & 10 & 1 \\ 10 & 5 & 0 & 2 & 0 \\ 0 & 2 & 0 & 0 & 3 \\ 3 & 3 & 4 & 0 & 6 \end{bmatrix}$$

Step 7: Repeat 3 to 6 to find a new solution. The new assignment is:

$$\begin{bmatrix} 7 & \boxed{0} & 8 & 12 & 11 \\ \boxed{0} & 4 & 13 & 10 & 1 \\ 10 & 5 & \cancel{0} & 2 & \boxed{0} \\ \cancel{0} & 2 & \boxed{0} & \cancel{0} & 3 \\ 3 & 3 & 4 & \boxed{0} & 6 \end{bmatrix}$$

Since the number of assignments (5) equals the number of rows (5), the solution is optimal. The optimum assignment is:

$$A \to \text{II},\ B \to \text{I},\ C \to \text{V},\ D \to \text{III},\ E \to \text{IV}$$

The minimum total time for this assignment schedule is $5 + 3 + 2 + 9 + 4 = 23$ hours.

Example 3

Solve the assignment problem represented by the matrix.

	1	2	3	4	5	6
A	9	22	58	11	19	27
B	43	78	72	50	63	48
C	41	28	91	37	45	33
D	74	42	27	49	39	32
E	36	11	57	22	25	18
F	3	56	53	31	17	28

Solution:

Step 1:

(i) Subtract the smallest element in each row from all the elements in that row.

(ii) Subtract the smallest element in each column from all the elements in that column.

The reduced matrix is as follows.

	1	2	3	4	5	6
A	0	13	49	0	0	13
B	0	35	29	5	10	0
C	13	0	63	7	7	0
D	47	15	0	20	2	0
E	25	0	46	9	4	2
F	0	53	50	26	4	20

Step 2: Make the zero-assignment as shown. Since row 2 and column 5 have no assignment, go to the next step. See the following Table.

	1	2	3	4	5	6
A	$\not{0}$	13	49	$\boxed{0}$	$\not{0}$	13
B	$\not{0}$	35	29	5	10	$\not{0}$
C	13	$\not{0}$	63	7	7	$\boxed{0}$
D	47	15	$\boxed{0}$	20	2	$\not{0}$
E	25	$\boxed{0}$	46	9	4	2
F	$\boxed{0}$	53	50	26	4	20

Step 3: Mark ✓ in row 2 as it is not having an assignment and in columns (1 and 6) as they are columns of ticked row 2 having zero's.

Next, mark ✓ to the rows (3 and 6) as these two rows contain assignment in marked columns (1 and 6).
Next, in marked row 3, there is zero in column 2. Assignment is made in this column which corresponds to row 5. So, mark ✓ to row 5 and column 2.

	1	2	3	4	5	6	
A	$\not{0}$	13	49	$\boxed{0}$	$\not{0}$	13	
B	$\not{0}$	35	29	5	10	$\not{0}$	✓
C	13	$\not{0}$	63	7	7	$\boxed{0}$	✓
D	47	15	$\boxed{0}$	20	2	$\not{0}$	
E	25	$\boxed{0}$	46	9	4	2	✓
F	$\boxed{0}$	53	50	26	4	20	✓
	✓	✓				✓	

Step 4: Draw straight lines through unmarked rows and marked columns as shown below.

	1	2	3	4	5	6
A	0	13	49	[0]	~~0~~	13
B	~~0~~	35	29	5	10	~~0~~
C	13	0	63	7	7	[0]
D	0	15	[0]	20	2	~~0~~
E	25	[10]	46	9	4	2
F	[0]	53	50	26	4	20

Since the number of lines is five which is less than the number of rows or columns, we move to the next step to get the optimal solution.

Step 5: The smallest element among all uncovered elements is 4. Subtract 4 from all elements not covered by lines and add 4 to all the elements that lie at the intersection of this lines. A reduced matrix is formed. Then, make zero assignments as shown below.

	1	2	3	4	5	6
A	4	17	49	[0]	~~0~~	17
B	[0]	35	25	1	6	~~0~~
C	13	~~0~~	59	3	3	[0]
D	51	19	[0]	20	2	4
E	25	[0]	42	5	~~0~~	2
F	~~0~~	53	46	22	[0]	20

The other solution is:

	1	2	3	4	5	6
A	4	17	49	[0]	~~0~~	17
B	~~0~~	35	25	1	6	[0]
C	13	[0]	59	3	3	~~0~~
D	51	19	[0]	20	2	4
E	25	~~0~~	42	5	[0]	2
F	[0]	53	46	22	~~0~~	20

Hence, the optimal assignments are:
(i) $A \to 4, B \to 1, C \to 6, D \to 3, E \to 2, F \to 5$
(ii) $A \to 4, B \to 6, C \to 2, D \to 3, E \to 5, F \to 1$.
with minimum cost z = Rs 142.

5.3 VARIATIONS OF THE ASSIGNMENT PROBLEM

1. **Non-square matrix (unbalanced assignment problem)**
The Hungarian method of assignment requires that the number of columns and rows in the assignment matrix must be equal. When the given cost matrix is not a square matrix, then the problem is called an *unbalanced problem*. In this case dummy row(s) or column(s) with zero cost is added to make it a square matrix. These cells are treated the same way as the real cells during the process. Then, adopt the Hungarian method to find the solution.
2. **Maximisation problem**
There may be problems of maximising the profit, revenue, and so on. Such problems may be solved by converting the given maximisation problem into a minimisation problem before the Hungarian method is applied. The transformation may be done in the following two ways:
(i) by subtracting all the elements from the highest element of the matrix.
(ii) by multiplying the matrix elements by – 1.
3. **Multiple optimal solutions**
While making an assignment in the reduced assignment matrix, it is possible to have two or more ways to strike off certain number of zeros. Such situation leads to multiple solutions with the same optimal value of objective function. In such cases the most suitable solution may be considered by the decision-maker.
4. **Restrictions on assignments (or) impossible assignment**
Cells in which assignments are not allowed are assigned a very heavy cost (written as M or ∞). Such cells are prohibited to enter into the final solution.

Example 1 Unbalanced Problem

In the modification of a plant layout of a factory four new machines M_1, M_2, M_3, M_4 are to be installed in a machine shop. There are five vacant places A, B, C, D and E that are available. Because of limited space, machine M_2 cannot be placed at C and M_3 cannot be placed at A. The cost of placing of machine i at place j (in rupees) shown below:

		Location				
		A	B	C	D	E
Machine	M_1	9	11	15	10	11
	M_2	12	9	–	10	9
	M_3	–	11	14	11	7
	M_4	14	8	12	7	8

Find the optimal assignment schedule.

Solution: As the cost matrix is unbalanced, add one dummy row with zero cost. Also, assign a high cost M to the pair (M_2, C) and (M_3, A). The cost matrix is shown below:

	A	B	C	D	E
M_1	9	11	15	10	11
M_2	12	9	M	10	9
M_3	M	11	14	11	7
M_4	14	8	12	7	8
M_5	0	0	0	0	0

Apply the Hungarian method to get the optimal solution. The optimal solution is shown below:

	A	B	C	D	E
M_1	[0]	2	6	1	2
M_2	3	[0]	M	1	⊠
M_3	M	4	7	4	[0]
M_4	7	1	5	[0]	1
M_5	⊠	⊠	[0]	⊠	⊠

The optimal solution is obtained. The schedule are:

$$M_1 \to A,\ M_2 \to B,\ M_3 \to E,\ M_4 \to D,\ M_5 \to C$$

and the total minimum cost is Rs : 9 + 9 + 7 + 7 + 0 = Rs 32.

Example 2

A department head has four task to be performed and three subordinates to perform the task. The subordinates differ in efficiency. The estimates of time, each subordinate would take to perform, is given below in the matrix. How should he allocate the tasks one to each person, so as to minimise the total man hours?

Task	Man		
	1	2	3
I	9	26	15
II	13	27	6
III	35	20	15
IV	18	30	20

Solution: Since the problem is unbalanced, we add a dummy column with all the entries as zero and use Hungarian method for optimum solution. The reduced cost matrix after applying the algorithm is:

$$\begin{array}{c} \\ \text{I} \\ \text{II} \\ \text{III} \\ \text{IV} \end{array} \begin{array}{c} \begin{array}{cccc} 1 & 2 & 3 & 4 \end{array} \\ \left[\begin{array}{cccc} \boxed{0} & 6 & 9 & \times \\ 4 & 7 & \boxed{0} & \times \\ 26 & \boxed{0} & 9 & \times \\ 9 & 10 & 14 & \boxed{0} \end{array}\right] \end{array}$$

The optimum assignment is I → 1, II → 3, III → 2, IV → 4. The total minimum time is [9 + 6 + 20 + 0] = 35 hours.

Example 3 Maximisation Problem

A company has four territories open, and four salesman available for the assignment. The territories are not equally rich in their sales potential; it is estimated that a typical salesman operating in each territory would bring in the following annual sales:

Territory :	I	II	III	IV
Annual Sales (Rs) :	60,000	50,000	40,000	30,000

Four salesman are also considered to differ in their ability. It is estimated that working under the same conditions, their yearly sales would be proportional as follows:

Salesman:	*A*	*B*	*C*	*D*
Proportion:	7	5	5	4

If the criterion is maximise expected total sales, then the intuitive answer is to assign the best salesman to the richest territory, the next best salesman to the second richest and so on. Verify this answer by the assignment technique.

Solution:

Step 1: *Construct the effectiveness matrix.*

To avoid the fractional values of annual sales of each salesman in each territory, for convenience consider the yearly sales as 21 (the sum of sale proportion = 7 + 5 + 5 + 4 = 21), taking Rs 10,000 as one unit. Now, divide the individual sales in each territory by 21 to obtain the required annual sales by each salesman.

Thus, the maximum sales matrix is obtained as follows:

Sales in 10,000 Rupees

		6	5	4	3
Sale proportion		I	II	III	IV
7	*A*	42	35	28	21
5	*B*	30	25	20	15
5	*C*	30	25	20	15
4	*D*	24	20	16	12

Step 2: *Convert maximisation into minimisation problem.*
Subtract from the highest element (i.e. 42) among all the elements of the given cost matrix. The resulting matrix is

	I	II	III	IV
A	0	7	14	21
B	12	17	22	27
C	12	17	22	27
D	18	22	26	30

Step 3: Apply Hungarian method to get the optimal solution.

(i) Subtract the smallest element in each row from every element in that row.

(ii) Subtract the smallest element in each column. The reduced matrix is obtained as follows:

	I	II	III	IV
A	0	3	6	9
B	0	1	2	3
C	0	1	2	3
D	0	0	0	0

Step 4: Assignment is made in row *A*. All zeros in the assigned column I are crossed out. Column II has only one zero in cell (*D*, II). Assignment is made in this column and other zeros are crossed in row *D*. The reduced matrix is shown below:

	I	II	III	IV
A	[0]	3	6	9
B	~~0~~	1	2	3
C	~~0~~	1	2	3
D	~~0~~	[0]	~~0~~	~~0~~

Step 5: Since rows *B* and *C* does not have any assignments, we tick this row. Since there is a zero in the I column of the ticked row, we tick I column. Further, there is an assignment in the first row of the ticked column, so we tick first row. Draw lines through all unmarked rows and marked columns.

	I	II	III	IV	
A	[0]	3	6	9	✓
B	~~0~~	1	2	3	✓
C	~~0~~	1	2	3	✓
D	~~0~~	[0]	~~0~~	~~0~~	
	✓				

The number of lines drawn is 2, which is less than order of cost matrix, therefore the current assignment is not optimal.

Step 6: Revised matrix is developed by selecting the minimum element (= 1) among all uncovered elements by the lines. Subtract 1 from each uncovered element and add it to the element at the intersection of two lines. The revised Table is:

	I	II	III	IV
A	[0]	2	5	8
B	~~0~~	[0]	1	2
C	~~0~~	~~0~~	1	2
D	1	~~0~~	~~0~~	[0]

Again, the solution is not optimal. Repeat Step 1 to 5. Two alternative optimal assignments we get are:

	I	II	III	IV
A	[0]	2	4	7
B	~~0~~	~~0~~	[0]	1
C	~~0~~	[0]	~~0~~	1
D	2	1	~~0~~	[0]

	I	II	III	IV
A	[0]	2	4	7
B	~~0~~	[0]	~~0~~	1
C	~~0~~	~~0~~	[0]	1
D	2	1	~~0~~	[0]

The two possible solutions are:

(i) $A \rightarrow 1, B \rightarrow \text{III}, C \rightarrow \text{II}, D \rightarrow \text{IV}$

(ii) $A \rightarrow \text{I}, B \rightarrow \text{II}, C \rightarrow \text{III}, D \rightarrow \text{IV}$

with maximum sales of Rs (42 + 20 + 25 + 12) = Rs 99

[or Rs (42 + 25 + 20 + 12) = Rs 99.]

Both solution show the best salesman *A* is assigned to the richest territory I, the worst salesman *D* to the poorest territory IV. Salesman *B* and *C* are equally good and they may be assigned VI or III.

Example 4 Maximisation Problem

A company has a team of four salesman and there are four districts where the company wants to start its business. After taking into account the capabilities of salesman and the nature of districts, the company estimates that the profit per day in rupees for each salesman in each district is as below:

	1	2	3	4
A	16	10	14	11
B	14	11	15	15
C	15	15	13	12
D	13	12	14	15

Find the assignment of salesman to various districts which will yield maximum profit.

Solution:

Step 1: Convert the maximisation problem into a minimisation problem by subtracting all the elements from the highest element 16. The reduced matrix is:

	1	2	3	4
A	0	6	2	5
B	2	5	1	1
C	1	1	3	4
D	3	4	2	1

Step 2: Apply Hungarian algorithm to get the optimal solution. The optimum assignment is as follows:

	1	2	3	4
A	[0]	6	2	5
B	1	4	[0]	~~0~~
C	~~0~~	[0]	2	3
D	2	3	1	[0]

Since the number of assignment = order of cost matrix, the current solution is optimal. The schedule is:

$$A \to 1, B \to 3, C \to 2, D \to 4$$

profit per day = Rs [16 + 15 + 15 + 15] = Rs 61.

Example 5 Multiple Solutions

Solve the minimal assignment problem whose cost matrix is given below:

	1	2	3	4
I	2	3	4	5
II	4	5	6	7
III	7	8	9	8
IV	3	5	8	4

Solution:

Step 1:

(i) Subtract the smallest element in the row from each element in that row.

(ii) Subtract the smallest element in the column from each element in that column.

The reduced matrix is:

	1	2	3	4
I	0	0	0	2
II	0	0	0	2
III	0	0	0	0
IV	0	1	3	0

Since single zeros do not exist either in the columns or in the rows, we get the following alternative solutions:

	1	2	3	4
I	$\not{0}$	$\boxed{0}$	$\not{0}$	2
II	$\not{0}$	$\not{0}$	$\boxed{0}$	2
III	$\not{0}$	$\not{0}$	$\not{0}$	$\boxed{0}$
IV	$\boxed{0}$	1	3	$\not{0}$

	1	2	3	4
I	$\boxed{0}$	$\not{0}$	$\not{0}$	2
II	$\not{0}$	$\boxed{0}$	$\not{0}$	2
III	$\not{0}$	$\not{0}$	$\boxed{0}$	$\not{0}$
IV	$\not{0}$	1	3	$\boxed{0}$

	1	2	3	4
I	$\not{0}$	$\not{0}$	$\boxed{0}$	2
II	$\not{0}$	$\boxed{0}$	$\not{0}$	2
III	$\not{0}$	$\not{0}$	$\not{0}$	$\boxed{0}$
IV	$\boxed{0}$	1	3	$\not{0}$

The possible optimal solutions with each of cost Rs 20 are:

$$\text{I} \rightarrow 2, \text{II} \rightarrow 3, \text{III} \rightarrow 4, \text{IV} \rightarrow 1$$
$$\text{I} \rightarrow 1, \text{II} \rightarrow 2, \text{III} \rightarrow 3, \text{IV} \rightarrow 4$$
$$\text{I} \rightarrow 3, \text{II} \rightarrow 2, \text{III} \rightarrow 1, \text{IV} \rightarrow 4$$
$$\text{I} \rightarrow 3, \text{II} \rightarrow 2, \text{III} \rightarrow 4, \text{IV} \rightarrow 1$$
$$\text{I} \rightarrow 2, \text{II} \rightarrow 3, \text{III} \rightarrow 1, \text{IV} \rightarrow 4$$

Example 6 Restrictions on Assignments

Four new machines M_1, M_2, M_3, M_4 are to be installed in a machine shop. There are five vacant places A, B, C, D, E that are available. Because of limited space, machine M_2 cannot be placed at C and M_3 cannot be placed at A. The cost matrix is shown below:

	A	B	C	D	E
M_1	4	6	10	5	4
M_2	7	4	–	5	4
M_3	–	6	9	6	2
M_4	9	3	7	2	3

Find the optimal assignment schedule.

Solution: As machine M_2 cannot be placed at C, M_3 cannot be placed at A, assign a large cost M in cells (M_2, C) and (M_3, A). The assignment problem is unbalanced. So, balance it by adding a dummy row with cost 0 as shown below:

	A	B	C	D	E
M_1	4	6	10	5	4
M_2	7	4	M	5	4
M_3	M	6	9	6	2
M_4	9	3	7	2	3
M_5	0	0	0	0	0

Applying Hungarian method, the following matrix is obtained.

	A	B	C	D	E
M_1	[0]	2	6	1	2
M_2	3	[0]	M	1	~~0~~
M_3	M	4	7	4	[0]
M_4	7	1	5	[0]	1
M_5	~~0~~	~~0~~	[0]	~~0~~	~~0~~

The optimal assignment is:
$M_1 \rightarrow A, M_2 \rightarrow B, M_3 \rightarrow E, M_4 \rightarrow D, M_5 \rightarrow C$ (C will remain vacant).
Total assignment cost = Rs (4 + 4 + 2 + 2) = Rs 12.

5.4 CREW ASSIGNMENT PROBLEM

The method of solution discussed in this section can be used to plan the assignment of crew members in different locations by a transport company.

Example 1

A trip from Chandigarh to Delhi takes six hours by bus. A typical time table of the bus service in both directions is given below:

Departure from Chandigarh	*Route Number*	*Arrival at Delhi*	*Arrival at Chandigarh*	*Route Number*	*Departure from Delhi*
6.00	a	12.00	11.30	1	5.30
7.30	b	13.30	15.00	2	9.00
11.30	c	17.30	21.00	3	15.00
19.00	d	01.00	00.30	4	18.30
00.30	e	6.30	06.00	5	00.00

The cost of providing this service by the transport company depends upon the time spent by the bus crew (driver and conductor) away from their places in addition to service time. There are five crews.

There is a constraint that every crew should be provided with more than 4 hours of rest before the return trip again and should not wait for more than 24 hours for the return trip. The company has residential facilities for the crew at Chandigarh as well as at Delhi. Find the optimal service line connections.

Solution: As the service time is constant for each line it does not appear directly in the computation. If the entire crew resides at Chandigarh then the waiting times in hours at Delhi for different route connections are given in the following Table:

	1	2	3	4	5
a	17.5	21	*M*	6.5	12
b	16	19.5	*M*	5	10.5
c	12	15.5	21.5	*M*	6.5
d	4.5	8	14	17.5	23
e	23	*M*	8.5	12	17.5

If route *a* is combined with route 1, the crew after arriving at Delhi at 12 noon start at 5.30 next morning. Thus, the waiting time is 17.5 hrs.

Some of the assignments are infeasible. Route 3 leaves Delhi at 15:00 hrs. Thus, the crew of route *a* reaching Delhi at 12 noon are unable to take rest of 4 hours if they are asked to leave by route 3. Hence, (*a*, 3) is an infeasible assignment. Its cost is thus, *M*, a large positive number.

Similarly, if the crew are assumed to reside at Delhi (so that they start from and came back to Delhi with halt for minimum time at Chandigarh), then waiting time at Chandigarh for different service line connections are given by the following Table.

	1	2	3	4	5
a	18.5	15	9	5.5	*M*
b	20	16.5	10.5	7	*M*
c	M	20.5	14.5	11	5.5
d	7.5	*M*	22	18.5	13
e	13	7.5	*M*	*M*	18.5

As the crew can be asked to reside at Chandigarh or Delhi, minimum waiting time from the above operation can be computed for different route combination by choosing the minimum of the two waiting time. These values of the waiting time are shown below:

	1	2	3	4	5
a	17.5	15	9	5.5	12
b	16	16.5	10.5	5	10.5
c	12	15.5	14.5	11	5.5
d	4.5	8	14	17.5	13
e	13	9.5	8.5	12	17.5

Hungarian method can now be applied for finding the optimal route connections which gives minimum over all waiting time and hence the minimum cost of bus service operations. It consists of the following steps.

Step 1: Find the first and second reduced cost matrix.

	1	2	3	4	5
a	12	9.5	3.5	0	6.5
b	11	11.5	5.5	0	5.5
c	6.5	10	9	5.5	0
d	0	3.5	9.5	13	8.5
e	4.5	1	0	3.5	9

	1	2	3	4	5
a	12	8.5	3.5	0	6.5
b	11	10.5	5.5	0	5.5
c	6.5	9	9	5.5	0
d	0	2.5	9.5	13	8.5
e	4.5	0	0	3.5	9

Step 2: Check for optimality.

	1	2	3	4	5
a	12	8.5	3.5	[0]	6.5
b	11	10.5	5.5	~~0~~	5.5
c	6.5	9	9	5.5	[0]
d	[0]	2.5	9.5	13	8.5
e	4.5	[0]	~~0~~	3.5	9

The solution is not optimal, as the number of assignments is less than the number of rows or columns.

Step 3: Mark ✓ in row *b* since no assignment was made in this row. Note that row *b* has zero in column 4, therefore mark ✓ in column 4. We then mark ✓ in row *a*, since column 4 has an assigned zero in row *a*. Draw straight lines through all unmarked rows and marked columns. The resulting matrix is:

	1	2	3	4	5	
a	12	8.5	3.5	[0]	6.5	✓
b	11	10.5	5.5	~~0~~	5.5	✓
c	6.5	9	9	5.5	[0]	
d	[0]	2.5	9.5	13	8.5	
e	4.5	[0]	~~0~~	3.5	9	
				✓		

Step 4: Develop a new revised matrix. Examine those elements that are not covered by the lines and among those elements take the smallest element. Here, 3.5 is the element. Subtract 3.5 from the uncovered elements and add 3.5 with the elements at the intersection of two lines, we get the new Table as:

8.5	5	0	0	3
7.5	7	2	0	2
6.5	9	9	9	0
0	2.5	9.5	16.5	8.5
4.5	0	0	7	9

Step 5: Check if optimal assignment can be made in the current feasible solution.

8.5	5	[0]	~~0~~	3
7.5	7	2	[0]	2
6.5	9	9	9	[0]
[0]	2.5	9.5	16.5	8.5
4.5	[0]	~~0~~	7	9

As the number of assignment is equal to the number of rows or columns, the current solution is optimal.

Therefore, the routes to be paired to achieve the minimum waiting time are $a - 3$, $b - 3$, $c - 5$, $d - 1$, and $e - 2$. We can obtain the waiting times of these assignments as well as the residence of the crew as

Crew	*Resident at*	*Service number*	*Waiting time hours*
1	Chandigarh	(*d* - 1)	4.5
2	Delhi	(2 - *e*)	9.5
3	Delhi	(3 - *a*)	9
4	Chandigarh	(*b* - 4)	5
5	Delhi	(5 - *c*)	5.5

Total minimum waiting time is thus 33.5 hours.

Example 2

A small aeroplane company, operating seven days a week, serves three cities *A*, *B*, *C* according to the schedule shown below. The layover cost per stop is roughly proportional to the square of the layover time. How should the planes be assigned the flights so as to minimise the total lay over cost?

Flight Route No.	From	Departure	To	Arrival
A_1B	A	09 A.M.	B	12 A.M.
A_2B	A	10 A.M.	B	01 P.M.
A_3B	A	03 A.M.	B	06 P.M.
A_4C	A	08 P.M.	C	12 P.M.
A_5C	A	10 P.M.	C	02 A.M.
B_1A	B	04 A.M.	A	07 A.M.
B_2A	B	11 A.M.	A	02 P.M.
B_3A	B	03 P.M.	A	06 A.M.
C_1A	C	07 A.M.	A	11 A.M.
C_2A	C	03 P.M.	A	07 P.M.

Solution: We make the following observations:

(1) Any plane flying from a station must come back within 24 hours for the scheduled trip.

(2) Any plane starting from *A* to *B* must avail the next opportunity to come back to *A*.

(3) A plane cannot make more than 2 trips, that is, going and coming back.

(4) Five planes are operating at a time.

From the given information, it is clear that there is no route between city *B* and city *C*. Therefore, the problem is divided into two sub-problems:

(i) Routes between *A* and *C*.

(ii) Routes between *A* and *B*.

Consider the cost matrix associated with the flights connecting cities *A* and *C*.

Any plane through C_1A must return back to *C* by evening route A_4C or A_5C. At station *A*, the lay over for A_4C is 9 hours (11 AM – 8 PM) while at station *C*, the lay over for C_1A is 7 hours (12 PM – 7 AM).

Thus, the layover cost for route $C_1A - A_4C$ is $9^2 + 7^2 = 130$ units. Similarly, the other route costs are as follows:

Flight	A_4C	A_5C
C_1A	130	146
C_2A	226	178

Next consider the cost matrix for the flights connecting *A* and *B*. Any plane through the route A_1B cannot return back by B_2A because reaching *B* at 12 AM the plane has to return the next day at 11 AM.

Thus, the route may be associated with infinte cost. Likewise, the layover costs for other routes can be computed and the following cost matrix is obtained.

	B_1A	B_2A	B_3A
A_1B	260	∞	234
A_2B	234	∞	260
A_3B	164	290	∞

Solving the problems by Hungarian method we get the following optimal assignment:

[0]	16
48	[0]

26	∞	[0]
[0]	∞	~~0~~
~~0~~	[0]	∞

Hence, we obtain the following optimal route schedule for the five planes:

Plane number:	1	2	3	4	5
Departure route:	A_1B	A_2B	A_3B	C_1A	C_2A
Arrival Route:	B_2A	B_1A	B_2A	A_4C	A_5C

5.5 TRAVELLING SALESMAN PROBLEM

Suppose a salesman wants to visit a certain number of cities starting from his headquarters. The distances (or cost or time) of journey between every pair of cities, denoted by c_{ij}, that is, distance from city i to city j is assumed to be known. The problem is:

Salesman starting from his home city visiting each city only once and returns to his home city in the shortest possible distance (or at the least cost or in the least time).

Given n cities and distance c_{ij}, the salesman starts from city 1, then any permutation of 2, 3, ..., n represents the number of possible ways for his tour. So, there are $(n-1)!$ possible ways for his tour. The problem is to select an optimal route that could achieve his objective.

The problem may be classified as:

(i) *Symmetrical*: If the distance between every pair of cities is independent of the direction of his journey.

(ii) *Asymmetrical*: For one or more pair of cities the distance changes with the direction.

5.5.1 Formulation of a Travelling-Salesman Problem as Assignment Problem

The distance (or cost or time) from city i to city j is c_{ij} and $x_{ij} = 1$, if the salesman goes directly from city i to city j, and $x_{ij} = 0$ otherwise. Then the objective function is

$$\text{Minimise } z = \sum_i \sum_j x_{ij}\, c_{ij}$$

with the additional restriction that x_{ij} must be so chosen that no city is visited twice before the tour is completed. In particular, he cannot go directly from city i to i. So, assign $c_{ii} = \infty$.

However, all $c_{ij} > 0$ and $c_{ij} + c_{jk} > c_{ik}$ for all i, j, k.

Omit the variable x_{ij} from the problem specification. It is also important to note that only single $x_{ij} = 1$ for each value of i and j. The cost matrix is shown below:

		To				
		1	2	3	⋯	n
	1	∞	c_{12}	c_{13}	⋯	c_{2n}
	2	c_{21}	∞	c_{23}	⋯	c_{2n}
From	⋮	⋮	⋮	⋮		⋮
	⋮	⋮	⋮	⋮		⋮
	n	c_{n1}	c_{n2}	c_{n3}	⋯	∞

Example 1

A machine operator processes five types of items on his machine each week, and must choose a sequence for them. The set-up cost per change depends on the item presently on the machine and the set-up to be made according to the following table.

	To item				
From item	A	B	C	D	E
A	∞	4	7	3	4
B	4	∞	6	3	4
C	7	6	∞	7	5
D	3	3	7	∞	7
E	4	4	5	7	∞

If he processes each type of item once and only once each week, how should he sequence the items on his machine in order to minimise the total set-up cost?

Solution:

Step 1: Reduce the cost matrix using Steps 1 and 2 of the Hungarian algorithm and then make assignments in rows and columns having single zeros as usual.

∞	1	3	[0]	1
1	∞	2	~~0~~	1
2	1	∞	2	[0]
~~0~~	[0]	3	∞	4
[0]	~~0~~	~~0~~	3	∞

Step 2: Note that row 2 is not assigned. So, Mark ✓ to row 2. Since there is a zero in the 4th column of the marked row, we tick 4th column. Further, there is an assignment in the first row of 4th column. So, tick first row. Draw lines through all unmarked rows and marked columns. We can find the number of lines is 4 which is less than the order of the matrix. So, go to next step (See table).

∞	1	3	[0]	1	✓
1	∞	2	~~0~~	1	✓
2	1	∞	2	[0]	
~~0~~	[0]	3	∞	4	
[0]	~~0~~	~~0~~	3	∞	
			✓		

Step 3: Subtract the lowest element from all the elements not covered by these lines and add the same with the elements at the intersection of two lines. Then we get the table as:

	1	2	3	4	5
1	∞	~~0~~	2	[0]	~~0~~
2	[0]	∞	1	~~0~~	~~0~~
3	2	1	∞	3	[0]
4	~~0~~	[0]	3	∞	4
5	~~0~~	~~0~~	[0]	4	∞

The optimum assignment is $1 \to 4, 2 \to 1, 3 \to 5, 4 \to 2, 5 \to 3$ with minimum cost as Rs 20.

This assignment schedule does not provide us the solution of the travelling salesman problem as it gives $1 \to 4, 4 \to 2, 2 \to 1$, without passing through 3 and 5.

Next, we try to find the next best solution which satisfies this restriction. The next minimum (non-zero) element in the cost matrix is 1. So, we bring 1 into the solution. But the element '1' occurs at two places. We consider all cases separately until we get an optimal solution.

We start with making an assignment at (2, 3) instead of zero assignment at (2, 1). The resulting assignment schedule is

$$1 \to 4, 4 \to 2, 2 \to 3, 3 \to 5, 5 \to 1$$

When an assignment is made at (3, 2) instead of zero assignment at (3, 5), the resulting assignment schedule is

$$1 \to 5, 5 \to 3, 3 \to 2, 2 \to 4, 4 \to 1$$

The total set-up cost in both the cases is 21.

Example 2

Solve the following travelling-salesman problem.

		To			
		A	*B*	*C*	*D*
From	*A*	∞	46	16	40
	B	41	∞	50	40
	C	82	32	∞	60
	D	40	40	36	∞

Solution: After subtracting the smallest element of each row from all the row elements of the corresponding row we obtain the first reduced matrix. Next, subtract the smallest element of each column from all the column elements of the corresponding column. We obtain the second reduced matrix. Then, make the assignment in rows and columns having single zero. We get

∞	30	[0]	24
[0]	∞	10	~~0~~
49	[0]	∞	28
3	4	~~0~~	∞

Since only three assignments are made, which is less than the order of the cost matrix, the current assignment is not optimal.

Proceeding as usual, we get the second iteration as:

∞	27	[0]	21
~~0~~	∞	13	[0]
49	[0]	∞	28
[0]	1	~~0~~	∞

which gives optimal solution. The optimum assignment is

$$A \to C, B \to D, C \to B, D \to A$$

that is,

$$A \to C, C \to B, B \to D, D \to A$$

with minimum cost = Rs 128.

Example 3

Solve the travelling salesman problem given by the following data:

$$c_{12} = 20,\ c_{13} = 4,\ c_{14} = 10,\ c_{23} = 5,\ c_{34} = 6$$
$$c_{25} = 10,\ c_{35} = 6,\ c_{45} = 20 \text{ where } c_{ij} = c_{ji}$$

and there is no route between cities i and j if the value for c_{ij} is not shown.

Solution: The cost matrix is:

∞	20	4	10	∞
20	∞	5	∞	10
4	5	∞	6	6
10	∞	6	∞	20
∞	10	6	20	∞

Repeating the steps as before using the Hungarian algorithm, the optimum table obtained is :

∞	12	[0]	~~0~~	∞
11	∞	~~0~~	∞	[0]
~~0~~	1	∞	[0]	1
[0]	∞	~~0~~	∞	9
∞	[0]	~~0~~	8	∞

The solution is

$$1 \to 3,\ 3 \to 4,\ 4 \to 1,\ 5 \to 2,\ 2 \to 5$$

which is not the solution of the travelling salesman problem as the sequence obtained is not in the cyclic order.

The next lowest number (other than 0) is 1. Therefore, make an assignment in the cell (3, 2) having the element 1. Consequently, make an assignment in the cell (5, 4) having element 8, instead of zero element in the cell (5, 2). The assignment Table is

∞	12	[0]	~~0~~	∞
11	∞	~~0~~	∞	[0]
~~0~~	[1]	∞	~~0~~	1
[0]	∞	~~0~~	∞	9
∞	~~0~~	~~0~~	[8]	∞

The shortest route for the travelling salesman is

$$1 \to 3 \to 2 \to 5 \to 4 \to 1.$$

EXERCISES

Section 5.1

1. What is an assignment problem? Give two areas of its applications.
2. Explain the conceptual justification that an assignment problem can be viewed, as a linear programming problem.
3. Explain the difference between a transportation problem and an assignment problem.
4. What is the dual of an assignment problem? What are the techniques used in solving an assignment problem.
5. Show that the assignment problem is a special case of the transportation problem. Why is the transportation method to find the optimal solution not preferred in the assignment problem.
6. Describe the mathematical formulation of an assignment problem.

Section 5.2

7. Explain the steps in the Hungarian method for solving assignment problems.
8. Find the optimal solution for the assignment problem with the following cost matrix.

	I	II	III	IV	V
A	11	17	8	16	20
B	9	7	12	6	15
C	13	16	15	12	16
D	21	24	17	28	26
E	14	10	12	11	15

[*Answer*: $A \to$ I, $B \to$ IV, $C \to$ V, $D \to$ III, $E \to$ II, minimum cost = 60.]

9. Show that the lines drawn in the assignment algorithm pass through all the zeros and have the property that every line passes through one and only one assignment.
10. Five men are available to do five different jobs. From the past records, the time (in 2 hour) that each man takes to do each job is known and given in the following Table.

	I	II	III	IV	V
A	2	9	2	7	1
B	6	8	7	6	1
C	4	6	5	3	1
D	4	2	7	3	1
E	5	3	9	5	1

Find the assignment of men to jobs that will minimise the total time taken.
[*Answer*: $A \to$ III, $B \to$ V, $C \to$ I, $D \to$ IV, $E \to$ II, optimal value = 13 hours.]

11. A company has four machines on which three jobs have to be done. Each job can be assigned to one and only one machine. The cost of each job on each machine is given in the following Table.

		Machine			
		P	*Q*	*R*	*S*
Job	A	18	24	28	32
	B	8	13	17	19
	C	10	15	19	22

What are the job assignments which will minimise the cost?
[*Answer*: $A \to P$, $B \to Q$, $C \to R$, $D \to S$, cost = 50.]

12. Solve the following assignment problems.

(i)

	1	2	3	4	5
A	8	4	2	6	1
B	0	9	5	5	4
C	3	8	9	2	6
D	4	3	1	0	3
E	9	5	8	9	5

[*Answer*: $A \to 5$, $B \to 1$, $C \to 4$, $D \to 3$, $E \to 2$ minimum cost = 9.]

Section 5.3

13. The owner of a small machine shop has four machinists available to assign jobs for the day. Five jobs are offered with the expected profit (in Rs) for each machinist on each job as follows:

		Job				
		A	*B*	*C*	*D*	*E*
	1	6.20	7.80	5.00	10.10	8.20
Machinist	2	7.10	8.40	6.10	7.30	5.90
	3	8.70	9.20	11.10	7.10	8.10
	4	4.80	6.40	8.70	7.70	8.00

Find the assignment of machinists to jobs that will result in a maximum profit. Which job should be declined?
[*Answer*: $1 \to D$, $2 \to B$, $3 \to C$, $4 \to E$, $5 \to A$, cost = Rs 37.60. Job A should be declined.]

14. How do you deal with the assignment problem where the objective function is to be maximised?
15. Explain how to modify an effectiveness matrix in an assignment problem if a particular assignment is prohibited.
16. What is an unbalanced assignment probl m? How is the Hungarian method applied for obtaining a solution if the matrix is rectangular?
17. In an assignment problem, how do you take care of an impossible assignment.
18. A department has four subordinates and four tasks to be performed. The subordinates differ in efficiency and the tasks differ in their intrinsic difficulty. The estimates of the profit in rupees each man would earn is given in the effectiveness matrix. How should the tasks be allocated, one to each man, so as to maximise the total earnings?

	A	*B*	*C*	*D*
1	5	40	20	5
2	25	35	30	25
3	15	25	20	10
4	15	5	30	15

[*Answer*: $1 \to$ B, $2 \to D$, $3 \to A$, $4 \to C$ with maximum earnings = Rs 110.]

19. Consider the problem of assigning five operators to five machines. The assignment costs are given below:

	M_1	M_2	M_3	M_4	M_5
A	7	7	–	4	8
B	9	6	4	5	6
C	11	5	7	–	5
D	9	4	8	9	4
E	8	7	9	11	3

Operator *A* cannot be assigned to machine M_3 and operator *C* cannot be assigned to machine M_4. Find the optimum assignment schedule.
[*Answer*: $A \to M_4$, $B \to M_3$, $C \to M_2$, $D \to M_1$, $E \to M_5$, minimum cost = Rs 25.]

20. A student has five days to revise the subject before examination. The course is divided into 4 sections. He decides to devote a whole day to the study of some section so that he may study a section for one day, two days, three days and so on, or not at all. The expected grade points he will get for alternative arrangements are as follows:

	Course Sections I	II	III	IV
0	1	1	0	0
1	2	1	0	1
2	2	2	1	2
3	3	3	2	2
4	4	3	3	3
5	4	4	3	4

How should he distribute the days to various sections to maximise his grade point average?

21. Assign the five jobs to the five men so as to maximize the sales.

		Jobs I	II	III	IV	V
	A	+ 2	− 1	+ 4	+ 1	− 6
Men	*B*	− 6	+ 4	− 7	+ 1	+ 4
	C	− 8	− 2	+ 5	− 2	− 4
	D	+ 5	− 3	+ 2	+ 3	+ 2
	E	− 1	+ 6	+ 1	+ 4	− 3

22. Solve the following unbalanced problem of assigning four jobs to three different men. The time to perform the job by different men is given in the following Table.

		Job J_1	J_2	J_3	J_4
	M_1	7	5	8	4
Men	M_2	5	6	7	4
	M_3	8	7	9	8

[*Answer*: $M_1 \to J_4$, $M_2 \to J_1$, $M_3 \to J_2$, $M_4 \to J_3$, minimum time = 16 hrs.]

Section 5.4

23. An airline that operates seven days a week has the following time-table. Crews must have a minimum lay over of 5 hours between flights. Obtain the pairing of flights that minimises layover time away from home. For any given pairing, the crew will be based at the city that results in smaller layover.

Delhi-Jaipur

Flight No.	Departure	Arrival
1	7:00 AM	8:00 AM
2	8:00 AM	9:00 AM
3	1:30 PM	2:30 PM
4	6:30 PM	7:30 PM

Jaipur-Delhi

Flight No.	Departure	Arrival
101	8:00 AM	9:15 AM
102	8:30 AM	9:45 AM
103	12:00 Noon	11:15 PM
104	5:30 PM	6:45 PM

For each pair also mention the town where the crew should be based.

[*Answer*: 3 → 101, 102 → 4, 2 → 104, 103 → 1, 2 → 104, minimum layover : 52.5 hrs].

Section 5.5

24. Explain the nature of a travelling salesman problem and give its mathematical formulation.
25. What is the difference between assignment problem and travelling salesman problem?
26. What is the objective of the travelling salesman problem?
27. Write short note on travelling salesman problem.
28. Solve the travelling salesman problem as to minimise the cost per cycle.

		To				
	City	1	2	3	4	5
	1	∞	10	25	25	10
From	2	1	∞	10	15	2
	3	8	9	∞	20	10
	4	14	10	24	∞	15
	5	10	8	25	27	∞

[*Answer*: 1 → 5 → 2 → 3 → 4 → 1, cost = Rs 62.]

29. Solve the travelling salesman problem, given

$c_{12} = 16, c_{13} = 4, c_{14} = 12, c_{23} = 6, c_{34} = 5, c_{25} = 8$
$c_{35} = 6, c_{45} = 20, c_{ij} = c_{ji}, c_{ij} = \infty,$

for all values of i and j not given in the data.

[*Answer*: 1 → 2 → 5 → 3 → 4 → 1, with minimum cost = Rs 47.]

30. Distances between six islands in the Indian Ocean are given in the Table.

		To					
		A	*B*	*C*	*D*	*E*	*F*
From	*A*	–	120	90	100	110	160
	B	120	–	190	200	210	270
	C	90	190	–	30	40	90
	D	100	200	30	–	30	80
	E	110	210	40	30	–	70
	F	160	270	90	80	70	–

It is proposed to lay submarine cable of smallest length to connect all the islands by telephone connections. Find the minimum spanning tree for the six islands mentioned.
[*Answer*: $A \to B$, $B \to A$, $C \to D$, $D \to C$, $E \to F$, $F \to E$, minimum distance = 440.]

31. Solve the travelling salesman problem.

	City	*A*	*B*	*C*	*D*	*E*
	A	∞	4	10	14	2
From	*B*	12	∞	6	10	4
	C	16	14	∞	8	14
	D	24	8	12	∞	10
	E	2	6	4	16	∞

[*Answer*: $A \to B \to C \to D \to E \to A$, minimum cost = Rs 30.]

6

Dynamic Programming

6.1 INTRODUCTION

The mathematical technique of optimising a sequence of interrelated decisions over a period of time is called *dynamic programming* (DP). It uses the idea of recursion to solve a complex problem, broken into a series of sub-problems. The word dynamic has been used because time is explicitly taken into consideration. The objective in dynamic programming is to select a decision policy so to optimise the returns that are in the form of costs or benefits.

Mathematically a dynamic programming problem DPP is a decision-making problem in n variables, the problem being sub-divided into n sub-problems and each such problem being a decision-making problem in one variable only. The solution to a DPP is achieved sequentially starting from one (initial) stage to the next till the final stage is reached. This dynamic programming technique was developed by Richard Bellman in the early 1950. Dynamic programming can be called as *recursive optimisation*.

To convert a verbal problem into a multi-stage structure is not always simple and sometimes it becomes very difficult, DP technique can be applied to problems of inventory control, production planning, chemical reactor design, business situation to take an optimal decision for investment and so on.

Some important concepts in DP are:

(i) *Stage*: The DPP can be decomposed or divided into a sequence of smaller sub-problems called *stages* of the original problem. At each stage there are a number of decision alternatives, and the best out of those is called the *stage decision*, which may not be optimal for the stage, but contributes to obtain the optimal decision policy. Stages very often represent different time periods associated with the planning period of the problem, places, people or other entities. For example, in the replacement problem, each year is a stage and in the salesman allocation problem, each territory is a stage.

(ii) *State*: The condition of the decision process at a stage is called its *state*. The variables which specify the condition of the decision process at a particular stage are called *State variables*. The number of state variables should be as small as possible, since larger the number of state variables, more complicated is the decision process.

(iii) *Principle of optimality*: An optimal policy (set of decisions) has the property that whatever be the initial state and initial decisions, the remaining decisions must constitute an optimal policy for the state resulting from the first decision.

The problem which does not satisfy the principle of optimality cannot be solved by the dynamic programming method. This principle implies that a wrong decision taken at a stage does not prevent from taking optimal decisions for the remaining stages.

(iv) *Return function*: At each stage, a decision is made which can effect the state of the system at the next stage and help in arriving at the optimal solution at the current stage. Every decision that is made has its merit which can represented in an algebraic equation form. This equation is called a *return function*. This return function generally depends on the state variable and the decision made at a particular stage. In optimal policy at a stage gives optimal return for a given value of the state variable.

(v) *Recursive relationship*: A mathematical relationship that connects the decision policy for each state at stage *n*, given the optimal policy for each state at stage $(n - 1)$, is called *recursive relationship*.

Let n = stage number, s_n = state input to stage n from stage $(n + 1)$, d_n = the decision variable at stage n independent of previous stages.

$$f_n = r_n (s_n, d_n) = \text{return (or objective) function for stage } n.$$

Suppose, there are n stages at which a decision is to be made. These n stages are interconnected by the relationship called transition function $s_{n-1} = s_n * d_n$, where * represents any mathematical operation like addition, multiplication, subtraction or division. The units of s_n, d_n, s_{n-1} are homogeneous.

At each stage of the problem, there are two inputs: one is, state variable s_n and the other is decision variable d_n. The state variable is the state input which relates the present stage back to the previous stage. The decision d_n is made at stage n for optimising the total return over the remaining $n - 1$ stages. The decision d_n which optimises the output at stage n produces two outputs:

(i) the return $r_n (s_n, d_n)$

(ii) the new state variable s_{n-1}.

The return function which is expressed as function of the state variable s_n and decision variable d_n indicates about the state of the process at the beginning of the next state (stage $n - 1$) and is denoted by $s_{n-1} = t_n (s_n, d_n)$ where t_n represents a state transformation function and its form depends on the particular problem to be solved.

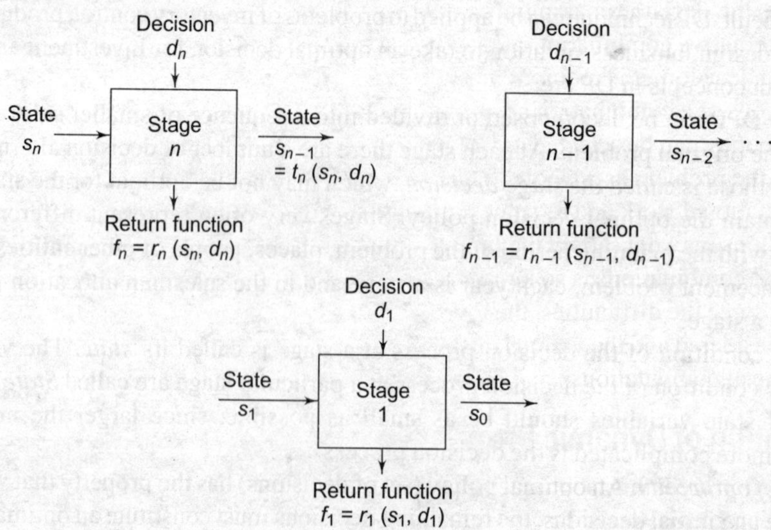

Based on the principle of optimality, the solution procedure starts by solving a one-stage problem and then sequentially adding a series of one-stage problems that are solved until the overall optimum of the initial problem is obtained. The solution procedure is based on *backward induction process* and *forward induction process.*

In backward process, the problem is solved from the last stage and then worked backwards towards the first stage, making optimal decision at each stage of the problem. In certain cases the second process is used to solve a problem by first solving the first stage and then keep moving to the last stage, making an optimal decision at each stage of the problem.

The recursion relationship would vary according to the nature of the DPP. The stage 1 return is given by

$$f_1 = r_1\,(s_1,\, d_1)$$

and the optimal value of f_1 under the state variable s_1 can be obtained by selecting a suitable decision variable d_1. That is,

$$f_1^*\,(s_1) = \underset{d_1}{\text{optimum}}\,\{r, (s_1, d_1)\}$$

The range of d_1 is determined by s_1, but s_1 is determined by what has happened in stage 2.

Then, in stage 2, the return function will take the form

$$f_2^*\,(s_2) = \underset{d_2}{\text{optimum}}\,\{r_2\,(s_2, d_2) * f_1^*\,(s_1)\}$$

where,
$$s_1 = t_2\,(s_2, d_2).$$

Recursively, we can define,

$$f_n^*\,(s_n) = \underset{d_n}{\text{optimum}}\,\{r_n\,(s_n, d_n) * f_{n-1}^*\,(s_{n-1})]$$

where, $s_{n-1} = t_n\,(s_n, d_n)$, '*' denotes addition, subtraction, multiplication and so on.

6.1.1 Need for Dynamic Programming Problem

In many situations, the decision-making process consists of selecting a combination of plans from a large number of alternative combinations.

Before making a decision, it is required that

(a) all the decisions of a combination are specified.

(b) the optimal policy can be selected only after all the combinations are evaluated.

While considering the problem as a whole, the problem has some important drawbacks like

(i) lot of computational work and too much time is involved.

(ii) all combinations may not satisfy the limitations and thus may be infeasible.

(iii) the number of combinations is so large.

So, in order to overcome the difficulties, the given problem is divided into sub-problems or stages. One stage at a time is considered and the various infeasible combinations are eliminated with the objective of reducing the volume of computations.

6.1.2 Application of Dynamic Programming Problem

1. In the production area, this technique has been used for production, scheduling and employment smoothening problems.
2. It has been used in inventory model, for formulating inventory recording.

3. It has been used in assignment of different resources to different destinations.
4. It has been used in replacement model to determine at which age the equipment is to be replaced for optimal return from the facilities.
5. It has been used in network analysis, job shop scheduling, capital budgeting, allocation of research and development funds, long-term corporate planning, solving linear programming problem, probabilistic decision and so on.

6.1.3 Characteristics of Dynamic Programming

The characteristics of DPP are as follows:

(i) The problem can be divided into stages, with a policy decision at each stage.
(ii) Each stage consists of a number of states associated with it.
(iii) Decision at each stage converts the current stage into a state associated with the next stage.
(iv) The state of the system at a stage is described by state variables.
(v) When the current state is known, an optimal policy for the remaining stages is independent of the policy of the previous ones.
(vi) The solution procedure begins by finding the optimal policy for each state to the last stage.
(vii) To identify the optimal policy for each state of the system a recursive relationship is formulated with n stages remaining, given the optimal policy for each state with $(n-1)$ stages left.
(viii) Using recursive equation approach, each time the solution procedure moves backward stage by stage for obtaining the optimum policy of each state for that particular stage, till it attains the optimum policy beginning at the initial stage.

6.1.4 Definition

Decision Tree

A multi-stage decision system, in which each decision and state variable can take only finite number of values, can be represented graphically by a *decision tree.*

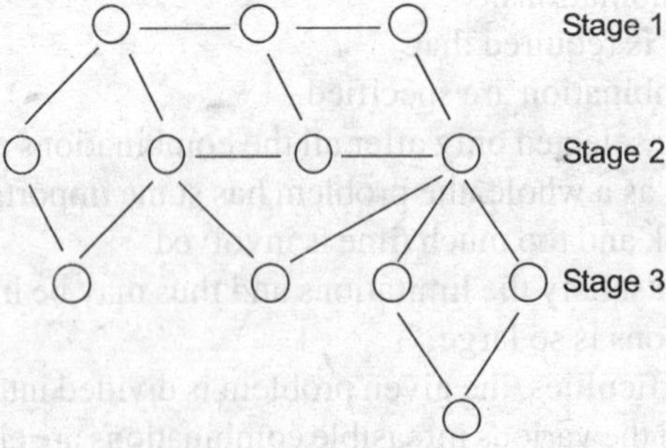

figure 6.1

In Fig 6.1, circles represents the nodes corresponding to stages and lines between circles denotes arcs corresponding to decisions.

6.1.5 Dynamic Programming Algorithm

The procedure for solving a problem by using the DPP approach can be summarised in the following steps.

Step 1: Identify the problem decision variables and specify objective function to be optimised under certain limitations, if any.

Step 2: Decompose the given problem into a number of smaller sub-problems (or stages). Identify the state variable at each stage and write down the transformation function.

Step 3: Write down a general recursive relationship for computing the optimal policy. Decide which method is to be followed either backward or forward.

Step 4: Construct appropriate stages to show the required values of the return function at each stage as shown in Table 6.1.

Step 5: Determine the overall optimal policy or decisions and its value at each stage. There may be more than one such optimal policy.

table 6.1

Possible decision.

$d_n \rightarrow$ / Entering states $s_n \downarrow$	$\dfrac{f_n(s_n, d_n)}{d_n}$	Optimum return $f_n^*(s_n)$	Optimal decision d_n^*

Example 1

Solve the following by dynamic programming

$$\text{Minimise } z = y_1 + y_2 + \cdots + y_n$$

subject to the constraints

$$y_1\, y_2 \cdots y_n = b$$

and,

$$y_1, y_2, \cdots y_n \geq 0 \text{ (or)}$$

Factorise a positive quantity b into n factors in such a way so that their sum is a minimum.

Solution: *To develop the recursive equation.*

Let $f_n(b)$ be the minimum attainable sum $y_1 + y_2 \cdots + y_n$, when the positive quantity 'b' is factorised in to n factors $y_1, y_2, \cdots y_n$.

One-stage problem: Let $n = 1$

Here b is factorised into one factor namely $y_1 = b$. Hence,

$$f_1(b) = \min_{y_1 = b} \{y_1\} = b \text{ (trivial case)} \tag{6.1.1}$$

Two-stage problem: Let $n = 2$.

Factorise b into two factors, $y_1 = x$ and $y_2 = \dfrac{b}{x}$ such that $y_1\, y_2 = b$. Then,

$$f_2(b) = \min_{0 \leq x \leq b} \{y_1 + y_2\} = \min_{0 \leq x \leq b} \left\{x + \frac{b}{x}\right\}$$

$$= \min_{0 \le x \le b} \left\{ x + f_1\left(\frac{b}{x}\right) \right\} \qquad (6.1.2)$$

Three-stage problem. Factorise b into three factors $y_1 = x$, $y_2 \cdot y_3 = \frac{b}{x}$ so that $y_1\, y_2\, y_3 = b$. Then,

$$f_3(b) = \min_{0 \le x \le b} \{y_1 + y_2 + y_3\}$$

$$= \min \left\{ x + f_2\left(\frac{b}{x}\right) \right\} \quad \text{[by (6.1.2)]} \qquad (6.1.3)$$

In general the recursive equation for the n-stage problem is

$$f_n(b) = \min_{0 \le x \le b} \left\{ x + f_{n-1}\left(\frac{b}{x}\right) \right\}$$

To solve the recursive equation.
For $n = 2$, equation (6.1.2) is

$$f_2(b) = \min_{0 \le x \le b} \left\{ x + f_1\left(\frac{b}{x}\right) \right\}$$

$$= \min_{0 \le x \le b} \left\{ x + \frac{b}{x} \right\}.$$

The function $x + \frac{b}{x}$ will attain its minimum when $\frac{dy}{dx} = 0$ and $\frac{d^2y}{dx^2} > 0$ (by using calculus).

That is, when $\frac{dy}{dx} = 1 - \frac{b}{x^2} = 0$, the function $x + \frac{b}{x}$ will become minimum.

That is, when $x = \sqrt{b}$ the function will become minimum. Hence,

$$f_2(b) = \min_{0 \le x \le b} \left\{ x + \frac{b}{x} \right\}$$

$$= \sqrt{b} + \frac{b}{\sqrt{b}} = 2\sqrt{b}$$

The optimum policy is $\left(\sqrt{b}, \sqrt{b}\right)$ and $f_2(b) = 2\sqrt{b}$

For $n = 3$, equation (6.1.3) becomes

$$f_3(b) = \min_{0 \le x \le b} \left\{ x + f_2\left(\frac{b}{x}\right) \right\}$$

$$= \min_{0 \le x \le b} \left\{ x + 2\sqrt{\frac{b}{x}} \right\}$$

The function $x + 2\sqrt{\frac{b}{x}}$ will attain its minimum, when $\frac{dy}{dx} = 1 + 2 \times \frac{1}{2}\left(\frac{b}{x}\right)^{-1/2} \left(\frac{-b}{x^2}\right) = 0$

That is, the function $x + 2\sqrt{\frac{b}{x}}$ will attain its minimum when $x = b^{1/3}$. Hence,

$$f_3(b) = \min_{0 \le x \le b} \left\{ x + 2\left(\frac{b}{x}\right)^{1/2} \right\}$$

$$= b^{1/3} + 2\left(\frac{b}{b^{1/3}}\right)^{1/2}$$

$$= 3b^{1/3}.$$

The optimal policy is $(b^{1/3}, b^{1/3}, b^{1/3})$ and $f_3(b) = 3b^{1/3}$. Let us assume that the optimal policy for $n = m$ is $\left(b^{\frac{1}{m}}, b^{1/m}, \ldots b^{\frac{1}{m}}\right)$ and $f_m(b) = mb^{1/m}$.

Now, for $n = m + 1$, equation (6.1.3) becomes

$$f_{m+1}(b) = \min_{0 \le x \le b} \left\{ x + f_m\left(\frac{b}{x}\right) \right\}$$

$$= \min_{0 \le x \le b} \left\{ x + m\left(\frac{b}{x}\right)^{\frac{1}{m}} \right\}$$

The function $y = x + m\left(\frac{b}{x}\right)^{1/m}$ will attain its minimum when $\frac{dy}{dx} = 1 + m \times \left(\frac{1}{m}\right)\left(\frac{b}{x}\right)^{\frac{1}{m}-1}\left(\frac{-b}{x^2}\right) = 0.$

That is, when $1 - \frac{b^{\frac{1}{m}}}{x^{1/m+1}} = 0$, the function $x + m\left(\frac{b}{x}\right)^{\frac{1}{m}}$ attains its minimum. That is, when $x = b^{\frac{1}{m+1}}$ the function $x + m\left(\frac{b}{m}\right)^{\frac{1}{m}}$ attains its minimum. Further, note that $\frac{d^2y}{dx^2} > 0$, for $x = b^{\frac{1}{m+1}}$. Hence,

$$f_{m+1}(b) = \min_{0 \le x \le b} \left\{ x + m\left(\frac{b}{x}\right)^{\frac{1}{m}} \right\}$$

$$= b^{\frac{1}{m+1}} + m\left[\frac{b}{b^{\frac{1}{m+1}}}\right]^{\frac{1}{m+1}}$$

$$= (m + 1)\, b^{\frac{1}{m+1}}.$$

The result is true for $n = m + 1$. Hence, by mathematical induction, the optimal policy is $(b^{1/n}, b^{1/n}, \cdots b^{1/n})$ and $f_n(b) = nb^{1/n}$.

6.2 SOME DYNAMIC PROGRAMMING TECHNIQUES

6.2.1 Single Additive Constraint, Multiplicatively Separable Return

Consider the problem:

To maximise $$z = \prod_{j=1}^{n} f_j(y_j)$$

subject to $\sum_{j=1}^{n} a_j\, y_j = b, \quad y_j \geq 0, \quad a_j \geq 0.$

First, introduce state variables, that is,

$$s_n = \sum_{j=1}^{n} a_j y_j = b;\ s_{j-1} = s_j - a_j\, y_j,\ j = 2, 3, \cdots n$$

Let $$f_j^*(s_j) = \max_{y_1 y_2, \cdots y_j} \prod_{1}^{j} f_j(y_j)$$

then the general recursion formula becomes

$$f_j^*(s_j) = \max_{y_j}\left[f_j(y_j)\, f_{j-1}^*(s_{j-1})\right], j = n, n-1, \cdots 2$$

or, $$f_1^*(s_1) = f_1(y_1)$$

6.2.2 Single Additive Constraint, Additively Separable Return

Consider the problem.

To minimise $z = \sum_{j=1}^{n} f_j(y_j)$ subject to the constraints

$$\sum_{j=1}^{n} a_j\, y_j \geq b,\ a_j \text{ and } b \text{ are real numbers}$$

$$a_j \geq 0, \quad y_j \geq 0, \quad b > 0$$

The return at the jth stage is the function $f_j(y_j)$, where y_j is the decision variable. Introduce state variables $s_0, s_1, s_2, \ldots, s_n$ as

$$\begin{aligned}
s_n &= a_1 y_1 + a_2 y_2 + \cdots + a_n y_n \geq b\\
s_{n-1} &= a_1 y_1 + a_2 y_2 + \cdots + a_{n-1}\, y_{n-1} = s_n - a_n\, y_n\\
s_{n-2} &= a_1 y_1 + \cdots + a_{n-2}\, y_{n-2} = s_{n-1} - a_{n-1}\, y_{n-1}\\
&\cdots \qquad \cdots \qquad \cdots\\
s_1 &= a_1 y_1 = s_2 - a_2\, y_2.
\end{aligned}$$

Also, $$s_{j-1} = t_j(s_j, y_j) \quad 1 \leq j \leq n$$

is the stage transformation function and indicates that each state variable is a function of next state and decision variables. $f_n^*(s_n)$ denotes the minimum value of z for any feasible value of s_n, where s_n being the function of all decision variables. Thus,

$$f_n^*(s_n) = \min_{y_1, y_2, \cdots y_n} \left[f_1(y_1) + f_2(y_2) + \cdots + f_n(y_n)\right]$$

where, $s_n \geq b.$

Now,

$$f_n^*(s_n) = \min_{y_1, y_2, \cdots y_{n-1}} \left[\sum_{j=1}^{n} f_j(y_j)\right] + f_n(y_n)$$

$$= \min_{y_n} \left[f_n(y_n) + f_{n-1}^*(s_{n-1})\right]$$

Recursively,

$$f_j^*(s_j) = \min_{y_j} \left[f_j(y_j) + f_{j-1}^*(s_{j-1})\right]$$

$1 \leq j \leq n$, and $f_1^*(s_1) = f_1(y_1)$.

6.2.3 Single Multiplicative Constraint, Additively Separable Return

Consider the problem.

$$\text{Minimise } z = f_1(y_1) + f_2(y_2) + \cdots + f_n(y_n)$$

subject to the constraints

$$y_1 y_2 \cdots y_n \geq p, \quad p > 0, y_i \geq 0 \text{ for all } i$$

State variables are defined as:

$$s_n = y_n y_{n-1} \cdots y_2 y_1 \geq p$$

$$s_{n-1} = \frac{s_n}{y_n} = y_{n-1} y_{n-2} \cdots y_2 y_1$$

$$\cdots \quad \cdots \quad \cdots \quad \cdots$$

$$s_2 = \frac{s_3}{y_3} = y_2 y_1$$

$$s_1 = \frac{s_2}{y_2} = y_1$$

These state variables are stage transformation of the type

$$s_{j-1} = t_j(s_j, y_j)$$

Let $f_n^*(s_n)$ be the minimum value of the objective function for any feasible s_n. Thus, proceeding as earlier obtain the recursive formula

$$f_j^*(s_j) = \min_{y_j} \left[f_j(y_j) + f_{j-1}^*(s_{j-1})\right], \quad 2 \leq j \leq n.$$

6.2.4 Systems Involving More than One Constraint

Dynamic programming models discussed so far involved only one constraint apart from non-negativity. In single constraint problems there has to be single state variable for each stage, whereas in multi-constraint problems there has to be one state variable per constraint per stage. Sometimes, it is possible to reduce the number of state variables. The procedure for this model will remain the same.

Problems

Example 1 Discrete Variables

A government space project is conducting research on a certain engineering problem that must be solved before man can fly to the moon safely.These research teams are currently trying three different approaches for solving this problem. The estimate has been made that, under present circumstances, the probability that the respective teams, call them *A*, *B* and *C* will not succeed are 0.40, 0.60 and 0.80, respectively. Thus, the current probability that all three teams will fail is (0.40) × (0.60) × (0.80) = 0.192. Since the objective is to minimise this probability, the decision has been made to assign two or more top scientists among the three teams in order to lower it as much as possible. The following Table gives the estimated probability that the respective teams will fail when 0, 1, or additional scientists are added to that team.

		Team		
		A	*B*	*C*
	0	0.40	0.60	0.80
Number of new	1	0.20	0.40	0.50
scientists	2	0.15	0.20	0.30

How should the additional scientists be allocated to the team?

Solution: In this problem, the research teams are corresponding to the *stages* in the DPP.

Let s denote the number of new scientists still available for assignment at that stage, x_j the number of additional scientists allocated to team (stage j), $p_j\,(x_j)$ denote the probability of failure for team j if it is assigned x_j additional scientists as prescribed in the Table.

The problem is: minimise $z = p_1\,(x_1)\,p_2(x_2)\,p_3(x_3)$

subject to the constraints

$$x_1 + x_2 + x_3 = 2$$

and, $x_1, x_2,. x_3 \geq 0,\quad x_1, x_2, x_3$ are integers.

To obtain Recursive Equations

Let $f_j(x_j)$ be the value of the optimal allocation for teams 1 through j both inclusive.

For $j = 1$

$$f_1\,(x_1) = p_1\,(x_1)$$

If $f_j\,(s, x_j)$ is the probability associated with the optimum solution $f_j^*\,(s)$, $j = 1, 2, \cdots n$, then

$$f_j(s, x_j) = p_j(x_j) \times \min\,[p_{j+1}\,(x_{j+1})\,p_{j+2}\,(x_{j+2})\,p_{j+3}\,(x_{j+3})]$$

such that $\sum_{i=j}^{3} x_i = s$ and x_i are non-negative integers, $j = 1, 2, 3$.

The recursive equations are:

$$f_j^*(s) = \min_{x_j \le s} f_j(s, x_j)$$

$$f_j(s, x_j) = p_j(x_j) f^*_{j+1}(s - x_j)$$

where,

$$f_j^*(s) = \min_{x_j \le s} \left[p_j(x_j) f_{j+1}^*(s - x_j) \right] \quad j = 1, 2, 3.$$

When $j = 3$

$$f_3^*(s) = \min_{x_3 \le s} p_3(x_3).$$

Solution of Recursive Equations

Since all the quantities in the recurrence equation are discrete, differential calculus method cannot be used. Let the optimal policy be denoted by x_j^*, $j = 1, 2, 3$. Now, proceed backward from $j = 3$ for each stage one by one.

Computation for one-stage problem 1: ($j = 3$)

s	$f_3^*(s)$	x_3^*
0	0.80	0
1	0.50	1
2	0.30	2

Computation for two-stage problem 2: ($j = 2$)

	$f_2(s, x_2) = p_2(x_2) \times f_3^*(s - x_2)$			Optimum solution	
s \ x_2	0	1	2	$f_2^*(s)$	x_2^*
0	(0.60) (0.80) = 0.48	–	–	0.48	0
1	(0.60) (0.50) = 0.30	(0.40) (0.80) = 0.32	–	0.30	0
2	(0.60) (0.30) = 0.18	(0.40) (0.50) = 0.20	(0.20) (0.80) = 0.16	0.16	2

Computation for three-stage problem 3: ($j = 1$)1

	$f_1(s, x_1) = p_1(x_1) \times f_2^*(s - x_1)$			Optimum solution	
s \ x_1	0	1	2	$f_1^*(s)$	x^*_1
0	(0.40) (0.48) = 0.192	–	–	0.192	0
1	(0.40) (0.30) = 0.120	(0.20) × (0.48) = 0.096	–	0.096	1
2	(0.40) (0.16) = 0.064	(0.20) × (0.30) = 0.060	(0.15) (0.48) = 0.072	0.060	

Therefore, optimum solution will have $x_1^* = 1$, making $s = 1$ at the second stage, so that $x_2^* = 0$ making $s = 1$ at the third stage, so that $x_3^* = 1$. Hence, first and third teams should receive one additional scientist. The new probability that all the three teams will fail is 0.060

Example 2

Divide a positive quantity c into n-parts in such away that their product is maximum.

(or)

$$\text{Maximise } z = y_1 y_2 \cdots y_n \text{ subject to the constraints}$$

$$y_1 + y_2 + \cdots + y_n = c$$

and, $$y_i \geq 0, i = 1, 2 \cdots n$$

Solution: Let y_i be the ith part of c and each i may be considered as a stage. Since y_i may assume any non-negative value which satisfies $y_1 + y_2 + \cdots + y_n = c$, the alternative at each stage are infinite. This means y_i is continuous. Hence, the optimal decision at each stage are obtained by using differentiation.

Let $f_n(c)$ be the maximum attainable product $y_1 y_2 \cdots y_n$ when c is divided into n parts $y_1, y_2, \cdots y_n$. Thus, $f_n(c)$ is a function of n.

One-stage problem 1: For $n = 1$, c is divided into one part only, then $y_1 = c \therefore f_1(c) = c$

Two-stage problem 2: For $n = 2$, c is divided into two parts $y_1 = x$, $y_2 = c - x$ such that $y_1 + y_2 = c$. Then,

$$f_2(c) = \max_{0 \leq x \leq c} \{y_1 y_2\}$$

$$= \max_{0 \leq x \leq c} \{x(c - x)\}$$

$$= \max_{0 \leq x \leq c} \{x \cdot f_1(c - x)\}$$

Three-stage problem 3: For $n = 3$, c is divided into three parts y_1, y_2, y_3. Let $y_1 = x$, $y_2 + y_3 = c - x$. So that $y_1 + y_2 + y_3 = c$

$$f_3(c) = \max_{0 \leq x \leq c} \{y_1 y_2 y_3\} = \max_{0 \leq x \leq c} \{x \cdot f_2(c - x)\}$$

Generalising, for n stages we get

$$f_n(c) = \max_{0 \leq x \leq c} \{x \cdot f_{n-1}(c - x)\}$$

To solve the recursive equation

$$f_2(c) = \max_{0 \leq x \leq c} \{x\, f_1(x - c)\}$$

$$= \max_{0 \leq x \leq c} \{x(c - x)\}$$

The function $x(c - x)$ will attain its maximum, when $\frac{d}{dx}(x(c - x)) = 0 \Rightarrow x(-1) + (c - x) = 0$

$$\Rightarrow \quad x = \frac{c}{2}.$$

Hence,

$$f_2(c) = \max_{0 \leq x \leq c} \{x(c - x)\}$$

$$= \left(\frac{c}{2}\right)\left(c - \frac{c}{2}\right) = \left(\frac{c}{2}\right)^2$$

The optimal policy is $\left(\frac{c}{2}, \frac{c}{2}\right)$ and $f_2(c) = \left(\frac{c}{2}\right)^2$.

For $n = 3$,

$$f_3(c) = \max_{0 \le x \le c} \{x\, f_2(c - x)\}$$

$$= \max_{0 \le x \le c} \left\{x\left(\frac{c - x}{2}\right)^2\right\}$$

The function $x\left(\frac{c-x}{2}\right)^2$ will attain its maximum value at $x = \frac{c}{3}$.

Hence,

$$f_3(c) = \left(\frac{c}{3}\right)\left(\frac{c - c/3}{2}\right)^2 = \left(\frac{c}{3}\right)^3$$

The optimal policy is $\left(\frac{c}{3}, \frac{c}{3}, \frac{c}{3}\right)$ and $f_3(c) = \left(\frac{c}{3}\right)^3$.

Assume the optimal policy for $n = m$. That is $\left(\frac{c}{m}, \frac{c}{m}, \cdots \frac{c}{m}\right)$ and $f_m(c) = \left(\frac{c}{m}\right)^m$. Then, for $n = m + 1$,

$$f_{m+1}(c) = \max_{0 \le x \le c} \{x\, f_m(c - x)\}$$

$$= \max_{0 \le x \le c} \left\{x\left(\frac{c - x}{m}\right)^m\right\}.$$

The function $x\left(\frac{c-x}{m}\right)^m$ attains its maximum value at $\frac{d}{dx}\left(x\left(\frac{c-x}{m}\right)^m\right) = 0$

$$\Rightarrow \quad \left(\frac{c-x}{m}\right)^m + xm\left(\frac{c-x}{m}\right)^{m-1}\left(\frac{-1}{m}\right) = 0$$

$$\Rightarrow \quad \left(\frac{c-x}{m}\right)^{m-1}\left[\frac{c-x}{m} - x\right] = 0 \Rightarrow c - x - mx = 0$$

$$\Rightarrow \quad c - x(1 + m) = 0 \Rightarrow x = \frac{c}{1+m}.$$

Hence,

$$f_{m+1}(c) = \max_{0 \le x \le c} \left\{x\left(\frac{c - x}{m}\right)^m\right\}$$

$$= \left(\frac{c}{1+m}\right)\left(\frac{c-\left(\frac{c}{1+m}\right)}{m}\right)^m = \left(\frac{c}{m+1}\right)^{m+1}$$

Hence, the result is true for $n = m + 1$. By induction the optimal policy is $\left(\frac{c}{n}, \frac{c}{n}, \cdots \frac{c}{n}\right)$ and $f_n(c) = \left(\frac{c}{n}\right)^n$.

Example 3 Discrete Variable

A student has to take examination in three courses x, y, z. He has three days available for study. He feels it would be best to devote a whole day to study the same course, so that he may study a course for one day, two days or three days or not at all. His estimates of grades he may get by studying are as follows:

Study days/course	x	y	z
0	1	2	1
1	2	2	2
2	2	4	4
3	4	5	4

How should he plan to study so that he maximises the sum of his grades?

Solution: Let n_1, n_2, n_3 be the number of days he should study the courses x, y, z, respectively. If $f_1(n_1), f_2(n_2), f_3(n_3)$ be the grades earned by such a study then the problem becomes,

maximise $z = f_1(n_1) + f_2(n_2) + f_3(n_3)$

subject to $n_1 + n_2 + n_3 \leq 3$

and, $n_1, n_2, n_3 \geq 0$ and integers.

Here, n_j are decision variables and $f_j(n_j)$ are the respective return functions for $j = 1, 2, 3$. Let us define the state variables s_j (s_j denote number of days available for jth stage) as follows:

$$s_3 = n_1 + n_2 + n_3 \leq 3$$
$$s_2 = n_1 + n_2 = s_3 - n_3$$
$$s_1 = n_1 = s_n - n_2$$

The state transformation functions are defined by

$$s_{j-1} = t_j(s_j, n_j), j = 2, 3.$$

The recursive equations are:

$$f_j^*(s_j) = \max_{n_j}\left\{f_j(n_j) + f_{j-1}^*(s_{j-1})\right\}, j = 2, 3$$

(or) $f_1^*(s_1) = f_1(n_1)$

where $$f_3^*(s_3) = \max_{n_1, n_2, n_3} \left[f_1(n_1) + f_2(n_2) + f_3(n_3)\right]$$

for any feasible value of s_3 the required solution would become $\max_{s_3} f_3(s_3)$.

Recursive equation gives the solution as follows.

Stage returns $f_j(n_j)$

j \ n_j	0	1	2	3
1	1	2	2	4
2	2	2	4	5
3	1	2	4	4

Stage transformation $s_{j-1}, j = 2, 3$

s_j \ n_j	0	1	2	3
0	0	–	–	–
1	1	0	–	–
2	2	1	0	–
3	3	2	1	0

Recursive operations

	$f_n(n_2)$				$f_1^*(s_1) = f_1(n_1)$				$f_2(n_2) + f_1^*(s_1)$				$f_2^*(s_2)$
s_2 \ n_2	0	1	2	3	0	1	2	3	0	1	2	3	
0	2	–	–	–	1	–	–	–	3	–	–	–	3
1	2	2	–	–	2	1	–	–	4	3	–	–	4
2	2	2	4	–	2	2	1	–	4	4	5	–	5
3	2	2	4	5	4	2	2	1	6	4	6	6	6

	$f_3(n_3)$				$f_2^*(s_2) = f_2(n_2)$				$f_3(n_3) + f_2^*(s_2)$				$f_3^*(s_3)$
s_3 \ n_3	0	1	2	3	0	1	2	3	0	1	2	3	
0	1	–	–	–	3	–	–	–	4	–	–	–	4
1	1	2	–	–	4	3	–	–	5	5	–	–	5
2	1	2	4	–	5	4	3	–	6	6	7	–	7
3	1	2	4	4	6	5	4	3	7	7	8	7	8

Proceeding backwards through enclosed type numbers the optimal policy is obtained as $n_3 = 2, n_2 = 0$, $n_1 = 1$, keeping in view that $n_1 + n_2 + n_3 \leq 5$.

The required maximum return is 5.

Example 4 Maximisation Problem

A truck can carry a total of 10 tonnes of product. Three types of product are available for shipment. Their weights and values are tabulated. Assuming that at least one of each type must be shipped determine the loading which will maximise the total value.

Type	*Value* Rs	*Weight* *(tonnes)*
A	20	1
B	50	2
C	60	2

Solution: In this problem, a decision is to be taken as to how many units of *A*, *B*, *C* should be transported. Let each stage represent the decision regarding the number of units of a given product to be transported. The problem is divided into three stages.

One-stage problem 1: It is a problem of dividing the amount of product *C* to be transported. The profit per unit of *C* (weight = 2 tonnes) is Rs 60. The restriction is that at least one limit of types *A* and *B* must be transported. Out of maximum of 10 tonnes, (1 + 2) tonnes are allotted to *A* and *B*. Hence, we can load the remaining 7 tonnes only. The total load varies from 2 tonnes to 7 tonnes represented by s_j. Let x_j be the states representing one, two, or three units ($j = 1, 2, 3$). Let $f_1^*(s_1)$ be the optimum profit in one stage problem. The results are tabulated as:

s_1 \ x_1	1	2	3	$f_1^*(s)$	x_1^*
2	$1 \times 60 = 60$	-	-	60	1
3	$1 \times 60 = 60$	-	-	60	1
4	$1 \times 60 = 60$	$2 \times 60 = 120$	-	120	2
5	$1 \times 60 = 60$	$2 \times 60 = 120$	-	120	2
6	$1 \times 60 = 60$	$2 \times 60 = 120$	$3 \times 60 = 180$	180	3
7	$1 \times 60 = 60$	$2 \times 60 = 120$	$3 \times 60 = 180$	180	3

Two-stage problem 2: Here, how much of product *B* and *C* is to be transported problem is to be decided. We take the decision to allot some space to transport product *B* and the remaining space is available for *C* for which the optimum values are taken from the results of one stage problem. The space to be used for *B* and *C* together varies from 4 tonnes to 9 tonnes.

Let x_2 represent the amount of units of product of type *B* to be transported. Then, the remaining $(s_2 - x_2)$ tonnes of space is allotted for *C*. The results are shown in the following Table.

s_2 \ x_2	1	2	3	$f_2^*(s)$	x_2^*
4	$1 \times 50 + 60 = 110$	-	-	110	1
5	$1 \times 50 + 60 = 110$	-	-	110	1
6	$50 + 120 = 170$	$2 \times 50 + 60 = 160$	-	170	1
7	$50 + 120 = 170$	$2 \times 50 + 60 = 160$	-	170	1
8	$50 + 180 = 230$	$2 \times 50 + 120 = 220$	$3 \times 50 + 60 = 210$	230	1
9	$50 + 180 = 230$	$2 \times 50 + 120 = 220$	$3 \times 50 + 60 = 210$	230	1

Three-stage problem: Here we consider all the three types of products. Let s_3 be the space available for transporting items, x_3 be the amount of product of type *A* to be transported, so that $(s_3 - x_3)$ is available for loading the products *B* and *C* together for which the optimum profit is taken from the two-stage problem. The space available (s_3) varies from 5 tonnes to 10 tonnes and x_3 varies from one tonne to 6 tonnes. The result of the three stage problem is as shown below:

s_3 \ x_3	1	2	3	4	5	6	f_3^*	x_3^*
5	20 + 110 = 130	–	–	–	–	–	130	1
6	20 + 110 = 130	40 + 110 = 150	–	–	–	–	150	2
7	20 + 170 = 190	40 + 110 = 150	60 + 110 = 170	–	–	–	190	1
8	20 + 170 = 190	40 + 170 = 210	60 + 130 = 190	–	–	–	210	2
9	20 + 230 = 250	40 + 170 = 210	60 +130 = 190	80 + 110 = 190	–	–	250	1
10	20 + 230 = 250	40 + 230 = 270	60 + 170 = 230	80 + 170 = 250	100 + 110 = 210	120 + 110 = 230	270	2

Thus, the optimal solution is given by

$$x_1^* = 3, x_2^* = 1, x_3^* = 2, \text{ with } f_3^*(s) = 270.$$

Example 5 Project Management

A construction contractor has four construction projects underway and wants to minimise the time required to complete all projects. The following Table gives the estimated time required to complete the project for a specified number of foremen assigned to the project.

Estimated time

Project	*No. of foremen*		*Assigned*
	1	2	3
A	15	13	12
B	17	15	13
C	19	18	17
D	21	18	18

Solution: The contractor has only six foremen and each project must be assigned to at least one foreman. By considering each project as a stage, we have 4 stages. Let x_i be the number of foremen assigned to project i and s_i be the number of foremen still available for allocation. Let $t(x_i)$ be the completion time for project i when x_i foremen are assigned. The state variables $s_{i-1} = s_i - x_i$, from which the values of s_4, s_3, s_2, s_1 can be computed. For the stage 4, six foremen are available for allocation. x_4 can take values 1, 2 or 3, since at least one foremen must be allocated to each project. Hence, the possible values for s_3 are $6 - 1 = 5$, $6 - 2 = 4$, $6 - 3 = 3$. Also, as x_3 assume values 1, 2, 3, s_2 takes the values as $5 - 1 = 4$, $5 - 2 = 3$, $5 - 3 = 2$. Note that s_2 cannot have value 1, since we need at least one foreman for the project A. Lastly, s_1 has the values as 3, 2, 1. The total return function is

$$f_i\,(s_i, x_i) = t(x_i) + f_{i-1}^*\,(s_i - x_i)$$

where, $f_{i-1}^*\,(s_i - x_i)$ is the minimum completion time for stage ($i = 1$).

Stage 1:

	$f_1(s_1, x_1) = t(x_1)$				
s_1 \ x_1	1	2	3	$f_1^*(s_1)$	x_1^*
1	15	–	–	15	1
2	–	13	–	13	2
3	–	–	12	12	3

Stage 2:

	$f_2(s_2, x_2) = t(x_2) + f_1^*(s_2 - x_2)$				
s_2 \ x_2	1	2	3	$f_2^*(s_2)$	x_2^*
2	17 + 15 = 32	–	–	32	1
3	17 + 13 = 30	15 + 15 = 30	–	30	1 or 2
4	17 + 12 = 29	15 + 13 = 28	13 + 15 = 28	28	2 or 3

Note that $f_2(3, 1) = t(1) + f_1^*(3 - 1) = t(1) + f_1^*(2) = 17 + 13 = 30$, where $f_1^*(2)$ is read from the table for stage 1, and $t(1)$ the estimated time for project *B*.

Stage 3:

	$f_3(s_3, x_3) = t(x_3) + f_2^*(s_3 - x_3)$				
s_3 \ x_3	1	2	3	$f_3^*(s_3)$	x_3^*
3	19 + 32 = 51	–	–	51	1
4	19 + 30 = 49	18 + 32 = 50	–	49	1
5	19 + 28 = 47	18 + 30 = 48	17 + 32 = 49	47	1

Stage 4:

	$f_4(s_4, x_4) = t(x_4) + f_3^*(s_4 - x_4)$				
s_4 \ x_4	1	2	3	$f_4^*(s_4)$	x_4^*
6	21 + 47 = 68	18 + 49 = 67	18 + 51 = 69	67	2

Since $x_4^* = 2$, two foremen are assigned to project *D*. Hence, $s_3 = 6 - 2 = 4$. Since $x_3^* = 1$; one foreman is assigned to project *C*. Now, $s_2 = 4 - 1 = 3$. As $x_2^* = 1$ or 2, we have two possibilities. If $x_2^* = 1$, one foreman is assigned to *B* and the remaining two foremen to *A*, otherwise, if $x_2^* = 2$, two foremen are assigned to *B* and the remaining one foreman is assigned to *A*. The minimum completion time is 67.

Example 6

Solve the following using dynamic programming.

$$\text{Minimise } z = y_1^2 + y_2^2 + \cdots + y_n^2$$

subject to the constraints

$$y_1 y_2 \cdots y_n = c \neq 0$$

and, $$y_j \geq 0, j = 1, 2, \cdots n$$

Solution: Let $f_n(c)$ be the minimum attainable value of z when c is divided into n factors.

For $n = 1$, we have $y_1 = c$ and therefore,

$$f_1(c) = \min_{y_1 = c} \{y_1^2\} = c^2$$

For $n = 2$, let $y_1 = x$, $y_2 = c/x$. Then,

$$f_2(c) = \min_{0 \leq x \leq c} \{y_1^2 + y_2^2\} = \min_{0 \leq x \leq c} \{x^2 + (c/x)^2\}$$

$$= \min_{0 \leq x \leq c} \{x^2 + f_1(c/x)^2\}$$

since $f_1(c) = c^2$, we have $f_1(c/x) = (c/x)^2$.

For $n = 3$, let $y_1 = x$, $y_2 y_3 = c/x$. Then,

$$f_3(c) = \min_{0 \leq x \leq c} \{y_1^2 + y_2^2 + y_3^2\}$$

$$= \min_{0 \leq x \leq c} \{x^2 + f_2(c/x)\}$$

Recursively, we have

$$f_n(c) = \min_{0 \leq x \leq c} \{x^2 + f_{n-1}(c/x)\}$$

Solution of recursive equation:

Let $f(x) = x^2 + (c/x)^2$. Then, $\dfrac{df}{dx} = 2x - \dfrac{2c^2}{x^3} = 0$ gives $x = c^{1/2}$, therefore $y_1 = c^{1/2}$ and $y_2 = c/x = c^{1/2}$.

Now, $$f_2(c) = \min_{0\ x \leq c} \{x^2 + (c/x)^2\}$$

$$= (c^{1/2})^2 + \left(\frac{c}{c^{1/2}}\right)^2 = 2c$$

Thus, $$f_2(c/x) = 2(c/x).$$

Hence, the optimal policy is $(c^{1/2}, c^{1/2})$ and $f_2(c) = 2c$.

$$f_3(c) = \min_{0 < x \leq c} \{x^2 + f_2(c/x)\}$$

$$= \min_{0 < x \leq c} \{x^2 + 2(c/x)\}$$

$$= (c^{1/3})^2 + 2\left(\frac{c}{c^{1/3}}\right) = 3c^{2/3}$$

as the minimum of $f_3(x) = x^2 + 2c/x$ was obtained at $x = c^{1/3}$. Hence, the optimal policy is $\{c^{1/3}, c^{1/3}, c^{1/3}\}$ and $f_3(c) = 3c^{1/3}$. Generalising, the optimal policy for an n-stage problem will be $\{c^{1/n}, c^{1/n}, \cdots, c^{1/n}\}$ and $f_n(c) = nc^{2/n}$.

Example 7

Suppose there are n machines which can perform two jobs. If x of them do the first job, then they produce goods worth $g(x) = 3x$, and if y of them perform the second job, then they produce goods worth $n(y) = (2.5)y$. Machines are subject to depreciation so that after performing the first job only $a(x) = x/3$ machines remain available and after performing the second job $b(y) = 2y/3$ machines remain available in the beginning of the second year. This process is repeated with the remaining machines. Obtain the maximum total return after three years and also find the optimal policy in each year.

Solution: Let us define the following notations:

j = stages, each year being viewed as a stage, $y = 1, 2, 3$.
x_j = number of machines devoted to job 1 in the year j.
y_j = number of machines devoted to job 2 in the year j
s = state variable, the number of machines at any stage (year)
$f_j(s)$ = maximum return function when initial available machines are s with j more stages (years) to go.

There are three stages in this problem.

Stage 1: When $j = 1$

Using backward approach, that is, $j = 1$ (third year) we have $s = s_3$ number of machines at the beginning of the year. The return function is

$$f_1(s_3) = \max_{x_3,\, y_3} [3x_3 + (2.5)y_3]$$

subject to the constraint

$$x_3 + y_3 \le s_3$$

and,

$$x_3, y_3 \ge 0.$$

Since the function $f_1(s_3)$ is a linear function in x_3 and y_3 its maximum can be obtained by the concept of extreme point solutions of linear programming problems as shown in Figure 6.3

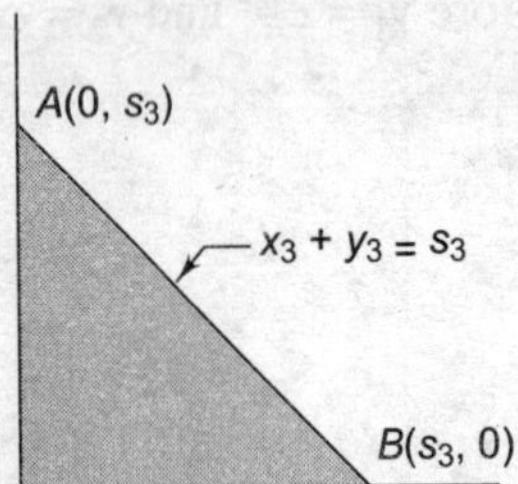

figure 6.3 **Extreme point solution (Stage 1)**

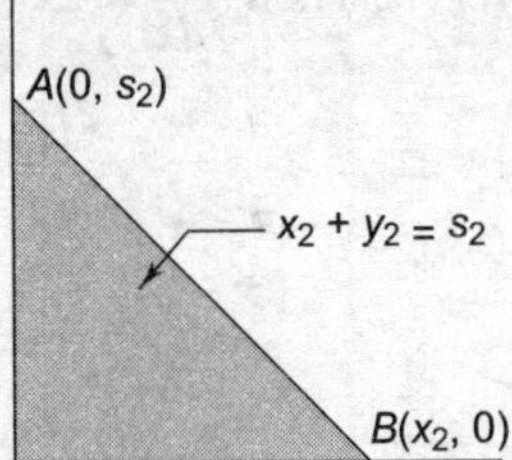

figure 6.4 **Extreme point solution (Stage 2)**

The maximum value of return function occurs at $B(s_3, 0)$, where

$$f_1(s_3) = 3x_3 + (2.5)\,y_3$$
$$= 3s_3 + (2.5)0 = 3s_3$$

Hence, optimal decision at this stage is

$$x_3^* = s_3,\ y_3^* = 0,\ f_1^*(s_3) = 3s_3$$

Stage 2: $j = 2$

$$f_2(s_2) = \max_{x_2,\, y_2} \left\{ f_1(s_3) + f_1^*\left(\frac{x_2}{3} + \frac{2y_2}{3}\right) \right\}$$

where $f_1(s) = 3x_2 + (2.5)\,y_2$ = Immediate return from stage 2, $f_1^*\left(\frac{x_2}{3} + \frac{2y_2}{3}\right)$ is the maximum known return from stage 1, subject to the constraint

$$x_2 + y_2 \le s_2$$
$$x_2,\ y_2 \ge 0$$

where, $x_2,\ y_2$ = number of machines devoted to job 1 and 2, respectively in the second year.

$\frac{x_2}{3}, \frac{2y_2}{3}$ = machines which will be available at the beginning of the next year.

Since $f_1^*\ (s_3) = 3s_3$, we have

$$f_2(s_2) = \max_{x_2,\,y_2} \left\{3x_2 + 2.5\,y_2 + 3\left(\frac{x_2}{3} + \frac{2y_2}{3}\right)\right\}$$
$$= \max_{x_2,\,y_2}\ \{4x_2 + (4.5)\ y_2\}$$

This is again a linear function in x_2 and y_2 whose maximum occurs at corner point $A(0, s_2)$ as shown in Fig. 6.4.

The maximum value of the return function at $A(0, s_2)$ is $f_2(s_2) = 4 \times 0 + (4\,.\,5)s_2 = (4.5)\ s_2$.

Hence, the optimal decision at this stage is

$$x_2^* = 0,\ y_2^* = s_2 \text{ and } f_2^*\ (s_2) = (4.5)s_2.$$

Stage 3: *(j = 3).* By backward approach, consider the first year as third stage. The return function at this stage is

$$= \max \begin{Bmatrix} \text{Immediate return} & + & \text{Maximum known} \\ \text{from stage 3} & & \text{return from stage 2} \end{Bmatrix}$$

$$f_3\ (s_1) = \max \left\{3x_1 + 2.5\ y_1 + f_2^*\left(\frac{x_1}{3} + \frac{2y_1}{3}\right)\right\}$$
$$= \max_{x_1, y_1} \left\{3x_1 + 2.5y_1 + (4.5)\left(\frac{x_1}{3} + \frac{2y_1}{3}\right)\right\}$$
$$= \max_{x_1,\,y_1}\{4.5x_1 + 5.5y_1\}$$

subject to the constraint

$$x_1 + y_1 \le s_1$$
$$x_1,\ y_1 \ge 0$$

The return function is a linear function in x_1 and y_1, whose maximum occurs at the corner point $A\ (0, s_1)$ as shown in Fig. 6.5. The maximum value of return function at $A\ (0, s_1)$ is

$$f_3\ (s_1) = 4.5 \times 0 + (5.5)s_1 = (5.5)s_1$$

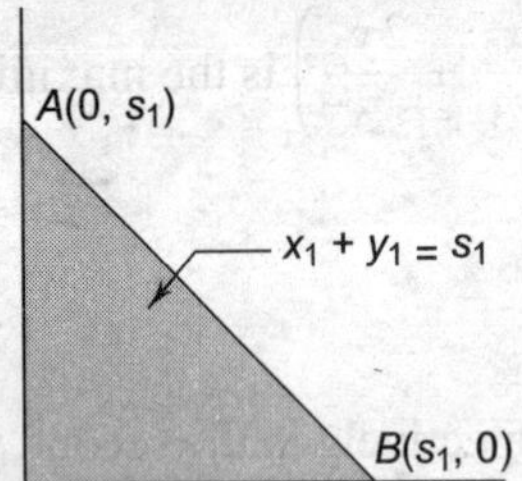

figure 6.5

Hence, the optimal decision at this stage is

$$x_1^* = 0,\ y_1^* = s_1 \text{ (= } m \text{ machines say), and } f_3^*\ (s_1) = (5.5)\ m.$$

The values of x_2, x_3, y_2, y_3 can be obtained as follows

$$y_2 = s_2 = 2y_1/3 = 2m/3 \text{ and } x_2 = 0$$

$$x_3 = s_3 = 2y_2/3 = \frac{2}{3}\left[\frac{2m}{3}\right] = \frac{4m}{9} \text{ and } y_3 = 0$$

$$x_1 = 0;\ y_1 = m.$$

Hence,

At first year (stage 3) : $x_1 = 0$, $y_1 = m$

At second year (stage 2) : $x_2 = 0$, $y_2 = 2m/3$

At third year (stage 1) : $x_3 = 4m/9$, $y_3 = 0$.

Example 8 Cargo Load Problem

A vessel is to be loaded with stocks of three items. Each unit of item *i* has a weight w_1 and value r_i. The maximum cargo weight the vessel can take is 5 and the details of the three items are as follows:

i	w_i	r_i
1	1	30
2	3	80
3	2	65

Develop the recursive equation for the above case and find the most valuable cargo load without exceeding the maximum cargo weight by using dynamic programming.

Solution: Number of items is 3. So, it is a three-stage problem. Let x_j (j = 1, 2, 3) be the decision variable, $f_j\,(x_j)$ denote the value of the optimal allocation for the three types of items. $f_j\,(s, x_j)$ be the value associated with the optimum solution $f_j^*\,(s)$ and,

$$f_1^*(s) = \max_{0 \le x_1 \le s} f_1\,(s, x_1)$$

$$f_j^*(s) = \max_{0 \le x_j \le s} \{p_j\,(x_j) + f_{j-1}^*\,(s - x_j)\}$$

where $p_j\,(x_j)$ denotes the expected value obtained from allocation of x_j units of weight to the item j.

One-stage problem: (One item cargo loading)

$$f_1^*\,(s) = \max_{x_1}\,(30\,x_1),$$

the largest value of x_1 is $\left[\dfrac{\text{maximum cargo weight}}{w_1}\right] = \left[\dfrac{5}{1}\right] = 5$

Value of 30 x_1

s \ x_1	0	1	2	3	4	5	$f_1^*(s)$	x_1^*
0	0	–	–	–	–	–	0	0
1	0	30	–	–	–	–	30	1
2	0	30	60	–	–	–	60	2
3	0	30	60	90	–	–	90	3
4	0	30	60	90	120	–	120	4
5	0	30	60	90	120	150	150	5

Two-stage problem: the largest value of x_2 is $\left[\dfrac{5}{3}\right] = 1$ and

$$f_2^*(s) = \max\,[80x_2 + f_1(s - 3x_2)]$$

Value of $80x_2 + f_1(s - 3x_2)$

s \ x_2	0	1	$f_2^*(s)$	x_2^*
0	0 + 0 = 0	–	0	0
1	0 + 30 = 30	–	30	0
2	0 + 60 = 60	–	60	0
3	0 + 90 = 90	80 + 0 = 80	90	0
4	0 + 120 = 120	80 + 30 = 110	120	0
5	0 + 150 = 150	80 + 60 = 140	150	0

For three-stage problem, the largest value of x_3 is $\left[\dfrac{5}{2}\right] = 2$, and

$$f_3^*(s) = \max_{x_3}\,[65x_3 + f_2^*(s - 2x_3)]$$

Value of $65x_3 + f_2^*(s - 2x_3)$

s \ x_3	0	1	2	$f_3^*(s)$	x_3^*
0	0 + 0	–		0	0
1	0 + 30	–		30	0
2	0 + 60	65 + 0 = 65		65	1
3	0 + 90	65 + 30 = 95		95	1
4	0 + 120	65 + 60 =125	130 + 0 = 130	130	2
5	0 + 150	65 + 90 =155	130 + 30 = 160	160	2

Given maximum weight is 5, the optimal solution is $x_3^* = 2$, $x_2^* = 0$, $x_1^* = 1$ and $f_3^*(s) = 160$.

Example 9 Application in Inventory Control

A man is engaged in buying and selling identical items. He operates from a warehouse that can hold 500 items. Each month he can sell any quantity that he chooses up to the stock at the beginning of the month. Each month, he can buy as much as he wishes for delivery at the end of the month so long as his stock does not exceed 500 items. For the next four months, he has the following error-free forecasts of cost sales prices:

Month:	i	1	2	3	4
Cost:	c_i	27	24	26	28
Sale Price:	p_i	28	25	25	27

If he currently has a stock of 200 units, what quantities should he sell and buy in next four months? Find the solution using dynamic programming.

Solution:

Let x_i = amount of items to be sold during the month i

y_i = amount of items to be ordered during the month i

b_i = stock level in the beginning of month i

p_i = sale price in the month i

c_i = purchase price in the ith month

H = warehouse capacity.

Let $f_n(b_n)$ be the maximum possible return when there are n months to precede and initial stock is b_n.

This is a four-stage problem. Thus,

$$f_1(b_n) = \max_{x_n, y_n} [p_1 x_n - c_n y_n]$$

where $b_n \geq x_n$, $b_n - x_n + y_n \leq H$. Also,

$$f_n(b_n) = \max_{x_n, y_n} [p_n x_n - c_n y_n + f_{n-1}(b_n - x_n + y_n)]$$

For $n = 1$,

$$f_1(b_1) = \max_{x_1, y_1} [p_1 x_1 - c_1 y_1]$$

Clearly, $y_1 = 1$, $x_1 = b_1$, hence $f_1(b_1) = p_1 b_1 = 27 b_1$

and, $b_1 = b_2 - x_2 + y_2$

For $n = 2$,

$$f_2(b_2) = \max_{x_2, y_2} [p_2 x_2 - c_2 y_2 + f_1(b_2 - x_2 + y_2)]$$

where $y_2 \leq H - b_2 + x_2 = 500 - b_2 + x_2$.

Therefore,

$$f_2(b_2) = \max_{x_2, y_2} [25x_2 - 26y_2 + f_1(b_2 - x_2 + y_2)]$$

$$= \max_{x_2, y_2} [25x_2 - 26y_2 + 27(b_2 - x_2 + y_2)]$$

$$= \max_{x_2, y_2} [27b_2 - 2x_2 + y_2]$$

$$= \max_{x_2} [27b_2 - 2x_2 + (500 - b_2 + x_2)]$$

$$= \max_{x_2} [26b_2 - x_2 + 500]$$

$$= 26b_2 + 500 \qquad \text{(taking } x_2 = 0 \text{ for maximum).}$$

and,

$$b_2 = b_3 - x_3 + y_3$$

For $n = 3$,

$$f_3(b_3) = \max_{x_3, y_3} [p_3x_3 - c_3y_3 + f_2(b_3 - x_3 + y_3)]$$

$$= \max_{x_3, y_3} [25x_3 - 24y_3 + 26b_2 + 500]$$

$$= \max_{x_3, y_3} [25x_3 - 24y_3 + 26(b_3 - x_3 + y_3) + 500]$$

$$= \max_{x_3, y_3} [25x_3 - 24y_3 + 26b_3 - 26x_3 + 26y_3 + 500)]$$

$$= \max_{x_3, y_3} [26b_3 - x_3 + 2y_3 + 500]$$

$$= \max_{x_3} [26b_3 - x_3 + 2[500 - b_3 + x_3] + 500]$$

$$= \max_{x_3} [24b_3 + x_3 + 1500]$$

$$= 25b_3 + 1500 \text{ (since } b_3 \geq x_3, \text{ therefore } b_3 = x_3 \text{ for maximum)}$$

As $b_3 = b_4 - x_4 + y_4$, taking $n = 4$,

$$f_4(b_4) = \max_{x_4, y_4} [p_4x_4 - c_4y_4 + f_3(b_4 - x_4 + y_4)]$$

$$= \max_{x_4, y_4} [28x_4 - 27y_4 + 25(b_4 - x_4 + y_4) + 1500]$$

$$= \max_{x_4, y_4} [25b_4 + 3x_4 - 2y_4 + 1500]$$

$$= [25b_4 + 3b_4 + 1500] \qquad \text{(since } y_4 = 0, x_4 = b_4 \text{ for maximum)}$$

$$= 28b_4 + 1500.$$

Given: $b_4 = 200$

which implies $b_3 = b_4 - x_4 + y_4$, since

$= 200 - 200 + 0 = 0$, $\quad x_4 = 200, y_4 = 0$

which implies $b_2 = b_3 - x_3 + y_3$

$= 0 - 0 + 500$, $\quad$ since $x_3 = 0, y_3 = 500$

$= 500$

which implies $b_1 = b_2 - x_2 + y_2$

$= 500 - 0 - 0 = 500$, $\quad$ since $x_2 = 0, y_2 = 0$,

and, $y_1 = 0, x_1 = b_1 = 500$.

Thus, the required solution is:

Month	:	1	2	3	4
Purchase	:	0	500	0	0
Sales	:	700	0	500	500

Maximum possible return = 28 × 200 + 1500 = 7100.

6.3 CAPITAL BUDGETING PROBLEM

Example 1

Suppose that a company is planning its capital improvement projects for the next few years. The company has budgeted Rs 100 lakh and is reviewing four possible projects for funding. The following Table shows the projects, together with their costs and net present value of return:

Project	1	2	3	4
Cost (lakhs)	50	70	30	30
Return	50	100	60	30

A project will be funded entirely or not at all.
Maximise the return on the projects.

Solution: Let x_i be 1 if *i*th project ($i = 1, 2, 3, 4$) is funded and 0 otherwise. Hence, we have to maximise $z = 50x_1 + 100x_2 + 60x_3 + 30x_4$

subject to

$$50x_1 + 70x_2 + 30x_3 + 30x_4 \leq 100$$

Each project is taken as stage and the state is the amount of capital available. Thus, s_i = the capital available at stage *i*.

The policy decision to be made is whether to fund a project *i* or not. Hence, the decision variable is

$$x_i = \begin{cases} 1 & \text{if } i\text{th project is funded} \\ 0 & \text{otherwise} \end{cases}$$

Let the cost of the project *i* be c_i. Then, the successive state variables are related as

$s_{i-1} = s_i - c_i x_i$
$= s_i - c_i$ if the project is selected
$= s_i$ if the project is not selected.

The recursive equation for the total return is

$$f_i(s_i, x_i) = \max \{p_i x_i + f^*_{i-1}(s_i - c_i x_i)\},$$

where p_i is the return from project *i*.

The available fund for stage 4 is 100 lakh. The available capital for stage *i* depends on the projects funded in stages $i + 1, \cdots, 4$. So various possible states are

Stage	*State*
4	100
3	70 if the project 4 is funded 100 if1 the project 4 is not funded
2	40 if projects 3 and 4 are funded 70 if any one of projects 3 or 4 is funded 100 if none of the projects are funded
1	0 if projects 2 & 3 are funded 30 if project 2 is funded 40 if projects 3 & 4 are funded 70 if any one of 3 or 4 is funded 100 if none of the projects are funded.

Stage 1:

States	$f_1(s_1, x_1)$		$f_1^*(s_1, x_1)$	x_1^*
	$x_1 = 0$	$x_1 = 1$		
0	0	–	0	0
30	0	–	0	0
40	0	–	0	0
70	0	50	50	1
100	0	50	50	1

Stage 2:

States s_2	$f_2(s_2, x_2) = 100x_2 + f_1^*(s_2 - 70x_2)$		$f_2^*(s_2)$	x_2^*
	$x_2 = 0$	$x_2 = 1$		
40	0	–	0	0
70	0 + 50 = 50	100 + 0 = 100	100	1
100	0 + 50 = 50	100 + 0 = 100	100	1

Stage 3:

States s_3	$f_3(s_3, x_3) = 60x_3 + f_2^*(s_3 - 30x_3)$		$f_3^*(s_3)$	x_3^*
	$x_3 = 0$	$x_3 = 1$		
70	0 + 100 = 100	60 + 0 = 60	100	0
100	0 + 100 = 100	60 + 100 = 160	160	1

Stage 4:

States s_4	$f_4(s_4, x_4) = 30x_4 + f_3^*(s_4 - 30x_4)$		$f_4^*(s_4)$	x_4^*
	$x_4 = 0$	$x_4 = 1$		
100	0 + 160 = 160	30 + 100 = 130	160	0

The optimal return is 160 lakh. Since $x_4 = 0$, project 4 is not selected for funding. Hence, the capital available at stage 3, $s_3 =$ is 100. In this case $x_3 = 1$. So project 3 is funded. The capital available for stage 2 is 100 – 30 = 70. For stage 2, $x_2 = 1$ for $s_2 = 70$. Thus, project 2 is selected for funding. Since project 2 costs 70 lakh, the capital available for stage 1 is 0. Here, $x_1 = 0$. Hence, the projects 2 and 3 are selected for funding.

6.4 RELIABILITY PROBLEM

Example 1

Using dynamic programming, solve the reliability problem with the following data.

M_i	$i = 1$		$i = 2$		$i = 3$	
	R	C	R	C	R	C
1	0.6	2	0.8	3	0.7	1
2	0.7	4	0.8	5	0.8	2
3	0.9	5	0.9	6	0.9	3

The total capital available is 10 (in units of thousand rupees).

Solution: Let R be the total reliability of a system having n components arranged in series and let there be k parallel units per component ($j = 1, 2, 3$).

The problem is

$$\text{Maximise } R = R_1\, R_2\, R_3$$

subject to

$$c_1 + c_2 + c_3 \leq c$$

where c is the total capital available.

Let $x_j =$ the capital allocated to the stage j, where $j = 1, 2, 3$.

$k_j =$ the number of parallel units assigned to main component j.

$f_j(x_j) =$ reliability of components (stages) 1 through j inclusive, given that $0 \leq x_j \leq c$.

The recursive equations are

$$f_1(x_1) = \max_{0 \leq c_1 \leq x_1} \{R_1(c_1)\}; \quad k_1 = 1, 2, 3.$$

$$f_j(x_j) = \max_{0 \leq c_j \leq x_j} \{R_j(c_j) \times f_{j+1}(x_j - c_j)\},$$

$$j = 1, 2, 3,\ k_j = 1, 2, 3.$$

Starting with stage 1, as the component 1 must include at least one (parallel) unit, we find that x_1 must at least equal $c_1(1) = 2$. By the same reasoning x_1 cannot exceed 10 – (3 + 1) = 6; otherwise, the remaining capital will not be sufficient to provide main components 2 and 3 with at least one (parallel) unit each. Following the same reasoning.

$$x_2 = 5, 6, 7, 8 \text{ or } 9 \text{ and } x_3 = 6, 7, 8, 9 \text{ or } 10.$$

Stage 1:

	$f_1(x_1) = \max R_1(c_1)$				
x_1	$k_1 = 1$ $R_1 = 0.6, c_1 = 2$	$k_2 = 2$ $R_1 = 0.7, c_1 = 4,$	$k_3 = 3$ $R_1 = 0.9, c_1 = 5$	$f_1^*(x_1)$	k_1^*
2	0.6	–	–	0.6	1
3	0.6	–	–	0.6	1
4	0.6	0.7	–	0.7	2
5	0.6	0.7	0.9	0.9	3
6	0.6	0.7	0.9	0.9	3

Stage 2:

	$f_2(x_2) = R_2(c_2) f_1(x_2 - c_2)$				
x_2	$k_2 = 1$ $R_2 = 0.8, c_2 = 3$	$k_2 = 2$ $R_2 = 0.8, c_2 = 5$	$k_2 = 3$ $R_2 = 0.9, c_2 = 6$	$f_2^*(x_2)$	k_2^*
5	$(0.8) \times (0.6) = .48$	–	–	0.48	1
6	$(0.8 \times 0.6) = 0.48$	–	–	0.48	1
7	$(0.8) \times 0.7 = 0.56$	$(0.8) \times 0.6 = 0.48$	–	0.56	1
8	$(0.8) \times (0.9) = 0.72$	$0.8 \times 0.6 = 0.48$	$(0.9) \times (0.6) = 0.54$	0.72	1
9	$(0.8) \times (0.9) = 0.72$	$0.8 \times 0.7 = 0.56$	$(0.9) \times (0.6) = 0.54$	0.72	1

Stage 3:

	$f_3(x_3) = R_3(c_3) f_2(x_3 - c_3)$				
x_3	$k_3 = 1$	$k_3 = 2$	$k_3 = 3$	$f_3^*(x_3)$	k_3^*
	$R_3 = 0.7, c_3 = 1$	$R_3 = 0.8, c_3 = 2$	$R_3 = 0.9, c_3 = 3$		
6	$0.7 \times 0.48 = 0.336$	–	–	0.336	1
7	$0.7 \times 0.48 = 0.336$	$0.8 \times 0.48 = 0.384$	–	0.384	2
8	$0.7 \times 0.56 = 0.392$	$0.8 \times 0.48 = 0.384$	$0.9 \times 0.48 = 0.432$	0.432	3
9	$0.7 \times 0.72 = 0.504$	$0.8 \times 0.56 = 0.448$	$0.9 \times 0.48 = 0.432$	0.504	1
10	$0.7 \times 0.72 = 0.504$	$0.8 \times 0.72 = 0.576$	$0.9 \times 0.56 = 0.504$	0.576	2

The optimal solution (given $c = 10$) is $k_3 = 2$, $k_2 = 1$, $k_1 = 3$ with the maximum reliability as 0.576.

Note: The reliability problem is similar to the shortest route problem or stage coach problem except that the recursive relation involves multiplication rather than addition, since the reliability over two components connected in series is given by the product of respective probabilities (or reliabilities).

Example 2

Solve the following reliability network:

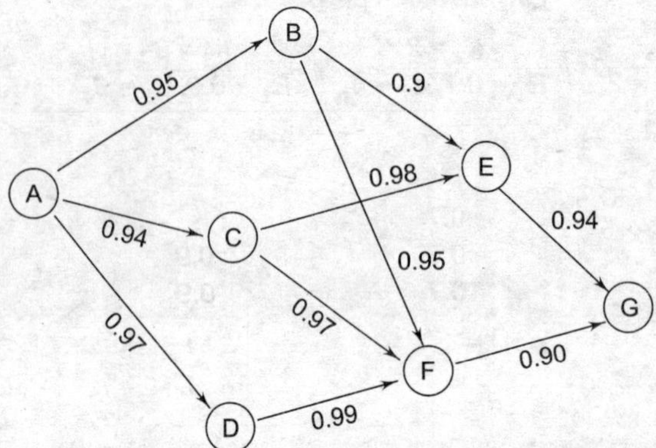

The letters *A*, *B*,*C*, *D*, *E*, *F*, *G* represent nodes of the network and the directed line segments show the possible flow of information from one node to another. The numbers assigned to each line represent the probabilities that the line will not have a failure during data transmission from one node to another node. The objective is to determine the most reliable path from *A* to *G*.

This problem has three stages.

Stage 1: Set of lines converging to *G*.

Stage 2: Corresponds to all the lines emanating {*B*, *C*, *D*} and converging to {*E*, *F*}. Stage 3 corresponds to all lines reaching {*B*, *C*, *D*}. The problem is to find the sequence of lines where total reliability is maximum. States s_i correspond to the node locations at a given stage i and the decision variable x_i represent the lines to be chosen in order to reach the next stage. Let the return function $r_i(s_i, x_i)$ represent the reliability at a given stage. So, $r_i(s_i, x_i)$ is the probability of the selected line. The total reliability function satisfies the recurrence relation.

$$f_n(s_n, x_n) = r_n(s_n, x_n) \times f^*_{n-1}(s_{n-1}, x_{n-1})$$
$$f_1(s_1, x_1) = r_1(s_1, x_1)$$

Stage 1:

s_1	$f_1(s_1, x_1) = r_1(s_1, x_1)$	$f_1^*(s_1, x_1)$	x_1^*
	$x_1 = G$		
E	0.94	0.94	G
F	0.90	0.90	G

Stage 2:

s_2	$f_2(s_2, x_2) = r_2(s_2, x_2)\, f_1^*(s_1, x_1)$		$f_2^*(s_2, x_2)$	x_2^*
	$x_2 = E$	$x_2 = F$		
B	(0.90) × (0.94) = 0.846	(0.95) × (0.90) = 0.855	0.855	F
C	0.98 × 0.94 = 0.9212	(0.97) × (0.90) = 0.873	0.9212	E
D	-	(0.99) × (0.90) = 0.891	0.891	F

Stage 3:

s_3	$f_3(s_3, x_3) = r_3(s_3, x_3) \times f_2^*(s_2, x_2)$			$f_3^*(s_3)$	x_3^*
	$x_3 = B$	$x_3 = C$	$x_3 = D$		
A	(0.95) × (0.855) = 0.81225	0.94 × 0.9212 = 0.8659	(0.97) × (0.891) = 0.86427	0.8659	C

The optimum reliability is 0.8659 and the path is $A \to C \to E \to G$.

6.5 STAGE COACH PROBLEM

Consider the network of routes from node A to B as below:

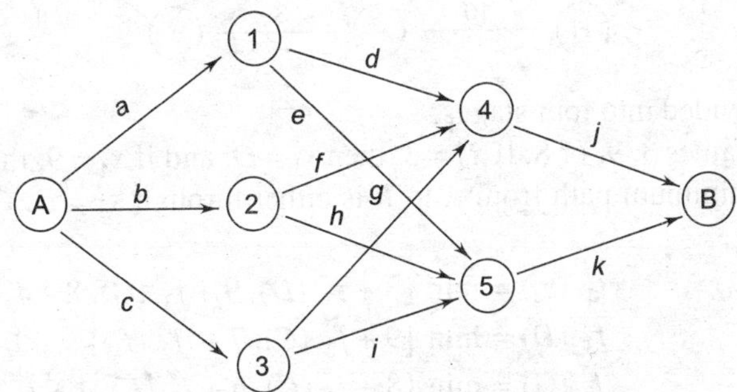

The objective is to find the shortest route from A to B. The salesman wants to find the shortest route. From city A he should visit any one of the cities 1, 2 or 3 and from there he select next tour to any of the cities 4 or 5, and then from 4 or 5 he can reach B. Each leg of his trip can be thought of as a stage. A stage number n can be defined to be the number of stages to be covered. When he is in city 4 or 5, he has to go one more stage to complete the journey. This, we call as one stage problem indicating there is one more stage to complete the journey. When he is in city 1, 2 or 3, he has to go two more stages to reach the final destination. Like this we have 3 stages from A to B.

Each move will change the state of the system denoted by s_j. Thus, s_o is the state in which the node A lies. Also, s_o has only single value, say $s_o = 1$. State s_2 has only 2 possible values corresponding to two nodes in stage 1, and so on. Possible alternative paths from one stage to the other will be called decision variables denoted by x_j. The return or profit function is $f_j(x_j)$. Note that x_j is identified with the length of the corresponding line. Here, $f_j(x_j) = x_j$. The minimum transportation cost (or path) from s_o to s_j will be denoted by $f_j^*(s_j)$.

The general recursive formula is given by

$$f_j^*(\text{sj}) = \underset{x_j}{\text{mix}}\ [x_j + f^*_{j-1}(s_{j-1})]$$

with $f_1^*(s_1) = x_1$.

Example 1

Find the shortest path from vertex A to B along arcs joining various vertices lying between A and B. Length of each path is given.

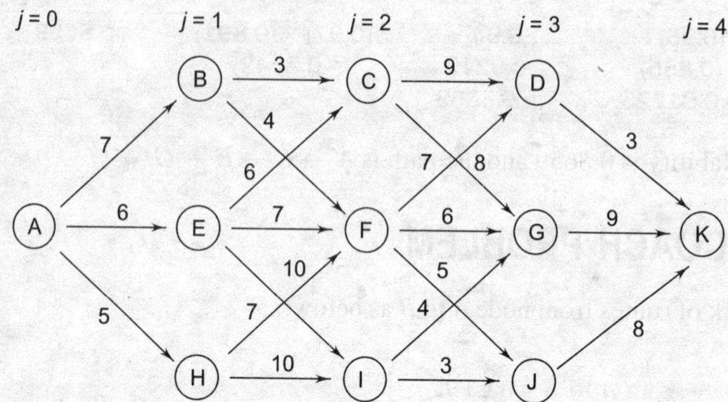

The problem can be divided into four stages.

At $j = 4$, x_4 has three values 3, 9, or 8. If $x_4 = 3$, then $s_3 = D$, and if $x_4 = 9$, then $s_3 = G$, and if $x_4 = 8$, then $s_3 = J$. Hence, the minimum path from A to B is either through $s_3 = D$, G or J corresponding to $x_4 = 3$, 9 or 8. Hence,

$$f_4^*(s_4) = \min\{3 + f_3^*(D), 9 + f_3^*(G), 8 + f_3^*(J)\}$$

Similarly,

$$f_3^*(D) = \min\{9 + f_2^*(C), 7 + f_2^*(F)\}$$

$$f_3^*(G) = \min\{8 + f_2^*(C), 6 + f_2^*(F), 4 + f_2^*(I)\}$$

$$f_3^*(J) = \min\{5 + f_2^*(F), 3 + f_2^*(I)\}$$

Similarly, we can calculate $f_2^*(C)$, $f_2^*(F)$, $f_2^*(I)$, and $f_1^*(B)$, $f_1^*(E)$, $f_1^*(H)$.

Stage 1:

State s_1	Decision variable x_1	$f_1^*(s_1)$
B	7	[7]
E	6	6
H	5	5

Stage 2:

s_2 \ x_2	$f_1^*(s_1)$ 3	4	6	7	10	$x_2 + f_1^*(s_1)$ 3	4	6	7	10	$f_2^*(s_2)$ minimum
C	7	–	6	–	–	10	–	12	–	–	10
F	–	[7]	–	6	5	–	11	–	13	15	[11]
I	–	–	–	6	5	–	–	–	13	15	13

Stage 3:

s_3 \ x_3	$f_2^*(s_2)$ 3	4	5	6	7	8	9	$x_3 + f_2^*(s_2)$ 3	4	5	6	7	8	9	$f_3^*(s_3)$
D	-	-	-	-	11	-	10	-	-	-	-	18	-	19	18
G	-	13	-	11	-	10	-	-	17	-	17	-	18	-	17
J	13	-	11	-	-	-	-	16	-	16	-	-	-	-	16

Stage 4:

s_4 \ x_4	$f_3^*(s_3)$ 3	9	8	$x_4 + f_3^*(s_3)$ 3	9	8	$f_3^*(s_4)$
K	18	17	16	21	26	24	21

The minimum path from A to B whose cost (or distance) is 21. By tracing the minimum path and decision backwards, successive distances are

$$f_4^*(s_4) = \min\{3 + f_3^*(D), 9 + f_3^*(G), 8 + f_3^*(J)\}$$
$$= 3 + f_3^*(D), \text{ gives the path } DK$$
$$f_3^*(D) = \min\{9 + f_2^*(C), 7 + f_2^*(F)\}$$
$$= 7 + f_2^*(F) \text{ gives the path } FD$$
$$f_2^*(F) = \min\{4 + f_1^*(B), 7 + f_1^*(E), 10 + f_1^*(H)\}$$
$$= 4 + f_1^*(B) \text{ gives the path } BF$$

The optimal route connecting A and K is $A \to B \to F \to D \to K$.

6.6 SOLUTION OF LINEAR PROGRAMMING PROBLEM BY DYNAMIC PROGRAMMING

The general linear programming problem is

$$\text{Maximise } z = c_1 x_1 + c_2 x_2 + \cdots + c_n x_n$$

subject to the constraints

$$a_{11} x_1 + a_{12} x_2 + \cdots + a_{1n} x_n \le b_1$$
$$a_{21} x_1 + a_{22} x_2 + \cdots + a_{2n} x_n \le b_2$$
$$a_{m1} + a_{m2} x_2 + \cdots a_{mn} x_n \le b_m$$
$$x_1, x_2, \cdots x_n \ge 0.$$

This problem can be formulated as a dynamic programming problem as follows.
Here, each activity j ($j = 1, 2, \ldots n$) may be regarded as a stage. The level of activity x_j (≥ 0) represents the aternatives (decision variables) at stage j. Since x_j is continuous, each stage possesses an infinite number of alternatives within the feasible region.

Since the linear programming problem is an allocation problem, states may be defined as the amounts of resources to be allocated to the current stage and the succeeding stages. Here, we shall use backward computational procedure. As there are m limited resources, the states must be represented by an m-dimensional factor. Let $(\beta_{1j}, \beta_{2j}, \ldots \beta_{mj})$ be the state of the system at stage j and $f_j(\beta_{1j}, \beta_{2j}, \cdots \beta_{mj})$ be the optimum value of the objective function for stages (activities) $j, j+1, \cdots n$, for given states $\beta_{1j}, \beta_{2j}, \cdots \beta_{mj}$.

Thus,

$$f_n(\beta_{1n}, \beta_{2n}, \cdots \beta_{mn}) = \max_{0 \le a_{in} x_n \le \beta_{in}} \{c_n x_n\}$$

$$f_j(\beta_{1j}, \beta_{2j}, \cdots, \beta_{mj}) = \max_{0 \le a_{ij} x_j \le \beta_{ij}} \{c_j x_j + f_{j+1}[\beta_{1j} - a_{1j} x_1, \beta_{2j} - a_{2j} x_2, \cdots \beta_{mj} - a_{mj} x_j]$$

$$i = 1, 2, \cdots m, j = 1, 2, \cdots n-1 \tag{6.6.1}$$

Here, $0 \le \beta_{ij} \le b_i$, for all i and j. The above recursive equation (6.6.1) is used to solve a LPP by dynamic programming approach.

Example 1

Solve the following LPP by dynamic programming approach.

$$\text{Maximise } z = 2x_1 + 5x_2$$

subject to

$$2x_1 + x_2 \le 43$$
$$2x_2 \le 46$$
$$x_1, x_2 \ge 0.$$

Solution: Since there are two resources (constraints), the states must be represented by a pair (u_j, v_j), $j = 1, 2$. Note that the problem has two variables only, so the problem has two stages.
$f_j((u_j, v_j)$ be the optimal value of the objective function for stage $j = 1, 2$ given (u_j, v_j).
Using backward procedure, we have

$$f_2(u_2, v_2) = \max_{\substack{0 \le x_2 \le u_2 \\ 0 \le 2x_2 \le v_2}} \{5x_2\}$$

$$= 5 \max_{\substack{0 \le x_2 \le u_2 \\ 0 \le x_2 \le \frac{v_2}{2}}} \{x_2\}$$

Since max $\{x_2\}$, with $0 \le x_2 \le u_2$, $0 \le x_2 \le \dfrac{v_2}{2}$ is the minimum of $\left\{u_2, \dfrac{v_2}{2}\right\}$, we have

$$f_2(u_2, v_2) = 5 \min\left\{u_2, \frac{v_2}{2}\right\} \tag{6.6.2}$$

and,

$$x_2^* = \min\left\{u_2, \frac{v_2}{2}\right\} \tag{6.6.3}$$

$$f_1(u_1, v_1) = \max_{\substack{0 \le 2x_1 \le u_1 \\ 0 \le 0x_1 \le v_1}} \{2x_1 + f_2(u_1 - 2x_1, v_1 - 0)\}$$

$$= \max_{0 \le 2x_1 \le u_1} \left\{2x_1 + 5\min\left\{u_1 - 2x_1, \frac{v_1}{2}\right\}\right\}$$

Since it is a two stage problem, at the first stage $u_1 = 43$, $v_1 = 46$. Thus,

$$f_1(u_1, v_1) = f_1(43, 46)$$

$$= \max_{0 \le x_1 \le \frac{43}{2} = 21.5} \left\{2x_1 + 5\min\left(43 - 2x_1, \frac{46}{2}\right)\right\}$$

$$= \max \left\{2x_1 + 5\begin{cases} 23 & \text{if } 0 \le x_1 \le 10 \\ 43 - 2x_1 & \text{if } 10 \le x_1 \le 21.5 \end{cases}\right\}$$

$$= \max \begin{cases} 2x_1 + 115 & \text{if } 0 \le x_1 \le 10 \\ 215 - 8x_1 & \text{if } 10 \le x_1 \le 21.5 \end{cases}$$

$= \max \{2(10) + 115, 218 - 8 \times 10\} = 135$, which is obtained for $x_1 = 10$. Hence,

$$f_1(u_1, v_1) = 135, \text{ achieved at } x_1^* = 10$$

Now,
$$x_2 = \min\left\{u_2, \frac{v_2}{2}\right\} = \min\left\{u_1 - 2x_1, \frac{v_1 - 0}{2}\right\}$$

$$= \min\left\{43 - 20, \frac{46}{2}\right\}$$

$$= \min\{23, 23\} = 23.$$

The optimal solution is given by

$$\text{maximum } z = 135,\ x_1 = 10,\ x_2 = 23.$$

Example 2

Solve the following LPP using dynamic programming approach.

$$\text{Maximum } z = 3x_1 + 5x_2$$

subject to

$$x_1 \le 4$$
$$x_2 \le 6$$
$$3x_1 + 2x_2 \le 18$$
$$x_1, x_2 \ge 0.$$

Solution: Since we have two decision variables and three resources, the problem has two stages and three states. Let the states be represented by (u_j, v_j, w_j) for $j = 1, 2$, and $f_j(u_j, v_j, w_j)$ be the optimal value of the objective function for stage $j = 1, 2$.

Using backward procedure, we have

$$f_1(u_1, v_1, w_1) = \max \begin{cases} 3x_1 + 30 & \text{if } 0 \le x_1 \le 2 \\ \dfrac{90 - 9x_1}{2} & \text{if } 2 \le x_1 \le 4 \end{cases}$$

$$= \max\left\{3 \times 2 + 30, \frac{90 - 9 \times 2}{2}\right\} = 36, \text{ which occurs at } x_1 = 2. \text{ Hence,}$$

$$f_1(u_1, v_1, w_1) = 36 \text{ which is achieved at } x_1^* = 2.$$

Now,

$$x_2^* = \min\left\{v_2, \frac{w_2}{2}\right\} = \min\left\{6, \frac{18-13x_1}{2}\right\} = \min(6.6) = 6$$

The optimal solution is maximum $z = 36$, $x_1 = 2$, $x_2 = 6$.

Advantages and Disadvantages of Dynamic Programming

The following are the major advantages and disadvantages of dynamic programming technique:

Advantages

1. For problems in areas such as inventory management, chemical engineering design, or control theory, dynamic programming is the only technique used to solve the problems.
2. It is well suited for multi-stage or multi-period or sequential decision process.
3. It is applicable to linear or nonlinear problems, discrete or continuous variables, deterministic or stochastic problems.
4. It can be implemented in a computer and also used to perform sensitivity analysis.

Disadvantages

1. No general formation of the dynamic programme is available. Each problem has its own design. So, we can model a problem by experience and insight through it.
2. The number of state variables must be kept very low in order to avoid excessive computational work.
3. In general, this approach is not efficient compared to other mathematical programming algorithms like the simplex method.

EXERCISES

Section 6.1

1. Explain the concept of dynamic programming.
2. State the principle of optimality in your own words.
3. Explain the relationship between stages and states in dynamic programming.
4. If both linear programming and dynamic programming could be used to model and solve a decision problem which method would you prefer to use and why?
5. What are the essential characteristics of dynamic programming.
6. Set up the recursive relation, using dynamic programming approach, when an n-stage objective function is to be maximised.
7. What sort of problems can be solved using dynamic programming?

Section 6.2

8. Use dynamic programming to find the value of

maximise $z = y_1 y_2 y_3$

subject to the constraints $y_1 + y_2 + y_3 = 5$

$y_1, y_2, y_3 \geq 0.$

[*Hint*: $f_1(x_1) = \max_{z_1 = x_1} (z_1), f_j(x_j) = \max_{0 \le z_j \le x_j} z_j f_{j-1}(x_j - z_j)$

Answer: $\left(\frac{5}{3}, \frac{5}{3}, \frac{5}{3}\right)$ with $f_3(5) = \left(\frac{5}{3}\right)^3$.]

9. Obtain the functional equation for maximising

$$z = g_1(x_1) + g_2(x_2) + \cdots + g_n(x_n)$$

subject to $x_1 + x_2 + \cdots x_n = c, x_j \ge 0, j = 1, 2, \cdots n$

[*Answer*: $f_n(c) = \max_{0 \le z \le c} \{.. \max \{g_1(z) + f_{n-1}(c - z)\}, f_1(c) = \max_{0 \le z \le c} g_1(z) = g_1(c)$.

10. Using dynamic programming, minimise $z = uvw$ subject to $u + v + w = 5, u, v, w \ge 0$
[*Answer*: $u = 0, v = 0, w = 5$, and min $z = 0$.]
11. Maximise $12x_1^2 + 27x_2^2 + 147x_3^2$ subject to the constraints $x_1 + 2x_2 + x_3 \le 8, x_1, x_2\, x_3 \ge 0$.
[*Answer*: $x_1 = x_2 = x_3 = 2$, minimum $z = 20$.]
12. Find the maximum value of $z = x_1^2 + 2x_2^2 + 4x_3$ subject to $x_1 + 2x_2 + x_3 \le 8, x_1, x_2, x_3 \ge 0$.
13. A manufacture of golf balls is trying to determine the most effective advertising and promotional strategy for the introduction of its new line of golf balls. The company has budgeted Rs 500000 for advertising. Listed below is the estimate of the potential sales revenue per Rs 100000 spent.

Medium	*Expenditure (in 100000s)*				
	1	2	3	4	5
Radio	0.90	1.80	2.70	3.60	4.50
TV	1.05	2.10	3.15	4.25	5.50
Magazine	1.10	2.05	3.20	4.20	5.10
Direct Mail	0.85	1.95	3.50	4.40	5.60

All expenditures should be in increments of Rs 100000. Determine the amount to be spent on the media.
[*Answer*: The optimal strategy is to allocate Rs 300000 for direct mailing, Rs 100000 to TV and magazine, advertisements, and nothing to radio.]
14. An oil company has 8 units of money available for exploration of three sites. If oil is present at a site, the probability of finding it depends upon the amount allocated for exploiting the site as given below.

	Units of money allocated								
	0	1	2	3	4	5	6	7	8
Site 1	0.0	0.0	0.1	0.2	0.3	0.5	0.7	0.9	1.0
Site 2	0.0	0.1	0.2	0.3	0.4	0.6	0.7	0.8	1.0
Site 3	0.0	0.1	0.1	0.2	0.3	0.5	0.8	0.9	1.0

The probability that oil exists at the sites 1, 2 and 3 is 0.4, 0.3, 0.2, respectively. Find the optimal allocation of money.
[*Answer*: All the 8 units of money are to be allocated to site 1, $x_1^* = 8, x_2^* = 0, x_3^* = 0$, and the maximum probability of obtaining oil is 40%.]

15. A drug manufacturing concern has ten medical representatives working in three sales areas. The profitability of each representative in the three sales areas is as follows:

No. of representatives		0	1	2	3	4	5	6	7	8	9	10
Profits	Area 1	15	22	30	38	45	48	54	60	65	70	70
(thousands of	Area 2	26	35	40	46	55	62	70	76	83	90	95
rupees)	Area 3	30	38	44	50	60	65	72	80	85	90	85

Determine the optimum allocation of the medical representatives in order to maximise the profits. What will be the optimum allocation if the number of representatives available at present is only six?

[*Answer*: (3, 6, 1,) : Rs 2,46,000; (4, 1, 1) ; Rs 1,18,000.]

16. A company has three media *A*,*B*, *C* available for advertising its product. The data collected over the past years about the relationship between the sales and frequency of advertisement in the different media is as follows:

Frequency/month	*Estimated Sale/month*		
	A	B	C
1	125	180	300
2	225	290	350
3	260	340	450
4	300	370	500

The cost of advertisement is Rs 5000 in medium *A*, Rs 10000 in medium *B* and Rs 20000 in medium *C*. The total budget allocated for advertising the product is Rs 40000. Determine the optimal combination of the advertising media and the frequency of advertisement.

[*Answer*: once in *C*, once in *B*, twice in *A*.]

17. A manufacture has entered into a contractor for the supply of the following number of units of a product at the end of each month.

Month	Jan	Mar	Aug	Oct	Nov	Dec
No. of units	10	5	20	3	6	30

The units manufactured during a month are available for supply at the end of the month or they may be kept in the storage at a cost of Rs 2 per unit per month. Each time the manufacture of a batch of units is undertaken, there is a set up cost of Rs 400. Determine the production schedule which will minimise the total cost.

[*Answer*: 15 units in Jan, 59 units in August with minimum z = Rs 1098.]

18. Explain capital budgeting problem with one example. Also, explain how a contractor determines the size of a labour force during a certain period from the dynamic programming stand point.

19. Maximise hydro electric power $P(s)$, $s = (s_1, s_2, s_3)$ produced by building dams on three different given basins, where

$$P(s) = f_1(s_1) + f_2(s_2) + f_3(s_3)$$

and, $f_i(s_i)$ is the power generated from the ith basin by investing Rs s_i. The total budgeting provision is $s_1 + s_2 + s_3 \le 3$. The functions f_1, f_2, f_3 are given in the following Table. Integer solution of the problem is required

s_i	0	1	2	3
f_1	0	2	4	6
f_2	0	1	5	6
f_3	0	3	5	6

Section 6.4

20. An electronic device consists of four components each of which must function for the system to function. The system reliability can be improved by installing parallel units in one or more components. The reliability of component (R) with one, two or three parallel units, and the corresponding cost (c) are given below. The maximum amount available for this device is 100. The problem is to determine the number of parallel units in each component.

No. of parallel units	Components							
	1		2		3		4	
	R	c	R	c	R	c	R	c
1	0.7	10	0.5	20	0.7	10	0.6	20
2	0.8	20	0.7	40	0.9	30	0.7	30
3	0.9	30	0.8	0	0.95	40	0.9	40

[*Answer*: The device should have 1, 2, 1, 3 units in components 1, 2, 3 and 4, respectively with maximum reliability of 0.308].

Section 6.5

21. A salesman has to travel between A and B indicated by the network shown in figure:

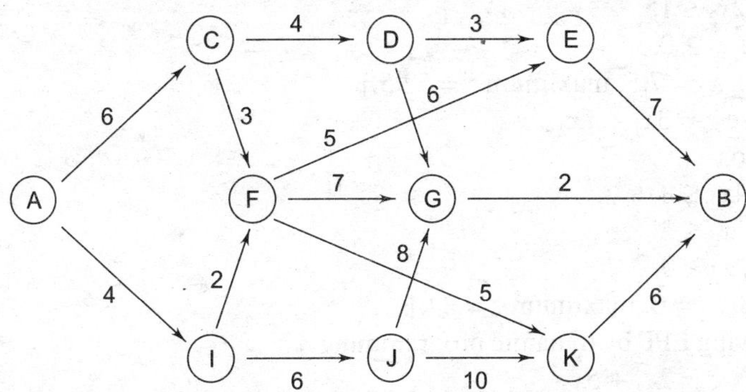

Find the minimum distance travelled by salesman from A to B.
[*Answer*: $A \to I \to F \to G \to B$ with minimum cost 15.]

22. A company has to transport some goods from city A to city J. The cost of transportation between the different cities is given in the following network. Find the optimal route connecting cities A and J.

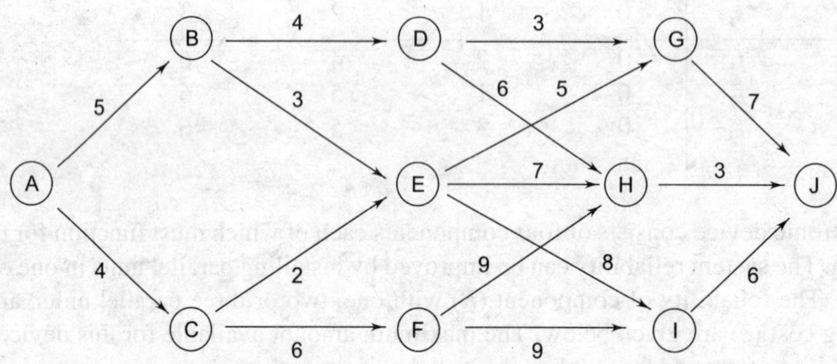

[*Answer*: $A \to C \to E \to H \to J$]

Section 6.6

23. Solve the following LPP by dynamic programming

(i) maximise $z = 2x_1 + 3x_2$
subject to
$$x_1 - x_2 \le 1$$
$$x_1 + x_2 \le 3$$
$$x_1, x_2 \ge 0.$$
[*Answer*: (i) $x_1 = 0, x_2 = 3$, max $z = 9$.]

(ii) minimise $z = x_1 + 3x_2 + 4x_3$
subject to
$$2x_1 + 4x_2 + 3x_3 \ge 60$$
$$3x_1 + 2x_2 + x_3 \ge 60$$
$$2x_1 + x_2 + 3x_3 \ge 90$$
$$x_1, x_2, x_3 \ge 0$$

24. Solve the following LPP by dynamic programming,

(i) Maximise $z = 8x_1 + 7x_2$
subject to
$$2x_1 + x_2 \le 8$$
$$5x_1 + 2x_2 \le 15$$
$$x_1, x_2 \ge 0.$$
[*Answer*: $x_1 = 0, x_2 = 7.5$, maximum $z = 52.5$.]

(ii) Maximise $z = 3x_1 + 7x_2$
subject to
$$x_1 + 4x_2 \le 8$$
$$x_2 \le 2$$
$$x_1, x_2 \ge 0.$$
[*Answer*: $x_1 = 8, x_2 = 0$, maximum $z = 24$.]

25. Solve the following LPP by dynamic programming

Maximise $z = 2x_1 + 5x_2$
subject to $2x_1 + x_2 \le 430$
$$2x_2 \le 460$$
$$x_1, x_2 \ge 0.$$
[*Answer*: $x_1 = 100, x_2 = 230$, max $z = 1350$.]

26. Solve the following LPP by dynamic programming
Maximise $z = 50x_1 + 100x_2$
subject to

$$2x_1 + 3x_2 \leq 48$$
$$x_1 + 3x_2 \leq 42$$
$$x_1 + x_2 \leq 21$$
$$x_1, x_2 \geq 0.$$

[*Answer*: $x_1 = 6$, $x_2 = 12$, maximum $z = 60$.]

7

Decision Theory and Games

7.1 INTRODUCTION TO DECISION ANALYSIS

Everyday we make many decisions and occasionally we make an important decision that can have immediate and/or long-term effects on our lives. Decisions like where to attend school, whether to rent or buy, whether your company should accept a merger proposal, whether to run a cool drink stall or a snacks stall and so on, are important decisions—for which we would prefer to make the correct choice. A decision, in general, may be defined as the selection by the decision-maker of an act, considered to be best according to some predesignated standard, from among the several options.

7.1.1 Steps in Decision Theory Approach

Step 1: Determine the various alternative courses of action (or choices or strategies) from which the final decision is to be made.

Step 2: Identify the possible outcomes, called the states of nature or events for the decision problem. The events are beyond the control of the decision-maker.

Step 3: *Construct a pay off table*

The decision-maker now constructs a pay off table (table representing profit, benefit and so on) for each possible combination of alternative course of action and state of nature.

If there are m alternative courses of action $A_1, ..., A_m$ and n states of nature $E_1, ..., E_n$, then the pay off matrix will be represented as

table 7.1

States of nature	Decision alternatives (courses of action)			
	A_1	A_2	...	A_m
E_1	a_{11}	a_{12}	...	a_{1m}
E_2	a_{21}	a_{22}	...	a_{2m}
$\vdots$	$\vdots$	$\vdots$	...	$\vdots$
E_n	a_{n1}	a_{n2}	...	a_{nm}

where a_{ij} — the pay off resulting from the combination of ith event and j^{th} action.

Step 4: The decision-maker will choose the criterion which results in largest pay off. The criterion may be economic, quantitative or qualitative (for example, market share, profit, fragrance of a perfume and so on)

Example

A firm manufactures three types of products. The fixed and variable costs are given below:

	Fixed cost (Rs)	Variable cost per unit (Rs)
Product *A*	25,000	12
Product *B*	35,000	9
Product *C*	53,000	7

The likely demand (units) of the products is given below:

Poor demand : 3,000
Moderate demand : 7,000
High demand : 11,000

If the sale price of each type of product is Rs 25 then prepare the pay off matrix.

Solution: Let D_1, D_2 and D_3 denote poor, moderate and high demands. Then, pay off is given by

Pay off = Sales revenue – Cost (fixed + variable).

The calculations for pay off (in thousand) for each pair of alternative demand (course of action) and the types of product (states of nature) are shown below:

$$D_1\,A = 3 \times 25 - 25 - 3 \times 12 = 14$$
$$D_1\,B = 3 \times 25 - 35 - 3 \times 9 = 13$$
$$D_1\,C = 3 \times 25 - 53 - 3 \times 7 = 1$$
$$D_2\,A = 7 \times 25 - 25 - 7 \times 12 = 66$$
$$D_2\,B = 7 \times 25 - 35 - 7 \times 9 = 77$$
$$D_2\,C = 7 \times 25 - 53 - 7 \times 7 = 73$$
$$D_3\,A = 11 \times 25 - 25 - 11 \times 12 = 118$$
$$D_3\,B = 11 \times 25 - 35 - 11 \times 9 = 141$$
$$D_3\,C = 11 \times 25 - 53 - 11 \times 7 = 145$$

The pay off table is

table 7.2

Events	Alternative demand		
	D_1	D_2	D_3
A	14	66	118
B	13	77	141
C	1	73	145

7.1.2 Decision-making Environments

Decision analysis is used to determine optimum strategies where a decision-maker is faced with several decision alternatives. There are four types of decision-making environment:

1. **Decisions under certainty**
   Whenever there exists only one outcome for a decision, we are dealing with this category.
   For example, the decision to purchase either a N.S.C. (National Saving Certificate), Indira Vikas Patra, or deposit in N.S.S. (National Saving Scheme) is one in which it is reasonable to assume the complete information about the future because there is no doubt that the Indian Government will pay the interest as it falls due and the principal at maturity.
   As other examples we can take linear programming transportation problem, assignment problem and sequencing and so on.
2. **Decisions under uncertainty**
   These refer to situations where more than one outcome can result from any single decision. That is, here more than one states of nature exist but the decision-maker lacks sufficient knowledge to allow him to assign probabilities to the various states of nature. For example, the probability that Mr X will be the Prime Minister of the country 15 years from now is not known.
3. **Decisions under risk**
   These refer to decision situations wherein the decision-maker chooses from among several possible outcomes where the probability of occurrence can be stated objectively from the past data.
4. **Decisions under conflict**
   In many situations, neither states of nature are completely known nor are they completely uncertain. Partial knowledge is available and therefore, it may be termed as decision-making under 'partial uncertainty'.
   As an example, we can consider the situation of conflict involving two or more competitors marketing the same product.

7.2 DECISION UNDER UNCERTAINTY

Under conditions of uncertainty, the decision-maker has a knowledge about the states of nature that happen but lacks the knowledge about the probabilities of their occurrence. Situations like launching a new product fall under this category. The insufficient data lead to a more complex decision model and perhaps, a less satisfactory solution. However, one uses scientific methods to exploit the available data to the fullest extent.

In decisions under uncertainty, situations exist in which two (or more) opponents with conflicting objectives try to make decisions with each trying to gain at the cost of the other(s). These situations are different since the decision-maker is working against an intelligent opponent. The theory governing these types of decision problems is called the theory of games and will be taken up later in this chapter.

Different persons have suggested several decision rules for making a decision under such situations.

7.2.1 Criterion of Pessimism (Minimax or Maximin)

The Maximin criterion is based upon the 'Conservative approach' to assume that the worst possible is going to happen. The decision-maker considers each strategy and locates the minimum pay off for each, and then selects that alternative which maximises the minimum pay off.

Thus, this criterion involves two steps.

Step 1: Find the minimum assured pay off for each alternative (course of action).

Step 2: Choose that alternative which corresponds to the maximum of the above minimum pay off.

When dealing with costs, the maximum cost associated with each alternative is considered and the alternative that minimises this maximum cost is chosen. This is known as Minimax criterion and it involves two steps.

Step 1: Determine the maximum possible cost for each alternative.

Step 2: Choose that alternative which corresponds to the minimum of the above costs.

7.2.2 Criterion of Optimism (Maximax or Minimin)

This criterion is based upon extreme optimism.

In this criterion the decision-maker ensures that he should not miss the opportunity to achieve the greatest possible pay off or lowest possible cost. In maximax criterion the decision-maker selects that particular strategy which corresponds to the maximum pay off for each strategy.

Thus, the maximax criterion consists of the following two steps.

Step 1: Determine the maximum possible pay off for each alternative.

Step 2: Select that alternative which corresponds to the maximum of the above maximum pay offs.

In decision problems dealing with costs, the minimum for each alternative is considered and then the alternative which minimises the above minimum cost is selected. This is termed as minimin principle.

7.2.3 Laplace Criterion or Equally Likely Decision Criterion

This criterion is based upon what is known as the principal of insufficient reason.

Since the probabilities associated with the occurrence of various events are unknown, there is not enough information to conclude that these probabilities will be different. Hence, it is assumed that all states of nature will occur with equal probability. That is, each state of nature is assigned an equal probability. As states of nature are mutually exclusive and collectively exhaustive, so the probability of each of these much be 1/(number of states of nature).

The working procedure is as follows.

Step 1: Assign equal probabilities 1/(number of states of nature) to each pay off of a strategy.

Step 2: Determine the expected pay off value for each alternative.

Step 3: Select that alternative which corresponds to the maximum (and minimum for cost) of the above expected pay offs.

7.2.4 Criterion of Realism (Hurwicz Criterion)

This criterion suggests that a rational decision-maker should be neither be completely optimistic nor pessimistic and therefore, must display a mixture of both. Hurwicz, who suggestes this criterion, introduced the idea of a coefficient of optimism (denoted by α) to measure the decision-maker's degree of optimism. This coefficient lies between 0 and 1, where 0 represents a completely pessimistic attitude about the future and 1 a completely optimistic attitude about the future. Thus, if α is the coefficient of optimism, then $(1 - \alpha)$ will represents the coefficient of pessimism.

The working procedure is summarised as follows.

1. Decide the coefficient of optimism α and then the coefficient of pessimising $1 - \alpha$.
2. Determine the maximum as well as minimum pay off for each alternative and obtain the quantities

$h = \alpha \times$ maximum for each alternative + $(1 - \alpha) \times$ minimum for each alternative.

3. Select an alternative with value of h as maximum.

7.2.5 Criterion of Regret (Savage Criterion) or Minimax Regret Criterion

This criterion is also known as opportunity loss decision criterion or minimax regret decision criterion because decision-maker feels regret after adopting a wrong course of action (or alternative) resulting in an opportunity loss of pay off.

The working procedure is as follows.

Step 1: From the given pay off matrix, develop an opportunity-loss (or regret) matrix.

(i) Find the best pay off corresponding to each state of nature (maximum for profit and minimum for cost)

(ii) ith regret $= \begin{cases} \text{(maximum pay off} - i\text{th pay off) for the } j\text{th event if the pay offs represent profits} \\ \text{(minimum pay off} - i\text{th pay off) for the } j\text{th event if the pay offs represent costs.} \end{cases}$

Step 2: Determine the maximum regret amount for each alternative.

Step 3: Choose that alternative which corresponds to the minimum regrets.

Remark: The reader may observe that while the other decision rules do not take into account the cost of opportunity loses by making the wrong decision, the minimax regret criterion does so.

Example 1

A farmer wants to decide which of the three crops he should plant on his 100-acre farm. The profit from each is dependent on the rainfall during the growing season. The farmer has categorised the amount of rainfall as high, medium and low. His estimated profit for each is shown in the Table below.

table 7.3

Rainfall	*Estimated conditional profit (Rs)*		
	Crop A	*Crop B*	*Crop C*
High	8000	3500	5000
Medium	4500	4500	5000
Low	2000	5000	4000

If the farmer wishes to plant only one crop, decide which should be his 'best crop' using

(a) Maximax Criterion
(b) Maximin Criterion
(c) Hurwicz criterion (farmer's degree of optimism being 0.6)
(d) Laplace criterion
(e) Minimax regret criterion.

Solution:

table 7.4

Rainfall	Estimated conditional profit (Rs)		
	Crop A	Crop B	Crop C
High	8000	3500	5000
Medium	4500	4500	5000
Low	2000	5000	4000
Maximum pay off	8000	5000	5000
Minimum pay off	2000	35000	4000

(a) *Maximax Criterion*: From Table 7.4 we observe that the maximum pay off for each alternative crop *A*, crop *B* and crop *C* are 8000, 5000 and 5000, respectively. Maximum among these is 8000 corresponding to crop *A*. Hence, best strategy under this criterion is crop *A*.

(b) *Maximin Criterion*: The minimum pay off for each alternative crop *A*, crop *B* and crop *C* are 2000, 3500 and 4000, respectively. The maximum among these is 4000 which corresponds to crop *C*. Hence, the best alternative according to this criterion is *C*.

(c) *Hurwicz Criterion*: Given that farmers degree of optimism $\alpha = 0.6$. Hence, $1 - \alpha = 0.4$

The quantity *h* is calculated in the following Table.

table 7.5

Alternative	Maximum pay off (i)	Minimum Pay off (ii)	$h = \alpha x(i) + (1 - \alpha)(ii)$
Crop A	8000	2000	5600
Crop B	5000	3500	4400
Crop C	5000	4000	4600

The maximimin value of *h* is 5600 corresponding to crop *A*. Hence, the best choice is crop *A*.

(d) *Laplace Criterion*: Assign equal probabilities (1/3) [∵ there are three outcomes or states of nature namely high, medium and low] to each pay off of a strategy. The expected pay offs are calculated for each alternative as follows.

E (Crop A) = 1/3 (8000) + 1/3 (4500) + 1/3 (2000) ≃ 4833

E (Crop B) = 1/3 (2000 + 3500 + 4000) ≃ 4333

E (Crop C) = 1/3 (5000 + 5000 + 4000) ≃ 4666

The maximum expected value is 4833 for Crop *A*. Hence, the best alternative under this criterion is *A*.

(e) *Minimax Regret Criterion*

table 7.6

Alternative	Pay off amounts			Regret pay off amount			Maximum
	High	Medium	Low	High	Medium	Low	regret
Crop A	8000	4500	2000	0	500	3000	3000
Crop B	3500	4500	5000	4500	500	0	4500
Crop C	5000	5000	4000	300	0	1000	3000
Maximum pay off	8000	5000	5000				

Since the pay offs are profits the regret value is called as follows.

*i*th regret = (maximum pay off–i^{th} pay off)

Since the minimum regret is 3000 corresponding to Crop *A* and Crop *C*, the best alternative under this category is either Crop *A* or Crop *C*.

Example 2

A manufacturer makes a product, of which the principal ingredient is a chemical *X*. At the moment, the manufacturer spends Rs 1000 per year on the supply of *X*, but there is a possibility that the price may soon increase to four times its present figure because of a worldwide shortage of the chemical. There is another chemical *Y*, which the manufacturer could use in conjunction with a third chemical *Z*, in order to give the same effect as chemical *X*. Chemicals *Y* and *Z* would together cost the manufacturer Rs 3000 per year; but their prices are unlikely to rise. What action should the manufacturer take? Apply the maximin and minimax regret criteria for decision-making and give two sets of solutions.

If the coefficient of optimism is 0.4, find the course of action that minimises the cost.

Solution: The data of the problem is summarised in the following Table (negative figures in the Table represents profit).

table 7.7

States of nature	Course of action	
	S_1 (Use Y and Z)	S_2 (use X)
N_1 (Price × increases)	- 3000	- 4000
N_2 (Price × does not increases)	- 3000	- 1000
Minimum pay off	- 3000	- 4000

(i) *Maximin Criterion*:

Since the maximum of the minimum pay offs is – 3000 which corresponds to S_1. Hence the best alternative is S_1.

(ii) *Minimax Regret Criterion*:

The opportunity loss Table is given below.

table 7.8

Course of action	Regret	Amount	Maximum regret
	N_1	N_2	
S_1	0	200	2000
S_2	1000	0	1000

Table 7.8 shows that the best criteria is S_2.

(iii) *Hurwicz Criterion*

Given that $\alpha = 0.4$. Hence, $1 - \alpha = 0.6$. The quantity h can be calculated as follows. $h = \alpha \times$ maximum pay off + $(1 - \alpha) \times$ minimum pay off.

table 7.9

Course of action	Maximum pay off	Minimum pay off	h
S_1	- 3000	- 3000	- 3000
S_2	- 1000	- 4000	- 2800

Since the pay offs or profits, the maximum h is – 2800, corresponding to S_2. Hence, the best alternative is S_2.

7.3 DECISION UNDER CERTAINTY

Since under this environment, only one state of nature exists, the decision-maker simply picks up the best pay off in that one column and chooses the associated alternative. Under conditions of certainty, the particular state of nature is associated with probability 1. Though the state of nature is only one, possible alternatives could be numerous.

Linear programming, transportation and assignment techniques, input out analysis, activity analysis and economic order quantity models are used for such situations. Few complex managerial decision making problems, however, ever enjoy the luxury of having complete information about the future and thus decision-making under certainty is of little consequential interest.

7.4 DECISION-MAKING UNDER RISK

Most business decisions may have to be made under conditions of risk. Here, more than one states of nature exist and the decision-maker has sufficient information to assign probabilities to each of these states. These probabilities could be obtained from the past records or from simply the subjective judgment of the decision-maker. Under conditions of risk, a number of decision criteria are available which could be of help to the decision-maker. The most popular criterion for evaluating the alternatives is the expected monetary value/expected opportunity loss of the expected pay off.

7.4.1 Expected Monetary Value (EMV) Criterion

The Expected Monetary Value (EMV) for a given course of action is the expected value of the conditional pay off for that action. The conditional pay offs are obtained for each action by considering various act-event combinations. The Expected Monetary Value (EMV) for a given course of action is the weighted average pay off, which is the sum of the pay offs for each course of action multiplied by the probabilities associated with each state of nature. Mathematically, EMV is stated as follows:

$$EMV\,(A_j) = \sum_{i=1}^{n} p_{ij} \times P_i$$

$n =$ number of states of nature
$P_i =$ probability of occurrence of state of nature i.
$p_{ij} =$ pay off for the ith state of nature and j^{th} course of action.

Procedure:

Step 1: List conditional profit for each act-event combinations, along with the corresponding event probabilities.
Step 2: For each act, determine the expected conditional profits.
Step 3: Determine EMV for each act.
Step 4: Choose the act which corresponds to the optimal EMV.

Example 1

A person has the choice of running a hot snack stall or an ice-cream and cool drink stall at Ooty. If the weather is cool and rainy, he can expect to make a profit of Rs 15000 and if it is warm he can expect to make a profit of Rs 3000 by running a hot snack stall. On the other hand, if his choice is to run an ice-cream and cool drink stall, he can expect to make a profit of Rs 18000 if the weather is warm and Rs 3000 if the weather is cool and rainy. There is 40% chance of weather being warm in the coming season. Should he opt for running the hot snack stall or an ice-cream stall?
Solution: The following pay off table shows the conditional profits resulting from the given act-event combinations.

table 7.10

Event	*Probability*	*Conditional pay off (Rs)*	
E_i (weather)	*$P(E_i)$*	*Hot snacks stall*	*Ice-cream stall*
Cool and rainy	0.6	15000	3000
Warm	0.4	3000	18000

The Expected Monetary Value (EMV) for the hot snack stall is = $0.6 \times 15000 + 0.4 \times 3000 =$ Rs 10200

The Expected Monetary Value (EMV) for cool drink stall is $= 0.6 \times 3000 + 0.4 \times 18000 =$ Rs 9000.
Since the EMV for hot snacks stall is maximum, the person should opt for running a hot-snack stall.

Example 2

A newspaper boy has the following probabilities of selling a magazine.

No. of copies sold	Probability
10	0.10
11	0.15
12	0.20
13	0.25
14	0.30

Cost of copy is 30 paise and sale price is 50 paise. He cannot return unsold copies. How many copies should he order?

Solution: The demand varies from 10 to 14. Hence, there is no reason for the newsboy to buy less than 10 or more than 14 copies. The following Table shows the profit resulting from any possible combination of supply and demand. Stocking 10 copies each day will always result in a profit of 200 paise irrespective of the demand. For instance, even if the demand on some day is 13 copies, he can sell only 10 and hence his conditional profit is 200 paise.

When he stocks 11 copies, his profit will be 220 paise on days when buyers request 11, 12, 13, or 14 copies. But on days he has 11 copies on stock and buyers buy only 10 copies, his profit decreases to 170 paise. The profit of 200 paise on the 10 copies sold must be reduced by 30 paise, the cost of one copy left unsold. The same will be true when he stocks 12, 13 or 14 copies. Thus, the conditional profit in paise is given by

$$\text{Pay off} = 20 \times \text{copies sold} - 30 \times \text{copies unsold}.$$

table 7.11 **Conditional Profit Table**

Possible demand (No. of copies)	*Probability*	*Possible stock action* 10 copies	11 copies	12 copies	13 copies	14 copies
10	0.10	200	170	140	110	80
11	0.15	200	220	190	160	130
12	0.20	200	220	240	210	180
13	0.25	200	220	240	260	230
14	0.30	200	220	240	260	280

Next, the expected value of each decision alternative is obtained by multiplying its conditional profit by the associated probability and adding the resulting values.

table 7.12 **Expected Profit Table**

Possible demand	*Probability*	*Expected profit from stocking (paise)* 10 copies	11 copies	12 copies	13 copies	14 copies
10	0.10	20	17	14	11	8
11	0.15	30	33	28.5	24	19.5
12	0.20	40	44	48	42	36
13	0.25	50	55	60	65	57.5
14	0.30	60	66	72	78	84
Total expected profit (paise)		200	215	222.5	220	205

The newsboy must, therefore, order 12 copies to earn the highest possible average daily profit of 222.5 paise. This stock will maximise the total profits over a period of time. Of course, there is no guarantee that he will make a profit of 222.5 paise the next day. However, if he stocks 12 copies each day under the conditions given, he will have a average profit of 222.5 paise per day. This is the best he can do because the choice of any one of the other four possible stock actions will result in a lower daily profit.

7.4.2 Expected Opportunity Loss (EOL) Criterion

An alternative approach to maximising Expected Monetary Value (EMV) is to minimise the Expected Opportunity Loss (EOL), also called expected value of regret.

The EOL is defined as the difference between the highest profit (or pay off) for a state of nature (event) and the actual profit obtained for the particular course of action taken. In other words, EOL is the amount of pay off that is lost by not selecting the course of action that has the greatest pay off for the state of nature that actually occured. The course of action for which EOL is minimum is recommended.

Since EOL is an alternative decision criterion for decision making under risk, therefore, the results will always be the same as those obtained by EMV criterion discussed in 7.4.1. Thus, only one of the two methods should be applied to reach a decision. Mathematically, it is stated as follows:

$$\text{EOL }(N_i) = \sum_{i=1}^{m} l_{ij}\, p_i$$

where, l_{ij} = opportunity loss due to state of nature N_i and course of action S_j,

p_i = probability of occurrence of state of nature N_i.

Procedure for calculating EOL

Step 1: Construct the conditional profit table for each act-event combination, along with the corresponding event probabilities.

Step 2: For each state of nature calculate the Conditional Opportunity Loss (COL) values by subtracting each pay off from the maximum pay off for that event.

Step 3: For each act, determine the expected COL values and add these values to get the Expected Opportunity Loss (EOL) for that act.

Step 4: Choose that act which corresponds to the minimum EOL value.

Example 1

Consider the newspaper boy problem given in example 2 of Section 7.4.1. Obtain the decision alternative using EOL criterion.

Solution: The conditional profit table is given in Table 7.11. The Conditional Opportunity Loss and EOL are calculated in Table 7.13.

table 7.13

Possible demand (Event)	Probability	COL (in paise) Possible stock action					EOL in (paise) Possible stock action				
		10	11	12	13	14	10	11	12	13	14
10	0.10	0	30	60	90	120	0	3	6	9	12
11	0.15	20	0	30	60	90	3	0	4.5	9	13.5
12	0.20	40	20	0	30	60	8	4	0	6	12
13	0.25	60	40	20	0	30	15	10	5	0	7.5
14	0.30	80	60	40	20	0	24	18	12	6	0
							50	35	27.5	30	45

Since the minimum EOL is 27.5 the best choice is to have a stock of 12 copies.

7.4.3 Expected Value of Perfect Information (EVPI)

In decision-making under risk each state of nature is associated with a probability of its occurrence. However, if the decision-maker can acquire perfect information about the occurrence of various states of nature, then he will be able to select a course of action that yields the desired pay off for whatever state of nature that actually occurs.

Expected profit with perfect information (EPPI) represents the expected profit, in the long run, if we have perfect information before a decision is made.

Expected value of perfect information (EVPI) represents the maximum amount of money the decision-maker has to pay to get this additional information about the occurrence of various states of nature before a decision has to be made

$$\text{EVPI} = \text{EPPI} - \text{Expected profit without information}$$

$$= \text{EPPI} - \text{EMV}^*$$

$$= \sum_{i=1}^{m} p_{ij}(N_i)\, p_i - \text{EMV}^*$$

where, p_{ij} – best pay off for state of nature, N_i

p_i – probability of states of nature, N_i

EMV* – maximum expected monetary value.

Example 1

A dairy firm wants to determine the quantity of butter it should produce to meet the demand past records have shown the following demand patterns:

Quantity required (kg)	15	20	25	30	35	40	50
No. of days demand occurred	6	14	20	80	40	30	10

The stock levels are restricted to the range 15 to 50 kg due to inadequate storing facilities. Butter costs Rs 40 per kg. and is sold at Rs 50 per kg.

(i) Construct a conditional profit table.
(ii) Determine the action alternative associated with maximisation of expected profit.
(iii) Determine EVPI.

Solution: The range is from 15 kg to 50 kg.

The probabilities are calculated as follows:

Quantity required (kg)	15	20	25	30	35	40	50
Probability	$\frac{6}{200}$	$\frac{14}{200}$	$\frac{20}{200}$	$\frac{80}{200}$	$\frac{40}{200}$	$\frac{30}{200}$	$\frac{10}{200}$

The conditional profit will be calculated as follows:

$$\text{Conditional Profit} = (50 - 40)\,S = 10\,S \text{ when } D \geq S$$
$$= 50\,D - 40\,S, \text{ when } D \leq S$$

where, D – Demand and S – stock.

table 7.14 **Conditional Profit Table**

Possible demand (event) (kg)	*Probability (i)*	*Possible stock action (kg)* 15	20	25	30	35	40	50
15	0.03 $\left(=\frac{6}{200}\right)$	150	-50	-250	-450	-650	-850	-1250
20	0.07	150	200	0	-200	-400	-600	-1,000
25	0.10	150	200	250	50	-150	-350	-750
30	0.40	150	200	250	300	100	-100	-500
35	0.20	150	200	250	300	350	150	-250
40	0.15	150	200	250	300	350	400	0
50	0.05	150	200	250	300	350	400	500

table 7.15 **Expected Profit Table**

Possible demand (event) (kg)	*Probability (i)*	*Possible stock action (kg)* 15	20	25	30	35	40	50
15	0.03	4.50	-1.50	-7.50	-13.50	-19.50	-25.50	-37.50
20	0.07	10.50	14.00	0.00	-14.00	-28.00	-42.00	-70.00
25.10	15.00	20.00	25.00	5.00	-150.00	-35.00	-75.00	
30	0.40	60.00	80.00	100.00	120.00	40.00	-40.00	-200.00
35	0.20	30.00	40.00	50.00	60.00	70.00	30.00	-50.00
40	0.15	22.50	30.00	37.50	45.00	52.50	60.00	0.00
50	0.05	7.50	10.00	12.50	15.00	17.50	20.00	25.00
	EMV (Rs)	150.00	192.50	217.50	217.50	117.50	-32.50	-407.50

Since the maximum EMV is Rs 217.50, the dairy firm may produce 25 kg or 30 kg of butter.

If perfect information is known then the stock level is kept equal to the demand level. Then, the EPPI is calculated as follows:

table 7.16

Market size	Conditional profit under certainty (Rs) (i)	Probability (ii)	Expected profit with perfect information (i) × (ii)
15	150	0.03	4.50
20	200	0.07	14.00
25	250	0.10	25.00
30	300	0.40	120.00
35	350	0.20	70.00
40	400	0.15	60.00
50	500	0.05	25.00
		EPPI	318.50

EVPI = EPPI – EMV*

= 318.50 – 217.50

= Rs 101

Example 2 EMV for Items that have a Salvage (Resale) Value

An ice-cream retailer buys ice-cream at a cost of Rs 5 per cup and sells it for Rs 8 per cup; any ice-cream remaining unsold at the end of the day can be disposed of at a salvage price of Rs 2 per cup. Past sales have ranged between 15 and 18 cups per day; there is no reason to believe that sales volume will take on any other magnitude in future. Find the EVPI, if the sale history has the following probabilities.

Market size	15	16	17	18
Probability	0.10	0.20	0.40	0.30

Solution: The market size ranges from 15 to 18 cups. Since the resale value of unsold ice-cream is Rs 2 per cup, the conditional profit is calculated as follows:

$$\text{Conditional profit} = (8 - 5)\,S = 3S \text{ when } D \geq S$$

$$= D - 5\,S + 2(S - D), \text{ when } D < S$$

where, D – Demand and S – Stock

The conditional profit and the expected conditional profit is calculated in Table 7.17.

table 7.17

Demand (event)	Probability	Conditional profit (Rs) Possible stock action				Expected conditional profit (Rs) Possible stock action			
		15	16	17	18	15	16	17	18
15	0.10	45	42	39	36	4.50	4.20	3.90	3.60
16	0.20	45	48	45	42	9.00	9.60	9.00	8.40
17	0.40	45	48	51	48	18.00	19.20	20.40	19.20
18	0.30	45	48	51	54	13.50	14.40	15.30	16.20
						45.00	47.40	48.60	47.40

Since the maximum EMV is 48.60, the best strategy is to have a stock of 17 cups. If the perfect information is known as which event would occur the retailer can keep the stock level equal to the demand level. Then, the EPPI is calculated as follows:

table 7.18

Demand (event)	Probability	Conditional profit with certainty	Expected profit with perfect information
15	0.10	45	4.50
16	0.20	48	9.60
17	0.40	51	20.40
18	0.30	54	16.02
		EPPI	50.52

$$\begin{aligned} \text{EVPI} &= \text{EPPI} - \text{EMV}^* \\ &= 50.52 - 48.60 \\ &= \text{Rs } 1.92. \end{aligned}$$

Example 3

A television dealer finds that the cost of a TV in stock for a week is Rs 30 and the cost of a unit shortage is Rs 70. For one particular model of TV the probability distribution of weekly sales is as follows:

Weekly sales	0	1	2	3	4	5	6
Probability	0.1	0.1	0.2	0.25	0.15	0.15	0.05

How many units per week should the dealer order? Also find EVPI.

Solution: Here, the cost of stocking = Rs 30 week and shortage cost = Rs 70. The conditional cost are calculated as follows:

$$\begin{aligned} \text{Conditional cost} &= 30 \times S \text{ when } D < S \\ &= 30\,S + 70\,(D - S) \text{ when } D \geq S. \end{aligned}$$

table 7.19 **Conditional Cost (in Rs)**

Weekly sales	Probability	Weekly stock action						
		0	1	2	3	4	5	6
0	0.10	0	30	60	90	120	150	180
1	0.10	70	30	60	90	120	150	180
2	0.20	140	100	60	90	120	150	180
3	0.25	210	170	130	90	120	150	180
4	0.15	280	240	200	160	120	150	180
5	0.15	350	310	270	230	190	150	180
6	0.05	420	380	340	300	250	220	180

The expected conditional cost is calculated by multiplying the conditional cost with the corresponding probabilities.

table 7.20 **Expected Cost (Rs)**

Weekly sales	Probability	Weekly stock action						
		0	1	2	3	4	5	6
0	0.10	0	3.0	6.0	9.0	12.0	15.0	18.0
1	0.10	7.0	3.0	6.0	9.0	12.0	15.0	18.0
2	0.20	28	20	12.0	18.0	24.0	30.0	36.0
3	0.25	5.25	42.5	32.5	22.5	30	37.5	45
4	0.15	42	36	30	24	18	22.5	27
5	0.15	52.5	42.5	40.5	34.5	28.5	22.5	27
6	0.05	21	19	17	15	13	11	9
		203	170	144	132	137.5	153.50	180

Since the minimal expected cost is Rs 132, the TV dealer should order 3 units/week.

If the perfect information is known the TV dealer makes the stock level is equal to the demand level. Then, the expected cost is calculated as follows:

table 7.21

Demand	Probability	Minimum cost (Rs) with certainty	Expected cost with perfect information (ECPI)
0	0.10	0	0.00
1	0.10	30	3.00
2	0.20	60	12.00
3	0.25	90	22.50
4	0.15	120	18.00
5	0.15	150	22.50
6	0.05	180	9.00
		ECPI	87.00

Now, EVPI = Minimum expected cost – ECPI

= 132 – 87

= Rs 45.

Thus, the TV dealer may give a maximum of Rs 45 to get the perfect information.

7.5 DECISION TREES

Decision-making problems discussed so far are referred to as single stage decision problems. Because the pay offs, states of nature, courses of action and probabilities associated with the occurrence of states of nature are not subject to change. However, situations may arise when decision-maker needs to revise his previous decisions on getting new information and make a sequence of other decisions. Thus, the problem becomes a multi-stage decision problem because the consequence of one decision affects future decisions. For example, in the process of marketing a new product, the first decision is often test marketing and the alternative courses of action might be either intensive testing or gradual testing. Given various possible consequences—favourable fair, or poor, the decision-maker may be required to decide between redesigning the product, and aggressive advertising campaign or complete withdrawal of product and so on. Given that decision, there will be an outcome which leads to another decision and so on.

Decision Tree is a graphical representation of various decision alternatives and the sequence of events as if they were branches of a tree.

A decision node is usually represented by a square and indicates places where a decision-maker must make a decision. Each branch leading away from a decision node represents one of the several possible courses of action available to the decision-maker. The chance node is represented by a circle and indicates a point at which the decision-maker will discover the response to his decision, that is, different possible outcomes which can result from a chosen course of action. The respective pay offs and the probabilities associated with alternative courses and the chance events are shown alongside these branches. At the terminal of the chance branch are shown the expected values of the outcome are shown.

The general approach used in decision tree analysis is to work backward through the tree from right to left, computing the expected value of each chance node. We then choose the particular branch leaving a decision node which leads to the chance node with the highest expected value. This is known as roll back or fold back process.

Steps in Decision Tree Analysis

Step 1: Identify the decision points and the alternative courses of action at each decision point systematically.

Step 2: At each decision point determine the probability and the pay off associated with each course of action.

Step 3: Commencing from the extreme right end, compute the expected pay offs (EMV) for each course of action.

Step 4: Choose the course of action that yields the best pay off for each decisions.

Step 5: Proceed backwards to the next stage of decision points.

Step 6: Repeat above steps till the first decision point is reached.

Step 7: Finally, identify the course of action to be adopted from the beginning to the end under different possible outcomes for the situation as a whole.

Advantages of Decision Tree Approach

1. It structures the decision process and helps decision-making in an orderly, systematic and sequential manner.
2. It requires the decision-maker to examine all possible outcomes, whether desirable or undesirable.
3. It communicates the decision-making process to others in an easy and clear manner, illustrating each assumption about the future.
4. It displays the logical relationship between the parts of a complex decision and identifies the time sequence in which various actions and subsequent events would occur.
5. It is especially useful in situations wherein the initial decision and its outcome affects the subsequent decisions. It can be applied in various fields such as introduction of a new product, marketing, make or buy decisions, investment decisions and so on.

Limitations of Decision Tree Approach

1. Decision tree diagrams become more complicated as the number of decision alternative increases and more variables are introduced.
2. It becomes highly complicated when inter-dependent alternatives and dependent variables are present in the problem.
3. It assumes that utility of money is linear with money.
4. It analyses the problem in terms of expected values and thus yields an 'average' valued solution.
5. There is often inconsistency in assigning probabilities for different events.

Example 1

An electronics company makes a profit of Rs 10000 per day at present. The company has the option to go in for licence from another company and make a profit of Rs 20000 per day gross but has to pay a royalty of Rs 6000 per day. It has another option to go in for research and development at a cost of Rs 10000. The company may make a profit of Rs 25000 with a probability of 70% success and 30% failure. Draw a decision tree and choose the appropriate action.

Solution:

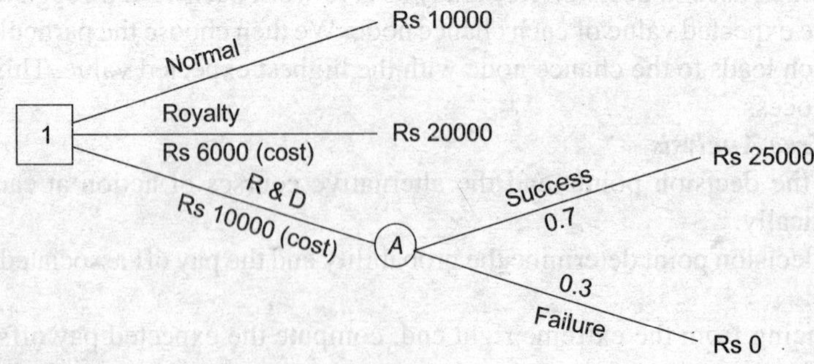

figure 7.1

$$\text{EMV}(A) = [0.7 \times 25000 + 0.3 \times 0] - 10000$$
$$= 7500$$

EMV (node 1) = Maximum [10000, 20000 – 6000, 7500]
= 14,000 (corresponding to loyalty)

Hence, the best choice is to go for licence by paying Rs 6000 as loyalty.

Example 2

Mr. Anish had to decide whether or not to drill a well on his farm. In his town, only 40% of the wells drilled were successful at 200 feet of depth. Some of the farmers who did not get water at 200 feet, drilled further upto 250 feet but only 20% struck water at 250 feet. Cost of drilling is Rs 50 per foot. Mr. Anish estimated that he would pay Rs 18000 during a 4-year period in the present value terms, if he continues to buy water from the neighbour rather than go for the well which would have a life of 4 years. Mr. Anish has three decisions to make (a) should he drill upto 200 feet and (b) if no water is found at 200 feet, should he drill upto 250 feet? (c) should he continue to buy water from his neighbour?

Solution:

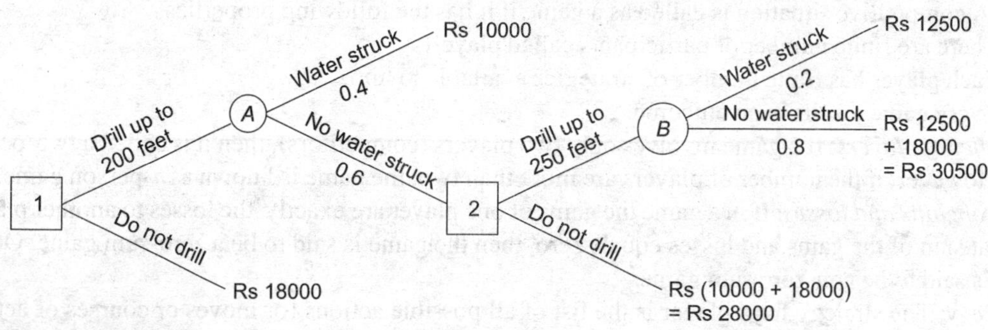

figure 7.2

EMV of node B = Rs [0.2 × 12500 + 0.8 × 30500]
= Rs (2500 + 24400)
= Rs 26900

∴ EMV of node 2 = Rs 26900 (Lesser of the two values of Rs 26900 and Rs 28000).

∴ EMV of node A = Rs [0.4 × 10000 + 0.6 × 26900]
= Rs [4,000 + 16140]
= Rs 20140

EMV of node 1 = Rs 18000 [Lesser of the two values of Rs 20140 and Rs 18000].

Hence, the optimal choice of Mr. Anish is not to drill the well and pay Rs 18,000 for water to his neighbour for five years.

7.6 GAME THEORY—AN INTRODUCTION

Many a decision is taken in a competitive situation in which the outcome depends not on that decision alone but rather on the interaction between the decision-maker and that of a competitor. The term 'games' now includes not only pleasurable activities of this kind, but also much more earnest competitive situations of war and peace. It was not coincidental that the classic work on the theory of games was first published during the Second World War. Many competitive situations are still too complex for the theory in its present state of development to solve. Other methods of which war games are long-established examples and business games of more recent origins, have been used. The availability of computers has allowed increasingly large scale operations to be represented with great realism. The theory of games has been represented with great realism. The theory of games has been developed alongside gaming techniques and a knowledge of the concepts involved, especially the importance of the role of chance, helps to clarify the issues in many decision making process.

7.6.1 Some Basic Terminologies

Game: A competitive situation is called as a game if it has the following properties:

(i) There are finite number of participants called playeRs

(ii) Each player has finite number of strategies available to him.

(iii) Every game results in an outcome.

Number of players: If a game involves only two players (competitors), then it is called a two-person game. However, if the number of players are more than two, the game is known as *n*-person game.

Sum of gains and losses: If in a game the gains of one player are exactly the losses to another player, such that sum of the gains and losses equals zero, then the game is said to be a zero-sum game. Otherwise it is said to be non-zero sum game.

Strategy: The strategy for a player is the list of all possible actions (or moves or courses of action) that he will take for every pay off (outcome) that might arise. It is assumed that the rules governing the choices are known in advance to the players. The outcome resulting from a particular choice is also known to the players in advance and is expressed in terms of numerical values. Here, it is not necessary that players have a definite information about each other strategies.

Optimal strategy: The particular strategy (or complete plan) by which a player optimises his gains or losses without knowing the competitor's strategies is called optimal strategy.

Value of the game: The expected outcome per play when players follow their optimal strategy is called the value of the game.

Pure strategy: It is a decision rule which is always used by the player to select the particular course of action. Thus, each player knows in advance of all the strategies out of which he always selects only one particular strategy irrespective of the strategy others may choose and the objective of the players is to maximise gains or minimise losses.

Mixed strategy: When both players are guessing as to which course of action is to be selected on a particular occasion with some fixed probability, it is a mixed strategic game. Thus, there is a probabilistic situation and objective of the players is to maximise expected gains or to minimise expected losses by making a solution among pure strategies with fixed probabilities.

Two person zero sum game: A game with only two persons is said to be two-person zero-sum game if the gain of one player is equal to the loss of the other.

Pay off matrix: The pay offs in terms of gains or losses, when players selected their particular strategies, can be represented in the form of a matrix, called the pay off matrix.

If the game is zero-sum, the gain of one player is equal to the loss of the other and vice versa. In other words, one player's pay off table would contain the same amounts in pay off table of the other player with the sign changed. If the player A has strategies $A_1, ..., A_m$ and the player B has strategies $B_1, ..., B_n$ and if a_{ij} represent the pay offs (gain for A) for i^{th} action of A and j^{th} action of B then the pay off matrix for player A is given by

Player B's strategies

$$\text{Player } A\text{'s strategies} \quad \begin{array}{c} \\ A_1 \\ A_2 \\ \vdots \\ A_m \end{array} \begin{array}{c} \begin{array}{cccc} B_1 & B_2 & \cdots & B_n \end{array} \\ \begin{bmatrix} a_{11} & a_{12} & \cdots & a_{1n} \\ a_{21} & a_{22} & \cdots & a_{2n} \\ \vdots & \vdots & \vdots & \vdots \\ a_{m1} & a_{m2} & \cdots & a_{mn} \end{bmatrix} \end{array}$$

7.7 TWO-PERSON ZERO-SUM GAME

Basic assumptions of the game:

1. Each player has available to him a finite number of possible courses of action. The list may not be same for each player.
2. Player A attempts to maximise gains and player B minimise losses.
3. The decisions of both players are made individually prior to the play with no communication between them.
4. The decisions are made simultaneously and also announced simultaneously so that neither player has an advantage resulting from direct knowledge of the other player's decision.
5. Both the players know not only the possible pay offs to themselves but also of each other.

7.7.1 Games with Saddle Point

Minimax and Maximin Principle: Consider the pay matrix of a game which represents pay off of player A. Now, the objective of the study is to know how these players must elect their respective strategies so that they may optimise their pay off. Such a decision-making criterion is referred to as the minimax-maximin principle.

For player A minimum value in each row represents the least gain (pay off) to him if he chooses his particular strategy. These are written in the matrix by row minima. He will then select the strategy that gives largest gain among the row minimum values. This choice of player A is called the maximin principle, and the corresponding gain is called the maximin value of the game denoted by v.

For player B (who is assumed to be the loser), the maximum value in each column represents the maximum loss to him if he chooses his particular strategy. These are written in the pay off matrix by column minima. He will then select the strategy that gives minimum loss among the column maximum values. This choice of player B is called the minimax principle, and the corresponding loss is the minimax value of the game denoted by $\overline{v}$.

Saddle point: A saddle point of a pay off matrix is that position in the pay off matrix where maximum of row minima coincides with the minimum of the column maxima. The saddle point need not be unique.

Value of the game: The pay off of the saddle point is called the value of the game denoted by v.

Fair game: A game is said to be fair if $v = 0 = \bar{v}$.

Strictly determinable game: A game is said to be strictly determinable if $\underline{v} = v = \bar{v}$.

Procedure to determine saddle point

Step 1: Select the minimum element in each row and enclose it in a rectangle (□)

Step 2: Select the maximum element in each column and enclose it in a circle (○)

Step 3: Find out the element which is enclosed by the rectangle as well as the circle. Such element is the value of the game and that position is called as the saddle point.

Example 1

Solve the game whose payoff matrix is given by

Player *B*

		I	II	III	IV	V
Player *A*	I	−2	0	0	5	3
	II	3	2	1	2	2
	III	−4	−3	0	−2	6
	IV	5	3	−4	2	−6

Solution: Select the row minimum and enclose it in a rectangle. Then, select the column maximum and enclose it in a circle.

Player *B*

		I	II	III	IV	V
Player *A*	I	[−2]	0	0	(5)	3
	II	3	2	([1])	2	2
	III	[−4]	−3	0	−2	(6)
	IV	(5)	(3)	−4	2	[−6]

It is clear that saddle point is (II, III) and the value of the game $v = 1$.

Player *A* uses his course of action II throughout.

Player *B* uses his course of action III throughout.

The value of the game is 1 (for player *A* and −1 for player *B*)

Example 2

A company management and the labour union are negotiating a new three-year settlement. Each of these has 4 strategies.

I : Hard and aggressive bargaining
II : Reasoning and logical approach
III : Legalistic strategy
IV : Conciliatory approach.

The costs to the company are given for every pair of strategy choice.

		Company strategies			
		I	II	III	IV
Union strategies	I	20	15	12	35
	II	25	14	8	10
	III	40	2	10	5
	IV	–5	4	11	0

What strategy will the two sides adopt? Also, determine the value of the game.

Solution: Select the row minimum and enclose it by a rectangle. Then, select the column maximum and enclose it in a circle.

		Company strategies			
		I	II	III	IV
Union strategies	I	20	(15)	[(12)]	(35)
	II	25	14	[8]	10
	III	(40)	[2]	10	5
	IV	[–5]	4	(11)	0

Hence, the saddle point is (I, III) and the value of the game is 12 (for union and – 12 for the company).

Hence, the company will always adopt strategy III – Legalistic strategy and union will always adopt strategy I – Hard and aggressive bargaining.

Example 3

Find the range of values of p and q which will render the entry (2, 2) a saddle point for the game.

		Player B	
	2	4	5
Player A	10	7	q
	4	p	6

Solution: First ignoring the values of p and q determine the maximin and minimax values of the pay off matrix.

	Player B		
	B_1	B_2	B_3
A_1	[2]	4	5
A_2	(10)	([7])	q
A_3	[4]	p	([6])

Maximin value $\underline{v} = 7$ = minimax value.

This imposes the condition on p as $p \leq 7$ and on q as $q \geq 7$. Hence, the range of p and q will be $p \leq 7$, $q \leq 7$.

Example 4

Determine which of the following two-person zero-sum games are strictly determinable and fair. Give optimum strategies for each player in the case of strictly determinable games.

(i)

	Player B	
Player A	1	1
	4	– 3

(ii)

	Player B	
Player A	– 5	2
	– 7	– 4

Solution: (i)

		Player *B*	
		I	II
Player *A*	I	1	1
	II	4	–3

Maximin value = minimax value = 1 = value of the game. Hence, the game is strictly determinable but not fair. The optimum strategies are I for player *A* and II for player *B*

(ii)

		Player *B*	
		I	II
Player *A*	I	–5	2
	II	–7	–4

Maximin value = Minimax value = 5 = value of the game. Hence, the game is strictly determinable but not fair. The optimum strategies are I for player *A* and I for player *B*.

7.7.2 Games without Saddle Point Mixed Strategies

There are some games for which no saddle point exists. In such cases both the players must determine an optimal combination of strategies to find a saddle (equilibrium) point. The optimal strategy combination for each player may be determined by assigning to each strategy its probability of being chosen. The strategies so determined are called mixed strategies because they are probabilistic combination of available choices of strategy.

The value of game obtained by the use of mixed strategies represents least pay off which player *A* can expect to win and the least which player *B* can lose. The expected pay off to a player in a game with arbitrary pay off matrix $[a_{ij}]$ of order m × n is defined as

$$E(p, q) = \sum_{i=1}^{m} \sum_{j=1}^{m} p_i \, a_{ij} \, q_j$$

$$= P^T A Q \quad \text{(in matrix notaion)}$$

where, $P = (p_1, p_2, ..., p_m)$ and $Q = (q_1, ..., q_n)$ denotes the mixed strategies for players *A* and *B*. Also, $p_1 + p_2 + ... + p_m = 1$ and $q_1 + ... + q_n = 1$. A particular strategy with particular probability a player chooses can also be interpreted as the relative frequency with which a strategy is chosen from the number of strategies of the game.

A mixed strategy game can be solved by different solution methods such as

1. Algebraic method
2. Analytical or calculus method

3. Matrix method
4. Graphical method, and
5. Linear programming method.

These methods will be discussed in detail in the next section.

Dominance Property of Reducing the Size of the Game

We can sometimes reduce the size of a game's pay off matrix by eliminating a course of action which is so inferior to another as never to be used. Such a course of action is said to be dominated by the other. The concept of dominance is especially useful for the evaluation of two-person zero-sum games where a saddle point does not exist.

General rule

1. If all the elements of a row, say k^{th}, are less than or equal to the corresponding elements of any other row, say r^{th}, then kth row is dominated by the rth row.
2. If all the elements of a column, say k^{th} are greater than or equal to the corresponding elements of any other column, say r^{th}, then k^{th} column is dominated by r^{th} column.
3. Omit dominated rows or columns.
4. If some linear combination of some rows dominates i^{th} row, then i^{th} row will be deleted. Similar argument follows for columns.

Example 1

Reduce the size of the game whose matrix is given by

		Player B		
		I	II	III
	I	– 4	6	3
Player A	II	– 3	– 3	4
	III	2	– 3	4

Solution:

	I	II	III
I	−4	6	3
II	−3	−3	4
III	2	−3	4

We observe that no saddle point exists. Consider Ist and IIIrd column's from the player B's point of view. We observe that each pay off in the IIIrd column is greater than the corresponding element in the 1st column regardless of player A's strategy. Evidently, the choice of IIIrd strategy by the player B will always result in the greater loss compared to that of selecting the 1st strategy. Column III is inferior to I and is never to be used. Hence, deleting the IIIrd column which is dominated by I, the reduced size pay off matrix is obtained.

		B I	*B* II
A	I	–4	6
	II	–3	–3
	III	2	–3

Again, if the reduced matrix is looked at from player *A*'s point of view, it is seen that the player *A* will never use the IInd strategy which is dominated by III. Hence, the size of the matrix can be reduced further by deleting the IInd row. Hence, the reduced matrix is

		B I	*B* II
A	I	–4	6
	II	2	–3

7.8 SOLUTION METHODS OF GAMES WITHOUT SADDLE POINT

Solution of 2 × 2 games: (*Standard method*)

For any 2 × 2 two-person zero-sum game without any saddle point having the pay off matrix for player *A*

	B_1	B_2
A_1	a_{11}	a_{12}
A_2	a_{21}	a_{22}

The optimum mixed strategies

$$S_A = \begin{bmatrix} A_1 & A_2 \\ p_1 & p_2 \end{bmatrix} \text{ and } S_B = \begin{bmatrix} B_1 & B_2 \\ q_1 & q_2 \end{bmatrix}$$

where,

$$p_i = \frac{a_{22} - a_{21}}{(a_{11} + a_{22}) - (a_{12} + a_{21})} \quad \text{and} \quad p_2 = 1 = p_1$$

$$q_1 = \frac{a_{22} - a_{12}}{(a_{11} + a_{22}) - (a_{12} + a_{21})} \quad \text{and} \quad q_2 = 1 - q_1$$

and the value of the game is

$$v = \frac{a_{11}a_{22} - a_{12}a_{21}}{(a_{11} + a_{22}) - (a_{12} + a_{21})}.$$

7.8.1 Algebraic Method

This method can be used to determine probability value by using different strategies by players A and B. This method becomes quite lengthy when number of strategies for both.

Consider a game where pay off matrix is $[a_{ij}]_{m \times n}$

Let $(p_1, p_2, ..., p_m)$ and $(q_1,..., q_n)$ be the probabilities with which players A and B adopt their mixed strategies $(A_1, ... A_m)$ and $(B_1, B_2, ..., B_n)$, respectively. If v is the value of game, then the expected gain to player A for this game when player B selects strategies $B_1, B_2, ..., B_n$ one by one is given by left hand side of simultaneous equations respectively. Since player A is the gainer and expects at least v, we must have

$$\begin{aligned} a_{11}\,p_1 + a_{12}\,p_2 + ... + a_{1m}\,p_m &\geq v \\ a_{21}\,p_1 + a_{22}\,p_2 + ... + a_{2m}\,p_m &\geq v \\ &\vdots \\ a_{n1}\,p_1 + a_{n2}\,p_2 + ... + a_{nm}\,p_m &\geq v \end{aligned}$$

where, $\quad p_1 + p_2 + ... + p_m = 1$ and $p_i \geq 0$ for all i.

Similarly, the expected loss to player B when player A adopts strategies $A_1, ..., A_m$ can be determined. Since player B is the loser, we must have,

$$\begin{aligned} a_{11}\,q_1 + a_{21}\,q_2 + ... + a_{n1}q_1 &\geq v \\ a_{12}\,q_2 + a_{22}\,q_2 + ... + a_{n2}\,q_2 &\geq v \\ &\vdots \\ a_{1m}\,q_1 + a_{2m}\,q_2 + ... + a_{nm}\,q_n &\geq v \end{aligned}$$

where, $\quad q_1 + ... + q_n = 1$ and $q_j \geq 0$ for all j

Now, to get the values of p_i's and q_j's from inequalities (1) and (2), these inequalities are considered as equations and are then solved for given unknowns. However, if the system of equations so obtained is inconsistent, then at least one of the inequalities must hold as strict inequality. The solution can now be obtained only by applying trial and error method.

Example 1

In a game of matching coins with two players, suppose A wins one unit of value when there are two tails and loses 1/2 unit of value when there is one head and one tail. Determine the pay off matrix, the best strategies for each player and the value of the game to A.

Solution: The pay off matrix for the given matching coin game is

		Player B	
		B_1	B_2
Player A	A_1	-1	$-\frac{1}{2}$
	A_2	$-\frac{1}{2}$	0

As the pay off matrix does not have a saddle point, the game can be solved by algebraic method.

For player A, let p_1 and p_2 be probabilities of selecting a strategy A_1 and A_2, respectively. Then, the expected gain to player A when player B uses its B_1, and B_2 strategies is given by

$$-p_1 - 1/2\, p_2 \geq v \qquad (7.8.1)$$

$$-1/2\, p_1 + 0 \cdot p_2 \geq v \qquad (7.8.2)$$

where,

$$p_1 + p_2 = 1 \qquad (7.8.3)$$

For obtaining values of p_1 and p_2 considering these inequalities as equations. From (7.8.1), (7.8.2) and (7.8.3) we get $p_1 = -2v$ and $p_2 = -6v$

Substituting these values of p_1 and p_2 in (7.8.3), we get $p_1 = 0.25$, $p_2 = 0.75$ and $v = -1/8$

For player B, let q_1 and q_2 be the probabilities of selecting strategies B_1 and B_2, respectively. Then, the expected loss to player B when player A uses its strategies A_1 and A_2 is given by

$$q_1 - 1/2\, q_2 \geq v \qquad (7.8.4)$$

$$-1/2 \,.\, q_1 + 0 \cdot q_2 \geq v \qquad (7.8.5)$$

$$q_1 + q_2 = 1 \qquad (7.8.6)$$

Considering (7.8.4), (7.8.5) as equations and then solving we get

$$q_1 = 2v \quad \text{and} \quad q_2 = -6v$$

Substituting the values of q_1 and q_2 in (7.8.6)

we get, $v = -1/8$, $q_1 = 0.25$ and $q_2 = 0.75$

Hence, the optimal strategies for players A and B are (0.25, 0.75) and (0.25, 0.75), respectively and the value of the game is $v = -1/8$.

7.8.2 Arithmetic Method

Arithmetic method provides an easy technique for obtaining the optimum strategies for each player in (2×2) games without saddle point. This method consists of the following steps:

Step 1: Find the difference of two numbers in column I and put it under the column II, neglecting the negative sign if occurs.

Step 2: Find the difference of two numbers in column II, and put it under the column I, neglecting the negative sign if occurs.

Step 3: Repeat the above two steps for the two rows also.

The values thus obtained are called the *oddments*.

These are the frequencies with which the players must use their courses of action in their optimum strategies.

Example 1

Two players A and B showing each other, put on a table a coin, with head or tail up. A wins Rs 8 when both the coins show head and Re 1 when both are tails. B wins Rs 3 when the coins do not match. Given the choice of being matching player (A) or non-matching player (B) which one would you choose and what would be your strategy?

Solution: The pay off matrix for player A is given by

		Player B *H*	Player B *T*		
Player A	*H*	8	−3	4	$\frac{4}{11+4} = 4/15$
	T	−3	1	11	$\frac{11}{11+4} = \frac{11}{15}$
		4	11		

Since no saddle point, is found the optimal strategies will be mixed strategies.

Step 1: Taking the difference of two numbers in column I, we find 8 – (– 3) = 11, and put it under column II.

Step 2: Taking the difference of two numbers in column II, we find (– 3 – 1) = – 4 and put the number 4 (neglecting the – ve sign) under column I.

Step 3: Repeat the above two steps for the two rows also.
Thus, for optimum gains, player *A* must use strategy *H* with probability 11/15 and strategy *T* with probability 4/15, while player *B* must use strategy *H* with probability 4/15 and strategy *T* with probability 11/15.

Step 4: To obtain the value of the game any of the following expressions may be used.

Using *B*'s oddments

$$B \text{ plays } H, \text{ value of the game, } v = \text{Rs } \frac{4\times 8+11\times(-3)}{11+4}$$
$$= \text{Rs } (-1/15)$$
$$B \text{ plays } T, \text{ value of the game, } v = \text{Rs } \frac{4\times(-3)+11\times 1}{11+4}$$
$$= \text{Rs } (-1/15)$$

Using A's oddments

$$A \text{ plays } H, \text{ value of the game, } v = \text{Rs } \frac{4\times 8+11\times(-3)}{4+11}$$
$$= \text{Rs } (-1/15)$$
$$A \text{ plays } T, \text{ value of the game, } v = \text{Rs } \frac{4\times(-3)+11\times 1}{4+11}$$
$$= \text{Rs } (-1/15).$$

The above values of v are equal only if the sum of the oddments vertically and horizontally are equal. Cases in which it is not so will be discussed later.

Thus, the complete solution of the game is:

(i) Optimum strategy for *A* is (4/15, 11/15)
(ii) Optimum strategy for *B* is (4/15, 11/15)
(ii) Value of the game to *A* is v = Rs (– 1/15)

Thus, player gains Rs (–1/15), ie he loses Rs (1/15) which *B*, is turn, gets.

Remark: Even though arithmetic method is easier than algebraic method but it cannot be applied to larger games.

7.8.3 Matrix Method

If the pay off matrix of a game is a square matrix, then optimal strategy mixture as well as value of the game obtained by the matrix method. The solution of a two-person zero-sum game with mixed strategies with a square pay off matrix may be found by using the following formulae:

$$\text{Player } A\text{'s optimal strategy} = \frac{[1 \quad 1]\, P_{adj}}{[1 \quad 1]\, P_{adj}\begin{bmatrix}1\\1\end{bmatrix}}$$

$$\text{Player } B\text{'s optimal strategy} = \frac{[1 \quad 1]\, P_{cof}}{[1 \quad 1]\, P_{adj}\begin{bmatrix}1\\1\end{bmatrix}}$$

Value of the game = (Player A's optimal strategies) × (Payoff matrix P_{ij}) × (Player B's optimal strategies)

where P_{adj} = adjoint matrix, P_{cof} = cofactor matrix. Player A's optimal strategies are in the form of a row vector and B's optimal strategies are in the form of a column vector.

This method can be used for finding a solution of a game with size more than 2×2. However, in rare cases, the solution violates the non-negative condition of probabilities that is $p_i \geq 0$, $q_i \geq 0$ although the requirement $p_1 + ... + p_m = 1$ or $q_1 + ... + q_n = 1$ is satisfied.

Example 1

Solve the following game after reducing it to a 2×2 game.

		Player B		
		B_1	B_2	B_3
Player A	A_1	1	7	2
	A_2	6	2	7
	A_3	5	1	6

Solution:

Reduction to 2×2 matrix.

In the given pay off matrix, the third row is dominated by second row and third column is dominated by the first column. Hence, by the dominance property, the matrix is reduced to

		Player B	
		B_1	B_2
Player A	A_1	1	7
	A_2	−6	2

Calculation of P_{adj} and P_{cof}

$$P_{adj} = \begin{bmatrix}2 & -7\\-6 & 1\end{bmatrix} \quad \text{and} \quad P_{cof} = \begin{bmatrix}2 & -6\\7 & 1\end{bmatrix}$$

$$\text{Player } A\text{'s optimal strategies} = \frac{[1 \quad 1]\begin{bmatrix}2 & -7\\-6 & 1\end{bmatrix}}{[1 \quad 1]\begin{bmatrix}2 & -7\\-6 & 1\end{bmatrix}\begin{bmatrix}1\\1\end{bmatrix}} = \frac{[-4 \quad -6]}{-10} = \frac{[4 \quad 6]}{10}$$

This solution can be broken into the optimal strategy combination for player A as $p_1 = 4/10 = 2/5$ and $p_2 = 6/10 = 3/5$, where p_1 and p_2 represent the probabilities of player A's using his strategies A_1 and A_2, respectively.

Similarly, the optimal strategy combination for player B is obtained as player B's optimal strategies

$$= \frac{[1 \quad 1]\begin{bmatrix} 2 & -6 \\ -7 & 1 \end{bmatrix}}{[1 \quad 1]\begin{bmatrix} 2 & -7 \\ -6 & 1 \end{bmatrix}\begin{bmatrix} 1 \\ 1 \end{bmatrix}} = \frac{[-5 \quad -5]}{-10} = \frac{[5 \quad 5]}{10}$$

This solution can also be broken down into the optimal strategy combination for player B as $q_1 = 5/10 = 1/2$ and $q_2 = 5/10 = 1/2$, where q_1 and q_2 represent the probabilities of player B's using his strategies B_1 and B_2, respectively.

Value of the game

$$v = [2/5 \quad 3/5]\begin{bmatrix} 1 & 7 \\ 6 & 2 \end{bmatrix}\begin{bmatrix} 1/2 \\ 1/2 \end{bmatrix} = 4.$$

7.8.4 Graphical Method (for $2 \times n$ or for $m \times 2$ Games)

The graphical method is useful for the game where the pay off matrix is of the size $2 \times n$ or $m \times 2$. That is, the game with mixed strategies that has only two pure strategies for one of the players in the two person zero-sum game. Optimal strategies for both the players assign non-zero probabilities to the same number of pure strategies. Therefore, if one player has only two strategies, the other will also use the same number of strategies. Hence, this method is useful in finding out which of the two strategies can be used.

Consider the $2 \times n$ pay off matrix of a game without a saddle point.

$$\begin{array}{cc} & \text{Player } B \\ & \begin{array}{cccc} B_1 & B_2 & \cdots & B_n \end{array} \\ \text{Player } A \begin{array}{c} A_1 \\ A_2 \end{array} & \begin{bmatrix} a_{11} & a_{12} & \cdots & a_{1n} \\ a_{21} & a_{22} & \cdots & a_{2n} \end{bmatrix} \end{array}$$

Let the mixed strategy for player A be given by $S_A = \begin{bmatrix} A_1 & A_2 \\ p_1 & p_2 \end{bmatrix}$, where $p_1 + p_2 = 1$ and $p_1 \geq 0, p_2 \geq 0$.

Now, for each of the pure strategies available to B, expected pay off for player A would be as follows:

B's pure move	A's expected pay off $E(p)$
B_1	$E_1(p) = a_{11}p_1 + a_{21}p_2$
B_2	$E_2(p) = a_{12}p_1 + a_{22}p_2$
$\vdots$	$\vdots$
B_n	$E_n(p) = a_{1n}p_1 + a_{2n}p_2$

The player B would like to choose that pure move B_j against S_A for which $E_j(p)$ is a minimum for $j = 1, ..., n$. Let us denote this minimum expected pay off for A by

$$v = \text{Min}\{E_j(p), j = 1, ..., n\}$$

The objective of player A is select p_1 and hence p_2 in such a way that v is as large as possible. This may be done by plotting the straight lines.

$E_j(p) = a_{ij}\,p_1 + a_{2j}\,p_2 = (a_{ij} - a_{2j})p_1 + a_{2j}$ $(j = 1, 2, ..., n)$ as linear function of p_1.

The highest point on the lower boundary of these lines will give maximum expected pay off among the minimum expected pay offs on the lower boundary (lower envelope) and the optimum value of probability p_1 and p_2.

Now the two strategies of player B corresponding to those lines which pass through the maximum point can be determined. It helps in reducing the size of the game to (2×2).

The $(m \times 2)$ games are also treated in the same way except that the upper boundary (upper envelope) of the straight lines corresponding to B's expected pay off will give the maximum expected pay off to player B and the lowest point on this boundary will then give the minimum expected payoff (minimax value) and the optimum value of probability q_1 and q_2.

Example 1

Solve the following 2×5 graphically

$$\begin{array}{cc} & \text{Player } B \\ & \begin{array}{ccccc} B_1 & B_2 & B_3 & B_4 & B_5 \end{array} \\ \text{Player } A \quad \begin{array}{c} A_1 \\ A_2 \end{array} & \begin{pmatrix} 2 & -1 & 5 & -2 & 6 \\ -2 & 4 & -3 & 1 & 0 \end{pmatrix} \end{array}$$

Solution: The game does not have a saddle point. Let the probability of player A playing A_1 and A_2 in the strategy combination is denoted by p_1 and p_2, respectively where $p_2 = 1 - p_1$. Then, the expected pay off (gain) to player A will be

B's pure strategies	*A's expected payoff (Ei)*
B_1	$2p_1 - 2p_2$
B_2	$-p_1 + 4p_2$
B_3	$5p_1 - 3p_2$
B_4	$-2p_1 + p_2$
B_5	$6p_1 + 0p_2$

These five expected pay off lines are plotted on the graph below.

Here, p_1 is measured on the x-axis. Since p_1 cannot exceed 1, the x-axis is cut-off at $p = 1$. The expected pay off of player A is measured along y-axis. From the game matrix, if player B plays B_1, the expected pay off of player A is 2 when A plays A_1 with $p_1 = 1$ and -2 when A plays A_2 with $p_1 = 0$. These two extreme points are connected by a straight line, which shows the expected pay off of A when B plays B_1. Four other straight lines are similarly drawn for B_2, B_3, B_4 and B_5.

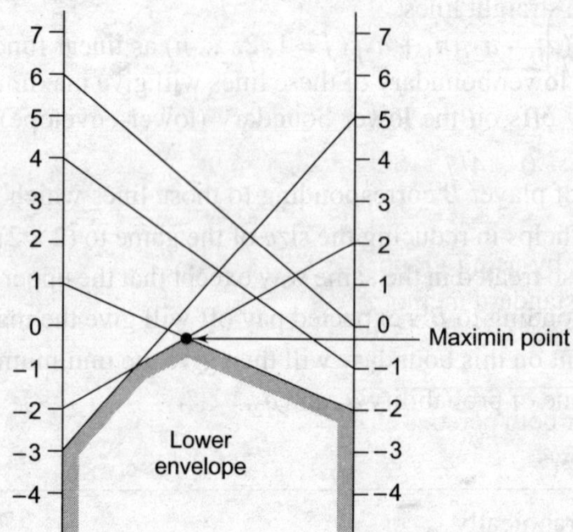

The maximin point shows that the reduced pay off matrix is for player A is

$$\text{Player } A \quad \begin{array}{c} \\ A_1 \\ A_2 \end{array} \overset{\text{Player } B}{\begin{array}{cc} B_1 & B_4 \end{array}} \begin{pmatrix} 2 & -2 \\ -2 & 1 \end{pmatrix} = \begin{pmatrix} a_{11} & a_{12} \\ a_{21} & a_{22} \end{pmatrix}$$

Let $S_A = \begin{pmatrix} A_1 & A_2 \\ p_1 & p_2 \end{pmatrix}$ be the mixed strategies for player A.

Then, $$p_1 = \frac{a_{22} - a_{21}}{(a_{11} + a_{22}) - (a_{12} + a_{21})} = \frac{1-(-2)}{(2+1)-(-2-2)} = \frac{3}{7}$$

and, $p_2 = 1 - p_1 = 1 - 3/7 = 4/7$

Let $S_B = \begin{pmatrix} B_1 & B_2 & B_3 & B_4 & B_5 \\ q_1 & 0 & 0 & q_2 & 0 \end{pmatrix}$ be the optimum strategies for player B.

Then, $$q_1 = \frac{a_{22} - a_{12}}{(a_{11} + a_{22}) - (a_{12} + a_{21})} = \frac{1-(-2)}{(2+1)-(-2-2)} = 3/7$$

and, $q_2 = 1 - q_1 = 4/7$

Value of the game $$v = \frac{a_{11}\,a_{22} - a_{21}\,a_{12}}{(a_{11} + a_{22}) - (a_{12} + a_{21})}$$

$$= \frac{2(1)-(-2)(-2)}{(2+1)-(-2-2)} = -2/7$$

Thus, optimum strategy are given by

$$S_A = \begin{pmatrix} A_1 & A_2 \\ 3/7 & 4/7 \end{pmatrix}$$

$$S_B = \begin{pmatrix} B_1 & B_2 & B_3 & B_4 & B_5 \\ 3/7 & 0 & 0 & 4/7 & 0 \end{pmatrix}$$

and value of the game is $v = -2/7$.

Remark: We observe that by using graphical method, the $2 \times n$ game is converted into a 2×2 game and then solved by using standard method.

Example 2

Obtain the optimal strategies for both persons and the value of the game for zero-sum two-person game whose pay off matrix is as follows:

		Player B	
		B_1	B_2
Player A	A_1	1	−3
	A_2	3	5
	A_3	−1	6
	A_4	4	1
	A_5	2	2
	A_6	−5	0

Solution: We observe that the given problem does not posses any saddle point. So, let the player B play the mixed strategy $S_B = \begin{bmatrix} B_1 & B_2 \\ q_1 & q_2 \end{bmatrix}$ with $q_2 = 1 - q_1$ against player A. Then, B's expected pay offs against A's pure moves are given by

A's pure move	*B's expected pay off E (q_1)*
A_1	$q_1 - 3q_2$
A_2	$3q_1 + 5q_2$
A_3	$-q_1 + 6q_2$
A_4	$4q_1 + q_2$
A_5	$2q_1 + 2q_2$
A_6	$-5q_1 + 0q_2$

The expected payoff equations are then plotted as functions of q_1 in the following graph.

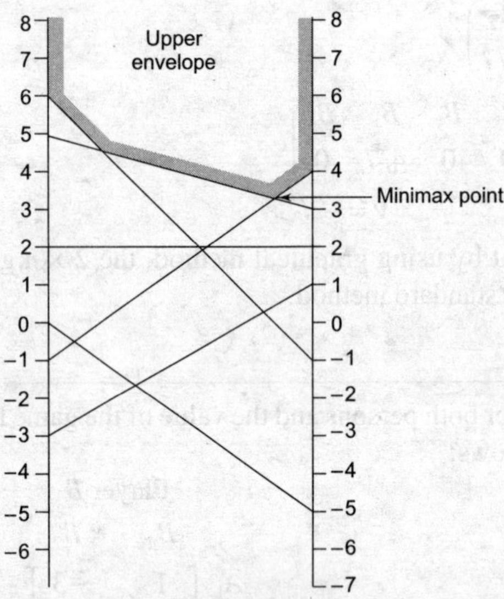

Since the player B wishes to minimise his maximum expected pay off, we consider the minimax point on the upper envelope of B's expected payoff equations. Hence, the given pay off matrix of the game is reduced to

$$\begin{array}{cc} & \text{Player } B \\ & \begin{array}{cc} B_1 & B_2 \end{array} \\ \text{Player } A \begin{array}{c} A_1 \\ A_4 \end{array} & \begin{bmatrix} 3 & 5 \\ 4 & 1 \end{bmatrix} = \begin{bmatrix} a_{11} & a_{12} \\ a_{21} & a_{22} \end{bmatrix} \end{array}$$

Let $S_A = \begin{bmatrix} A_1 & A_2 & A_3 & A_4 \\ 0 & p_1 & 0 & p_2 \end{bmatrix}$ be the optimal strategy of player A.

Then, $\quad p_1 = \dfrac{a_{22} - a_{21}}{(a_{11} + a_{22}) - (a_{12} + a_{21})} = \dfrac{(1-4)}{(3+1)-(4+5)} = \dfrac{3}{5}$

Hence, $\quad p_2 = 1 - p_1 = 2/5$

Let $S_B = \begin{bmatrix} B_1 & B_2 \\ q_1 & q_2 \end{bmatrix}$ be the optimal strategies of the player B.

Then, $\quad q_1 = \dfrac{a_{22} - a_{21}}{(a_{11} + a_{22}) - (a_{12} + a_{21})} = \dfrac{1-5}{(3+1)-(4+5)} = \dfrac{4}{5}$

and so $\quad q_2 = 1 - 4/5 = 1/5$

and the value of the game

$$v = \frac{a_{11}a_{22} - a_{12}a_{21}}{(a_{11}a_{22} - a_{12}a_{21})} = \frac{3\times1 - 4\times5}{(3+1)-(4+5)} = \frac{17}{5}$$

Remark: Using dominance property a $m \times n$ game can be converted into a $2 \times n$ or a $m \times 2$ game and the graphical method may be applied to convert it into a 2×2 game. Finally, standard method for a 2×2 game can be applied.

Example 3

Use the notion of dominance to simplify the rectangular game with the following pay off, and solve it graphically.

		Player *B*			
		I	II	III	IV
	1	18	4	6	4
Player *A*	2	6	2	13	7
	3	11	5	17	3
	4	7	6	12	2

Solution: Column I is dominated by column II. Column III is dominated by column IV. Hence, omit column I and column III.

The given pay off matrix is reduced to a 4×2 matrix

		Player *B*	
		II	IV
	1	4	4
Player *A*	2	2	7
	3	5	3
	4	6	2

Now, the given problem can be solved by graphical method.

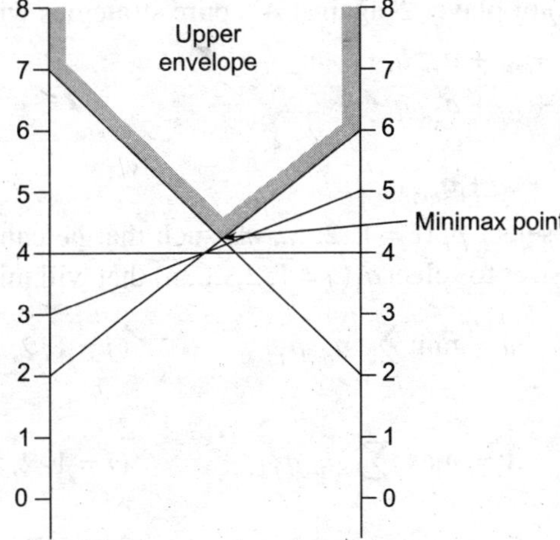

From the graph it is clear that the reduced 2×2 matrix is

$$\begin{array}{c} \text{Player } B \\ \text{II} \quad \text{IV} \end{array}$$

$$\text{Player } A \begin{array}{c} 2 \\ 4 \end{array}\begin{pmatrix} 2 & 7 \\ 6 & 2 \end{pmatrix} = \begin{pmatrix} a_{11} & a_{12} \\ a_{21} & a_{22} \end{pmatrix}$$

Let $S_A = \begin{pmatrix} 1 & 2 & 3 & 4 \\ 0 & p_1 & 0 & p_2 \end{pmatrix}$ and $S_B = \begin{pmatrix} \text{I} & \text{II} & \text{III} & \text{IV} \\ 0 & q_1 & 0 & q_2 \end{pmatrix}$ be the optimum strategies for player *A* and *B*, respectively. By applying the standard method we can get $p_1 = 4/9, p_2 = 5/9, q_1 = 5/9, q_2 = 4/9$ and the value of the game $v = 38/9$.

7.8.5 Linear Programming Method

A two person zero-sum game can also be solved by linear programming approach. The major advantage of using linear programming technique is that it solves mixed strategy of any size.

To illustrate the connection between a game problem and linear programming let us consider $(m \times n)$ pay off matrix (a_{ij}) for player *A*.

$$\text{Let } S_A = \begin{bmatrix} A_1 & \cdots & A_m \\ p_1 & \cdots & p_m \end{bmatrix} \text{ and } S_B = \begin{bmatrix} B_1 & \cdots & B_n \\ q_1 & \cdots & q_n \end{bmatrix}$$

be the optimum strategies for player *A* and player *B* respectively.

Then, $\sum_{i=1}^{m} p_i = \sum_{j=1}^{n} q_i = 1$

Then, the expected gains g_j $(j = 1, ..., n)$ of player *A* against *B*'s pure strategies will be

$$g_1 = a_{11}\, p_1 + a_{21}\, p_2 + ... + a_{m1}\, p_m$$
$$g_2 = a_{12}\, p_1 + a_{22}\, p_2 + ... + a_{m2}\, p_m$$
$$\vdots$$
$$g_m = a_{1n}\, p_1 + a_{2n}\, p_2 + ... + a_{mn}\, p_m$$

and the expected loss l_i $(i = 1, ... n)$ of player *B* against *A*'s pure strategies will be

$$l_1 = a_{11}\, q_1 + a_{12}\, q_2 + ... + a_{1n}\, q_n$$
$$l_2 = a_{12}\, q_1 + a_{22}\, q_2 + ... + a_{2n}\, q_n$$
$$\vdots$$
$$l_n = a_{m1}\, q_1 + a_{m2}\, q_2 + ... + a_{mn}\, q_n$$

The objective of player *A* is to select p_i $(i = 1, 2, ..., m)$ such that he can maximise his minimum expected gains and the player *B* desires to select q_j $(j = 1, 2, ..., n)$ that will minimise his expected loss.

Thus, if we let $\quad u = \min_j \sum_{i=1}^{m} a_{ij}\, p_j \quad (j = 1, 2, ... n)$

and, $\quad v = \max_i \sum_{j=1}^{m} a_{ij}\, q_j \quad (i = 1, 2, ... m)$

the problem of two players could be written as

Player A: Maximise u = minimise $\frac{1}{u} = \sum_{i=1}^{m} \frac{p_i}{u}$

subject to the constraints

$$\sum_{i=1}^{m} a_{ij}\, p_i \geq u \text{ and } \sum p_i = 1$$

$$p_i \geq 0 \quad (i = 1, 2, ..., m)$$

Player B: Minimise v = maximise $\frac{1}{v} = \sum_{j=1}^{n} \frac{q_i}{v}$

subject to constraints

$$\sum_{i=1}^{m} a_{ij}\, q_j \leq v \text{ and } \sum q_j = 1$$

$$q_i \geq 0 \qquad (j = 1, 2, ..., n)$$

Assuming $u > 0$, $v > 0$, introduce a new variable defined by $p_i^1 = \frac{p_i}{u}$ and $q_i^1 = \frac{q_j}{v}$

$$(i = 1, 2, ..., m, j = 1, 2, ..., n)$$

Then, the pair of linear programming problem can be re-written as

Player A: Minimise $p_0 = p_1^1 + p_2^1 + \cdots + p_m^1$

subject to $\quad a_{1j}\, p_1^1 + a_{2j'}\, p_2^1 + \cdots + a_{mj}\, p_m^1 \geq 1$

$$p_i^1 \geq 0 \quad (i = 1 \text{ to } m \text{ and } j = 1 \text{ to } n)$$

Player B: Maximise $q_0 = q_1' + \cdots + q_n^1$

subject to the constraints $\quad a_{i1}\, q_1^1 + q_{i2}\, q_2^1 + \cdots + a_{in}\, q_n^1 \leq 1$

$$q_j^1 \geq 0 \quad (i = 1 \text{ to } m, j = 1 \text{ to } n)$$

It is easy to note that the LPPs of the 2 players represent a primal dual pair. Therefore, by fundamental theorem of duality one can read the optimum solution of one player, just from the optimum simplex table of the opponent. That is, solve one player's LPP.

Remark: In case there are negative elements in the pay off matrix add a suitable constant, then value of the game = value of the game – constant.

Example 1

Solve the following game by using simplex method

$$\text{Player } A \quad \overset{\text{Player } B}{\begin{bmatrix} 1 & -1 & 3 \\ 3 & 5 & -3 \\ 6 & 2 & -2 \end{bmatrix}}$$

Solution: Since some of the entries in the pay off matrix are negative, we add a suitable constant, say $c = 4$ to each element.

Player B

$$\text{Player } A \begin{bmatrix} 5 & 3 & 7 \\ 7 & 9 & 1 \\ 10 & 6 & 2 \end{bmatrix}$$

Let the strategies for 2 players be

$$S_A = \begin{bmatrix} A_1 & A_2 & A_3 \\ p_1 & p_2 & p_3 \end{bmatrix}$$

$$S_B = \begin{bmatrix} B_1 & B_2 & B_3 \\ q_1 & q_2 & q_3 \end{bmatrix}$$

where $p_1 + p_2 + p_3 = 1$, $q_1 + q_2 + q_3 = 1$

The linear programming problem for B is

$$\text{Minimise } v = \text{maximise } \frac{1}{v} = y_1 + y_2 + y_3$$

subject to

$$5y_1 + 3y_2 + 7y_3 \leq 1,$$
$$7y_1 + 9y_2 + y_3 \leq 1,$$
$$10y_1 + 6y_2 + 2y_3 \leq 1,$$

$y_j \geq 0\ j = 1, 2, 3$ where $y_j = q_j/v, j = 1, 2, 3$

Introduce slack variables $s_1 \geq 0$, $s_2 \geq 0$, $s_3 \geq 0$

Starting Table:

table 7.22

	c_j		1	1	1	0	0	0	
C_B	Y_B	X_B	y_1	y_2	$y_3 \downarrow$	s_1	s_2	s_3	*Min ratio*
0	s_1	1	5	3	[7]	1	0	0	1/7
0	s_2	1	7	9	1	0	1	0	1/1
0	s_3	1	10	6	2	0	0	1	1/2
			–1	–1	–1	0	0	0	

From Table 7.22 we observe that y_3 enters into the basis and s_i leaves the basis.

First iteration: Introduce y_3 and drop s_1.

table 7.23

	c_j		1	1	1	0	0	0	
C_B	Y_B	X_B	y_1	$y_2 \downarrow$	y_3	s_1	s_2	s_3	*Min ratio*
–1	y_3	1/7	5/7	3/7	1	1/7	0	0	1/3
0	$s_2 \leftarrow$	6/7	44/7	[60/7]	0	– 1/7	1	0	1/10
0	s_3	5/7	60/7	36/7	0	– 2/7	0	1	5/36
$z_j - c_j$			–2/7	– 4/7	0	1/7	0	0	

From Table 7.23, we observe that y_2 enters into the basis and s_1 leaves the basis.

Second iteration:

			1	1	1	0	0	0
C_B	Y_B	X_B	y_1	y_2	y_3	s_1	s_2	s_3
1	y_3	1/10	2/5	0	1	3/20	-1/20	0
1	y_2	1/10	11/15	1	0	-1/60	7/60	0
0	s_3	1/5	24/5	0	0	-1/5	-3/5	1
	$z_j - c_j$	1/5	2/5	0	0	2/15	1/15	0

Since all $z_j - c_j \geq 0$ the current solution is optimum.

$$\frac{1}{v} = \frac{1}{5} \quad \therefore \quad v = 5$$

The value of the game $v = v - 4$

$= 5 - 4$

$= 1$

Optimum strategies for B are

$$q_1 = 0, \; q_2^1 = \frac{1}{10} \times 5 = \frac{1}{2}$$

$$q_3^1 = \frac{1}{10} \times 5 = \frac{1}{2}$$

Making use of duality, the optimum strategies for player A are obtained as

$$p_1^1 = \frac{2}{15} \times 5 = \frac{2}{3}, \quad p_2^1 = \frac{1}{10} \times 5 = \frac{1}{3}$$

$$p_3^1 = 0.$$

EXERCISES

Section 7.1

1. Discuss the difference between decision-making under certainty and risk.
2. Discuss the difference between decision-making under uncertainty and risk.
3. What are the steps in decision-making analysis?

Section 7.2

4. Explain various quantitative methods which are useful for decision-making under uncertainty.
5. What techniques are used to solve decision-making problems under uncertainty? Which technique results in an optimistic decision? Which technique results in a pessimistic decision?
6. Give an example of a good decision you made that resulted in a bad outcome. Also, give an example of a good decision you made that had a good outcome. Why was each decision good or bad?

7. A Super Bazar must decide on the level of supplies it must stock to meet the needs of its customers during Diwali days. The exact number of customers is not known, but it is expected to be in one of the four categories, 300, 350, 400 or 450 customeRs For levels of supplies are thus suggested with level *j* being ideal (from the view point of incurred costs) if the number of customers falls in category *j*. Deviations from the ideal levels results in additional costs either because extra supplies are stocked needlessly or because demand cannot be satisfied. The Table below provides these costs in thousands of rupees.

Customer category	*Supplies level*			
	A_1	A_2	A_3	A_4
E_1	7	12	20	27
E_2	10	9	10	25
E_3	23	20	14	23
E_4	32	24	21	17

What is the best alternative under

(i) Laplace criterion
(ii) Minimax criterion
(iii) Maximum criterion
(iv) Savage criterion
(v) Hurwicz criterion

[*Answer*: (i) A_2 or A_3 (ii) A_3 (iii) A_4 (iv) A_2 (v) A_3]

8. A food product company is contemplating the introduction of a revolutionary new product with new packaging to replace the existing product at much higher price (S_1), or a moderate change in the composition of the existing product with a new packaging at a small increase in price (S_2), or a small change in the composition of the existing except the word 'New' with a negligible increase in price (S_3). The three possible states of nature or events are: (i) high increase in sales (N_1), (ii) no change in sales (N_2) and (iii) decrease in sales (N_3). The marketing department of the company worked out the pay offs in terms of yearly net profits for each of the strategies of three events (expected sales). This is represented in the following Table.

Strategies	*States of nature*		
	N_1	N_2	N_3
S_1	7,00,000	3,00,000	1,50,000
S_2	5,00,000	4,50,000	0
S_3	3,00,000	3,00,000	3,00,000

which strategy should the concerned executive choose on the basis of

(i) Maximin criterion
(ii) Maximax criterion
(iii) Minimax regret criterion
(iv) Laplace criterion

[*Answer*: (i) S_3 (ii) S_1 (iii) S_1 (iv) S_1.]

9. Dr. Vivekanandan has been thinking about starting his own independent nursing home. The problem is to decide how large the nursing home should be. The annual returns will depend on both the size of nursing home and a number of marketing factors. After a careful analysis, Dr. Vivekanandan developed the following Table:

Size of nursing home	*Good market (Rs)*	*Fair market (Rs)*	*Poor market (Rs)*
Small (*S*)	50,000	20,000	–10,000
Medium (*M*)	70,000	35,000	–25,000
Large (*L*)	90,000	35,000	–45,000
Very large (*VL*)	2,00,000	25,000	–1,20,000

(a) What is the maximax decision?
(b) What is the maximin decision?
(c) What is equally likely decision?
(d) What is the criterion of realism decision? Use $\alpha = 0.8$
(e) Develop an opportunity loss table and determine the minimax decision.

[*Answer*: (a) *VL* (b) *S* (c) *VL* (d) *VL* (e) *L* or *VL*.]

10. In the Anish Toy manufacturing company, suppose the product acceptance probabilities are not known, but the following data is known.

Product acceptance	*Anticipated first year profit (Rs 1,000s) Product line*		
	Full	*Partial*	*Minimal*
Good	8	70	50
Fair	50	45	40
Poor	– 25	– 10	0

Determine the optimal decision under each of the following criteria and show how you arrived at it.

(a) Maximax
(b) Maximin
(c) Equally likelihood
(d) Minimax regret

[*Answer*: (a) Full (b) Minimal (c) Full or partial (d) partial.]

11. Mr Nishok has Rs 10,000 to invest in one of three options. *A*, *B* or *C*. The return on his investment depends on whether the economy experiences inflation recession, or no change at all. His possible returns under each economic condition are given below:

Strategy	*States of nature*		
	Inflation	*Recession*	*No change*
A	2000	1200	1500
B	3000	800	1000
C	2500	1000	1800

What should he decide using the pessimistic criterion, optimistic criterion, equally likely criterion and regret criterion?

[*Answer*: Persimistic: *A* Rs 120, optimistic: *B* Rs 300, equally likely: *C* Rs 176.6, regret: *C* Rs 50.]

Section 7.4

12. A physician purchases a particular vaccine on Monday of each week. The vaccine must be used within the following week, otherwise it becomes worthless. The vaccine costs Rs 2 per dose and the physician charges Rs 4 per dose. In the past 50 weeks, the physician had administered the vaccine in the following quantities.

Dose per week	20	25	50	60
Number of weeks	5	15	25	5

Determine how many does the physician should buy every week.
[*Answer*: EMV = Rs 54; purchase 40 doses.]

13. Mr. Navin quite often flies from town *A* to town *B*. He can use the airport bus which costs Rs 25 but if he takes it, there is a 0.08 chance that he will miss the flight. A hotel limousine costs Rs 270 with a 0.96 chance of being on time for the flight. For Rs 350 he can use a taxi which will make 99 of 100 flights. If Mr. Navin catches the plane on time, he will conclude a business transaction which will produce a profit of Rs 10000, otherwise he will lose it. Which mode of transport should Mr. Navin use? Answer on the basis of the EMV criterion.
[*Answer*: EMV : 9550 Best choice is: Taxi.]

14. The Blossom flower shop promises its customers delivery within four hours on all flower ordeRs All flowers are purchased on the previous day and delivered to Blossom by 8.00 AM the next morning. Parker's daily demand for roses is as follows.

Dozens of roses	7	8	9	10
Probability	0.1	0.2	0.4	0.3

Blossom purchases roses for Rs 10.00 per dozen and sells them for Rs 30.00. All unsold roses are donated to a local hospital. How many dozens of roses should Blossom order each evening to maximise its profits? What is the optimum expected profit?
[*Ans*wer: EMV = Rs 168. Best course of action is 9.]

15. The probability of the demand for lorries for hiring on any day in a given district is as follows:

No. of lorries demanded	0	1	2	3	4
Probability	0.1	0.2	0.3	0.2	0.2

Lorries have a fixed cost of Rs 90 each day to keep the daily hire charges (net of variable costs of running) Rs 200. If the lorry-hire company owns four lorries, what is its daily expectation? If the company is about to go into business and currently has no lorries, how many lorries should it buy?
[*Answer*: *EMV* = Rs 140. Best course of action : 2.]

16. You are given the following pay offs of three acts A_1, A_2 and A_3 and the events E_1, E_2, E_3.

States of nature	Three acts		
	A_1	A_2	A_3
E_1	25	−10	−125
E_2	400	440	400
E_3	650	740	750

The probabilities of the states of nature are respectively 0.1, 0.7 and 0.2. Calculate and tabulate EMV and conclude which of the acts can be choosen as the best.
[*Answer*: EMV (A_1) = 412.5; EMV (A_2) = 455, EMV (A_3) = 417.5 Best choice is A_2.]

17. The marketing staff of a certain industrial organisation has submitted the following pay off table, giving profits in million rupees, concerning a certain proposal depending upon the rate of technology advance.

Technological advance	Decision	
	Accept	Reject
Much	2	3
Little	5	2
None	−1	4

The probabilities are 0.2, 0.5 and 0.3 for much, little and none technological advance, respectively. What decision should be taken?
[*Answer*: EMV (accept) = 2.6, EMV (reject) = 2.8, Best choice is reject.]

18. A company buys certain items which sells for Rs 16 and costs Rs 12. A tabulation of recent demand for the product appears as follows:

Quantity sold	80	81	82	83	84	85
Number of days	14	36	70	30	20	10

Future demands for this product during the next 30 days should be comparable to past demand.
(a) What is the expected profit and loss from stocking 83 units?
(b) What is the expected profit under uncertainty?
(c) What would be the maximum expected profits certainly if this were possible ?
[*Answer*: (a) Profit = Rs 328.72,loss = Rs 11.28
(b) *EMV* = Rs 328.72
(c) *EPPI* = Rs 32.72]

19. A motor-parts dealer finds that the cost of a particular item in stock for a week is Rs 20 and the cost of a unit shortage is Rs 50. The probability distribution of weekly sales (in thousand items) is as follows:

Weekly sales (in thousands)	0	1	2	3	4	5	6
Probability	0.10	0.10	0.20	0.20	0.20	0.15	0.05

How many units per week should the dealer order? Also, find the EVPI.

[*Answer*: 3000 or 4000 units; EVPI = Rs 33.50.]

20. A bicycle repairman has an opportunity to purchase a stock of discontinued bicycles. They were originally supposed to be sold for Rs 400 each. The repairman is offered all five bicycles for Rs 500, which makes his cost for each bicycle Rs 100. If he sells them he believes he can get Rs 250 for each bicycle, thereby making a profit of Rs 150. He has two options: either to buy all the discontinued bicycles or not to buy at all. There are six states of nature; these being the demand for 0, 1, 2, 3, 4 and 5 bicycles.
    (i) Prepare the pay off as well as regret tables for the problem.
    (ii) If the repairman has the option of buying any number of bicycles (0 to 5), find the average expected pay off and average expected regret for each stock action.

    [*Answer*: EMV = Rs 200 or EOL = Rs 175 and the optimum stock level is 300 discontinued bicycles.]

21. A certain piece of equipment can be purchased for a construction project at a remote location. This equipment contains an expensive part which is subject to random failure. Spare of this part can be purchased at the same time the equipment is purchased. Their unit cost is Rs 1,500 and they have no scrap value. If the part fails on the job and no spare is available, the part will have to be manufactured on a special order basis. If this is required, the total cost including down time of the equipment is estimated as Rs 9000 for each such occurrence. Based on previous experience with similar parts, the following probability estimates of the number of failures expected over the duration of the project are provided as given below.

Failure	0	1	2
Probability	0.80	0.15	0.05

    (a) Determine optimal EMV and optimal number of spares to purchase initially.
    (b) Based on opportunity loss, determine the optimal course of action and optimal value of EOL.
    (c) Determine expected profit with perfect information and expected value of perfect information.

    [*Answer*: EMV (one spare purchased) = 1950. Hence, the best choice is to purchase one spare part.
    (b) Minimum EOL is (one spare purchased) = 1575. Hence, the best choice is one spare purchased.
    (c) EPPI = – 375; EVPI = Rs 1575.]

22. A producer of boats has estimated the following distribution of demand for a particular kind of boat.

Number demanded	0	1	2	3	4	5	6
Probability	0.14	0.27	0.27	0.18	0.09	0.09	0.01

    Each boat costs him Rs 7000 and he sells them for Rs 10000 each. Boats left unsold at the end of the season must be disposed off for Rs 6000 each. How many boats should be in stock so as to maximise his expected profit?

[*Hint*: Conditional profit = $\begin{cases} 3000\,D & \text{when } D \geq S \\ 3000\,D - 1000(S-D) & \text{when } D < S \end{cases}$

Maximum EMV = 4080, stock 3 boats.]

23. A small industry finds from the past data, and the cost of making an item is Rs 25, the selling price of the item is Rs 30, if it is sold within a week, and it could be disposed off at Rs 20 per item at the end of the week.

Weekly sales	< 3	4	5	6	7	< 8
No. of sales	0	10	20	40	30	0

Find the optimum number of items per week the industry should produce.

[*Answer*: Conditional profit = $\begin{cases} SD & \text{when } D \geq S \\ 10\,D - 5\,D & \text{when } D < S \end{cases}$ < D – Demand S – Stock

Maximum EMV = Rs 48.60, produce 6 items.]

24. A firm makes pastries which it sells at Rs 8 per dozen in special boxes containing one dozen each. The direct cost of pastries for the firm is Rs 4.50 per dozen. At the end of the week the stale pastries are sold off for a lower price of Rs 2.50 per dozen. The overhead expenses attributable to pastry production are Rs 1.25 per dozen. Fresh pastries are sold in special boxes which cost 50 paise each and the stale pastries are sold wrapped in ordinary paper. The probability distribution of demand per week is as under:

Demand (in dozen)	0	1	2	3	4	5
Probability	0.01	0.14	0.2	0.5	0.1	0.05

Find the optimal production level of pastries per week.

[*Hint*: Conditional profit = $\begin{cases} 1.75\,D & \text{when } D \geq S \\ 2D - 0.25S & \text{when } D < S \end{cases}$

Maximum EMV = Rs 4.28, produce 4 dozen pastries.]

25. The probability distribution of monthly sales of an item as follows:

Monthly sales (units)	0	1	2	3	4	5	6
Probabilities	0.01	0.06	0.25	0.30	0.22	0.10	0.06

The cost of carrying inventory (unsold during the month) is Rs 30 per unit per month and cost of unit shortage is Rs 70. Determine optimum stock to minimise expected cost.

[*Answer*: Cost function = $\begin{cases} 70\,(D - S);\, D \geq S \\ 200\,(S - D);\, D < S \end{cases}$

where D – demand S – stock

Since expected cost, Rs 46 is minimum for course of action 4, optimum stock to minimise the cost is 4 units per month.]

26. A modern home appliances dealer finds that the cost of holding a mini cooking range in stock for a month is Rs 200 (insurance, minor deterioration, interest on borrowed capital and others). Customers who cannot obtain a cooking range immediately tend to go to other dealers and the estimates that for every customer who cannot get immediately delivery, he loses an average of Rs 500. The probabilities of a demand of 0, 1, 2, 3, 4, 5 mini cooking ranges in a month are 0.05, 0.10, 0.20, 0.30, 0.20 and 0.15, respectively. Determine the optimal stock of mini cooking ranges. Also, find EVPI.

[*Answer*: Cost function = $\begin{cases} 500\,(D - S) & \text{when } D \geq S \\ 200\,(S - D) & \text{when } D < S \end{cases}$

Since expected cost, Rs 315 is minimum for course of action 4, optimum stock to minimise the cost is four mini cooking ranges.]

Section 7.5

27. Nishok Company is currently working with a process which after paying for materials, labour and so on, brings a profit of Rs 12000. The following alternatives are made available to the company:
   (i) The company can conduct research (R_1) which is expected to cost Rs 10000 having 90% chances of success. If it proves a success, the company gets a gross income of Rs 25000.
   (ii) The company can conduct research (R_2) which is expected to cost Rs 8000 having a probability of 60% success, the gross income will be Rs 25000.
   (iii) The company can pay Rs 6000 as royalty for a new process which will bring a gross income of Rs 20000.
   (iv) The company continues the current process. Due to limited resources it is assumed that only one of the two types of research can be carried out at a time. Use decision tree analysis, locate the optimal strategy for the company.

   [*Answer*: Optimal decision is to procure new process on royalty basis.]
28. A manager has a choice between (i) A risky contract promising Rs 7 lakhs with probability 0.6 and Rs 4 lakhs with probability 0.4, and (ii) A diversified portfolio consisting of two contracts with independent outcomes each promising Rs 3.5 lakhs with probability 0.6 and Rs 2 lakhs with probability 0.4.

   Construct a decision tree for using EMV Criteria. Can you arrive at the decision using EMV criteria ?

   [*Answer*: EMV = Rs 5.8 lakhs.]
29. A client asks an estate agent to sell three properties *A*, *B* and *C* for him and agrees to pay him 5% commission on each sale . He specifies certain conditions. The estate agent must sell *A* first, and this he must do within 50 days. The agent receives his 5% commission on that sale only after *A* is sold. He can then either back out at this stage or nominate and try to sell one of the two remaining properties within 60 days. If he does not succeed in selling the nominated property in that period, he is not given the opportunity to sell the other. If he does sell it in the period, he is given the opportunity to sell the third property on the same conditions. The following Table summarises the prices, selling costs and the estate agent's estimated probability of making a sale.

Property	Price of property	Selling cost	Probability of sale
A	Rs 12000	Rs 400	0.7
B	Rs 25000	Rs 225	0.6
C	Rs 50000	Rs 450	0.5

(i) Draw up an appropriate decision tree for the estate agent.
(ii) What is the estate agent's best strategy under EMV approach?
[*Answer*: Accept the offer and try to sell *A*, then *B* and then *C*.]

30. An industry has three alternatives namely:
    (a) Start production commercially.
    (b) Build a pilot plant and then decide about the commercial production.
    (c) Stop production.

    It is estimated that the pilot plant has a 0.8 chance of high yield and 0.2 chance of low yield. If the yield is high, the commercial production will have a 0.85 chance of high profit and 0.15 chance of low profit. If the yield is low the commercial production will have 0.1 chance of high profits and 0.9 chance of low profits. However, direct commercial production is expected to have 0.7 chance of high profit and 0.3 chance of low profit.

    Draw a decision tree to illustrate the situation for decision-making. Expected values of high profits and low profits are Rs 12 lakh and Rs 15 lakh. The cost of the pilot plant is Rs 3.5 lakh. What should be your decision ?

31. A company is currently working with a process, which, after paying for materials, labour and so on, brings a profit of Rs 12000. The company has the following alternatives.
    (i) The company can conduct research R_1 which is expected to cost Rs 10000 and having 90% probability of success. If successful, the gross income will be Rs 26000.
    (ii) The company can conduct research R_2, which is expected to cost Rs 6000 and having a probability of 60% success. If successful, the gross income will be Rs 24000.
    (iii) The company can pay Rs 5000 as royalty of a new process which will bring a gross income of Rs 20000.
    (iv) The company may continue the current process.

    [*Answer*: The company should pay Rs 5000 as royalty of the new process to earn maximum expected profit of Rs 15000.]

32. Raman Industries Ltd., has a new product which they expect has great potential. At the moment they have two courses of action open to them. To test market (S_1) and to drop product (S_2). If they test it, it will cost Rs 50000 and the response could be positive or negative with probabilities 0.70 and 0.30, respectively. If it is positive, they could either market it with full scale or drop the product. If they market with full scale, then the result might be low, medium or high demand and the respective net pay offs would be Rs 100000, Rs 100000 or Rs 500000. These outcomes have probabilities of 0.25, 0.55 and 0.20, respectively. If the result of the test marketing is negative they have decided to drop the product. If, at any point, they drop the product there is a net gain of Rs 25000 from the sale of scrap. All financial values have been discounted to the present. Draw a decision tree for the problem and indicate the most preferred decision.
    [*Answer*: The company should test market rather than drop the product. If test market result is positive, company should market the product otherwise drop the product.]

33. *XYZ* company manufactures guaranteed tennis balls. At present, approximately 10% of the tennis balls are defective. A defective ball leaving the factory costs the company Re 0.50 to honour its guarantee. Assume that all defective balls are returned. At a cost of Re 0.10 per ball, the company can conduct a test, which always correctly identifies both good and bad tennis balls.
    (*a*) Draw a decision tree and determine the optimal course of action and its expected cost.
    (*b*) At what test cost the company should be indifferent to testing?
    [*Answer*: (a) Do not test, Re 0.05 (b) Re 0.05].

Section 7.6

34. Give a comprehensive explanation of the term "game theory".
35. Define with the help of an example, the following:
    (a) Players (b) Pay off matrix (c) Outcome
36. Define the following:
    (a) Strategy (b) Pure strategy (c) Mixed strategy
37. Explain the difference between a pure strategy and a mixed strategy.
38. What are the assumptions made in game theory?
39. State the major limitations of game theory.

Section 7.7

40. Explain two-person zero-sum game by giving suitable example.
41. Explain: Minimax and maximin principle used in the theory of games.
42. What is a game in game theory? What are the properties of a game? Explain the 'best strategy' on the basis of minimax criterion of optimality.
43. Define 'saddle point'. Is it necessary that a game should always possess a saddle point.
44. Game theory provides a systematic quantitative approach for analysing competitive situations in which the competitors make use of logical processes and techniques in order to determine an optimal strategy for winning. Comment.
45. Define value of the game.
46. Define the following terms:
    (a) Fair game (b) Strictly determinable game
47. For the game with pay off matrix

$$\text{Player } A \quad \begin{array}{c} \\ A_1 \\ A_2 \end{array} \overset{\text{Player } B}{\begin{array}{ccc} B_1 & B_2 & B_3 \\ \end{array}} \begin{bmatrix} -1 & 2 & 2 \\ 6 & 4 & -6 \end{bmatrix}$$

Determine the best strategies for players *A* and *B*. Also, determine the value of the game. Is the game
    (i) fair (ii) strictly determinable
    [*Answer*: Best strategy for player *A* is A_1 and that for *B* is B_3, value of the game = –2. The game is strictly determinable but not fair.]

48. Consider the game with the following pay off

$$\begin{array}{cc} & \text{Player } B \\ & \begin{array}{cc} B_1 & B_2 \end{array} \\ \text{Player } A \begin{array}{c} A_1 \\ A_2 \end{array} & \begin{bmatrix} 2 & 6 \\ -2 & \lambda \end{bmatrix} \end{array}$$

(i) Show that game is strictly determinable, whatever λ may be.
(ii) Determine the value of the game.
[*Answer*: (ii) Value of the game $v = 2$.]

49. Determine whether the following game is strictly determinable and fair or not.

$$\begin{array}{cc} & \text{Player } B \\ \text{Player } A & \begin{bmatrix} 1 & 2 \\ -4 & -3 \end{bmatrix} \end{array}$$

[*Answer*: Not fair value of game = 1.]

50. Solve the following games whose pay off matrix is given by

(a)

$$\begin{array}{cc} & \text{Player } B \\ & \begin{array}{cccc} B_1 & B_2 & B_3 & B_4 \end{array} \\ \text{Player } A \begin{array}{c} A_1 \\ A_2 \\ A_3 \end{array} & \begin{bmatrix} 1 & 7 & 3 & 4 \\ 5 & 6 & 4 & 5 \\ 7 & 2 & 0 & 3 \end{bmatrix} \end{array}$$

(b)

$$\begin{array}{cc} & \text{Player } B \\ & \begin{array}{cccc} B_1 & B_2 & B_3 & B_4 \end{array} \\ \text{Player } A \begin{array}{c} A_1 \\ A_2 \\ A_3 \end{array} & \begin{bmatrix} -3 & -5 & 0 & 6 \\ -4 & -2 & 1 & 2 \\ 5 & 4 & 2 & 3 \end{bmatrix} \end{array}$$

(c)

$$\begin{array}{cc} & \text{Player } B \\ & \begin{array}{cccc} B_1 & B_2 & B_3 & B_4 \end{array} \\ \text{Player } A \begin{array}{c} A_1 \\ A_2 \\ A_3 \end{array} & \begin{bmatrix} 3 & 3 & 1 & 10 \\ 5 & 5 & 4 & 6 \\ 4 & -2 & 0 & -5 \end{bmatrix} \end{array}$$

[*Answers*: (a) Optimal strategy for $A : A_2$, $B : B_3$ value of the game = 4, (b) optimal strategy for $A : A_3$, $B : B_3$, $v = 1$, (c) optimal strategy for $A : A_2$, $B : B_3$, $v = 1$]

51. Find out whether there is any saddle point in the following problem

Player B

$$\text{Player } A \quad \begin{bmatrix} -3 & 1 \\ 3 & -1 \end{bmatrix}$$

[*Answer*: Saddle point does not exist.]

52. For the game with pay off matrix:

Player B

$$\text{Player } A \quad \begin{matrix} \\ \text{I} \\ \text{II} \end{matrix} \begin{matrix} \text{I} & \text{II} & \text{III} \\ \end{matrix} \begin{bmatrix} -1 & 2 & -2 \\ 6 & 4 & -6 \end{bmatrix}$$

Determine the best strategies for players A and B and also the value of the game for them. Is the game (i) fair? (ii) strictly determinable?

[*Answer*: Saddle point (I, III), $v = -2$. Game is strictly determinable and not fair.]

53. Find the saddle point (or points) and hence solve the following games

(a)

Player B

$$\text{Player } A \quad \begin{matrix} A_1 \\ A_2 \\ A_3 \end{matrix} \begin{bmatrix} 15 & 2 & 3 \\ 6 & 5 & 7 \\ -7 & 4 & 0 \end{bmatrix}$$

(b)

B

$$A \quad \begin{matrix} \\ \text{I} \\ \text{II} \\ \text{III} \end{matrix} \begin{matrix} \text{I} & \text{II} & \text{III} & \text{IV} \end{matrix} \begin{bmatrix} -5 & 2 & 1 & 20 \\ 5 & 5 & 4 & 6 \\ 4 & -2 & 0 & -5 \end{bmatrix}$$

(c)

Player B

$$\text{Player } A \quad \begin{matrix} \\ \text{I} \\ \text{II} \end{matrix} \begin{matrix} \text{I} & \text{II} & \text{III} \end{matrix} \begin{bmatrix} 6 & 8 & 6 \\ 4 & 12 & 2 \end{bmatrix}$$

[*Answers*: (a) Saddle point (A_2, B_2), $v = 5$,
(b) Saddle point (II, III), $v = 4$,
(c) Saddle point (I, I), $v = 6$]

54. Solve the game whose pay off matrix is given by

Player B

$$\text{Player } A \quad \begin{matrix} & B_1 & B_2 & B_3 \\ A_1 & 1 & 3 & 1 \\ A_2 & 0 & -4 & -3 \\ A_3 & 1 & 5 & -1 \end{matrix}$$

[*Answer*: Saddle point is (A_1, B_1) or (A_1, B_3) and $v = 1$.]

55. The following games have saddle point solutions. Determine the saddle point and the optimum strategies for each player.

(a)

Player Y

$$\text{Player } X \quad \begin{matrix} & \text{I} & \text{II} & \text{III} \\ \text{I} & 4 & 6 & 4 \\ \text{II} & 2 & 10 & 0 \end{matrix}$$

(b)

Player B

$$\text{Player } A \quad \begin{matrix} & \text{I} & \text{II} \\ \text{I} & 4 & 4 \\ \text{II} & 4 & 4 \end{matrix}$$

[*Answer*: (a) Saddle point: $(x_1, y_1,)$ or (x_1, y_3), $v = 4$
(b) Saddle point: all the points are saddle points, $v = 4$.]

56. Determine the optimum strategies and values of the following game:

Player B

$$\text{Player } A \quad \begin{matrix} & \text{I} & \text{II} & \text{III} & \text{IV} \\ \text{I} & -3 & 4 & 2 & 9 \\ \text{II} & 7 & 8 & 6 & 10 \\ \text{III} & 6 & 2 & 4 & -1 \end{matrix}$$

[*Answer*: Saddle point : (A_2, B_3), $v = 6$.]

Section 7.8

57. Solve the following 2×2 games without saddle points:

(a)

Player B

$$\text{Player } A \quad \begin{matrix} & \text{I} & \text{II} \\ \text{I} & 5 & 1 \\ \text{II} & 3 & 4 \end{matrix}$$

(b)

$$\begin{array}{cc} & \text{Player } B \\ & \begin{array}{cc} \text{I} & \text{II} \end{array} \\ \text{Player } A \begin{array}{c} \text{I} \\ \text{II} \end{array} & \begin{bmatrix} 2 & 5 \\ 7 & 3 \end{bmatrix} \end{array}$$

(c)

$$\begin{array}{cc} & \text{Player } B \\ & \begin{array}{cc} \text{I} & \text{II} \end{array} \\ \text{Player } A \begin{array}{c} \text{I} \\ \text{II} \end{array} & \begin{bmatrix} -4 & 6 \\ 2 & -3 \end{bmatrix} \end{array}$$

[*Answers*: (a) $S_A : \begin{bmatrix} \text{I} & \text{II} \\ 1/5 & 4/5 \end{bmatrix}; S_B : \begin{bmatrix} \text{I} & \text{II} \\ 3/5 & 2/5 \end{bmatrix}; v = \dfrac{17}{5}$

(b) $S_A : \begin{bmatrix} \text{I} & \text{II} \\ 4/7 & 3/7 \end{bmatrix}; S_B : \begin{bmatrix} \text{I} & \text{II} \\ 2/7 & 5/7 \end{bmatrix}; v = \dfrac{29}{7}$

(c) $S_A : \begin{bmatrix} \text{I} & \text{II} \\ 1/3 & 2/3 \end{bmatrix}; S_B : \begin{bmatrix} \text{I} & \text{II} \\ 3/5 & 2/5 \end{bmatrix}; v = 0$]

58. Two players A and B match coins. If the coin matches, then A wins one unit of value, if the coins do not match, then B wins one unit of value. Determine optimum strategies for the players and the value of the game.

[*Answers* Strategy for player A is $S_A : \begin{pmatrix} \text{I} & \text{II} \\ 1/2 & 1/2 \end{pmatrix}$

Strategy for player B is $S_B : \begin{pmatrix} \text{I} & \text{II} \\ 1/2 & 1/2 \end{pmatrix}$ and $v = 0$.]

59. Solve the following game using dominance property

$$\begin{array}{cc} & \text{Player } B \\ & \begin{array}{ccc} \text{I} & \text{II} & \text{III} \end{array} \\ \text{Player } A \begin{array}{c} \text{I} \\ \text{II} \\ \text{III} \end{array} & \begin{bmatrix} 1 & 7 & 2 \\ 6 & 2 & 7 \\ 5 & 2 & 6 \end{bmatrix} \end{array}$$

[*Answer*: $S_A : \begin{bmatrix} \text{I} & \text{II} & \text{III} \\ 2/5 & 3/5 & 0 \end{bmatrix} S_B : \begin{bmatrix} \text{I} & \text{II} & \text{III} \\ 1/2 & 1/2 & 0 \end{bmatrix}, v = 4$]

60. Using dominance property solve the following games

(a)

		Player B			
		I	II	III	IV
Player A	I	3	2	4	0
	II	3	4	2	4
	III	4	2	4	0
	IV	0	4	0	8

(b)

		Player B		
		I	II	III
Player A	I	2	3	1/2
	II	3/2	2	0
	III	1/2	1	1

[*Answer*: (a) $S_A : \begin{bmatrix} \text{I} & \text{II} & \text{III} & \text{IV} \\ 0 & 0 & 2/3 & 1/3 \end{bmatrix}$ $S_B : \begin{bmatrix} \text{I} & \text{II} & \text{III} & \text{IV} \\ 0 & 0 & 2/3 & 1/3 \end{bmatrix}$ and $v = 8/3$

(d) $S_A = S_B : \begin{pmatrix} \text{I} & \text{II} & \text{III} \\ 1/4 & 0 & 3/4 \end{pmatrix}$ and $v = 34/9$.]

61. *A* and *B* play a game in which each has three coins, a 5 *p*, a 10 *p*, and a 20 *p*. Each selects a coin without the knowledge of the others choice. If the sum of the coins is odd amount, *A* wins *B*'s coin, if the sum is even *B* wins *A*'s coin. Find the best strategy for each player and the value of the game.

[*Answer*: The pay off matrix is

		Player B		
		5*p*	10*p*	20*p*
Player A	5*p*	– 5	10	20
	10*p*	5	– 10	– 10
	20*p*	5	– 20	– 20

$S_A : \begin{pmatrix} 5p & 10p & 20p \\ 1/2 & 1/2 & 0 \end{pmatrix}$; $S_B \begin{pmatrix} 5p & 10p & 20p \\ 2/3 & 1/3 & 0 \end{pmatrix}$, $v = 0$.]

62. Solve the following games by reducing them to 2×2 games by graphical method

(a)

		Player B				
		I	II	III	IV	V
Player A	I	3	0	6	–1	7
	II	–1	5	–2	2	1

(b)

$$\text{Player } A \begin{array}{c} \text{I} \\ \text{II} \end{array} \overset{\begin{array}{ccccccc} \text{I} & \text{II} & \text{III} & \text{IV} & \text{V} & \text{VI} & \text{VII} \end{array}}{\begin{bmatrix} -1 & 3 & -1 & 4 & 2 & 2 & -5 \\ -3 & 5 & 6 & 1 & 2 & 2 & 0 \end{bmatrix}}$$

Player B

(c)

Player B

$$\text{Player } A \begin{array}{c} \text{I} \\ \text{II} \end{array} \overset{\begin{array}{ccc} \text{I} & \text{II} & \text{III} \end{array}}{\begin{bmatrix} 3 & -3 & 4 \\ -1 & 1 & -3 \end{bmatrix}}$$

(d)

Player B

$$\text{Player } A \begin{array}{c} \text{I} \\ \text{II} \end{array} \overset{\begin{array}{ccc} \text{I} & \text{II} & \text{III} \end{array}}{\begin{bmatrix} 4 & -1 & 0 \\ -1 & 4 & -2 \end{bmatrix}}$$

[*Answer*: (a) $S_A : \begin{pmatrix} \text{I} & \text{II} \\ 3/7 & 4/7 \end{pmatrix}$; $S_B : \begin{pmatrix} \text{I} & \text{II} & \text{III} & \text{IV} & \text{V} \\ 3/7 & 0 & 0 & 4/7 & 0 \end{pmatrix}$; $v = \frac{5}{7}$

(b) $S_A : \begin{pmatrix} \text{I} & \text{II} \\ 4/5 & 1/5 \end{pmatrix}$; $S_B : \begin{pmatrix} \text{I} & \text{II} & \text{III} & \text{IV} & \text{V} & \text{VI} \\ 0 & 3/5 & 0 & 2/5 & 0 & 0 \end{pmatrix}$; $v = \frac{17}{5}$

(c) $S_A : \begin{pmatrix} \text{I} & \text{II} \\ 4/11 & 7/11 \end{pmatrix}$; $S_B : \begin{pmatrix} \text{I} & \text{II} & \text{III} \\ 0 & 7/11 & 4/11 \end{pmatrix}$; $v = -\frac{5}{11}$

(d) $S_A : \begin{pmatrix} \text{I} & \text{II} \\ 3/7 & 4/7 \end{pmatrix}$; $S_B : \begin{pmatrix} \text{I} & \text{II} & \text{III} \\ 2/7 & 0 & 5/7 \end{pmatrix}$; $v = \frac{8}{7}$]

63. Solve the following games by reducing them to 2×2 games by graphical method:

(a)

Player B

$$\text{Player } A \begin{array}{c} \text{I} \\ \text{II} \\ \text{III} \end{array} \overset{\begin{array}{cc} \text{I} & \text{II} \end{array}}{\begin{bmatrix} 2 & 7 \\ 3 & 5 \\ 11 & 2 \end{bmatrix}}$$

(b)

Player B

$$\text{Player } A \begin{array}{c} \text{I} \\ \text{II} \\ \text{III} \\ \text{IV} \\ \text{V} \end{array} \overset{\begin{array}{cc} \text{I} & \text{II} \end{array}}{\begin{bmatrix} -6 & 7 \\ 4 & -5 \\ -1 & -2 \\ -2 & 5 \\ 7 & 6 \end{bmatrix}}$$

(c) Player B

$$\text{Player } A \begin{array}{c} \text{I} \\ \text{II} \\ \text{III} \\ \text{IV} \\ \text{V} \end{array} \overset{\begin{array}{cc} \text{I} & \text{II} \end{array}}{\begin{bmatrix} 2 & 3 \\ 6 & 7 \\ -6 & 10 \\ -3 & -2 \\ 3 & 2 \end{bmatrix}}$$

[*Answer*: (a) $S_A : \begin{pmatrix} \text{I} & \text{II} & \text{III} \\ 9/14 & 0 & 5/4 \end{pmatrix}; S_B : \begin{pmatrix} \text{I} & \text{II} \\ 5/14 & 9/14 \end{pmatrix}; v = \frac{73}{14}$

(b) $S_A : \begin{pmatrix} \text{I} & \text{II} & \text{III} & \text{IV} & \text{V} \\ 0 & 0 & 0 & 13/20 & 7/20 \end{pmatrix}; S_B : \begin{pmatrix} \text{I} & \text{II} \\ 11/20 & 9/20 \end{pmatrix}; v = \frac{23}{20}$

(c) Saddle point (A_2, B_1), $v = 6$]

64. Solve (3×3) game by the simplex method of linear programming whose pay off matrix is given as follows

Player B

$$\text{Player } A \begin{array}{c} 1 \\ 2 \\ 3 \end{array} \overset{\begin{array}{ccc} 1 & 2 & 3 \end{array}}{\begin{bmatrix} 3 & -1 & -3 \\ -3 & 3 & -1 \\ -4 & -3 & 3 \end{bmatrix}}$$

$$\left[\textit{Answer: } S_A : \begin{bmatrix} 1 & 2 & 3 \\ 20/45 & 11/45 & 14/45 \end{bmatrix}; S_B : \begin{bmatrix} 1 & 2 & 3 \\ 14/45 & 11/45 & 20/45 \end{bmatrix}; v = \frac{29}{45}. \right]$$

65. Two companies A and B are competing for the same product. Their different strategies are given in the following pay off matrix

		Company A		
		A_1	A_2	A_3
Company B	B_1	2	−2	3
	B_2	−3	5	−1

Use linear programming to determine the best strategies for both the players.

$$\left[\textit{Answer: } S_A : \begin{pmatrix} \text{I} & \text{II} & \text{III} \\ 2/3 & 1/3 & 0 \end{pmatrix}; S_B : \begin{pmatrix} \text{I} & \text{II} \\ 7/12 & 5/12 \end{pmatrix}; v = \frac{1}{3}. \right]$$

66. Solve the following games by linear programming

(a)

$$\text{Player } A \begin{array}{c} \text{I} \\ \text{II} \\ \text{III} \end{array} \overset{\begin{array}{ccc} & \text{Player } B & \\ \text{I} & \text{II} & \text{III} \end{array}}{\begin{bmatrix} -1 & 2 & 1 \\ 1 & -2 & 2 \\ 3 & 4 & -3 \end{bmatrix}}$$

(b)

$$\text{Player } A \begin{array}{c} \text{I} \\ \text{II} \\ \text{III} \end{array} \overset{\begin{array}{ccc} & \text{Player } B & \\ \text{I} & \text{II} & \text{III} \end{array}}{\begin{bmatrix} 3 & -2 & 4 \\ -1 & 4 & 2 \\ 2 & 2 & 6 \end{bmatrix}}$$

[*Answer*: (a) $S_A : \begin{pmatrix} \text{I} & \text{II} & \text{III} \\ 17/46 & 20/46 & 9/46 \end{pmatrix}$; $S_B : \begin{pmatrix} \text{I} & \text{II} & \text{III} \\ 7/23 & 6/23 & 10/23 \end{pmatrix}$; $v = \dfrac{15}{23}$

(b) $S_A : \begin{pmatrix} \text{I} & \text{II} & \text{III} \\ 0 & 0 & 1 \end{pmatrix}$; $S_B : \begin{pmatrix} \text{I} & \text{II} & \text{III} \\ 4/5 & 1/5 & 0 \end{pmatrix}$; $v = 2$.]

67. For the following pay off table, transform the zero-sum-game into an equivalent linear programming problem and solve it by simplex method.

$$\text{Plyaer } A \begin{array}{c} A_1 \\ A_2 \\ A_3 \end{array} \overset{\begin{array}{ccc} & \text{Player } B & \\ B_1 & B_2 & B_3 \end{array}}{\begin{bmatrix} 9 & 1 & 4 \\ 0 & 6 & 3 \\ 5 & 2 & 8 \end{bmatrix}}$$

[*Answer*: $S_A : \begin{bmatrix} A_1 & A_2 & A_3 \\ 3/8 & 13/24 & 1/12 \end{bmatrix}$; $S_B : \begin{bmatrix} B_1 & B_2 & B_3 \\ 7/24 & 5/9 & 11/72 \end{bmatrix}$; $v = \dfrac{91}{24}$]

8

Sequencing Models

8.1 INTRODUCTION AND BASIC ASSUMPTION

This chapter deals with situations in which the effectiveness measure (time, cost, distance, and so on) is a function of the order or sequence of performing a series of jobs (tasks). The selection of the appropriate order in which waiting customers may be served is called sequencing. A practical situation may correspond to an industry producing a number of products, each of which is to be processed through different machines, of course, finite in number.

Suppose there are n jobs to perform, each of which requires processing on some or all of m different machines. The effectiveness (that is, cost, time, mileage, and henceforth) can be measured for any given sequence of job at each machine, and the most suitable sequence is to be selected (which optimises the effectiveness measure) among all $(n!)^m$ theoretically possible sequences.

Although theoretically, it is always possible to select the best sequence by testing each one, in practice, it is impossible because of the large number of computations involved. For example, if there are 4 jobs to be processed at each of the 5 machines (i.e. $n = 4$ and $m = 5$), the total number of theoretically possible different sequences will be $(4!)^5 = 7,962,624$. So, easier methods of dealing with such problems are needed. That is, a technique which helps us arrive at an optimal sequence without trying all or most of the possibilities is needed.

8.1.1 Definition

Suppose there are n jobs (1, 2, 3, ..., n) each of which has to be processed one at a time at each of m machines $A, B, C, \ldots$. The order of processing each job through machines is given (for example, job 1 is processed through machines, in the order A, C, B). The time that each job must require on each machine is known. The problem is to find a sequence $(n!)^m$ number of all possible sequences (or combinations or order) for processing the jobs so that the total elapsed time for all the jobs will be minimum.

Mathematically, let

A_i = time for job i on machine A,
B_i = time for job i on machine B, and so on
T_i = time from start of first job to completion of the last job.

Then, the problem is to determine for each machine a sequence of jobs $(i_1, i_2, \ldots, i_n)$ where $(i_1, \ldots, i_n)$ is the permutation of integers which will minimise T.

8.1.2 Terminology and Notations

***Notation*:**

t_{ij} = processing time (time required) for job i on machine j
T = total elapsed time for processing all the jobs. This includes idle time, if any.
I_{ij} = Idle time on machine j from the end of job $(i - 1)$ to the start of job i.

***Terminology*:**

Number of machines: It means the service facilities through which a job must pass before it is completed.

For example, a book to be published has to be processed through composing, printing, binding, and so on. In this example, the book constitutes the job and the different processes constitute the number of machines.

Processing order: It refers to the order (sequence) in which given machines are required for completing the job.

Processing time: It is the time required by a job on each machine.

Idle time on a machine: It is the time for which a machine does not have a process; it is idle time from the end of job $(i - 1)$ to the start of job i.

Total elapsed time: It is the time interval between starting the first job and completing the last job including the idle time (if any) in a particular order by the given set of machines.

No passing rule: It refers to the rule of maintaining the order in which jobs are to be processed on given machines. For example, if n jobs are to be processed on two machines M_1 and M_2 in the order $M_1 M_2$, then each job should go to machine M_1 first and then to M_2.

8.1.3 Assumptions

1. No machine can process more than one operation at a time.
2. Each operation, once started, must be performed till completion.
3. A job in an entity is even though it represents a lot of individual parts; no lot may be processed by more than one machine at a time.
4. Each operation must be completed before any other operation, which following it, can begin.
5. Time intervals for processing are independent of the order in which the operations are performed.
6. There is only one of each type of machine.
7. A job is processed as soon as possible subject to ordering requirements.
8. All jobs are known and are ready to start processing before the period under consideration begins.
9. The time required to transfer jobs between machines is negligible.

8.1.4 Solution of Sequencing Problems

At present, a solution of following cases are available:

1. n jobs and two machines A and B, all jobs processed in the order AB.
2. n jobs and three machines A, B and C, all jobs processed in the order ABC.
3. Two jobs and m machines. Each job is to be processed through the machines in a prescribed order (which is not necessarily the same for both the jobs).
4. Problem with n jobs and m machines.

We will discuss these cases in the following sections.

8.2 PROCESSING *n* JOBS THROUGH TWO MACHINES

Let there be n jobs, each of which is to be processed through two machines A and B in the order AB. That is, each job has to pass through the same sequence of operations. In other words, a job is assigned on machine A first and after it has been completely processed, it is assigned to machine B. If machine B is not free, then the job has to wait in a waiting line for its turn, that is, passing is not allowed. Therefore, machine A will remain busy in processing all the n jobs one-by-one while machine B may remain idle after completion of one job and before starting of another job. Thus, the objective is to minimise the idle time of the second machine. This can be achieved only by determining sequence of n jobs which are to be processed on the two machines A and B.

Johnson's algorithm for n jobs through 2 machines

Step 1: List the jobs along with their processing times in a table as follows:

		Job number				
		1	2	3	...	n
Processing time	A	t_{11}	t_{12}	t_{13}	...	t_{1n}
Processing time	B	t_{21}	t_{22}	t_{23}	...	t_{2n}

Step 2: Examine the columns for processing time on machines A and B, and find the smallest processing time in each column, that is, find out

min (t_{1j}, t_{2j}) for all j

Step 3: *(a)*: If the smallest processing time is for the first machine A, then place the corresponding job in the first available position in the sequence. If it is for the second machine, then place the corresponding job in the last available position in the sequence.

Step 3: *(b)*: If there is a tie in selecting the minimum of all the processing times, then there may be three situations:

(i) Minimum among all processing times is same for the machine, that is, min $(t_{1j}, t_{2j}) = t_{1k} = t_{2r}$, then process the kth job first and the rth job last.

(ii) If the tie for the minimum occurs among processing times t_{1j} on machine A only, then select the job corresponding to the smallest job subscript first.

(iii) If the tie for the minimum occurs among processing times t_{2j} on machine B, then select the job corresponding to the largest job subscript last.

Step 4: Remove the assigned jobs from the table. If the table is empty, stop and go to step 4. Otherwise, go to step 2.

Step 5: Calculate idle time for machines A and B.

(a) Idle time for machine A =
(Total elapsed time) – (Time when the last job in a sequence finishes in machine A)

(b) Idle time for machine B =
Time at which the first job in a sequence finishes on machine A +

$$\sum_{j=2}^{n} \{\text{time when the } j^{\text{th}} \text{ job in a sequence starts on machine } B) -$$

(Time when the $(j-1)^{\text{th}}$ j on in a sequence finishes on machine B)

Step 6: The total elapsed time to process all jobs through two machines is given by

Total elapsed time = Time when n^{th} job in a sequence finishes on machine B.

$$= \sum_{j=1}^{n} B_{2j} + \sum_{j=1}^{n} I_{2j}$$

where, B_{2j} = Time required for processing j^{th} job on machine B.

I_{2j} = Time for which B remains idle after processing $(j-1)^{th}$ job and before starting work on the jth job.

Example 1

There are five jobs, each of which is to be processed through two machines M_1, M_2 in the order M_1, M_2, processing hours are as follows:

Job	1	2	3	4	5
Machine *A*	3	8	5	7	4
Machine *B*	4	10	6	5	8

Determine the optimum sequence for the five jobs, and minimum elapsed time. Also, find the idle time of machines *A* and *B*.

Solution:

Table 8.1

Machine/ Job No.	*A*	*B*	*Order of cancellation*
1	3	4	(1)
2	8	10	(5)
3	5	6	(3)
4	7	5	(4)
5	4	8	(2)

Optimal sequence

1	5	3	4	2

Explanation: The minimum element (among all the elements) is 3 corresponding to job 1 occurring under machine *A*.

Next minimum element is 4 against the job 5 under machine *A*; so next, do job 5. Now, the minimum element is 5. But there is a tie. 5 occurs against jobs 3 and 4. But job 3 is chosen since the entry corresponding to 5 under machine *B* is 6 smaller than 7 found for job 4 under machine *A*.

Obviously, job 4 is chosen next and lastly the job 2 is chosen.

To find the minimum total elapsed time:

table 8.2

Job No.	Machine A Time in	Machine A Time out	Machine B Time in	Machine B Time out
1	10	13	13	17
5	13	17	17	15
3	17	12	15	21
4	12	19	21	26
2	19	27	27	37

Therefore, minimum elapsed time = 37 hr.
Idle time for machine $A = 37 - 27 = 10$ hr.
Idle time for machine $B = 3 + 1 = 4$ hr.

Example 2

A book binder has one printing press, one binding machine, and the manuscripts of a number of different books. The time required to perform the printing and binding operations for each book are shown below. Determine the order in which books should be processed, in order to minimise the total time required to turn out all the books.

Books	1	2	3	4	5	6
Printing time (hr)	30	120	50	20	90	110
Binding time (hr)	80	100	90	60	30	10

Solution: Here, the books will first go to the printing press and then on the binding machine. If P_i ($i = 1, 2, .., 6$) denotes the time in hours on the printing press and B_i ($i = 1, 2, ..., 6$) the binding time for books, then since min $\{P_i, B_i\} = 10$, corresponding to B_6, book 6 will be processed in the last. Now, min $\{P_i, B_i\} = 20$, which corresponds to P_4. Hence, book 4 will be processed in the beginning. Now, the minimum of P_i and B_i is 30 which corresponds to P_1 and B_5, that is, there is a tie for the minima. So, we place book 1 next to the first and book 5 next to last. Since smallest time is 50 for book 3, we place book 3 in the 3rd cell and the remaining book 2 in the 4th cell getting the optimal sequence.

4	1	3	2	5	6

To find the minimum elapsed time:

table 8.3

Book	Printing machine Time in	Printing machine Time out	Binding machine Time in	Binding machine Time out	Idle time of binding machine
4	0	20	20	80	20
1	20	50	80	160	0
3	50	100	160	250	0
2	100	220	250	350	0
5	220	310	350	380	0
6	310	420	420	430	40

From Table 8.3, it is clear that the minimum elapsed time is 430 hours. Idle time for the printing machine is 10 hours (from 420 hours to 430 hours) and that for binding machine is 20 + 40 = 60 hours.

Remark It may be noted that the total elapsed time is equal to the sum of the idle time of the binding machine and the total processing time on it.

8.3 PROCESSING *n* JOBS THROUGH 3 MACHINES

The problem can be described as

(i) Only three machines *A, B* and *C* are involved.
(ii) Each job is processed in the prescribed order *ABC*.
(iii) Transfer of jobs is not permitted, that is, adhere strictly the order over the machine.
(iv) Exact or expected processing time are given in Table 8.4.

table 8.4

Job	Machine A	Machine B	Machine C
1	A_1	B_1	C_1
2	A_2	B_2	C_2
⋮	⋮	⋮	⋮
n	A_n	B_n	C_n

There is no general procedure available for obtaining optimal sequence in this case. Johnson's method can be extended to cover special cases when either one or both of the following conditions hold:

1. The minimum time on machine $A \geq$ the maximum time on machine B.
2. The minimum time on machine $C \geq$ the maximum time on machine B.

The method is to replace the problem with an equivalent problem, involving n jobs and 2 imaginary machines G and H, and corresponding time G_i and H_i are defined by $G_i = A_i + B_i$, $H_i = B_i + C_i$.

If this problem with the prescribed ordering *GH* is solved, the resulting optimal sequence will also be optimal for the original problem.

Example 1

Find the sequence that minimises the total elapsed time required to complete the following tasks:

Tasks	A	B	C	D	E	F	G
Time on Machine I	3	8	7	4	9	8	7
Time on Machine II	4	3	2	5	1	4	3
Time on Machine III	6	7	5	11	5	6	12

Solution: We are given 7 jobs each of which is to be processed through 3 machines I, II and III in the order I, II, III.

Now, Minimum time on machine I = 3
Maximum time on machine II = 5
Minimum time on machine III = 5

Since maximum time on machine II = minimum time on machine III, the given problem can be reduced to a problem of processing n jobs through two machines.

Thus, if G and H are two imaginary machines such that

$$G_i = M_{i1} + M_{i2}$$
$$H_i = M_{i2} + M_{i3} \text{ for } i = 1, 2, .., 7$$

then the problem can be rewritten as the following 7 jobs and 2 machines problems.

Job	A	B	C	D	E	F	G
G	3 + 4 = 7	8 + 3 = 11	7+ 2 = 9	4 + 5 = 9	9 + 1 = 10	8 + 4 = 12	7 + 3 = 10
H	4 + 6 = 10	3 + 7 = 10	2 + 5 = 7	5 + 11 = 16	1 + 5 = 6	4 + 6 = 10	3 + 12 = 15

Using the optimal sequence algorithm, the following optimal sequence can easily be obtained.

A	B	G	B	F	C	E

To find the minimum elapsed time:

table 8.5

Job sequence	Machine I Time in	Machine I Time out	Machine II Time in	Machine II Time out	Machine III Time in	Machine III Time out
A	0	3	3	7	7	13
D	3	7	7	12	13	24
G	7	14	14	17	24	36
B	14	22	22	25	36	43
F	22	30	30	34	43	49
C	30	37	37	39	49	54
E	37	46	46	47	54	59

From Table 8.5, we observe that, the minimum total elapsed time is 59 hours.
Idle time is 13 hours for machine I,
37 hours for machine II, and
7 hours for machine III.

8.4 PROCESSING *n* JOBS THROUGH *m* MACHINES

Suppose there are n jobs to be processed through m machines, say $M_1, M_2, ..., M_m$ in the order $M_1 M_2 ...M_m$, and let T_{ij} denote the time taken by the ith machine to complete the jth job. There is no general method available by which we can obtain optimal sequence(s) in problems of this type. This problem can be converted to a problem of processing n jobs through 2 machines if no passing of jobs is permissible and if either or both of the conditions given in step 2 of the procedure that follows is satisfied:

Procedure for obtaining optimal sequence

Step 1: Find (i) $\min_j (T_{1j})$, (ii) $\min_j (T_{mj})$, and

(iii) $\max_j (T_{2j}, T_{3j}, ... T_{(m-1)j})$ for $j = 1, 2, ...n$

Step 2: Check whether

(i) $\min_j (T_{1j}) \geq \max_j (T_{ij})$ for $i = 2, 3, ..., m - 1$

or, (ii) $\min_j (T_{mj}) \geq \max_j (T_{ij})$ for $i = 2, 3, ..., m - 1$

Step 3: If inequations of Step 2 are not satisfied, this method fails. Otherwise, go to next step.

Step 4: Convert the m machine problem into a 2-machine problem considering two imaginary machines G and H, so that

$$T_{Gj} = T_{1j} + T_{2j} + ... + T_{(m-1)j} \text{ and}$$
$$T_{Hj} = T_{2j} + T_{3j} + ... + T_{Mj}$$

Now, determine the optimal sequence on n jobs through 2 machines by using the optimal sequence algorithm.

Step 5: In addition to conditions given in Step 4, if $T_{2j} + T_{3j} + ... + T_{(m-1)} = C$ (a fixed positive constant) for all $j = 1, 2, ..., n$, then determine the optimal sequence for n jobs and two machines M_1 and M_m in the order $M_1 M_m$ by using the Jhonson's algorithm.

Remark

1. If in addition to the condition given in Step 4, $T_{ij} = T_{mj}$ and $T_{Gj} = T_{Hj}$, for $j = 1, 2, ..., n$, then $n!$ number of optimal sequence will exist.
2. This procedure for sequencing n jobs is not a general procedure. It is applicable to only to such those sequencing problems in which minimum time or cost of processing the jobs through first and/or last machine is greater than or equal to the time or cost of processing the jobs through the mediocre machine.

Example 1

There are 4 jobs each of which has to go through the machines M_i, $i = 1, 2, ..., 6$ in the order $M_1 M_2 ... M_6$. Processing time are given.

table 8.6

			Machine (i)			
Job (j)	M_1	M_2	M_3	M_4	M_5	M_6
A	20	10	9	4	12	27
B	19	8	11	8	10	21
C	13	7	10	7	9	17
D	22	6	5	6	10	14

Determine a sequence for these four jobs which minimises the total elapsed time T.

Solution: Here, $\min_j T_{1j} = 13$, $\min_j T_{6j} = 14$ and

$$\max_j T_{2j} = 10, \max_j T_{3j} = 11, \max_j T_{4j} = 8, \max_j T_{5j} = 12.$$

Hence, the conditions,

$$\min_j T_{1j} \geq \max_j T_{ij} \text{ and } \min_j T_{6j} \geq \max_j T_{ij}$$

for $i = 2, 3, 4, 5$

are satisfied. The problem can be converted into 4-job and 2-machine problem. Thus, if G and H are two imaginary machines such that $T_{Gj} = \sum_{i=1}^{m-1} T_{ij}$ and $T_{Hj} = \sum_{i=2}^{m} T_{ij}$, then the problem can be reformulated as

table 8.7

Machines \ Jobs	A	B	C	D
G	55	56	46	49
H	62	58	50	41

Using Johnson's algorithm, an optimal sequence is obtained as *C–A–B–D*.

To find the minimum elapsed time:

table 8.8

Optimal sequence	M_1 In	M_1 Out	M_2 In	M_2 Out	M_3 In	M_3 Out	M_4 In	M_4 Out	M_5 In	M_5 Out	M_6 In	M_6 Out
C	0	13	13	20	20	30	30	37	37	46	46	53
A	13	33	33	43	43	52	52	56	56	68	68	95
B	33	52	52	60	60	71	71	79	79	89	95	116
D	52	74	74	80	80	85	85	91	91	101	116	130

From Table 8.8, we observe that the minimum elapsed time is 130 hours.

Example 2

Four jobs 1, 2, 3 and 4 are to be processed on each of the four machines *A, B, C* and *D* in the order *ABCD*. The processing time in minutes are given in Table 8.9. Find for no passing, the minimum elapsed time.

table 8.9

Jobs \ Machines	A	B	C	D
1	58	14	14	48
2	30	10	18	32
3	28	12	16	44
4	64	16	12	42

Solution: Min $A_i = 28$, min $D_i = 32$, max $B_i = 16$ and max $C_i = 18$

Since min $A_i \geq$ Max B_i;

min $A_i \geq$ max C_i; and

min $D_i \geq$ max B_i;

min $D_i \geq$ max C_i,

the problem can be converted to a 2-machine and 4-job problem.

Further, since $B_1 + C_1 = B_2 + C_2 = B_3 + C_3 = B_4 + C_4 = 28$, a fixed constant, the given problem reduces to that of finding the optimal sequence for 4 jobs and 2 machines *A* and *D* in the order *AD*. Machines *B* and *C* do not have any effect on the optimality of the sequence.

Examining columns A and D only, the optimal sequence is given by

$$3 - 2 - 1 - 4$$

To find the minimum elapsed time:

Use the individual processing time given in the original problem.

table 8.10

Optimal sequence	A		B		C		D	
	Time in	Time out	Time in	Time out	Time in	Time out	Time in	Time out
3	0	28	28	40	40	56	56	100
2	28	58	56	68	68	86	100	132
1	58	116	116	130	130	144	144	192
4	116	180	180	196	196	208	208	250

Thus, the minimum elapsed time is 250 minutes.

8.5 PROCESSING TWO JOBS ON *n* MACHINES

Suppose there are two jobs, job 1 and job 2, each of which is to be processed on n machines M_1, M_2, ...M_n.

General assumptions

1. The technological ordering of each of the two jobs through n machines is known in advance (ordering may not be same for both the jobs).
2. Each machine can process only one job at a time.
3. The exact processing time on all the n machines are known.

Graphical method is used to determine the minimum total elapsed time.

Graphical method

Algorithm:

Step 1: Use the x-axis to represent the processing time for job 1 and the y-axis to represent the process time for job 2.

Step 2: Mark the machine time for the two jobs and their corresponding axes as per the given technological order.

Note that in such a graph moving horizontally will imply that job 1 is being processed while job 2 remains idle. Moving vertically will imply that job 2 is processed while job 1 remains idle. Diagonal movement along 45° line to the horizontal (line with slope 1) will imply that both the jobs 1 and 2 are being processed simultaneously. Since each machine can process only one job at a time, overlapping region for the machines should be determined first and movement across them should be avoided.

Step 3: An optimal path is the shortest one (that minimises the idle time for job 1 in the horizontal movement or the shortest one that minimises the idle time for job 2 in the vertical movement) consisting of horizontal, vertical and 45° lines from the origin to the end. Obviously, we must choose such a combination in which diagonal movement is maximum possible.

Step 4: Compute the total elapsed time by adding processing time of job 1 + idle time of job 1.

Or

Processing time of job 2 + idle time of job 2, and the optimal sequence of processing from the graph drawn.

Example 1

Determine the minimum time needed to process two jobs on four machines M_1, M_2, M_3 and M_4. The technological order for these jobs as given below:

Job 1	M_1	M_2	M_3	M_4
Job 2	M_4	M_2	M_1	M_3

Processing time (in hours) are as given under:

	M_1	M_2	M_3	M_4
Job 1	5	7	8	4
Job 2	5	8	6	9

Solution: The rectangular blocks in Figure 8.1 represent the overlaps (for clashes) which are to be avoided.

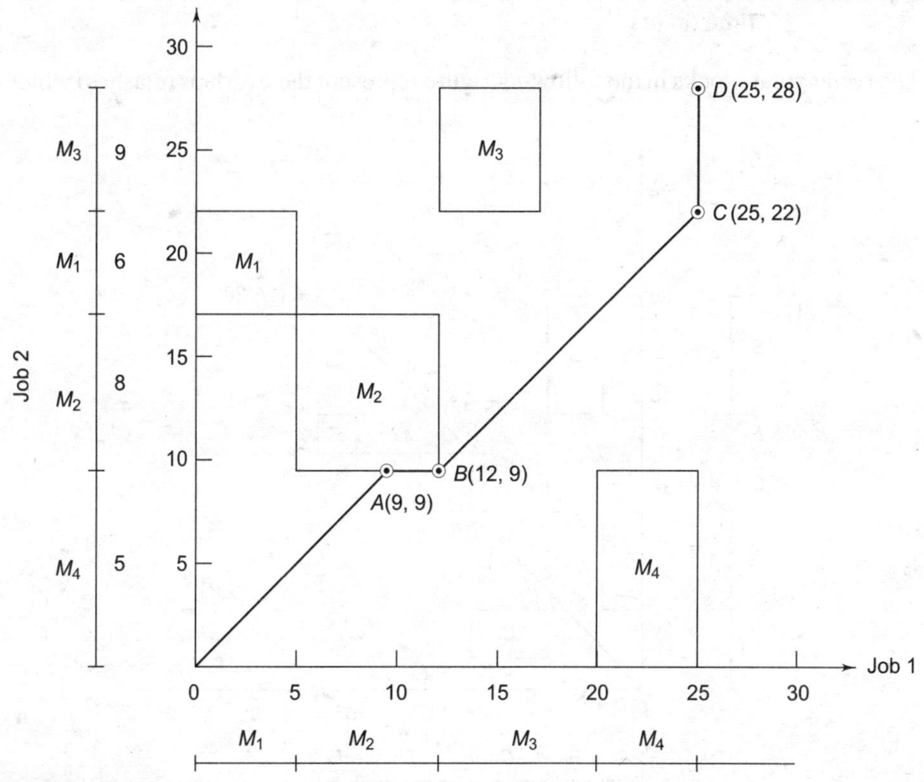

figure 8.1

An optimal path is one that minimises the idle time for job 1 (horizontal movement). Similarly, an optimal path is one that minimises the idle time for job 2 (vertical movement).

Hence, minimum elapsed time = processing of job 1 + idle time for job 1
= 25 + 6 = 31 hours.

Or

Minimum elapsed time = processing of job 2 + idle time for job 2
= 28 + 3 = 31 hours.

Sequence: Process both the jobs for 9 hours; process job 1 for 3 hours, process both the jobs for the next 13 hours; process job 2 only for 6 hours to complete all the jobs and their processing. Total elapsed time = 31 hours.

Example 2

Use graphical method to minimise the time needed to process the following jobs on the machines shown below, that is, for each machine find the job which should be done first. Also, calculate the total time needed to complete both the jobs.

table 8.11

Job 1	Sequence of machines	*A*	*B*	*C*	*D*	*E*
	Time (in hrs)	2	3	4	6	2
Job 2	Sequence of machines	*C*	*A*	*D*	*E*	*B*
	Time (in hrs)	4	5	3	2	6

Solution: The rectangular blocks in the following figure represent the overlaps (clashes) which are to be avoided.

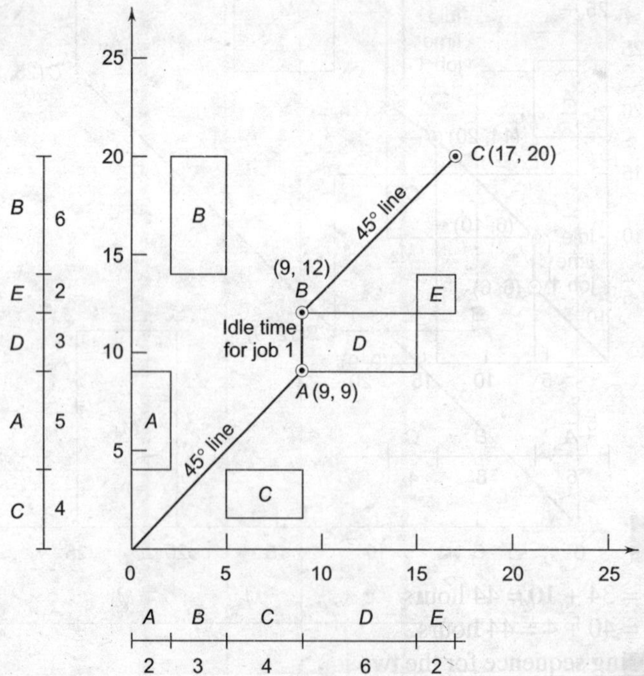

figure 8.2

Minimum total elapsed time = processing of job 1 + idle time for job 1 = 17 + 3 = 20 hours.

Or, Minimum total elapsed time = processing of job 2 + idle time on job 2 = 20 + 0 = 20 hours.

Idle time for job 1 is 3 hours and that for job 2 is 0 hours. Process both the jobs for 9 hours and then perform job 2 for 3 hours and then perform both the jobs for 8 hours.

Example 3

Use graphical method to minimise the time required to process the following jobs on the machines, that is, for each machine specify the job which should be done first. Also, calculate the total elapsed time to complete both the jobs.

Job 1	**Sequence of machines**	*A*	*B*	*C*	*D*	*E*
	Time (in hrs)	6	8	4	12	4
Job 2	**Sequence of machines**	B	C	A	D	E
	Time (in hrs)	10	8	6	4	12

Solution: The rectangular blocks in Fig. 8.3 represent the overlaps (clashes) which are to be avoided.

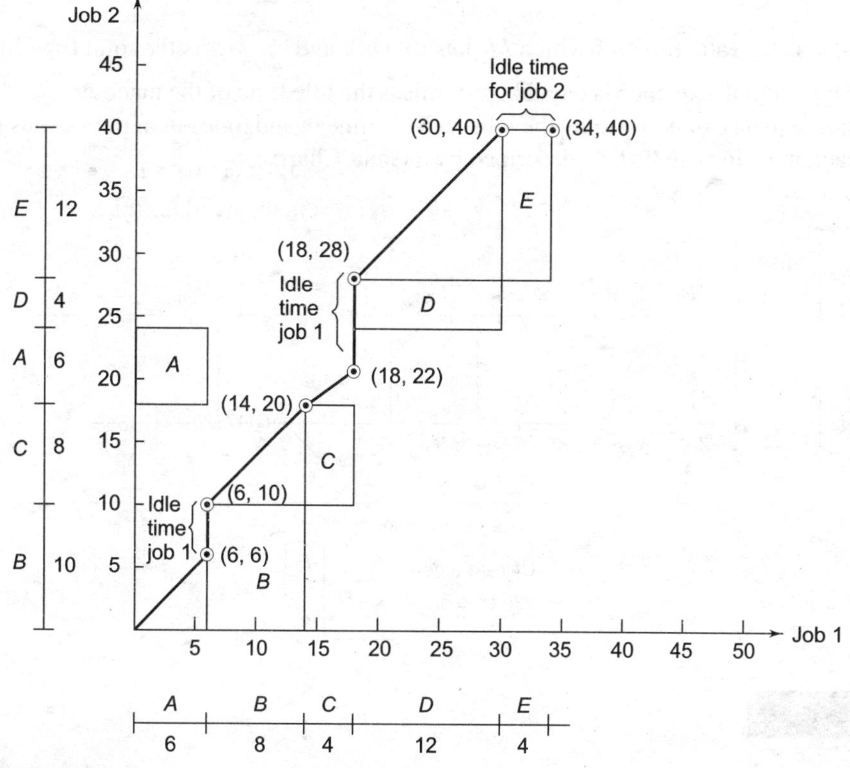

figure 8.3

Total elapsed time = 34 + 10 = 44 hours

Or, = 40 + 4 = 44 hours

The optimal processing sequence for the two jobs on various machines is as follows:

Job 1 precedes job 2 on machine *A*,
Job 2 precedes job 1 on machine *B*,
Job 2 precedes job 1 on machine *C*,
Job 1 precedes job 2 on machine *D*, and
Job 2 precedes job 1 on machine *E*.

8.6 GANTT CHART

Consider the problem of processing *n* jobs through two machines M_1 and M_2 in the order $M_1\,M_2$. Let X_{i2} be the time for which machine M_2 remains idle after finishing $(i-1)^{\text{th}}$ job and before starting the processing of ith job ($i = 1, 2, ..., n$). We know that total elapsed time is given by $T = \sum_{i=1}^{n} M_{i2} + \sum_{i=1}^{n} X_{i2}$

Or,

$$T = \sum_{i=1}^{n} M_{i1} + \sum_{i=1}^{n} X_{i1}$$

where, $\sum M_{i2}$ is the total time for which M_2 has to work and $\sum M_{i1}$ is the total time for which M_1 has to work. The optimal sequence is one that minimises the idle time of the machines.

The optimum sequence of performing the jobs, and the time in and time out of the various jobs and the idle time on each machine can be best described by a Gantt Chart:

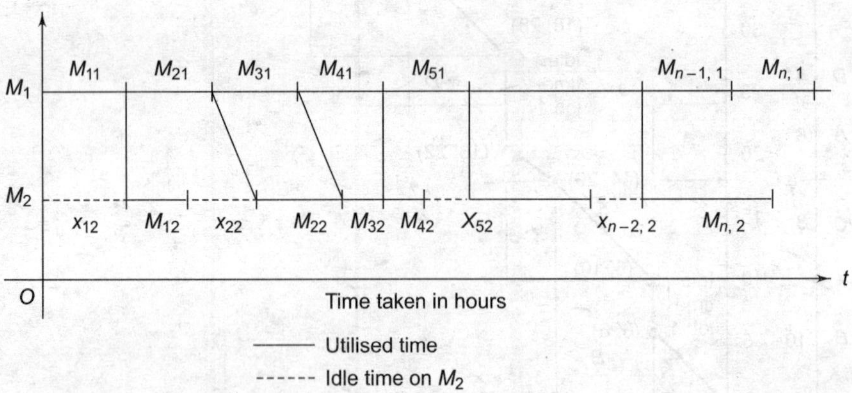

figure 8.4

Example 1

Consider the sequencing problem given in Example 1 under Section 8.2. The optimal sequence of the job is 1–5–3–4–2, the total minimum elapsed time is 37 hours, idle time for machine *A* is 10 hours, and the idle time for machine *B* is 4 hours. Gantt Chart representation of the above problem is given in Fig. 8.5.

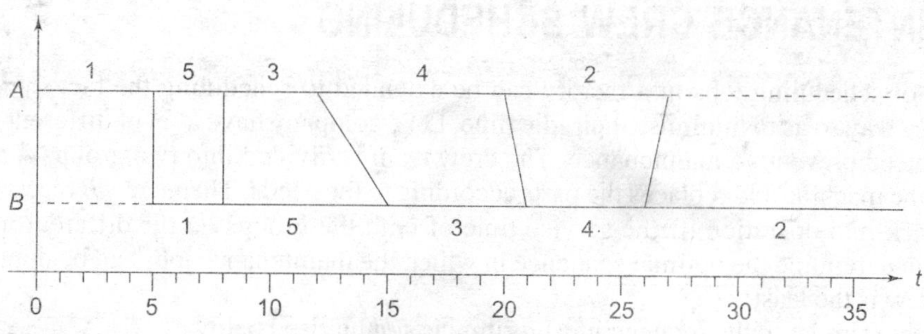

figure 8.5 Gantt Chart

Remark: Gantt Chart can also be used to solve sequencing problem of processing 2 jobs on n machines.

Example 2

Two jobs are to be processed through four machines A, B, C, D with the following technological ordering:

Job 1	*A*	*B*	*C*	*D*
Job 2	*D*	*B*	*A*	*C*

Processing time (in hours) are given in the following table:

	Machine			
	A	*B*	*C*	*D*
Job 1	20	40	50	10
Job 2	20	50	30	60

Find the minimum elapsed time for both jobs and also find the idle time for both jobs using a Gantt Chart.

Solution: In this method we draw the time taken for both jobs on different machines in the prescribed order. Care should be taken that no machines are busy simultaneously. The solution is given below:

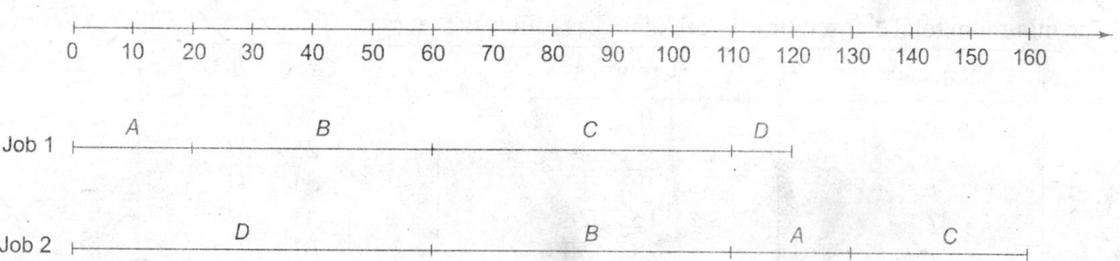

figure 8.6 Gantt Chart

Hence, minimum elapsed time is 160 hours, idle time for the chosen path is 40 for job 1 and 0 for job 2.

8.7 MAINTENANCE CREW SCHEDULING

The problem of scheduling jobs in a factory can be extended to scheduling the factory maintenance crews in such a way so as to minimise their idle time. Let a company have a set of different machines in its plant that need preventive maintenance. The crew team is divided into two groups *A* and *B*. First, crew *A* takes the machine and replaces the parts according to the needs. Then, crew *B* resets the machine and puts it back into operation. If the service time of both the groups on the different machines are known, we can determine the optimal sequence in which the maintenance jobs can be done so that idle time of the crew is the least.

The algorithm similar to the sequencing algorithm is summarised below:

(i) Choose the smallest of all values that appear in the two rows.

(ii) If the value is in the first row then the machine corresponding to the value shall be serviced by crew *A* in the beginning.

(iii) If it is in the second row, that machine will be serviced by crew *A* in the end.

(iv) If there is a tie in the same row, either of the machines involved is selected and assigned the first, or the last place accordingly as the tie is in the first or the second row.

(v) After assigning, cross out the machine involved.

(vi) Repeat steps (i) to (v) until all the assignments are made.

Example 1

A company has eight large machines which receive preventive maintenance. The maintenance team is divided into two crews *A* and *B*. Crew *A* takes the machine 'power' and replaces parts according to a given maintenance schedule. The second crew resets the machine and puts it back into operation. At all times the no passing rule is considered to be in effect. The servicing time for each machine are given below:

Machine	*a*	*b*	*c*	*d*	*e*	*f*	*g*	*h*
Crew *A*	5	4	22	16	15	11	9	4
Crew *B*	6	10	12	8	20	7	2	21

Determine the optimal sequence of scheduling the factory maintenance crews to minimise their idle time and represent it on a chart.

Solution: Treating the machines as 8 items and crew as the servicing facilities, the optimal sequence can be obtained as

$$h - a - b - e - c - d - f - g \text{ (using Johnson's algorithm)}$$

The minimum total elapsed time is calculated as follows:

Sequence	Crew *A*		Crew *B*	
h	0	4	4	25
a	4	9	25	31
b	9	13	31	41
e	13	28	41	61
c	28	50	61	73
d	50	66	73	81
f	66	77	81	88
g	77	86	88	90

Minimum total elapsed time is 90 time units.

Idle time for Crew A and Crew B is 4 time unit each. The chart representation is given in Fig. 8.7.

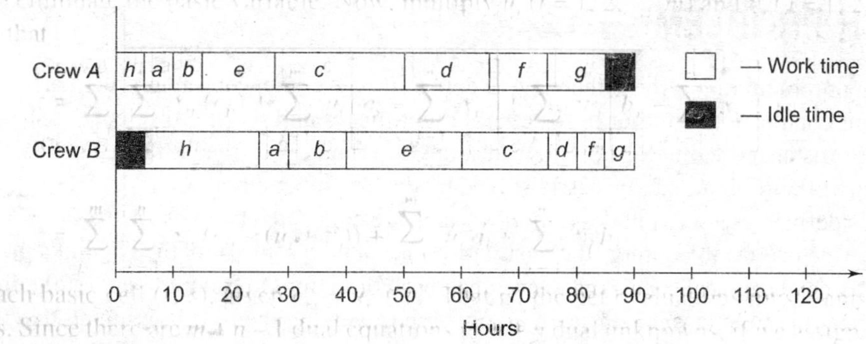

figure 8.7

8.8 SHORTEST CYCLIC ROUTE MODELS (TRAVELLING SALESMEN PROBLEM)

A salesman has to visit a number of cities. The problem of finding the shortest route (in distance or time or cost) to visit all the cities (each city exactly once) is known as the Travelling Salesman Problem (TSP).

If the distance (time or cost) between every pair of cities is independent of the direction, the problem is said to be symmetrical. Otherwise, it is known as asymmetrical.

Example 1 It takes more time to go up a hill between two stations than come down the hill between them.

Example 2 A flight may take more time against the wind direction compared to that in the direction of the wind.

If there are n cities then there will be $(n-1)!$ possible routes. It may be noted that since the salesman has to visit all the n cities, the shortest route will be independent of the selection of the starting city.

Unfortunately, there is no analytical method which can be used satisfactorily to find the best route. However, there are some computational techniques.

Some of the areas in which this type of problem arises consists of

1. postal deliveries
2. inspection
3. school bus routing
4. television relays
5. assembly lines.

TSP appears to be related to the sequencing problem but actually is more similar to assignment problem with the difference that there is an additional constraint specifying that the solution should be in cyclic order. The solution methodologies of TSP are discussed in detail in Chapter 5.

8.9 SHORTEST ACYCLIC ROUTE MODELS (MINIMAL PATH PROBLEM)

TSP is a routing problem involving rather severe constraints. Another routing problem arises when we wish to go from one place to another or to several other places and we are to select the shortest route (involving least distance or time or cost) out of many alternatives to reach the desired station. Such acylic route network problems can be easily solved by the graphic method.

A network is defined as a set of nodes or nodes which are connected by lines or links. A way of going from one node (the origin) to another (the destination) is called a route or path. The links in a network may be one way (in either direction) or two way (in both directions). The numbers on the links in the network represent the time, cost or distance involved in traversing them. It is assumed that the way in which we enter a node has no effect on the way of leaving it—an assumption which does not hold good in TSP.

Example 1

Consider the network given below. A person wishes to reach the destination k while starting from the station a in the network. Links work in both directions unless marked otherwise. The numbers on the links represent cost (in rupees) of going from one node to another. Find the route that involves the least cost.

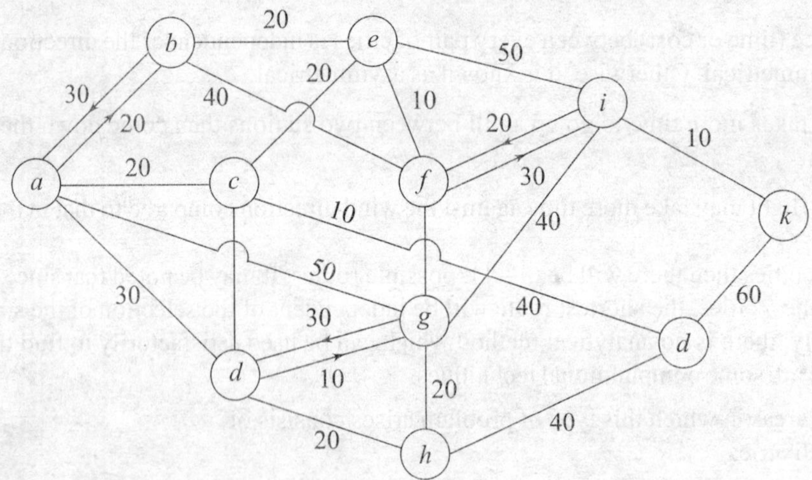

figure 8.8

Solution: The given problem is not a TSP, but a shortest route problem. It is a acylic route problem as it is not required to come back to the starting point a.

In order to find the shortest route between a and k we must find the shortest route from a to every other point in the network. The graphic method used for this purpose consists of the following steps:

Step 1: Identify the nodes having direct links with a and represent the cost from a on each of these nodes.

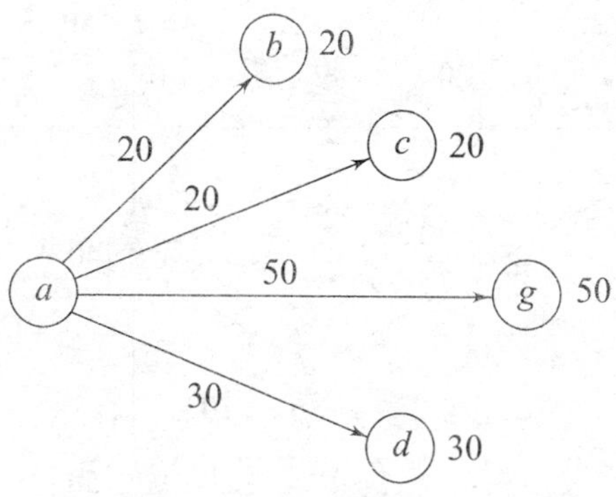

figure 8.9

Step 2: In case there are links between any of the nodes obtained in Step 1, determine for each of these links if the indirect route from a is shorter than the direct route. Draw the shorter route as a solid line and the longer route as a dotted line. Insert the shortest cost found on each such node.

For example, the cost of going from a to g through d is lower than the cost of going from a to g directly. Hence, link ag is drawn as a dotted line. In case of a tie both links are represented by solid lines. Thus, one can go from a to d directly or through c at the same cost. Hence, the links connecting them are drawn solid as in Fig. 8.10.

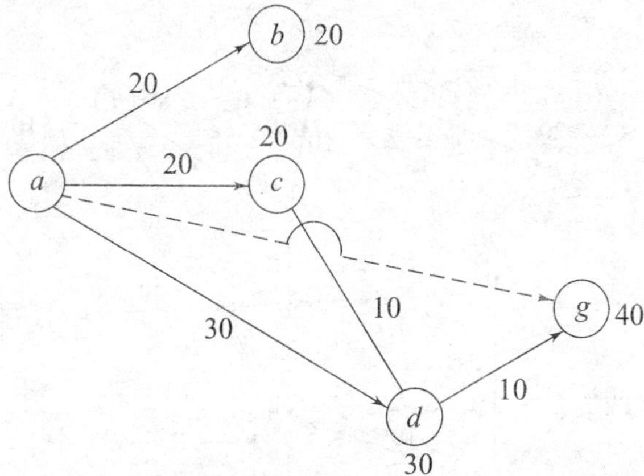

figure 8.10

Step 3: Add nodes to which one can go grom the nodes represented in Step 2 and repeat Step 2. This is given in Fig. 8.11.

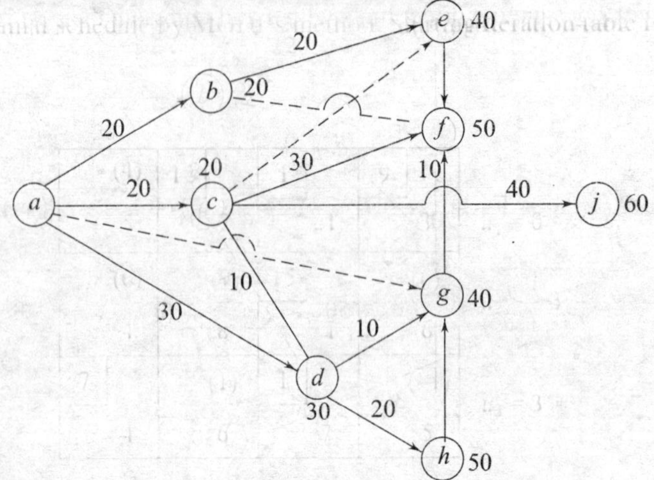

figure 8.11

Step 4: Continue till completed. This is shown in Fig. 8.12. Solid lines show the routes that can be taken from a to every other node.

Evidently, there are a number of alternative paths giving least cost. They are:

(i)	$a-b-e-f-i-k$	(involving 6 nodes)
(ii)	$a-c-f-i-k$	(involving 5 nodes)
(iii)	$a-c-d-g-f-i-k$	(involving 7 nodes)
(iv)	$a-d-g-f-i-k$	(involving 6 nodes)
(v)	$a-d-g-i-k$	(involving 5 nodes)
(vi)	$a-c-e-f-i-k$	(involving 6 nodes)

All the routes have the same cost (Rs 90) of travelling from a to k.

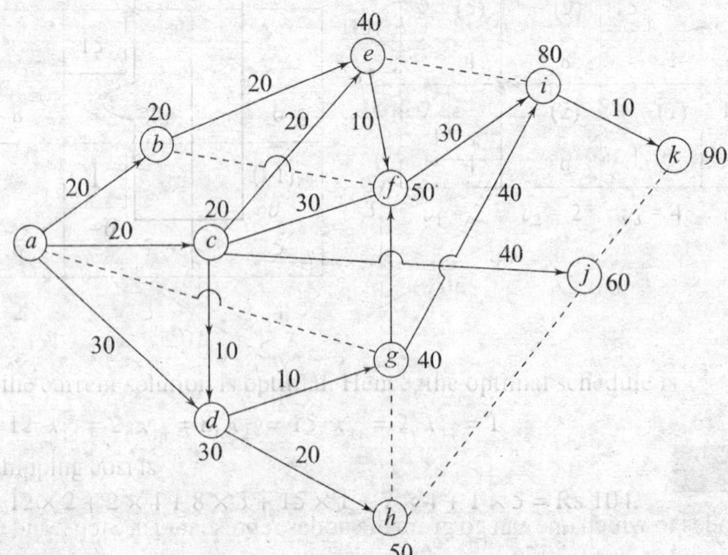

figure 8.12

If, however, an additional constraint is imposed, for example, a person is to visit a minimum number of stations before reaching k, the number of alternative optimum (shortest) cost routes decreases to only two:

(i) $a - c - f - i - k$
(ii) $a - d - g - i - k$

EXERCISE

Section 8.1

1. Define a sequencing problem.
2. Define the following terms:
   (a) Number of machines
   (b) Processing order
   (c) Processing time
   (d) Idle time on a machine
   (e) Total elapsed time
3. What are the general assumptions of a sequencing model?

Section 8.2

4. Explain the Johnson's algorithm of processing n jobs through two machines.
5. What is no passing rule in a sequencing algorithm?
6. Find the sequence that minimises the total elapsed time required to complete the following tasks on two machines:

Task	*A*	*B*	*C*	*D*	*E*	*F*	*G*	*H*	*I*
Machine *I*	2	5	4	9	6	8	7	5	4
Machine *II*	6	8	7	4	3	9	3	8	11

[*Answer*: The optimal sequences are *A C I B H F D G E* and *A C I H B F D G E*. Minimum elapsed time is 61 hours and the idle time for machine II is 2 hours.]

7. Give three different examples of sequencing problems from your daily life.
8. Write a short note on the sequencing decision problem for n jobs on two machines.
9. We have five jobs, each of which must be processed on the two machines *A* and *B* in the order *AB*. Processing times in hours are given in the Table below:

Job	1	2	3	4	5
Machine A	5	1	9	3	10
Machine B	2	6	7	8	4

Determine a sequence for the five jobs that will minimise the elapsed time T.
[*Answer*: Optimal sequence is 2 – 4 – 3 – 5 – 1, Elapsed time = 30 hours. Idle time for machine A = 2 hours, and for machine B = 3 hours.]

10. In a factory, there are six jobs to be performed, each of which should go through two machines *A* and *B*, in the order *AB*. The processing time (in hours) for the jobs are given. You are required to determine the sequence for performing the jobs that would minimise the total elapsed time T. What is the value of T?

Job	J_1	J_2	J_3	J_4	J_5	J_6
Machine *A*	1	3	8	5	6	3
Machine *B*	5	6	3	2	2	10

[*Answer*: The optimal sequence is $J_1 - J_6 - J_2 - J_3 - J_4 - J_5$; minimum elapsed time = 29 hours. Idle time for machine *A* is 3 hours and that for machine *B* is 1 hour.]

11. A company has 3 jobs on hand. Each of these must be processed through two departments, the sequential order for which is
Department *A* : Press shop Department *B* : Finishing
The Table below lists the number of days remquired by each job in each department.

	Job I	*Job II*	*Job III*
Department *A*	8	6	5
Department *B*	8	3	4

Find the sequence in which the 3 jobs should be processed so as to take minimum time to finish all the 3 jobs.
[*Answer*: Optimal sequence is Job I – Job III – Job II, minimum elapsed time = 23 days.]

12. A company has six jobs on hand coded *A* to *F*. All the jobs have to go through 2 machines M_1 and M_2. The time required for the jobs on each machine, in hours, is given below:

	A	*B*	*C*	*D*	*E*	*F*
M_2	1	4	6	3	5	2
M_3	3	6	8	8	1	5

Draw a sequence table scheduling the six jobs on the two machines.
[*Answer*: Optimal sequence : $A \to F \to D \to B \to C \to E$, Minimum elapsed time is 32 hours.]

13. We have 5 jobs each of which must go through 2 machines in the order *AB*. Processing time are given in the Table below:

Job No.	1	2	3	4	5
Machine *A*	10	2	18	6	20
Machine *B*	4	12	14	16	8m

Determine a sequence for the 5 jobs that will minimise the total elapsed time.
[*Answer*: 2 – 4 – 3 – 5 –1, Total elapsed time (minimum) is 60 hours.]

14. Find the sequence that minimises the total elapsed time required to complete the following jobs:

No. of jobs	1	2	3	4	5	6
Machine *A*	4	8	3	6	7	5
Machine *B*	6	3	7	2	8	4

[*Answer*: 3 – 1 – 5 – 6 – 2 – 4, Total time = 35 hours.]

15. Five jobs are performed, first on machine *X* and then on machine *Y*. The time taken in hours by each job on each machine is given below:

Job	*A*	*B*	*C*	*D*	*E*
Time on Machine X	12	4	20	14	22
Time on Machine Y	6	14	16	18	10

Determine the optimum sequence of jobs that minimises the total elapsed time to complete the jobs. Also, compute the minimum time.
[*Answer*: Optimal sequence $B - D - C - E - A$, minimum elapsed time = 48 hours, idle time is 12 hours each for machines X and Y.]

16. A manufacturing company processes 6 different jobs on two machines A and B. The number of units of each job and its processing times on A and B are given below:

Job	*Units of each job*	*Processing Time*	
		Machine A (minutes)	*Machine B (minutes)*
1	3	5	8
2	4	16	7
3	2	6	5
4	5	3	5
5	2	9	7.5
6	3	6	14

Find the optimal sequence, the total minimum elapsed time and the idle time for each machine.
[*Answer*: Optimal sequence: 4(5 units) – 1 (3 units) – 3 (2 units) – 6 (3 units) – 5 (2 units) – 2 (4 units);elapsed time = 159 minutes; idle time for machine A = 17 minutes, and for machine B = 3 minutes.]

17. A company plans to fill 6 positions. Since the positions are known to vary considerably with respect to skill and responsibility, different types of aptitude tests and interviews are required for each. While the aptitude tests are conducted by people from the clerical positions, the job interviews are held by the personnel from the management cadre. The job interviews immediately follow the aptitude test. The time required (in minutes) by each of the positions is given below:

Position	P_1	P_2	P_3	P_4	P_5	P_6
Aptitude Test	140	180	150	200	170	100
Job Interview	70	120	110	80	100	90

In order to minimise the waiting time of the management personnel, in what order should the position filling be handled?
[*Answer*: Optimal sequence is $P_2 - P_3 - P_5 - P_6 - P_4 - P_1$; minimum elapsed time is 7 hours and 20 minutes.]

18. A company has 8 large machines which receive preventive maintenance. The maintenance team is divided into two crews A and B. Crew A takes the machine 'power' and replaces parts according to a given maintenance schedule. The second crew resets the machine and puts it back into operation. At all times the no passing rule is considered to be in effect. The servicing times for each machine are given below:

Machine	*a*	*b*	*c*	*d*	*e*	*f*	*g*	*h*
Crew *A*	5	4	22	16	15	11	9	4
Crew *B*	6	10	12	8	20	7	2	21

Determine the optimal sequence of scheduling the factory maintenance crews to minimise their idle time and represent it on a chart.

[*Answer*: $b - a - e - c - d - f - h - g$.]

Section 8.3

19. Given the following data:

Job	1	2	3	4	5	6
Machine *A*	12	10	9	14	7	9
Machine *B*	7	6	6	5	4	4
Machine *C*	6	5	6	4	2	4

(a) Order of processing Job : *ACB*.

(b) Sequence suggested: Jobs 5, 3, 6, 2, 1, 4

(i) Determine the total elapsed time for the sequence suggested.

(ii) Is the given sequence optimal?

(iii) If your answer to (ii) is no determine the optimal sequence and the total elapsed time associated with it.

[*Answer*: *b*(i) 73 hours; (ii) No; (iii) Optimal sequence is 1 – 3 – 2 – 4 – 6 – 5 and elapsed time is 67 hours.]

20. We have 5 jobs, each of which must go through machines *A*, *B* and *C* in the order *ABC*. Processing times (in hours) are given in the following Table.

Job	1	2	3	4	5
Machine *A* (A_i)	8	10	6	7	11
Machine *B* (B_i)	5	6	2	3	4
Machine *C* (C_i)	4	9	8	6	5

Obtain the optimal sequence

[*Answer*: Optimal sequence is 3 – 2 – 4 – 1 – 5, minimum elapsed time is 51 hours.]

21. (a) A book binder has one printing press, one binding machine and the manuscripts of a number of different books. The times required to perform the printing and binding operation for each book are known. Determine the order in which the books should be processed in order to minimise the total time required to process all the books. Also, find the total time required.

	Processing time (in minutes)				
Book	1	2	3	4	5
Printing Time	40	90	80	60	50
Binding Time	50	60	20	30	40

(b) Suppose that an additional operation is added to the process described in (a), namely, finishing. The time required for the operations are given below:

	Finishing time (in minutes)				
Book	1	2	3	4	5
Finishing time	80	100	60	70	100

What is the order in which the books should be processed? Also, find the minimal total elapsed time.

[*Answer*: (a) Optimal sequence is 1 – 2 – 5 – 4 – 3; minimum elapsed time is 5 hours and 40 minutes.

(b) Optimal sequence is 5 – 1 – 4 – 2 – 3; minimum elapsed time is 8 hours and 30 minutes.]

22. A readymade garments manufacturer has to process 7 items through 2 stages of production, namely, cutting and sewing. The time taken for each of these at the different stages is given below in appropriate units:

	Processs time						
Item	1	2	3	4	5	6	7
Cutting time	5	7	3	4	6	7	12
Sewing time	2	6	7	5	9	5	8

(a) Find the order in which these items are to be processed through these stages so as to minimise the total processing time.

(b) Suppose a third stage of production is added, namely, pressing and packing with the following processing time:

Item	1	2	3	4	5	6	7
Processssing time for pressing and packing	10	12	11	13	12	10	11

Find the order in which these 7 items are to be processed so as to minimise the time taken to process all the items through all the three stages.

[*Answer*: (a) 3 – 5 – 7 – 2 – 6 – 1; minimum time is 46 hours.

(b) 1 – 4 – 3 – 6 – 2 – 5 – 7; minimum time is 86 hours.]

23. Find the sequence for the following 8 jobs that will minimise the total elapsed time for the completion of all jobs. Each job is processed in the order *CAB*.

Time for	*Job*							
machines	1	2	3	4	5	6	7	8
A	4	6	7	4	5	3	6	2
B	8	10	7	8	11	8	9	13
C	5	6	2	3	4	9	15	11

The entries give the time in hours on the machines.

[*Answer*: Optimal sequence is 4 – 5 – 3 – 1 – 2 – 8 – 7 – 6; minimum total elapsed time is 61 hours.]

24. A firm works 40 hours a week and has a capacity of overtime work to the extent of 20 hours in a week. It has received 7 orders to be processed on 3 machines *A*, *B* and *C* in the order *ABC*, to be delivered in a week's time. The process times (in hours) are recorded in the following Table:

Job	1	2	3	4	5	6	7
Machine *A*	7	8	6	6	7	8	5
Machine *B*	2	2	1	3	3	2	4
Machine *C*	6	5	4	4	2	1	5

The manager, who, in fairness, insists on performing the jobs in the sequence in which they are received is refusing to accept an eighth order, which requires 7, 2 and 5 hours, respectively on *A*, *B* and *C* machines, because according to him the eighth job would require a total of 61 hours for processing, which extends the firm's capacity. What is your advice to him?
[*Answer*: Optimal sequence is 7 – 1 – 4 – 8 – 2 – 3 – 5 – 6 or, 7 – 1 – 8 – 4 – 2 – 3 – 5 – 3 and minimum time is 57 hours.]

25. A company has 6 jobs which go through 3 machines *X*, *Y* and *Z* in the order *XYZ*. The processing time in minutes for each job on each machine is as follows:

Job	1	2	3	4	5	6
Machine *X*	18	12	29	36	43	37
Machine *Y*	7	12	11	2	6	12
Machine *Z*	19	12	23	47	28	36

What should be the sequence of jobs?
[*Answer*: Optimal sequence is 2 – 1 – 4 – 6 – 5 – 3 minimum time = 209 min.; idle time: 34 min. for *X*, 159 min. for *Y*, and 44 min. for *Z*.]

26. Find the sequence that minimises the total elapsed time (in hours) required to complete the follwing jobs on three machines M_1, M_2 and M_3 in the order M_1 M_2 M_3.

Job	*A*	*B*	*C*	*D*	*E*
Machine M_1	4	9	8	6	5
Machine M_2	5	6	2	3	4
Machine M_3	8	10	6	7	11

[*Answer*: Optimal sequence *A* – *D* – *E* – *B* – *C* or, *A* – *E* – *D* – *B* – *C* or, *D* – *A* – *E* – *B* – *C* or, *D* – *E* – *A* – *B* – *C* or, *E* – *D* – *A* – *B* – *C* or, *E* – *A* – *D* – *B* – *C*.
Minimum time: 51 hours; idle time: 9 hours for I, 31 hours for II, 19 hours for III.

27. There are 5 jobs, each of which is to be processed through 3 machines *A*, *B* and *C* in the order *ABC*. Processing time in hours are:

Job	*A*	*B*	*C*
1	3	4	7
2	8	5	9
3	7	1	5
4	5	2	6
5	4	3	10

Determine the optimum sequence for the 5 jobs and the minimum elapsed time.
[*Answer*: The optimal sequence : 4 – 1 – 5 – 2 – 3 or, 4 – 5 – 1 – 2 – 3 or, 1 – 4 – 5 – 2 – 3 or, 1 – 5 – 4 – 2 – 3 or, 5 – 1 – 4 – 2 – 3 or, 5 – 4 – 1 – 2 – 3. The minimum elapsed time is 44 hours; idle time: 17 hours for machine *A*, 29 hours for machine *B* and 7 hours for machine *C*.]

Section 8.4

28. Four jobs 1, 2, 3 and 4 are to be processed on each of the five machines *A*, *B*, *C*, *D* and *E* in the order *ABCDE*. Find the total minimum elapsed time if no passing of jobs is permitted.

Jobs \ *Machine*	*A*	*B*	*C*	*D*	*E*
1	7	5	2	3	9
2	6	6	4	5	10
3	5	4	5	6	8
4	8	3	3	2	6

[*Answer*: Optimal sequence is 1 – 3 – 2 – 4. The minimum elapsed time is 51 time units; Machines *A*, *B*, *C*, *D* and *E* remain idle for 25, 33, 37, 35 and 18 time units, respectively.]

29. Solve the following sequencing problem giving an optimal solution when no passing is allowed.

Machine \ *Job*	1	2	3	4	5
A	14	7	12	8	10
B	5	6	4	7	3
C	3	2	4	1	5
D	10	12	8	15	16

[*Ans*wer: Optimal sequence is 2 – 4 – 5 – 1 – 3. Minimum elapsed time is 76 time units; idle time: 25, 51, 61, 15 time units, respectively.]

30. Solve the following problem giving an optimal solution when passing is not allowed.

(a)

Job \ *Machine*	M_1	M_2	M_3	M_4	M_5
A	9	7	4	5	11
B	8	8	6	7	12
C	7	6	7	8	10
D	10	5	5	4	8

(b)

Job \ *Machine*	M_1	M_2	M_3	M_4
A	15	5	4	15
B	12	2	10	12
C	16	3	5	16
D	17	3	4	17

[*Answer*: (a) $A-C-B-D$, minimum time = 67 hours
(b) $D-A-B-C$, minimum time =25 hours
(c) 4! = 24 sequences, minimum time = 84 hours.]

31. Find the optimal sequence for processing the jobs J_1, J_2, J_3, J_4 on four machines M_1, M_2, M_3 and M_4, in that order. The processing time are given as under.

Job	*Machine*			
	M_1	M_2	M_3	M_4
J_1	25	15	14	24
J_2	22	12	20	22
J_3	23	13	16	25
J_4	26	10	13	29

[*Answer*: $J_4-J_3-J_2-J_1$; minimum elapsed time is 149 hours.]

32. Six jobs $J_1, J_2, \ldots, J_6$ are required to be processed on 4 machines M_1, M_2, M_3 and M_4, in that order. Using the processing time given below determine the optimal sequence(s) of the job performance.

job	*Machine*			
	M_1	M_2	M_3	M_4
J_1	12	6	6	8
J_2	8	3	6	12
J_3	13	4	6	9
J_4	9	2	4	14
J_5	10	6	7	12
J_6	12	7	2	10

[*Answer*: Optimal sequence is $J_4-J_2-J_5-J_1-J_3-J_6$ or, $J_4-J_2-J_5-J_1-J_6-J_3$.]

33. Solve the following sequencing problem giving an optimal solution if passing is not allowed.

		Machines			
		M_1	M_2	M_3	M_4
Jobs	A	13	8	7	14
	B	12	6	8	19
	C	9	7	8	15
	D	8	5	6	15

[*Answer*: Optimal sequence is $D-C-B-A$; minimum total elapsed time = 82 time units; idle time: 40 time units for M_1; 56 time units for M_2; 53 time units for M_3; and 19 time units for M_4.]

34. Find the optimal solution for the following sequencing problem:

	Jobs			
Machine	1	2	3	4
A	15	5	5	15
B	12	2	10	12
C	16	2	4	16
D	18	3	4	18

[*Answer*: Optimal sequence $C - B - D - A$ or, $C - B - D$; minimum total elapsed time = 86 time units. Idle time = 25 time units for M_1, 74 time units for M_2, 63 time units for M_3, 25 time units for M_4.]

Section 8.5

35. Use the graphical method to minimise the time added to process the following jobs on the machines shown, that is, for each machine find the job which should be done first. Also, calculate the total time elapsed to complete both the jobs.

Job 1	Sequence	*A*	*B*	*C*	*D*	*E*
	Time in hrs.	3	4	2	6	2
Job 2	Sequence	*B*	*C*	*A*	*D*	*E*
	Time in hrs.	5	4	3	2	6

[*Answer*: Minimum elapsed time = 22 hours; idle time for Job 1 = 5 hours; idle time for Job 2 = 2 hours.]

36. There are 2 jobs to be processed through 4 machines *A*, *B*, *C*, *D*. The technological ordering of the jobs is as follows:

Job 1 : *ABCD*
Job 2 : *DBAC*

Processing time in hours is in the following Table.

Job \ *Machine*	*A*	*B*	*C*	*D*
1	2	4	5	1
2	6	4	2	3

Calculate the total minimum elapsed time to complete the two jobs.
[*Answer*: Total minimum elapsed time = 21 hours.]

37. Two jobs are to be processed on four machines *A*, *B*, *C* and *D*. The technological order for these machines is as follows:

Job 1	:	*A*	*B*	*C*	*D*
Job 2	:	*D*	*B*	*A*	*C*

Processing time in hours are given in the following Table.

	A	B	C	D
Job 1	4	6	7	3
Job 2	4	7	5	8

Find the optimal sequence of jobs on each of the machines.
[*Answer*: Total minimum elapsed = 26 hours.]

38. A machine job has four machines *A*, *B*, *C* and *D*. Two jobs must be processed through each of these machines. The time (in hours) taken on each of the machines and the necessary sequence of jobs through the shop are given below:

Job 1	Sequence	A	B	C	D	E
	Time in hrs.	2	4	5	1	2
Job 2	Sequence	D	B	A	C	B
	Time in hrs.	6	4	2	3	6

Use graphic method to obtain the total minimum elapsed time.
[*Answer*: Minimum total elapsed time is = 15 hours.]

39. Two jobs are to be processed through four machines *A*, *B*, *C*, *D* with the following technological ordering.

Job 1 : A B C D
Job 2 : D B A C

Processing time (in hours) is given in the following Table:

	Machine			
	A	B	C	D
Job 1	20	40	50	10
Job 2	20	50	30	60

Find the minimum processing time and the idle time for both the jobs.
[*Answer*: Minimum elapsed time = 160 hours. Idle time for the chosen path is 40 for Job 1 and 0 for Job 2.]

Section 8.7

40. The maintenance crew of a company is divided into two groups, C_1 and C_2. Crew C_1 is responsible for replacement of parts which are worn out while Crew C_2 oils and resets the machines for operation. The time (in hours) required by crews C_1 and C_2 on different machines which need working on are follows:

Machine	M_1	M_2	M_3	M_4	M_5	M_6	M_7
Crew C_1	8	6	10	11	10	14	4
Crew C_2	5	3	7	12	8	6	7

In what order should the machines be handled by crews C_1 and C_2 so that the total time is minimised?
[*Answer*: The optimal sequence is M_7 M_4 M_5 M_3 M_6 M_1 M_2; the total minimum elapsed time is 66 hours.]

Section 8.9

41. Find the shortest route from a to i by using graphic method. The numbers on the links represent the distance in kilometres.

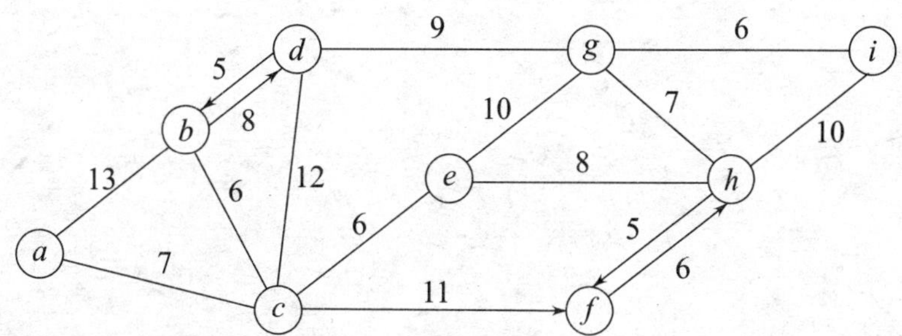

[*Answer*: $a - c - e - g - i$, 29 km.]

42. Find the shortest route from 0 to 8 by using the graphic method. The numbers on the links represent the distance in kilometres.

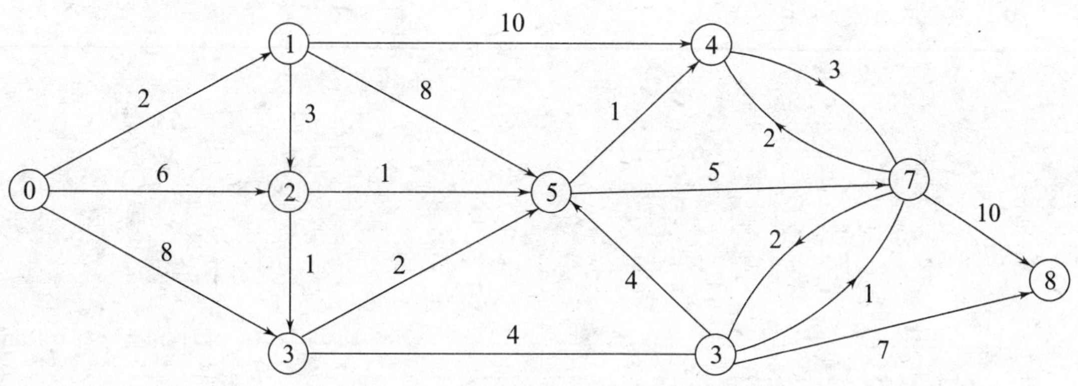

[*Answer*: Best route is 0 – 1 – 2 – 3 – 6 – 8; and the minimum distance is 17 km.]

9

Replacement Models

9.1 INTRODUCTION

The problem of replacement occurs when the job performing units like men, machines, equipment, and others become less effective or useless due to either sudden or gradual deterioration in their efficiency or failure or complete breakdown. For example, an electric light bulb fails all of a sudden, pipeline get blocked, parts of machines become faulty. These are some situations that need most economic replacement policy for replacing faulty units or to take some remedial to restore efficiency.

Suppose an item goes on performing and with decreasing efficiency, then it requires more money to be spent in order to increase the operating cost, repair cost and so on. In such a situation the replacement of an old item with new one is the only alternative. Thus, the problem of replacement is to decide the best policy to determine an age at which the replacement is most economical instead of continuous increase in cost. The need for replacement arises in many situations like:

(i) We may decide either to wait for complete failure of the item or to replace earlier due to high expense.

(ii) The expensive items may be considered individually to decide whether we should replace now or when it should be next replaced.

(iii) Whether the replaced item is of same type or of different type (latest model) of the item.

The main objective of replacement is to direct the organisation for maximising its profit (or minimising its cost).

Replacement problems can be classified as:

(i) When the equipment deteriorates with time due to constant use and needs increased operating and maintenance cost.

(ii) When equipments such as light bulbs, tubes, radio and television parts, and so on, do not give any indication of deterioration with time, but fail completely all of a sudden.

(iii) Existing working staff in an organisation reduces gradually due to retirement, death, and so on.

Failure Mechanism of Items

There are two types of failures:

(i) *Gradual failure*: It is progressive in nature, that is, the lifetime increases, but its efficiency deteriorates causing

(a) increased maintenance and operating costs,

(b) decreased productivity, and
(c) decrease in the value of the equipment, that is, resale or salvage value.

(ii) *Sudden failure*: This type of failure occurs after some period of service rather than deterioration while in service. The period of giving desired service is not constant but follows some frequency distribution which may be progressive, retrogressive or random in nature.

(a) *Progressive failure*: If the probability of failure increases with the increase in its life, then the failure is said to be progressive.
For example: electric light bulbs, automobile tubes.

(b) *Retrogressive failure*: If the probability of failure in the beginning of the life of an item is more and due to change of time, the chances of failure decreases, then the failure is said to be retrogressive.

(c) *Random failure*: The constant probability of failure is associated with items that fail from random causes like physical shocks not related to age. In such a case, virtually all items fail before ageing has any effect.

9.2 REPLACEMENT OF ITEMS THAT DETERIORATE WITH TIME

Generally, the cost of maintenance and repair of certain items increases with time. When years go by, these costs become so high that it is more economical to replace the item by a new one. In this situation, a replacement is necessary.

9.2.1 Value of Money Does Not Change with Time

Theorem

The cost of maintenance of a machine is given as a function increasing with time and its scrap value is constant.

(a) If time is a continuous variable, then the average annual cost will be minimised by replacing the machine when the average cost to date becomes equal to the current maintenance cost.

(b) If time is a discrete variable, then the average annual cost will be minimised by replacing the machine when the next period's maintenance cost becomes greater than the current average cost.

Proof: The aim is to find the optimum replacement age of an item whose running or maintenance cost increases with time and the value of money remains constant during the period. Let

C: capital cost of the item
S: scrap value of the item
n: number of years that the equipment would be in use
$f(t)$: maintenance cost function
$A(n)$: Average total annual cost

(a) *When t is a continuous variable*: If the item is used for n years, then the total cost used during this period is:

$$\text{Total cost} = \text{Capital cost} - \text{Scrap value} + \text{Maintenance cost}$$

$$= C - S + \int_0^n f(t)\,dt$$

Average annual total cost is given by

$$A(n) = \frac{\text{Total cost}}{n} = \frac{C-S}{n} + \frac{1}{n}\int_0^n f(t)\,dt$$

Now, we find such time n for which $A(n)$ is minimum. Therefore, differentiating $A(n)$ with respect to 'n', we get

$$\frac{d\,A(n)}{dn} = \frac{1}{n}f(n) - \frac{1}{n^2}\int_0^n f(t)\,dt - \frac{C-S}{n^2}$$

$$= 0 \text{ for minimum of } A(n)$$

which implies

$$f(n) = \frac{C-S}{n} + \frac{1}{n}\int_0^n f(t)\,dt$$

$$= A(n)$$

Note that $\frac{d^2}{dn^2}[A(n)] \geq 0$ at $f(n) = A(n)$.

(b) *When time t is a discrete variable*: Since the time t is taken as discrete, it can take the values 1, 2, 3, ...

Then,

$$A(n) = \frac{C-S}{n} + \frac{1}{n}\sum_{t=1}^{n} f(t)$$

By using finite differences $A(n)$ will be minimum if the relationship is satisfied.

$A(n+1) - A(n) \geq 0$ and $A(n) - A(n-1) \leq 0$

For this, we write

$$A(n+1) - A(n) = \left[\frac{C-S}{n+1} + \frac{1}{n+1}\sum_{t=1}^{n+1} f(t)\right] - \left[\frac{C-S}{n} + \frac{1}{n}\sum_{t=1}^{n} f(t)\right]$$

$$= \frac{1}{n+1}\left[C-S+\sum_{t=1}^{n} f(t) + f(n+1)\right] - \frac{1}{n}\left[C-S+\sum_{t=1}^{n} f(t)\right]$$

$$= \frac{f(n+1)}{n+1} + \sum_{t=1}^{n} f(t)\left[\frac{1}{n+1} - \frac{1}{n}\right] + (C-S)\left[\frac{1}{n+1} - \frac{1}{n}\right]$$

$$= \frac{f(n+1)}{n+1} - \frac{1}{n(n+1)}\left[\sum_{t=1}^{n} f(t) + (C-S)\right]$$

since $A(n+1) - A(n) \geq 0$,

$$\frac{f(n+1)}{n+1} \geq \frac{1}{n(n+1)}\left(\sum_{t=1}^{n} f(t) + (C-S)\right)$$

$$\Rightarrow \qquad f(n+1) \geq A(n).$$

Similarly, it can be shown that

$$A(n) - A(n-1) \leq 0 \Rightarrow f(n) \leq A(n-1)$$

This suggests the optimal replacement policy as:

Replace the machine at the end of n years when the maintenance cost in the $(n+1)$th year is more than the average total cost in the nth year and the nth year maintenance cost is less than the previous year's average total cost.

Example 1

The cost of a machine is Rs 6,100 and its scrap value is only Rs 100. The maintenance costs are found from experience to be:

Year	1	2	3	4	5	6	7	8
Maintenance cost in Rs	100	250	400	600	900	1,250	1,600	2,000

When should machine be replaced?

Solution: Find an average cost per year during the life of the machine. That is,

Total cost in a year

= Capital (or purchase) cost – Scrap value + Maintenance (running) cost.

We calculate the average cost per year during the life of the machine, which are given in Table 9.1.

Table 9.1

Replace at the end of year (n)	Maintenance cost f(n)	Total maintenance cost $\Sigma f(n)$	Difference between price and scrap value C – S	Total cost $\Sigma f(n) + C - S$	Average cost $\frac{\Sigma f(n)}{n} + \frac{C-S}{n}$
1	100	100	6,000	6,100	6,100
2	250	350	6,000	6,350	3,175
3	400	750	6,000	6,750	2,250
4	600	1,350	6,000	7,350	1,837
5	900	2,250	6,000	8,250	1,650
6	1,250	3,500	6,000	9,500	1,583
7	1,600	5,100	6,000	11,100	1,586
8	2,000	700	6,000	13,100	1,638

From Table 9.1, it may be observed that the average cost per year is minimum in the sixth year (Rs 1,583) and the maintenance cost in the seventh year (Rs 1,600) becomes greater than the average cost for six years. Hence, the machine should be replaced at the end of the 6th year.

Example 2

(a) Machine A costs Rs 9,000. Annual operating costs are Rs 200 for the first year, and then increase by Rs 2,000 every year. Determine the best age at which to replace the machine. If the optimum replacement policy is followed, what will be the average yearly cost of owning and operating the machine?

(b) Machine B costs Rs 10,000. Annual operating costs are Rs 400 for the first year, and then increase by Rs 800 every year. You now have a machine of type A which is one year old. Should you replace it with B; if so when?

Solution: (a) Let the machine have no scrap (resale) value when replaced. For Machine *A*, the average total cost is computed as follows:

Replace at the end of year (n)	Maintenance cost f(n)	Σ f(n)	C – S	Σ f(n) + (C – S)	$\frac{\Sigma f(n)}{n}+\frac{C-S}{n}$
1	200	200	9,000	9,200	9,200
2	2,200	2,400	9,000	11,400	5,700
3	4,200	6,600	9,000	15,600	5,200
4	6,200	12,800	9,000	21,800	5,450
5	8,200	21,000	9,000	30,000	6,000

From the Table, it follows that the Machine *A* may be replaced at the end of the 3rd year.

(b) For Machine *B*, we compute the average cost per year.

Replace at the end of year (n)	f(n)	Σ f(n)	C – S	f(n) + (C – S)	$\frac{\Sigma f(n)+C-S}{n}$
1	400	400	10,000	10,400	10,400
2	1,200	1,600	10,000	11,600	5,800
3	2,000	3,600	10,000	13,600	4,533
4	2,800	6,400	10,000	16,400	4,100
5	3,600	10,000	10,000	20,000	4,000
6	4,400	14,400	10,000	24,400	4,066

From the two Tables, we observe that the average cost for Machine *B* is lower than the average cost for Machine *A*. So Machine *B* should be replaced by Machine *A*.

Next, to decide the time of replacement we compare the minimum average cost for *B* (Rs 4,000) with yearly cost of maintaining and using Machine *A*. Since there is no resale value, we consider only the maintenance cost. Machine *A* is kept as long as the yearly maintenance cost is lower than Rs 4,000 and replace when it exceeds Rs 4,000.

On a one-year old Machine *A*, Rs 2,200 would be required for maintenance in the next year, and Rs 4,200 is needed for the following year. Since the maintenance cost in the third year is more than the lowest in average cost (Rs 4,000) of Machine *B*, Machine *A* should be replaced by Machine *B* after two years. As Machine *A* is one year old, the replacement should be one year from now.

Example 3

A taxi owner estimates from his past records that the cost per year for operating a taxi whose purchase price when new is Rs 60,000. These are given below.

Age in years	1	2	3	4	5
Operating cost (Rs)	10,000	12,000	15,000	18,000	20,000

After 5 years, the operating cost is Rs 6,000 K where K = 6, 7, 8, 9, 10, age in years. If the resale value decreases by 10% of purchase price each year, what is the optimum replacement policy? Cost of the money is zero.

Solution: 10% of purchase price = Rs 60,000 × $\frac{10}{100}$ = Rs 6,000.

Thus, the resale value decreases by Rs 6,000 every year. The average annual cost of the taxi is computed as follows.

Years (n)	Operating cost f (n)	Resale value S	C – S	Σ f(n) + (C – S)	$\frac{\Sigma f(n) + C - S}{n}$
1	10,000	54,000	6,000	16,000	16,000
2	12,000	48,000	12,000	34,000	17,000
3	15,000	42,000	18,000	55,000	18,333
4	18,000	36,000	24,000	79,000	19,750
5	20,000	30,000	30,000	1,05,000	21,000
6	36,000	24,000	36,000	1,47,000	24,500
7	42,000	18,000	42,000	1,95,000	27,857
8	48,000	12,000	48,000	2,49,000	31,125
9	54,000	6,000	54,000	3,09,000	34,333
10	60,000	0	60,000	3,75,000	37,500

The optimum period for replacement is one year.

Example 4

An auto owner finds from his past records that the cost per year of an auto whose purchase price is Rs 60,000 are:

Year	1	2	3	4	5	6	7	8
Running price	10,000	12,000	14,000	18,000	23,000	28,000	34,000	40,000
Resale price	30,000	15,000	7,500	3,750	2,000	2,000	2,000	2,000

(a) Determine at what age is its replacement due.

(b) Let the owner have 3 auto two of which are two years old and the third, is one year old. The cost price, running costs and resale value of these autos is same as given in Part (a). Now, the owner is considering a new type of autos with 50% more capacity than one of the old one at a unit price of Rs 80,000. He estimates that the running costs and resale price for the new auto will be as follows:

Year	1	2	3	4	5	6	7
Running cost	12,000	15,000	18,000	24,000	31,000	40,000	50,000
Resale price	40,000	20,000	20,000	5,000	3,000	3,000	3,000

Assuming the loss of flexibility due to the new autos is of no importance, the owner will continue to have sufficient work for three of the old autos, what should his policy be?

Solution: The computation of average cost per year during the life of the auto are:

Year (n)	Running cost Rs f (n)	Σ f(n)	C – S	Σ f(n) + (C – S)	$\frac{\Sigma f(n) + C - S}{n}$
1	10,000	10,000	30,000	40,000	40,000.00
2	12,000	22,000	45,000	67,000	33,500.00
3	14,000	36,000	52,500	88,500	29,500.00
4	18,000	54,000	56,250	1,10,250	27,562.00
5	23,000	77,000	58,000	1,35,000	27,000.00
6	28,000	1,05,000	58,000	1,63,000	27,166.00
7	34,000	1,39,000	58,000	1,97,000	28,142.85
8	40,000	1,79,000	58,000	2,37,000	29,625.00

It may be noted that the average cost per year is minimum in the 5th year. Hence, the autos should be replaced after 5th year.

Average Cost of New Auto

Year (n)	Running cost Rs f (n)	Σ f(n)	C – S	Σ f(n) + (C – S)	$\frac{\Sigma f(n) + C - S}{n}$
1	12,000	12,000	40,000	52,000	52,000.00
2	15,000	27,000	60,000	87,000	43,500.00
3	18,000	45,000	70,000	1,15,000	38,333.33
4	24,000	69,000	75,000	1,44,000	36,000.00
5	31,000	1,00,000	77,000	1,77,000	35,400.00
6	40,000	1,40,000	77,000	2,17,000	36,166.66
7	50,000	1,90,000	77,000	2,67,000	38,142.85
8	61,000	2,51,000	77,000	3,28,000	41,000.00

The new auto should be replaced after every 5th year. The average yearly cost would be Rs 35,400. Since the capacity of the new auto is 50% more than that of the old auto, therefore, capacity of two new autos is equivalent to three smaller autos. Thus, the minimum average yearly cost for the new auto will be equal to $35{,}400 \times \frac{2}{3}$ = Rs 23,600, which is less than average yearly cost Rs 27,000 for one of the old autos. Hence, the old, smaller auto should be replaced by a new auto.

The 3 old autos should be replaced with 2 new autos as the running cost for the next year of the 3 old autos exceeds the average yearly cost of 2 new autos.

The total cost of old autos in the subsequent years are as follows:

Years	Total cost of old autos
1	40,000
2	67,000 - 40,000 = 27,000
3	88,500 - 67,000 = 21,500
4	1,10,250 - 88,500 = 21,750
5	1,35,500 - 1,10,250 = 25,250
6	1,63,000 - 1,35,500 = 27,500

The average cost for 2 new autos will be 35,400 × 2 = Rs 70,800, for 3 old autos 27,000 × 3 = Rs 81,000. Further, it is given that two autos are two years old and one is one year old. Therefore, total average cost during first year will be 21,500 × 2 + 27,000 = Rs 70,000 as two of the autos would be running in the third year and the third one in the second year. Similarly, the average cost in the subsequent years will be

2 × 21,750 + 21,500 = Rs 65,000 (2nd year)

2 × 25,250 + 21,750 = Rs 72,250 (3rd year)

2 × 27,500 + 25,250 = Rs 80,250 (4th year)

The total cost of three old autos up to the third year is Rs 65,000 which is less than total average cost for new autos. Total cost for three old autos during the fourth year is Rs 72,250 which is more than the total average cost for new autos (Rs 70,800). Hence, all the three old autos should be replaced after two years of service without waiting for the replacement age of five years.

9.2.2 Value of Money Changes with Constant Rate During the Period

As money value changes with time, we calculate the *present value* or *present worth* of the money to be spent in a few years. One rupee a year from now is equivalent to $(1.1)^{-1}$ rupee at the interest rate of 10 per cent per year. One rupee spent two years from now is equivalent to $(1.1)^{-2}$ today. Hence, one rupee spent n years from now is equivalent to $(1.1)^{-n}$ today. The quantity $(1.1)^{-n}$ is called the *present worth factor* of one rupee spent in n years from now.

Generally, if r is the rate of interest per year, then $(1 + r)^{-n}$ is called the *present worth factor* of one rupee spent in n years time from now. The expression $(1 + r)^{n}$ is known as the *payment compound amount factor* of one rupee spent in n years time.

Discount rate or Depreciation value: The present worth factor of unit amount to be spent after one year is given by $v = (1 + r)^{-1}$, where r is the interest rate. Then, v is called *discount rate* or *depreciation value*.

Theorem

The maintenance cost increases with time and money value decreases at a constant rate, that is, depreciation value is given. Then, replacement policy will be:

(i) Replace the equipment if the next period's operating cost is greater than the weighted average of previous costs.

(ii) Do not replace the equipment if the next period's operating cost is less than the weighted average of previous costs.

Solution: Assume that

(i) the equipment in question has no resale value at the time of replacement, (ii) the maintenance costs are incurred in the beginning of different time periods.

Let

C = purchase price of new equipment

R_n = operating cost at the beginning of nth year ($R_{n+1} > R_n$)

r = annual interest rate

$v = (1 + r)^{-1}$ is the depreciation value per unit of money during a year

The present worth of all future costs of purchasing and operating the item with the policy of replacing the equipment at the end of each n years is given by

$$V_n = \{(C+R_1) + vR_2 + v^2R_3 + \ldots + v^{n-1}R_n\} \quad \text{(for 1 to } n \text{ years)}$$
$$+ \{(C+R_1)v^n + v^{n+1}R_2 + v^{n+2}R_3 + \ldots + v^{2n-1}R_n\} \quad \text{(for } n+1 \text{ to } 2n \text{ years)}$$
$$+ \{(C+R_1)v^{2n} + v^{2n+1}R_2 + v^{2n+2}R_3 + \ldots + v^{3n-1}R_n\} \quad \text{(for } 2n+1 \text{ to } 3n \text{ years)}$$
$$+ \ldots$$
$$= (C+R_1)[1 + v^n + v^{2n} + \ldots] + R_2v[1 + v^n + v^{2n} + \ldots] + \ldots + R_nv^{n-1}[1 + v^n + v^{2n} + \ldots]$$
$$= [C + R_1 + R_2v + \ldots + R_nv^{n-1}]\left[\frac{1}{1-v^n}\right] \quad \text{since } v < 1$$

Let $F(n) = C + R_1 + R_2v + \ldots + R_nv^{n-1}$

Then, $V_n = \dfrac{F(n)}{1-v^n}$ and $V_{n+1} = \dfrac{F(n+1)}{1-v^{n+1}}$

Now, V_n will be a minimum for that value of n, for which $V_{n+1} - V_n > 0$ and $V_{n-1} - V_n > 0$. For this, we write,

$$V_{n+1} - V_n = \frac{F(n+1)}{1-v^{n+1}} - V_n$$
$$= \frac{\left[C + \sum_{k=1}^{n+1} v^{k-1}R_k\right]}{1-v^{n+1}} - V_n$$
$$= \frac{v^n\,[R_{n+1} - (1-v)\,V_n]}{1-v^{n+1}}$$

and similarly,

$$V_n - V_{n-1} = \frac{v^{n-1}\,[R_n - (1-v)\,V_n]}{1-v^{n-1}}$$

Since v is the depreciation value of money, it will always be less than 1 and, therefore, $1 - v$ will always be positive. Hence, $\dfrac{v^n}{1-v^{n+1}} \geq 0$. Hence,

$$V_{n+1} - V_n > 0 \Rightarrow R_{n+1} > (1-v)\,V_n$$

and $\quad V_n - V_{n-1} < 0 \Rightarrow R_n < (1-v)\,V_n$

Thus, $\quad R_n < (1-v)V_n < R_{n+1}$

or,
$$R_n < \frac{C + R_1 + R_2v + R_3v^2 + \cdots + R_nv^{n-1}}{1 + v + v^2 + \cdots + v^{n-1}} < R_{n+1} \qquad (9.2.1)$$

as $\quad (1-v^n)(1-v)^{-1} = \sum_{k=0}^{n-1} v^k.$

The expression which lies between R_n and R_{n+1} is called the 'weighted average cost' of all the previous n years with weights $1, v, v^2, \ldots, v^{n-1}$, respectively. The value of n satisfying the relationship (9.2.1) will be the best replacement age of the equipment.

When the time value of money is not taken into consideration, the rate of interest becomes zero and hence v approaches unity. In this case,

$$R_{n-1} < \frac{C + R_1 + R_2 + \cdots R_n}{(1+1+\cdots+1)\ n \text{ times}} < R_n$$

Remarks: In case salvage is considered, then $W(n) = \dfrac{C + \Sigma R_n v^{n-1} - s_n v^n}{\Sigma v^{n-1}}$

Selection of Best Equipment: Suppose two machines M_1 and M_2 are given. The data required for determining the best replacement age of each machine is also given from past experience.

Step 1: First, find the best replacement age for machines M_1 and M_2 by using the relationship

$$R_n < (1 - v)V_n < R_{n+1}$$

Let the optimum replacement age for M_1 and M_2 be n_1 and n_2. Then,

Step 2: Compute the fixed annual payment (or weighted average cost) for each equipment by using the formula

$$W(n) = \frac{C + R_1 + R_2 v + \cdots + R_n v^{n-1}}{1 + v + v^2 + \cdots + v^{n-1}}$$

Substitute $n = n_1$ for equipment A, and $n = n_2$ for equipment B in $W(n)$.

Step 3: (i) If $W(n_1) < W(n_2)$, choose equipment A

(ii) If $W(n_1) > W(n_2)$, choose equipment B

(iii) If $W(n_1) = W(n_2)$, both equipments are equally good.

Example 1

Let the value of money be assumed to be 10 per cent per year and suppose that machine A is replaced after every 3 years whereas machine B is replaced after every six years. The yearly costs of both the machines are given below.

Year	1	2	3	4	5	6
Machine A	1,000	200	400	1000	200	400
Machine B	1,700	100	200	300	400	500

Determine which machine should be purchased.

Solution: Since money carries the rate of interest, the present worth factor is given by

$$v = \frac{100}{100 + 10} = \frac{10}{11}$$

Total discount cost (present worth) of A for 3 years is

$$= \text{Rs}\left[1{,}000 + 200 \times \frac{10}{11} + 400 \times \left(\frac{10}{11}\right)^2\right]$$

$$= \text{Rs } 1{,}512 \text{ (appx.)}$$

The total discount cost of B for six years is

$$= \text{Rs}\left[1{,}700 + 100 \times \left(\frac{10}{11}\right) + 200 \times \left(\frac{10}{11}\right)^2 + 300 \times \left(\frac{10}{11}\right)^3 + 400 \times \left(\frac{10}{11}\right)^4 + 500 \times \left(\frac{10}{11}\right)^5\right]$$

$$= \text{Rs } 2{,}765$$

$$\text{Average yearly cost of } A = \frac{1{,}513}{3} = \text{Rs } 504$$

$$\text{Average yearly cost of } B = \frac{2{,}765}{6} = \text{Rs } 461$$

This shows the apparent advantage with B, but the comparison is unfair because the periods of consideration are different. So, if we consider 6-year period for machine A, then the total discounted cost of A is

$$= 1{,}000 + 200 \times \left(\frac{10}{11}\right) + 400 \times \left(\frac{10}{11}\right)^2 + 1{,}000 \times \left(\frac{10}{11}\right)^3 + 200 \times \left(\frac{10}{11}\right)^4 + 400 \times \left(\frac{10}{11}\right)^5$$

$$= \text{Rs } 2{,}647$$

Hence, the average yearly cost of A is $\frac{2{,}647}{6} = 441$ which is lesser than the average yearly cost of B (461). Hence, machine A should be purchased.

Example 2

The initial cost of an item is Rs 15,000 and maintenance or running costs (in Rs) for different years are given below.

Year	1	2	3	4	5	6	7
Running cost	2,500	3,000	4,000	5,000	6,500	8,000	10,000

What is the replacement policy to be adopted if the capital is worth 10 per cent and there is no salvage value?

Solution: The present worth factor is given by

$$v = \frac{100}{100+10} = \frac{10}{11} = 0.9091$$

The following Table gives $W(n)$ for one rupee to be spent in n years time

Year (n)	Running cost R_n	v^{n-1} (pwf)	$R_n v^{n-1}$	$\Sigma R_n v^{n-1}$	$C + \Sigma R_n v^{n-1}$	Σv^{n-1}	$W(n)$
1	2,500	1000	2,500	2,500	17,500	1.000	17,500
2	3,000	0.9091	2,727	5,227	20,227	1.9091	10,595
3	4,000	$(0.9091)^2 = 0.8264$	3,306	8,533	23,533	2.7355	8,603
4	5,000	$(0.9091)^3 = .7513$	3,756	12,289	27,289	3.4868	7,826
5	6,500	$(0.9091)^4 = .6830$	4,440	16,729	31,729	4.1698	7,609
6	8,000	$(0.9091)^5 = 0.6209$	4,967	21,696	36,696	4.7907	7,660
7	10,000	$(0.9091)^6 = 0.5645$	5,645	27,341	42,341	5.3552	7,906

From the Table it is clear that $R_4 < W(5) < R_5$.

That is, Rs 6,500 < 7,609 < 8,000. The optimum replacement period is the 5th year.

Example 3

The cost pattern for two machines A and B, when money value is not considered is given below.

Year	Machine A	Machine B
1	900	1,400
2	600	100
3	700	700

Find the cost pattern for each machine when money is worth 10 per cent per year, and hence find which machine is less costly.

Solution: The total outlay for three years for machine $A = 900 + 600 + 700 =$ Rs 2,200. The same for machine $B = 1{,}400 + 100 + 700 =$ Rs 2,200. The total outlay for machines A and B are same (i.e. Rs 2,200) when money value is not considered. Hence, both the machines will appear to be equally good in this case.

Considering the money value at the rate of 10 per cent per year, the discount cost pattern for each machine for three years is:

Year	1	2	3	Total
Machine A	900.00	$600 \times \frac{100}{110} = 545.45$	$700 \times \left(\frac{100}{110}\right)^2 = 578.50$	Rs 2,023.95
Machine B	1,400.00	$100 \times \frac{100}{110} = 90.90$	$700 \times \left(\frac{100}{110}\right)^2 = 578.50$	Rs 2,069.40

The total outlay for machine A is less than that of machine B, hence machine A is preferred.

Example 4

The cost of a new machine is Rs 5,000. The running cost of nth year is given by $R_n = 500\,(n - 1)$, $n = 1, 2, \ldots$ Suppose that money is worth 5 per cent per year. After how many years will it be economical to replace the machinery with a new one?

Solution: The present worth factor for one year is given by

$$v = \left(1 + \frac{5}{100}\right)^{-1} = (1.05)^{-1} = 0.9523.$$

The following Table gives the values of $W(n)$.

Year (n)	R_n	v^{n-1}	$R_n v^{n-1}$	$\Sigma R_n v^{n-1}$	$C + \Sigma R_n v^{n-1}$	Σv^{n-1}	$W(n)$
1	0	1.000	0	0	5,000	1.000	5,000
2	500	0.9523	476	476	5,476	1.9523	2,805
3	1,000	0.9070	907	1,383	6,383	2.8593	2,232
4	1,500	0.8638	1,296	2,679	7,679	3.7231	2,063
5	2,000	0.8227	1,645	4,324	9,324	4.5458	2,061
6	2,500	0.7835	1,959	6,283	11,283	5.3293	2,117

Since $R_4 = 1{,}500 < W_5 = 2{,}061 < R_6 = 2{,}500$, it is economical to replace the machine with new one at the end of the 5th year.

Example 5

A pipeline is due for repair. It will cost Rs 10,000 and lasts for 3 years. Alternatively, a new pipeline can be laid at a cost of Rs 30,000 and that will last for 10 years. Assuming the cost of capital to be 10 per cent and ignoring salvage value, which alternative should be chosen?
Solution: Consider the two pipelines for infinite replacement cycles of 10 years for the new pipeline, and 3 years for the existing pipeline. The present worth factor is given by

$$v = \frac{100}{100+11} = 0.9091.$$

Let D_n denote the discounted value of all future costs associated with the policy of replacing equipment after n years. Suppose C is the initial outlay; then,

$$D_n = C + v^n C + v^{2n} C + \ldots.$$

$$= C\,[1 + v^n + v^{2n} \ldots] = \frac{C}{1-v^n}$$

Substitute the values of C, v, and n for the two types of pipelines. The discounted value for the existing pipeline is given by

$$D_3 = \frac{10{,}000}{1-(0.9091)^3} = \text{Rs } 4{,}021$$

For the new pipeline it is

$$D_{10} = \frac{30{,}000}{1-(0.9091)^{10}} = \text{Rs } 48{,}820$$

Since $D_3 < D_{10}$, the existing pipeline should be continued. Alternatively, the comparison may be made over $3 \times 10 = 30$ years.

Example 6

A manual stamper currently valued at Rs 1,000 is expected to last 2 years and costs Rs 4,000 per year to operate. An automatic stamper which can be purchased for Rs 3,000 will last 4 years and can be operated at an annual cost of Rs 3,000. If money carries the rate of interest 10 per cent per annum, determine which stamps should be purchased.

Solution: The present worth factor is $v = \frac{100}{110} = 0.9091$. The given two stampers have different expected lives. So, we shall consider a span of four years during which we have to purchase either two manual stampers or one automatic stamper. The present worth of investments on the two manual stampers used in 4 years is (nearly)

$$1{,}000\,(1 + v^2) + 4{,}000\,(1 + v + v^2 + v^3)$$

$$= 1{,}000\,[1 + (0.9091)^2] + 4{,}000\,[1 + (0.9091) + (0.9091)^2 + (0.9091)^3]$$

$$= \text{Rs } 15{,}773.$$

Also, the pwf on the automatic stamper for the next four years

$$= \text{Rs } 3{,}000 + 3{,}000\,(1 + v + v^2 + v^3)$$

$$= \text{Rs } 3{,}000 + 10{,}460$$

$$= \text{Rs } 13{,}460$$

Since the pwf of future costs of the automatic stamper is less than that of the manual stamper, it will be more profitable to purchase the former.

Example 7

A production machine has initial investment of Rs 30,000 and its salvage value at the end of i years of its use is estimated as Rs $\frac{30{,}000}{i+1}$ The annual operating and maintenance cost in the first year is Rs 15,000 and it increases by Rs 1,000 each subsequent years for the first five years and increases by Rs 5,000 each year thereafter. The replacement policy is to be planned over a period of 7 years. During this period, the cost of capital may be taken as 10 per cent per year. Solve the problem for optimal replacement.

Solution: Given $C = 30{,}000$ and pwf for one year as $\frac{1}{1+\frac{10}{100}} = 0.9091$. The following Table gives the value of $W(n) = \frac{C - s_n v^n + \Sigma R_n v^{n-1}}{\Sigma v^{n-1}}$

Year (n)	1	2	3	4	5	6	7
R_n	15,000	16,000	17,000	18,000	19,000	24,000	29,000
v^{n-1}	1.0000	0.9091	0.8264	0.7512	0.6829	0.6208	0.5643
$R_n v^{n-1}$	15,000	14,545.6	14,049.8	13,521.6	12,975.1	14,899.2	16,364.7
$\Sigma R_n v^{n-1}$	15,000	29,545.50	43,595.3	57,116.9	70,092.0	84,991.2	10,1355.9
Σv^{n-1}	1.0000	1.9091	2.7355	3.4867	4.1696	4.7904	5.3547
s_n	15,000	10,000	7,500	6,000	5000	4286	3750
$s_n v^n$	13,636.50	8,264.00	5,634.00	4,097.40	3,104.00	2,418.58	1923.76
$\frac{C - s_n v^n + \Sigma R_n v^{n-1}}{\Sigma v^{n-1}}$	31,363.50	26,861.61	24,844.19	23,810.33	23,260.74	23,499.62	24,171.68

Optimum replacement plan is after 5 years.

9.3 REPLACEMENT OF ITEMS THAT FAIL COMPLETELY AND SUDDENLY

It is not easy to predict when a particular equipment will fail and its time of failure. This difficulty can be overcome by determining the probability distribution of failures. Assume that the failures occur only at the end of the period, say, t. The objective is to find the value of t for which the total cost after replacement of an equipment is minimum.

We have the following replacement policies:

Individual Replacement Policy: Under this policy, an item is replaced immediately on its failure.

Group Replacement Policy: It is concerned with those items that either work or fail completely. It happens that a system contains a large number of items that are increasingly liable to failure with age. In this case it is advisable to replace all items irrespective of the fact that the items have failed or not failed, with a provision that if any item fails before optimal time, it may be individually replaced. Such a policy is called group replacement policy and is best when the value of any individual item is so small that the cost of keeping records of individual ages cannot be justified.

9.3.1 Theorem (Mortality)

A large population is subject to a given mortality law for a very long period of time. All deaths are immediately replaced by births and there are no other entries or exits. Then, the age distribution ultimately becomes stable and that the number of deaths per unit time becomes constant (which is equal to the size of the total population divided by the mean age at death).

Proof: Assume that death (or failure) occurs just before the age of $(k + 1)$ years, where k is an integer. That is, the lifespan of any item is between $t = 0$ to $t = k$. Define,

$f(t) =$ number of births (replacement) at time t

$p(x) =$ probability of death (failure) just before the age $x + 1$, that is, failure at the age x.

It is easy to see that $\sum_{x=0}^{k} p(x) = 1$.

If $f(t - x)$ represents the number of births at time $t - x$, $t = k, k + 1, k + 2, \ldots$, then the new born attains the age x at time t. For easy observation see the following diagram.

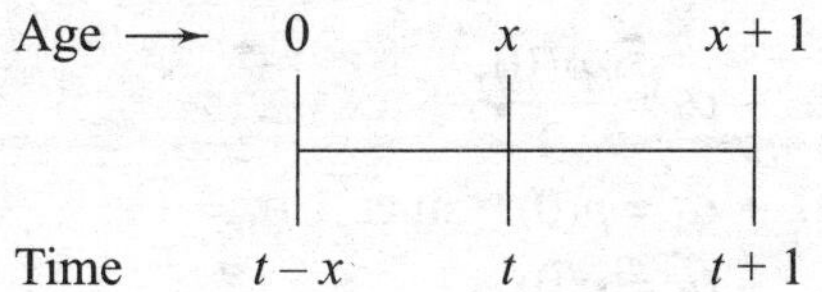

The expected number of deaths of such new borns before reaching the time $t + 1$ (i.e. at time t) is =

$\sum_{x=0}^{k} p(x) f(t - x)$, $t = k, k + 1, \ldots$

Since all deaths at time t (failures) are immediately replaced by births (replacements) at time $t + 1$, we have

$$f(t + 1) = \sum_{x=0}^{k} p(x) f(t - x), \quad t = k, k + 1, \ldots \tag{9.3.1}$$

The differenctial equation (9.3.1) in t may be solved by substituting

$$f(t) = A\alpha^t, \quad A \text{ is a constant.}$$

Then, equation (9.3.1) becomes,

$$A\alpha^{t+1} = A \sum_{x=0}^{k} p(x)\, \alpha^{t-x} \tag{9.3.2}$$

Dividing both sides of (9.3.2) by $A\alpha^{t-k}$, we get

$$\alpha^{k+1} = \sum_{x=0}^{k} p(x)\ \alpha^{k-x} = \alpha^k \sum_{x=0}^{k} p(x)\, \alpha^{-x}$$

$$= \alpha^k \{p(0) + p(1)\alpha^{-1} + p(2)\alpha^{-2} + \ldots + p(k)\, \alpha^{-k}\}$$

$$= \alpha^{k+1} - \{\alpha^k p(0) + \alpha^{k-1} p(1) + \ldots + p(k)\} = 0 \tag{9.3.3}$$

This is a linear homogeneous difference equation of degree $k+1$ and thus has exactly $k+1$ roots. Let the roots be $\alpha_0, \alpha_1, ..., \alpha_k$.

For $\alpha = 1$, the equation (9.3.3) becomes

$$1 - \{p(0) + p(1) + ... + p(k)\} = 0$$

$$\Rightarrow \quad 1 - 1 = 0 \quad \text{since } \sum_{x=0}^{k} p(x) = 1$$

Hence, $\alpha = 1$ is a root of equation (9.3.3). Let us denote it by $\alpha_0 = 1$. The general solution of equation (9.3.3) will take the form

$$f(t) = A_0\, \alpha_0^t + A_1 \alpha_1^t + ... + A_k\, \alpha_k^t$$

$$= A_0 + A_1 \alpha_1^t + ... + A_k\, \alpha_k^t \qquad (9.3.4)$$

where $A_0, A_1, ... A_k$ are constants whose values are to be determined.

Since one of the roots of equation (9.3.3), $\alpha_0 = 1$, is positive, by Descarte's sign rule which states, 'The number of positive roots of the polynomial equation $f(x) = 0$ cannot exceed the number of changes of sign (from +ve to –ve or from –ve to +ve) in the terms occurring in $f(x)$' the remaining roots $\alpha_1, \alpha_2, ..., \alpha_k$ will be negative.

By theory of equations, sum of all the roots of (9.3.3) is given by

$$\alpha_0 + \alpha_1 + \alpha_2 + ... + \alpha_k = \frac{p(0)}{1}$$

or, $$1 + \alpha_1 + \alpha_2 + ... + \alpha_k = p(0) \quad \text{since} \quad \alpha_0 = 1$$

or, $$\alpha_1 + \alpha_2 + ... + \alpha_k = p(0) - 1$$

$$|\alpha_1 + \alpha_2 + ... + \alpha_k| = |p(0) - 1| < 1 \quad \text{since} \quad 0 \le p(0) \le 1$$

$$\Rightarrow \quad |\alpha_1| + |\alpha_2| + ... + |\alpha_k| < 1$$

$\Rightarrow$ for each i, $$|\alpha_i| < 1, \quad i = 1, 2, ... k$$

Hence, $\alpha_1^t, \alpha_2^t, ... \alpha_k^t \to 0$ as $t \to \infty$. Consequently, equation (9.3.4) becomes $f(t) = A_0$, which shows that the number of deaths per unit time (as well as the number of births) becomes constant.

Next, we find the value of A_0. Let

$g(x)$ = probability of survivor for more than x years.

or, $g(x) = 1 - p$ {survivor will die before attaining the age x}

$$= 1 - \{p(0) + p(1) + ... + p(x-1)\}$$

Obviously, it is assumed that $g(0) = 1$

Since the number of births as well as deaths have become constant, each equal to A_0, the expected number of survivors of age x is given by $A_0 g(x)$.

As deaths are immediately replaced by births, the size N of the population remains constant. That is,

$$N = A_0 \sum_{x=0}^{k} g(x)$$

or, $$A_0 = \frac{N}{\sum_{x=0}^{k} g(x)} \qquad (9.3.5)$$

Now, the number of survivors aged 0, 1, 2, 3 ... can be computed from equation (9.3.5) as $A_0, A_0P(1)$, $A_0P(2)$, and so on.

Finally, if the denominator in (9.3.5) became the mean age at death, then the age distribution will become stable. For this,

$$\sum_{x=0}^{k} g(x) = \sum_{x=0}^{k} g(x)\,\Delta x \qquad \text{[since } \Delta(x) = (x+1) - x \text{ by finite differences]}$$

$$= [g(x)x]_0^{k+1} - \sum_{x=0}^{k} (x+1)\,\Delta g(x)$$

$$= g(k+1)\,(k+1) - g(0)0 - \sum_{x=0}^{k} (x+1)\,\Delta g(x)$$

$$= g(k+1)\,(k+1) - \sum_{x=0}^{k} (x+1)\,\Delta g(x) \qquad (9.3.6)$$

But, $\quad g(k+1) = 1 - \{p(0) + p(1) + \ldots + p(k)\}$

$$= 1 - 1 = 0.$$

Since no one can survive for more than $(k + 1)$ years of age and

$$\Delta g(x) = g(x+1) - g(x)$$

$$= \{1 - p(0) - p(1) - \ldots - p(x)\} - \{1 - p(0) - p(1) - \ldots - p(x-1)\}$$

$$= -\,p(x)$$

Now, equation (9.3.6) becomes,

$$\sum_{x=0}^{k} g(x) = 0 + \sum_{x=0}^{k} (x+1)\,p(x)$$

$$= \sum_{y=1}^{k+1} y\,p(y-1) \text{ by setting } x+1 = y$$

$$= \sum_{y=1}^{k+1} yp \quad \text{[age at death is } y\text{]}$$

$$= \text{Mean (expected) age at death}$$

Hence, $\quad A_0 = \dfrac{N}{\text{Mean age at death}}$

This completes the proof.

The rate of replacement and total cost involved in group replacement is based on the following theorem.

9.3.2 Theorem (Group Replacement Policy)

(a) Group replacement should be made at the end of the period t if the cost of individual replacements for the period t is greater than the average cost per period through the end of period, t.

(b) Group replacement is not advisable at the end of period t if the cost of individual replacements at the end of period $t - 1$ is less than the average cost per period through the end of tth period.

Proof: Let

N = total number of items in the system

$f(t)$ = number of items failing during time t

$C(t)$ = total cost of group replacement after time period t

C_1 = cost of replacing an item in a group

C_2 = cost of replacing an individual item

Then, clearly

$$C(t) = NC_1 + C_2\,[f(0) + f(1) + \ldots + f(t-1)]$$

The average cost of group replacement per unit period of time during a period t is

$$A(t) = \frac{C(t)}{t} = \frac{NC_1 + C_2\,[f(0) + f(1) + \cdots + f(t-1)]}{t} \tag{9.3.6}$$

Now, in order to determine the replacement age t, the average cost per unit period should be minimum. The condition for minimum of $A(t)$ is

$$\Delta A(t-1) < 0 < \Delta A(t)$$

Now, $\Delta A(t) = A(t+1) - A(t)$

$$= \frac{C(t+1)}{t+1} - \frac{C(t)}{t}$$

$$= \frac{C(t) + C_2 f(t)}{t+1} - \frac{C(t)}{t}$$

$$= \frac{tC_2 f(t) - C(t)}{t(t+1)}$$

$$= \frac{C_2 f(t) - \dfrac{C(t)}{t}}{(t+1)}$$

which must be greater than zero for minimum $A(t)$. That is,

$$C_2 f(t) > \frac{C(t)}{t}$$

Similarly, from $\Delta A(t-1) < 0$, we get

$$C_2 f(t-1) < \frac{C(t)}{t}$$

Thus, $$C_2 f(t-1) < \frac{C(t)}{t} < C_2 f(t)$$

Note that $$C_2 f(t) > \frac{C(t)}{t}$$

$$\Rightarrow \quad \frac{C_2 f(t)}{t+1} > \frac{C(t)}{t(t+1)}$$

$$\Rightarrow \quad \frac{C_2 f(t)}{t+1} + \frac{C(t)}{t+1} > \frac{C(t)}{t(t+1)} + \frac{C(t)}{t+1}$$

$$\Rightarrow \quad \frac{C(t) + C_2 f(t)}{t+1} > \frac{C(t)}{t+1}\left[\frac{1}{t} + 1\right]$$

$$\Rightarrow \quad \frac{C(t+1)}{t+1} > \frac{C(t)}{t}$$

Or, average cost for $(t + 1)$ periods > average cost for t periods.

Example 1

The following failure rates have been observed for a certain type of light bulbs:

Week	1	2	3	4	5
% failed by the end of week	10	25	50	80	100

There are 1,000 bulbs in use, and it costs Rs 2 to replace an individual bulb which has burnt out. If all bulbs were replaced simultaneously it would cost 50 paise per bulb. It is proposed to replace all bulbs at fixed intervals of time, whether or not they have burnt out, and to continue replacing burnt out bulbs as and when they fail. At what interval should all the bulbs be replaced? At what group replacement price per bulb would a policy of strictly individual replacement become preferable to the adopted policy?

Solution: Let p_i be the probability of a bulb which was new when placed in position for use, fails during the ith week of its life. Given 1,000 bulbs are in use. Then, we have

$$p_1 = \text{the prob. of failure in 1st week} = \frac{10}{100} = 0.10$$

$$p_2 = \text{the prob. of failure in 2nd week} = \frac{25-10}{100} = 0.15$$

$$p_3 = \text{the prob. of failure in 3rd week} = \frac{50-25}{100} = 0.25$$

$$p_4 = \text{the prob. of failure in 4th week} = \frac{80-50}{100} = 0.30$$

$$p_5 = \text{the prob. of failure in 5th week} = \frac{100-80}{100} = 0.20$$

Note that $\sum_{i=1}^{5} p_i = 1$, so p_6, p_7, p_8 ... equal zero.

Thus, a bulb that has already lasted four weeks is sure to fail during the fifth week.

Let N_i denote the number of replacements at the end of the ith week, if all 1,000 bulbs are new initially. Thus,

$$N_0 = \text{number of items in the beginning} = 1{,}000.$$

$$N_1 = N_0 p_1 = 1{,}000 \times 0.10 = 100$$

$$N_2 = N_0 p_2 + N_1 p_1 = 1{,}000 \times 0.15 + 100 \times 0.10 = 160$$

$$N_3 = N_0 p_3 + N_1 p_2 + N_2 p_1$$

$$= 1{,}000 \times 0.25 + 100 \times 0.15 + 160 \times 0.10 = 281$$

$$N_4 = N_0 p_4 + N_1 p_3 + N_2 p_2 + N_3 p_1 = 377$$

$$N_5 = N_0 p_5 + N_1 p_4 + N_2 p_3 + N_3 p_2 + N_4 p_1 = 350$$
$$N_6 = 0 + N_1 p_5 + N_2 p_4 + N_3 p_3 + N_4 p_2 + N_5 p_1 = 230$$
$$N_7 = 0 + 0 + N_2 p_5 + N_3 p_4 + N_4 p_3 + N_5 p_2 + N_6 p_1 = 286$$
$$\ldots \quad \ldots \quad \ldots$$

Observe that the expected number of bulbs burnt out in each week increases until the 4th week, then decreases until the 6th week and again starts increasing. Thus, the number will continue to oscillate and ultimately the system settles down to a steady state in which the proportion of bulbs failing each week is the reciprocal of their average life.

The expected life of each item

$$= \sum_{i=1}^{5} i p_i$$
$$= 1 \times 0.10 + 2 \times 0.15 + 3 \times 0.25 + 4 \times 0.30 + 5 \times 0.20$$
$$= 3.35 \text{ weeks.}$$

The number of failures in each week in steady state becomes $= \frac{1{,}000}{3.35} = 299$ (by mortality theorem). The cost of replacing bulbs only on failure = 2 × 299 (at the rate of Rs 2 per bulb) = Rs 598.

Since the replacement of all the 1,000 items simultaneously costs Rs 0.50 per bulb and replacement of an individual bulb on failure costs Rs 2, the average cost for different group replacement polices is shown below.

End of Week	Cost of individual replacement	Total cost of group replacement		Average cost per week
1	100 × 2 = 200	1,000 × 0.50 + 100 × 2	= Rs 700	Rs 700.00
2	160 × 2 = 320	700 + 160 × 2	= Rs 1,020	Rs 510.00
3	281 × 2 = 562	1,020 + 281 × 2	= Rs 1,582	Rs 527.30

The cost of individual replacement in the third week exceeds the average cost for two weeks. Hence, it is advisable to replace all the bulbs after every two weeks; otherwise the average cost will start increasing. Further, since the group replacement after one week costs Rs 700 and the individual replacement after one week costs Rs 598, individual replacement is preferable.

Example 2

Let $p(t)$ be the probability that a machine in a group of 30 machines would breakdown in period t. The cost of repairing a broken machine is Rs 200.00. Preventive maintenance is performed by the servicing team on all the 30 machines at the end T units of time. Preventive maintenance cost is Rs 15 per machine. Find optimal T which will minimise the expected total cost per period of servicing, given

$$p(t) = \begin{cases} 0.03 & \text{for } t = 1 \\ p(t-1) + 0.01 & \text{for } t = 2, 3, \ldots, 10 \\ 0.13 & \text{for } t = 11, 12, \ldots \end{cases}$$

Solution: Given

t	1	2	3	4	5	6	7	8	9	10	11	12
$p(t)$	0.03	0.04	0.05	0.06	0.07	0.08	0.09	0.10	0.11	0.12	0.13	0

as $p(1) + p(2) + \ldots\, p(11) = 0.88 < 1$, and if

we add $p(12) = 0.13$, then $p(1) + ... + p(12) = 1.01 > 1$, where the sum of all probabilities can never be greater than 1, so consider $p_{12} = 0$, $p_{13} = 0$, and so on.

This means that a machine which has already lasted up to the 11th period is sure to fail in the 12th period. Let N_i be the number of machines at the end of ith period. Then,

$$N_0 = 30$$
$$N_1 = N_0 p_1 = 30 \times 0.03 = 0.9 \approx 1$$
$$N_2 = N_0 p_2 + N_1 p_1 = 30 \times 0.04 + 1 \times 0.03 \approx 1.23 \approx 1$$
$$N_3 = N_0 p_3 + N_1 p_2 + N_2 p_1 = 30 \times 0.05 + 1 \times 0.04 + 1 \times 0.03 = 1.57 \approx 2$$

Similarly,

$$N_4 = 2, \quad N_5 = 2, \quad N_6 = 3, \quad N_7 = 3, \quad N_8 = 4, \quad N_9 = 4, \quad N_{10} = 5, \quad N_{11} = 6.$$

As the expected life of each machine, $\sum_{i=1}^{11} i p_i = 6.41$ time units, the average number of machines failed per period is $\frac{30}{6.41} = 5$ (nearly). Hence, cost of individual replacement = Rs 5×200 = Rs 1,000.

Group maintenance cost is computed as:

End of period	Cost of group maintenance (in Rs)	Average cost of group maintenance per period (in Rs)
1	$(30 \times 15) + 1 \times 200$ = 650	650
2	$(30 \times 15) + 2 \times 200$ = 850	425
3	$(30 \times 15) + 4 \times 200$ = 1,250	417
4	$(30 \times 15) + 6 \times 200$ = 1,650	412
5	$(30 \times 15) + 8 \times 200$ = 2,050	410
6	$(30 \times 15) + 11 \times 200$ = 2,650	442

Since the minimum cost occurs in the 5th period it is optimal to maintain all the machines up to the 5th period.

Example 3

There are a large number of electric bulbs, all of which must be kept in working order. If an electric bulb fails in service, it costs Re 1 to replace it, but if all the electric bulbs are replaced in the same operation, it costs only 35 paise an electronic bulb. If the proportion of bulbs failing in successive time intervals is known, decide on the best replacement policy and give reasons. The following mortality rates for light bulbs have been observed:

Proportion failing during the first week = 0.09
Proportion failing during the second week = 0.16
Proportion failing during the third week = 0.24
Proportion failing during the fourth week = 0.36
Proportion failing during the fifth week = 0.12
Proportion failing during the sixth week = 0.03

Solution: Let the number of an electronic bulbs initially be N_0 = 10,000 (say). Let p_i denote the probability of failure of bulbs during ith week. Then,

$p_1 = 0.09$, $p_2 = 0.16$, $p_3 = 0.24$, $p_4 = 0.36$, $p_5 = 0.12$, $p_6 = 0.03$.

Let N_i denote the number of replacements at the end of ith week. Then,

$$N_0 = 10{,}000$$
$$N_1 = N_0 p_1 = 10{,}000 \times 0.09 = 900$$
$$N_2 = N_0 p_2 + N_1 p_1 = 10{,}000 \times 0.16 + 900 \times (0.09) = 1{,}681$$
$$N_3 = N_0 p_3 + N_1 p_2 + N_2 p_1 = 10{,}000 \times (0.24) + 900 \times (0.16) + 1{,}681 \times (0.09) = 2{,}695$$

Similarly,

$$N_4 = 4{,}327;\ N_5 = 2{,}747;\ N_6 = 2{,}599$$

As the expected life of each bulb is $\sum_{i=1}^{6} i\,p_i = 3.35$, the average number of failures per week $= \dfrac{10{,}000}{3.35}$ = 2,985 (nearly).

The cost of individual replacement = 2,985 × 1 = Rs 2,985

Group replacement cost is computed as:

End of week	*Cost of individual replacement*	*Total cost of group replacement*	*Average cost Rs*
1	900	(10,000 × 0.35) + (900 × 1) = 4,400	4,440
2	900 + 1,681 = 2,581	(10,000 × 0.35) + (2581 × 1) = 6,081	3,040.50
3	900 + 1,681 + 2,695 = 5,276	(10,000 × 0.35) + (5276 × 1) = 8,776	2,925.33
4	900 + 1,681 + 2,695 + 4,327 = 9,603	(10,000 × 0.35) + (9603 × 1) = 13,103	3,275.75
5	12,350	(10,000 × 0.35) + (12350 × 1) = 15,850	3,170.00

From the Table, it follows that the optimal policy is to have group replacement for every 3rd week. Also, average cost is Rs 2,925, which is less than Rs 2,985 for individual replacements. Therefore, the policy of group replacement is preferred.

9.4 OTHER REPLACEMENT PROBLEMS

Like industrial replacement problems, we can see a few others, like problems of recruitment and promotion of staff and equipment renewal.

9.4.1 Recruitment and Promotion Problems

The principles of replacement may be applied to formulate some useful recruitment and promotion polices for the staff working in an organisation. We assume the life distribution for the service of staff in the organisation is already known.

Example 1

An airline requires 200 assistant hostesses, 300 hostesses, and 50 supervisors. Girls are recruited at age 21 and if still in service, retire at the age of 60. Given the following life table, determine:

(i) How many girls should be recruited each year?
(ii) At what age should promotion take place?

Airline Hostesses' Life Record

Age	21	22	23	24	25	26	27	28
No. of service	1,000	600	480	384	307	261	228	206
Age	29	30	31	32	33	34	35	36
No. of service	190	181	173	167	161	155	150	146
Age	37	38	39	40	41	42	43	44
No. of service	141	136	131	125	119	113	106	99
Age	45	46	47	48	49	50	51	52
No. of service	93	87	80	73	66	59	53	46
Age	53	54	55	56	57	58	59	–
No. of service	39	33	27	22	18	14	11	

Solution: The total number of girls recruited at the age of 21 and those serving up to the age of 59 is 6,480. The total number of women recruited in the airline are 200 + 300 + 50 = 550.

The recruitment of girls every year is 1,000 when total number of girls are 6,480 after 59 years. Therefore, in order to maintain a strength of 550 hostesses, the recruitment policy should be

$550 \times \dfrac{1{,}000}{6{,}480} = 85$ (nearly) every year.

If the assistant hostesses are promoted at the age of x, then up to age $(x - 1)$, 200 assistant hostesses will be required. Among 550, there are 200 assistant hostesses. Therefore, out of 1,000 hostesses there will be $200 \times \dfrac{1{,}000}{550} = 364$ assistant hostesses. From the Table we observe that the number is available up to the age of 24 years. Hence, the promotion of assistant hostesses will be due in the 25th year. Also, out of 550 recruitments, we need only 300 hostesses. Therefore, if 1,000 girls are recruited, then we require $1{,}000 \times \dfrac{300}{550} = 545$ (nearly) hostesses. Hence, the number of hostesses and assistant hostesses in a recruitment of 1,000 will be 545 + 364 = 909. So, only 1,000 – 909 = 91 supervisors are needed. But from the life table we can see this number is available up to the age of 46 years. Thus, promotion of hostesses to supervisors will be due in the 47th year.

Example 2

A research team is planned to be raised to a strength of 50 chemists and then to remain at that level. The wastage of recruits depends on their length of service which is as follows:

Year	1	2	3	4	5	6	7	8	9	10
Total % who have left upon the end of year	5	36	55	63	68	73	79	87	97	100

What is the recruitment per year necessary to maintain the required strength? There are 8 senior posts for which the length of service is the main criterion. What is the average length of service after which a new entrant expects promotion to one of the senior posts.

Solution: We can calculate the probability of a person p_i in service at the end of ith year as follows:

Year	No. of persons who left at the end of the year	No. of persons in service at the end of the year (q_n)	Prob. of leaving at the end of the year	Prob. of in service at the end of year (p_n)
0	0	100	0	1.00
1	5	95	0.05	0.95
2	36	64	0.36	0.64
3	56	44	0.56	0.44
4	63	37	0.63	0.37
5	68	32	0.68	0.32
6	73	27	0.73	0.27
7	79	21	0.79	0.21
8	87	13	0.87	0.13
9	97	3	0.97	0.03
10	100	0	1.00	0

From the Table, we observe that if 100 chemists are appointed each year, then the total number of chemists serving the organisation is 436.

Thus, in order to maintain a strength of 50 chemists, we must recruit $100 \times \frac{50}{436} = 12$ chemists every year.

Out of 12 recruits, the total number of survivals at the end of year n will be $12p_n$. Now,

Years (n)	0	1	2	3	4	5	6
p_n	1.00	0.95	0.64	0.44	0.37	0.32	0.27
No. of chemists $12\,p_n$	12	11	8	5	4	4	3
Years (n)	7	8	9	10			
p_n	0.21	0.13	0.03	0			
No. of chemists $12\,p_n$	2	2	0	0			

Given there are 8 senior posts for which the length of service is the main criterion. From the Table we observe that there are 3 persons in service during 6th year, 2 persons in 7th year, and 2 persons in the 8th year. Altogether, there are 7 persons in service from the 6th year to the 8th year, which is less than the total number of senior posts. Hence, the promotions of the new entrants will start by the end the 5th year.

Example 3

An airline, whose staff are subject to the same survival rates as in Example 1, presently has a staff whose ages are distributed in the following Table. It is estimated that for the next two years staff requirement will increase by 10 per cent per year for supervisors. If girls are to be recruited at the age of

21, how many should be recruited for next year and at what age will promotions take place ? How many should be recruited for the following year and at what age will promotions take place?

Assistant											
Age	:	21	22	23	24	25					
Number	:	90	50	30	20	10	(Total 200)				
Hostesses	:										
Age	:	26	27	28	29	30	31	32	33	34	
Number	:	40	35	35	30	28	26	20	18	16	
Age	:	35	36	37	38	39	40	41			
Number	:	12	10	8	–	8	8	6	(Total 300)		
Supervisors											
Age	:	42	43	44	45	46	47	48	49	50	
Noumber	:	5	4	5	3	3	3	6	2	–	
Age	:	51	52	53	54	55	56	57	58	59	
Number	:	–	4	3	5	–	3	2	–	2	(Total 50)

Solution: Compute the expected number of survivals from supervisors for one year by making use of the probability of survival for one year at each age given in the following Table. Note that the probability of being in service for one year and more at the age of 42 is calculated as

$$\frac{\text{No. in service at age 42}}{\text{No. in service at age 41}} = \frac{113}{119} = 0.949$$

Therefore, expected survivals after one year = 5 × 0.949 = 4.745. Hence, we form the Table.

Present age	No. of supervisors	Prob. of being in service one year hence	Expected survivals after one year	New age
42	5	$\frac{113}{119} = 0.949$	4.745	43
43	4	$\frac{106}{113} = 0.938$	3.752	44
44	5	$\frac{99}{106} = 0.933$	4.669	45
45	3	$\frac{93}{99} = 0.939$	2.818	46
46	3	$\frac{87}{93} = 0.935$	2.806	47
47	3	$\frac{80}{87} = 0.919$	2.758	48
48	6	$\frac{73}{80} = 0.912$	5.475	49

Present age	No. of supervisors	Prob. of being in service one year hence	Expected survivals after one year	New age
49	2	$\frac{66}{73} = 0.904$	1.808	50
50	0	$\frac{59}{66} = 0.893$	0	51
51	0	$\frac{53}{59} = 0.898$	0	52
52	4	$\frac{46}{53} = 0.868$	3.472	53
53	3	$\frac{39}{46} = 0.847$	2.541	54
54	5	$\frac{33}{39} = 0.846$	4.23	55
55	0	$\frac{27}{33} = 0.818$	0	56
56	3	$\frac{22}{27} = 0.814$	2.442	57
57	2	$\frac{18}{22} = 0.818$	1.636	58
58	0	$\frac{14}{18} = 0.777$	0	59
59	2	$\frac{11}{14} = 0.785$	1.57	60
	50		$44.7 \simeq 45$	

From this Table we observe that only 45 supervisors will be in service after one year. Also, there is 10 per cent increase in the posts. That is, there will be $50 \times 10\% = 5$ new posts of supervisors after one year and hence 55 supervisors are needed after one year. But only 45 will remain in service. Hence, 10 hostesses are to be promoted on the basis of their age. Since there are 6 hostesses of age 41 and their probability of survival for one year is $\frac{119}{125} = 0.952$, the expected number of hostesses aged 41 who will be in service for one year more (i.e. up to age of 42 years) $= 6 \times 0.952 = 5.712$.

Similarly, there are 8 hostesses of age 40 years with probability of survival for one year being 0.954; the expected number of hostesses aged 40 who will be in service for one year more is $8 \times 0.954 = 7.633$.

Thus, after one year from present age (5.712 + 7.633) = 13.3 senior hostesses will be of age 42 and 41 years and we want to promote 10 out of them. Hence, all hostesses of age 41 and all but 3 of age 40 will be promoted as supervisors after one year.

EXERCISES

Section 9.1

1. What is a replacement problem? Describe some important replacement situations.
2. What are the situations which make the replacement of items necessary?
3. Explain with examples the failure mechanism of items.
4. Enumerate various replacement problems.
5. Discuss in brief the replacement procedure for items that deteriorate with time.
6. Give a brief account of the situations in which replacement problems arise. What does the theory of replacement establish?

Section 9.2

7. A firm is considering replacement of a machine whose cost price is Rs 12,200 and scrap value is Rs 200. The maintenance costs are found from experience to be as follows:

Year	1	2	3	4	5	6	7	8
Maintenance cost (Rs)	200	500	800	1,200	1,800	2,500	3,200	4,000

When should the machine be replaced?
[*Answer*: Replace at the end of the 6th year.]

8. A machine costs Rs 10,000. Operating costs are Rs 50 per year for the first five years. In the sixth and succeeding years operating costs increase by Rs 100 per year. Find the optimum length of time to hold the machine before replacing it.
[*Answer*: After 15 years.]

9. A truck with first cost of Rs 80,000 has depreciation and service pattern as shown below:

Year	1	2	3	4	5	6
Depreciation during year	28,000	20,000	14,000	5,000	4,000	4,000
Annual service cost	18,000	21,000	25,000	29,000	34,000	40,000

Assume no interest charges are necessary for the evaluation. How many years should the truck be kept in service before replacement?
[*Answer*: 6 years]

10. Machine *A* costs Rs 3,600. Annual operating costs are Rs 40 for the first year and then increase by Rs 360 every year. Assuming that machine A has no resale value, determine the best replacement age.
Another machine *B*, which is similar to machine *A*, costs Rs 4,000. Annual running cost are Rs 200 for the first year and then increase by Rs 200 every year. It has resale value of Rs 1,500, Rs 1,000 and Rs 500 if replaced at the end of first, second, third years, respectively. It has no resale value from the fourth year onwards.
Which machine would you prefer to purchase? Future costs are not to be discounted.
[*Answer*: Replace *A* at the end of 5th year, *B* at the end of 6th year; machine *B* should be purchased.]

11. A truck owner from his past experience estimated that the maintenance cost per year of a truck whose purchase price is Rs 1,50,000 and the resale value of truck will be as follows.

Year	1	2	3	4	5	6	7	8
Maintenance cost (Rs)	10,000	50,000	20,000	25,000	30,000	40,000	45,000	50,000
Resale value (Rs)	1,30,000	1,20,000	1,15,000	1,05,000	90,000	75,000	60,000	50,000

Determine at which time it is profitable to replace the truck.
[*Answer*: Though the minimum value of the average cost occurs at the end of the first year, it is impractical to replace the truck every year. So, the next minimum occurs at the end of the 4th year. Hence, it is profitable to replace the truck after every four years.]

12. Madras Cola Inc. uses a bottling machine that cost Rs 50,000 when new. The Table below gives the expected operating costs per year, the annual expected production per year and the salvage value of the machine. The wholesale price for a bottle of drink is Re 1.00.

Age	1	2	3	4	5
Operating cost (Rs)	7,000	8,000	10,000	14,000	20,000
Production cost (Rs)	2,08,000	2,08,000	2,00,000	1,90,000	1,75,000
Resale cost (Rs)	30,000	19,000	15,000	12,000	10,000

When should the machine be replaced?
[*Answer*: At the end of the 4th year.]

13. Explain the terms (a) present worth factor, (b) discount rate.

14. The cost of a new machine is Rs 5,000. The maintenance cost of nth year is given by $R_n = 500\ (n-1), n = 1, 2, \ldots$ Suppose that the discount rate per year is 0.5. After how many years will it be economical to replace the machine by a new one?

[*Answer*: $d = 0.9523 = \dfrac{1}{1+0.05}$; Replace the machine at the end of 5th year, $W(n) =$ Rs 2051.]

15. A scooter costs Rs 6,000 when new. The running cost and salvage value at the end of different years are as follows:

Year	1	2	3	4	5	6	7
Running cost	1,200	1,400	1,600	1,800	2,000	2,000	2,400
Salvage	4,000	2,666	2,000	1,500	1,000	600	600

If the interest rate is 10 per cent per year and running costs are assumed to have occurred mid-year, find when the scooter should be replaced.
[*Hint*: Discount the running cost to the start of the year using $\sqrt{v} = 0.95346$.]
[*Answer*: End of the 6th year.]

16. A truck has been purchased at a cost of Rs 1,60,000. The value of the truck is depreciated in the first three years by Rs 20,000 each year and Rs 16,000 per year thereafter. Its maintenance and operating costs for the first three years are Rs 16,000, Rs 18,000 and Rs 20,000 in that order and increase by Rs 4,000 every year. Assuming an interest rate of 10 per cent find the economic life of the truck.
[*Answer*: End of the 11th year.]

17. The yearly cost of two machines A and B, when money value is neglected is shown below. Find their cost patterns if money is worth 10 per cent per year and hence find which machine is more economical.

Year	1	2	3
Machine A (Rs)	1,800	1,200	1,400
Machine B (Rs)	2,800	1,200	1,400

[*Answer*: Machine A is more economical.]

18. If we wish to have a minimum rate of return of 10 per cent per annum on our investment, which of the following two plans should we prefer?

	Plan A	Plan B
First cost	75,000	75,000
Estimated scrap value after 20 years	37,500	6,000
Excess of annual receipts over annual disbursements	7,500	9,000

Pwfs at 10 per cent for 20 years = 8.514.

Section 9.3

19. Find the cost per period of individual replacement policy of an installation of 300 electric bulbs given the following:
(i) Cost of replacing individual bulb is Rs 3.
(ii) Conditional probability of failure is given below:

Week	0	1	2	3	4
Conditional prob. of failure	0	1/10	1/3	2/3	1

[*Answer*: Rs 206.]

20. A computer contains 10,000 resistors. When any resistor fails it is replaced. The cost of replacing a resistor individually is Re 1. If all the resistors are replaced at the same time, the cost per resistor would be reduced to 35 paise. The percentage of surviving resistors, say, $s(t)$ at the end of month t and $p(t)$, the probability of failure during the month t are:

t	0	1	2	3	4	5	6
$s(t)$	100	97	90	70	30	15	0
$p(t)$	–	0.03	0.07	0.20	0.40	0.15	0.15

What is the optimum replacement plan?
[*Answer*: Group replacement at the end of 3rd month.]

21. A factory has a large number of bulbs, all of which must be in working condition. The mortality of bulbs is given in the following table:

Week	1	2	3	4	5	6
Proportion of bulbs failing	0.10	0.15	0.25	0.35	0.12	0.03

If a bulb fails in service, it costs Rs 3.50 to be replaced; but if all the bulbs are replaced at a time it costs Rs 1.20 each, find the optimum group replacement policy.
[*Answer*: After every third week.]

22. It has been suggested by a data processing firm that a company adopt the policy of periodically replacing all the tubes in a certain piece of equipment. A given type of tube is known to have mortality distribution as shown in the following table.

Tube failures/week	1	2	3	4	5
Prob. of failure	0.3	0.1	0.1	0.2	0.3

The cost of replacing the tubes on an individual basis is estimated to be Re 1.00 per tube and the cost of a group replacement policy average Re 0.30 per tube. Compare the cost of preventive replacement with that of remedial replacement.
[*Answer*: It is optimal to have a group replacement after every fourth week along with individual replacements.]

23. An electric company which generates and distributes electricity conducted a study on the life of poles. The appropriate data are given in the following Table:

Years after Installation	1	2	3	4	5	6	7	8	9	10
Percentage poles failing	1	2	3	5	7	12	20	30	16	4

(i) If the company now installs 5,000 poles and follows a policy of replacing poles only when they fail, how many poles are expected to be replaced each year during the next 10 years? To simplify the computation assume that failures occur and replacements are made only at the end of a year.

(ii) The cost of replacing individual poles is Rs 160. If we have a common group replacement policy, it costs Rs 80 per pole. Find the optimal period for group replacement.

[*Answer*: (i) 5,533 poles (ii) Replace all poles after 6th year.]

Section 9.4

24. A survey agency requires 150 investigators, 225 senior investigators, and 40 supervisors. The persons recruited must be at least 18 years old. The retirement age is 58 years. Given the following Table, determine the,

(i) Number of persons to be recruited each year.

(ii) Age at which promotions should take place.

Age :	18	19	20	21	22	23	24	25			
No. in service :	500	300	240	192	154	130	114	103			
Age :	26	27	28	29	30	31	32	33	34	35	
No. in service :	95	90	87	83	80	77	75	73	70	68	
Age :	36	37	38	39	40	41	42	43	44	45	46
No. in service :	65	62	60	67	53	50	47	44	40	37	33
Age :	47	48	49	50	51	52	53	54	55	56	57
No. in service :	30	26	23	19	16	14	11	9	7	6	3

[*Answer*: (i) 64 persons (ii) Promotion of investigator will be due in 22nd year. Promotion of supervisor will be due in 42nd year.]

25. Suppose the life X of electric bulbs follows the gamma distribution

$$P(x \leq X \leq x + \Delta x) = \frac{a^p}{\Gamma(p)} e^{-ax} x^{p-1} \Delta x$$

$$a > 0, 0 \leq x < 0$$

Determine the renewal rate for one point at the end of time period $(0, t)$.

$$\left[\textit{Answer: } h(t) = e^{-at} \sum_{r=1}^{\infty} \frac{a^{rp} t^{rp-1}}{\Gamma(rp)} \right]$$

26. Suppose the life of electric bulbs follows the following distribution:

$$f(x) = \lambda e^{-\lambda x}$$

Determine the renewal density $h(x)$ after the end of the period $(0, t)$. If there are N points in the house, how many bulbs would you expect to be replaced within a period (t_1, t_2) where $t_2 > t_1$?

$$\left[\textit{Answer: } h(x) = \lambda \quad h(t_2 - t_1) = N \int_{t_1}^{t_2} h(y)dy \right]$$

27. Write short note on staffing problem.
28. Name any two uses of the replacement model.
29. What is the optimal replacement policy when value of money changes and maintenance and repair costs increases with time?

10

Inventory Models

10.1 INTRODUCTION

Inventory is defined as any idle resource of an enterprise. It is a stock of goods, commodities or economic resources that are kept for the purpose of future business affairs. Inventory may be kept in any form—raw materials in process, finished product, packaging, spares and others—stocked in order to meet the expected demand. It is important to maintain some inventories for the smooth functioning of an enterprise. Suppose a firm is not maintaining inventories; if a sales order comes, it has to purchase the raw materials required, wait till these arrive and then start production. This increases the waiting time of customers to get the delivery goods.

The disadvantages of not maintaining inventories are: raw materials are purchased at a high price because of piecemeal buying; production costs would be high because of not being able to take advantage of batching, and so on.

Meaning of Inventory Control

Inventory control may be defined as 'the function of directing the movement of goods through the entire manufacturing cycle from the requisitioning of raw materials to the inventory of finished goods in an orderly manner to meet the objectives of maximum customer-service with minimum investment and low-cost plant operation.'

There are two basic functions.

(i) Maintaining an accounting record to handle the inventory transactions concerning each inventory item. For inventory transactions, a record-keeping system called *Kardex file* is maintained for each inventory item.

(ii) Deciding inventory replenishment decisions. There are two basic replenishment decisions:

  (a) When is it necessary to place an order (or produce) to replenish inventory?
  If the demand of an item is independent of that of other items, then the recorder point technique can be used to know the time of replenishment.

  (b) How much is to be ordered (or produced) in each replenishment?

The decision about the number of units to order (or produce) for replenishment depends on the types of inventory costs.

Reasons for Carrying Inventories

It is essential for any firm to have inventory because of the following reasons:

1. It provides adequate service to customers.
2. It helps in smooth and efficient running of business.
3. It reduces the possibility of duplicating orders.
4. Timely shipment of customers' orders will improve cash flow.
5. It takes care of economic fluctuations.
6. It helps in minimising the loss due to deterioration, obsolescence, damage, and so on.
7. It acts as a buffer stock when raw materials are received late and shop rejections are too many.
8. It takes advantages of price discounts by bulk purchasing.
9. It improves the manpower, equipment and facility utilisation because of better planning and scheduling.

Maintenance of inventories costs money by means of expenses on stores, equipment, personnel, and so on. So excess inventories are undesirable. This calls for controlling the inventories in the most profitable manner.

Types of Inventories

Basically, there are five types of inventories.

(i) *Fluctuation Inventories (Buffer Inventories)*: These are inventories to meet uncertainties of demand and supply. Buffer inventories in excess of those necessary to meet the average demand during lead time (the time lapsing between placing an order and having the goods in stock ready for use) and held for protecting against the fluctuations in demand and lead time, are termed as safety stocks or reserve stocks.

(ii) *Anticipation Inventories*: These are built up in advance for the season of large sales, a promotion programme, or a plant shut down period. It keeps men and machine ready for future needs. For example, keeping crackers well before Diwali or air coolers or air conditioners before summer.

(iii) *Cycle (Lot Size) Inventories*: These are built up in advance because the purchases are usually made in lots rather than for the exact amounts needed at a point of time.

(iv) *Transportation Inventories (or Pipeline Inventories)*: Such inventories exist because of the transportation of inventory items to various distribution centres and customers from the production centres. This type of inventory is also called *process inventory* where the significant amount of time is consumed in the trans-shipment of items from one location to another. To meet the demand it is essential to hold extra stock at various work stations. The amount of transportation inventory depend on the time taken in transportation and the nature of demand.

(v) *Decoupling Inventories*: If various production stages operate successively, then the breakdown of one or many may affect the entire system. This kind of interdependence is not only expensive but also disruptive for the entire system. The inventories used to reduce the interdependence of various stages of the production system are known as decoupling inventories. These inventories may be classified as:

(a) *Raw materials and component parts*: It is used to decouple the producer from the suppliers. That is, raw materials and component parts inventory could

(i) act as a buffer to take care of delays on the part of suppliers.

(ii) guard against seasonal variations in the demand of final product.

(b) *Work-in-process inventory*: As it takes time to convert raw material into finished product, work-in-process inventory is developed. This inventory takes the form of orders waiting to be transported between machines or of orders waiting to be processed on a particular machine. The level of such inventory can be increased by changing the production process, lot sizes, and so on.

(c) *Finished goods inventory*: It is the inventory of final products which could be released for sale to the customers. The size of this inventory depends upon the demand and the ability of the firm to sell its products.

(d) *Spare parts inventory*: These are the parts which are used in the production process but do not become part of the product.

Operating Constraints The stock level of various items in the inventory is governed by various constraints such as limited warehouse space, limited budget available for inventory, customer service level to be achieved, and management attitude about the individual items in the inventory.

Operating Decision Rules Two types of decisions need to be made in managing inventories:

(i) How much (size) is to order when the inventory of that item is to be replenished?

(ii) When to place an order (or set up production) to replenish inventory?

These may be classified as follows:

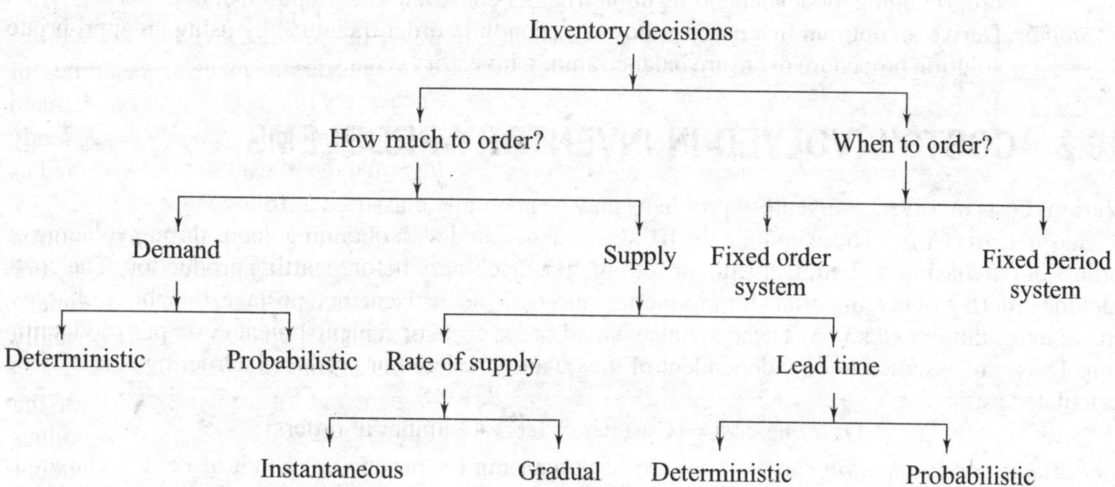

The size and timing of replenishment of orders for an item can be decided using any one of the four inventory control decision rules given below:

Inventory Decision Rules

Order frequency	Order quantity: Fixed (Q)	Order quantity: Variable (S)
Variable (R)	(Q, R)	(S, R)
Fixed (t)	(Q, t)	(S, t)

The above involves placing orders for either a fixed or a variable quantity (size) with either a fixed or a variable time between successive orders. For example, under the (Q, R) rule, an order for a fixed quantity (Q) for replenishment is placed when the inventory balance drops to a certain point (reorder level R). This policy is called a *fixed order quantity* or *reorder-point policy*. Similarly, under (S, t) rule, order of replenishment are placed after every fixed interval (t) for an amount equal to the difference between the current stock level in hand and a predetermined maximum level. This policy is known as the *fixed-interval policy* or *order-period policy* or *periodic review policy*.

The success of any inventory decision rule depends on the accurate determination of its parameters Q, R.

Steps of Inventory Model Building

The steps to build up a suitable inventory model and then to derive decision rules are as follows:

Step 1: First take physical stock of all the inventory items in an organisation.

Step 2: Classify the stock of items into various categories.

Step 3: After classification of inventories, each item should be assigned a suitable code.

Step 4: Build up a mathematical model to achieve the objective function of minimising the total inventory costs subject to changes in inventory reorder policy and constraints of limited resources. The model would çither be an unconstrained optimisation model or a mathematical programming model depending upon whether constraints are imposed or not.

Step 5: Derive an optimal inventory policy (i.e economic order quantity) by using an appropriate solution procedure to ensure balance among inventory costs.

10.2 COSTS INVOLVED IN INVENTORY PROBLEMS

Various costs involved in inventory problems may be generally classified as follows:

Set-up Cost (C_S): These include the fixed cost associated with obtaining goods through placing of orders or purchasing or manufacturing or setting up a machinery before starting production. The costs include ordering of raw materials for production purposes, advertisements, postage, telephone charges, travel expenditure and so on. These are also called order costs or replenishment costs per production run. These are assumed to be independent of the quantity ordered or produced. Ordering cost may be calculated as:

$$\text{Ordering cost} = (\text{Cost per order}) \times (\text{Number of orders})$$

Purchase or Production Cost: The cost of purchasing (or producing) a unit of item is known as purchase (or production) cost. Purchase cost per unit item is affected by the quantity purchased due to *quantity discounts* or *price breaks*.

$$\text{Purchase cost} = (\text{Price per unit item}) \times (\text{Demand per unit time})$$

When price break or quantity discounts are available for bulk purchase of a specified quantity, the unit price becomes smaller as size of order Q exceeds a specified quantity level. In such cases the purchase cost becomes variable and depends on the size of the order. In this case purchase cost is given by

$$\text{Purchase cost} = \begin{pmatrix}\text{Price per unit when}\\ \text{order size is } Q\end{pmatrix} \times \begin{pmatrix}\text{Demand per}\\ \text{unit time}\end{pmatrix}$$

Carrying or Holding Cost (C_1): The cost associated with carrying or holding goods in stock is called holding or carrying cost per unit of item for a unit of time. It is assumed to vary directly with the size of inventory as well as the time the item is held in stock. This cost generally includes:

(i) *Invested Capital Cost*: This is the interest charged on the capital investment.

(ii) *Record-keeping and Administrative Cost*: This shows that there is a need of funds for maintaining the records and necessary administration.

(iii) *Handling Costs*: These include all costs associated with movement of stock, cost of labour, and so on.

(iv) *Storage Costs*: These involve the rent for storage space or depreciation and interest even if own space is used.

(v) *Depreciation, Deterioration, Obsolescence Costs*: These costs arise due to the items in stock being out of fashion or the items undergoing chemical changes during storage, date expiring and so on.

(vi) *Taxes and Insurance Costs*: These costs require careful study and generally amounts to 1 per cent to 2 per cent of the invested capital.

Shortage (or Stock out) Cost: The shortage of items occurs when actual demand cannot be fulfilled from the existing stock. These costs arise due to shortage of goods, and may cause loss of sales. Goodwill may be lost either by a delay in meeting the demand or being unable to meet the demand. The shortage can be looked at in two different ways:

(i) The supply of items is awaited by the customers, that is, the items are back ordered and therefore there is no loss in sale.

(ii) Customers are not ready to wait and, therefore, there, is loss of sale. In the case of shortage, cost shall be measured in terms of goodwill lost and lost profit on the unit which were demanded but were not available.

Shortage cost may be calculated as:

$$\text{Shortage cost} = \begin{pmatrix}\text{Cost of being one unit}\\ \text{short in the inventory}\\ \text{planning period}\end{pmatrix} \times \begin{pmatrix}\text{Average number of units short in the}\\ \text{inventory planning period}\end{pmatrix}$$

Average number of units short

$$= \frac{\begin{pmatrix}\text{Minimum shortance in the} \\ \text{inventory planning period}\end{pmatrix} + \begin{pmatrix}\text{Maximum shortage in} \\ \text{the inventory planning period}\end{pmatrix}}{2} \times \begin{pmatrix}\text{Time for which} \\ \text{shortage occurs}\end{pmatrix}$$

Salvage Cost (or Selling Price): When the demand for commodity is affected by the quantity stocked, the decision model of the problem depends upon the profit maximisation criterion and includes revenue from selling. Generally, salvage value may be combined with the cost of storage and not considered independently.

Revenue Cost: When it is assumed that both the price and the demand of the product are not under the control of the organisation, the revenue from the sales is independent of the company's inventory policy. It may be neglected expect when the organisation cannot meet the demand and the sale is lost. The revenue cost may or may not be included in the study of inventory.

Total Inventory Cost: If the unit price of an item depends on the quantity purchased, that is, price discount is available, then we formulate an inventory policy that considers the purchase cost of the items held in stock. The total inventory cost is given as

$$\text{Total inventory cost} = \text{Ordering cost} + \text{Carrying cost} + \text{Shortage cost}$$

Total Variable Inventory Cost: When price discounts are not offered, the purchase cost remains constant and is independent of the quantity purchased. Hence,

$$\text{Total variable inventory cost} = \text{Ordering cost} + \text{Carrying cost} + \text{Shortage cost.}$$

Other Factors in Inventory Analysis: The factors which play an important role in the study of inventory problems are:

Demand: Demand is the number of units required per period and may be known exactly or in terms of probabilities or be completely unknown. If demand is known, it may be either fixed or variable per unit time. Problems in which the demand is known and fixed are called *deterministic* problems. If the demand is assumed to be a random variable, then those problems are called *stochastic* or *probabilistic problems.*

Lead time: The time gap between placing of an order and its actual arrival in the inventory is known as lead time. The level of inventory of an item depends upon the length of its lead time. The longer the lead time, the higher is the average inventory. Lead time has two components, namely *administrative lead time*—time from the initiation of procurement action to the placing of an order, and *delivery lead time*—time from placing of an order to the delivery of the ordered material.

Order cycle: The time period between placement of two successive orders is referred to as an order cycle. The order may be placed on the basis of following two types of inventory review systems.

(a) *Continuous review*: The record of the inventory level is checked continuously until a specified point is reached where a new order is placed.

(b) *Periodic review*: The inventory levels are viewed at equal time intervals and orders are placed at such intervals. The quantity ordered each time depends on the available inventory level at the time of review.

Stock replenishment: Actually, the replacement of stock may occur instantaneously or uniformly. Instantaneous replenishment occurs in case the stock is purchased from outside sources whereas uniform replenishment occurs when the product is manufactured by the company.

Time horizon: The time period over which the inventory level will be controlled is called the time horizon.

Reorder level: The level between maximum and minimum stock, at which purchasing (or production) activities start for replenishment is called *reorder level.*

Variables in Inventory Problem: The variables used in any inventory model are of two types.

(a) Controlled variables
(b) Uncontrolled variables

(a) *Controlled Variables*: The following variables are controlled separately or in combination.

(i) How much quantity to buy (purchase, production, so on).
(ii) The frequency or timing of acquisition—how often or when to replenish the inventory?
(iii) The completion stage of stocked items.

(b) *Uncontrolled Variables*: These include the holding costs, shortage or penalty costs, set up costs, demand, lead time, and supply of goods.

10.3 EOQ MODELS

10.3.1 Economic Order Quantity (EOQ)

The concept of economic ordering quantity was first developed by F. Harries in 1916. The inventory problems in which the demand is assumed to be fixed and completely predetermined are usually referred to as economic order quantity. When the size of order increases, the ordering costs (cost of purchase, etc.) will decrease, whereas the carrying charges (cost of storage, insurance etc.) will increase. Hence, there are two opposite costs in the production process, one encourages increase in the order size and the other discourages. Hence, Economic Ordering Quantity (*EOQ*) is that size of order which minimises total annual (or any other time period as determined by individual firms) costs of carrying inventory and cost of ordering.

Note that by order quantity we mean the quantity produced or procured during one production cycle.

10.3.2 Determination of EOQ by Tabular Method

This method involves the following steps.

Step 1: Select a number of possible lot sizes to purchase.

Step 2: Determine the total cost for each lot size chosen.

Step 3: Finally, select the ordering quantity which minimises total cost.

10.3.3 Determination of EOQ by Graphical Method

The data calculated in tabular method can be graphed as below:

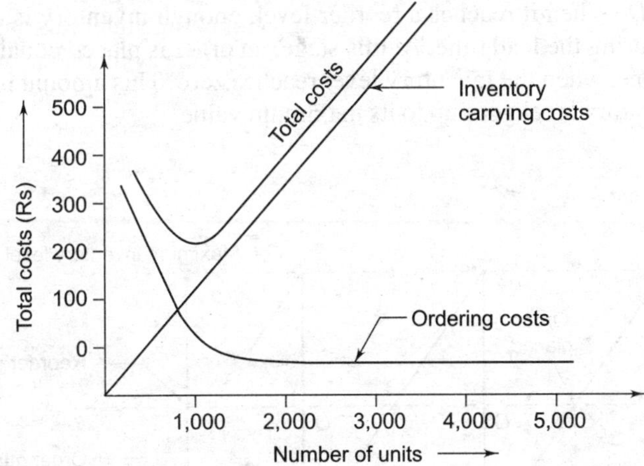

figure 10.1

The minimum total cost occurs at the point where the ordering costs and inventory carrying costs are equal.

A disadvantage of the graphical method is that without specific costs and values, an accurate plotting of the carrying costs, ordering costs, and total costs is not feasible.

10.3.4 Deterministic Inventory Models

In this section we study the inventory models in which demand is assumed to be fixed and completely predetermined, that is, static demand. Such models are usually referred to as the economic lot size models or economic order quantity (*EOQ*) models.

Case 1: *EOQ problem without shortages with uniform demand (Purchasing problem with no shortages).*

Here, we derive the *EOQ* for the optimum production quantity Q per cycle (i.e. per production run) of a single product so as to minimise the total inventory cost. The inventory control system in this case can be described in terms of the following assumptions:

(i) Demand is known and uniform

(ii) Lead time (LT) is zero (or known exactly)

(iii) Production rate is infinite, that is, production is instantaneous (the entire order quantity Q is received at one time as soon as the order is placed).

(iv) Shortages are not allowed.

(v) Q denotes the lot size in each production run and D denotes the total number of units purchased/produced or supplied per time period.

(vi) Set up cost per production run is C_S per set up.

(vii) Holding cost is C_1 per quantity per unit time, that is, $C_1 = IC$, where C is the unit cost and I is inventory carrying charge expressed as a percentage of the value of the average inventory. Fig. 10.2 shows the behaviour of an inventory system which operates on the assumptions listed above. At the beginning of the cycle time, let us start with maximum amount of inventory equal to the order quantity Q. As this amount is used, the level of inventory drops at a fixed rate equal to the demand rate D. When it reaches a reorder level, enough inventory is available to cover expected demand during the lead time. At this stage, an order is placed equal to Q which arrives at the end of lead time, when the inventory level reaches zero. This amount is placed in stock all at once and the inventory level goes up to its maximum value.

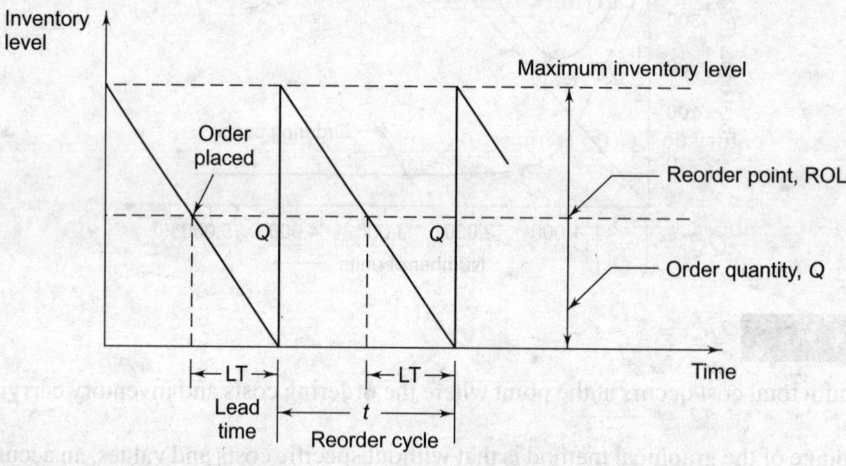

figure 10.2 **EOQ Problem with Uniform Demand**

Rise and fall in inventory level for a particular item over time reflects the periodic cycles of depletion and replenishment. Since the actual consumption of inventory varies constantly, the concept of average inventory is applicable here. With a constant rate of demand,

$$\text{Average inventory} = \frac{1}{2}\,[\text{Maximum level} + \text{Minimum level}]$$

$$= \frac{(Q+0)}{2} = \frac{Q}{2}$$

Total inventory carrying cost per unit

$$= \begin{pmatrix}\text{Average no. of units}\\ \text{in inventory}\end{pmatrix}\begin{pmatrix}\text{Cost of}\\ \text{one unit}\end{pmatrix}\begin{pmatrix}\text{Inventory carrying}\\ \text{cost percentage}\end{pmatrix}$$

$$= (Q/2) \times C \times I = \frac{1}{2}\ QCI,$$

where the term (CI) can be written simply as C_1 (the holding or carrying cost per unit for a unit time)

Total annual ordering costs

$$= \begin{pmatrix}\text{No. of orders}\\ \text{per year}\end{pmatrix}\begin{pmatrix}\text{Ordering}\\ \text{cost per order } C_S\end{pmatrix}$$

$$= \left(\frac{D}{Q}\right) C_S$$

$$\text{Total inventory cost} = \begin{pmatrix}\text{Total inventory}\\ \text{carrying cost}\end{pmatrix}\begin{pmatrix}\text{Total annual}\\ \text{ordering cost}\end{pmatrix}$$

$$C(Q) = \frac{1}{2} QC_1 + \left(\frac{D}{Q}\right) C_S$$

But the total inventory cost $C(Q)$ is minimum when the inventory carrying cost becomes equal to the total annual ordering cost.

Therefore, $$\frac{1}{2} QC_1 = \left(\frac{D}{Q}\right) C_S$$

or, $$QC_1 = \frac{2D}{Q} C_S$$

or, $$Q^2 = \frac{2C_S D}{C_1}$$

or, $$Q = \sqrt{\frac{2C_S D}{C_1}}$$

Hence,

$$\text{optimal } (EOQ) Q^\circ = \sqrt{\frac{2 \times \text{Set-up cost} \times \text{Demand rate}}{\text{Carrying cost}}}$$

This is known as the *economic (optimum) lot size formula.*

The above EOQ formula can also be expressed in terms of the economic order value as:

$$Q^\circ = \sqrt{\frac{2C_S D}{IC}}$$

Characteristics

Minimum total annual inventory cost

$$C(Q^\circ) = \frac{1}{2}Q^\circ C_1 + \frac{DC_S}{Q^\circ}$$

$$= \frac{1}{2}\sqrt{\frac{2C_S D}{C_1}} \times C_1 + DC_S \times \sqrt{\frac{C_1}{2C_S D}}$$

$$= \frac{1}{2}\sqrt{2C_1 C_S D} + \sqrt{\frac{C_1 DC_S}{2}}$$

$$= \sqrt{\frac{C_1 C_S D}{2}} + \sqrt{\frac{C_1 C_S D}{2}}$$

$$= \sqrt{2C_1 C_S D}$$

Hence, minimum total annual inventory cost

$$= \sqrt{[2 \times (\text{Holding cost}) \times (\text{Set-up cost}) \times (\text{Demand rate})]}$$

Optimum length of time between orders

$$t^\circ = \text{interval of ordering} = \frac{\text{Economic ordering quantity}}{\text{Demand rate}}$$

or,
$$t^\circ = \frac{Q^\circ}{D} = \frac{1}{D}\sqrt{\frac{2C_S D}{C_1}} = \sqrt{\frac{2C_S}{DC_1}}$$

That is, optimum ordering interval (t°)

$$= \sqrt{\frac{2 \times \text{Set-up cost}}{\text{Demand rate} \times \text{Holding cost}}}$$

Optimum number of orders placed is

$$n^\circ = \frac{D}{Q^\circ} = \sqrt{\frac{DC_1}{2C_S}}$$

Hence, optimum number of orders (n°)

$$= \frac{\text{Total annual quantity requirement } (D)}{\text{Economic order quantity } (Q^\circ)}$$

Number of days supply (d)

$$= \frac{365}{\text{Optimum number of orders } (N)}$$

Note: If the carrying cost is given as a percentage of average value of inventory held, then total annual carrying cost may be expressed as

$$C_1 = C \times I \quad \text{or,} \quad P \times I$$

Total inventory cost then becomes

$$C(Q) = (\text{Total inventory carrying costs}) + (\text{Total ordering cost})$$

$$= \frac{1}{2} Q \times C \times I + \frac{DC_S}{Q}$$

Optimum order quantity $Q^\circ = \sqrt{\dfrac{2DC_S}{CI}}$

Corollary: In the *EOQ* problem discussed above, if the set-up is $C_S + bQ$ instead of being fixed, (where b is set-up cost per unit item produced) then there is no change in the optimum order quantity produced due to change in the set-up cost.

Proof: The annual inventory cost $C(Q) = \dfrac{1}{2} QC_1 + \dfrac{D}{Q}(C_S + bQ)$

The optimum value of Q which minimises $C(Q)$ can also be obtained by using differential calculus (concepts of maxima and minima) as follows:

$$\frac{d}{dQ}(C(Q)) = 0 \Rightarrow + \frac{1}{2} C_1 - \frac{D}{Q^2} C_S = 0$$

$$\Rightarrow Q = \sqrt{\frac{2DC_S}{C_1}}$$

and, $$\frac{d^2}{dQ^2}(C(Q)) = \frac{2D}{Q^3} C_S > 0 \text{ for } Q > 0$$

Hence, there is no change in Q° if the set-up cost varies.

Example 1

An aircraft company uses rivets at an approximate customer rate of 2,500 kg per year. Each unit costs Rs 30 per kg and the company personnel estimate that it costs Rs 130 to place an order, and that the carrying cost of inventory is 10 per cent per year. How frequently should orders for rivets be placed? Also, determine the optimum size of each order.

Solution:

Given $D = 2{,}500$ per year

$C_1 = (\text{Cost of each unit}) \times (\text{Inventory carrying cost})$

$= \text{Rs } 30 \times (0.1)$

$= \text{Rs } 3.$

$C_S = \text{Rs } 130$ per order

Then, $$Q^\circ = \sqrt{\frac{2DC_S}{C_1}} = \sqrt{\frac{2 \times 2500 \times 130}{3}} = 465.47 \text{ units}$$

$$t^\circ = \frac{Q^\circ}{D} = \frac{465.47}{2500} = 0.186 \text{ year} = 2.232 \text{ months}$$

$$n = \text{number of orders} = \frac{D}{Q^\circ} = 5.37 \text{ orders per year.}$$

Example 2

The production department for a company requires 3,600 kg of raw material for manufacturing a particular item per year. It has been estimated that the cost of placing an order is Rs 36 and the cost of carrying inventory is 25 per cent of the investment in the inventories. The price is Rs 10 per kg. The purchase manager wishes to determine an ordering policy for the raw material.

Solution: Given, D = 3,600 kg year

C_S = Rs 36 per order

C_1 = 25 per cent of the investment in inventories

$$= \text{Rs } 10 \times \frac{25}{100}$$

$$= \text{Rs } 2.5 \text{ per kg/year}$$

So, $$Q^\circ = \sqrt{\frac{2DC_S}{C_1}} = \sqrt{\frac{2 \times 3600 \times 36}{2.5}} = 321.99 \text{ kg per order}$$

$$t^\circ = \frac{Q^\circ}{D} = \frac{321.99}{3,600} = 0.0894 \text{ year}$$

Minimum total annual inventory cost is

$$C(Q^\circ) = \sqrt{2C_1C_SD} = \sqrt{2 \times 3,600 \times 36 \times 2.5} = \text{Rs } 804.98 \text{ per year.}$$

Minimum yearly total cost

$$= C(Q^\circ) + DC$$

$$= \text{Rs } 804.98 + (3,600 \text{ kg}) (\text{Rs } 10/\text{kg})$$

$$= \text{Rs } 36,804.98 \text{ per year.}$$

10.3.5 Case 2: EOQ Problem with No Shortages and Several Production Runs of Unequal Lengths

Inventory system in this case is based on all assumptions as in Case 1, except that the demand is uniform and the production runs differ in units.

Let $t_1, t_2, \ldots t_n$ denote the times of successive production runs, such that $t_1 + t_2 + \ldots + t_n$ = Total period T. The behaviour of the system which operates by the above assumption is shown in Fig. 10.3.

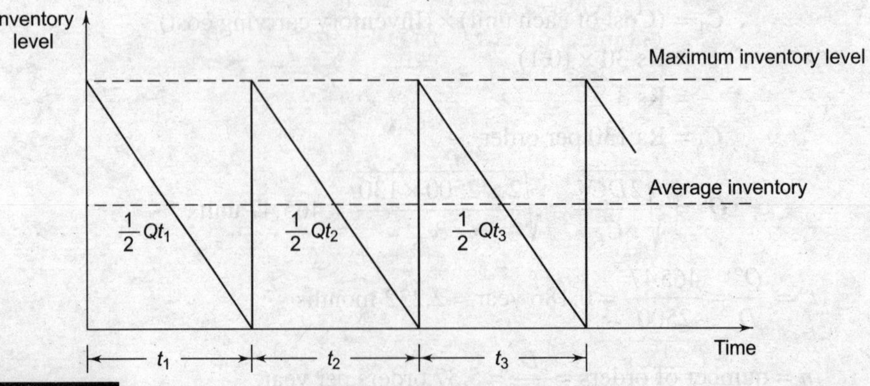

figure 10.3 Inventory level with different rates of demand in different cycles.

The inventory carrying cost and set-up cost for the time period T will be

$$\text{Carrying cost} = \left(\frac{1}{2}Qt_1\right)C_1 + \left(\frac{1}{2}Qt_2\right)C_1 + \cdots + \left(\frac{1}{2}Qt_n\right)C_1$$

$$= \frac{1}{2}QC_1\,(t_1 + t_2 + \cdots + t_n)$$

$$= \frac{1}{2}QC_1T$$

Set-up cost associated with runs of size Q are $\frac{D}{Q}\,C_S$ since $nQ = D$.

Hence, total yearly variable inventory cost is given by

$$C(Q) = \frac{1}{2}\,QC_1\,T + \frac{D}{Q}\,C_S$$

For optimum value of Q which minimises $C(Q)$ equating set-up cost and carrying cost, we have

$$\frac{1}{2}\,QC_1T = \frac{D}{Q}\,C_S$$

On simplification, we have

$$Q = \sqrt{\frac{2DC_S}{TC_1}}$$

The minimum value of $C(Q)$ can be obtained as

$$C(Q^\circ) = \sqrt{2C_1C_S\left(\frac{D}{T}\right)}$$

where D/T denote the average rate of demand.

Example 1

A manufacturing company purchases 9,000 parts of a machine for its annual requirements, ordering one month's requirement at a time. Each part costs Rs 20. The ordering cost per order is Rs 15 and the carrying charges are 15 per cent of the average inventory per year. You have been assigned to suggest a more economical purchasing policy for the company. What advice would you offer, and how much would it save the company per year?

Solution:

Given, $D =$ 9,000 parts per year

$C_S =$ Rs 15 per order

$C_1 =$ 15 per cent of the average inventory per year

$$= \text{Rs } 20 \times \frac{15}{100} = \text{Rs 3 each part per year}$$

$$\text{Optimal lot size } Q^\circ = \sqrt{\frac{2DC_S}{C_1}} = \sqrt{\frac{2 \times 9{,}000 \times 15}{3}} = 300 \text{ units}$$

Optimal order cycle = $t^\circ = \dfrac{300}{9{,}000} = \dfrac{1}{30}$ years = 12 days (approx.)

Minimum average yearly cost = $\sqrt{2C_1C_SD}$

$= \sqrt{2 \times 9{,}000 \times 15 \times 3}$

= Rs 900

But, if the company follows the policy of ordering every month, then the annual ordering cost is Rs 12 × 15 = Rs 180, and lot size of inventory each month = $\dfrac{9{,}000}{12}$ = 750 parts.

Average inventory at any time = $Q/2 = \dfrac{750}{2}$ = 375 parts

Storage cost at any time = $375C_1$ = 375 × 3 = Rs 1,125.

Total annual cost = Rs 1,125 + Rs 180

= Rs 1,305.

Hence, if the company orders 300 parts after every 12 days instead of 750 parts after every month, there will be a net saving of Rs (1,305 – 900) = Rs 405 per year.

Example 2

A chemical company is considering the optimal batch size for reorder of sulphur. The management accountant has supplied the following information: The purchase price of sulphur is Rs 150 per kg. The clerical and data processing costs are Rs 500 per order. All the transport is done by rail. A charge of Rs 2,000 is made each time the special line to the factory is opened. A charge of Rs 20 per kg is also made. The company uses 40,000 kg per year. Maintenance costs of stock are Rs 400 per kg per year. Each kg requires 0.5 sq.ft. of storage space. If warehouse space is not used, it can be rented out to another company at Rs 200 per sq.ft. per annum. Available warehouse space is 1,000 sq.ft., the overhead costs being Rs 5,000 per annum.

Assume that all free warehouse space can be rented out.

(i) Calculate the economic reorder size.

(ii) Calculate the minimum total annual cost of holding and reordering stock.

Solution:

Given ordering cost = Rail transport + Clerical and data processing costs.

= Rs 2,000 + 500 = Rs 2,500

Demand = 40,000 kg per year

Carrying cost = Maintenance cost + Rented cost

$= \text{Rs } 400 + \left(\dfrac{200}{2}\right)$

$$= \text{Rs } 500$$

$$Q^\circ = \sqrt{\frac{2DC_S}{C_1}} = \sqrt{\frac{2 \times 40{,}000 \times 2{,}500}{500}}$$

$$= \text{Rs } 632.45$$

Hence, the minimum total annual costs are

$$\text{Ordering: } \frac{DC_S}{Q^\circ} = \frac{40{,}000}{632.45} \times 2500 = \text{Rs } 1{,}58{,}115.26$$

$$\text{Carrying: } \frac{Q^\circ C_1}{2} = \frac{632.45}{2} \times 500 = \text{Rs } 1{,}58{,}112.50$$

Total inventory cost : Ordering + Carrying

= Rs 1,58,115.26 + Rs 1,58,112.50

= Rs 3,16,227.76

Rail transport

$40{,}000 \times 20$ = Rs 8,00,000.00

Storage overhead cost = Rs 5, 000.00

Purchase cost

$40{,}000 \times 150$ = Rs 60,00,000.00

Total = Rs 71,21,227.70.

10.3.6 Case 3: Economic Lot Size with Finite Rate of Replenishment (Production Problem with No Shortage)

All the assumptions are same as in Case 1, except that of instantaneous replenishment (supply). This is because, in many situations, the amount ordered is not delivered all at once, but is made available at a finite rate (i.e. given supply rate) per unit of time. That is, order is produced internally or will be supplied at some uniform rate k. This situation arises when an order is being filled by a machine having finite production rate.

Assume that each production run of length t consists of two parts, say t_1 and t_2 such that:

(i) The inventory is building up at a constant rate of $(k - r)$ units, per unit of time during t_1; $k > r$, where r is the number of items required per unit time (i.e. demand rate).

(ii) There is no replenishment (or production) during time t_2 and the inventory is decreasing at the rate of r per unit of time.

Since the inventory does not pile up in one shot but happens continuously over a time period and it is also consumed simultaneously, the average inventory level would be determined not only by the lot size Q, but also be affected by production rate k and depletion (demand) r. The graphical representation is given in Fig. 10.4.

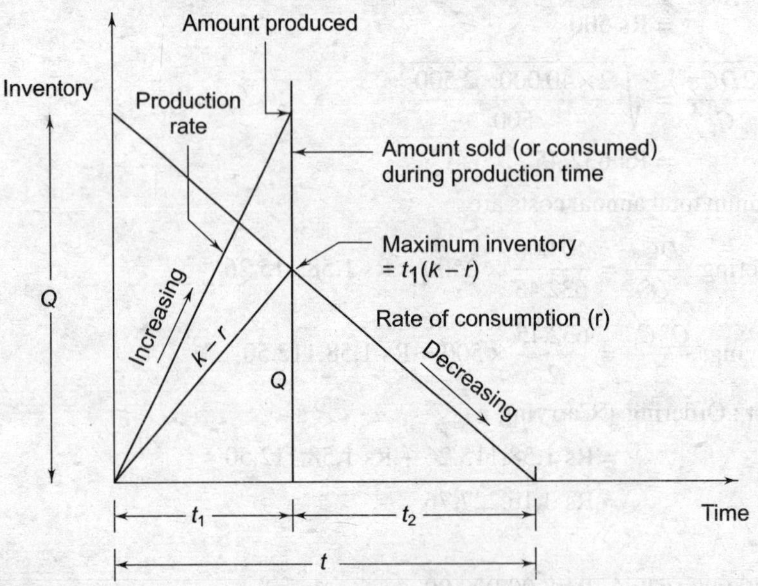

figure 10.4 **Inventory level with finite rate of replenishment**

Since t is the time required to produce Q at a rate k, we have

$$Q = kt_1 \text{ or } t_1 = \frac{Q}{k}$$

During production period t_1, inventory is increasing at the rate of k and simultaneously decreasing at the rate of r. Thus, inventory accumulates at the rate of $(k - r)$ units. Therefore, the maximum inventory level is $t_1(k - r)$. Hence, average inventory $= \frac{1}{2} t_1 (k - r) = \frac{1}{2}\frac{Q}{k}(k - r) = \frac{Q}{2}\left(1 - \frac{r}{k}\right)$.

Let C_1 be the holding cost per unit per year. Then, the total annual holding cost is

$$\frac{1}{2} QC_1 \left(1 - \frac{r}{k}\right)$$

Annual set-up (or ordering) cost is $\frac{D}{Q} C_S$, since D is the total demand in a year. Hence,

total inventory cost $C(Q) = \frac{1}{2} QC_1 \left(\frac{1-r}{k}\right) + \left(\frac{D}{Q}\right) C_S$.

Now, $\frac{d}{dQ}(C(Q)) = 0 \Rightarrow Q = \sqrt{\frac{2DC_S}{C_1\left(1 - \frac{r}{k}\right)}}$

or, $$Q = \sqrt{\frac{2DC_S}{C_1}} \sqrt{\frac{k}{k - r}}.$$

Note that $\frac{d^2}{dQ^2}(C(Q)) > 0$ for $Q > 0$. Thus

$$Q^\circ = \sqrt{\frac{2DC_S}{C_1}}\sqrt{\frac{k}{k-r}}.$$

Characteristics of case 3

1. Total minimum inventory cost

$$C(Q^\circ) = \frac{D}{Q^\circ} + \frac{1}{2}Q^\circ\left(1-\frac{r}{k}\right)C_1$$

$$= \sqrt{2DC_SC_1\left(1-\frac{r}{k}\right)}$$

2. Optimal length of each lot size production run

$$t_1^\circ = \frac{Q^\circ}{k} = \frac{1}{k}\sqrt{\frac{2DC_S}{C_1}}\sqrt{\frac{k}{k-r}}$$

$$= \sqrt{\frac{2DC_S}{C_1k(k-r)}}$$

3. Optimum number of production runs per year

$$n^\circ = D/Q^\circ = D\frac{1}{\sqrt{\frac{2DC_S}{C_1}\left(\frac{k}{k-r}\right)}}$$

$$= \sqrt{\frac{DC_1}{2C_S}\left(\frac{k-r}{k}\right)}$$

4. Optimal production cycle time

$$t^\circ = \frac{Q^\circ}{D} = \sqrt{\frac{2C_S}{DC_1}\left(\frac{k}{k-r}\right)}$$

Note: If $k \to r$, then there will be no holding cost and no set-up cost. If $k \to \infty$, then the above problem reduces to the one in case 1, as the ratio $\left(1-\frac{r}{k}\right)$ tends to be 1, the holding cost per unit of time is reduced from the cost discussed in case 1 and the set-up cost remains the same.

Example 1

A contractor has to supply 10,000 bearings per day to an automobile manufacturer. He finds that when he starts a production run, he can produce 25,000 bearings per day. The cost of holding a bearing in stock for one year is 2 paise and the set-up cost of a production run is Rs 18. How frequently should the production run be made?

Solution:

Given, C_1 = Re 0.02 per bearing per year

= 0.000055 per bearing per day

C_S = Rs 18.00 per production run

r = 10,000 bearings per day

k = 25,000 bearings per day

D = 10,000 bearings per day

Therefore,

$$Q° = \sqrt{\frac{2DC_S}{C_1}}\sqrt{\frac{k}{k-r}}$$

$$= \sqrt{\frac{2 \times 10{,}000 \times 18}{0.000055}}\sqrt{\frac{25{,}000}{25{,}000 - 10{,}000}}$$

$$= \sqrt{\frac{2 \times 18}{0.000055}}\sqrt{\frac{10{,}000 \times 25{,}000}{25{,}000 - 10{,}000}}$$

= 1,04,447 bearings

$$t° = \frac{Q°}{D} = \frac{1}{10{,}000} \times 1{,}04{,}447 \text{ days} = 10.4 \text{ days}$$

Length of production cycle = $\frac{1{,}04{,}447}{25{,}000}$ = 40 days (approx.)

Thus, the production cycle starts at an interval of 10.4 days and production continues for 4 days so that in each cycle a batch of 1,04,447 bearings is produced.

Example 2

(a) At present a company is purchasing an item X from outside suppliers. The consumption is 10,000 units/year. The cost of the item is Rs 5 per unit and the ordering cost is estimated to be Rs 100 per order. The cost of carrying inventory is 25%. If the consumption rate is uniform, determine the economic purchasing quantity.

(b) In the above problem assume the company is going to manufacture the item with the equipment that is estimated to produce 100 units per day. The cost of the unit thus produced is Rs 3.50 per unit. The set-up cost is Rs 150 per set up and the inventory carrying charge is 25%. How has your answer changed?

Solution: (a) Given D = 10,000 units per year

C = Rs 5 per unit

C_S = Rs 100 per order

C_1 = 25% of Rs 5

= Rs 1.25 per year

$$\text{EOQ } Q^\circ = \sqrt{\frac{2\,DC_S}{C_1}} = \sqrt{\frac{2\times 10{,}000\times 100}{1.25}}\ 1{,}265 \text{ units}$$

(b) Given, k = 100 units per day

C = Rs 3.50 per unit.

C_S = Rs 150 per set up

C_1= Rs 25% of Rs 3.50 = Re 0.875 per year

$r = \dfrac{10{,}000}{250}$ = 40 units per day (assuming 250 working days in the year).

EOQ for each production run is

$$Q^\circ = \sqrt{\frac{2DC_S}{C_1}\left(\frac{k}{k-r}\right)}$$

$$= \sqrt{\frac{2\times 10{,}000\times 150}{0.875}\times\frac{100}{100-40}}$$

$$= 2{,}391 \text{ units}$$

The increase in EOQ in case (b) may be due to the following reasons:

(i) Increased item cost (c) and procurement cost (i.e. set-up cost C_S)

(ii) Consumption of inventory simultaneous with production reduces average inventory from $Q/2$ to $\left(1-\dfrac{r}{k}\right)\dfrac{Q}{2}$.

10.3.7 Case 4: EOQ Problem with Shortages (Purchasing Problem with Shortages)

This model is an extension of the previous cases by allowing shortages. We discuss this case in three parts, the assumptions in each part being slightly different.

Two types of situations occur when an inventory system runs out of stock:

(a) Customers are not likely to keep buying inventory items, and therefore any sale that would have resulted is lost.

(b) Customer leave order(s) with the supplier and this back order is filled on stock availability.

Thus, all the costs (cost of keeping backlog recorders, cost of shipping the items to the customers, loss or goodwill, etc.) of placing an ordinary order are involved in placing a back order. The back order cost depends upon how long a customer waits to receive an order. It is expressed in rupees per unit of time.

Fig. 10.5 describes the situation.

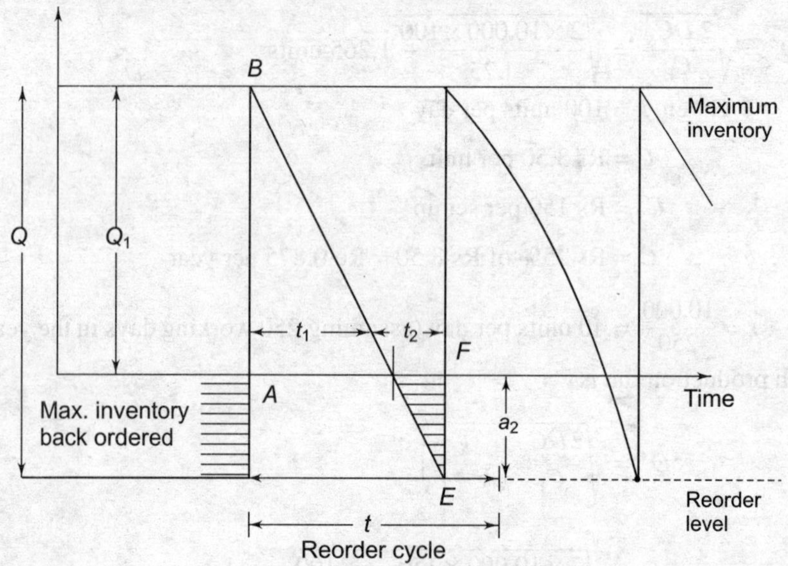

figure 10.5 **EOQ problem with shortages**

Sub-case 1: Production is instantaneous

The problem is same as that of Case 1 with addition that the shortages are allowed. So, total inventory cost is $C(Q)$ = ordering cost + carrying cost + shortage cost. Let C_2 be the shortage cost per unit of time per unit quantity. Let the total time period t be divided into equal parts say t_1 and t_2 such that $t = t_1 + t_2$.

During the interval t_1, the items are drawn from the inventory as needed and during t_2, orders for the item are being accumulated but not filled. Then, at the end of the interval t, an amount Q is produced (or delivered). The amount Q has been divided into Q_1 and Q_2 such that $Q = Q_1 + Q_2$, where Q_1 denotes the amount which goes into the inventory and Q_2 denotes the amount which is immediately taken to satisfy past orders or unfilled demand.

As we want to calculate an optimal value of Q and Q_1, we have to express t_1 and t_2 in terms of Q and Q_1. Now,

$$\text{Time period in days} = \frac{\text{Total units over time period}}{\text{Demand in units per day}}$$

Hence, $t_1 = \dfrac{Q_1}{D}$ and $t = \dfrac{Q}{D}$ and $t_2 = \dfrac{Q_2}{D}$

Thus, the average inventory level over the reorder cycle time t can be determined by dividing the area of the triangle ABC by the cycle time, that is,

$$\text{Average inventory level} = \frac{(\text{Average level over } t_1) + (\text{Average level over } t_2)}{t}$$

$$= \frac{\left(\dfrac{Q_1}{2}\right)t_1 + 0.t_2}{t} = \frac{\dfrac{Q_1}{2}\left(\dfrac{Q_1}{D}\right)}{(Q/D)} = \frac{Q_1^2}{2Q}$$

and, therefore, carrying cost $= \dfrac{Q_1^2}{2Q} \times C_1$

Similarly, the average back order level can be determined by dividing the area under the triangle *CFE* by the cycle time *t*, that is,

$$\text{Average number of units on back order} = \frac{(\text{Average level over } t_1) + (\text{Average level over } t_2)}{t}$$

$$= \frac{(0.t_1) + \left(\frac{Q_2}{2}\right) t_2}{t}$$

$$= \frac{\left(\frac{Q_2}{2}\right)\left(\frac{Q_2}{D}\right)}{(Q/D)} = \frac{Q_2^2}{2Q}$$

and, therefore,

$$\text{Shortage cost} = \frac{Q_2^2}{2Q} C_2$$

Hence, the total yearly variable inventory cost

$$C(Q) = \frac{D}{Q} C_S + \frac{Q_1^2}{2Q} C_1 + \frac{Q_2^2}{2Q} C_2.$$

For determining the optimum values of Q_1 and Q so as to optimise $C(Q)$, we have

$$\frac{\partial}{\partial Q_1}(C(Q)) = 0 \Rightarrow Q_1 = \frac{C_2 Q}{(C_1 + C_2)}$$

$$\frac{\partial}{\partial Q}(C(Q)) = 0 \Rightarrow Q = \sqrt{\frac{2C_S D + C_1 Q_1^2}{C_2} + Q_1^2}$$

and, $\dfrac{\partial^2 (C(Q))}{\partial Q_1^2} > 0, \quad \dfrac{\partial^2 (C(Q))}{\partial Q^2} > 0$

Thus, the optimum quantities are given after simplification as

$$Q^\circ = \sqrt{\frac{2C_S D}{C_1}} \sqrt{\frac{(C_1 + C_2)}{C_2}}$$

$$Q_1^\circ = \left(\frac{C_2}{C_1 + C_2}\right) Q^\circ = \sqrt{\frac{C_2}{C_1 + C_2}} \sqrt{\frac{2C_S D}{C_1}}$$

Characteristics of Case 4

1. Time between receipt of orders

$$t^\circ = \frac{Q^\circ}{D} = \sqrt{\frac{2C_S}{DC_1}}\sqrt{\frac{C_1 + C_2}{C_2}}$$

2. Total optimum inventory cost

$$C(Q^\circ) = \sqrt{2DC_SC_1}\sqrt{\frac{C_2}{(C_1 + C_2)}}$$

3. Reordered level $= Q^\circ - Q_1{}^\circ = Q^\circ\left(1 - \dfrac{C_2}{C_1 + C_2}\right)$.

Note 1: If the penalty cost $C_2 = \infty$, then

$$Q^\circ = \sqrt{\frac{2DC_S}{C_1}\left(\frac{C_1 + C_2}{C_2}\right)} = \sqrt{\frac{2DC_2}{C_1}\left(1 + \frac{C_1}{C_2}\right)}$$

$$= \sqrt{\frac{2DC_2}{C_1}}$$

(2) If $C_1 = \infty$, $C_2 > 0$, then $Q_1^\circ = 0$ and

$$Q^\circ = \sqrt{\frac{2C_SD}{C_2}\cdot\frac{C_1 + C_2}{C_1}}$$

$$= \sqrt{\frac{2C_SD}{C_2}\left(1 + \frac{C_2}{C_1}\right)}$$

$$= \sqrt{\frac{2C_SD}{C_2}}.$$

(3) If penalty costs are negligible, that is, $C_2 \to 0$, with $C_1 > 0$, then $Q_1 \to 0$ and $Q^\circ \to \infty$.

(4) If inventory costs are negligible, then $C_1 \to 0$ and $C_2 > 0$, In this case $Q^\circ \to \infty$, and $Q_1^\circ \to \infty$. That is, $Q_1 \to Q^\circ$. Thus, as inventory costs become very small, increasingly large batches should be produced and used entirely as inventory for future demands.

(5) When inventories and shortages are equally costly, that is, when $C_1 = C_2$, then $\dfrac{C_2}{C_1 + C_2} = \dfrac{1}{2}$. In this case $Q^\circ = \sqrt{2}\sqrt{\dfrac{2C_SD}{C_1}} = (1.414)\sqrt{\dfrac{2C_SD}{C_1}}$.

Sub-case 2: Production is instantaneous and time is fixed

Let the reorder cycle time t be fixed. That is, inventory is to be supplied after every time period t at the rate of r units per unit of time during this fixed time period.

Also, let $Q = D.t$, where D is the demand rate per unit time, Q is the fixed lot size to meet the demand for the period t.

As shown in Fig. 10.5, the amount Q_1 ($< Q$) is planned to meet the demand during time $t_1 = \frac{Q_1}{D}$. Since the reording (or set-up) cost and time t are constant, therefore, total variable inventory cost $C(Q)$ is given by carrying cost + shortage cost

$$= \frac{Q_1^2}{2Q} C_1 + \frac{(Q - Q_1)^2}{2Q} C_2$$

Now, the optimum value Q_1° is obtained as follows:

$$\frac{\partial}{\partial Q_1}(C(Q)) = 0 \Rightarrow \frac{C_1 Q_1}{Q} + \frac{1}{Q}\cdot(Q - Q_1)C_2(-1) = 0$$

$$\Rightarrow C_1 Q_1 - C_2 (Q - Q_1) = 0$$

$$\Rightarrow Q_1 = \left(\frac{C_2}{C_1 + C_2}\right) Q = \left(\frac{C_2}{C_1 + C_2}\right) Dt$$

and $\frac{\partial^2 (C(Q))}{\partial Q_1^2} > 0$ for all values of Q_1, therefore,

$$Q_1^\circ = Dt \frac{C_2}{C_1 + C_2}.$$

The set-up cost is not considered because of it being fixed as the time of one production run is fixed.

Sub-case 3: Production is finite

This model is based on the same assumptions as in case 3 (10.3.6) except that shortages (back orders) are allowed.

The inventory system is shown in Fig. 10.6.

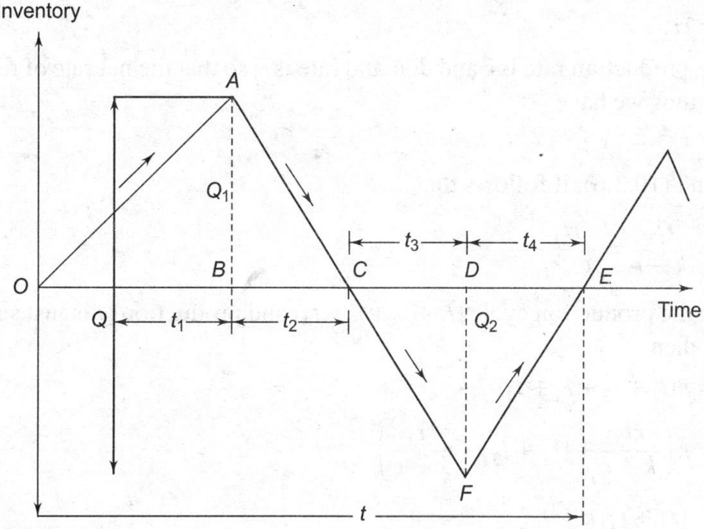

figure 10.6 Inventory cycle-production is finite

Stocks start at zero and increase for a period t_1 and then decrease for a period t_2 until they again reach zero at the point where a backlog piles up for the period t_3. At the end of t_3, production starts and backlog diminishes for the time t_4 when the backlong reaches zero. The cycle repeats itself after total time $(t_1 + t_2 + t_3 + t_4)$. Now,

$$\text{Holding cost} = C_1 \times \Delta OAC = C_1 \times \frac{1}{2}\, Q_1\,(t_1 + t_2)$$

$$\text{Shortage cost} = C_2 \times \Delta EFC = C_2 \times \frac{1}{2}\, Q_2\,(t_3 + t_4)$$

and the set-up cost is C_S. Hence, average cost per unit time

$$C = \frac{\frac{1}{2}\left[C_1 Q_1\,(t_1 + t_2) + C_2 Q_2\,(t_3 + t_4)\right] + C_S}{t_1 + t_2 + t_3 + t_4} \qquad (10.3.1)$$

The inventory is zero at O and during the period t_1 and amount kt_1 is produced. But because orders are filed at a rate r, the net increase Q_1 in inventory during t_1 is given by

$$Q_1 = kt_1 - rt_1 = t_1\,(k - r) \qquad (10.3.2)$$

After time t_1, the production is stopped and the stock Q_1 is used up during t_2, and because the rate of use is r, we have

$$Q_2 = rt_2 \qquad (10.3.3)$$

From (10.3.2) and (10.3.3), we have

$$t_1 = \frac{Q_1}{k - r} = \frac{rt_2}{k - r} \qquad (10.3.4)$$

During period t_3, shortages accumulate at the rate r,

So,

$$Q_2 = rt_3 \qquad (10.3.5)$$

During period t_4, production rate is k and demand rate is r, so that the net rate of reduction of shortage becomes $k - r$, and thus we have

$$Q_2 = t_4(k - r) \qquad (10.3.6)$$

From (10.3.5) and (10.3.6), it follows that

$$t_4 = \frac{Q_2}{k - r} = \frac{rt_3}{k - r} \qquad (10.3.7)$$

Finally, because the production cycle $(t_1 + t_2 + t_3 + t_4)$ and production Q is just sufficient to meet the demand at a rate r, then

$$Q = r(t_1 + t_2 + t_3 + t_4)$$

$$= r\left(\frac{rt_2}{k - r} + (t_2 + t_3) + \frac{rt_3}{k - r}\right)$$

$$= \frac{(t_2 + t_3)\,k}{k - r} \qquad (10.3.8)$$

Now, eliminating t_1, t_4, Q_1 and Q_2 from (10.3.1) by using relations (10.3.4), (10.3.7), we have

$$C = \frac{\frac{1}{2}\left\{C_1(rt_2)\left[\frac{rt_2}{k-r}+t_2\right]+C_2(rt_3)\left[t_3+\frac{rt_3}{k-r}\right]\right\}+C_S}{\frac{rt_2}{k-r}+t_2+t_3+\frac{rt_3}{k-r}}$$

$$= \frac{\frac{1}{2}\left\{\frac{C_1t_2^2rk}{k-r}+\frac{C_2t_3^2rk}{k-r}\right\}+C_S}{(t_2+t_3)\left[1+\frac{r}{k-r}\right]}$$

$$= \frac{\frac{1}{2}\left[C_1t_2^2rk+C_2t_3^2rk\right]+C_S(k-r)}{k(t_2+t_3)}$$

The optimum values t_2°, t_3° of t_2 and t_3 after differentiating partially with respect to t_2 and t_3 and setting the results equal to zero is

$$t_2^{\circ} = \sqrt{\frac{2C_SC_2\left(1-\frac{r}{k}\right)}{r(C_1+C_2)C_1}}$$

$$t_3^{\circ} = \sqrt{\frac{2C_SC_1\left(1-\frac{r}{k}\right)}{r(C_1+C_2)C_2}}$$

and from (10.3.5), (10.3.8), the optimum order quantity is obtained as:

$$Q^{\circ} = \sqrt{\frac{2rC_S(C_1+C_2)}{C_1C_2}}\cdot\sqrt{\frac{1}{\left(1-\frac{r}{k}\right)}}$$

$$Q_2^{\circ} = \sqrt{\frac{2r_SC_1C_S\left(1-\frac{r}{k}\right)}{(C_1+C_2)C_2}}$$

The minimum cost is given by

$$C^{\circ} = \sqrt{\frac{2rC_1C_2C_S\left(1-\frac{r}{k}\right)}{C_1+C_2}}$$

Characteristics

Production cycle time between set-ups

$$t^{\circ} = \frac{Q^{\circ}}{r} = \frac{1}{r}\times\sqrt{\frac{2rC_S}{C_1}\left(\frac{k}{k-r}\right)\left(\frac{C_1+C_2}{C_2}\right)}$$

$$= \sqrt{\frac{2C_S}{rC_1}\left(\frac{k}{k-r}\right)\left(\frac{C_1+C_2}{C_2}\right)}$$

Optimum inventory level

$$Q_1^\circ = \left(\frac{k-r}{k}\right)Q^\circ - Q_2^\circ = \sqrt{\frac{2rC_2}{C_1}\left(1-\frac{r}{k}\right)\left(\frac{C_2}{C_1+C_2}\right)}$$

Optimum production time$= \dfrac{Q^\circ}{k} = \sqrt{\dfrac{2rC_S}{kC_1}\left(\dfrac{1}{k-r}\right)\left(\dfrac{C_1+C}{C_2}\right)}$

$$C(Q^\circ) = \sqrt{2rC_SC_1\left(1-\frac{r}{k}\right)\left(\frac{C_2}{C_1+C_2}\right)}$$

Note 1: If $k = \infty$, then

$$Q^\circ = \sqrt{\frac{2rC_S}{C_1}\left(\frac{C_1+C_2}{C_2}\right)}$$

$$Q_1^\circ = \sqrt{\frac{2rC_S}{C_1}\left(\frac{C_2}{C_1+C_2}\right)}$$

$$C(Q^\circ) = \sqrt{2rC_SC_1\left(\frac{C_2}{C_1+C_2}\right)}$$

Example 1

A commodity is to be supplied at a constant rate of 200 units per day. Supplies of any amounts can be had at any required time, but each ordering costs Rs 50.00. Cost of holding the commodity in inventory is Rs 2.00 per unit per day while delay in the supply of the items induces a penalty of Rs 10.00 per unit per delay of 1 day. Find the optimal policy (q, t) where t is the reorder cycle period and q is the inventory level after reorder. What would be the best policy, if the penalty cost became ∞?

Solution: Here,

$D = 200$ units per day
$C_S =$ Rs 50 per order
$C_1 =$ Rs 2.00 per unit per day
$C_2 =$ Rs 10.00 per unit per day

$$Q^\circ = \sqrt{\frac{2\times 50\times 200}{2}\left(\frac{10+2}{10}\right)} = 109.5 \text{ units}$$

$$t^\circ = \frac{Q^\circ}{D} = \frac{109.5}{200} = \frac{1}{2} \text{ day}$$

Thus, an optimal order quantity of 109.5 units must be supplied after every 1/2 day.
If the penalty cost $C_2 = \infty$, then,

$$Q^\circ = \sqrt{\frac{C_S D}{C_1}} = \sqrt{\frac{2 \times 50 \times 200}{2}} = 100 \text{ units}$$

and, $$t^\circ = \frac{Q^\circ}{D} = \frac{100}{200} = \frac{1}{2} \text{ day.}$$

Example 2

A commodity is to be supplied at a constant rate of 25 units per day. A penalty cost is being charged at the rate of Rs 10 per unit per day late. The cost of carrying the commodity in inventory is Rs 16 per unit per month. The production process is such that each month (30 days) a batch of items is started and is available for delivery any time after the end of the month. Find the optimal level of inventory at the beginning of each month.

Solution: Given,

D = 25 units/day
C_S = Rs 16/30 = 0.53 per unit\per day
C_2 = Rs 10 per unit/per day
t = 30 days

The optimal inventory level is given by

$$Q_1^* = \left(\frac{10}{0.53 + 10}\right)(25)(30) = 712 \text{ units.}$$

Example 3

The demand for an item in a company is 18,000 units per year, and the company can produce the item at a rate of 3,000 per month. The cost of one set-up is Rs 500 and the holding cost of 1 unit per month is 15 paise. The shortage cost of one unit is Rs 20 per month. Determine (i) optimum production batch quantity and the number of strategies; (ii) optimum cycle time and production time; (iii) maximum inventory level in the cycle; and (iv) total associated cost per year if the cost of the item is Rs 20 per unit.

Solution: Given, C_1 = Re 0.15 per month, C_2 = Rs 20.00
C_S = Rs 500.00, k = 3,000 units per month,
r = 18,000 units per year or 1,500 units per month.

(i) Optimum production batch quantity is given by

$$Q^\circ = \sqrt{\frac{2C_S(C_1 + C_2)}{C_1 C_2}\left(\frac{kr}{k - r}\right)}$$

$$= \sqrt{\frac{2 \times 500(0.15 + 20)}{(0.15) \times 20}\left(\frac{1{,}500 \times 3{,}000}{3000 - 1{,}500}\right)}$$

= 4,489 units (approx.)

Number of shortages is given by

$$Q_2^\circ = \sqrt{\frac{2C_S C_1 r}{(C_1 + C_2)C_2}\left(\frac{k - r}{k}\right)}$$

$$= \left(\frac{C_1}{C_1 + C_2}\right)Q^\circ\left(\frac{k - r}{k}\right)$$

$$= \left(\frac{0.15}{0.15+20}\right)(4{,}489)\left(\frac{3{,}000-1{,}500}{3000}\right)$$

$= 17$ units (approx.)

$$\text{Optimum production time} = \frac{Q^\circ}{k} = \frac{4{,}489}{3{,}000} = 1.5 \text{ months}$$

$$\text{Optimum cycle time between set-ups} = \frac{Q^\circ}{r} = \frac{4{,}489}{1{,}500} = 3 \text{ months}$$

Maximum inventory level is

$$Q_1^\circ = \frac{k-r}{k}\, Q^\circ - Q_2^\circ = \frac{1{,}500}{3000} \times 4489 - 17 = 2227 \text{ units (approx.)}$$

Total associated cost is given by

$$C(Q^\circ)) = \sqrt{2rC_S C_1\left(1-\frac{r}{k}\right)\left(\frac{C_2}{C_1+C_2}\right)}$$

$$= \sqrt{2\times 1{,}500\times 500\left(1-\frac{1500}{3000}\right)\left(\frac{20}{0.15+20}\right)}$$

$=$ Rs 334.

10.4 EOQ PROBLEMS WITH PRICE BREAKS

Nowadays, discounts are offered for the purchase of large quantities of products, and discounts take the form of price breaks. If discount is not available, then

$$Q^\circ = \sqrt{\frac{2DC_S}{C_1}}$$

If a discount (price break) is available, then total cost per unit of the inventory is

$$C(Q^\circ) = DK_1 + \frac{D}{Q}C_S + \frac{1}{2}QK_1 \times I$$

where, the cost of manufacturing (or purchasing) is K_1 per unit and I denotes the holding cost per unit. The optimum value of Q will be obtained by

$$\frac{d}{dQ}(C(Q^\circ)) = 0 \Rightarrow Q = \sqrt{\frac{2C_S D}{K_1 I}}, \text{ with } \frac{d^2}{dQ^2}(C(Q) > 0$$

$$\text{Hence,}\quad Q^\circ = \sqrt{\frac{2C_S D}{K_1 I}}$$

Case 1: *EOQ problem with one price break*

When there is only one price break (one quantity discount), and if

Range of quantity	*Purchase*	*Cost per unit*
$0 \le Q_1 < b$	k_{11}	
$b \le Q_2$	k_{12}	

where b is that quantity at and beyond which the quantity discount applies and $k_{12} < k_{11}$.

The procedure to obtain EOQ is:

Step 1: Compute Q_2°. That is, find the optimum order quantity for the highest discount and compare it with the quantity b.

If $Q_2^\circ \geq b$, then place orders for quantities of size Q_2° and obtain discount, otherwise go to the next step.

Step 2: If $Q_2^\circ < b$, then compare the total inventory for $Q = Q_1^\circ$ (for price k_{11}) with $Q = b$.

The values of $C(Q_1^\circ)$ and $C(b)$ are:

$$C(Q_1^\circ) = Dk_{11} + \frac{D}{Q_1^\circ} C_S + \frac{1}{2} Q_1^\circ \times k_{11} \times I$$

$$C(b) = Dk_{12} + \frac{D}{b} C_S + \frac{1}{2} b \times k_{12} \times I$$

If $C(Q_1^\circ) > C(b)$, then $Q^\circ = b$, otherwise $Q^\circ = Q_1^\circ$.

Case 2: *EOQ problems with two price breaks*

Where there are two price breaks (two quantity discounts), and if

Range of quantity	*Purchase*	*Cost per unit*
$0 \leq Q_1 < b_1$	k_{11}	
$b_1 \leq Q_2 < b_2$	k_{12}	
$b_2 \leq Q_3$	k_{13}	

where b_1 and b_2 are those quantities which determine the price breaks.

The procedure to obtain EOQ is:

Step 1: Compute Q_3° and compare it with b_2.

Step 2: If $Q_3^\circ \geq b_2$, the optimum order quantity is Q_3°.

Step 3: If $Q_3^\circ < b_2$, compute Q_2°. As $Q_3^\circ < b_2$, we have $Q_2^\circ < b_2$, as $Q_1^\circ < Q_2^\circ < Q_3^\circ < ... < Q_n^\circ$
So, either $Q_2^\circ < b_1$ or $b_1 \leq Q_2^\circ < b_2$.

Step 4: If $Q_3^\circ < b_2$, $b_1 \leq Q_2^\circ < b_2$, then proceed as in the case of only one price break.

Step 5: If $Q_2^\circ < b_2$, $Q_2^\circ < b_1$, then compute Q_1°.

Now, $Q_1 < b_1$ compare $C(Q_1^\circ)$ with $C(b_1)$ and $C(b_2)$ in order to get optimum purchase quantity.

Case 3: *EOQ problem with n price breaks*

When there are n price breaks, we have

Range of quantity	*Purchase cost per unit*
$0 \leq Q_1 < b_1$	k_{11}
$b_1 \leq Q_2 < b_2$	k_{12}
$\vdots$	$\vdots$
$b_{n-1} \leq Q_n$	k_{1n}

where $b_1, b_2, ..., b_{n-1}$, are those quantities which determine the price breaks.

The procedure to obtain EOQ is:

Step 1: Compute Q_n° and compare it with b_{n-1}.

Step 2: If $Q_n^\circ \geq b_{n-1}$, the optimum order quantity is Q_n°.

Step 3: If $Q_n^\circ < b_{n-1}$, compute Q_{n-1}°. If $Q_{n-1}^\circ \geq b_{n-2}$, proceed as in the case of one price break.

Step 4: If $Q_{n-1}^{\circ} < b_{n-2}$, compute Q_{n-2}°. If $Q_{n-2}^{\circ} \geq b_{n-3}$, proceed as in the case of one price break.

Step 5: If $Q_{n-2}^{\circ} < b_{n-2}$, $Q_{n-2}^{\circ} < b_{n-3}$, then compute Q_{n-3}°. If $Q_{n-3}^{\circ} \geq b_{n-4}$, then compare $C(Q_{n-3}^{\circ})$ with $C(b_{n-3})$, $C(b_{n-2})$ and $C(b_{n-1})$

Compute this process until $Q_{n-j}^{\circ} \geq b_{n-(j+1)}$; $0 \leq j \leq n-1$ and then compare $C(Q_{n-j}^{\circ})$ with $C(b_{n-j-2})$, ... $C(b_{n-1})$.

Example 1

Find the optimum order quantity for a product for which the prices are as follows:

Quantity	*Unit Cost (Rs)*
$0 \leq Q_1 < 500$	10.00
$500 \leq Q_2$	9.25

The monthly demand for the product is 200 units, the cost of storage is 2% of the unit cost and the cost of ordering is Rs 350.00

Solution: Given C_S = Rs 350.00, $D = 200$, I = Re 0.02

k_{11} = Rs 10.00, k_{12}= Rs 9.25

Step 1: $$Q_2^{\circ} = \sqrt{\frac{2C_S D}{k_{12} I}} = \sqrt{\frac{2 \times 350 \times 200}{(9.25) \times (0.02)}} = 870 \text{ units}$$

Since $Q_2^{\circ} = 870 > b_2 = 500$, the optimum purchase quantity is $Q^{\circ} = Q_2^{\circ} = 570$ units.

Example 2

The annual demand of a product is 10,000 units. Each unit costs Rs 100 if the order placed is in quantities below 200 units. But for orders of 200 or above the price is Rs 95. The annual inventory holding cost is 10% of the value of the item and the ordering cost is Rs 5 per order. Find the economic lot size.

Solution: Given D = 10,000 units, C_S = Rs 5,00, I = 10% of price of an item = Re 0.10, k_{11} = Rs 100, $k_{12} = 95$.

The optimum order quantity is based on price k_{12} = Rs 95.

Now,

$$Q_2^{\circ} = \sqrt{\frac{2DC_S}{k_{12} \times I}} = \sqrt{\frac{2 \times 10{,}000 \times 5}{95 \times 0.10}} = 103 \text{ units}$$

Since $Q_2^{\circ} < b$ (=200), calculate Q_1°

$$Q_1^{\circ} = \sqrt{\frac{2DC_S}{k_{11} \times I}} = \sqrt{\frac{2 \times 10{,}00 \times 5}{100 \times 0.10}} = 100 \text{ units}$$

As $Q_1^{\circ} < b$ (=200), therefore, comparing $C(Q_1^{\circ})$ and $C(b = 200)$, we have

$$C(Q_1^{\circ}) = Dk_{11} + \frac{D}{Q_1^{\circ}} C_S + \frac{1}{2} Q_1^{\circ} \times k_{11} \times I$$

$$= 10{,}000 \times 100 + \frac{10{,}000}{100} \times 5 + \frac{100}{2} \times 100 \times 0.10$$

$$= \text{Rs } 10{,}01{,}000$$

$$C(b) = Dk_{12} + \frac{D}{b}C_S + \frac{1}{2}\, b \times k_{12} \times I$$
$$= \text{Rs } 10{,}01{,}250$$

Since $C(Q_1^\circ) < C(b)$, the optimal order quantity is $Q^\circ = Q_1^\circ = 100$ units.

Example 3

Annual demand for an item is 6,000 units. Ordering cost is Rs 600 per order. Inventory carrying cost is 18% of the purchase price/unit/year. The price break up is as shown below:

Quantity	*Price*
$0 \le Q_1 < 2{,}000$	20
$2{,}000 \le Q_2 < 4000$	15
$4{,}000 \le Q_3$	9

Find the optimal order size.

Solution: Given $D = 6{,}000$/year, $C_S =$ Rs 600, $I = 0.18$, $k_{11} = 20$, $k_{12} = 15$, $k_{13} = 9$.

Step 1:

$$Q_3^\circ = \sqrt{\frac{2DC_S}{k_{13} \times I}} = \sqrt{\frac{2 \times 6{,}000 \times 600}{9 \times 0.18}} = 2{,}109 \text{ units.}$$

Since $Q_3^\circ < b_2$ (= 4,000), proceed to next step.

Step 2: Compute Q_2°. Here, the highest discount available is $k_{12} = 15$. Now,

$$Q_2^\circ = \sqrt{\frac{2DC_S}{k_{12} \times I}} = \sqrt{\frac{2 \times 6{,}000 \times 600}{15 \times 0.18}}$$
$$= 1{,}633 \text{ units.}$$

Here, $Q_2^\circ < b_1$ (= 2000), so go to next step.

Step 3: Compute Q_1°. Here, $k_{11} =$ Rs 20. Now,

$$Q_1^\circ = \sqrt{\frac{2DC_S}{k_{11} \times I}} = \sqrt{\frac{2 \times 6{,}000 \times 600}{20 \times 0.18}} = 1{,}415 \text{ units}$$

Since $Q_1^\circ < b_1$, compare $C(Q_1^\circ)$ with $C(b_1)$ and $C(b_2)$ to get the optimum purchase quantity.

$$C(Q_1^\circ) = 6{,}000 \times 20 + \frac{600 \times 6{,}000}{1{,}415} + \frac{(0.18) \times (20) \times (1{,}415)}{2}$$
$$= \text{Rs } 1{,}25{,}091$$

$$C(b_1) = 6{,}000 \times 15 + \frac{600 \times 6{,}000}{2{,}000} + \frac{(0.18) \times (15) \times (2000)}{2}$$
$$= \text{Rs } 94{,}500$$

$$C(b_2) = 6{,}000 \times 9 + \frac{600 \times 6{,}000}{4{,}000} + \frac{(0.18) \times (9) \times (4{,}000)}{2}$$
$$= \text{Rs } 58{,}140$$

The least cost is Rs 58,140, hence optimal order quantity is $b_2 = 4000$ units.

Example 4

The annual demand for a product is 500 units. The cost of storage per unit per year is 10% of the unit cost. The ordering cost is Rs 180 for each order. The unit cost depends upon the amount ordered.

The range of amount ordered and the unit cost price are as follows:

Quantity	*Price*
$0 \le Q_1 < 500$	Rs 25
$500 \le Q_2 < 1{,}500$	Rs 24.80
$1{,}500 \le Q_3 < 3{,}000$	Rs 24.60
$3{,}000 \le Q_4$	Rs 24.40

Find the optimal order quantity.

Solution: Given $D = 500$ units, $C_S =$ Rs 180, $I = 0.10$, $b_1 = 500$, $b_2 = 1{,}500$, $b_3 = 3{,}000$, $k_{11} = 25$, $k_{12} = 24.80$, $k_{13} = 24.60$, $k_{14} = 24.40$.

Step 1: $$Q_4^\circ = \sqrt{\frac{2DC_S}{k_{14} \times I}} = \sqrt{\frac{2 \times 500 \times 180}{(24.40) \times (0.10)}} = 272 \text{ units}$$

Since $Q_4^\circ < b_3$, we compute Q_3°.

Step 2: $$Q_3^\circ = \sqrt{\frac{2DC_S}{k_{13} \times I}} = \sqrt{\frac{2 \times 500 \times 180}{(24.60) \times (0.10)}} = 270 \text{ units}$$

Since $Q_3^\circ < b_2$ (= 1,500), we calculate Q_2°.

Step 3: $$Q_2^\circ = \sqrt{\frac{2DC_S}{k_{12} \times I}} = \sqrt{\frac{2 \times 500 \times 180}{(24.80) \times (0.10)}} = 269 \text{ units}$$

Since $Q_2^\circ < b_1$, we go to next step.

Step 4: $$Q_1^\circ = \sqrt{\frac{2DC_S}{k_{11} \times I}} = \sqrt{\frac{2 \times 500 \times 180}{25 \times 0.10}} = 268 \text{ units.}$$

Now, compute $C(Q_1^\circ)$ and compare with $C(b_1)$, $C(b_2)$ and $C(b_3)$ in order to get optimal order quantity.

$$C(Q_1^\circ) = Dk_{11} + \frac{D}{Q_1^\circ} C_S + \frac{1}{2} Q_1^\circ \times k_{11} \times I$$

$$= 500 \times 25 + \frac{500 \times 180}{268} + \frac{268 \times 25 \times 0.10}{2}$$

$$= \text{Rs } 13{,}170.82$$

$$C(b_1) = Dk_{12} + \frac{DC_S}{b_1} + \frac{1}{2} b_1 \times k_{12} \times I$$

$$= 500 \times (24.80) + \frac{500 \times 180}{500} + \frac{500 \times (24.80) \times (0.10)}{2}$$

$$= \text{Rs } 13{,}200.00$$

$$C(b_2) = Dk_{13} + \frac{DC_S}{b_2} + \frac{1}{2} b_2 \times k_{13} \times I$$

$$= 500 \times (24.60) + \frac{500 \times 180}{1{,}500} + \frac{1{,}500 \times (24.60) \times (0.10)}{2}$$

$$= \text{Rs } 14{,}205.00$$

$$C(b_3) = Dk_{14} + \frac{DC_S}{b_3} + \frac{1}{2}\, b_3 \times k_{14} \times I$$

$$= 500 \times (24.40) + \frac{500 \times 180}{3{,}000} + \frac{3{,}000 \times (24.40) \times (0.10)}{2}$$

$$= \text{Rs } 15{,}890.00.$$

Since $C(Q_1^\circ) < C(b_1) < C(b_2) < C(b_3)$, $Q_1^\circ = 268$ units is the optimum order quantity.

Example 5

(a) Mini Computer Company purchases a component of which it has a steady usage of 1,000 units per year. The ordering cost is Rs 50 per order. The estimated cost of money invested is 25% per year. The unit cost of the component is Rs 40. Calculate the optimal ordering policy and total cost of inventory system, including purchase cost of the components.

(b) If in the above, the component supplier agrees to offer price discounts of minimum lot supplies as per schedule given below, reassess the decision on the optimal ordering policy and the total cost.

Lot Size	*Price*
Upto 149	Rs 40
150 – 499	Rs 39
500 or more	Rs 38

Solution: Given $D = 1000$ units, $C_S =$ Rs 50, $I = 0.25$, $k_1 =$ Rs 40

Now,
$$Q^\circ = \sqrt{\frac{2C_S D}{k_1 I}} = \sqrt{\frac{2 \times 1{,}000 \times 50}{(0.25) \times 40}} = 100 \text{ units.}$$

Total inventory cost

$$C(Q^\circ) = 1{,}000 \times 40 + \frac{50 \times 1{,}000}{100} + \frac{100 \times 40 \times 0.25}{2}$$

$$= \text{Rs } 41{,}100.$$

(b) When quantity discounts are allowed

$$Q_1^\circ = \sqrt{\frac{2 \times 50 \times 1{,}000}{40 \times 0.25}} = 100 \text{ units}$$

Since $Q_1^\circ < 149$ units, it is not feasible.
Total cost

$$C(Q_1^\circ) = 1{,}000 \times 40 + \frac{1{,}000}{100} \times 50 + \frac{100 \times 50 \times 0.25}{2}$$

$$= \text{Rs } 41{,}000.00$$

Also, $$Q_2^\circ = \sqrt{\frac{2\times 50\times 1{,}000}{39\times 0.25}} = 101 \text{ units}$$

Since $Q_2^\circ < 150$ units, it is not feasible.

Now, $$C(150) = 1{,}000\times 39 + \frac{1{,}000\times 50}{150} + \frac{150\times 39\times 0.25}{2}$$

$$= \text{Rs } 40{,}064.58$$

Since $C(150) < C(Q_1)$, the optimum purchase quantity = 150 units.

Saving $= \text{Rs } 41{,}000.00 - \text{Rs } 40{,}064.58$

$= \text{Rs } 935.42.$

10.5 MULTI-ITEM DETERMINISTIC MODELS (EOQ WITH CONSTRANTS)

When the inventories consist of several items under some limitations, then it is not possible to consider each item separately, since there exists a relation among the items. After constructing the cost equation in such models we use Lagrange's multipliers technique to minimise the cost in simple cases. We first solve the problem by ignoring the limitations and then consider the effect of limitations.

Let us suppose

n = number of items with instantaneous production and no lead time.
D_i = the demand rate which is known and is uniform for the ith item ($i = 1, 2, \ldots n$)
C_{1i} = the holding (or carrying) cost per unit of the quantity of ith item
C_{si} = the set-up cost per production run for the ith term
Q_i = quantity of ith item in inventory at the beginning of the period

The total cost per unit of time is given by

$$C = \sum_{i=1}^{n}\left\{\frac{1}{2}C_{1i}Q_i + C_{Si}\frac{D_i}{Q_i}\right\} \tag{10.5.1}$$

(as calculated in case 1, 10.3.4)

To determine the optimum values for Q_i, $i = 1, 2, \ldots n$ so as to minimise the total cost C, we have the necessary condition $\frac{\partial C}{\partial Q_i} = 0$, and $\frac{\partial^2 C}{\partial Q_i^2} > 0$ for all Q_i.

$$\frac{\partial C}{\partial Q_i} = \frac{1}{2}C_{1i} - \frac{C_{Si}D_i}{Q_i^2}$$

$$\frac{\partial C}{\partial Q_i} = 0 \Rightarrow Q_i = \sqrt{\frac{2C_{Si}D_i}{C_{1i}}}$$

$$\frac{\partial^2 C}{\partial Q_i^2} = \frac{2C_{Si}D_i}{Q_i^3} > 0$$

Hence the optimum Q_i° of Q_i is given by

$$Q_i^\circ = \sqrt{\frac{2C_{Si}D_i}{C_{1i}}}, \; i = 1, 2, ..., n \tag{10.5.2}$$

Limitation on Inventories

In this case, there is an upper limit, say K (in Rs) on the amount to be invested on inventory. Let C_{ui} be the unit price of ith item. Then,

$$\sum_{i=1}^{n} C_{ui}Q_i \le K \tag{10.5.3}$$

For the given problem, if the additional constraint (10.5.3) is satisfied, Q_i obtained as in (10.5.2) is optimal. If the additional constraint is not satisfied, optimal values of Q_i are found using Lagrange multiplier technique.

Let us formulate the Lagrangian function.

$$L = \sum_{i=1}^{n}\left(\frac{1}{2}C_{1i}Q_i + \frac{C_{Si}D_i}{Q_i}\right) + \lambda\left[\sum_{i=1}^{n} C_{ui}Q_i - K\right]$$

where λ is a Lagrange multiplier.

The necessary condition for L to be minimum is

$$\frac{\partial L}{\partial Q_i} = \frac{\partial L}{\partial \lambda} = 0 \; (i = 1, 2, ... n).$$

Therefore, $$\frac{\partial L}{\partial Q_i} = \frac{1}{2}C_{1i} - \frac{C_{Si}D_i}{Q_i^2} + \lambda C_{ui} = 0 \; (i = 1, 2, ..., n)$$

and,

$$\frac{\partial L}{\partial \lambda} = \sum_{i=1}^{n} C_{ui}Q_i - K = 0$$

These equations give

$$Q_i^\circ = \sqrt{\frac{2C_{Si}D_i}{C_{1i} + 2\lambda C_{ui}}}$$

and, $$\sum_{i=1}^{n} C_{ui}Q_i^\circ = M$$

The second equation implies that Q_i° must satisfy the investment constraint in equality sense.

Limitation of the Floor (Storage) Space

If instead of limitation of the inventories there is a limit of space in terms of square feet, then the additional constraint becomes

$$\sum_{i=1}^{n} f_iQ_i \le f$$

where f_i denotes the floor space required per unit per square feet for the ith item and f is the upper limit on the storage space.

The optimum values of Q_i are given by

$$Q_i^\circ = \sqrt{\frac{2C_{Si}D_i}{C_{1i} + 2\lambda f_i}}\ , i = 1, 2, ..., n$$

and,
$$\sum_{i=1}^{n} f_i Q_i^\circ = f$$

Limitation on Capital Investment

If the constraint is on the capital invested, the constraint is of the form

$$\sum_{i=1}^{n} C_i Q_i \le C$$

where C_i denotes the cost per unit of the ith item and C the total capital available for inventory investment. The optimum values of Q_i are given by

$$Q_i^\circ = \sqrt{\frac{2C_{Si}D_i}{C_{1i} + 2\lambda C_i}}]$$

Example 1

Consider a shop which produces three items. The items are produced in lots. The demand rate for each item is constant and can be assumed to be deterministic. No back orders are allowed. The pertinent data for the items is given below:

Item	1	2	3
Holding cost (Rs)	20	20	20
Set-up cost (Rs)	50	40	60
Cost per unit (Rs)	6	7	5
Yearly demand rate	10,000	12,000	7,500

Determine approximately the economic order quantities when the total value of average inventory levels of these items is Rs 1,000.

Solution: First, compute the optimal value Q°, without considering the restriction. Then,

$$Q_1^\circ = \sqrt{\frac{2 \times 50 \times 10{,}000}{20}} = 223 \text{ (approx.)}$$

$$Q_2^\circ = \sqrt{\frac{2 \times 40 \times 12{,}000}{20}} = 216 \text{ (approx.)}$$

$$Q_3^\circ = \sqrt{\frac{2 \times 60 \times 7{,}500}{20}} = 210 \text{ (approx.)}$$

Since the average optimal inventory at any time is $\frac{1}{2}Q_i^\circ$, the investment over the average inventory becomes

$$= \sum_{i=1}^{n} C_{ui}\left(\frac{1}{2}Q_i^\circ\right)\}$$

$$= \text{Rs}\left[6 \times \frac{223}{2} + 7 \times \frac{216}{2} + 5 \times \frac{210}{2}\right]$$

$$= \text{Rs } 1{,}950.00.$$

This amount is greater than the available amount of Rs 1,000. Therefore, we try to find the suitable value of λ by trial and error method for computing Q_1°. If we set λ = 4, then,

$$Q_1^\circ = \sqrt{\frac{2 \times 50 \times 10{,}000}{20 + 2 \times 4 \times 6}} = 121$$

$$Q_2^\circ = \sqrt{\frac{2 \times 40 \times 12{,}000}{20 + 2 \times 4 \times 7}} = 112$$

$$Q_3^\circ = \sqrt{\frac{2 \times 60 \times 7{,}500}{20 + 2 \times 4 \times 5}} = 123$$

and hence, the cost of average inventory

$$= \text{Rs}\left[6 \times \frac{121}{2} + 7 \times \frac{112}{2} + 5 \times \frac{123}{2}\right]$$

$$= \text{Rs } 1{,}112.50$$

This is also greater than Rs 1,000.

If we set λ = 5, then we get

$$Q_1^\circ = 111,\ Q_2^\circ = 102,\ Q_3^\circ = 113$$

and the cost of average inventory is Rs 972.50 which is less than Rs 1,000.

From this we conclude that the most suitable value of λ lies between 4 and 5.

To find the most suitable value of λ, we draw a graph between cost of average inventory and the value of λ as shown.

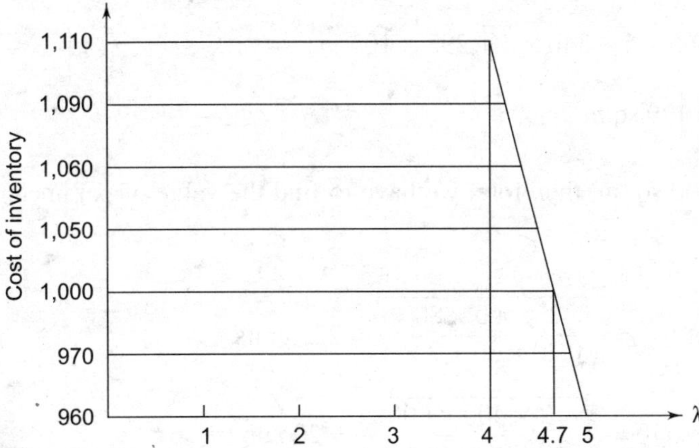

figure 10.7

From Fig. 10.7, the value λ = 4.7 is the most suitable value corresponding to which the cost of inventory is Rs 999.50 which is sufficiently close to Rs 1,000.

For λ = 4.7, the optimal values of the three items are $Q_1^\circ = 114$, $Q_2^\circ = 105$, $Q_3^\circ = 116$.

Example 2

A machine shop produces three products 1, 2 and 3 in lots. The shop has a warehouse whose total floor area is 4,000 sq. m. The relevant data for the three items is given below.

Item	1	2	3
Annual demand (units/year)	500	400	600
Cost per unit (Rs)	30	20	70
Set-up cost per lot (Rs)	800	600	1,000
Floor area required (sq. metres)	5	4	10

The inventory carrying charges for the shop are 20% of the average inventory valuation per annum for each item. If no stock-outs are allowed and at no time can the warehouse capacity be exceeded, determine the optimal lot size for each item.

Solution: First, ignoring the restriction of total warehouse capacity,

$$Q_1^\circ = \sqrt{\frac{2 \times 500 \times 800}{(0.20) \times 30}} = 365 \text{ units (nearly)}$$

$$Q_2^\circ = \sqrt{\frac{2 \times 400 \times 600}{(0.20) \times 20}} = 346 \text{ units (nearly)}$$

$$Q_3^\circ = \sqrt{\frac{2 \times 600 \times 1{,}000}{(0.20) \times 70}} = 292 \text{ units (nearly)}$$

Using these values, we calculate the floor area as

$$\sum_{i=1}^{3} Q_i^\circ f_i = 365 \times 5 + 346 \times 4 + 292 \times 10$$

$$= 6{,}129 \text{ sq.m}$$

As $\sum_{i=1}^{3} Q_i^\circ f_i > 4{,}000$ sq. m, therefore, we have to find the value of parameter λ by trial and error substitution.

For $\lambda = 1$, we get $Q_1^\circ = \sqrt{\dfrac{2 \times 500 \times 800}{0.20 \times 30 + 2 \times 5}} = 223$ units

$$Q_2^\circ = \sqrt{\frac{2 \times 400 \times 600}{(0.20) \times 20 + 2 \times 4}} = 200 \text{ units}$$

$$Q_3^\circ = \sqrt{\frac{2 \times 600 \times 1{,}000}{(0.20) \times 70 + 2 \times 10}} = 187 \text{ units}$$

and $\sum_{i=1}^{3} Q_i^\circ f_i = 223 \times 5 + 200 \times 4 + 187 \times 10 = 3{,}785$ sq.m, which is less than 4,000 sq.m.

Trying for other values of λ in the range 0.8 to 1.0, where for $\lambda = 0.8$, we get $\sum_{i=1}^{3} Q_i^\circ f_i = 4{,}051$ which is slightly greater than 4,000 sq. m. If we choose $\lambda = 0.835$, we get $Q_1^\circ = 236$ units, $Q_2^\circ = 211$ units, $Q_3^\circ = 197$ units and $\sum_{i=1}^{3} Q_i^\circ f_i = 3{,}994$ units which is nearer to 4,000. But, if we choose $\lambda = 0.83$, $\sum_{i=1}^{3} Q_i^\circ f_i =$ 4,008 which exceeds 4,000, which violate the constraint $\sum_{i=1}^{3} Q_i^\circ f_i \le f$. Hence, $\lambda = 0.835$ is the required value of λ, and the optimal quantity is given by

$$Q_1^\circ = 236 \text{ units}, Q_2^\circ = 211 \text{ units}, Q_3^\circ = 197 \text{ units}.$$

10.6 INVENTORY PROBLEMS WITH UNCERTAIN DEMAND

In this section we study inventory models in which demand cannot be completely predetermined. It is uncertain and can fluctuate either way. In many situations, it is found that neither the consumption rate nor the lead time is constant throughout the year. In order to overcome these difficulties, extra stock is maintained to meet the demands, if any. This extra stock is termed as *Buffer Stock* or *Safety Stock*.

Determining Buffer Stock and Reorder Level

To determine the buffer stock, we approximate the estimated maximum lead time and normal lead time for a particular item. Let B denote the buffer stock, L_d denote the difference between the maximum and normal lead times, r the consumption rate during the lead time. Then,

$$B = L_d \times r$$

If we do not maintain a buffer stock, then the total requirements for inventory during lead time L will become Lr. Thus, as soon as the stock reaches the level Lr, a quantity Q should be ordered. This point is called *reorder level* (ROL). However, due to uncertainty in supply, the policy of ROL may result in shortages. To avoid this, a buffer stock B is added and an order placed when stock level reaches $B + Lr$. Hence, the reorder level is given by

$$\begin{aligned} \text{ROL} &= \text{Lead time demand} + \text{Buffer stock} \\ &= B + Lr \\ &= L_d r + Lr = (L_d + L)r \end{aligned}$$

For example, if the monthly consumption rate for items is 150 units, the normal lead time is 15 days, and the buffer stock is 200 units, Then, $\text{ROL} = 200 + \frac{1}{2} \times 150 = 275$ units.

If we take t days for reviewing the reorder system, then on the assumption of uniform consumption rate, we get

$$\text{ROL} = B + Lr + \frac{rt}{2}$$

Determining optimum Buffer Stock

When the buffer stock maintained is very low, the inventory holding cost is low but shortages will occur frequently and the cost of shortages will be high. Suppose the buffer stock maintained is very high, then shortages would be rare, and the cost of shortages would be low, but the inventory costs would be high.

Hence, it becomes necessary to strike a balance between the cost of shortages and the cost of inventory holding to arrive at the optimum buffer stock.

Example 1

The average monthly consumption for an item is 300 units and the normal lead time is one month. If maximum consumption has been up to 370 units per month and maximum lead time is $1\frac{1}{2}$ months, what should be the buffer stock for the item?

Solution: Max. lead time demand = (Max. lead time) × (Max. demand rate) = $\frac{3}{2} \times 370 = 555$ units.

Normal lead time demand = $1 \times 300 = 300$ units
Buffer stock = $555 - 300 = 255$ units

Example 2

A firm uses 12,000 units of a raw material every year costing Rs 1.25 per unit. Ordering cost is Rs 15 per order and the holding cost is 5% per year of average inventory.

(i) Find the economic order quantity.

(ii) The firm follows EOQ purchasing policy. It operates for 300 days per year. Procurement time is 14 days and safety stock is 400 units. Find the reorder point, the maximum inventory and the average inventory.

Solution: (i) EOQ = $Q^\circ = \sqrt{\dfrac{2DC_S}{C_1}} = \sqrt{\dfrac{2 \times 15 \times 12{,}000}{5/100 \times 1.25}} = 2{,}400$ units

(ii) ROL = Buffer stock + Consumption during normal lead time

$$= 400 + 14 \times \frac{12{,}000}{300} = 960 \text{ units}$$

(iii) Maximum inventory $= Q^\circ + B = 2{,}400 + 400 = 2{,}800$ units

(iv) Minimum inventory $= B = 400$ units

(v) Average inventory $= \dfrac{2{,}800 + 400}{2} = 1{,}600$ units

Example 3

The following information is provided for an item:

Annual usage = 12,000 units, ordering costs = Rs 60 per order, carrying cost 10%, unit cost of item = Rs 10 and lead time = 10 days.

There are 300 working days a year. Determine the EOQ and the number of orders per year. In the past two years the use rate has gone as high as 70 units per day. For a reordering system based on the inventory level, what should be the safety stock? What should be the reorder level at this safety stock? What would be the carrying costs for a year?

Solution: Given C_S = Rs 60/order, D = 12,000 units/year,

$$C_1 = CI = \text{Re. } 10 \times \frac{10}{100} = \text{Re } 1\text{/unit/year}$$

Therefore, EOQ = $Q^\circ = \sqrt{\dfrac{2DC_S}{CI}} = \sqrt{\dfrac{2 \times 60 \times 12{,}000}{1}}$ = 1,200 units

No. of orders/year = $\dfrac{12{,}000}{1{,}200} = 10$

Average usage $= \dfrac{12{,}000}{300} = 40$ units/day

Maximum usage = 70 units/day

Therefore, safety stock = (70 – 40) × 10 = 300 units

ROL = Average lead time demand + Safety stock

= 40 × 10 + 300 = 700 units

Average inventory $= \dfrac{Q^\circ}{2} + B = \dfrac{1{,}200}{2} + 300 = 900$ units

Hence, inventory carrying cost/year = Rs. 900 × $\left(10 \times \dfrac{10}{100}\right)$ = Rs 900.

Example 4

An automobile has determined that 16 spare engineers will result into a stock-out risk of 25% while 20 will reduce the risk to 15% and 24 to 10%. If the lead time is 3 months and the average usage is 6 engines per month, what should be the ROL to maintain 85% service level?

Solution: Lead time demand = 3 × 6 = 18 engines

Safety stock for 85% service or 15% disservice = 20 engines

ROL = 18 + 20 = 38 engines.

Example 5

A firm has normally distributed forecast of usage with mean absolute deviation (MAD) = 60 units. It desires a service level which limits stock outs to one order cycle per year.

(a) How much safety stock should be kept if the order quantity is normally a week's supply?

(b) What will be the safety stock if the order quantity is 4 weeks' supply?

Solution: (a) No. of orders/year = 52. Since there is one stock out per year, service level $= \dfrac{52-1}{52} \times 100 = 98\%$.

Therefore, safety (service) factor Z = 2.05 (by normal table), $Z = \dfrac{x-\mu}{\sigma}$

$$\text{MAD} = \sqrt{\frac{2}{\pi}} \times \text{standard deviation.}$$

$$= (0.8) \times \text{SD}$$

Therefore, $\text{SD} = \dfrac{\text{MAD}}{0.8} = \dfrac{60}{0.8} = 75$. Now,

$$Z = \frac{\text{Safety stock} - 0}{75} \Rightarrow \text{Safety stock} = Z \times 75 = (2.05) \times 75 = 154 \text{ units.}$$

(b) No. of orders per year $= \dfrac{52}{4} = 13$

Therefore, service level = $\frac{13-1}{13} \times 100 = 92.3\%$

Safety factor Z from normal table = 1.43

$$Z = \frac{SS - 0}{S \cdot D} \Rightarrow \text{Safety stock} = 1.43 \times 75 = 107 \text{ units.}$$

Example 6

The average demand for an item is 120 units/year. The lead time is 1 month and the demand during lead time follows normal distribution with average of 10 units and standard deviation of 2 units. If the item is ordered once in 4 months and the policy of the company is that there should not be more than 1 stock out every two years, determine the reorder level.

Solution: No. of orders in 2 years = $\frac{1}{4}$ (2 × 12) = 6 %. Service = $\frac{6-1}{6} \times 100 = 83.33$. Therefore, Z (from normal table) = 0.96. Now $Z = \frac{\text{Safety stock} - 0}{SD} \Rightarrow \text{Safety stock} = Z \times SD = (0.96) \times 2 = 1.96.$

Also, ROL = 1 × 10 + 1.92 = 11.92 units

Example 7

A look at the past records gives the following distribution for lead time and daily demand during lead time.

Lead time distribution											
Lead time (days)	0	1	2	3	4	5	6	7	8	9	10
Frequency	0	0	1	2	3	4	4	3	2	2	1

Demand rate distribution								
Demand/day (units)	0	1	2	3	4	5	6	7
Frequency	3	5	4	5	2	3	2	1

What should be the buffer stock?

Solution: The computation of average lead time.

Lead time (1)	Frequency (2)	(1) × (2)
1	0	0
2	1	2
3	2	6
4	3	12
5	4	20
6	4	24
7	3	21
8	2	16
9	2	18
10	1	10
	22	129

Average lead time = $\dfrac{129}{22}$ = 5.56 days

Computation of average demand rate

Demand (1)	Frequency (2)	(1) × (2)
0	3	0
1	5	5
2	4	8
3	5	15
4	2	8
5	3	15
6	2	12
7	1	7
	25	70

Average demand rate = $\dfrac{70}{25}$ = 2.8

Average lead time demand = 5.86 × 2.8 = 16.4 units
Max. lead time demand = Max. lead time × Max. demand rate
= 10 × 7 = 70 units
Buffer stock = 70 – 16.4 = 53.6 units

Example 8

(a) A company uses 24,000 units of a raw material annually which costs Rs 1.25 per unit. Placing each order costs Rs 22.50 and the carrying cost is 5.47 per year of the average inventory, find the EOQ and total inventory cost.

(b) Should the company accept the offer made by the supplier of a discount of 5% on the cost price on a single order of 24,000 units?

(c) Suppose the company works for 300 days a year. If the procurement time is 12 days and safety stock is 400 units, find the reorder point, the minimum, maximum and average inventory.

Solution:

(a) Give D = 24000 units/year, C = Rs 1.2/unit, C_S = Rs 22.50/order

$$C_1 = \text{CI} = \text{Re}\left(1.25 \times \frac{5.4}{100}\right) = \text{Re } 0.0675\text{/unit/year. Hence,}$$

$$Q^{\circ} = \sqrt{\frac{2C_S D}{C_1}} = \sqrt{\frac{2 \times 22.50 \times 24{,}000}{0.0675}} = 4{,}000 \text{ units}$$

$$C(Q^{\circ}) = \text{Rs}\left[24{,}000 \times 1.25 + \sqrt{2 \times 0.0675 \times 22.50 \times 24{,}000}\right]$$

$$= \text{Rs } 30{,}270.$$

(b) Under the discount offer, annual fixed cost (capital cost) = Rs (1.25 × 0.95 × 24,000) = Rs 28,500

inventory carrying cost = $\dfrac{q}{2}\,C_1 = \dfrac{1}{2} \times 24{,}000 \times (0.0675 \times 0.95)$ = Rs 769.50

ordering cost = Rs 22.50. Total annual cost = Rs (28,500 + 769.50 + 22.5) = Rs 29,292. Since the total annual cost under discount conditions is lower than that under optimal order size, discount offer should be accepted.

(c) Since the company works for 300 days, the daily demand = $\frac{24000}{300}$ = 80 units. $t_0 = \frac{d_0}{80} = \frac{4,000}{80}$

= 50 days lead time = 12 days since t_0 > lead time,

ROL = 80 × 12 + 400 = 1,360 units.

Minimum inventory = 400 units, Maximum inventory = 400 × $Q°$ = 4,400 units

Average inventory = $\frac{1}{2}$ (400 + 4,400) = 2,400 units.

Example 9

A company uses 50,000 units of an item annually each costing Rs 1.20. Each order costs Rs 45 and inventory carrying costs are 15% of the annual average inventory value.

(i) Find EOQ.

(ii) If the company operates 250 days a year, the procurement time is 10 days and safety stock is 500 units, find reorder level, maximum, minimum and average inventory.

Solution: Given D = 50,000 units, C_S = Rs 45 per order

C = Rs 1.20 per unit

I = Re 0.15

(i) $Q° = \sqrt{\dfrac{2 \times 45 \times 50{,}000}{(1.20) \times (0.15)}}$ = 5,000 units

(ii) The company operates 250 days a year, requirement per day = $\frac{50,000}{250}$ = 200 units.

Lead time demand = r = 10 × 200 = 2,000 units
Safety Stock = B = 500 units
Hence, ROL = 2,000 + 500 = 2,500 units
Maximum inventory = 5,000 + 500 = 5,500 units
Minimum inventory = 5,000 units

Average inventory = $\frac{1}{2}$ × 5,000 + 500

= 3,000 units.

10.7 INVENTORY CONTROL SYSTEM

10.7.1 Fixed Order Quantity System (Q-System)

In this system, a reorder is made when the stock falls to a reorder level, which is equal to the lead time requirements plus buffer stocks. The procurement and consumption cycle is shown in Fig. 10.8.

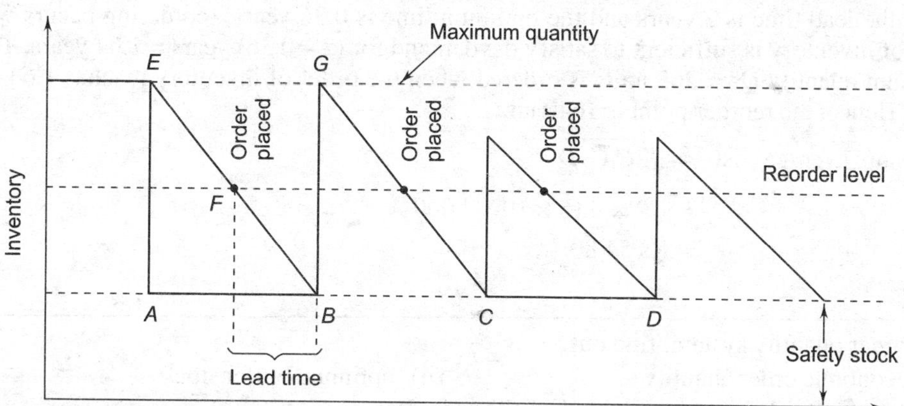

figure 10.8

From Fig. 10.8 it is clear that a supply equal to EOQ is received at the point A and the quantity in stock reaches a point E. The items are then issued and at time F, when the stock reaches ROL, an order is placed for quantity Q = EOQ and the issues continue. At point B the supplies of order placed at point F are received and the stock reaches G. At G, there is a delay in receiving the supplies due to lead time and so we issue from the buffer stock.

Thus, for this system of ordering, we fix the size of the order, that is, every time the same quantity is ordered, but the time of ordering is allowed to vary depending on the actual demand (or usage). This is named as 'fixed order quantity' or 'ordering level' or 'two-bin system'.

This is called the two-bin system because the system is separated into two bins. One bin contains sufficient stocks to meet the demand of time between the arrival of an order and placing the next. The second bin contains sufficient stock to meet any probable demand during the period of replenishment. An order is placed as soon as the first bin is exhausted.

In this system of ordering we have to decide three parameters of an item before placing the order. These are:

(a) $\text{EOQ} = \sqrt{\dfrac{2DC_S}{C_1}}$

(b) Optimum buffer stock

(c) ROL.

Example 1

Consider the inventory system with the following data in usual notations:

D = 1,000 units/year, I = 0.30, P = Re 0.50 per unit

C_S = Rs 10, L = 2 years (lead time)

Find (i) EOQ, (ii) reorder point, (iii) minimum average cost

Solution: (i) $$Q^\circ = \sqrt{\frac{2C_SD}{IP}} = \sqrt{\frac{2 \times 10 \times 1{,}000}{0.15}} = 365 \text{ units}$$

$$t^\circ = \frac{Q^\circ}{D} = \frac{365}{1{,}000} = 0.36 \text{ years.}$$

(ii) Since the lead time is 2 years and the optimum time is 0.36 years, reordering occurs when the level of inventory is sufficient to satisfy the demand for (2 – 0.36) years = 1.64 years. Thus, the optimum quantity $Q^\circ = 365$ units is ordered when the order of inventory reaches $1.64 \times 1{,}000$ units. Hence, the reorder point is 1640 units.

(iii) Minimum average cost $= \sqrt{2C_1C_SD}$

$= \sqrt{2 \times 0.15 \times 10 \times 1{,}000}$

= Rs 54.80.

Example 2

For a fixed order quantity system, find out

(i) economic order quantity (ii) optimum buffer stock

(iii) reorder level, for an item with the following data

Annual consumption (D) = 10,000 units

Cost of one unit = Re 1.00

C_s = Rs 12 per production run

C_1 = Re 0.24 per unit.

Past lead times: 15 days, 25 days, 13 days, 14 days, 30 days, 17 days.

Solution: (i) EOQ $= \sqrt{\dfrac{2C_SD}{C_1}} = \sqrt{\dfrac{2 \times 12 \times 10{,}000}{0.24}}$ = 1,000 units.

(ii) Optimum buffer stock = Max. lead time – Normal lead time × Monthly consumption

$= \left(\dfrac{30-15}{30}\right) \times \left(\dfrac{10{,}000}{12}\right)$

= 416.66 units (since normal lead time is 15 days)

≅ 420.

(iii) Normal lead time consumption

= Normal lead time × Monthly consumption

$= \left(\dfrac{15}{30}\right) \times \left(\dfrac{10{,}000}{12}\right)$ = 417 units

Therefore, the reorder level = safety stock + normal lead time consumption

= 420 + 417

= 837 units

≅ 840 units. (approx.)

(iv) Maximum inventory level

= Safety stock + Economic order quantity

= 420 + 1000

= 1,420 units.

Since inventory would fluctuate from a maximum of 1,420 units to a minimum of 420 units, the average inventory

$$= \frac{1}{2} \text{ [Safety stock + Max. inventory]}$$

$$= \frac{1}{2} [420 + 1{,}420]$$

$$= 920 \text{ units.}$$

10.7.2 Periodic Review System (P-System)

This system is also known as the fixed internal system or replenishment inventory system or cyclic review system.

In this system, the size of order quantity may vary with the fluctuations in demand, but the ordering interval is fixed. The system is specified for any item by

(i) Review period t

(ii) Replenishment level (R).

In this system, the inventory position is periodically reviewed—once, weekly, monthly, quarterly or half-yearly. At each review period, an order is placed for an amount equal to the difference between a fixed replenishment level and the actual inventory level. The calculation of R is based on the formula

Replenishment level = Average consumption during a review period + Lead time + Safety stock

Order quantity = Replenishment level – Stock available

The following diagram gives the way in which the periodic review system operates.

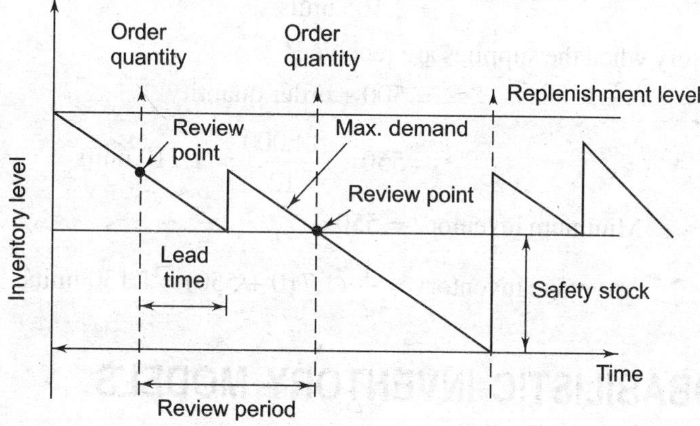

figure 10.9 **Periodic review system**

The order quantity is variable in size from one review date to another.

Example 3

For a periodic review system, find out the various parameters for an item with the following data:

Annual consumption = 14,000 units,

Cost of one unit = Rs 10

Supplier's minimum quantity = 1,000 units

Normal lead time = 10 days
Maximum lead time = 15 days
Maximum consumption = 1,020 (average consumption)

Solution: The maximum number of orders to cover the annual requirement is $\frac{1,40,000}{1,000} = 14$ orders.

Therefore, the review period should be $\frac{1}{14}$th of an year or 26 days. We consider the period of review as 1 month instead of 26 days. Now,

Safety stock = Max. consumption rate (review period + max. lead time) – Normal consumption rate (review period + normal lead time)

$$= (1.20) \times \frac{14,000}{12}\left(1+\frac{15}{30}\right)-\frac{14,000}{12}\left(1+\frac{10}{30}\right)$$

$$= 545 \text{ units}$$

$$\approx 550 \text{ units}$$

Replenishment level = Average rate of consumption × (Review period + normal lead time) + Safety stock

$$= \frac{14,000}{12}\left(1+\frac{10}{30}\right) + 550$$

$$\approx 2,105 \text{ units}$$

Maximum inventory when the supplies are received

$$= 500 + \text{order quantity}$$

$$= 550 + \frac{14,000}{12} \approx 1,710 \text{ units}$$

Minimum inventory = 550

$$\text{Average inventory} = \frac{1}{2}(1,710 + 550) = 1,130 \text{ units.}$$

10.8 PROBABILISTIC INVENTORY MODELS

The inventory models discussed in earlier sections seems to be unrealistic. In some reality, the demand of an item is not known exactly but it will follow some probability distribution. In probabilistic systems we minimise the total expected costs rather than actual costs.

Consider the case of a perishable item, for example, newspaper. Assume that the demand of the newspaper does not follow a fixed pattern. A probability distribution can be obtained for the same. The probability distribution may be a discrete distribution, a uniform distribution or continuous distribution.

For example, for each unsold newspaper, there will be a penalty which is given by the formula

Marginal cost of surplus/unit C_1 = Purchase price/unit – Salvage value/unit

For each shortage unit, there will be a penalty which is given by the formula

Marginal cost of shortage/unit C_2 = Selling price/unit – purchase price/unit.

Let the generalised probability distribution of the demand of the item be a discrete distribution as shown below:

Observation	i	1	2	3	...	i	...	n
Demand	D_i	D_1	D_2	D_3	...	D_i	...	D_n
Probability	p_i	p_1	p_2	p_3	...	p_i	....	p_n

The optimal order size D_i^o is determined by the relation

$$P_{i-1} < \frac{C_2}{C_1 + C_2} < P_i$$

where P_i is the cumulative probability of having demand up to D_i^o.

Example 1

The daily demand of bread at a bakery follows a discrete distribution as follows:

S.No.	1	2	3	4	5	6	7	8	9	10	11
Demand D_i	25	26	27	28	29	30	31	32	33	34	35
Probability p_i	0.2	0.11	0.10	.09	0.08	0.12	0.14	0.05	0.04	0.04	0.03

The purchase price of the bread is Rs 8 per packet. The selling price is Rs 11 per packet. If the bread packets are not sold within the day of purchase, they are sold at Rs 4 per packet to hotels for secondary use. Find the optimal order size of the bread.

Solution:

Given purchase price/packet = Rs 8

Selling price/packet = Rs 11

Salvage price/packet = Rs 4

Marginal cost of surplus $C_1 = 8 - 4 =$ Rs 4

Marginal cost of shortages $C_2 = 11 - 8 =$ Rs 3

$$\text{Therefore, cumulative probability } P = \frac{C_2}{C_1 + C_2} = \frac{3}{4+3} = 0.43$$

Next, we find the cumulative probability of demand

S.No.	Demand D_i	Probability p_i	Cumulative probability P_i
1	25	0.2	0.2
2	26	0.11	0.31
3	27	0.10	0.41
4	28	0.09	0.50
5	29	0.08	0.58
6	30	0.12	0.70
7	31	0.14	0.84
8	32	0.05	0.89
9	33	0.04	0.93
10	34	0.04	0.97
11	35	0.03	1.00

From the Table it follows that

$$P_3 < \frac{C_2}{C_1 + C_2} = 0.43 < P_4$$

That is,

$$0.41 < 0.43 < 0.50$$

Therefore, the optimal order size is D_4, which is equal to 28 breads.

10.8.1 Single Period Problem with Instantaneous Demand

The following assumptions should be made in this case:

(i) t is the constant interval between orders (t may be considered as unity, e.g. daily, weekly, monthly, etc.).
(ii) Q is the stock (in discrete units) for time t.
(iii) r is the estimated (random) demand at a discontinuous rate with probability $p(r)$.
(iv) C_1 is the holding cost per item per t time unit.
(v) C_2 is the shortage cost per item per t time unit.
(vi) Lead time is zero.

In this model, it is assumed that the total demand is filled at the beginning of the period. Thus, depending on the amount r demanded, the inventory position just after the demand occurs as a surplus or shortage. These two cases are shown in Fig. 10.10(a) and 10.10(b).

Case 1: When demand r does not exceed the stock Q, that is, $r \le Q$

Case 2: When demand r exceeds the stock Q, that is, $r > Q$.

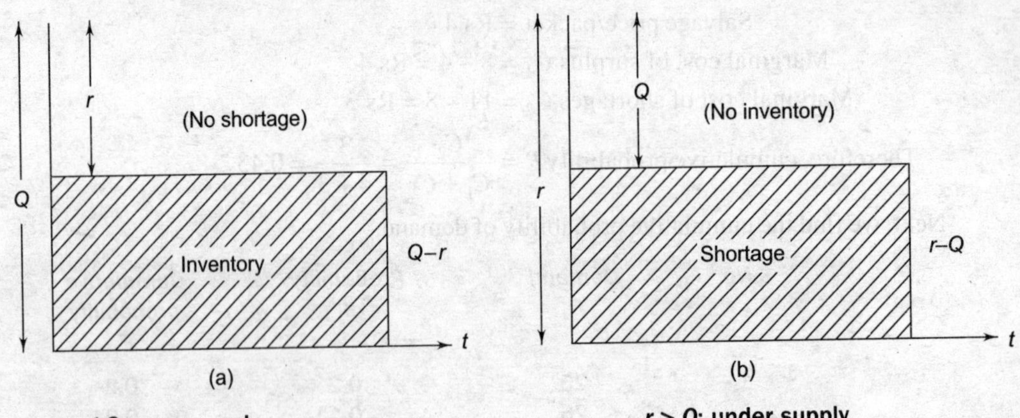

figure 10.10

Suppose $r \le Q$; the holding cost becomes

$$= \begin{cases} (Q-r)\,C_1, & \text{for } r \le Q \\ C_1 \times 0 & \text{for } r > Q \quad \text{(no stock)} \end{cases}$$

Suppose $r > Q$; the shortage cost per unit time is

$$= \begin{cases} (C_2 \times 0) & \text{for } r \le Q \quad \text{(no shortage)} \\ (r-Q)C_2 & \text{for } r > Q \end{cases}$$

To get the expected cost, multiply the cost by given probability $p(r)$. To get the total expected cost, sum over all expected costs, that is, the costs associated with each possible value of r. Hence, the total expected cost per unit time is given by the cost equation

$$C(Q) = \sum_{r=0}^{Q}(Q-r)\,C_1\,p(r) + \sum_{r=Q+1}^{\infty} C_1 \times 0 \times p(r) + \sum_{r=0}^{Q} C_2 \times 0 \times p(r) + \sum_{r=Q+1}^{\infty}(r-Q)\,C_2\,p(r)$$

$$= \sum_{r=0}^{Q}(Q-r)C_1 p(r) + \sum_{r=Q+1}^{\infty}(r-Q)\,C_2\,p(r)$$

To find the optimum value of Q so as to minimise $C(Q)$, the following conditions must hold

$$C(Q+1) - C(Q) > 0 \qquad \text{and} \qquad C(Q) - C(Q-1) < 0$$

i.e. $\qquad \Delta C(Q-1) < 0 < \Delta C(Q)$

Now,

$$\Delta C(Q) = C(Q+1) - C(Q)$$

$$= C_1 \sum_{r=0}^{Q}(Q-r)\,p(r) + C_1 \sum_{r=0}^{Q} p(r) + C_2\left(\sum_{r=Q+1}^{\infty}(r-Q)p(r) - \sum_{r=Q+1}^{\infty} p(r)\right) - C(Q)$$

$$= \sum_{r=0}^{Q}(Q-r)\,C_1 p(r) + \sum_{r=Q+1}^{\infty}(r-Q)p(r)\,C_2 + C_1 \sum_{r=0}^{Q} p(r) - C_2 \sum_{r=Q+1}^{\infty} p(r) - C(Q)$$

$$= C_1 \sum_{r=0}^{Q} p(r) - C_2 \sum_{r=Q+1}^{\infty} p(r)$$

$$= C_1 \sum_{r=0}^{Q} p(r) - C_2\left[\sum_{r=0}^{\infty} p(r) - \sum_{r=0}^{Q} p(r)\right]$$

$$= C_1 \sum_{r=0}^{Q} p(r) - C_2 + C_2 \sum_{r=0}^{Q} p(r), \text{ since } \sum_{r=0}^{\infty} p(r) = 1$$

$$= (C_1 + C_2)\sum_{r=0}^{Q} p(r) - C_2$$

For minimum of $C(Q)$, $\Delta C(Q) > 0$. Hence,

$$(C_1 + C_2)\sum_{r=0}^{Q} p(r) - C_2 > 0$$

$$\Rightarrow \qquad \sum_{r=0}^{Q} p(r) > \frac{C_2}{C_1 + C_2}.$$

Similarly, $\Delta C(Q-1) = C(Q) - C(Q-1)$ implies

$$\sum_{r=0}^{Q-1} p(r) < \frac{C_2}{C_1 + C_2}$$

Hence ,
$$\sum_{r=0}^{Q-1} p(r) < \frac{C_2}{C_1 + C_2} < \sum_{r=0}^{Q} p(r)$$

Instantaneous Demand and Continuous Units

Here, the instantaneous demand is a continuous random variable rather than a discrete one. The probability distribution $p(r)$ are replaced by the probability density function $f(r)$. In this case $\int_0^{\infty} f(r)dr = 1$ and $f(r) \geq 0$. The cost matrix for this model bcomes

$$C(Q) = C_1 \int_0^{Q} (Q-r) f(r)\, dr + C_2 \int_{r=Q}^{\infty} (r-Q) f(r)\, dr \qquad (10.8.1)$$

The optimal value of Q is obtained by equating to zero the first derivative of $C(Q)$.

That is,
$$\frac{dC}{dQ} = 0.$$

We know that if
$$C(Q) = \int_{r=a(Q)}^{b(Q)} f(r, Q)\, dr, \text{ then,}$$

$$\frac{dC}{dQ} = \int_{r=a(Q)}^{b(Q)} \frac{\partial f(r,Q)}{\partial Q} dr + \left[f(r,Q) \frac{dr}{dQ} \right]_{r=a(Q)}^{b(Q)}$$

$$r = a(Q)$$

Differentiating the equation (10.8.1), we get

$$\frac{dC}{dQ} = C_1 \int_{r=0}^{a} (1-0) f(r)\, dr + C_1 \left[(Q-r)\, f(r) \frac{dr}{dQ} \right]_{r=0}^{Q}$$

$$+ C_2 \int_{r=Q}^{\infty} (0-1)\, f(r)\, dr + C_2 \left[(r-Q) f(r) \frac{dr}{dQ} \right]_{r=Q}^{\infty}$$

$$= C_1 \int_0^{Q} f(r)\, dr - C_2 \int_Q^{\infty} f(r)\, dr$$

$$= C_1 \int_0^{Q} f(r)\, dr - C_2 \left[\int_0^{\infty} f(r) dr - \int_0^{Q} f(r)\, dr \right]$$

$$= C_1 \int_0^{Q} f(r)\, dr - C_2 \left[1 - \int_0^{Q} f(r)\, dr \right]$$

$$= (C_1 + C_2)\int_0^Q f(r)\,dr - C_2$$

In order that $\frac{dC}{dQ} = 0$, we have

$$(C_1 + C_2)\int_0^Q f(r)\,dr - C_2 = 0$$

$$\Rightarrow \qquad \int_0^Q f(r)\,dr = \frac{C_2}{C_1 + C_2}$$

Further,
$$\frac{d^2}{dQ^2}C(Q) = (C_1 + C_2)\left[f(r)\frac{dr}{dQ}\right]_0^Q$$
$$= (C_1 + C_2)f(Q) > 0$$

Hence, the optimum value of Q satisfies

$$\int_0^Q f(r)\,dr = \frac{C_2}{C_1 + C_2}$$

Example 2

A baking company sells cake by the kilogram. It makes a profit of 50 paise a kilo on every kilogram sold on the day it is backed. It disposes of all cakes not sold on the date it is baked at a loss of 12 paise a kilogram. If demand is known to be rectangular, between 2,000 and 3,000 kg, determine the optimum daily amount baked.

Solution: Given, C_1 = Re 0.12 (sale of left overs);
C_2 = Re 0.50 (profit on sale)

Demand r is rectangular between 2,000 and 3,000 kg means the distribution $f(r)$ is continuous, given by

$$f(r) = \begin{cases} \frac{1}{b-a} & \text{if} \quad a \le r \le b \\ 0 & \text{otherwise} \end{cases}$$

$$= \begin{cases} \frac{1}{3{,}000 - 2{,}000} & \text{if} \quad 2{,}000 \le r \le 3{,}000 \\ 0 & \text{otherwise} \end{cases}$$

$$= \begin{cases} \frac{1}{1{,}000} & \text{for} \quad 2{,}000 \le r \le 3{,}000 \\ 0 & \text{otherwise} \end{cases}$$

$$\int_0^Q f(r)\,dr = \int_0^{2000} f(r)\,dr + \int_{2000}^Q f(r)\,dr$$

$$= 0 + \int_{2000}^Q \frac{1}{1{,}000}\,dr$$

$$= \int_{2000}^Q \frac{1}{1{,}000}\,dr$$

We know $$\int_0^Q f(r)\,dr = \frac{C_2}{C_1 + C_2}.$$

So, $$\int_{2000}^{Q} \frac{1}{1{,}000}\,dr = \frac{0.50}{0.62}$$

$$\Rightarrow \quad \left[\frac{r}{1{,}000}\right]_{r=2000}^{Q} = \frac{50}{62}$$

$$\Rightarrow \quad \frac{1}{1{,}000}[Q - 2{,}000] = \frac{50}{62}$$

$$\Rightarrow \quad Q = 2{,}806.45 \text{ kg}.$$

10.8.2 Single Period Problem with Uniform Demand

The problem has the following assumptions:

1. Demand is uniform over a period. Let it be r units per time period.
2. Reorder time is fixed and known. Thus, the set-up cost is not included in the total cost.
3. Q is the stock level to which the stock is raised at the end of every period t.
4. The production is instantaneous.
5. C_1 is the carrying cost per quantity per unit time.
6. C_2 is the shortage cost per quantity per unit time.
7. Lead time is zero.
8. $f(r)$ is the probability distribution for demand r which is known.

The demand occurs uniformly rather than instantaneously during the period as shown in Fig. 10.11(a) and Fig. 10.11(b)

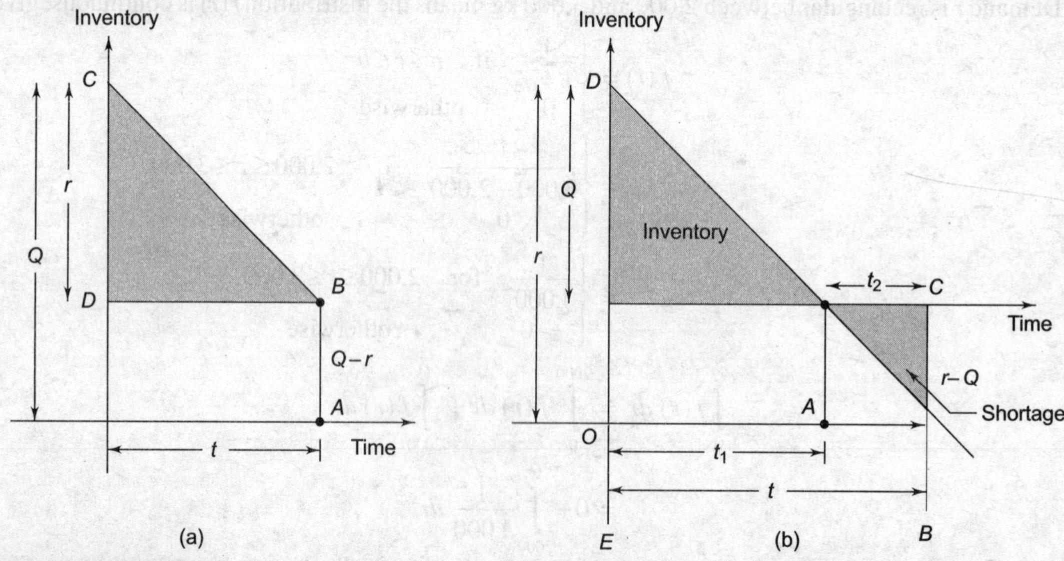

figure 10.11 **Uniform demand no set up cost model**

The problem is to determine the optimal order level α where $r \le Q$ or $r \le Q$. For $r > Q$, no shortage occur. For $r > Q$, shortage occurs.

Now, for $r \le Q$: No shortage and only carrying cost is involved (see Fig. 10.11(a).

$$= C_1 \text{ (inventory area } OABC]$$

$$= C_1 \cdot \frac{1}{2} [AB + OC] \times OA \text{ (by the area of trapezium)}$$

$$= C_1 \cdot \frac{1}{2} [Q - r + Q]t$$

$$\text{Total carrying cost} = \frac{1}{2} \cdot C_1 t \sum_{r=0}^{\alpha} (2Q - r)\, p(r)$$

$r > Q$: Both costs are involved (see Fig. 10.11 (b)).

= The carrying cost is the area of ΔOAD and the shortage cost is the area of ΔABC

$$\Delta OAD = \frac{1}{2} Q t_1 C_1 \text{ and } \Delta ABC = \frac{1}{2}(r - Q)\, C_2 t_2$$

By the property of similar triangles of ΔDEB and ΔDOA, we get $\dfrac{DO}{DE} = \dfrac{OA}{OB}$ or $\dfrac{Q}{r} = \dfrac{t_1}{t}$

Similarly, we get $$\frac{t_2}{t} = \frac{r-Q}{r}.$$

Hence, total shortage cost $$= \sum_{r=Q+1}^{\infty} \left[\frac{1}{2} QC_1\left(\frac{Q}{r}t\right) + \frac{1}{2}(r-Q)\, C_2\left(\frac{r-Q}{r}t\right)\right] p(r)$$

The total expected cost is given by the cost equation

$$C\,(Q) = C_1 t \sum_{r=0}^{Q} \left(\frac{2Q-r}{2}\right) p(r) + \sum_{r=Q+1}^{\infty} (C_1 Q^2 + C_2 (r-Q)^2) \frac{p(r)}{2r}$$

The optimum value of Q is obtained by minimising (CQ). In that case, we have

$$C\,(Q + 1) - C(Q) > 0 \quad \text{and} \quad C(Q) - C(Q - 1) < 0$$

i.e. $$\Delta\, C(Q - 1) < 0 < \Delta C\,(Q)$$

Now, $$\Delta C(Q) = C\,(Q + 1) - C(Q)$$

$$= C_1 \sum_{r=0}^{Q+1} \left(Q + 1 - \frac{r}{2}\right) p(r) + \sum_{r=Q+2}^{\infty} C_1 (Q+1)^2 + C_2\,(r - Q - 1)^2$$

$$\frac{p(r)}{2r} - C(Q)$$

$$= (C_1 + C_2) \sum_{r=0}^{Q} p(r) + (C_1 + C_2)\left(Q + \frac{1}{2}\right) \sum_{r=Q+1}^{\infty} \frac{p(r)}{r} - C_2$$

But $\Delta C(Q) > 0$, for minimum $C\,(Q)$. Hence,

$$\sum_{r=0}^{Q} p(r) + \left(Q + \frac{1}{2}\right) \sum_{r=Q+1}^{\infty} \frac{p(r)}{r} > \frac{C_2}{C_1 + C_2}$$

By using the condition $\Delta C\,(Q-1) < 0$ for minimum $C\,(Q)$, we obtain the relation

$$\sum_{r=0}^{Q-1} p(r) + \left(Q - \frac{1}{2}\right) \sum_{r=Q}^{\infty} \frac{p(r)}{r} < \frac{C_2}{C_1 + C_2}$$

Hence,

$$\sum_{r=0}^{Q-1} p(r) + \left(Q - \frac{1}{2}\right) \sum_{r=Q}^{\infty} \frac{p(r)}{r} < \frac{C_2}{C_1 + C_2} < \sum_{r=0}^{Q} p(r) + \left(Q + \frac{1}{2}\right) \sum_{r=Q+1}^{\infty} \frac{p(r)}{r}$$

Using the above relation we can find the range of optimum value of Q.

Uniform Demand and Continuous Units

The cost equation for this model is similar to the above model. Replace $p(r)$ by $f(r)$ and the summation by integration. Hence, the total expected cost is

$$C\,(Q) = C_1 \int_0^Q \left(Q - \frac{r}{2}\right) f(r)\,dr + C_1 \int_Q^{\infty} \frac{Q^2}{2r} f(r)dr + C_2 \int_Q^{\infty} (r-Q)^2 \frac{f(r)}{2r}\,dr$$

Now,

$$\frac{dC(Q)}{dQ} = C_1 \int_0^Q (1-0) f(r)\,dr + C_1 \int_Q^{\infty} \frac{Q}{r} f(r)\,dr - C_2 \int_Q^{\infty} \left(\frac{r-Q}{r}\right) f(r)\,dr$$

$$= C_1 \int_0^Q f(r)dr + C_1 \int_Q^{\infty} \frac{Q}{r} f(r)\,dr - C_2 \int_Q^{\infty} f(r)\,dr + C_2 \int_Q^{\infty} \frac{Q}{r} f(r)\,dr$$

$$= C_1 \int_0^Q f(r)dr + C_1 Q \int_Q^{\infty} \frac{1}{r} f(r)\,dr - C_2 \left[\int_0^{\infty} f(r)\,dr - \int_0^Q f(r)\,dr\right] + C_2 \int_Q^{\infty} \frac{Q}{r} f(r)\,dr$$

$$= C_1 \int_0^Q f(r)dr + C_1 Q \int_Q^{\infty} \frac{1}{r} f(r)dr - C_2 + C_2 \int_0^Q f(r)dr + C_2 \int_Q^{\infty} \frac{Q}{r} f(r)\,dr$$

$$\left[\text{Since } \int_0^{\infty} f(r)\,dr = 1\right]$$

$$= (C_1 + C_2) \left[\int_0^Q f(r)\,dr + \int_Q^{\infty} \frac{Q}{r} f(r)\,dr\right] - C_2$$

For minimum cost, $\frac{dC(Q)}{dQ} = 0$, therefore,

$$(C_1 + C_2) \left[\int_0^Q f(r)\,dr + \int_Q^{\infty} \frac{Q}{r} f(r)\,dr\right] = C_2$$

$$\Rightarrow \qquad \int_0^Q f(r)\,dr + \int_Q^\infty \frac{Q}{r} f(r)\,dr = \frac{C_2}{C_1 + C_2}$$

Further, $$\frac{d^2}{dQ^2} C(Q) = (C_1 + C_2) \int_Q^\infty \frac{f(r)}{r}\,dr > 0$$

Hence, $\int_0^Q f(r)dr + \int_Q^\infty \frac{Q}{r} f(r)\,dr$ gives the condition for finding the optimum value of Q.

Example 3

The probability distribution of monthly sales of a certain item is as follows:

Monthly sales	0	1	2	3	4	5	6
Probability	0.02	0.05	0.30	0.27	0.20	0.10	0.06

The cost of carrying inventory is Rs 10.00 per unit per month. The current policy is to maintain a stock of four items at the beginning of each month. Assuming that the cost of shortage is proportional to both time and quantity short, obtain the imputed cost of a shortage of one item for one unit of time.

Solution: Given, C_1 = Rs 10 per unit per month

Q = Optimum stock = 4 units

Since the demand is uniformly distributed over the month, the least value of shortage cost C_2 can be determined using the relation

$$\sum_{r=0}^{Q-1} p(r) + \left(Q - \frac{1}{2}\right) \sum_{r=Q}^{\infty} \frac{p(r)}{r} < \frac{C_2}{C_1 + C_2} < \sum_{r=0}^{Q} p(r) + \left(Q + \frac{1}{2}\right) \sum_{r=Q+1}^{\infty} \frac{p(r)}{r}$$

Now,

$$\sum_{r=0}^{4-1} p(r) + \left(4 - \frac{1}{2}\right) \sum_{r=4}^{6} \frac{p(r)}{r} = (0.64) + \frac{7}{2}\left[\frac{0.20}{4} + \frac{0.10}{5} + \frac{0.06}{6}\right]$$

$$= 0.92$$

The least value of C_2 is given by

$$\frac{C_2}{C_1 + C_2} = 0.92 \quad \text{or,} \quad \frac{C_2}{10 + C_2} = 0.92$$

or, $\quad C_2(1 - 0.92) = 9.2 \quad$ or, $\quad C_2 = 115$

Similarly, the greatest value of C_2 is obtained by considering

$$\frac{C_2}{C_1 + C_2} = \sum_{r=0}^{4} p(r) + \left(4 + \frac{1}{2}\right) \sum_{r=5}^{6} \frac{p(r)}{r}$$

$$= (0.84) + \frac{9}{2}\left[\frac{0.10}{5} + \frac{0.06}{6}\right]$$

$$= 0.975$$

$$\Rightarrow \qquad C_2 = 0.975\,(10 + C_2) \Rightarrow C_2 = \text{Rs } 390$$

Hence, the range of shortage is given by

$$\text{Rs } 115 < C_2 < \text{Rs } 390.$$

Example 4

Let the probability density of demand of a certain item during a week be

$$f(r) = \begin{cases} 0.1 & 0 \le r \le 10 \\ 0 & \text{otherwise} \end{cases}$$

This demand is assumed to occur with a uniform pattern over the week. Let the unit carrying cost of the item in inventory be Rs 2.00 per week and unit shortage cost be Rs 8.00 per week. How will you determine the optimal order level of the inventory?

Solution: Since $f(r) = 0.1$, $0 \le r \le 10$, $C_1 = \text{Rs } 2.00$, $C_2 = \text{Rs } 8.00$, the optimum value of Q can be obtained by the relation

$$\int_0^Q (0.1)\, dr + \int_Q^{10} Q.\frac{(0.1)}{r}\, dr = \frac{8}{10}$$

or, $$(0.1)\,[Q + Q\,[\log r]_Q^{10}] = 0.8$$

or, $$(0.1)\,[Q + Q\,[\log 10 - \log Q] = 0.8$$

or, $$(0.1)\,[Q - Q \log Q + (2.3)Q] = 0.8$$

or, $$3.3Q - Q \log Q - 8 = 0$$

The solution of this equation $Q = 4.5$ is obtained by trial and error method.

10.9 SELECTION INVENTORY CONTROL TECHNIQUES

Inventories in an organisation cannot be controlled with equal importance. Some inventory items may be very important and some may not be that important. Efficient management calls for an understanding of the nature of inventories and accordingly they give priorities and extent of control required in respect of each item. For this purpose, items may be classified into groups depending upon their utility and importance. Such a classification is termed the *principle of selective control*, and is used to control the inventories. Now, we see techniques like ABC, VED, HML, FSN, XYZ, SOS, SDE of selective control.

ABC Analysis (Always, Better, Control)

An ABC analysis consists of separating the inventory items into three groups A, B, C, where A–high consumption value items, B–moderate consumption value items, C–low consumption value items. Note that

(i) *Category 'A' items*: Most costly and valuable items are classified as 'A'. Such items have large investment but not much in number. The items of this category are ordered frequently but in small number. A periodic review policy should be followed to minimise the shortage percentage of such items. Closed control is required for these category.

(ii) *Category 'B' items*: The items having average consumption value are classified as 'B'. These items have less importance than 'A' class items, but are much costly to pay more attention on their use. These items cannot be overlooked and require lesser degree of control than those in category 'A'. Statistical sampling is generally useful to control these.

(iii) *Category 'C' items*: The items having low consumption value are classified as 'C'. Nearly 75 per cent of inventory items account only for 10 per cent of the total invested capital. Such items can be stocked at an operative place where people can help themselves with any requisition formality. Not much control is needed for the 'C' category. This will increase their investment cost.

After dividing the items into various categories, the usage value of each item is plotted in a graph. Such a graph is known as the ABC distribution curve.

The following procedure of classifying in A, B, C categories may be observed.

Step 1: Determine the number of units sold or used in the past one year period.

Step 2: Determine the unit cost standard for each item.

Step 3: Compute the annual consumption value (in rupees) for each consumed item by multiplying annual consumption (of units) with the unit price.

Step 4: Arrange the items in descending order according to their usage value computed in Step 3.

Step 5: Prepare a Table showing unit cost, annual consumption and annual usage value for each item.

Step 6: Calculate the cumulative sum of the number of items and the usage value for each item obtained in Step 3.

Step 7: Find the percentage of the values obtained in Step 6 with respect to the grand total of the corresponding columns.

Step 8: Draw a graph by taking percentage of items on tbe *X*-axis and the corresponding usage value on the *Y*-axis. After plotting the various points on the graph draw a curve.

Step 9: Identify cut-off points *X* and *Y* where the curve sharply changes its shape. This provides three segments *A*, *B* and *C*. (See Fig. 10.12)

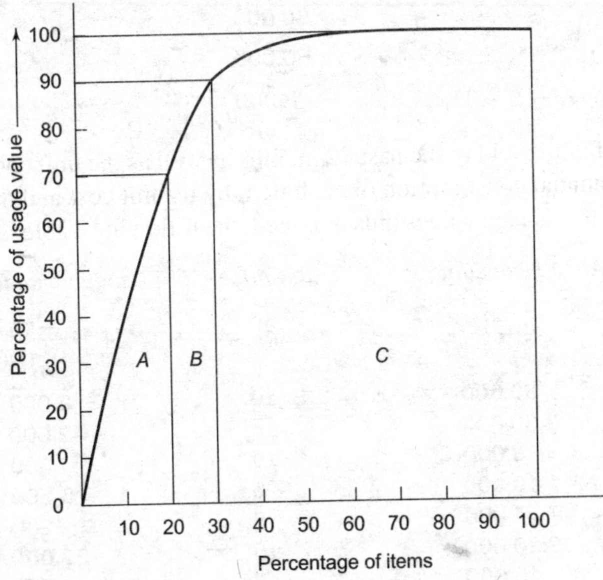

figure 10.12 **ABC distribution curve**

Limitation of ABC Analysis

ABC analysis does not permit precise consideration of all relevant problems of inventory control. For example, a never ending problem in inventory management is that of adequately handling thousands of low value '*C*' items.

Low value purchases frequently require more items, and consequently reduce the time allowance available and purchasing personnel for value analysis, vendor investigation and other '*B*' items.

(i) If *ABC* analysis is not updated and reviewed periodically, the real purpose of control may be defeated. For example, '*C*' items like diesel oil will become most high-value items during a power crisis and, therefore, it require more attention.

(ii) *ABC* classification can lead to overlooking the needs of spare parts whose criticality is high but consumption value is low.

Example 1

The following information is known about a graph of items. Classify the items as A, B, and C.

Model number	Annual consumption in pieces	Unit price (in paise)
501	30,000	10
502	2,80,000	15
503	3,000	10
504	1,10,000	5
505	4,000	5
506	2,20,000	10
507	15,000	5
508	80,000	5
509	60,000	15
510	8,000	10

Solution: Number of items sold in the past 12 months as well as the unit cost standard for each item are given. Multiplying annual consumption of each item by its unit cost and then ranking the items in the descending order of the usage values thus obtained, the following Table is obtained:

Model number (1)	Annual consumption (2)	Unit price (3)	Usage value (in Rs) (4) = $\frac{(2)\times(3)}{100}$	Ranking (5)
501	30,000	10	3,000	6
502	2,80,000	15	42,000	1
503	3,000	10	300	9
504	1,10,000	5	5,500	4
505	4,000	5	200	10
506	2,20,000	10	22,000	2
507	15,000	5	750	8
508	80,000	5	4,000	5
509	60,000	15	9,000	3
510	8,000	10	800	7

Compute the cumulative total number of items and their usage values and convert the accumulated total into percentage of the grand total.

The following classification is thus obtained.

Rank (1)	Model (2)	No. of items (3)	% of items (4)	Cumulative usage value (Rs) (5)	Cumulative % of (6)	% of Total value	Category (7)
1	502			42,000	48		
2	506	3	30%	64,000	73		A
3	509			73,000	83	83%	
4	504			78,500	90		
5	508			82,500	94		
6	501	4	40%	85,500	98	15.6%	B
7	510			86,300	98.6		
8	507			87,050	94.4		
9	503	3	30%	87,350	99.6		
10	505			87,550	100	1.4%	C
Total		10	100%			100%	

The cut-off points have been determined from the graph

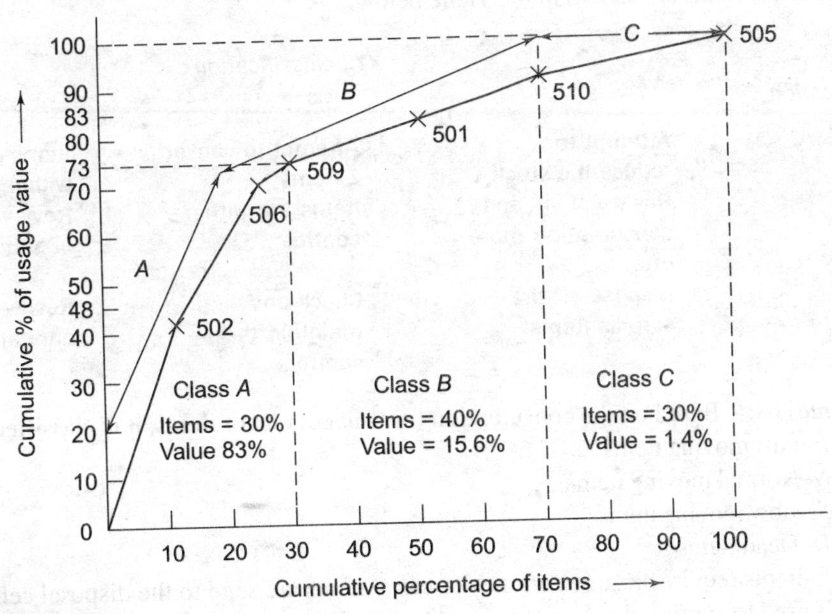

figure 10.13

VED Analysis: It is based on the criticality of items. If the items are arranged in the descending order of their criticality, then the more attention is to these type of items.

V—Vital items: Those items which when required are not available, which make the whole system inoperative (e.g. clutch-wire used for scooters, motor cycles, etc.)

E—Essential items: Items which are not available when demanded, reduce the efficiency of the system, e.g., telephone.

D—Desirable item: The items which neither stop the system nor reduce its efficiency, but it will be good if they are present in the system.

This analysis is useful in controlling the inventory of spare parts.

Both *ABC* and *VED* are combined to control the stocking of spare parts. The control actions are:

ABC classification	VED classification		
	V	E	D
A	Regular stock with constant control	Medium stock	No stock
B	Medium stock	Medium stock	Very low stock
C	High stock	Medium stock	Low stock

XYZ Analysis: The classification is based on the closing value of items in storage. Items whose inventory values are high and moderate are classified as *X* items and *Y* items, while *Z* items are those having low inventory value.

XYZ classification is usually performed once in a year during annual stock-taking. It helps us in identifying items which are being stocked extensively. It can also be combined with *ABC* analysis and the controls of the items are shown in the Table below:

ABC classification	XYZ classification		
	X	Y	Z
A	Attempt to reduce the stock	Attempt to convert *Z* items	Items are within control
B	Review stock and consumption more often	Items are within control	Review bi-annually
C	Dispose off the surplus items	Check and maintain the control.	Review annually

FNSD Analysis: Based on the consumption (or usuage) rate of items, it is classified as

F–Fast moving items
N–Normal moving items
S–Slow moving items
D–Dead items

F–items require close attention whereas *D* items are send to the disposal cell. This method is used to control obsolescence in all types of inventories.

XYZ and *FNSD* can be combined to control obsolete items which are useful in the timely prevention of obsolescence.

SDE Analysis: It is based on the nature of procurement (or availability) of items. Here, *S* represents scarce items, *D* represents difficult items which are available and but not always traceable and *E* represents easy to obtain items.

HML Analysis: Based on the unit price of items, *HML* classification separates inventory items, such as high price, medium price and low price. This controls over-consumption of various items of inventory.

The suitability of selective control techniques *ABC*, *VED*, *XYZ*, and *FNSD* depends on the nature of inventories carried by an organisation. The following Table gives the classification and their uses.

Selective control techniques	*Basis of classification*	*Main uses*
ABC	Consumption value	Control raw material components and work-in-progress inventories.
VED	Critically of the items	Determining the inventory levels of spare parts.
XYZ	Value of items in storage	Reviewing the inventories and other uses.
FNSD	Consumption rate (or movement) of the item	Controlling obsolescence.

EXERCISES

Sections 10.1 and 10.2

1. What are the types of inventory? Why are they maintained? Explain the various costs related to inventory.
2. Explain in detail what constitutes ordering cost and carrying cost.
3. What are the categories of costs associated within developing a sound inventory model? What are the components of cost under each of them?
4. What are the advantages and disadvantages of increase in inventory? Briefly explain the objectives of holding inventory.
5. Describe the basic characteristics of an inventory system.
6. Distinguish between deterministic and stochastic models.
7. Explain the significance of lead time and safety stock in inventory control.
8. Explain the terms lead time, reorder point, stock-out cost and set-up cost.

Section 10.3

9. An oil engine manufacturer purchases lubricants at the rate of Rs 42 per piece from a vendor. The requirement of these lubricants is 1,800 per year. What should be the order quantity per order, if the cost per placement of an order is Rs 16 and inventory carrying charge per rupee per year is only 20 paise?
   [*Answer*: Optimum inventory quantity of lubricant is 83.]
10. A manufacturer has to supply his customers 600 units of his product per year. Shortages are not allowed and the shortage cost amounts to Re 0.60 per unit per year. The set-up cost per run is Rs 80. Find the optimum run size and the minimum average yearly cost.
    [*Answer*: $Q = 346$ units, $C(Q) =$ Rs 173, $t^{\circ} = 11.5$ days.]
11. A shopkeeper has uniform demand of an item at the rate of 600 items per year. He buys from a supplier at a cost of Rs 8 per item and the cost of ordering is Rs 12 each time. If the stockholding

costs are 20% per year of stock value, how frequently should he replenish his stocks and what is the optimal order quantity?
[*Answer*: Q° =95 units, t° = 57 days.]

12. For an item, the production is instantaneous. The storage cost of one item is Re 1 per month and the set-up cost is Rs 25 per run. If the demand is 200 units per month, find the optimum quantity to be produced per set-up and hence determine the total cost of storage and set-up per month.
[*Answer*: Q° = 100 units, t° = 15 days, $C(Q)$ = Rs 125.]
13. An item is produced at the rate of 50 items per day. The demand occurs at the rate of 25 items per day. If the set-up cost is Rs 100 and holding cost is Re 0.01 per unit of item per day, find the economic lot size for one run, assuming that shortages are not permitted. Also, find the time of cycle and minimum total cost for one run.
[*Answer*: Q° = 1,000 units, t° = 40 days, $C(Q)$ = Rs 200.]
14. A contractor has to supply 20,000 units per day. He can produce 30,000 units per day. The cost of holding a unit in stock is Rs 3 per year and the set-up cost per run is Rs 50. How frequently, and of what size, should the production runs be made?
[*Answer*: Q° = 1414 units, t° = 1.68 hours.]
15. The demand for a certain item is 16 units per period. Unsatisfied demand causes a shortage cost of Re 0.75 per unit per short period. The cost of initiating purchasing action is Re 1 per purchase and the holding cost is 15% of average inventory valuation per period. Item cost is Rs 8.00 per unit (Assume that shortages are being back ordered at the above mentioned cost). Find the minimum cost of purchase quantity.
[*Answer*: Q° = 33.3 units, minimum cost = Rs 14.88 (nearly)].
16. A commodity is to be supplied at a constant rate of 200 units per day. Supplies for any amounts can be had at any required time, but each ordering costs Rs 50.00; costs of holding the commodity in inventory is Rs 2.00 per unit per day while the delay in the supply of the item induces a penalty of Rs 10.00 per unit per delay of 1 day. Formulate the average cost function of this situation and find the optimal policy (Q t) where t is the reorder cycle-period and Q is the inventory level after reorder. What should be the best policy if the penalty cost becomes infinite?

$$\left[\textit{Answer}\text{: } Q^\circ = 109.5 \text{ units}, t^\circ = \frac{1}{2} \text{ day; when } C_2 \to \infty, Q^\circ = 100 \text{ units.}\right]$$

17. A dealer supplies you the following information with regard to a product handled by him: Annual demand = 5,000 units, buying cost = Rs 250 per order, inventory carrying cost = 30% per year, price = Rs 100 per unit.
The dealer is considering the possibility of allowing some back-orders to occur for the product. He has estimated that the annual cost of back-ordering (allowing shortage) of the product that will be Rs 10 per unit.
(i) What quantity of the product should he allow to be back-ordered?
(ii) What should be the optimum number of units of the product he should buy in one lot?
(iii) How much additional cost will he have to incur on inventory if he does not permit back ordering?
[*Answer*: (i) 577.35 (ii) 433 units (iii) Rs 4,330.13.]
18. A company has a demand of 12,000 units/year for an item and it can produce 2,000 such items per month. The cost of one set-up is Rs 400 and the holding cost per unit month is Re 0.15. Th

shortage cost of one unit is Rs 20 per year. Find the optimum lot size and the total cost per year, assuming the cost of 1 unit as Rs 14. Also, find the maximum inventory, manufacturing time and total time.
[*Answer*: $Q° = 3{,}413$ units, $C(Q) =$ Rs 51,336.]

19. The demand of an item is uniform at a rate of 25 units per month. The fixed cost is Rs 15 each time a production run is made. The production cost is Re 1 per item, and the inventory carrying cost is Re. 0.30 per item per month. If the shortage cost is Rs 1.50 per item per month, determine how often to make a production run and of what size it should be?

$$\left[\textit{Answer}: Q° = \sqrt{\frac{2DC_S}{C_1}\left\{\frac{C_1 + C_2}{C_2}\right\}} = 54 \text{ items}, t° = \frac{Q°}{D} = 2.16 \text{ months.}\right]$$

20. An automobile factory manufactures a particular type of gear within the factory. This gear is used in the final assembly. The particulars of this gear are: demand rate $r = 14{,}000$ units/year, production rate $k = 35{,}000$ units/year.
Set-up cost $C_s =$ Rs 500 per set-up and carrying cost $C_1 =$ Rs 15/unit/year. Find the economic batch quantity and cycle time.

$$\left[\textit{Answer}: Q° = \sqrt{\frac{2C_S D}{C_1\left(1 - \frac{r}{k}\right)}} = \sqrt{\frac{2 \times 500 \times 14{,}000}{15\left(1 - \frac{14{,}000}{35{,}000}\right)}}\right.$$

$$= 1{,}248 \text{ (approx.)}$$

$$t_1° = \frac{Q°}{k} = 13 \text{ days (approx.)}$$

$$t_2° = \frac{Q°}{r}\left(1 - \frac{r}{k}\right) = 20 \text{ days (approx.)}$$

Therefore, the cycle time is $t = t_1° + t_2° = 13 + 20 = 33$ days

$$\left.\text{The number of set-ups per year} = \frac{D}{Q°} = 11.22.\right]$$

Section 10.4

21. Find the optimum order quantity for the following: annual demand = 3,600 units, ordering cost = Rs 50, cost of storage = 20% of the unit cost

Price break Quantity	*Unit cost (Rs)*
$0 \le Q_1 < 100$	20.00
$100 \le Q_2$	18.00

[*Answer*: $Q° = 316.23$ units.]

22. Anil's company buys 2,000 bats annually. A fixed cost of Rs 50 is incurred each time an order is placed. Inventory carrying cost is estimated at 20%. Supplier offers a 10% discount in price per bat of Rs 100 if orders are placed for more than or equal to 150 bats at a time. In what order size should the company purchase?
[*Answer*: $Q° = 150$ bats.]

23. Find the optimal order quantity for a product for which the price breaks are as follows:

Quantity	*Unit cost (Rs)*
$0 \le Q_1 < 500$	10.00
$500 \le Q_2 < 750$	9.25
$750 \le Q_3$	8.75

The monthly demand for the product is 200 units, the cost of storage is 2% of the unit cost and the cost of ordering is Rs 350.

[*Answer*: $Q_3^\circ = \sqrt{\dfrac{2 \times 350 \times 200}{(8.57)(0.02)}}$ = 894 units > b_2 = (750). Hence, optimum order is $Q^\circ = Q_3^\circ$ = 894 units.]

24. The consumption of an item is known to be fixed at 4,800 units per year. The cost of processing an order for purchase of this item is Rs 400 and the inventory carrying charges work out to 24% per annum of the cost of the item. The cost of the item depends on the purchase lot size as per the schedule given below. Determine the optimum ordering policy.

Quantity per order	*Cost per unit (Rs)*
up to 999	20.00
1,000 to 1499	18.50
1,500 and over	17.00

[*Answer*: Q° = 1,000 units.]

25. A shopkeeper has a uniform demand of an item at the rate of 50 items per month. He buys from supplier at a cost of Rs 6 per item and the cost of ordering is Rs 10 each time. If the stockholding costs are 20% per year of stock value, how frequently should he replenish his stocks? Now, suppose the supplier offers a 5% discount on orders between 200 and 999 items, and a 10% discount on orders exceeding or equal to 1,000. Can the shopkeeper reduce his costs by taking advantage of either of these discounts?

[*Answer*: (i) Q° = 100 items $C(Q)$ = Rs 3,720.00
(ii) Q° = 200 items $C(Q)$ = Rs 3,564.00
Saving by adopting EOQ is Rs 156.00.]

26. Annual demand for an item is 500 units, ordering cost is Rs 18 per order. Inventory carrying cost is Rs 15 per unit per year. Relationship between price and quantity ordered is as follows:

Quantity ordered	1 to 15	16 to 149	150 to 549	550 & over
Price per unit (Rs)	10	9	8.75	8.50

Specify optimal order quantity and the corresponding price of this item.

[*Answer*: Q° = 16 units, $C(Q)$ = Rs 6,142.50.]

Section 10.5

27. Consider a shop which produces and stocks three items. The management desires never to have an investment in inventory of more than Rs 15,000. The items are produced in lots. The demand rate for each item is constant and can be assumed to be deterministic. No back-orders are to be allowed. The relevant data for the items are given below. Determine the optimal lot size for each item.

	Data for problem		
Item	1	2	3
Demand rate (unit per year)	1,000	500	2,000
Cost per unit (Rs)	20	100	50
Set-up cost per lot (Rs)	50	75	100
Holding cost (Rs)	20	20	20

[*Answer*: $Q_1^\circ = 70.71$ units, $Q_2^\circ = 61.23$ units, $Q_3^\circ = 141.40$ units.]

28. Four products *A*, *B*, *C* and *D* are to be manufactured successively in batches on the same machine. The consumption and production rates, batch set-up and item holding costs for each product are shown in the following Table:

Product	*Consumption per year*	*Production rate per day*	*Holding cost per item per day (Rs)*	*Set-up cost (Rs)*
A	10,000	250	0.85	42
B	5,000	100	1.20	18
C	8,000	200	2.00	60
D	12,000	300	0.65	20

If there are 250 working days per year, calculate the production batch sizes for each product and how many complete production run of all four products should be made per year?
[*Answer*: The optimal lot size for each product *A, B, C, D*, are 146, 149, 153, 178 respectively, per year. The complete production run of all four products is 626 per year.]

29. The uniform annual demands for two bulky items are 90 tonne and 160 tonne, respectively. The carrying costs are Rs 250 and Rs 200 per tonne per year, and set-up costs Rs 50 and Rs 40 per production, respectively. No shortages are allowed. Space considerations restrict the average amount inventory of items to 4,000 sq.ft. A tonne of the first item occupies 1,000 sq.ft., and a tonne of second item 500 sq.ft. Find the optimal lot size.
[*Answer*: $Q_1^\circ = 2$ tonne, $Q_2^\circ = 12$ tonne.]

Section 10.6

30. A factory uses Rs 32,000 worth of a raw material per year. The ordering cost per order is Rs 50 and the carrying cost is 20% per year of the average inventory.
If the company follows the EOQ purchasing policy, calculate the reorder point, the maximum inventory, the minimum inventory and the average inventory, given that the factory works for 360 days a year, the replenishment time is 9 days and the safety stock is worth Rs 300.
[*Answer*: *ROL* = Rs 1,100, maximum inventory = Rs 4,300, minimum inventory = Rs 300, average inventory = Rs 2,300.]

Section 10.7

31. Calculate the various parameters for putting an item with following data on EOQ system:
Annual consumption is 12,000 units at the cost of Rs 7.50 per unit. Set-up cost is Rs 6, and the average inventory holding cost is Re. 0.12 per unit. Normal lead time is 15 days and maximum

lead time is 20 days.
[*Answer*: ROL = 367 units, buffer stock = 200 units.]

32. A company uses 10,000 units of an item per year. The purchase price is Re 1 per item, ordering cost Rs 25 per order and carrying cost per year is 12% of the inventory value.
    (i) Find the EOQ.
    (ii) Find the number of orders/year.
    (iii) If the lead time is 4 weeks and assuming 50 working weeks per year, find the reorder point.
    [*Answer*: (i) 2,083 units, (ii) 5 orders, (iii) 800 units.]

Section 10.8

33. A newspaper boy buys paper for 60 paise each and sells them for Rs 1.40 each. He cannot return unsold newspapers. Daily demand has the following distribution:

No. of customers	23	24	25	26	27	28	29	30	31	32
Probability	0.01	0.03	0.06	0.10	0.20	0.25	0.15	0.10	0.05	0.05

If each days's demand is independent of the previous day's demand, how many papers should be ordered each day?
[*Answer*: No. of papers to be ordered is 28.]

34. Some of the spare parts of a ship cost Rs 1,00,000 each. These spare parts can only be ordered together with the ship. If not ordered at the time when the ship is constructed, these parts cannot be available on need. Suppose that a loss of Rs 1,00,00,000 is suffered for each spare that is needed as replacement during the life term of the class of the ship discussed are:

Spare parts required	0	1	2	3	4	5
Probability	0.9488	0.0400	0.0100	0.0010	0.0002	0.000

How many spare parts should be procured?
[*Answer*: 3 spare parts should be procured.]

35. A baking company sells one type of cake by weight. It makes a profit of Rs 9.50 on every kg of cake sold on the day it is baked. It disposes of all cakes not sold on the date it is baked at a loss of Rs 1.50 per kg. If demand known to be rectangular between 300 and 400 kg, determine the optimum amount to be baked?
[*Answer*: $Q° = 386.4$ kg (nearly).]

36. An ice-cream company sells one type of ice-cream by weight. If the product is not sold on the day it is prepared, it can be sold at a loss of 50 paise per pound. But there is an unlimited market for one-day old ice-creams. On the other hand, the company makes a profit of Rs 3.20 on every pound of ice-cream sold on the day it is prepared. Past daily orders form as distribution with

$$f(x) = 0.02 - 0.0002\,x, \; 0 \leq x \leq 100.$$

How many pounds of ice-cream should the company prepare every day?
[*Answer*: $Q° = 63.3$ pounds.]

Section 10.9

37. Perform ABC analysis for the following inventory

Item	*Annual consumption*	*Price per unit (in Paise)*
A	300	10
B	2,800	15
C	30	10
D	1,100	5
E	40	5
F	220	100
G	1,500	5
H	800	5
I	600	15
J	80	10

Answer:

Class	*Name of items*	*% of usage value*	*% of items*
A	*B* & *F*	67.49	20
B	*I* & *G*	17.58	20
C	*D*, *H*, *A*, *J*, *C*, *E*	14.63	60

38. A company purchases three items *A*, *B*, *C*. Their annual demand and unit prices are given in the following Table.

Item	Annual demand (Units)	Unit price (Rs)
A	1,00,000	3
B	80,000	2
C	600	96

If the company wants to place 40 orders per year for the three items what is the optimal number of orders for each item?
[*Answer*: 18 orders of *A*, 14 orders of *B*, and 8 orders of *C*.]

39. What is selective inventory control?
40. The following 30 numbers represents the annual value in thousand of rupees of some 30 items of materials selected at random. Carry out an ABC analysis and list out the values of '*A*' items only

1	2	4	9	75	4	25	3	6
13	2	4	12	30	100	2	7	40
15	55	1	11	15	8	19	1	20
1	3	5						

11

Queuing Models

Queue is a common word that means a waiting line or the act of joining a line. It is formed when the number of customers arriving is greater than the number of customers being served during a period of time. Various queuing situations are:

Examples	*Members of Queue*	*Server(s)*
Airport runways	Planes	Runways
Petrol bunk	Vehicles	Bank employee
Library	Students	Counter clerk
Bank counter	Account holders	Counter clerk
Hospital	Patient	Doctor
Telephone booth	Customers	Telephone attendant
Traffic signal	Vehicles	Signal point
Barber shop	Customers	Barber
Ration shop	Ration card holders}	Shop clerk

Depending on the server status, the incoming customer either waits at the queue or gets served. If the server at the counter is free at the time of arrival, the customer can get served without a waiting time. In this process, over a period of time, the system may experience 'customer waiting' and/or 'server idle time'. In any service system involving queuing situation, the objective is to design the system in such a manner that the average waiting time of the customers is minimised and the percentage utilisation of the server is maintained above a desired level.

A queue system can be described by

1. the input (or arrival pattern)
2. the service mechanisms (or service pattern)
3. the queue discipline
4. customer behaviour

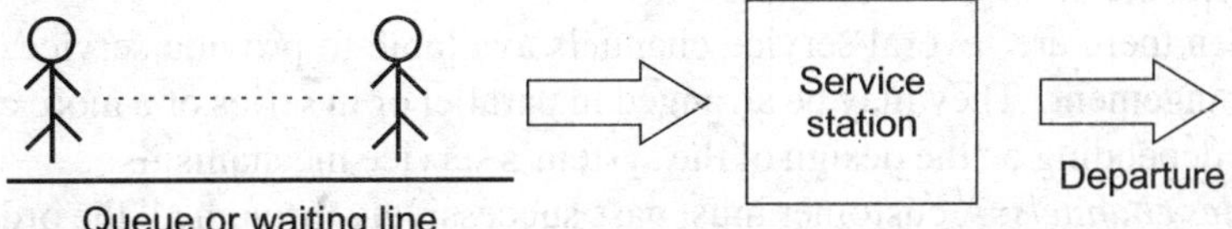

figure 11.1 **Queuing system**

11.1 CHARACTERISTICS OF QUEUING MODELS

A queuing system consists of one or more servers, an arrival pattern of customers, service pattern, queue disciple, the order in which the service is provided and customer behaviour. The word 'queue' is sometimes used to describe the whole system, but it is, in fact, part of the system that holds the excess customers who cannot be immediately served. Hence, the total number of customers in the system at any given time will be equal to the number of customers in the queue plus the number of customers being served. Of course, these numbers will vary with time due to customer's arrival and departure, so they are random process.

1. *The input* (or *arrival pattern*): It represents the pattern in which customers arrive at the system. Arrivals may also be represented by the *inter-arrival time*, which is the time period between two successive arrivals. Arrivals may be separated by equal intervals of time, or unequal but definitely known intervals of time, or by unequal intervals of time whose probabilities are known; these are called *random arrivals*. The rate at which customers arrive at the service station, that is, the number of customers arriving per unit of time is called *arrival rate*.

   The assumption regarding the distribution of arrival rate has a great impact on the mathematical model. If the number of customers is very large, the probability of an arrival in the next interval of time does not depend upon the customers already in the system. Hence, the arrival is completely random and it follows the Poisson process with mean equals the average number of arrivals per unit time, represented by λ.
2. *The service mechanism* (or *service pattern*): The service pattern is similar to the arrival pattern, but there are some important differences. Service time may be a constant or a random variable. Distributions of service time which we are following are 'negative exponential distribution', which is characterised by a single parameter, the mean rate μ or its mean service time $\dfrac{1}{\mu}$.

   The servicing system in which the customers may be served in batches of fixed size or of variable size by the same server is termed as *bulk service system*. The system in which service depends on the number of waiting customers is termed as *state-dependent system*.
3. *Capacity of the system*: Some of the queuing processes admit physical limitations to the amount of waiting room, so that when the waiting line exceeds a fixed length, no further customers are allowed to enter until space becomes available by a service completion. This type of situation is termed as *finite source queues*.
4. *Service arrangements*: For providing service to the incoming customers, one or more service points are established. The number depends on the number of customers, rate of arrivals, time taken for providing service to a single customer, and so on. Depending on these variables, a service channel is single or multiple.

   When there are several service channels available to provide service, much depends upon their arrangement. They may be arranged in parallel or in series or a more complex combination of both, depending on the design of the system's service mechanism.

   Series channels: A customer must pass successively through all the ordered channels before service is completed, for example, in public offices where parts of the service are done at different service counters.

Parallel channels: A number of channels providing identical service facilities so that several customers may be serviced simultaneously.

figure 11.2 (a) **Queuing system with single queue and single service station**

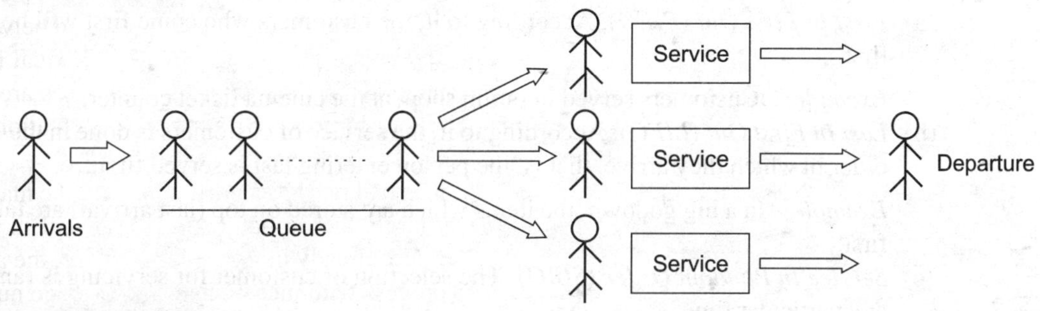

figure 11.2 (b) **Queuing system with single queue and several service stations, parallel servers**

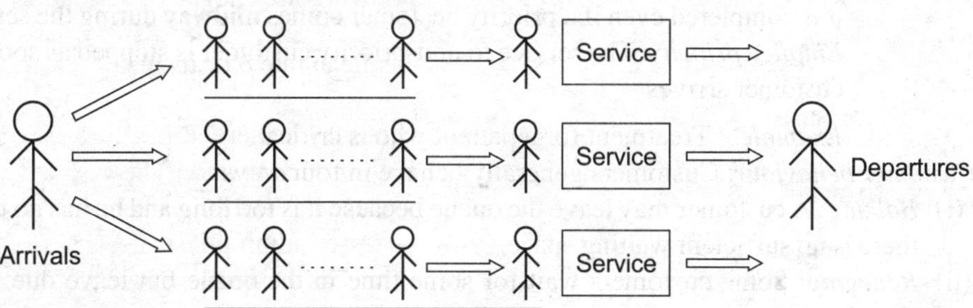

figure 11.2 (c) **Queuing system with several queues and several service stations**

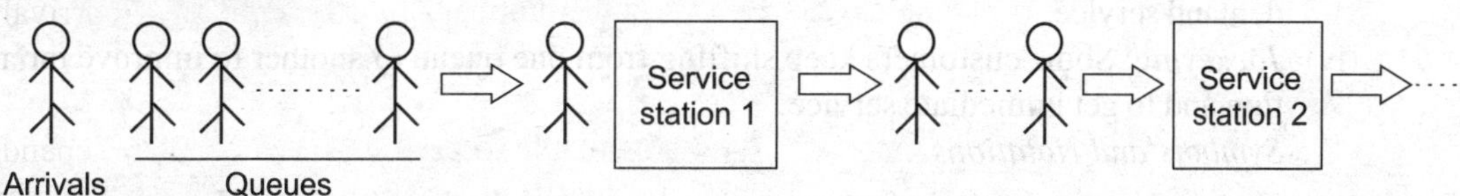

figure 11.2 (d) **Queuing system with series servers**

A queuing system is called a *one-server model* when the system has one server only, and a *multiple-server model* when the system has a number of parallel channels with one server.

5. *Service time*: The time required for servicing a customer is called service time. Service time may be constant or it may vary with the customer. However, for sake of simplicity, it is assumed that the time required for servicing for all customers is constant. Moreover, since the arrival pattern is assumed to be random, the service time is also taken as random. Hence, the service time follows exponential distribution with mean equal to reciprocal of the mean rate of service. In cases where the assumption of service time following exponential distribution fails, Erlang distribution is applied.
6. *Queue discipline*: It is a rule determining the formation of the queue, the manner of customer's behaviour while waiting, and the manner is which they are chosen for service. The following are queue discipline.
   (a) *First In First Out* (*FIFO*): According to it, the customers who come first will be served first.

   Example: Customers served in ration shop, at the cinema ticket counter.

   (b) *Last In First Out* (*LIFO*): According to it, the service of customers is done in the reverse order in which they arrive, that is, the person entering last is served first.

   Example In a big godown, the items which are stored on top (last arrival) are taken out first.

   (c) *Service In Random Order* (*SIRO*): The selection of customer for servicing is random at any particular time.

   (d) *Service in Priority* (*SIP*): Certain customers are given priority over others in selection for service. It is of two types: non-emptive priority and, emptive priority.

   Non-emptive priority: The customer already getting service is allowed to continue till it is completed even if a priority customer comes midway during the service.

   Emptive priority: The service to non-priority customer is stopped as soon as a priority customer arrives.

   Example Treatment for a patient who is critical.
7. *Customer behaviour*: Customers generally behave in four ways:
   (i) *Balking*: A customer may leave the queue because it is too long and he has no time to wait or there is no sufficient waiting space.

   (ii) *Reneging*: Some customers wait for some time in the queue but leave due to impatience without getting the service.

   (iii) *Collusion*: Some of the customers join together and only one of them, instead of all, stay in the queue. However, when their turn comes for service, the customers who were in collusion demand service.

   (iv) *Jockeying*: Some customers keep shifting from one queue to another to improve their position and to get immediate service.

   Symbols and Notations.

n = number of customers in the system, both waiting and in service.

λ = average number of customers arriving per unit of time.

μ = average number of customers being served per unit of time.

$$\rho = \frac{\lambda}{\mu} = \text{traffic intensity}$$

C = number of parallel service channels
$E(n)$ = average number of customs in the system, both waiting and in service
$E(m)$ = average number of customers waiting in the queue.
$E(v)$ = average waiting time of a customer in the system, both waiting and in service.
$E(w)$ = average waiting time of a customer in the queue.
$P_n(t)$ = probability that there are n customers in the system at any time t, both waiting and in service.

General Relationships Among System Characteristics

By definition of various measures of performance (operating characteristics), we have

$$E(n) = \sum_{n=0}^{\infty} n P_n$$

$$E(m) = \sum_{n=s}^{\infty} (n-s) P_n$$

The relationship between the system characteristics are:

1. The expected number of customers in the system is equal to the expected number of customers in queue + in service. That is,

$$E(n) = E(m) + E(\text{number of customers in service})$$
$$= E(m) + \frac{\lambda}{\mu}$$

The value of E (number of customers in service) should not be confused with the number of service channels, and we denote $\rho = \frac{\lambda}{\mu}$ for all models, except finite queue cases.

2. Expected waiting time of the customer in the system is equal to the expected waiting time in queue plus the expected service time.

$$E(v) = E(W) + \frac{1}{\mu}$$

3. Expected number of customers served per busy period is given by

$$E\left(n/_{n>0}\right) = \frac{E(n)}{P(n>0)} = \frac{\mu}{\mu - \lambda}$$

where $E(n)$ is the expected number of customers in the system and $P(n>0)$ is the probability of the system being busy.

4. Expected length of a busy period is given by

$$E\left(W/_{W>0}\right) = \frac{E(\omega)}{P(n>0)} = \frac{1}{\mu - \lambda}$$

where $E(w)$ is the expected waiting time in the queue.

5. $E(v) = \frac{1}{\lambda} E(n)$ or $E(n) = \lambda E(v)$

6. $E(W) = \frac{1}{\lambda} E(m)$ or $E(m) = \lambda E(W)$

For applying (5) and (6) for systems with finite queues, instead of using λ, its effective value $\lambda(1 - P_N)$ must be used.

7. The probability P_n of n customers in the queuing system at any time can be used to determine all the basic measures of performance, as follows:

$$E(n) = \sum_{n=0}^{\infty} n P_n \Rightarrow E(v) = \frac{1}{\lambda} E(n) \Rightarrow E(W) = E(v) - \frac{1}{\mu} \Rightarrow E(m) = \lambda E(W)$$

Hence, if we know one characteristic, then the remaining characteristics can be calculated.

Kendall's Notation for Representing Queuing Model

Different models in queuing theory are classified by using special (or standard) notations described as

$$(a|b|c) : (d|e)$$

where,
a = arrival (inter-arrival) distribution
b = departure (or service time) distribution
c = number of channels
d = capacity of the system
e = queue discipline

It is important to note the first three characteristics ($a|b|c$) in the above notation were introduced by D. Kendall (1953), and the fourth (d), and fifth (e) characteristics were introduced by A. Lee (1966). The descriptive notations used for arrivals and service time distributions in place of a and b are as follows:

a(or b) = M = inter-arrival times (or service times) having exponential distributions, where M stands for Markovian property of the exponential.

a(or b) = D = inter-arrival times (or service times) that are constant or deterministic.

b = G = service time distributions of general type.

Definition of Transient and Steady States

A system is said to be in *transient state* when its operating characteristics (like input, output, mean queue length, etc.) are dependent upon time.

A system is said to be in *steady state* when the behaviour of the system is independent of time t. In the steady state case,

$$\lim_{t \to \infty} P_n(t) = P_n \qquad \text{(independent of } t\text{)}$$

This implies

$$\lim_{t \to \infty} \frac{d}{dt} P_n(t) = 0.$$

11.2 POISSON PROCESS AND EXPONENTIAL DISTRIBUTION

In this section we derive the Poisson distribution and show that assuming the number of occurrences in some time interval to be a Poisson random variate is equivalent to assuming the time between successive occurrences to be an exponentially distributed random variable.

Given an arrival process $\{N(t) : t \geq 0\}$, where $N(t)$ denotes the total number of arrivals up to time t, $N(0) = 0$, we can make the following assumptions (or axioms):

1. The number of arrivals in non-overlapping intervals are statistically independent, that is, the process has independent increments.
2. Assuming that there are n units in the system at time t, the probability that exactly one arrival (birth) will occur during small time interval Δt is given by $\lambda \Delta t + O(\Delta t)$, where λ is independent of t and $O(\Delta t)$ includes the terms of higher order of Δt.
3. Δt is considered to be so small that the probability of more than one arrival in time Δt is almost zero.

Distribution of Arrivals (Pure Birth Process)

The model in which only arrivals are counted and no departures take place are called *pure birth models*. We prove the number of arrivals in a time interval $(0, t)$ is a random variable which follows a Poisson distribution with parameter λt.

Let $P_n(t)$ be the probability of n arrivals in a time interval of length t, where $n \geq 0$ is an integer.

Case 1: When $n > 0$, there may be two mutually exclusive ways of having n units at time $t + \Delta t$.

(i) There are n units in the system at time t and no arrival takes place during the time interval Δt. The probability of the event is

$$= \text{Probability of } n \text{ units at time } t \times \text{Probability of no arrival during } \Delta t$$
$$= P_n(t)(1 - \lambda \Delta t)$$

(Since probability of exactly one arrival in the time Δt is $\lambda \Delta t$, and probability of no arrival is $1 - \lambda \Delta t$)

(ii) There are $(n - 1)$ units in the system at time t and one arrival takes place during time Δt. The probability of the event is $= P_{n-1}(t) \times \lambda \Delta t$.

Hence, the probability of n arrivals at time $t + \Delta t$ is:

$$P_n(t + \Delta t) = P_n(t)(1 - \lambda \Delta t) + P_{n-1}(t)\, \lambda \Delta t. \tag{11.2.1}$$

Case 2: When $n = 0$,

$$P_0(t + \Delta t) = \text{Prob. [no unit at time } t] \times \text{Prob. [no arrival in time } \Delta t]$$
$$= P_0(t)(1 - \lambda \Delta t) \tag{11.2.2}$$

Rewriting (11.2.1) and (11.2.2) and dividing throughout by Δt, we get

$$\frac{P_n(t+\Delta t) - P_n(t)}{\Delta t} = -\lambda P_n(t) + \lambda P_{n-1}(t) \tag{11.2.3}$$

$$\frac{P_0(t+\Delta t) - P_0(t)}{\Delta t} = -\lambda P_0(t) \tag{11.2.4}$$

Taking limit $\Delta t \to 0$ on both sides of (11.2.3), and (11.2.4), we get

$$\lim_{\Delta t \to 0} \frac{P_n(t+\Delta t) - P_n(t)}{\Delta t} = -\lambda P_n(t) + \lambda P_{n-1}(t) \tag{11.2.5}$$

and,

$$\lim_{\Delta t \to 0} \frac{P_0(t+\Delta t) - P_0(t)}{\Delta t} = -\lambda P_0(t) \tag{11.2.6}$$

(11.2.5) implies

$$P'_n(t) = -\lambda P_n(t) + \lambda P_{n-1}(t) \qquad (11.2.7)$$

(11.2.6) implies

$$P'_0(t) = -\lambda P_0(t) \qquad (11.2.8)$$

(11.2.7) and (11.2.8) are called *differential difference equations.*

Equation (11.2.8) can be written as

$$\frac{P'_0(t)}{P_0(t)} = -\lambda$$

Integrating both sides with respect to t, we get

$$\log P_0(t) = -\lambda t + A \qquad (11.2.9)$$

The constant A can be calculated by using the boundary conditions:

$$P_n(0) = \begin{cases} 1, & n = 0 \\ 0, & n > 0 \end{cases}$$

Substituting $n = 0$, $P_0(0) = 1$ in (11.2.9) gives $A = 0$.

Thus, $\log P_0(t) = -\lambda t$ or $P_0(t) = e^{-\lambda t}$

Substituting $n = 1$, $P_1(0) = 0$ in (11.2.7) gives

$$P'_1(t) = -\lambda P_1(t) + \lambda P_0(t)$$

or, $$P'_1(t) + \lambda P_1(t) = \lambda e^{-\lambda t} \qquad (11.2.10)$$

Since this is a linear differential equation of the first order, it can be solved by multiplying both sides by the integrating factor $e^{\int \lambda\, dt} = e^{\lambda t}$ [since $y' + P(x)y = Q(x)$ can be solved by multiplying the integrating factor $e^{\lambda x}$ and we get $ye^{\int P dx} = \int Q(x) e^{\int P dx} dx + C$.

Then, (11.2.10) becomes

$$e^{\lambda t}[P'_1(t) + \lambda P_1(t)] = \lambda$$

$$\Rightarrow \qquad \frac{d}{dt}[e^{\lambda t} P_1(t)] = \lambda$$

Integrating both sides w.r. to t, we get

$$e^{\lambda t} P_1(t) = \lambda t + B, \ B \text{ is a constant of integration.}$$

Put $t = 0$, then $e^{\circ} P_1(0) = B \Rightarrow e^{\circ} \times 0 = B \Rightarrow B = 0$.

Hence, $$e^{\lambda t} P_1(t) = \lambda t$$

or, $$P_1(t) = \frac{\lambda t e^{-\lambda t}}{1!}$$

Similarly, putting $n = 2$, in (11.2.7) we get

$$P'_2(t) + \lambda P_2(t) = \lambda P_1(t)$$

$$\Rightarrow \qquad P'_2(t) + \lambda P_2(t) = \lambda \left[\frac{\lambda t e^{-\lambda t}}{1!}\right]$$

Multiplying on both sides by its integrating factor $e^{\lambda t}$ gives

$$\frac{d}{dt}\left[e^{\lambda t} P_2(t)\right] = \frac{\lambda(\lambda t)}{1!}$$

Integrating w.r. to t, $$e^{\lambda t} P_2(t) = \frac{(\lambda t)^2}{2!} + C$$

Put $t = 0$, then since $P_2(0) = 0$, we get $C = 0$. Hence,

$$P_2(t) = \frac{(\lambda t)^2 e^{-\lambda t}}{2!}$$

In general for $n = m$, $$P_m(t) = \frac{(\lambda t)^m e^{-\lambda t}}{m!} \qquad (11.2.11)$$

We prove by mathematical induction, that for any n the result is true. Assume $n = m$ the result, is true we prove for $n = m + 1$ the result is true. Now by (11.2.7),

$$P'_{m+1}(t) + \lambda P_{m+1}(t) = \lambda \frac{(\lambda t)^m e^{-\lambda t}}{m!}$$

or, $$\frac{d}{dt}[e^{\lambda t} P_{m+1}(t)] = \frac{\lambda(\lambda t)^m}{m!}$$

Integrating w.r. to t, on both sides we get,

$$e^{\lambda t} P_{m+1}(t) = \frac{(\lambda t)^{m+1}}{(m+1)!}$$

or, $$P_{m+1}(t) = \frac{(\lambda t)^{m+1} e^{-\lambda t}}{(m+1)!}$$

Hence, in general $$P_n(t) = \frac{(\lambda t)^n \cdot e^{-\lambda t}}{n!}$$

which is a well-known Poisson distribution with mean λt. Thus, if we consider the random variable defined as the number of arrivals to a system in time t, this random variable has the Poisson distribution with a mean of λt arrivals, or a mean arrival rate of λ.

Distribution of Inter-arrival Time (Exponential Process) Let T be the time between two consecutive arrivals (called the inter-arrival time), and $f(t)$ denotes the probability density function for inter-arrival times. Then,

Distribution of Departures (or Pure Death Process) In this process assume there are N customers in the system at time t. Assume no arrivals (births) can occur in the system. Departures occur at a rate μ per unit time. Assume

1. Prob. [one departure during Δt] = $\mu\Delta t + O(\Delta t)^2 = \mu\Delta t$ (since $O(\Delta t)^2$ is negligible).
2. Prob. [more than one departure during Δt] = $O(\Delta t)^2 \equiv 0$.
3. The number of departures in non-overlapping intervals are statistically independent and identically distributed random variables, that is, the process $N(t)$ has independent increments.

We obtain the differential difference equation in three mutually exclusive ways:

Case 1: When $0 < n < N$. As in case of pure birth process we have

$$P_n(t+\Delta t) = P_n(t)\,[1-\mu\Delta t] + P_{n+1}(t)\,\mu\Delta t \tag{11.2.12}$$

Case 2: When $n = N$

Since there are exactly N units in the system, $P_{n+1}(t) = 0$

$$P_n(t+\Delta t) = P_N(t)\,[1-\mu\Delta t] \tag{11.2.13}$$

Case 3: When $n = 0$

$$P_0(t+\Delta t) = P_0(t) + P_1(t)\,.\,\mu\Delta t \tag{11.2.14}$$

Since there is no unit in the system at time t, the question of any departure Δt does not arise. Therefore, probability of no departure is unity in the case.

Dividing throughout by Δt and taking limit $\Delta t \to 0$ in equations (11.2.12), (11.2.13), (11.2.14) we get,

$$\underset{\Delta t\to 0}{\text{lt}} \frac{P_n(t+\Delta t)-P_n(t)}{\Delta t} = -\mu P_n(t) + \mu P_{n+1}(t),\ 0<n<N$$

$$\underset{\Delta t\to 0}{\text{lt}} \frac{P_N(t+\Delta t)-P_N(t)}{\Delta t} = -\mu P_N(t), \quad n = N$$

$$\underset{\Delta t\to 0}{\text{lt}} \frac{P_0(t+\Delta t)-P_0(t)}{\Delta t} = \mu P_1(t), \quad n = 0$$

(or)

$$P_n'(t) = -\mu P_n(t) + \mu P_{n+1}(t), \quad 0<n<N \tag{11.2.15}$$

$$P_N'(t) = -\mu P_N(t), \quad n = N \tag{11.2.16}$$

$$P_0'(t) = \mu P_1(t), \quad n = 0 \tag{11.2.17}$$

From (11.2.16), we get $\dfrac{P_N'(t)}{P_N(t)} = -\mu$. Integrating w.r.t. to t on both sides, $\log P_N(t) = -\mu t + A$.

The constant A can be eliminated by using the boundary condition $P_N(0) = 1$, and so $\log 1 = A$ which implies $A = 0$ (since $\log 1 = 0$). Hence,

$$\log P_N(t) = -\mu t$$

or,

$$P_N(t) = e^{-\mu t}$$

In equation (11.2.15), put $n = N - 1$, then

$$P'_{N-1}(t) = -\mu P_{N-1}(t) + \mu P_N(t)$$

$$= -\mu P_{N-1}(t) + \mu e^{-\mu t}$$

$$\Rightarrow \quad P'_{N-1}(t) + \mu P_{N-1}(t) = \mu e^{-\mu t}$$

The solution of this equation is obtained as

$$P_{N-1}(t)\cdot e^{\int \mu dt} = \int \mu\, e^{-\mu t}\,(e^{\int \mu dt})dt + C$$

[since for the equation $y' + P(x)\,y = Q(x)$, the solution is $ye^{\int P dx} = \int Q(x)\,(e^{\int P dx})\,dx + C$]

That is,

$$P_{N-1}{}^{(t)} \cdot e^{\mu t} = \int \mu\, dt + C$$

$$\Rightarrow \qquad P_{N-1}(t) = \mu t\, e^{-\mu t} + C\, e^{-\mu t}$$

To determine C, put $t = 0$, then $P_{N-1}(0) = 0$, and so $C = 0$. Hence,

$$P_{N-1}(t) = \frac{\mu t\, e^{-\mu t}}{1!}$$

Putting $n = N - 2$ in (11.2.15), we get

$$P_{N-2}(t) = \frac{e^{-\mu t}(\mu t)^2}{2!}$$

In general, for $n = N - i$, we get

$$P_n(t) = P_{N-i}(t) = \frac{e^{-\mu t}(\mu t)^i}{i!} = \frac{e^{-\mu t}(\mu t)^{N-n}}{(N-n)!} \qquad \text{for } n = 1, 2, \ldots, N \quad (11.2.18)$$

To find $P_0(t)$, we know $\sum_{n=0}^{N} P_n(t) = 1$,

so,
$$1 = \sum_{n=0}^{N} P_n(t) = P_0(t) + \sum_{n=1}^{N} P_n(t)$$

or,
$$P_0(t) = 1 - \sum_{n=1}^{N} P_n(t)$$

$$= 1 - \sum_{n=1}^{N} \frac{(\mu t)^{N-n} e^{-\mu t}}{(N-n)!} \qquad (11.2.19)$$

Combination of (11.2.18) and (11.2.19) gives

$$P_n(t) = \begin{cases} \dfrac{(\mu t)^{N-n} e^{-\mu t}}{(N-n)!} & \text{for } n = 1, 2, \ldots, N \\ 1 - \sum_{n=1}^{N} \dfrac{(\mu t)^{N-n} . e^{-\mu t}}{(N-n)!} & \text{for } n = 0 \end{cases}$$

The number of departures in time t follows the *truncated Poisson distribution.*

Markovian Queuing Models

For a Markovian queuing system is, both inter-arrival times and service times must be negative exponentially distributed, so that it is fully characterised by only one parameter, which is related to the mean. When the queuing model is Markovian, we can solve a set of linear equations whose solution provides the steady-state distribution of the number of customers in the system. The simplest Markovian model to set up and solve is the single server infinite capacity model.

Example 1

Prove the Markovian property of inter-arrival times, which states that at any instant the time until the next arrival occurs is independent of the time that has elapsed since the occurrence of the last arrival. That is,

$$P\left(T \geq t_1 \big/ T \geq t_0\right) = P(0 \leq T \leq t_1 - t_0),$$

where, $P(T \geq t_1 | T \geq t_0)$ denotes the conditional probability given by

$$P(T \geq t_1 | T \geq t_0) = \frac{P[(T \geq t_1) \text{ and } (T \geq t_0)]}{P(T \geq t_0)}$$

Solution:
$$P(T \geq t_1 | T \geq t_0) = \frac{\int_{t_0}^{t_1} \lambda e^{-\lambda t} dt}{\int_{t_0}^{\infty} \lambda e^{-\lambda t} dt} = \frac{[e^{-\lambda t_1} - e^{-\lambda t_0}]}{e^{-\lambda t_0}} = 1 - e^{-\lambda (t_1 - t_0)}$$

and,
$$P(0 \leq T \leq t_1 - t_0) = \int_0^{t_1 - t_0} \lambda e^{-\lambda t} dt = -[e^{-\lambda t}]_0^{t_1 - t_0}$$
$$= 1 - e^{-\lambda (t_1 - t_0)}$$

Hence,
$$P(T \geq t_1 | T \geq t_0) = P[0 \leq T \leq t_1 - t_0]$$

Classification of Queuing Models

I–Probabilistic Queuing Models: When both arrival rate and service rate are unknown and are assumed to be random variables, the probability distribution obtained is based on past experience.

Model I: (Erlang Model) (M|M|1): (∞/F1FO). This denotes Poisson arrival (exponential inter-arrival), Poisson departure (exponential service time), single server, infinite capacity, First in, First Out service discipline.

Model II: (General Erlang Model) (M/M/1): (∞/F1 FO). This is a general queuing model in which the rate of arrival and the service depend on the length *n* of the line.

Model III: (M/M/1): (N/F1 FO). In this model capacity is finite, say N.

Model IV: (M/M/C): (∞/F1 FO), In this the, number of service stations is *C* which are in parallel.

Model V: ($M/E_K/1$): (∞/F1 FO), That is, Poisson arrivals, Erlangian service time for *k* phases, and a single server.

Model VI: (Machine Servicing Model) (M/M/R): (K/GD), K > R. That is, Poisson arrivals, exponential service time, R repairmen, K machines in the system, and general queue discipline.

Model VII: Power supply model.

Model VIII: Cost profit model.

Model IX: (M/G/1): (∞/GD), where G is the general output distribution, and GD represents general service discipline.

II–Mixed Queuing Model. When either the arrival rate or the service rate is exactly known and the other is not known.

Model X: (M/D/1): (∞/F1 FO), where D stands for deterministic service time.

III–Deterministic Queuing Model. When both arrival rate and service rate are exactly known.

Model XI: (D/D/1): (K – 1/FIFO), where D–Deterministic arrivals, D–Deterministic service time distribution.

11.3 POISSON QUEUES

The models which have Poisson arrivals (exponential inter-arrival time) and Poisson services (exponential service time) are called Poisson queuing models.

(M/M/1) Queuing System: This queuing system deals with the process in which arrivals and services occur randomly over time. Arrivals can be considered as *births* to the system, since, if the system is in state E_n and an arrival occurs, the state is changed to E_{n+1}. If departure occurs the state E_n is changed down to E_{n-1} and can be looked up on as a death. This type of process is referred to as *a birth death process.*

***Model I*: (M/M/1): (∞/*F*1 *FO*)**. This model deals with a queuing system having single server, Poisson arrival, exponential service is on the basis of 'First in, First out'.

Step 1: *Obtain the system of differential–difference equations.*

Let $P_n(t)$ be the probability that there are n customers in the system at time t. To be in state E_n at time $t + \Delta t$

(i) the system could have been in state E_n at time t and have no arrivals or service completions in the time Δt.

(ii) the system could have been in E_{n-1} at time t and during Δt, there was one arrival and no service completions.

(iii) the system could have been in state E_{n+1} at time t and during Δt time, one service was completed and there were no arrivals. Hence, as arrivals and service are both independent of each other, for $n \geq 1$, we have

$$
\begin{aligned}
P_n(t+\Delta t) &= P_n(t).\ P(\text{no arrival in } \Delta t).\ P(\text{no service completed in } \Delta t) \\
&\quad + P_n(t).\ P(\text{one arrival in } \Delta t).\ P(\text{one service completed in } \Delta t) \\
&\quad + P_{n+1}(t).\ (\text{one service completed in } \Delta t)\ P(\text{no arrival in } \Delta t) \\
&\quad + P_{n-1}(t).\ P(\text{one arrival in } \Delta t)\ P(\text{no service in } \Delta t). \\
&= P_n(t)\,[1-\lambda\Delta t]\,[1-\mu\Delta t] + P_n(t)\,(\lambda\Delta t)\,(\mu\Delta t) \\
&\quad + P_{n+1}(t)\,\mu\Delta t\,(1-\lambda\Delta t) + P_{n-1}(t)\,\lambda\Delta(t)\,(1-\mu\Delta t) \\
&= P_n(t)\,1[1-(\lambda+\mu)\Delta t] + P_{n-1}(t)\,\lambda\Delta t(1-\mu\Delta t) + P_{n+1}(t)\,\mu\Delta t(1-\lambda\Delta t)
\end{aligned}
$$

This implies,

$$\lim_{\Delta t \to 0} \frac{P_n(t+\Delta t) - P_n(t)}{\Delta t} = (\lambda+\mu)P_n(t) + P_{n-1}(t)\,\lambda + P_{n+1}(t).\,\mu \qquad (11.3.1)$$

Next, we find $P_0(t + \Delta t)$ separately as $P_{n-1}(t)$ does not exist for $n = 0$. The system can be in state E_0 at time $t + \Delta t$ if it were in E_0 at t and there was no arrival during Δt (no service is possible since the system is empty) or the system can be in E_1 at time t and there are no arrivals but one service completed in Δt. Thus,

$$P_0(t+\Delta t) = P_0(t)\,[1-\lambda\Delta t] + P_1(t)\,\mu\Delta t(1-\lambda\Delta t)$$

$$\Rightarrow \quad \lim_{\Delta t \to 0} \frac{P_0(t+\Delta t) - P_0(t)}{\Delta t} = -\lambda P_0(t) + \mu P_1(t) \qquad (11.3.2)$$

(11.3.1) and (11.3.2) are called *differential difference equations.*

Step 2: *Obtain steady state equations.*

To get the steady state solution for P_n, the probability of n customers in the system at an arbitrary point of time after steady state is reached, we take the limit as $t \to \infty$ in (11.3.1) and (11.3.2)

The steady state solution exists only when $\frac{\lambda}{\mu} < 1$ as $t \to \infty$. If $\lambda = \mu$ there is no queue, and if $\frac{\lambda}{\mu} > 1$, then the state is called the explosive state. Throughout this chapter we assume that the steady state conditions prevails. In the steady state, $P_n(t) \to P_n$ and $\frac{d}{dt} P_n(t) \to 0$ as $t \to \infty$.

Now, (11.3.1) and (11.3.2) become

$$0 = -(\lambda + \mu) P_n + \mu P_{n+1} + \lambda P_{n-1},\ n \geq 1$$
$$0 = -\lambda P_0 + \mu P_1, \quad n = 0 \tag{11.3.3}$$

Step 3: *Solution of difference equations.*
Using (11.3.3) we get,

$$P_1 = \frac{\lambda}{\mu} P_0$$

$$P_2 = \left(\frac{\lambda + \mu}{\mu}\right) P_1 - \frac{\lambda}{\mu} P_0$$
$$= \left(\frac{\lambda}{\mu}\right)^2 P_0$$

$$P_3 = \left(\frac{\lambda + \mu}{\mu}\right) P_2 - \frac{\lambda}{\mu} P_1$$
$$= \left(\frac{\lambda}{\mu}\right)^3 P_0$$

In general, $$P_n = \left(\frac{\lambda}{\mu}\right)^n P_0, \text{ for } n \geq 0.$$

To obtain the value of P_0, we make use of the boundary condition $\sum_{n=0}^{\infty} P_n = 1$.

That is, $$P_0 + P_1 + P_2 + \ldots + P_n + \ldots = 1$$

$$\Rightarrow \quad P_0 + \left(\frac{\lambda}{\mu}\right) P_0 + \left(\frac{\lambda}{\mu}\right)^2 P_0 + \ldots \left(\frac{\lambda}{\mu}\right)^n P_0 + \ldots = 1$$

$$\Rightarrow \quad P_0 \left[1 + \left(\frac{\lambda}{\mu}\right) + \left(\frac{\lambda}{\mu}\right)^2 + \ldots\right] = 1$$

$$\Rightarrow \quad P_0 \left[\frac{1}{1 - \left(\frac{\lambda}{\mu}\right)}\right] = 1, \quad \text{since } \frac{\lambda}{\mu} < 1$$

$$\Rightarrow \quad P_0 = 1 - \left(\frac{\lambda}{u}\right)$$

Hence, $$P_n = \left(\frac{\lambda}{\mu}\right)^n P_0 = \left(\frac{\lambda}{\mu}\right)^n \left(1 - \frac{\lambda}{\mu}\right) = \rho^n(1-\rho),$$

where, $$\rho = \frac{\lambda}{\mu} < 1, \text{ and } n \geq 0.$$

This expression gives the probability distribution of queue length.

Note 1. An important measure of a simple queue ($M/M/1$) is its traffic intensity ρ, where

$$\rho = \frac{\text{mean arrival rate}}{\text{mean service rate}} = \frac{\lambda}{\mu}$$

i.e. $$\rho = \frac{1/\mu}{1/\lambda} = \frac{\text{mean service time}}{\text{mean inter-arrival time}}$$

The unit of traffic intensity is Erlang.

2. $$P[\text{queue size} \geq N) = \sum_{n=N}^{\infty} P_n = \sum_{n=0}^{\infty} P_n - \sum_{n=0}^{N-1} P_n$$

$$= 1 - [P_0 + P_1 + \ldots + P_{N-1}]$$

$$= 1 - \left[P_0 + \left(\frac{\lambda}{\mu}\right) P_0 + \ldots + \left(\frac{\lambda}{\mu}\right)^{N-1} P_0\right]$$

$$= 1 - P_0 \left[\frac{1 - \left(\frac{\lambda}{\mu}\right)^N}{1 - \left(\frac{\lambda}{\mu}\right)}\right]$$

$$= 1 - \left(1 - \left(\frac{\lambda}{\mu}\right)\right)\left(\frac{1 - \left(\frac{\lambda}{\mu}\right)^N}{1 - \left(\frac{\lambda}{\mu}\right)}\right)$$

$$= \left(\frac{\lambda}{\mu}\right)^N$$

$$= \rho^N.$$

Probability that the queue size exceeds N is $P(n > N) = \rho^{N+1}$.

Step 4: To obtain the probability density function of waiting time excluding service time distribution.

In the steady state, each customer has the same waiting time distribution. This is a continuous distribution with probability density function $\psi(\omega)$ and we denote by $\psi(\omega)\,\Delta\omega$ the probability that a customer begins to be served in the interval $(\omega, \omega + \Delta\omega)$ where ω is measured from the time of his arrival.

Suppose the customer arrives at time $\omega = 0$ and service begins in the interval $(\omega, \omega + \Delta\omega)$.

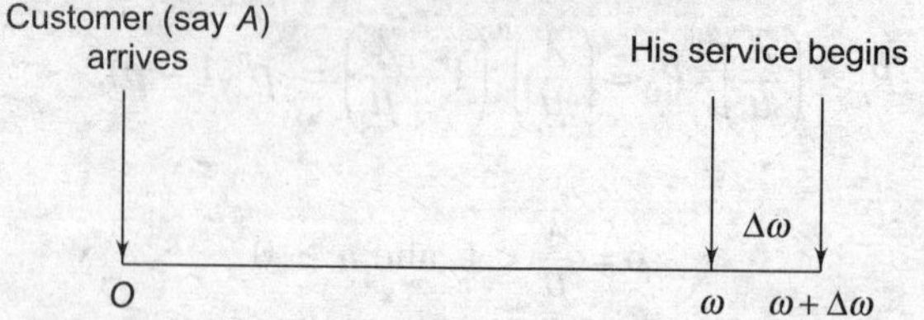

figure 11.3

Then,

(i) There is a finite probability that the waiting time is zero.

(ii) If there are n customers already in the system when the customer A arrives, n must leave before the service of A begins. More precisely, $(n - 1)$ customers must leave in the time interval $(0, \omega)$ and the nth customer in $(\omega, \omega + \Delta\omega)$.

The service mean rate is μ in the unit time and hence $\mu\omega$ in time ω, and the probability of $(n - 1)$ departures in time ω, during which the surver is busy, follows the Poisson distribution $\dfrac{(\mu\omega)^{n-1} e^{-\mu\omega}}{(n-1)!}$.

Let there be n units in the system. Then,

$$\psi_n(\omega)\Delta\omega = P[(n-1) \text{ units are served at time } \omega] \times P[\text{one served in time } \Delta\omega]$$

or,
$$\psi_n(\omega)\Delta\omega = \frac{(\mu\omega)^{n-1}\rho^{-\mu\omega}}{(n-1)!} \times \mu\Delta\omega.$$

Let W be the waiting time of a unit who has to wait such that $\omega \leq W \leq \omega + \Delta\omega$. Then, the probability

$$\psi(\omega)\Delta\omega = P[\omega \leq W \leq \omega + \Delta\omega]$$

$$= \sum_{n=1}^{\infty} P_n\, \psi_n(\omega)\Delta\omega,$$

[(the probability of n customers in the system when A arrives) $\times$ (the probability that exactly $n - 1$ customers leave in (o, ω) $\times$ (the probability that n^{th} customer leaves in $(\omega, \omega + \Delta\omega)$] summed over all n from 1 to ∞]

Now $\sum_{n=1}^{\infty} P_n\, \psi_n(\omega)\, \Delta\omega$

$$= \sum_{n=1}^{\infty}\left(\frac{\lambda}{\mu}\right)^n \left(1-\frac{\lambda}{\mu}\right)\frac{(\mu\omega)^{n-1} e^{-\mu\omega}}{(n-1)!}\,\mu\Delta\omega$$

$$= \left(\frac{\lambda}{\mu}\right)\left(1-\frac{\lambda}{\mu}\right)\mu e^{-\mu\omega}\sum_{n=1}^{\infty}\frac{\left[\left(\frac{\lambda}{\mu}\right)(\mu\omega)\right]^{n-1}}{(n-1)!}\,\Delta\omega$$

$$= \lambda\left(1-\frac{\lambda}{\mu}\right)e^{-\mu\omega}\sum_{n=1}^{\infty}\frac{(\lambda\omega)^{n-1}}{(n-1)!}\Delta\omega$$

$$= \lambda\left(1-\frac{\lambda}{\mu}\right)e^{-\mu\omega}\left(1+\frac{\lambda\omega}{1!}+\frac{(\lambda\omega)^2}{2!}+\ldots\right)\Delta\omega$$

$$= \lambda\left(1-\frac{\lambda}{\mu}\right)e^{-\mu\omega}e^{\lambda\omega}\;\Delta\omega$$

Hence, $$\psi(\omega)\,\Delta\omega = \lambda\left(1-\frac{\lambda}{\mu}\right)e^{-(\mu-\lambda)\omega}\;\Delta\omega$$

$\Rightarrow$ $$\psi(\omega) = \lambda\left(1-\frac{\lambda}{\mu}\right)e^{-(\mu-\lambda)\omega},\ \text{for } w \neq 0$$

For, $$w = 0$$

$$P[w = 0] = P[\text{no unit in the system}] = P_0$$

$$= 1 - \frac{\lambda}{\mu}.$$

Sum of all probabilities of waiting time

$$= \int_0^\infty P(\omega \le W \le \omega + \Delta\omega) + P(w = 0)$$

$$= \int_0^\infty \lambda\left(1-\frac{\lambda}{\mu}\right)e^{-(\mu-\lambda)\omega}d\omega + \left(1-\frac{\lambda}{\mu}\right)$$

$$= \lambda\left(1-\frac{\lambda}{\mu}\right)\left[\frac{e^{-(\mu-\lambda)\omega}}{-(\mu-\lambda)}\right]_0^\infty + \left(1-\frac{\lambda}{\mu}\right)$$

$$= \left(\frac{\lambda}{\mu}\right) + \left(1-\frac{\lambda}{\mu}\right)$$

$$= 1$$

Hence, it is concluded that the complete distribution for waiting time is partly continuous and partly discrete, that is, for $\omega \le W \le \omega + \Delta\omega$, the waiting time follows continuous distribution and for $W = 0$, it follows a discrete distribution,

$$P(W \ge \omega) = \int_\omega^\infty \psi(\omega)\,d\omega$$

$$= \int_\omega^\infty \lambda\left(1-\frac{\lambda}{\mu}\right)e^{-(\mu-\lambda)\omega}d\omega$$

$$= \lambda\left(1-\frac{\lambda}{\mu}\right)\left[\frac{e^{-(\mu-\lambda)\omega}}{-(\mu-\lambda)}\right]_\omega^\infty$$

$$= \frac{\lambda}{\mu}\,e^{-(\mu-\lambda)\,\omega}$$

$$= \rho.\,e^{-(\mu-\lambda)\,\omega}$$

Step 5: *To obtain busy period distribution.*

To find the probability density function for the distribution of total time (waiting + service) an arrival spends in the system. Let

$\psi\left(\omega/_{\omega>0}\right)$ = probability density function for waiting time such that a person has to wait (busy period)

By condition probability definition,

$$\psi(\omega/\omega{>}0)\,\Delta\omega = \frac{\psi(\omega)\Delta\omega}{P(\omega>0)} = \frac{\psi(\omega)\,\Delta\omega}{\int_0^\infty \psi(\omega)\,d\omega}$$

$$= \frac{\lambda\left(1-\frac{\lambda}{\mu}\right)e^{-(\mu-\lambda)\omega}\Delta\omega}{\int_0^\infty \lambda\left(1-\frac{\lambda}{\mu}\right)e^{-(\mu-\lambda)\omega}\,d\omega}$$

$$= \frac{\lambda\left(1-\frac{\lambda}{\mu}\right)e^{-(\mu-\lambda)\omega}\Delta\omega}{\frac{\lambda}{\mu}}$$

or $$\psi\left(\omega/_{\omega>0}\right) = (\mu-\lambda)\,e^{-(\mu-\lambda)\,\omega} \tag{11.3.4}$$

$$\int_0^\infty \psi\left(\omega/_{\omega>0}\right)\,d\omega = \int_0^\infty (\mu-\lambda)\,e^{-(\mu-\lambda)\,\omega} = 1$$

Hence, (11.3.4) is the probability density function for the busy period. If service time is included, then,

$$P\,(w \ge \omega) = \int_{\mathrm{w}}^{\infty} (\mu-\lambda)_e{}^{-(\mu-\lambda)\omega}\,d\omega$$

$$= \rho\,e^{-(\mu-\lambda)\omega}$$

Characteristics of Model I

(i) *Average (or expected) number of customers E (n) in the system is*

$$E\,(n) = \sum_{n=0}^{\infty} nP_n = \sum_{n=1}^{\infty} n\left(\frac{\lambda}{\mu}\right)^n\left(1-\frac{\lambda}{\mu}\right)$$

$$= \left(1-\frac{\lambda}{\mu}\right)\left(\frac{\lambda}{\mu}\right)\sum_{n=1}^{\infty} n\left(\frac{\lambda}{\mu}\right)^{n-1}$$

$$= \left(1-\frac{\lambda}{\mu}\right)\left(\frac{\lambda}{\mu}\right)\left[1+2\left(\frac{\lambda}{\mu}\right)+3\left(\frac{\lambda}{\mu}\right)^2+\ldots\right]$$

$$= \left(1-\frac{\lambda}{\mu}\right)\left(\frac{\lambda}{\mu}\right)\left(\frac{1}{\left(1-\frac{\lambda}{\mu}\right)^2}\right)$$

$$= \frac{\frac{\lambda}{\mu}}{1-\frac{\lambda}{\mu}} = \frac{\rho}{1-\rho}, \qquad \text{where, } \rho = \frac{\lambda}{\mu} < 1$$

Hence, $$\boxed{E(n) = \frac{\rho}{1-\rho} = \frac{\lambda}{\mu-\lambda}}$$

(ii) *Average (or expected) queue length E (m) is*

$$E(m) = \sum_{m=0}^{\infty} m\, P_n$$

where, $m = n - 1$ being the number of customers in the queue, excluding the customer being serviced. Therefore,

$$E(m) = \sum_{n=1}^{\infty} (n-1)P_n = \sum_{n=1}^{\infty} nP_n - \sum_{n=1}^{\infty} P_n$$

$$= \sum_{n=0}^{\infty} nP_n - \left[\sum_{n=0}^{\infty} P_n - P_0\right]$$

$$= \frac{\rho}{1-\rho} - [1-(1-\rho)]$$

$$= \frac{\rho}{1-\rho} - \rho = \frac{\rho^2}{1-\rho} = \frac{\lambda^2}{\mu(\mu-\lambda)}$$

Hence, $$\boxed{E(m) = \frac{\rho^2}{1-\rho} = \frac{\lambda^2}{\mu(\mu-\lambda)}}$$

(iii) *Mean (or expected) waiting time in the queue*

Expected time an arrival spends in the queue is

$$E(\omega) = \int_0^{\infty} \omega\, \psi(\omega)\, d\omega$$

$$= \int_0^{\infty} \omega \lambda \left(1 - \frac{\lambda}{\mu}\right) e^{-(\mu-\lambda)\omega} d\omega$$

$$= \int_0^{\infty} \omega \mu \rho (1-\rho)\, e^{-\mu(1-\rho)\omega} d\omega, \qquad \text{where } \rho = \frac{\lambda}{\mu}$$

$$= \rho \cdot \int_0^{\infty} \frac{xe^{-x}}{\mu(1-\rho)} dx, \text{ where } x = \mu(1-\rho)\omega$$

$$= \frac{\rho}{\mu(1-\rho)} \Gamma(2)$$

$$= \frac{\rho}{\mu(1-\rho)} 1!, \quad \text{since } \boxed{(2)} = (2-1)! = 1!$$

Hence, $$\boxed{E(\omega) = \frac{\rho}{\mu(1-\rho)} = \frac{\lambda}{\mu(\mu-\lambda)}}$$

(iv) *Average waiting time of the customer who has to wait* $\left(w/_{w>0}\right)$

The expected length of busy period is given by

$$E\left(w/_{w>0}\right) = \int_0^\omega \omega\psi(\omega>0)\, d\omega$$

$$= \int_0^\omega \omega(\mu-\lambda)\, e^{-(\mu-\lambda)\omega}\, d\omega$$

Hence, $$E(w/w>0) = \frac{1}{\mu-\lambda} = \frac{1}{\mu(1-\rho)}$$ by using integration by parts.

(v) *Average (expected) waiting time in the system (including service time), E (v)*

Since expected waiting time in the system $E(v)$

= Expected waiting time in queue + Expected service time (or mean service time)

$$= E(\omega) + \frac{1}{\mu}$$

$$= \frac{\lambda}{\mu(\mu-\lambda)} + \frac{1}{\mu} = \frac{1}{\mu-\lambda}$$

Hence, $$\boxed{E(v) = \frac{1}{\mu-\lambda}}$$

(vi) *Expected (average) length of non-empty queue (E (n/n > 0))*

$$E(n/n>0) = \frac{E(n)}{P(n>0)} = \frac{E(n)}{1-P_0}$$

$$= \frac{\lambda}{\mu-\lambda}\cdot\frac{\mu}{\lambda}, \quad \text{since } P_0 = 1 - \frac{\lambda}{\mu}$$

$$= \frac{\mu}{\mu-\lambda}$$

Hence, $$\boxed{E(n/n>0) = \frac{\mu}{\mu-\lambda}}$$

(vii) *The fluctuation (variance) of queue length is given by*

$$V(n) = \sum_{n=1}^{\infty} n^2 P_n - \left(\sum_{n=1}^{\infty} n P_n\right)^2$$

$$= \sum_{n=1}^{\infty} n^2 P_n - [E(n)]^2$$

$$= \sum_{n=1}^{\infty} n^2 (1-\rho)\rho^n - \left(\frac{\rho}{1-\rho}\right)^2$$

$$= (1-\rho)\,[1^2 \times \rho + 2^2\rho^2 + 3^2\rho^3 + 4^2\rho^4 + \ldots] - \frac{\rho^2}{(1-\rho)^2}$$

$$= (1-\rho)\,\rho[1 + 2^2\rho + 3^2\rho^2 + \ldots] - \frac{\rho^2}{(1-\rho)^2}$$

$$= (1-\rho)\rho\left[\frac{1+\rho}{(1-\rho)^3}\right] - \frac{\rho^2}{(1-\rho)^2}$$

$$= \frac{\rho}{(1-\rho)^2} = \frac{\lambda\mu}{(\mu-\lambda)^2}$$

where, $1 + 2^2\rho + 3^2\rho^2 + \ldots = S$ (say)

Integrating both sides within the limit 0 to ρ gives

$$\int_0^{\rho} S d\rho = \int_0^{\rho} 1 + 2^2\rho + 3^2\rho^2 + \ldots)\, d\rho = \rho + 2\rho^2 + 3\rho^3 + \ldots$$

$$= \rho\,[1 + 2\rho + 3\rho^3 + \ldots]$$

$$= \frac{\rho}{(1-\rho)^2}$$

Now, differentiating w.r. to ρ on both sides, we get

$$S = \frac{1}{(1-\rho)^2} + \frac{2\rho}{(1-\rho)^3} = \frac{1+\rho}{(1-\rho)^3}$$

Hence, $$V(n) = \frac{\rho}{(1-\rho)^2} = \frac{\lambda\mu}{(\mu-\lambda)^2}$$

(viii) *Probability of arrivals during the service time of any given customer*

Since the arrivals are Poisson and service times are exponential, the probability of r arrivals during the service time of any given customer is given by

$$K_r = \int_0^{\infty} \rho_r(t)\, s(t)\, dt$$

$$= \int_0^{\infty} \frac{e^{-\lambda}(\lambda t)^r}{r!}\, \mu\, e^{-\mu t}\, dt$$

$$= \frac{\lambda^r \mu}{r!} \int_0^\infty e^{-(\lambda+\mu)t} t^r dt$$

$$= \frac{\lambda^r \mu}{r!} \frac{\Gamma_{(r+1)}}{(\lambda+\mu)^{r+1}} \qquad \left[\text{using} \int_0^\infty e^{-at} t^n dt = \frac{\Gamma_{(n+1)}}{a^{n+1}}\right]$$

$$= \left(\frac{\lambda}{\lambda+\mu}\right)^r \cdot \frac{\mu}{\lambda+\mu} \qquad \text{since } \Gamma_{(r+1)} = r\,!$$

Relation between Average Queue Length and Average Waiting Time—Little's Formula

$$E\,(n) = \frac{\lambda}{\mu-\lambda} = \frac{\lambda}{\mu-\lambda} = \lambda E\,(v)$$

$$E\,(m) = \frac{\lambda^2}{\mu\,(\mu-\lambda)} = \lambda \frac{\lambda}{\mu(\mu-\lambda)} = \lambda \cdot E\,(\omega)$$

and, $$E\,(v) = E\,(\omega) + \frac{1}{\mu} \quad \text{or,} \quad E\,(\omega) = E\,(v) - \frac{1}{\mu}$$

Thus, multiplying both sides by λ and substituting the values gives

$$E\,(m) = E\,(n) - \frac{\lambda}{\mu}$$

This means that one of the four expected values (together with λ and μ) should yield the remaining three values.

Example 1

A self-service store employs one cashier at its counter. An average of nine customers arrive every 5 minutes while the cashier can serve 10 customers in 5 minutes. Assuming Poisson distribution for arrival rate and exponential distribution for service rate, find

1. Average number of customers in the system.
2. Average number of customers in queue or average queue length.
3. Average time a customer spends in the system.
4. Average time a customer waits before being served.

Solution: Arrival rate $\lambda = \frac{9}{5} = 1.8$ customers/minute

Service rate $\mu = \frac{10}{5} = 2$ customers/minute

1. Average number of customers in the system

$$E\,(n) = \frac{\lambda}{\mu-\lambda} = \frac{1.8}{2-1.8} = 9$$

2. Average number of customers in the queue

$$E\,(n) = \frac{\lambda^2}{\mu(\mu-\lambda)} = \frac{\lambda}{\mu}\frac{\lambda}{\mu-\lambda} = \frac{1.8}{2} \times \frac{1.8}{2-1.8} = 8.1$$

3. Average time a customer spends in the system $E(v) = \dfrac{1}{\mu - \lambda} = \dfrac{1}{2 - 1.8} = 5$ minutes

4. Average time a customer spends in the queue

$$E(\omega) = \frac{\lambda}{\mu(\mu - \lambda)} = \frac{1.8}{2(2 - 1.8)} = 4.5 \text{ minutes.}$$

Example 2

At what average must a clerk at a supermarket work in order to ensure a probability of 0.90 that the customer will not have to wait longer than 12 minutes? It is assumed that there is only one counter to which customers arrive in a Poisson fashion at an average rate of 15 per hour. The length of service by the clerk has an exponential distribution.

Solution: Given $\lambda = \dfrac{15}{60} = \dfrac{1}{4}$ customer/minute

P [waiting time ≥ 12] = 1 – (= 0.10

Therefore,

$$\int_{12}^{\infty} \lambda\left(1 - \frac{\lambda}{\mu}\right) e^{-(\mu-\lambda)\omega}\, d\omega = 0.10$$

or,

$$\lambda\left(1 - \frac{\lambda}{\mu}\right)\left[\frac{e^{-(\mu-\lambda)\omega}}{-(\mu-\lambda)}\right]_{12}^{\infty} = 0.10$$

$$\left(\frac{\lambda}{\mu}\right) e^{-12(\mu-\lambda)} = 0.10$$

$$e^{(3-12\mu)} = 0.4\,\mu$$

$$\frac{1}{\mu} = 2.48 \text{ minute per service}$$

$$\mu = \frac{1}{2.48} \text{ minute/service.}$$

Example 3

Customers arrive at a one-window drive-in-bank according to Poisson distribution with mean 10 per hour. Service time per customer is exponential with mean 5 minutes. The space in front of the window including that for the serviced care can accommodate a maximum of 3 cars. Others can wait outside this space.

(i) What is the probability that an arriving customer can drive directly to the space in front of the window?

(ii) What is the probability that an arriving customer will have to wait outside the indicated space?

(iii) How long is the arriving customer expected to wait before starting service?

Solution: We know

$$P_n = \rho^n (1 - \rho)$$

(i) The probability that an arriving customer can drive directly to the space in front of the window = $P_0 + P_1 + P_2$

$$= \left(1 - \frac{\lambda}{\mu}\right)\left[1 + \frac{\lambda}{\mu} + \left(\frac{\lambda}{\mu}\right)^2\right]$$

$$= \left(1 - \frac{10}{12}\right)\left[1 + \frac{10}{12} + \frac{100}{144}\right]$$
$$= 0.42$$

(ii) Probability that an arriving customer will have to wait outside the indicated space = 1 – 0.42 = 0.58

(iii) Average waiting time of a customer in the queue $= \frac{\lambda}{\mu} \cdot \frac{1}{\mu - \lambda} = \frac{10}{12} \cdot \frac{1}{12 - 10} = \frac{5}{12} = 0.417$ hours.

Example 4

On an average 96 patients per 24-hour day require the service of an emergency clinic. Also, on an average, a patient requires 10 minutes of active attention. Assume that the facility can handle only one emergency at a time. Suppose that it costs the clinic Rs 100 per patient treated to obtain an average servicing time of 10 minutes, and that each minute of decrease in this average time would cost Rs 10 per patient treated. How much would have to be budgeted by the clinic to decrease the average size of the queue from $1\frac{1}{3}$ patients to $\frac{1}{2}$ patient.

Solution: Given $\lambda = \frac{96}{24 \times 60} = \frac{1}{15}$ patient/minute

$\mu = \frac{1}{10}$ patient/minute

Expected number of patients in the waiting line

$$E\,(m) = \frac{\lambda^2}{\mu(\mu - \lambda)} = \frac{\left(\frac{1}{15}\right)^2}{\frac{1}{10}\left(\frac{1}{10} - \frac{1}{15}\right)} = 1\frac{1}{3} \text{ patients}$$

But $E\,(m) = 1\frac{1}{3}$ is reduced to $E\,(m)' = \frac{1}{2}$. Then, substituting $E\,(m)' = \frac{1}{2}$, $\lambda' = \lambda = \frac{1}{15}$ in the formula,

$$E\,(m)' = \frac{\lambda'^2}{\mu'(\mu' - \lambda')} \Rightarrow \frac{\left(\frac{1}{15}\right)^2}{\mu'\left(\mu' - \frac{1}{15}\right)} = \frac{1}{2}$$

$\Rightarrow$ $\mu' = \frac{2}{15}$ patient/minute

Hence, the average rate of treatment required is $\frac{1}{\mu'} = 7.5$ minutes. Consequently, the decrease in the average rate of treatment $= 10 - \frac{15}{2} = \frac{5}{2}$ minutes and the budget per patient $= 100 + \frac{5}{2} \times 10 =$ Rs 125.

So, in order to get the required size of the queue the budget should be increased from Rs 100 to Rs 125.

Example 5

If for a period of 2 hours in a day (8–10 a.m.) trains arrive at a yard every 20 minutes but the service time continues to remain 36 minutes, then calculate for this period

(a) the probability that the yard is empty
(b) average queue length
on the assumption that the line capacity of the yard is limited to 4 trains only.

Solution: $\rho = \dfrac{36}{20} = 1.8 > 1$, and $N = 4$.

(a) $$P_0 = \frac{\rho - 1}{\rho^5 - 1} = 0.04$$

(b) Average queue size $= P_0 \sum_{n=0}^{4} n\rho^n$

$$= 0.04\,\{\rho + 2\rho^2 + 3\rho^3 + 4\rho^4\}$$
$$= 0.04 \times 72$$
$$= 2.9$$
$$\cong 3 \text{ trains.}$$

Example 6

In a railway marshalling yard, goods trains arrive at a rate of 30 trains per day. Assume that the inter-arrival time follows an exponential distribution and the service time distribution is also exponential with an average of 36 minutes. Calculate
(a) The probability that the yard is empty.
(b) Average queue length assuming that the line capacity of the yard is 9 trains.

Solution: Given $\rho = \dfrac{\lambda}{\mu} = \dfrac{\left(\dfrac{30}{24 \times 60}\right)}{\dfrac{1}{36}} = \dfrac{36}{48} = 0.753/\text{minute}$

(a) The probability that the queue size is zero is given by

$$P_0 = \frac{1-\rho}{1-\rho^{N+1}}, \quad \text{where } N = 9.$$

$$= \frac{1-0.753}{1-(0.753)^{10}} = \frac{0.247}{0.941} = 0.2625$$

(b) Average queue length is given by the formula

$$E(m) = \frac{\rho^2\,[1 - N\rho^{N-1} + (N-1)\,\rho^N]}{(1-\rho)(1-\rho^{N+1})}$$

$$= \frac{(0.753)^2\,[1 - 9\,(0.753)^8 + 8(0.753)^9\,]}{[0.247)\,(1-(0.753)^{10}\,]}$$

$$= \frac{0.3878}{(0.247)\,(0.9414)} = \frac{0.3878}{0.2325} = 1.668$$

$$\simeq 2 \text{ trains.}$$

Example 7

A barber shop has space to accommodate 10 customers. The barber can service only one person at a time. If a customer comes to his shop and finds it full, he goes to the next shop. Customers random arrive at an average rate $\lambda = 10$ per hour and the barber's service time is negative exponential with an average of $\frac{1}{\mu} = 5$ minutes per customer. Find P_0 and P_n.

Solution: Given $N = 10$, $\lambda = \frac{10}{60}$, $\mu = \frac{1}{5}$, $\rho = \frac{\lambda}{\mu} = \frac{5}{6}$

$$P_0 = \frac{1-\rho}{1-\rho''} = \frac{1-5/6}{1-(5/6)''} = 0.1926$$

$$P_n = \left(\frac{1-\rho}{1-\rho''}\right)\rho^n = (0.1926)\,(5/6)^n, \quad n = 0, 1, 2, ..., 10$$

Model IV (Birth-Death Process) (General Erlang Queuing Model)

Assume that the system is in E_n. Then, the probability of a birth occurring in a small time interval Δt is considered as $\lambda_n \Delta t + O(\Delta t)$ and the death occuring Δt is considered as $\mu_n \Delta t + O(\Delta t)$, $n \geq 1$.

If the system is in E_n at time t, then it will remain in E_n at time $t + \Delta t$ provided there is no birth and no death or one birth and one death or the system was in E_{n-1} and had a birth, or in E_{n+1} and had a death. Let arrival rate $\lambda = \lambda_n$, and service rate $\mu = \mu_n$ (depending on n). Then, by the same arguments as in Model I

$$P_n(t + \Delta t) = P_n(t)\,(1 - \lambda_n \Delta t - O(\Delta t))\,(1 - \mu_n \Delta t - O(\Delta t)) + P_{n+1}(t)\,(\mu_{n+1} \Delta t + O(\Delta t))$$
$$(1 - \lambda_{n+1}\Delta t - O(\Delta t)) + P_{n-1}(t)\,(\lambda_{n-1}\Delta t + O(\Delta t))\,(1 - \mu_{n-1}\Delta t - O(\Delta t))$$
$$= P_n(t)\,[1 - (\lambda_n + \mu_n)\,\Delta t] + \mu_{n+1} P_{n+1}(t)\,\Delta t + \lambda_{n-1} P_{n-1}(t)\,\Delta(t) + O(\Delta t)),\ n \geq 1.$$

$$P_0(t + \Delta t) = P_0(t)\,[1 - \lambda_0 \Delta t - O(\Delta t)] + P_1(t)\,[\mu_1 \Delta(t) + O(\Delta t)]\,\{1 - \lambda_1, \Delta t + O(\Delta t)\} + O(\Delta t),$$

Dividing these equations by Δt and taking $\lim \Delta t \to 0$, we get

$$\lim_{\Delta t \to 0} \frac{P_n(t+\Delta t) - P_n(t)}{\Delta t} = -(\lambda_n + \mu_n)\,P_n(t) + \mu_{n+1} P_{n+1}(t) + \lambda_{n-1} P_{n-1}(t)$$

and,
$$\lim_{\Delta t \to 0} \frac{P_0(t+\Delta t) - P_0(t)}{\Delta t} = -\lambda_0 P_0(t) + \mu_1 P_1(t)$$

That is,
$$P'_n(t) = -(\lambda_n + \mu_n)\,P_n(t) + \mu_{n+1} P_{n+1}(t) + \lambda_{n-1} P_{n-1}(t)$$

and,
$$P'_0(t) = -\lambda_0 P_0(t) + \mu_1 P_1(t)$$

In the case of steady state, $P'_n(t) = 0$, $P'_0(t) = 0$, as $P_n(t)$ is independent of t, we have $P_n(t) = P_n$, $P_0(t) = P_0$

$$0 = -(\lambda_n + \mu_n)\,P_n + \mu_{n+1} P_{n+1} + \lambda_{n-1} P_{n-1} \tag{11.3.11}$$

$$0 = -\lambda_0 P_0 + \mu_1 P_1 \tag{11.3.12}$$

From (11.3.12), we have

$$P_1 = \left(\frac{\lambda_0}{\mu_1}\right) P_0$$

Substituting $n = 1$ in (11.3.11), we get

$$P_2 = \frac{(\lambda_1 + \mu_1)}{\mu_2} P_1 - \frac{\lambda_0}{\mu_2} P_0 = \frac{\lambda_1 \lambda_0}{\mu_2\, \mu_1} P_0$$

Substituting $n = 2$ in (11.3.11) gives

$$P_3 = \frac{(\lambda_2 + \mu_2)}{\mu_3} P_2 - \frac{\lambda_1}{\mu_3} P_1 = \frac{\lambda_2 \lambda_1 \lambda_0}{\mu_3 \mu_2\, \mu_1} P_0$$

$$\ldots \quad \ldots \quad \ldots \quad \ldots$$

In general, we can write the following formula

$$P_n = \frac{\lambda_{n-1} \lambda_{n-2} \ldots \lambda_0}{\mu_n\, \mu_{n-1} \ldots \mu_1} P_0$$

$$= P_0 \prod_{i=0}^{n-1} \frac{\lambda_i}{\mu_{i+1}},\ n \ge 1.$$

In order to find P_0, consider

$$\sum_{n=0}^{\infty} P_n = 1$$

$$\Rightarrow \qquad P_0 + P_1 + P_2 + \ldots + \infty = 1$$

$$\Rightarrow \qquad P_0 \left[1 + \frac{\lambda_0}{\mu_1} + \frac{\lambda_0 \lambda_1}{\mu_1 \mu_2} + \ldots \right] = 1$$

$$\Rightarrow \qquad P_0 = \frac{1}{S},$$

where, $S = 1 + \dfrac{\lambda_0}{\mu_1} + \dfrac{\lambda_0 \lambda_1}{\mu_1 \mu_2} + \ldots + \infty$

The series s is summable and meaningful only when it is a convergent series.

Special Cases

Case 1: $(\lambda_n = \lambda,\ \mu_n = \mu)$, then

$$S = 1 + \frac{\lambda}{\mu} + \left(\frac{\lambda}{\mu}\right)^2 + \ldots + \infty = \frac{1}{1 - (\lambda/\mu)},\ \text{when } \lambda/\mu < 1$$

Hence,

$$P_0 = \frac{1}{S} = 1 - \frac{\lambda}{\mu} = 1 - \rho,$$

and so,

$$P_n = \rho^n (1 - \rho),\ \text{for } n > 0$$

Case 2: $\lambda_n = \dfrac{\lambda}{n+1},\ \mu_n = \mu$. In this case,

$$S = 1 + \frac{\lambda}{\mu} + \frac{\lambda^2}{2\mu^2} + \frac{\lambda^3}{2.3\,\mu^3} + \ldots \infty$$

$$= 1 + \rho + \frac{\rho^2}{2!} + \frac{\rho^3}{3!} + \ldots \infty,\ \rho = \frac{\lambda}{\mu}$$

$$= e^{\rho}$$

So,

$$P_0 = \frac{1}{S} = e^{-\rho}$$

$$P_n = \frac{\lambda}{\mu} P_0 = \rho e^{-\rho}$$

$$P_2 = \frac{\lambda^2}{2\mu^2} P_0 = \frac{\rho^2 e^{-\rho}}{2!}$$

$$\ldots \quad \ldots \quad \ldots$$

$$P_n = \frac{\lambda^n}{n!\,\mu^n} P_0 = \frac{\rho^n e^{-\rho}}{n!} \text{ for all } n = 0, 1, \ldots\infty.$$

This case indicates that P_n follows the Poisson distribution, where $\frac{\lambda}{\mu} = \rho$. However, $\rho > 1$ or $\rho < 1$ must be finite.

Case 3: When $\lambda_n = \lambda$, and $\mu_n = n\mu$. In this case the arrival rate does not depend on n, but the service rate μ_n increases as n increases. In this case,

$$S = 1 + \frac{\lambda}{\mu} + \frac{\lambda^2}{2.1.\mu^2} + \frac{\lambda^3}{3.2.1.\mu^3} + \ldots + \infty$$

$$= 1 + \rho + \frac{\rho^2}{2!} + \frac{\rho^3}{3!} + \ldots + \infty$$

$$= e^{\rho}$$

$\Rightarrow$ $P_0 = e^{-\rho}$ and $P_n = \frac{e^{-\rho}\rho^n}{n!}$, which again follows Poisson law.

Example 8

Problems arrive at a computing centre in a Poisson fashion at an average rate of five per day. The rules of the computing centre are that any man waiting to get his problem solved must aid the man whose problem is being solved. If the time to solve a problem with one man has an exponential distribution with mean time of 1/3 day, and if the average solving time is inversely proportional to the number of people working on the problem, approximate the expected time in the centre for a person entering the line.

Solution: Given $\lambda = 5$ problems/day $\mu = 3$ problems /day.

Then, the expected number of persons working at any specified instant is

$$E(n) = \sum_{n=0}^{\infty} n P_n = \sum_{n=0}^{\infty} n . \frac{1}{n!}\left(\frac{\lambda}{\mu}\right)^n e^{-\lambda/\mu}$$

$$= e^{-\lambda/\mu} \sum_{n=0}^{\infty} \frac{n}{n!}\left(\frac{\lambda}{\mu}\right)^n$$

$$= e^{-\lambda/\mu}\left[\frac{\lambda}{\mu}+\frac{2}{2!}\left(\frac{\lambda}{\mu}\right)^2+\frac{3}{3!}\left(\frac{\lambda}{\mu}\right)^3+\ldots\right]$$

$$= e^{-\lambda/\mu}\left(\frac{\lambda}{\mu}\right)\left[1+\frac{\lambda}{\mu}+\frac{1}{2!}\left(\frac{\lambda}{\mu}\right)^2+\ldots\right]$$

$$= e^{-\lambda/\mu}\left(\frac{\lambda}{\mu}\right)e^{\lambda/\mu}$$

$$= \frac{\lambda}{\mu}$$

Substituting the values of λ and μ gives

$$E\,(n) = 5/3 \text{ persons.}$$

Now, the average solving time, which is inversely proportional to the number of people working on the problem, is given by $\frac{1}{5}$ day per problem.

Therefore Expected time for a person entering the line is given by

$$\frac{1}{5}\,E\,(n) = \frac{1}{5}\times\frac{5}{3}\text{ days} = \frac{1}{3}\text{ day or 8 hours.}$$

Example 9

A bank has two tellers working on savings accounts. The first teller handles withdrawals only. The second teller handles deposits only. It has been found that the service time distribution for both deposits and withdrawals is exponential with mean service time 3 minutes per customer. Depositors are found to arrive in a Poisson fashion throughout the day with mean arrival rate of 16 per hour. Withdrawers also arrive in a Poisson fashion with a mean arrival rate of 14 per hour. What would be the effect on the average waiting time for depositors and withdrawers, if each teller could handle both withdrawals and deposits. What would be the effect on average waiting time for depositors and withdrawers, if each teller could handle both the functions. What would be the effect of the average waiting time of the queue if the service time is increased to 3.5 minutes?

Solution: Initially there are two independent queuing systems: withdrawers and depositors, where arrivals follow Poisson distribution and service time follows exponential distribution.

For withdrawers: Given λ = 14/hour, μ = 3/minute or 20/hour

Average waiting time in queue

$$E(m) = \frac{\lambda}{\mu(\mu-\lambda)} = \frac{14}{20\,(20-14)} = \frac{7}{60}\text{ hours or 7 minutes.}$$

For depositors: Given λ = 16/hour, μ = 3/minute or 20/hour

Average waiting time in queue

$$E(m) = \frac{\lambda}{\mu(\mu-\lambda)} = \frac{16}{20\,(20-16)} = \frac{1}{5}\text{ hour or 12 minutes}$$

Combined case: In this case, there will be a common queue with two servers (tellers). Thus, we have

$$\lambda = 14 + 16 = 30/\text{hour},\ M = 20/\text{hour},\ C = 2,$$

$$P = \frac{\lambda}{C\mu} = \frac{3}{4}$$

Now,

$$P_0 = \left[\sum_{n=0}^{C-1} \frac{1}{n!}\left(\frac{\lambda}{\mu}\right)^n + \frac{1}{C!}\left(\frac{\lambda}{\mu}\right)^C \left(\frac{C\mu}{C\mu - \lambda}\right)\right]^{-1}$$

$$= \left[\sum_{n=0}^{C-1} \frac{1}{n!}\left(\frac{3}{2}\right)^n + \frac{1}{2!}\left(\frac{3}{2}\right)^2 \left(\frac{40}{40-30}\right)\right]^{-1}$$

$$= \left[1 + \frac{3}{2} + \frac{1}{2}\left(\frac{9}{4}\right)\cdot 4\right]^{-1} = \frac{1}{7}$$

Average waiting time of arrivals in the queue

$$E(\omega) = \frac{E(m)}{\lambda} = \frac{1}{(C-1)!}\left(\frac{\lambda}{\mu}\right)^C \frac{\mu}{(C\mu - \lambda)^2}.P_0$$

$$= \left(\frac{3}{2}\right)^2 \frac{20}{(40-30)^2} \times \frac{1}{7}$$

$$= \frac{9}{140} \text{ hour or 3.86 minutes}$$

Combined waiting time with increase service time, when $\lambda = 30/\text{hour}$, $\mu = 60 \times \frac{1}{3.5} = \frac{120}{7}$ per hour, then,

$$P_0 = \left[\sum_{n=0}^{1} \frac{1}{n!}\left(\frac{21}{12}\right)^n + \frac{1}{2!}\left(\frac{21}{12}\right)^2 \frac{2 \times \left(\frac{120}{7}\right)}{\left(2\left(\frac{120}{7}\right) - 30\right)}\right]^{-1}$$

$$= 1 + \frac{7}{4} + \frac{49}{4}$$

$$= \frac{1}{15}$$

Average waiting time of arrivals in the queue

$$E(\omega) = \frac{E(m)}{\lambda} = \frac{1}{(C-1)!}\left(\frac{\lambda}{\mu}\right)^C \frac{\mu}{(C\mu - \lambda)^2}.\ P_0$$

$$= \left(\frac{7}{4}\right)^2 \frac{\left(\frac{120}{7}\right)}{\left(2 \times \frac{120}{7} - 30\right)^2} \times \frac{1}{15}$$

$$= \frac{343}{30} \times 60 \text{ hours}$$
$$= 11.43 \text{ minutes.}$$

Example 10

A supermarket has two girls ringing up sales at the counters. If the service time for each customer is exponential with mean 4 minutes, and if people arrive in a Poisson fashion at the rate of 10 an hour, calculate the

(a) probability of having to wait for service?
(b) expected percentage of idle time for each girl?
(c) if a customer has to wait, what is the expected length of his waiting time?

Solution: Given, $\lambda = \frac{1}{6}$ per minute, $\mu = \frac{1}{4}$ per minute,

$$C = 2, \rho = \frac{\lambda}{C\mu} = \frac{\frac{1}{6}}{2 \times \frac{1}{4}} = \frac{1}{3}. \text{ Then,}$$

$$P_0 = \left[\sum_{n=0}^{2-1} \frac{1}{n!}\left(\frac{4}{6}\right)^n + \frac{1}{2!}\left(\frac{4}{6}\right)^2 \cdot \frac{2 \times \frac{1}{4}}{\left(2 \times \frac{1}{4} - \frac{1}{6}\right)}\right]^{-1}$$

$$= \left[1 + \frac{2}{3} + \frac{1}{3}\right]^{-1} = \frac{1}{2}$$

(a) Probability of having to wait for service

$$P(n \geq 2) = P(\omega > 0) = \sum_{n=2}^{\infty} P_n = \sum_{n=2}^{\infty} \frac{1}{2!2^{n-2}}\left(\frac{2}{3}\right)^n P_0$$

$$= \frac{1}{2!}\left(\frac{2}{3}\right)^2 P_0 \sum_{n=2}^{\infty}\left(\frac{2}{3 \times 2}\right)^{n-2}$$

$$= \frac{1}{9}\sum_{n=2}^{\infty}\left(\frac{1}{3}\right)^{n-2}$$

$$= \frac{1}{9}\left[1 + \frac{1}{3} + \left(\frac{1}{3}\right)^2 + \ldots\right]$$

$$= \frac{1}{9}\left[\frac{1}{1 - \frac{1}{3}}\right]$$

$$= \frac{1}{9} \times \frac{3}{2}$$

$$= \frac{1}{6}$$

(b) The fraction of time the servers are busy, $\rho = \frac{\lambda}{C\mu} = \frac{1}{3}$. Therefore, the expected idle time for each girl is $1 - \frac{1}{3} = \frac{2}{3} = 0.67$. Hence, the expected percentage of idle time for each girl is 67%.

(c) Expected length of the customer' waiting time in the system

$$E(v) = E(\omega) + \frac{1}{\lambda}$$

$$= \frac{1}{(C-1)!}\left(\frac{\lambda}{\mu}\right)^C \frac{\mu}{(C\mu - \lambda)^2} P_0 + \frac{1}{\mu}$$

$$= \left(\frac{4}{6}\right)^2 \frac{\frac{1}{4}}{\left(\frac{1}{2} - \frac{1}{6}\right)^2} \times \frac{1}{2} + 4 = 4.5 \text{ minutes}$$

Model VI: (M/M/C): (N/F1F0) Multi Server, Limited Queuing Model

This model is essentially the same as Model V, with maximum number of customers in the system is limited to N(>C). Then

$$\lambda_n = \begin{cases} \lambda & ; \quad 0 \le n \le N \\ 0 & ; \quad n \ge N \end{cases}$$

and

$$\mu_n = \begin{cases} n\mu & ; \quad 0 \le n < C \\ C\mu & ; \quad C \le n \le N \end{cases}$$

Probability of n customers in the system in the steady state condition is given by

$$P_n = \begin{cases} \frac{1}{n!}\left(\frac{\lambda}{\mu}\right)^n P_0 & ; \le n < C \\ \frac{1}{C!C^{n-C}}\left(\frac{\lambda}{\mu}\right)^n . P_0 ; C \le n \le N \end{cases}$$

The condition

$$\sum_{n=0}^{N} P_n = \sum_{n=0}^{C-1} P_n + \sum_{n=C}^{N} P_n = 1$$

$\Rightarrow$

$$P_0 = \left[\sum_{n=0}^{C-1} \frac{1}{n!}\left(\frac{\lambda}{\mu}\right)^n + \sum_{n=C}^{N} \frac{1}{C!C^{n-C}}\left(\frac{\lambda}{\mu}\right)^n\right]^{-1}$$

$$= \begin{cases} \left[\sum_{n=0}^{C-1} \frac{1}{n!}\left(\frac{\lambda}{\mu}\right)^n + \frac{1}{C!}\left(\frac{\lambda}{\mu}\right)^C \left\{1-\left(\frac{\lambda}{C\mu}\right)^{N-C+1}\right\} \frac{C\mu}{C\mu-\lambda}\right]^{-1}, & \frac{\lambda}{C\mu} \neq 1 \\ \left[\sum_{n=0}^{C-1} \frac{1}{n!}\left(\frac{\lambda}{\mu}\right)^n + \frac{1}{C!}\left(\frac{\lambda}{\mu}\right)^C (N+C+1)\right]^{-1}, & \frac{\lambda}{C\mu} = 1 \end{cases}$$

Remark: If we take $N \to \infty$, and consider $\frac{\lambda}{C\mu} < 1$, then the reduced result corresponds to that of Model III. Also, if we take $C = 1$, then the reduced result corresponds to that of Model II.

Characteristics of Model VI

(i) Expected number of customers in the queue

$$E(m) = \sum_{n=C}^{N} (n-C) P_n = \sum_{n=C}^{N} (n-C) \frac{\left(\frac{\lambda}{\mu}\right)^n}{C!C^{n-C}} P_0$$

$$= \frac{1}{C!}\left(\frac{\lambda}{\mu}\right)^C .P_0.\rho \sum_{j=0}^{N-C} j\rho^{j-1}, \text{ where } j = n - C, \rho = \frac{\lambda}{C\mu}$$

$$= \frac{1}{C!} \frac{(C\mu\rho)^C.P_0.\rho}{\mu C} \sum_{j=0}^{N-C} \frac{d}{d\rho}(\rho^j)$$

$$= \frac{1}{C!}(C\rho)^C P_0 P.\frac{d}{d\rho}\left[\sum_{j=0}^{N-C} \rho^j\right]$$

$$= \frac{1}{C!}(C\rho)^C P_0 P.\frac{d}{d\rho}[1 + \rho + \rho^2 ... + \rho^{N-C}]$$

$$= \frac{1}{C!}(C\rho)^C P_0\, \rho \frac{d}{d\rho}\left[\frac{1-\rho^{N-C+1}}{1-\rho}\right]$$

(GP of $N - C + 1$ terms)

$$= \frac{(C\rho)^C.\rho.P_0}{C!(1-\rho)^2} [1 - \rho^{N-C+1} - (1 - \rho)(N - C + 1) \rho^{N-C}]$$

(ii) Expected number of customers in the system

$$E(n) = \sum_{n=0}^{N} nP_n = \sum_{n=0}^{C-1} nP_n + \sum_{n=C}^{N} nP_n$$

$$= \sum_{n=0}^{C-1} nP_n + E(m) + C\sum_{n=C}^{N} P_n$$

$$= E(m) + C\left[\sum_{n=0}^{N} P_n - \sum_{n=0}^{C-1} P_n\right] + \sum_{n=0}^{C-1} nP_n$$

$$= E(m) + C - P_0 \sum_{n=0}^{C-1} \frac{(C-n)(\rho C)^n}{n!}$$

(iii) Expected waiting time in the system

$$E(v) = \frac{E(n)}{\lambda(1 - P_N)}$$

(iv) Expected waiting time in the queue

$$E(w) = E(v) - \frac{1}{\mu} = \frac{E(m)}{\lambda(1 - P_N)}.$$

Example 11

A barber shop has two barbers and three chairs for customers. Assume that the customers arrive in poisson fashion at a rate of 5 per hour and that each barber services customers according to an exponential distribution with mean of 15 minutes. Further if a customer arrives and there are no empty chairs in the shop, he will leave. What is the probability that the shop is empty? What is the expected number of customers in the shop.

Solution: Given $C = 2, N = 3, \lambda = \frac{5}{60} = \frac{1}{12}$ customer/minute $\mu = \frac{1}{15}$ per minute.

$$P_0 = \left[\sum_{n=0}^{2-1} \frac{1}{n!}\left(\frac{\lambda}{\mu}\right)^n + \sum_{n=2}^{3} \frac{1}{2^{n-2} 2!}\left(\frac{\lambda}{\mu}\right)^n\right]^{-1}$$

$$= \left[1 + \frac{5}{4} + \frac{1}{2}\left(\frac{5}{4}\right)^2 + \frac{1}{4}\left(\frac{5}{4}\right)^3\right]^{-1}$$

$$= \frac{256}{901} = 0.28$$

$$P_n = \begin{cases} \frac{1}{n!}\left(\frac{5}{4}\right)^n \times 0.28, & 0 \le n < 2 \\ \frac{1}{2^{n-2}\, 2!}\left(\frac{5}{4}\right)^n \times 0.28, & 2 \le n \le 3 \end{cases}$$

$$= \begin{cases} \frac{1}{n!}(1.25)^n \times 0.28, & 0 \le n < 2 \\ \frac{1}{2^{n-2}\, 2!}(1.25)^n \times 0.28, & 2 \le n \le 3 \end{cases}$$

Expected number of customers in the system

$$E(n) = E(m) + C - P_0 \sum_{n=0}^{C-1} \frac{(C-n)(\rho C)^n}{n!},$$

where,

$$E(m) = \sum_{n=2}^{3} (n-2) P_n = P_3 = \frac{1}{2^{3-2}.2!}(1.25)^3 \times 0.28$$
$$= 0.1367$$

So,

$$E(n) = 0.1367 + 2 - (0.28) \sum_{n=0}^{2-1} \frac{(2-n)(1.25)^n}{n!}$$
$$= 0.1367 + 2 - (0.28)\,(3.25)$$
$$= 1.2267 \text{ customers.}$$

Example 12

There are 5 machines, each of which when running suffers breakdown at an average rate of 2 per hour. There are 2 serviceman and only 1 man can work on a machine at a time. If n machines are out of order, where $n > 2$, then $(n - 2)$ of them wait until a serviceman is free. Once a serviceman starts work on a machine, the time to complete the repair has an exponential distribution with mean 5 minutes. Find the distribution of the number of machines out of action at a given time. Find also the average time an out-of-action machine has to spend waiting for the repairs to start.

Solution: Given, K = total number of machines in the system = 5

R = number of servicemen = 2

λ = 2/hour

$$\mu = \frac{60}{5} = 12\text{/hour}$$

n = number of machines out of order

So,

$$P_n = \begin{cases} 5C_n \left(\frac{2}{12}\right)^n P_0 & 0 \le n < 2 \\ 5C_n \frac{n!}{2^{n-2} \cdot 2!} \left(\frac{2}{12}\right)^n P_0, & 2 \le n \le 5 \end{cases}$$

where,

$$P_0 = \left[\sum_{n=0}^{1} 5C_n \left(\frac{2}{12}\right)^n + \sum_{n=2}^{5} 5C_n \frac{n!}{2^{n-2} \cdot 2!} \left(\frac{2}{12}\right)^n \right]^{-1}$$
$$= 0.455$$

Average number of machines out of action is given by

$$E(m) = \sum_{n=2}^{5} (n - R)P_n = \sum_{n=2}^{5} (n-2)P_n$$

$$= P_3 + 2P_4 + 3P_5$$

$$= 5C_3 \frac{3!}{2^{3-2} \cdot 2!}\left(\frac{2}{12}\right)^3 P_0 + 2 \times 5C_4 \frac{4!}{2^{4-2} \cdot 2!}\left(\frac{2}{12}\right)^4 P_0 + 3 \times 5C_5 \frac{5!}{2^{5-2} \cdot 2!}\left(\frac{2}{12}\right)^5 P_0$$

$$= 0.379 + 0.0105 + .00132$$

$$= 0.391 \text{ hours}$$

Average time an out-of-action machine has to spend waiting for the repairs to start is

$$E(\omega) = \frac{E(m)}{\lambda(K - E(n))},$$

where, $\lambda(K - E(n)) = \lambda K \sum_{n=0}^{K} P_n - \lambda \sum_{n=0}^{K} nP_n$

$$= \lambda \sum_{n=0}^{K} (K - n)\lambda P_n$$

$$= \lambda \sum_{n=0}^{5} (5 - n)P_n$$

$$= \lambda(5P_0 + 4P_1 + 3P_2 + 2P_3 + P_4)$$

$$= \lambda\left[5 \times 0.455 + 4 \times 5C_1 \times \left(\frac{2}{12}\right)^1 P_0 + 3 \times 5C_2 \times \frac{2!}{2^0 \cdot 2!}\left(\frac{2}{12}\right)^2 P_0\right.$$

$$\left. + 2 \times 5C_3 \times \frac{3!}{2 \times 2!}\left(\frac{2}{12}\right)^3 P_0 + 5C_4 \frac{4!}{2^{4-2} \cdot 2!}\left(\frac{2}{12}\right)^4 P_0\right]$$

$$= \lambda\,[2.275 + 0.379 + 0.063 + 0.0052$$

$$= 2 \times 2.722 = 5.444$$

$$= 5 \text{ machines}$$

$$E(\omega) = \frac{0.391}{5.444} = 0.072 \text{ hours.}$$

11.4 Non-Poisson Queues

Queues in which arrivals and/or departures may not follow the Poisson axioms are called non-Poisson queues. The following techniques are employed while studying the non-Poisson queues:

(a) Phase technique
(b) The technique of imbedded Markov chain
(c) The supplementary variable technique.

The phase technique is used when an arrival demands phases of service, say k in number.

The technique by which non-Markovian queues are reduced to Markovian is called Imbedded Markov chain technique. When one or more random variables are added to convert a non-Markovian process into a Markovian one, the technique involved in this conversion is called the supplementary variable technique. This technique is applied to: GI|G|C, M|G|I, GI|M|S, GI|E_K|1, D|E_K|1, models.

We now consider a queuing system in which the service time has an Erlang type K-distribution with average service time as $\frac{1}{K\mu}$, for each exponential phase.

Model I (M|E_K|1): (∞|FIFO)

This model consists of a single service channel queuing system in which there are n phases in the system (waiting or in service). Sinc each customer is served in K-phases one-by-one and a new service does not start until all K-phases have been completed, therefore each arrival increases the number of phases by K in the system. Hence, a new arrival creates K-phases of service and departure of one customer reduces k-phases of service.

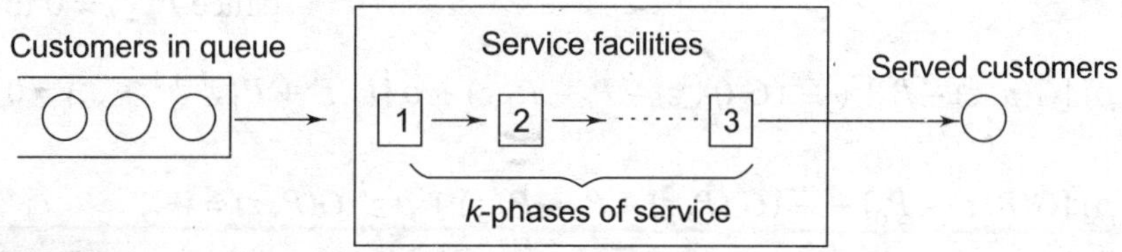

figure 11.4 **Service Channels in Series**

If $P_n(t)$ denotes the probability of $n(>0)$ phases in the system at time t. At time $t + \Delta t$,

$$P_n(t + \Delta t) = P_n(t)\,(-\lambda\Delta t - O(\Delta t))\,(1 - K\mu\Delta t - O(\Delta t))$$
$$+ P_n(t)\,(\lambda\Delta t)\,(K\mu\Delta t) + P_{n+1}(t)\,(1 - \lambda\Delta t - O(\Delta t))$$
$$+ P_{n-K}(t)\,(\lambda\Delta t)\,[1 - K\mu\Delta t - O(\Delta t)],\ n \geq 1$$

$$P_0(t + \Delta t) = P_0(t)\,(1 - \lambda\Delta t + O(\Delta t) + P_1(t)\,(K\mu\Delta t),\ n = 0$$

Here, a negative subscript indicates the term is zero. As usual, the steady state difference equations are:

$$0 = -(\lambda + k\mu)P_n + k\mu\, P_{n+1} + P_{n-k},\ n \geq 1$$
$$0 = -\lambda P_0 + k\mu P_1$$

Put $\frac{\lambda}{k\mu} = \rho$. Then, we have

$$0 = -(1 + \rho)P_n + P_{n+1} + \rho\, P_{n-k},\ n \geq 1 \quad (11.4.1)$$

$$0 = -\rho P_0 + P_1 \quad (11.4.2)$$

To solve these difference equations, we can use generating function concepts. By this definition of generating function (GF) for a sequence $\{P_n\}_{n=0}^{\infty}$

$$\text{G.F.} = G(p, z) = \sum_{n=0}^{\infty} P_n z_n \tag{11.4.3}$$

where, z is a formal variable.

Multiply both sides of (11.4.1) by z^n and taking summation over 1 to ∞ (this equation is true only when $n \geq 1$), we have

$$-\sum_{n=0}^{\infty} (1+\rho) P_n z^n + \sum_{n=1}^{\infty} P_{n+1} z^n + \sum_{n=1}^{\infty} \rho P_{n-k} z^n = 0$$

$$\Rightarrow \quad -(1+\rho)[P_1 z + P_2 z^2 + \ldots] + [P_2 z + P_3 z^2 + \ldots] + \sum_{n=1}^{\infty} \rho P_{n-k} z^n = 0$$

$$\Rightarrow \quad -(1+\rho)[G(p, z) - P_0] + \frac{1}{z}(G(P, z) - P_0 - P_1 z) + \rho \sum_{n=k}^{\infty} P_{n-k} z^n = 0$$

[since $P_{n-k} = 0$ for all $n - k < 0$]

$$\Rightarrow \quad -(1+\rho)[G(p, z) - P_0] + \frac{1}{z}(G(p, z) - P_0 - P_1 z) + \rho [P_0 z^k + P_1 z^{k+1} + \ldots] = 0$$

$$\Rightarrow \quad -(1+\rho)[G(P, z) - P_0] + \frac{1}{z}(G(P, z) - P_0 - P_1 z) + \rho z^k G(P, z) = 0$$

$$\Rightarrow \quad G(P, z)\left[-(1+\rho) + \rho z^k + \frac{1}{z}\right] = \frac{P_0}{z} + P_1 - P_0(1+\rho)$$

$$\Rightarrow \quad G(P, z)\left[\frac{-(1+\rho)z + \rho z^{k+1} + 1}{z}\right] = P_0\left[\frac{1}{z} - 1\right] \text{ by (11.4.2)}$$

$$\Rightarrow \quad G(p, z)[1 + Pz^{k+1} - (1+\rho)z] = P_0(1-z)$$

$$G(p, z) = \frac{P_0(1-z)}{1+\rho z^{k+1} - (1+\rho)z}$$

$$= \frac{P_0(1-z)}{(1-z) - \rho z(1-z^k)}$$

$$= \frac{P_0}{1 - \rho z\left(\frac{1-z^k}{1-z}\right)}$$

$$= P_0\left[1 - \rho z\left(\frac{1-z^k}{1-z}\right)\right]^{-1}$$

Using binomial thereom

$$G(P, z) = P_0\left[1 + \rho z\left(\frac{1-z^k}{1-z}\right) + (\rho z)^2\left(\frac{1-z^k}{1-z}\right)^2 + \cdots\right]$$

[since $(1 - x)^{-1} = 1 + x + x^2 + ...$]

$$= P_0 \sum_{n=0}^{\infty} (\rho z)^n \left(\frac{1-z^k}{1-z}\right)^n \qquad (11.4.4)$$

Since the sum of the series $1 + z + z^2 + ... + z^{k-1} = \dfrac{1-z^k}{1-z}$, it follows that

$$G(P, z) = P_0 \sum_{n=0}^{\infty} (\rho z)^n (1 + z + z^2 + ... + z^{k-1})^n \qquad (11.4.5)$$

Now, we find P_0 and P_n from equation (11.4.5), Put $z = 1$, then,

$$G(P, 1) = P_0 \sum_{n=0}^{\infty} \rho^n k^n, \text{ since } [1 + 1 + ... k \text{ times}]^n = k^n$$

$$\Rightarrow \quad G(P, 1) = \frac{P_0}{1-\rho k}\left[\text{since } \sum_{n=0}^{\infty} (\rho k)^n = \frac{1}{1-\rho k} \text{ for sum of infinite G.P.}\right] \qquad (11.4.6)$$

On the other hand, putting $z = 1$ in equation (11.4.3), we get

$$G(P, 1) = \sum_{n=0}^{\infty} P_n = 1 \text{ (by the law of probability)} \qquad (11.4.7)$$

Comparing the equations (11.4.6) and (11.4.7),

$$\frac{P_0}{1-\rho k} = 1$$

Which gives $\qquad P_0 = 1 - \rho k$

Substituting the value of P_0 in (11.4.4), and letting $n = m$ in equation (11.4.4)

$$G(P, z) = (1 - \rho k) \sum_{n=0}^{\infty} (\rho z)^m (1 - z^k)^m (1 - z)^{-m} \qquad (11.4.8)$$

But,

$$(1 - z^k)^m = 1 - mc_1 (z^k) + mc_2 (z^k)^2 + ... + (-1)^m mC_m(z^k)^m$$

$$= \sum_{i=0}^{m} (-1)^i mC_i z^{ik}$$

and,

$$(i - z)^{-m} = \sum_{j=0}^{\infty} (-1)^j \left(-m_{C_j}\right) z^j$$

$$= \sum_{j=0}^{\infty} (-1)^j (-1)^j (m + j - 1)_{C\,j}\, z^j, \text{ since } -mC_j = (m + j - 1)C_j$$

Therefore, (11.4.8) becomes

$$G(P, z) = (1 - \rho k) \sum_{m=0}^{\infty} (\rho z)^m \left[\sum_{i=0}^{m} (-i)^i\, mC_i z^{ki} \sum_{j=0}^{\infty} (m + j - 1) C_j z^j \right]$$

or,

$$G(P, z) = (1 - \rho k) \sum_{m=0}^{\infty} \rho^m \sum_{j=0}^{\infty} \sum_{i=0}^{m} (-1)^i\, mC_i \left(m + j - 1\right) C_j . z^{m + ik + j}$$

$$\text{or, } \sum_{n=0}^{\infty} P_n z^n = (1 - \rho k) \sum_{m=0}^{\infty} \rho^m \sum_{j=0}^{\infty} \sum_{i=0}^{m} (-1)^i\, mC_i \left(m + j - 1\right) C_j . z^{m + ik + j} \qquad (11.4.9)$$

Comparing the coefficients of z^n from both sides of equation (11.4.9) we get

$$P_n = (1 - \rho k) \sum_{i, j, m} \rho^m (-1)^i\, mC_i\, (m + j - 1) C_j$$

where, the summation is taken over all combinations of i, j and m satisfying $n = m + ik + j$, for given k.

Characteristics of Model I

(i) Average number of phases in the system $E(n_p)$ is given by

$$E(n_p) = \sum_{n=0}^{\infty} nP_n$$

$$= \frac{k(k+1)}{2} \frac{\frac{\lambda}{k\mu}}{1 - \frac{k\lambda}{k\mu}} = \left(\frac{k+1}{2}\right) \frac{\lambda}{\mu - \lambda},$$

by using the equation $(1 + \rho) P_n = \rho P_{n-k} + P_{n+1}$, $n \geq 1$

(ii) Average waiting time of the phases in the system is given by

$$E(\omega_p) = \frac{E(n_p)}{\mu} = \left(\frac{k+1}{2\mu}\right) \frac{\lambda}{\mu - \lambda}$$

(iii) Average waiting time of an arrival is given by

$$E(\omega) = \frac{E(\omega_p)}{k} = \left(\frac{k+1}{2k}\right) \frac{\lambda}{\mu(\mu - \lambda)}$$

(iv) Average time an arrival spends in the system is given by

$$E(n) = E(\omega) + \frac{1}{\mu}$$

$$= \left(\frac{k+1}{2k}\right)\frac{\lambda}{\mu(\mu-\lambda)} + \frac{1}{\mu}$$

(v) Average number of units in the system is given by

$$E(n) = \lambda E(v) = \left(\frac{k+1}{2k}\right)\frac{\lambda^2}{\mu(\mu-\lambda)} + \frac{\lambda}{\mu}$$

(vi) Average queue length is given by

$$E(m) = E(n) - \frac{\lambda}{\mu}$$

$$= \frac{k+1}{2k}\,\frac{\lambda^2}{\mu(\mu-\lambda)}$$

Model II (M|E$_K$|1): (1|FIFO)

This model differs from Model I in the sense that the capacity of the system is unity. That is, there is no queue and the system contains only one customer.

Let the customer be in nth phase of service such that $1 < n < k$.

Like model I, the following set of steady-state difference equations govern this model:

$$0 = -\lambda P_0 + k\mu P_1, \quad n = 0$$

$$0 = -\,k\mu\, P_k + \lambda P_0, \quad n = k$$

$$0 = -\,k\mu\, P_n + k\mu\, P_{n+1}, \quad 1 \le n < k$$

These equations, provide us the following relations:

$$P_1 = \frac{\lambda}{k\mu}\,P_0, \qquad \text{for } n = 0$$

$$P_k = \frac{\lambda}{k\mu}\,P_0, \qquad \text{for } n = k$$

and,

$$P_{n+1} = P_n\,,\ \text{for } 1 \le n < k$$

Now, we observe that $P_1 = P_2 = \cdots = P_{k-1} = P_k$, we have

$$P_1 = \frac{\lambda}{k\mu}\,P_0, \text{ for } i = 1, 2, \ldots k$$

To obtain the value of P_0, we make use of the fact that $\sum_{i=0}^{k} P_i = 1$. That is,

$$P_0 + \sum_{i=1}^{k} P_i = 1 \quad \text{or} \quad P_0 + \sum_{i=1}^{k} \frac{\lambda}{k\mu} P_0 = 1$$

or,

$$P_0 = \left[1 + \frac{\lambda}{\mu} \times \frac{1}{k} \times k\right]^{-1}$$

$$= \left[1 + \frac{\lambda}{\mu}\right]^{-1}$$

Hence,

$$P_n = \frac{\lambda}{k\mu}\left(1 + \frac{\lambda}{\mu}\right)^{-1} = \frac{\lambda}{k\mu} \times \frac{\mu}{\lambda + \mu} = \frac{1}{k} \frac{\lambda}{\lambda + \mu}.$$

Example 1

Prove that for the Erlang distribution with parameters *m* and *k*, the mode is at $\left(1 - \frac{1}{k}\right)\left(\frac{1}{\mu}\right)$

Solution: The probability density function of the Erlangian service time distribution is defined by

$$s(t, \mu, k) = \frac{(k\mu)^k \, t^{k-1} e^{-k\mu t}}{(k-1)!}, \; k = 1, 2, 3...$$

Mode:

$$\frac{d}{dt} (s(t, \mu, k) = 0$$

$$\Rightarrow \quad \frac{(k\mu)^k}{(k-1)!} [(k-1)t^{k-2} e^{-k\mu t} - t^{k-1} (k\mu) e^{-k\mu t}] = 0$$

$$\Rightarrow \quad (k-1)t^{k-2} = k\mu \, t^{k-1}$$

$$\Rightarrow \quad t = \frac{k-1}{k\mu}$$

Since $\frac{d^2}{dt^2}$ (s (t, μ, k)) is negative for $t = \frac{k-1}{k\mu}$, the mode lies at $t = \frac{k-1}{k\mu} = \left(1 - \frac{1}{k}\right)\left(\frac{1}{\mu}\right)$.

Example 2

In a factory cafeteria customers have to pass through three counters. The customers buy coupons at the first counter, select and collect snacks at the second counter and collect tea at the third. The server at each counter takes on an average 1.5 minutes although the distribution of service time is approximately Poisson at an average rate of 6 per hour. Calculate

(a) the average time a customer spends waiting in the cafeteria.

(b) the average time of getting the service.

(c) the most probable time in getting the service.

Solution: Given $\lambda = 6$ customers/hour,

Service time per phase = 1.5 minutes

$$\mu = 1.5 \times 3 \text{ (phases)}$$
$$= 4.5 \text{ customers/minute}$$
$$= 13.34/\text{hour}$$
$$k = 3$$

(a) Average time a customer spends waiting in the cafeteria

$$E(v) = E(\omega) + \frac{1}{\mu}$$
$$= \left(\frac{k+1}{2k}\right)\frac{\lambda}{\mu(\mu-\lambda)} + \frac{1}{\mu}$$
$$= \left(\frac{3+1}{2\times 3}\right)\left(\frac{6}{13.34(13.34-6)}\right) + \frac{1}{13.34}$$
$$= \left[\frac{9}{220} + \frac{1}{13.34}\right] \text{hour}$$
$$= [2.43 + 4.50] \text{ minutes}$$
$$= 6.93 \text{ minutes}$$

(b) Average time of getting the service is the average time when it is following the third phase of service. Thus, average time spent in getting the service is $\frac{1}{\mu} = \frac{1}{13.34}$ hour or 4.50 minutes.

(c) The most probable time spent in getting the service is the model value of service time for the third phase of service. Thus, most probable time spent is

$$\frac{k-1}{k\mu} = \frac{3-1}{3\times 13.34} = \frac{1}{20} \text{ hour or 3 minute.}$$

11.5 QUEUING CONTROL

Queuing theory is mainly used to determine the queue length, waiting time in queue, and so on. Majority of queuing literature involves *prescriptive models* rather than *descriptive models*. Prescriptive models are viewed as static optimisation models.

Static (Design) optimisation models are those in which steady state conditions are set up for the system and some long-run average criterion (such as cost and/or profit) is determined.

When the queuing systems are allowed to depend upon time and are controlled, then these systems are known as *dynamic control systems*. Some optimisation models are the mixture of static and dynamic categories. But, if the state-dependent system is controlled, then it is called under-dynamic control. Following are the objectives of *dynamic control*:

Arrival Process Control
(a) To accept or reject control
(b) To adjust mean arrival rate
(c) Customer exercised control
(d) Self versus social optimisation
(e) Projection times
Service Process Control
(a) Varying a number of servers
(b) Varying the service times
Control of Queue Discipline
(a) Priority models
(b) Scheduling models
(c) Allocation of customers to multi-server channels.

EXERCISES

Section 11.1

1. What is a queue? Give an example. Explain the basic elements of queues.
2. Explain Poisson input, exponential service.
3. Give some applications of queuing theory.
4. State three applications of waiting line theory in business enterprises.
5. Explain (i) queue discipline, (ii) capacity of the system, (iii) holding time, (iv) balking, (v) jockeying.
6. Briefly explain the important characteristics of queuing system.
7. What do you understand by
   (i) traffic intensity (ii) service channels
   (iii) steady and transient state (iv) utilisation factor.
8. Show that if the inter-arrival times is exponentially distributed, the number of arrivals in a period of time is a Poisson process and the converse is also true.
9. Draw a diagram showing the physical layout of a queuing system with multi-server multi-channel service facility.
10. Briefly discuss the application of queuing theory in industrial management.
11. Explain the basic queuing process. What are the important random variates in the queuing system to be investigated?
12. Explain (i) expected waiting time in queue, (ii) expected waiting time in system.

Section 11.2

13. Show that the distribution of the number of births up to time T in a simple birth process follows the Poisson law.
14. State and prove the Markovian property of inter-arrival times.
15. Establish the probability distribution formula for pure-death process. If the intervals between successive arrivals are random variables which follow the negative exponential distribution with mean $\frac{1}{\lambda}$, then show that the arrivals form a Poisson distribution with mean λ.

16. In a single server, Poisson arrival and exponential service time queuing system show that probability P_n of n customers in steady state satisfies the following equations:

$$\lambda P_0 = \mu P_1 \qquad n = 0$$
$$(\lambda + \mu) P_1 = \mu P_2 \qquad n = 1$$
$$(\lambda + \mu) P_n = \mu P_{n+1} + \lambda P_{n-1}, \qquad n \geq 2$$

17. Define the cumulative probability distribution of waiting time for a customer who has to wait and show that in an (M | M | 1|)(∞ | FIFO) queue it is given by $(1 - \rho)e^{-\mu/1 - \rho}$, where $\rho = \dfrac{\lambda}{\mu}$.
18. Write a short note on M | M | 1 queue and its applications.
19. What are the various queuing models available?
20. Explain the following queuing systems:
    (a) (M | M | R) : (GD | K), R < K
    (b) (M | G | 1) : (∞ | GD)

Section 11.3

Model I

21. The mean arrival rate to a service centre is 3 per hour. The mean service time is found to be 10 minutes per service. Assuming Poisson arrival and exponential service time, find
    (a) utilisation factor for this service facility.
    (b) probability of two units in the system.
    (c) expected number of units in the queue.
    (d) expected time in minutes that a customer has to spend in the system.

    $\left[\textit{Answer}: \text{(a) } \dfrac{1}{2}, \text{(b) } \dfrac{1}{8}, \text{(c) } 1, \text{(d) } \dfrac{1}{3}\right]$

22. At a one-man barber shop, customers arrive according to Poisson distribution with a mean arrival rate of 5 per hour and the hair cutting time is exponentially distributed, with an average hair cut taking 10 minutes. It is assumed that because of his excellent reputation, customers are always willing to wait. Calculate the following:
    (i) Average number of customers in the shop and the average number of customers waiting for a hair cut.
    (ii) The per cent of time an arrival can walk right in without having to wait.
    (iii) The percentage of customers who have to wait prior to getting into the barber's chair.
    [*Answer*: (i) 5 customers in the shop and approximately 4 customers waiting for hair cut. (ii) 16.7%, (iii) 83.3%.
23. A supermarket has a single cashier. During peak hours, customers arrive at a rate of 20 customers per hour. The average number of customers that can be processed by the cashier is 24 per hour. Calculate
    (i) the probability that the cashier is idle.
    (ii) the average number of customers in the queuing system.
    (iii) the average time a customer spends in the system.
    (iv) the average number of customers in the queue.
    (v) the average time a customer spends in the queue waiting for service.

[*Answer*: (i) 0.167, (ii) 5 customers, (iii) 15 minutes, (iv) 4.167 or 4 customers, (v) 12.5 minutes.]

24. At a public telephone booth in a post office, arrivals are considered to be Poisson with an average inter-arrival time of 12 minutes. The length of a phone call may be assumed to be distributed exponentially with an average of 4 minutes.
    (a) What is the probability that a fresh arrival will not have to wait for the phone?
    (b) What is the probability that an arrival will have to wait more than 10 minutes before the phone is free?
    (c) What is the average length of queues that form time to time?
    [*Answer*: (a) 67%, (b) 0.063, (c) 1.5 customers.]
25. A warehouse has only one loading dock manned by a 3 person crew. Trucks arrive at the loading dock at an average rate of 4 trucks per hour and the arrival rate is Poisson distributed. The loading of a truck takes 10 minutes on an average and can be assumed to be exponentially distributed. The operating cost of a truck is Rs 20 per hour and the members of the loading crew are paid @ Rs 6 each hour. Would you advise the truck owner to add another crew of three persons?
    [*Answer*: Total cost (present crew) = Rs 58 per hour,
    Total cost (proposed crew) = Rs 46 per hour
    Therefore, add a crew of 3 loaders.]
26. A repairman is to be hired to repair machines which breakdown at an average rate of 3 per hour. Breakdowns are distributed in time in a manner that may be regarded as Poisson. Non-productive time on a machine is considered to cost Rs 5 per hour. The company has narrowed the choice to 2 repairmen—one slow, but cheap, the other fast, but expensive.
    The slow-cheap repairman asks Rs 2 per hour and he will service machines at the rate of Rs 4 per hour. The fast-expensive repairman demands Rs 8 per hour, and will repair machines at the rate of 5 per hours. Which repairman should be hired?
    [*Answer*: Fast-expensive repairman to be hired.]

Model III

27. At a railway station, only one train is handled at a time. The railway yard is sufficient only for two trains to wait while the other is given the signal to leave the station. Trains arrive at the station at an average rate of 6 per hour and the railway station can handle them on an average of 12 per hour. Assuming Poisson arrivals and exponential service distribution, find the steady state probabilities for the various number of trains in the system. Also, find the average waiting time of a new train coming into the yard.
    [*Answer*: $P_0 = 0.53$, $P_1 = 0.27$, $P_2 = 0.13$, $P_3 = 0.07$; average number of trains in the system = 0.74; average waiting time of a new train coming into the yard = 3.5 minutes.]
28. Consider a single server queuing system with Poisson input and exponential service times. Suppose the mean arrival rate is 3 calling units per hour, the expected service time is 0.25 hour and the maximum permissible calling units in the system is 2. Derive the steady state probability distribution of the number of calling units in the system, and then calculate the expected number in the system.
    [*Answer*: $P_n = (0.43)\,(0.75)^n$, $P_0 = 0.431$, $E\,(n) = 0.81$.]
29. A car park contains 5 c ars. The arrival of car is Poisson at a mean rate of 10 per hour. The length of time each car spends in the car park has negative exponential distribution with mean of 5 hours.

How many cars are in the car park and what is the probability of a newly arriving customer finding the car park full and having to park elsewhere?

$$\left[\textit{Hint: } N = 5,\ \lambda = \frac{10}{60},\ \mu = \frac{1}{2\times 60},\ \rho = \frac{\lambda}{\mu} = 20.\ P_0 = \frac{1-\rho}{1-\rho^{N+1}},\ L_S = P_0 \sum_{n=0}^{N} n\rho_n.\right]$$

30. Customers arrive at a one-window drive-in-bank according to a Poisson distribution with a mean of 10 per hour. Service time per customer is exponential with a mean of 5 minutes. The car space in front of the window, including that for the serviced car, can accommodate a maximum of 3 cars. Other cars wait outside this space.
    (a) What is the probability that an arriving customer can drive directly to the space in front of the window?
    (b) What is the probability that an arriving customer will have to wait outside the indicated space?
    (c) How long is an arriving customer expected to wait before starting service?
    (d) How many spaces should be provided in front of the window so that all the arriving customers can wait in front of the window at least 20% of the time.

    [*Hint*: $\lambda = 10,\ \mu = \frac{60}{5} = 12,\ P_0 = 1 - \frac{\lambda}{\mu} = \frac{1}{6}$

    (a) $P_0 + P_1 + P_2 = \left[1 + \frac{\lambda}{\mu} + \left(\frac{\lambda}{\mu}\right)^2\right] P_0 = 0.42$

    (b) $1 - (P_0 + P_1 + P_2 + P_3) = 1 - 0.42 - P_3 = 0.58 - \left(\frac{\lambda}{\mu}\right)^3 P_0 = 0.48$

    (c) $\frac{\lambda}{\mu(\mu - \lambda)} = 0.417$

    (d) $P_0 + P_1 = 0.30$. There should be atleast one car space for waiting at least 20% of the time.]

31. A stenographer has 5 persons for whom she peforms stenographic work. Arrival rate is Poisson and service times are exponential. Average arrival rate is 4 per hour with an average service time of 10 minutes. Cost of waiting is Rs 8 per hour while the cost of servicing is Rs 2.50 each. Calculate the average
    (i) waiting time of an arrival
    (ii) length of the waiting line
    (iii) time which an arrival spends in the system
    [*Answer*: (i) 12.4 minutes, (ii) 0.8 ≃ 1 stenographer, (iii) 22.4 minutes.]

32. At a one-man barber shop, the customers arrive following Poisson process at an average rate of 5 per hour and they are served according to exponential distribution with an average service rate of 10 minutes. Assuming that only 5 seats are available for waiting customers, find the average time a customer spends in the system.

    [*Answer*: $E(n) = P_0 \sum_{n=0}^{5} n\rho^n = 0.166 \sum_{n=0}^{5} \left(\frac{5}{6}\right)^n = 0.166 \times 7.8$

    $E(\omega) = 0.26$ hours ≃ 16 minutes.]

Model IV

33. An international airport operates with three check-out counters. The sign by the check-out area advises the customers that an additional counter will be opened any time the number of customers in the queue exceeds 3. This means that for fewer than 4 customers, only one counter will be in operation. For 4 to 6 customers, two counters will be open. For more than 6 customers, all three counters will be open. The customers arrive at the counters area according to a Poisson distribution with a mean of 10 customers per hour. The average check-out time per customer is exponential with a mean of 12 minutes. Determine the following:
    (a) steady-state probability P_n of n customers in the check-out area
    (b) average number of busy counters
    (c) average number of idle counters.
    [*Hint*: $\lambda_n = \lambda = 10$ customers per hour, $n = 0, 1, 2, ...$

$$\mu_n = \begin{cases} \dfrac{60}{12} \simeq 5 \text{ customers per hour } (n = 0, 1, 2, 3), \\ 2 \times 5 = 10 \text{ customers per hours } (n = 4, 5, 6) \\ 3 \times 5 = 15 \text{ customers per hour } (n \geq 7) \end{cases}$$

Model V

34. A petrol pump station has two pumps. The service time follows exponential distribution with a mean of 4 minutes and cars arrive for service in a Poisson process at the rate of 10 cars per hour. Find the probability that a customer has to wait for service. For what proportion of time do the pumps remain idle?
    [*Answer*: 0.167, or 67% for each pump.]
35. Four counters are being run on the frontier of a country to check the passports and necessary papers of the tourists. The tourists choose a counter at random. If the arrival at the frontier is Poisson at the rate λ and the service time is exponential with parameter $\frac{\lambda}{2}$, what is the steady state average queue at each counter?
    $\left[\textit{Answer}: P_0 = \dfrac{3}{23}, E(m) = \dfrac{4}{23}. \right]$
36. Given an average arrival rate of 20 per hour, is it better for a customer to get service at a single channel with mean service rate of 22 customers or at one of two channels in parallel, with mean service rate of 11 customers for each of the two channels? Assume that both queues are of M | M | C type.
    [*Answer*: It is better to get service at two channels in parallel.]
37. A two-channel waiting line with Poisson arrival has a mean arrival rate of 50 per hour and exponential service with a mean service rate of 75 per hour for each channel. Find the
    (i) probability of an empty system
    (ii) probability that an arrival in the system will have to wait.
    [*Answer*: (i) 0.83, (ii) 0.167.]

38. Assume that customers arrive at a small post office in Poisson fashion and service is exponentially distributed. Determine the minimum number of parallel service channels needed in each of the following situations to guarantee that the operation of the queuing situation will be stable (i.e. the queue length will not grow indefinitely).
    (i) Customers arrive every 5 minutes and are served at the rate of 10 customers per hour.
    (ii) The average inter-arrival time is 2 minutes, and the average service time is 6 minutes.
    (iii) The arrival rate is 30 customers per hour, and the service rate per server is 40 customers per hour.

    [*Answer*: (i) $C \geq 2$, (ii) $C \geq 4$, (iii) $C \geq 1$.]
39. For a queuing system with k service stations, each having exponential service time distribution with mean service rate μ feed by a queue with built-up arrival rate λ, find (i) average number of customers in the system, (ii) average waiting time of a customer in the system if $k = 2$, $\mu = 5$ per minute, $\lambda = 8$ per minute.
    [*Answer*: (i) 4.44, (ii) 0.56 minutes.]
40. A telephone company is planning to instal telephone booths in a new airport. It has established the policy that a person should not have to wait more than 10% of the time to use the phone. The demand for use is estimated to be Poisson with an average rate of 30 per hour. The average phone call has an exponential distribution with mean time of 5 minutes. How many phone booths should be installed.
    [*Answer*: 6 phone booths.]
41. Ships arrive at a port at the rate of one in every 4 hours with exponential distribution of inter-arrival times. The time a ship occupies a berth for unloading has exponential distribution with an average of 10 hours. If the average delay of ships waiting for berths is to be kept below 14 hours, how many berths should be provided at the port?
    [*Answer*: Four berths must be provided at the port.]
42. A general insurance company has three claims adjusters in its branch office. People with claims against the company are found to arrive in Poisson fashion, at an average rate of 20 per 8 hour-day. The amount of time that an adjuster spends with a claimant is found to have a negative exponential distribution, with mean service time 40 minutes. Claimants are processed in the order of their appearance.
    (a) How many hours a week can an adjuster expect to spend with claimants?
    (b) How much time, on the average, does a claimant spend in the branch office?

    [*Answer*: (a) 22.2 hours, (b) 49 minutes.]

    Model VI
43. A car servicing station has two boys who offer service simultaneously. Due to space limitation, only four cars are accepted for servicing. The arrival pattern is Poisson with 12 cars per day. The service time of both the boys is exponentially distributed with $\mu = 8$ cars per day per boy. Find the average number of cars in the service station, the average number of cars waiting to be serviced and the average time a car spends in the system.
    [*Answer*: $P_0 = 0.196$, $E(n) = 6.28$ cars, $E(m) = 5.26$ cars, $E(v) = 35.8$ minutes.]

Model VIII

44. Two repairmen are attending to 5 machines in a workshop. Each machine breaks down according to a Poisson distribution with a mean of 3 per hour. The repair time per machine is exponential with mean 15 minutes.

(a) Find the probability that the two repairmen are idle, that one repairman is idle.
(b) What is the expected number of idle machines not being serviced?
[*Answer*: (a) 0.0405, 0.16144, (b) 0.911.]

45. A group of engineers has two terminals available to aid in their calculations. The average computing job requires 20 minutes of terminal time, and each engineer requires some computation about once every 0.5 hours, that is, the mean time between a call for service is 0.5 hours. Assume these are distributed according to an exponential distribution. If there are 6 engineers in the group find the
(a) expected number of engineers waiting to use one of the terminals
(b) total lost time per day.
[*Answer*: (a) 1.40 or 1 engineer approx., (b) 11.2 hours per day.]

Section 11.4

46. At a certain airport it takes exactly 5 minutes to load an aeroplane, once it is given the signal to land. Although incoming planes have scheduled arrival times, the wide variability produces an effect which makes the incoming planes appear to arrive in a Poisson fashion at an average rate of 6 per hour. This produces occasional stock-ups at the airport which can be dangerous and costly. Under these circumstances, how much time will a pilot expect to spend circuling the field waiting to land?
[*Answer*: $E(\omega) = \lim_{k\to\infty} \dfrac{k+1}{2k} \cdot \dfrac{\lambda}{\mu(\mu-\lambda)} = \dfrac{5}{2} = 2.5$ minutes.]

47. In a car manufacturing plant, a loading crane takes exactly 10 minutes to load a car into a wagon and again come back to position to load another car. If the arrival of cars is a Poisson stream at an average of 1 every 20 minutes, calculate the average waiting time of a car.
[*Answer*: $E(\omega) = 5$ minutes.]

48. Describe the use of Erlangian distribution in queuing process.

49. Repairing a certain type of machine which breaks down in a given factory consists of 5 basic steps that must be performed sequentially. The time taken to perform each of the 5 steps is found to have an exponential distribution with a mean of 5 minutes and is independent of the other steps. If these machines breakdown in Poisson fashion at an average rate of 2 per hour and if there is only 1 repairman, what is the average idle time for each machine that has broken down?
[*Answer*: $E(v) = \dfrac{k+1}{2k} \cdot \dfrac{\lambda}{\mu(\mu-\lambda)} + \dfrac{1}{\mu} = 100$ minutes.]

50. A barber takes exactly 25 minutes to complete one hair cut. If customers arrive in a Poisson fashion at an average rate of 1 every 40 minutes, how long, on an average, must a customer wait for service?
[*Answer*: 20.8 minutes.]

51. A one-man tailoring shop takes exactly 1 day to stitch a suit. Customers' arrival follow a Poisson pattern with mean rate of arrival of 1 in every 2 days. How long, on an average is a customer expected to wait in such a situation?
[*Answer*:]

12

Scheduling by PERT and CPM

12.1 INTRODUCTION

Critical Path Method (CPM) was developed by Morgan R. Walker for the E.I. DuPont de Nemours Company to solve project scheduling problems. It was later extended and enhanced by Mauchly Associates. During the same time, a team of engineers working on the Polar's Missile programme of the US Navy developed PERT.

A project is defined as a combination of interrelated activities all of which must be executed in a certain order to achieve a set goal. In a large and complex project involving a number of interrelated activities requiring men, machines and material, it is impossible for the management to make and execute an optimum schedule just by intution, based on organisational capabilities and work experience. A systematic scientific approach has become a necessity. So, a number of methods applying network scheduling techniques have been developed. PERT and CPM are two of the many network techniques which are widely used for planning, scheduling and controlling large, complex projects.

These two methods are essentially network-oriented techniques using the same principle. PERT and CPM are basically time-oriented methods in the sense that they both lead to the determination of a time schedule of a project. The significant difference between two approaches is that the time estimates for the different activities in CPM are assumed to be deterministic while in PERT these are described probabilistically. Nowadays, PERT and CPM comprise one technique and the differences, if any, are only historical. Hence, these techniques are referred to as Project Scheduling techniques.

12.1.1 Phases of Project Management

The three main managerial functions for any project are:

1. Planning
2. Scheduling
3. Control.

Planning: This phase involves listing of tasks or jobs (work breakdown structure or (WBS) that must be performed to complete a project under consideration. In this phase, men, machines and material required for the project in addition to the estimates of costs and duration of various activities of the project are also determined.

Scheduling: This phase involves the laying out of actual activities of the projects in a logical sequence of time in which they have to be performed. Also, start and finish times for each activity, critical path on which activities require special attention and slack (or) float for the non-critical paths are determined.

Control: This consists of reviewing the progress of the project and whether the actual performance is according to the planned schedule and finding the reasons for the difference, if any, between the schedule and the performance. The basic aspect of control is to analyse and correct this difference by taking remedial action wherever possible.

PERT and CPM are useful for these functions.

12.1.2 Differences between PERT and CPM

PERT

1. Since PERT was developed in connection with an R&D work, therefore, it had to cope with the uncertainties which are associated with such activities. In PERT, the total project is regarded as a random variable, and therefore, associated probabilities are calculated so as to characterise it.
2. It is an event-oriented network because in the analysis of network, emphasis is given to important stages of completion of task rather than to the activities required to be performed to reach a particular event or task.
3. PERT is normally used for projects involving activities of non-repetitive nature in which time estimates are uncertain.
4. It helps in pinpointing critical areas in a project so that necessary adjustments can be made to meet the scheduled completion date of the project.

CPM

1. Since CPM was developed in connection with a construction project which consisted of routine tasks whose resource requirements and duration were known with certainty, it is basically deterministic.
2. CPM is suitable for establishing a trade-off for optimum balancing between schedule time and cost of the project.
3. CPM is used for projects involving activities of a repetitive nature.

12.2 NETWORK CONSTRUCTION

12.2.1 Some Basic Definitions

Activity: An activity is a task or an item of work to be done in a project. It consumes resources like time, money and labour.

An activity is represented by an arrow with a node at the beginning and a node at the end indicating the start and termination (finish) of the activity. Nodes are denoted by circles. Since this is a logical diagram, length or shape has no meaning. The direction indicates the progress of the activity. Initial and terminal nodes are denoted by $i–j$ ($j > i$), respectively. For example, if A is the activity whose initial node is i and the terminal node is j, then it is denoted by

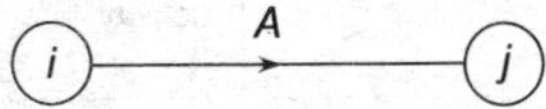

figure 12.1

The diagram in which arrow represents an activity is called an *arrow diagram.*

Predecessor activity: Activities that must be completed immediately prior to the start of another activity are called predecessor activities.

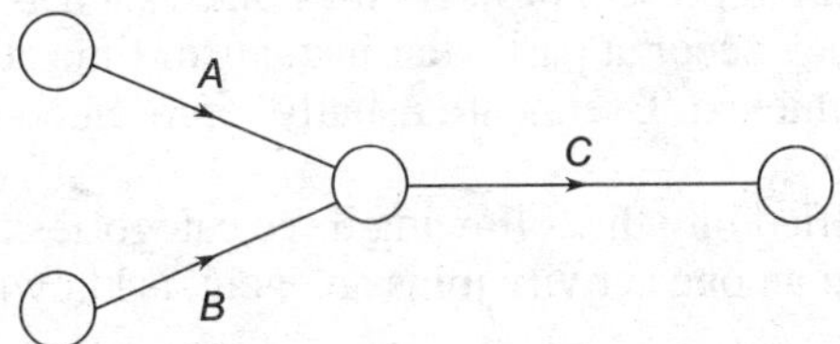

figure 12.2

Here, A and B are predecessors of C and these are denoted by $A < C$; $B < C$.

Successor activity: Activities that cannot be started until one or more of other activities are completed, but immediately succeed them are called successor activities. In Fig. 12.2, C is a successor of A and B. This is denoted by $C > A$ and $C > B$.

Concurrent or parallel activities: Activities which can be accomplished concurrently are known as concurrent activities. It may be noted that an activity can be a predecessor or a successor to an event or it may be concurrent with one or more of the other activities.

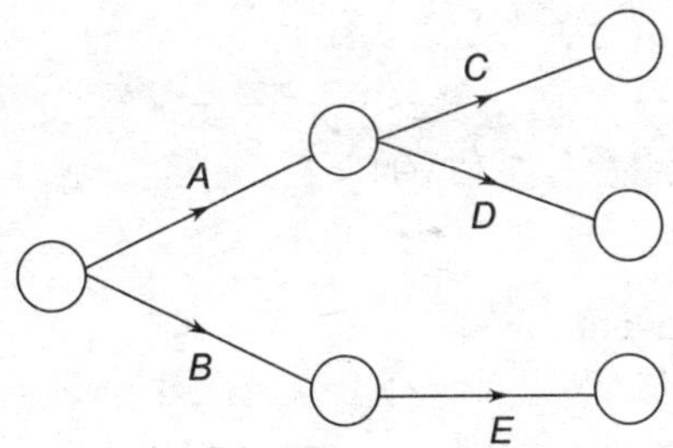

figure 12.3

In Fig. 12.3, A and B are parallel activities; C and D are parallel activities.

Dummy activity: An activity which does not consume any kind of resource but merely serves the purpose of indicating the predecessor or successor relationship clearly in network is called as a dummy activity.

Let A and B be the predecessors of C and B is the predecessor of D. This will be denoted by

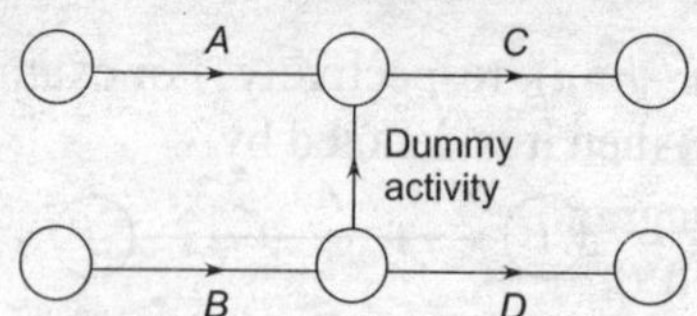

figure 12. 4

Activities which have no predecessors are called start activities of the project. All the start activities can be made to the same initial node. Activities which have no successors are called terminal activities of the project. These can be made to have the same terminal node (end node) of the project.

Events: Events of a network represent project milestones, such as the start or the completion of an activity (task) or activities. They occur at particular instants of time at which some specific part of the project has been or is to be achieved. Events are usually represented by a circle O in a network, also called a node or connector.

Events can be further classified into the following three categories:

Merge event: When more than one activity joins an event, is known as it merge event.

Example

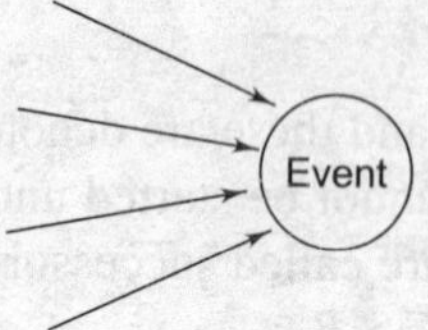

figure 12.5 **Merge event**

Burst event: When more than one activity leaves an event, such an event is known as a burst event.

Example

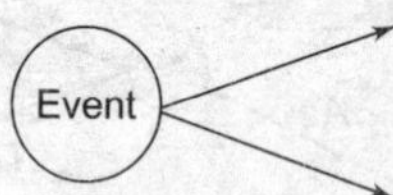

figure 12.6 **Burst event**

Merge and burst event: An activity may be a merge and burst event at the same time; as with respect to some activities it can be a merge event and with respect to some other activities it may be a burst event.

Example

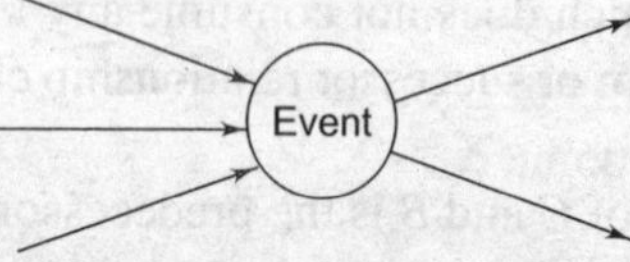

figure 12.7 **Merge and burst event**

Network: An arrow diagram denoting all the activities of a project taking into account the technological sequence of the activities is called a project network. It is represented by activities on an arrow diagram or simply an arrow diagram.

Remark: There is another method of depicting a project network—by representing activities on nodes. This is called an AON diagram. To avoid confusion we use only activity on arrow diagrams throughout the text.

12.2.2 Rules of Network Construction

1. There must be no loops. (That is, a loop is not possible in any real project network.)

Example

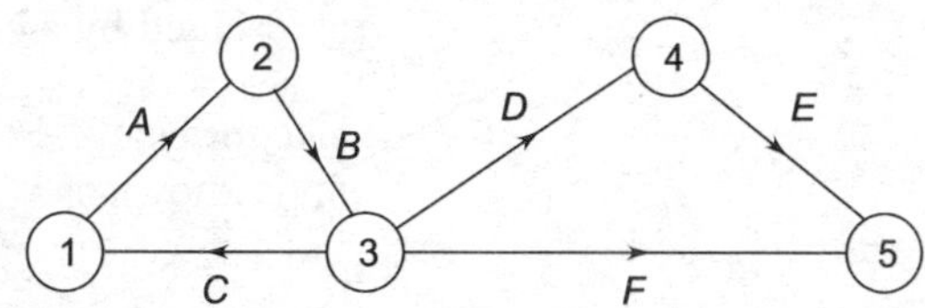

figure 12.8 **Loop in a network**

In the above network the activities, *A*, *B* and *C* forms a loop which is not allowed.

2. Each activity is represented by one and only one arrow in a network.
3. In order to ensure correct precedence relationship in an arrow diagram, the following questions must be asked whenever addition is made to the network.
   (i) What activity must be completed immediately before this activity starts?
   (ii) What activity follows this?
   (iii) What activities must occur simultaneously with this activity?
4. The logical sequence (or interrelationship) between activities must follow the following rules:
   (i) An event cannot occur until all the incoming activities into it have been completed.
   (ii) An activity cannot start unless all the preceding activities on which it depends have been completed.
   (iii) Though dummy activities do not consume either any resource or time, it still has to follow the rules 4(i) and 4(ii).

Some Important Tips for Drawing a Good Network

1. Avoid crossing of arrows.
2. Use straight arrows.
3. Do not attempt to represent the duration of an activity by its arrow length.
4. Use arrows from left to right (or right to left). Avoid mixing two directions; vertical and horizontal arrows may be used if necessary.
5. Use dummies freely in rough draft but final network should not have any redundant dummies.
6. The network has only one entry point, called the start event, and one point of emergence, called the end event.

Errors in a Network

1. Looping
2. *Dangling*: To disconnect an activity before the completion of all activities in a network diagram is known as dangling.

Example

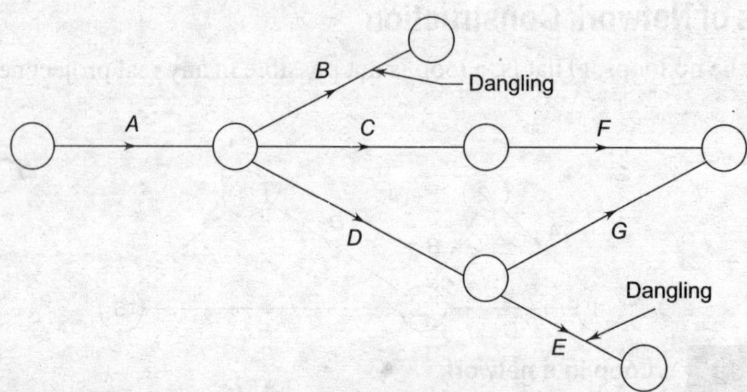

figure 12.9

3. *Redundancy*: Unnecessarily inserting a dummy activity in a network logic is known as the error of redundancy.

12.2.3 Fulkerson's Rule (i–j rule) of Numbering Events

1. Number the start node which has no predecessor activity as 1.
2. Delete all the activities emanting from this node 1.
3. Number all the resulting start nodes without any predecessor as 2, 3, ...
4. Delete all the activities originating from the start nodes without any predecessor next to the last number in step (3).
5. Number all the resulting new start nodes without any predecessor next to the last number in step (3).
6. Repeat the process until the terminal node without any successor activity is reached and number this terminal node suitably.

Example 1

Draw a network for the following project and number the events according to Fulkerson's rule.

(a) *A* is the start event and *k* is the end event.
(b) *J* is the successor event to *F*.
(c) *C* and *D* are successor events to *B*.
(d) *D* is the preceding event to *G*.
(e) *E* and *F* occur after event *C*.
(f) *E* precedes *F*.
(g) *C* restrains the occurrence of *G* and *G* precedes *H*.

(h) *H* precedes *J*.
(i) *F* restrains the occurrence of *H*.
(j) *K* succeeds event *J*.

Solution:

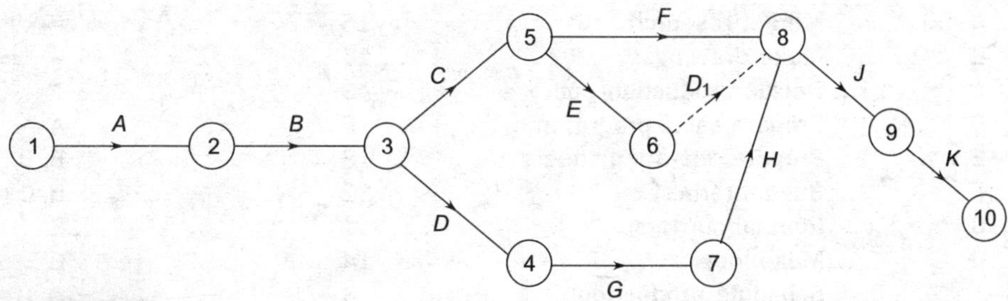

figure 12.10

Example 2

Consider a project with the following WBS for building a site preparation. Determine the precedence relationship, draw the network and number the events.

A. Clear the site
B. Survey and layout
C. Rough grade
D. Exavacate the sewer
E. Exavacate for electrical manholes
F. Instal sewer and backfill
G. Instal electrical manhones
H. Construct boundary wall.

Solution: Looking at the list of activities we can fix the following precedence order:

B succeeds *A* and *C* succeeds *B*, that is, $B > A$, $C > B$.

D and *E* can start together after the completion of *C*, that is, $D, E > C$.

F will follow *D* and *G* will follow *E*, that is, $F > D$, $G > 1$

H can start after *C*, that is, $H > C$.

Thus the precedence relationships are:

$B > A$; $C > B$; $D, E, H > C$; $F > D$; and $G > E$.

The project can be represented in the form of a network as shown below:

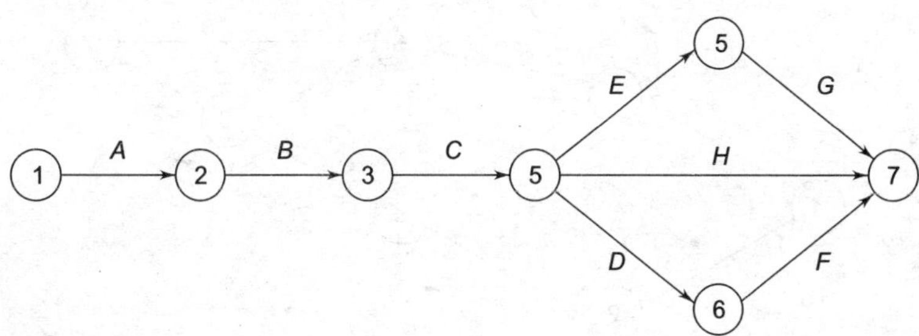

figure 12.11

Example 3

Draw a network for jobbing production and indicate the critical path from the following:

Activity	*Description*	*Time (weeks)*	*Preceded by*
A	Market research	15	–
B	Make drawings	15	–
C	Decide production policy	3	A
D	Prepare sales programme	5	A
E	Prepare operation sheets	8	B, C
F	Buy materials	12	B, C
G	Plan labour force	1	E
H	Make tools	14	E
I	Schedule production	3	D, G
J	Produce product	14	F, H, I

Solution: The network of the given situation is as follows:

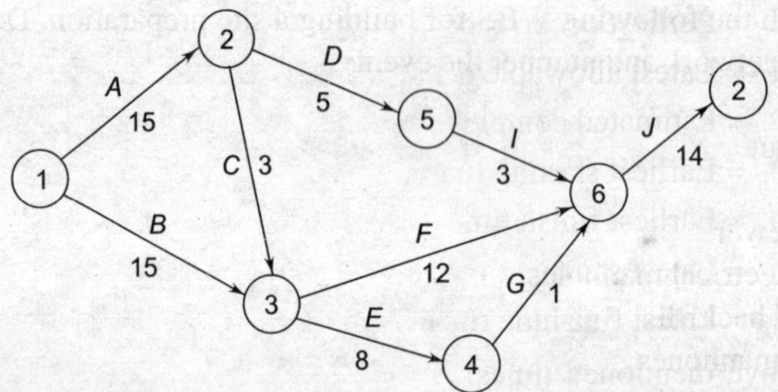

figure 12.12

Example 4

Draw an event-oriented network for the following data:

Event No.	1	2	3	4	5	6	7
Immediate predecessors:	–	1	1	2, 3	3	4, 5	5, 6

Solution:

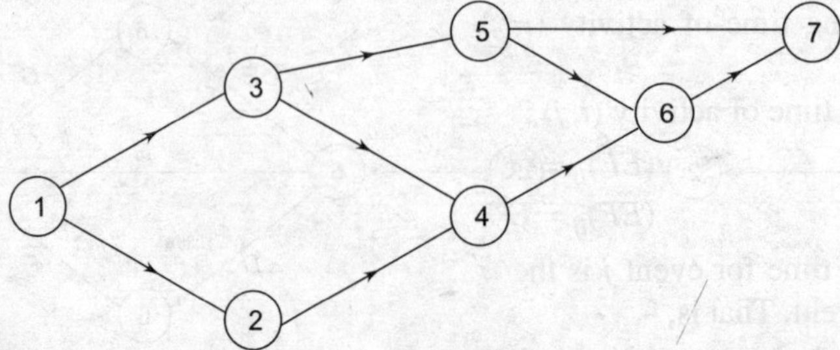

figure 12.13

12.3 CRITICAL PATH METHOD (CPM)

After the construction of a project's network its time analysis becomes essential for planning various activities of the project. The duration of individual activities may be uniquely determined (in case of CPM) or may involve three time estimates (in case of PERT) out of which the expected duration of an activity is computed.

The main objective of time analysis is to prepare the planning schedule of a project. The planning schedule should include the following factors:

(i) Total completion time for the project.
(ii) Earliest time when each activity can start.
(iii) Latest time when each activity can be started without delaying the total project.
(iv) Float for each activity, that is, the amount of time by which the completion of an activity can be delayed without delaying the total project completion.
(v) Identification of critical activities and critical path.

Notations: We shall use the following notations for basic scheduling computations:

(ij) or i–j = Activity (i, j) with tail event i and head event j.

T_E or E_i = Earliest occurrence time of event i

T_L or L_j = Latest allowable occurrence time of event j.

D_{ij} = Estimated completion time (duration of activity (i, j))

ES_{ij} (or) $(ES)_{ij}$ = Earliest starting time of activity (i, j)

EF_{ij} (or) $(EF)_{ij}$ = Earliest finish time of activity (i, j)

LS_{ij} (or) $(LS)_{ij}$ = Latest starting time of activity (i, j)

LF_{ij} or $(LF)_{ij}$ = Latest finishing time of activity (i, j)

For calculating the above mentioned times, we shall discuss two methods, namely forward pass computation and backward pass computation.

12.3.1 Forward Pass Computation (for Earliest Event Time)

In this method, calculations begin from the initial event, proceed through the network visiting events in an increasing order of event number and end at the final event. At each event we calculate earliest occurrence event time (T_E or E_i) and earliest start and finish time for each activity that begins at that event. When calculations end at the final event, its earliest occurrence time gives the earliest possible completion time of the entire project.

(i) Set the earliest occurrence time of initial event as zero.
(ii) The earliest starting time of activity (i, j) is the earliest event of the tail end event, that is, $(ES)_{ij}$, $= E_i$.
(iii) The earliest finish time of activity (i, j) is the earliest starting + the activity duration. That is,

$$(EF)_{ij} = (ES)_{ij} + D_{ij}$$

or,

$$(EF)_{ij} = E_i + D_{ij}$$

(iv) The earliest event time for event j is the maximum of the earliest finish times of all activities ending into that event. That is,

$$E_j = \max_i \, [(Ef)_{ij} \text{ for all immediate predecessor of } (i, j)]$$

$$E_j = \max_i \, [E_i + D_{ij}].$$

12.3.2 Backward Pass Computation (for Latest Allowable Time)

In this method calculations begin from the final event, proceed through the network visiting events in the decreasing order of event numbers and end at the initial node. At each event, we calculate the latest occurrence event time T_L or L_j for the corresponding event, latest finish and start time for each activity that is terminating at the event, such that the earliest finish time for the project remains the same.

(i) For ending event assume $E = L$. Remember that all *Es* have been computed by forward pass computation.

(ii) The latest finish time for activity (i, j) is equal to the latest event time of event j, $(Lf)_{ij} = L_j$

(iii) The latest starting time activity (i, j) = the latest completion time of (i, j) – the activity time (duration)

or, $$(LS)_{ij} = (Lf)_{ij} - D_{ij}$$

or, $$(LS)_{ij} = L_j - D_{ij}$$

(iv) Latest event time for event i is the minimum of the latest start time of all activities originating from that event.

That is,

$$L_i = \min_j \, [(LS)_{ij} \text{ for all immediate successors of } (i, j)]$$

$$= \min_j \, [(LF)_{ij} - D_{ij}]$$

$$= \min_j \, [L_j - D_{ij}].$$

12.3.3 Computation of Float and Slack Time

After calculating the earliest and latest occurrence time the next step is to calculate the floats as defined below:

Total float (TF or $(TF)_{ij}$) of an activity is defined as the difference between the latest finish and the earliest finish of the activity or the difference between the latest start and the earliest start of the activity.

i.e. $$TF = (TF)_{ij} = (LS)_{ij} - (ES)_{ij}$$

or, $$(TF)_{ij} = (LF)_{ij} - (EF)_{ij}$$

This is the most important type of float because it concerns with the overall project duration.

Total float of an activity is the amount of time by which that particular activity may be delayed without affecting the duration of the project.

Free Float (FF or $(FF)_{ij}$) of an activity is that portion of the total float which can used for rescheduling that activity without affecting the succeeding activity.

That is, $FF = FF_{ij}$ = Total float of i–j – $(T_L - T_E)$ of the event j

where, T_L = Latest occurrence Time

T_E = Earliest occurrence Time

Independent Float (IF or IF_{ij}) of an activity is the amount of time by which an activity can be rescheduled without affecting the preceding or succeeding activities of that activity.

i.e. $$IF = IF_{ij} = \text{Free float of } i\text{–}j - (T_L - T_E) \text{ of the event } i.$$

The negative value of independent float is considered as zero.

Remarks

1. The relation between the three floats is given by Independent float ≤ Free float ≤ Total float.
2. The concept of floats is useful to the management in representing under-utilised resources and flexibility of the schedule and the extent to which the resources will be utilised on different activities.
3. The floats can be used for redeployment of resources to level the same or to reduce project duration. However, one should bear in mind that whenever the float in a particular activity is utilised, the float of not only that activity but that of other activities would also change.

Event slacks: For any given event, event slack is defined as the difference between the latest event and earliest event times.

For a given activity $i - j$,

$$\text{Head event slack} = L_j - E_j$$
$$\text{Tail event slack} = L_i - E_i$$

The floats can be represented in terms of head and tail event slacks also.

$$\text{Total float} = L_j - E_i - D_{ij}$$
$$\text{Free float} = E_j - E_i - D_{ij} = (L_j - E_i - D_{ij}) - (L_j - E_j)$$
$$= \text{Total float} - \text{Head event slack}$$
$$\text{Independent float} = E_j - L_i - D_{ij}$$
$$= (E_j - E_i - D_{ij}) - (L_i - E_i)$$
$$= \text{Free float} - \text{Tail event slack.}$$

Time Scale Representation of Floats and Slacks

The various floats and slacks for an activity i–j can be represented on a time scale as follows:

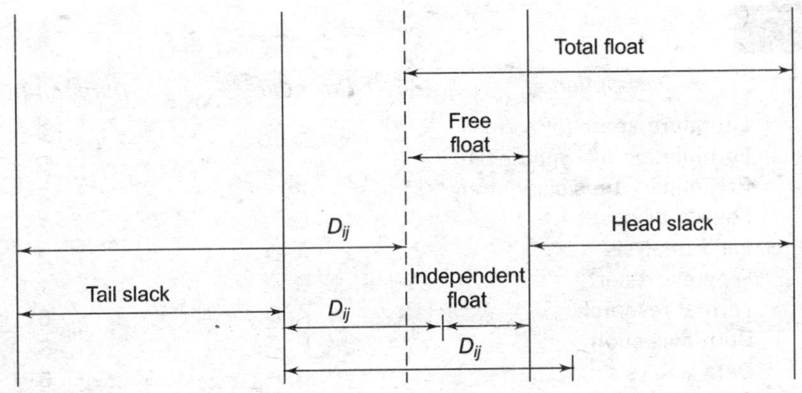

figure 12.14

Conclusions drawn from total values

From the value of a total float we can draw the following conclusion:

If the total float is positive then it may indicate that the resources for the activity are more than adequate. If the total float of an activity is zero it may indicate that the resources are just adequate for that activity. If the total float is negative, it may indicate that the resources for that activity are inadequate.

12.3.4 Critical Path

Critical event: Since the slack of an event is the difference between the latest and earliest event times, that is, slack $(i) = L_i - E_i$, the events with zero slack times are called critical events.

That is, the event (i) is said to be critical if $E_i = L_i$.

Critical activity: Since the difference between the latest start time and earliest start time of an activity is usually called as total float, the activities with zero total float are known as critical activities.

That is, an activity is said to be critical if a delay in its start will cause a further delay in the completion date of the entire project.

Critical path: Path connecting the first initial node to the very last terminal node, of longest duration in any project network is called the critical path.

That is, the sequence of critical activities in a network is called the critical path. Double or darker lines may be used to denote the critical path.

Main Features of Critical Path

1. If a project has to be shortened, then some of the activities on the critical path must also be shortened. The application of additional resources on other activities will not give the desired result unless the critical path is shortened first.
2. The variation in actual performance from the expected activity duration time will be completely reflected in a one-to-one fashion in the anticipated completion of the whole project.

Example 1

Listed in the Table are the activities and sequencing requirements necessary for the completion of a research project.

table 12.1

Activity	Description	Procedure	Duration (week)
A	Literature search	–	6
B	Formulation of hypothesis	–	5
C	Preliminary feasibility study	B	2
D	Formal disposal	C	2
E	Field analysis	A, D	2
F	Progress report	D	1
G	Formal research	A, D	6
H	Data collection	E	5
I	Data analysis	G, H	6
J	Conclusions	I	2
K	Rough draft	G	4
L	Final copy	J, K	3
M	Preparation of oral presentation	L	1

(a) Draw a network diagram for this project.
(b) Find the critical path and duration of the project.
(c) Find the total, free and independent floats of various activities.

Solution:

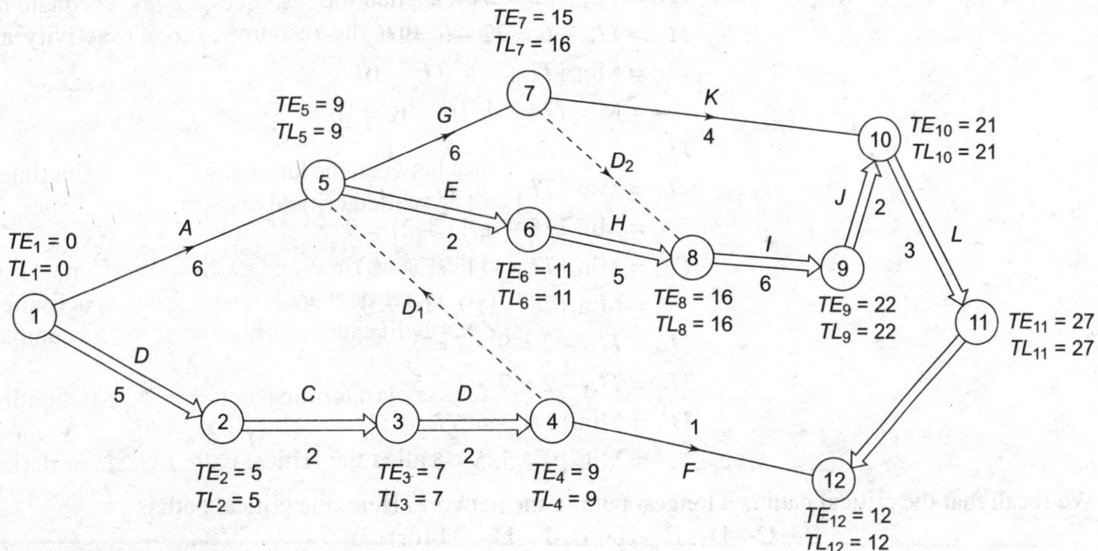

figure 12.15

The network of the given project is shown in Fig. 12.15. Using forward pass computations TE_i or E_i for each node i is as follows:

$$TE_1 = 0$$
$$TE_2 = TE_1 + 5 = 0 + 5 = 5$$
$$TE_3 = TE_2 + 2 = 5 + 2 = 7$$
$$TE_4 = TE_3 + 2 = 7 + 2 = 9$$
$$TE_5 = \text{Max. } \{TE_1 + 6, TE_4 + 0\} = \text{Max. } \{0 + 6, 9 + 0\} = 9$$
$$TE_6 = TE_5 + 2 = 9 + 2 = 11$$
$$TE_7 = TE_5 + 6 = 9 + 6 = 15$$
$$TE_8 = \text{Max } \{TE_7 + 0, TE_6 + 5\}$$
$$= \text{Max } \{15 + 0, 11 + 5\} = 16$$
$$TE_9 = TE_8 + 6 = 16 + 6 = 22$$
$$TE_{10} = \text{Max } \{TE_7 + 4, TE_9 + 2\}$$
$$= \text{Max } \{15 + 4, 22 + 2 = 24\}$$
$$TE_{11} = TE_{10} + 3 = 24 + 3 = 27$$
$$TE_{12} = \text{Max } \{TE_{11} + 1, TE_4 + 1\}$$
$$= \text{Max } \{27 + 1, 9 + 1\} = 28$$

Now, by using backward pass computations the values TL_j for each node j is computed.

$$TL_{12} = 28$$
$$TL_{11} = TL_{12} - 1 = 28 - 1 = 27$$
$$TL_{10} = TL_{11} - 3 = 27 - 3 = 24$$
$$TL_9 = TL_{10} - 2 = 24 - 2 = 22$$
$$TL_8 = TL_9 - 6 = 22 - 6 = 16$$
$$TL_7 = \text{Min}\ \{TL_{10} - 4, TL_8 - 0\}$$
$$= \text{Min}\ \{24 - 4, 16 - 0\} = 16$$
$$TL_6 = TL_8 - 5 = 16 - 5 = 11$$
$$TL_5 = \text{Min}\ \{TL_5 - 6, TL_6 - 2\}$$
$$= \text{Min}\ \{15 - 6, 11 - 2\} = 9$$
$$TL_4 = \text{Min}\ \{TL_{12} - 1, TL_5 - 0\}$$
$$= \text{Min}\ \{28 - 1, 9 - 0\} = 9$$
$$TL_3 = TL_4 - 2 = 9 - 2 = 7$$
$$TL_2 = TL_3 - 2 = 7 - 2 = 5$$
$$TL_1 = \text{Min}\ \{TL_5 - 6, TL_2 - 5\}$$
$$= \text{Min}\ \{9 - 5, 5 - 5\} = 0$$

We recall that the critical path is a longest path in the network. Here, the critical path is

B—C—D—E—H—I—J—L— M (or)

1—2—3— 4 —5—6—8—9—10—11— 12

It is shown by double lines in Fig. 12.15.

We observe that the critical path is the sequence of those activities for which *TE* and *TL* values are equal for the starting as well as the ending nodes. There is another way of identifying the critical path. It is the sequence of activities with total float equal to zero.

The various floats of each activity are calculated in Table 12.2.

table 12.2

Activity (i) i–j	*Duration (in weeks) (2)*	*Earliest*		*Latest*		*Total float (7) = (5) – (3)*	*Free float (8)*	*Independent float (9)*
		Start (3)	*Finish (4) = (3) + (2)*	*Start (5) = (6) – (2)*	*Finish (6)*			
A(1 – 5)	6	0	6	3	9	3	3	3
B(1 – 2)	5	0	5	0	5	0	0	0
C(2 – 3)	2	5	7	5	7	0	0	0
D(3 – 4)	2	7	9	7	9	0	0	0
E(5 – 6)	2	9	11	9	11	0	0	0
G(5 – 7)	6	9	15	10	16	1	0	0
H(6 – 8)	5	11	16	11	16	0	0	0
I(8 – 9)	6	16	22	16	22	0	0	0
(J(9 – 10)	2	22	24	22	24	0	0	0
K(7 – 10)	4	15	19	20	24	5	5	4
L(10 – 11)	3	24	27	24	27	0	0	0
M(11 – 12)	1	27	28	27	28	0	0	0

We observe that total float can also be calculated using (6) – (4). In the above Table free float is calculated by using the formula:

Free float of activity i–j = Total float – $(T_L - T_E)$ of event j

For example, Free float of 1 – 5 = 3 – $(T_L - T_E)$ of node 5

= 3 – 0 = 3

Independent float of each activity is calculated by using the formula:

Independent float of i–j = Total float of its i–j – $(T_L - T_E)$ of node i

For example, float of 1 – 5 = Free float of 1 – 5 – $(T_L - T_E)$ of node 1

= 3 – 0 = 3.

Example 2

Consider the following data for the activities of a project:

Activity	:	A	B	C	D	E	F
Immediate predecessors	:	–	A	A	B, C	–	E
Duration (days)	:	2	3	4	6	2	8

Draw the network and find the critical path and various floats.

Solution: The network is shown in Fig. 12.16 and the numbering of the nodes is according to Fulkerson's rule:

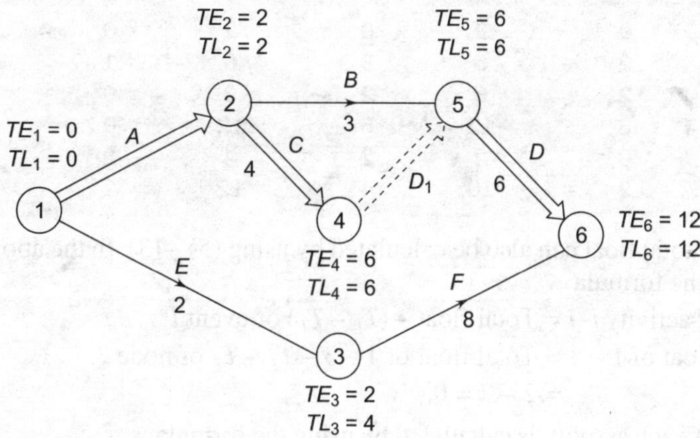

figure 12.16

Using forward pass computations, the earliest occurrence time TE_i or E_i is calculated for each node E_i as follows:

$$\text{Set } TE_1 = 0$$
$$TE_2 = TE_1 + 2 = 0 + 2 = 2$$
$$TE_3 = TE_1 + 2 = 0 + 2 = 2$$
$$TE_4 = TE_2 + 4 = 2 + 4 = 6$$
$$TE_5 = \text{Max }\{TE_2 + 3, TE_4 + 0\}$$

$$= \text{Max } \{2 + 3, 6 + 0\} = 6$$
$$TE_6 = \text{Max } \{TE_5 + 6, TE_3 + 8\} = 12$$

Now, by using backward pass computations, the latest occurrence time TL_i or L_i for each node i is calculated as follows:

$$\text{Set } TL_6 = TE_6 = 12$$
$$TL_5 = TL_6 - 6 = 6$$
$$TL_3 = TL_6 - 8 = 4$$
$$TL_4 = TL_5 - 0 = 6 - 0 = 6$$
$$TL_2 = \text{Min } \{TL_5 - 3, TL_4 - 4\} = 2$$
$$TL_1 = \text{Min } \{TL_2 - 2, TL_3 - 2\} = 0$$

Hence, the critical path is 1—2—4—5— 6.

The critical path is represented by double lines in the network. The project duration is 12 days. The various floats for each activity are calculated and represented in the following Table.

table 12.3

Activity (i–j) (1)	Duration (day) (2)	Earliest Start (3)	Earliest Finish (4) = (3) + (2)	Latest Start (5) = (6) – (2)	Latest Finish (6)	Total float (7) = (6) – (4)	Free float (8)	Independent float (9)
A(1 – 2)	2	0	2	0	2	0	0	0
B(2 – 5)	3	2	5	3	6	1	1	1
C(2 – 4)	4	2	6	2	6	0	0	0
D(5 – 6)	6	6	12	6	12	0	0	0
E(1 – 3)	2	0	2	2	4	2	0	0
F(3 – 6)	8	2	10	4	12	2	2	0

We observe that the total float can also be calculated by using (5) – (3). In the above Table free float is calculated by using the formula

Free float of activity i–j = Total float – $(T_L - T_E)$ of event j

For example, free float of 1 – 3 = Total float of 1 – 3 – $T_L - T_E$ of node 3.

= 2 – 2 = 0.

Independent float of each activity is calculated by using the formula:

Independent float of i–j = Total float of $i - j - (T_L - T_E)$ of node i

For example,

Independent float of 3 – 6 = Total float of 3 – 6 – $T_L - T_E$ of node 3

= 2 – 2 = 0.

Example 3

A project schedule has the following characteristics:

Activity	:	1 – 2	1 –3	2 – 4	3 – 4	3 – 5	4 – 9
Time (days)	:	4	1	1	1	6	5
Activity	:	5 – 6	5 – 7	6 – 8	7 – 8	8 – 10	9 –10
Time (days)	:	4	8	1	2	5	7

(i) Construct a network diagram.

(ii) Compute the total float, free float and independent float for each activity.

(iii) Find the critical path and total project duration.

Solution: The network is shown in Fig. 12.17. The earliest occurrence time TE_i and the latest occurrence time TL_i are calculated as explained in examples 1 and 2.

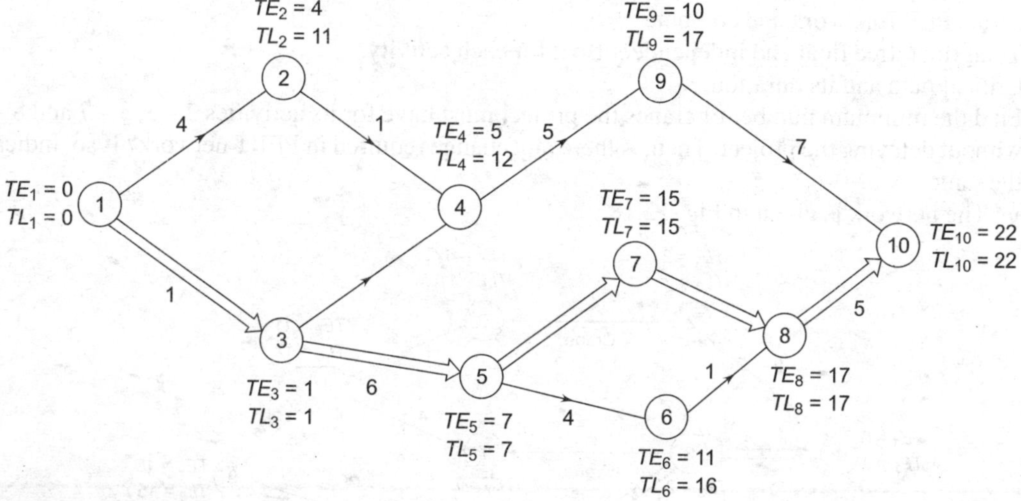

figure 12.17

Critical path is 1 – 3 – 5 – 7 – 8 – 10 and the project duration is 22 days.

table 12.4

Activity	Duration	Earliest		Latest		Total	Free	Independent
		Start	Finish	Start	Finish	float	float	float
(1)	(2)	(3)	(4)	(5)	(6)	(7)	(8)	(9)
1 - 2	4	0	4	7	11	7	0	0
1 - 3	1	0	1	0	1	0	0	0
2 - 4	1	4	5	11	12	7	2	0
3 - 4	1	1	2	11	12	10	5	5
3 - 5	6	1	7	1	7	0	0	0
5 - 7	8	7	15	7	15	0	0	0
5 - 6	4	7	11	12	16	5	0	0
7 - 8	2	15	17	15	17	0	0	0
6 - 8	1	11	12	16	17	5	5	0
4 - 9	5	5	10	12	17	7	0	0
9 - 10	7	10	17	15	22	5	5	0
8 - 10	5	17	22	17	22	0	0	0

Example 4

A project has the following time schedule:

Activity :	1 – 2	1 – 3	1 – 4	2 – 5	3 – 6	3 – 7	4 – 6	5 – 8	6 – 9	7 – 8	8 – 9
Duration : (in months)	2	2	1	4	8	5	3	1	5	4	3

Construct PERT network and compute

(a) Total float, free float and independent float for each activity.
(b) Critical path and its duration.
(c) Find the minimum number of cranes the project must have for its activities 2 – 5, 3 – 7 and 8 – 9 without delaying the project. Then, is there any change required in PERT network? If so, indicate the same.

Solution: The network is given in Fig. 12.18.

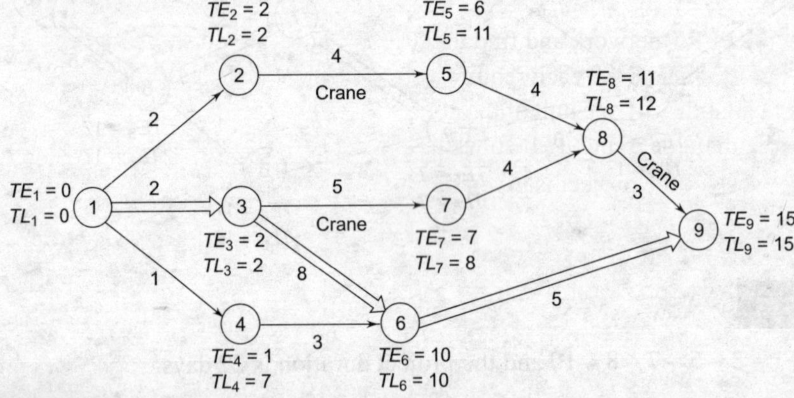

figure 12.18

The TE_i and TL_i are calculated using forward pass computations and backward pass computations, respectively. The various floats are calculated in the Table 12.5.

table 12.5

Activity (i–j) (1)	Duration (in months) (2)	Earliest Start (3)	Earliest Finish (4)	Latest Start (5)	Latest Finish (6)	Total float (7)	Free float (8)	Independent float (9)
1 - 2	2	0	2	4	6	4	4	4
1 - 3	2	0	2	0	2	0	0	0
1 - 4	1	0	1	6	7	6	0	0
2 - 5	4	2	6	7	11	5	0	0
3 - 6	8	2	10	2	10	0	0	0
3 - 7	5	2	7	3	8	1	0	0
4 - 6	3	11	4	7	10	6	6	0
5 - 8	1	6	7	11	12	5	4	3
6 - 9	5	10	15	10	15	0	0	0
7 - 8	4	7	11	8	12	1	0	0
8 - 9	3	11	14	12	15	1	1	0

(b) Critical path is 1 – 3 – 6 – 9 with duration 15 months.

(c) Minimum number of cranes.

Finish 3 – 7 at 7 with one crane.

Finish 2 – 5 at 7 + 4 = 11 with the same crane.

Finish 5 – 8 at 11 + 1 = 12 without the crane.

Finish 8 – 9 at 12 + 3 = 15 with the same crane.

Therefore, one crane will be sufficient, 2 – 5, 5 – 8, and start at 7, 11, and 12, respectively.

Example 5

A project schedule has the following characteristics:

Activity :	1 – 2	1 – 4	1 – 7	2 – 3	3 – 6	4– 5	4 – 8	5 – 6	6 – 9	7 – 8	8 – 9
Time : (in days)	2	2	1	4	1	5	8	4	3	3	5

(a) Construct the PERT network and find the critical path and time duration of the project.

(b) Activities 2 – 3, 4 – 5, 6 – 9 each requires one unit of the same key equipment to complete it. Do you think availability of one unit of the equipment in the organisation is sufficient for completing the project without delaying it? If so what is the schedule of these activities?

Solution: The network of the project is given in Fig. 12.19.

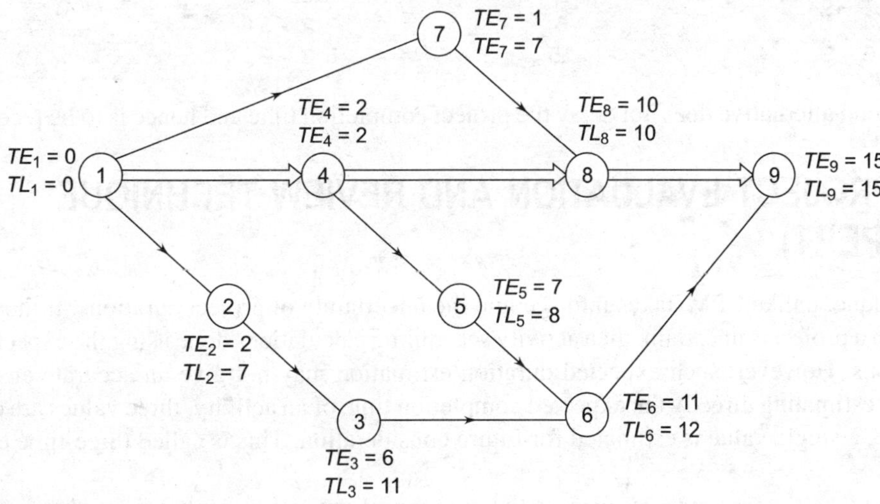

figure 12.19

TE_i and TL_i are calculated using forward pass computations and backward pass computations, respectively. The critical path is 1– 4 – 8 – 9 with a duration of 15 days. The total float of various activities are calculated as follows:

table 12.6

Activity (i, j)	Normal Duration	Earliest Start	Earliest Finish	Latest Start	Latest Finish	Total Float
1 - 2	2	0	2	5	7	5
1 - 4	2	0	2	0	2	0
1 - 7	1	0	1	6	7	6
2 - 3	4	2	6	7	11	5
3 - 6	1	6	7	11	12	5
4 - 5	5	2	7	3	8	1
4 - 8	8	2	10	2	10	0
5 - 6	4	7	11	8	12	1
6 - 7	3	11	14	12	15	1
7 - 8	3	1	4	7	10	6
8 - 9	5	10	15	10	15	0

(b) Activity 6 – 9 can only be undertaken when both 2 – 3 and 4 – 5 and their following activities are over. Thus, 2 – 3 and 4 – 5 contend for the equipment. The two alternative schedules for these are:

Activity	Start	Activity	Start
2 - 3	2	4 - 5	2
4 - 5	6	2 - 3	7
6 - 9	15	6 - 9	12

The second alternative does not delay the project completion time and hence is to be recommended.

12.4 PROJECT EVALUATION AND REVIEW TECHNIQUE (PERT)

This technique, unlike CPM, takes into account the uncertainty of project durations. If the duration of activities in a project is uncertain, then activity scheduling calculations done using the expected value of the durations. However, such expected duration estimation may not give an accurate answer. Thus, rather than estimating directly the expected completion time of an activity, three values are considered. From these, a single value is estimated for future consideration. This is called three-time estimates in PERT.

1. *Optimistic time estimate (t_o or a)* is the duration of any activity when everything goes on well during the project. That is, labourers are available and come in time, machines are working properly, money is available whenever needed, there is no scarcity of raw material needed, and so on.
2. *Pessimistic time estimate (t_p or b)* is the duration of an activity when almost everything goes against our will and a lot of difficulties are faced while doing a project.
3. *Most likely time estimate (t_m or m)* is the duration of an activity when some things go on well and some thing go wrong while doing a project.

These three time values are shown in Fig. 12.20.

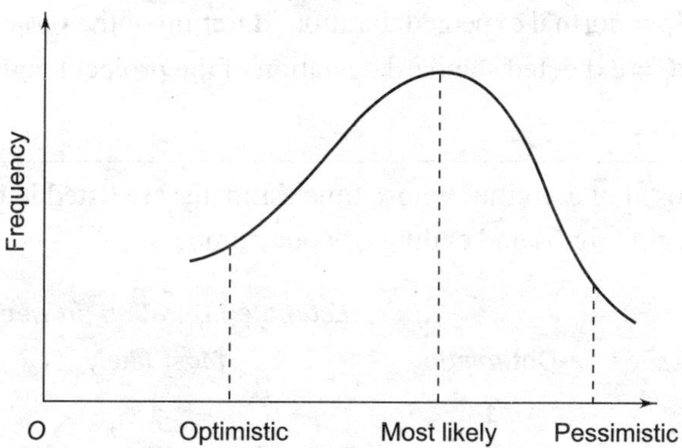

figure 12.20 **Time distribution curve**

Two main assumptions made in PERT calculations are:

1. The activity durations are independent. That is, the time required to complete an activity will have no bearing on the completion time of any other activity of the project.
2. The activity durations follow β distribution. β distribution is a probability distribution with density function $K(t - a)^{\alpha} (b - t)^{\beta}$ with mean $t_e = \frac{1}{3}[2t_m + \frac{1}{2}(t_o + t_p)]$ and standard deviation $\sigma_t = \frac{t_p - t_o}{6}$.

12.4.1 PERT Procedure

Step 1: Draw the project network.

Step 2: Compute the expected duration of each activity $t_e = \frac{t_o + 4t_m + t_p}{6}$.

Step 3: Compute the expected variance σ^2 of each activity.

Step 4: Compute the earliest start, earliest finish, latest start, latest finish and total float of each activity.

Step 5: Determine the critical path and identify critical activities.

Step 6: Compute the expected variance of the project length (also called the variance of the critical path) σ_c^2 which is the sum of the variances of all the critical activities.

Step 7: Compute the expected standard deviation of the project length σ_c and calculate the standard normal deviate $\frac{T_s - T_E}{\sigma_c}$

where, T_S = specified or scheduled time to complete the project
T_E = normal expected duration (duration of the project)
σ_c = expected standard deviation of the project length

Example 1

A small project is composed of activities whose time estimates are listed in the Table below. Activities are identified by their beginning (i) and ending (j) node numbers.

Activity	Estimated Duration (in weeks)		
i - j	*Optimistic*	*Most likely*	*Pessimistic*
1 - 2	1	1	7
1 - 3	1	4	7
1 - 4	2	2	8
2 - 5	1	1	1
3 - 5	2	5	14
4 - 6	2	5	8
5 - 6	3	6	15

(a) Draw the project network.
(b) Find the expected duration and variance for each activity. What is the expected project length?
(c) Calculate the variance and standard deviation of the project length. What is the probability that the project will be completed
 (i) at least 4 weeks earlier than expected?
 (ii) no more than 4 weeks later than expected time?
(d) If the project is 19 due in weeks, what is the probability of meeting the due date?

Given:	Z	0.5	0.67	1.00	1.33	2.00
	p	0.1915	0.2486	0.3413	0.4082	0.4772

Solution: The expected time and variance of each activity is calculated in the following Table:

table 12.7

Activity *i - j*	*Optimistic* *time* t_o	*Most likely* *time* t_m	*Pessimistic* *time* t_p	$t_e = \dfrac{(t_o + 4tm + t_1)}{6}$	*Variance* $\left(\dfrac{t_p - t_o}{6}\right)^2$
1 - 2	1	1	7	2	1
1 - 3	4	7	4	4	1
1 - 4	2	2	8	3	1
2 - 5	1	1	1	1	0
3 - 5	2	5	14	6	4
4 - 6	2	5	8	5	1
5 - 6	3	6	15	7	1

The network is drawn with the help of t_e.

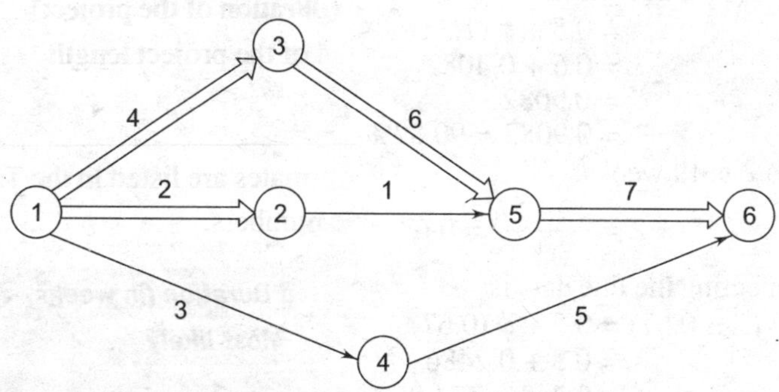

figure 12.21

(b) By examining all the paths we see that the critical path is 1 – 3 – 5 – 6. (Critical path can also be obtained by calculating TE_i and TL_i of each event or node.)

Duration of project = 17 days

Variance of project length = Sum of the variances of the critical activities

$$\sigma_c^2 = 1 + 4 + 4 = 9 \quad \text{or} \quad \sigma_c = 3$$

(c) The standard normal deviate is

$$z = \frac{\text{Due date} - \text{Expected date of completion}}{\sqrt{\text{Variance}}}$$

(i) $T_S = 13$

$$\therefore \quad Z = \frac{T_S - T_E}{\sigma_c} = \frac{13-17}{3} = -1.33$$

i.e $P(T_S \leq 13) = P(Z \leq -1.33)$

$= 0.5 - \phi(1.33)$

$= 0.5 - 0.4082$

$= 0.918 = 9.18\%$

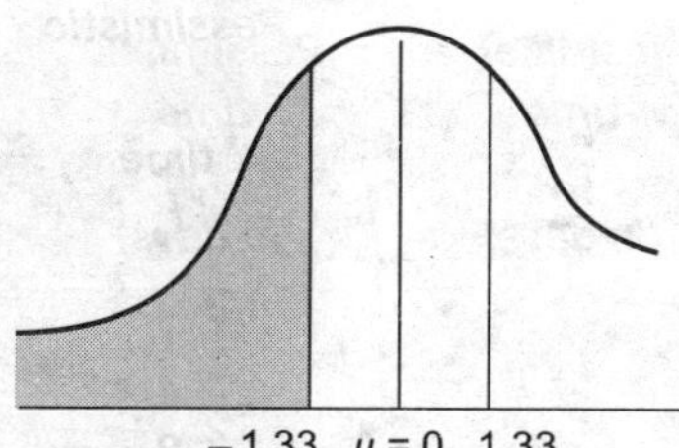

figure 12.22

(ii) $T_S = 21$

$$Z = \frac{T_S - T_E}{\sigma_c} = \frac{21-17}{3} = 1.33$$

$$
\begin{aligned}
P(T_S \le 21) &= P(Z \le 1.33) \\
&= 0.5 + \phi(1.33) \\
&= 0.5 + 0.4082 \\
&= 0.9082 \\
&= 0.9082 \approx 90.82\%
\end{aligned}
$$

(d) When the due date is 19 weeks,

$$Z = \frac{19-17}{3} = 0.67$$

$\therefore$ Probability of meeting the due date is

$$
\begin{aligned}
P(Z \le 0.67) &= 0.5 + \phi(0.67) \\
&= 0.5 + 0.2486 \\
&= 0.7486 \approx 74.86\%
\end{aligned}
$$

Thus, the probability of not meeting the due date is 1 – 0.7486 is 0.2514 or 25.14%.

Example 2

For the network below, the estimate t_o, t_m and t_p are given in this order for each of the activities. Find the probability of completing the project in 25 days.

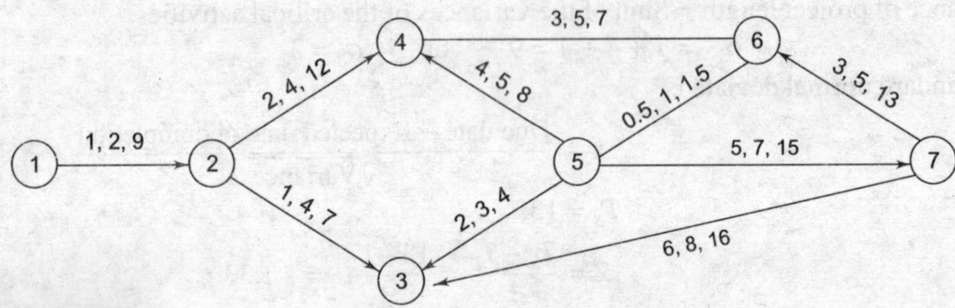

figure 12.23

Solution: The expected time and variance of each activity are calculated in Table 12.8.

table 12.8

Activity $i-j$	Optimistic time t_o (in days)	Most likely time t_m (in days)	Pessimistic time t_p (in days)	$t_e = \frac{t_o + 4t_m + t_p}{6}$	σ_c^2 $\sigma_x = \left(\frac{t_p - t_o}{6}\right)^2$
1 - 2	1	2	9	3	1.78
2 - 3	1	4	7	4	1
2 - 4	2	4	12	5	0.69
3 - 5	2	3	4	3	0.11
4 - 5	4	5	8	5.3	0.44
4 - 6	3	5	7	5	0.44
5 - 6	0.5	1	1.5	1	0.03
3 - 7	6	8	16	9	2.78
6 - 7	3	5	13	6	2.78
5 - 7	5	7	15	8	2.78

The t_e is represented in the network as follows:

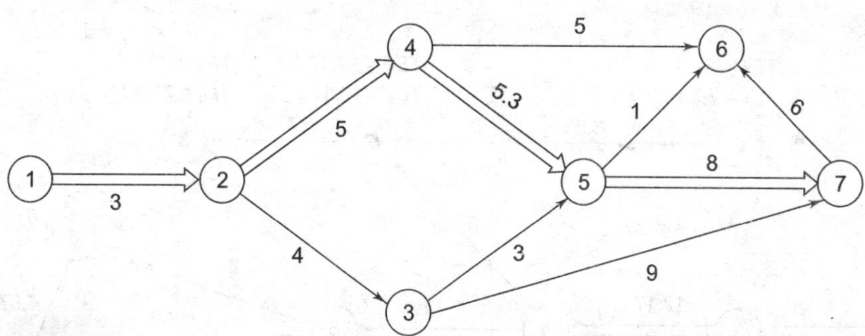

figure 12.24

The critical path is calculated as follows:

Path	Duration
1 – 2 – 4 – 6 – 7	19 days
1 – 2 – 4 – 5 –6 – 7	20.3 days
1 – 2 – 4 – 5 – 7	21.3 days
1 – 2 – 3 – 7	16 days
1 – 2 – 3 – 5 – 7	18 days
1 – 2 – 3 – 5 – 6 – 7	17 days

Hence, the critical path is 1 – 2 – 4 – 5 –7 and the expected project duration $T_E = 21.3$ days.

$$\sigma_c^2 = \text{sum of variances of the critical activities}$$
$$= 1.78 + 0.69 + 0.44 + 2.78$$
$$= 5.69$$

$$T_S = 25 \qquad T_E = 21.3 \qquad \sigma_c = \sqrt{5.69} = 2.385$$

$$Z = \frac{T_S - T_E}{\sigma_c} = \frac{25 - 21.3}{2.385} = 1.55$$

$$P(T_S \le 25) = P(Z \le 1.55)$$
$$= 0.5 + \phi(1.55)$$
$$= 0.5 + 0.4394 = 0.9394 \approx 93.94\%$$

Here, we note that the critical path can also be determined by calculating TE_i and TL_i for each node.

Example 3

The following Table gives optimistic time (a), the most likely time (m) and the pessimistic time (b). Draw the network of the project and calculate the slack for each event. Find the critical path and the probability of completing the project in 35 days.

Activity	1 – 2	1 – 3	2 – 5	3 – 4	4 – 5	5 – 8	4 – 6	4 – 7	6 – 9	8 – 9	7–10	9 – 10
a	3	1	6	8	0	5	6	3	1	3	8	2
m	5	2	8	12	0	7	9	6	2	5	15	5
b	7	3	12	17	0	9	12	8	3	8	20	6

Solution: By using t_e in Table 12.19 the network is drawn.

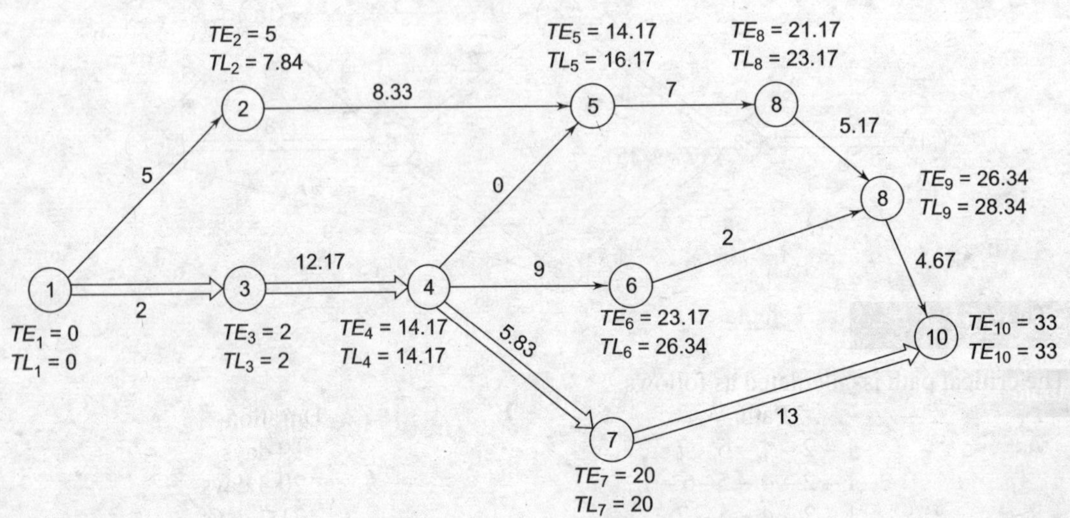

figure 12.25

The expected time for each activity and variance of each activity are calculated in Table 12.9.

table 12.9

Activity $i-j$	Optimistic time a	Most likely time m	Pessimistic time b	Expected time $t_e = (a+4m+b)/6$	Variance $\sigma_e^2 = \left(\frac{b-a}{6}\right)^2$
1 - 2	3	5	7	5	0.44
1 - 3	1	2	3	2	0.11
2 - 5	6	8	12	8.33	1
3 - 4	8	12	17	12.17	2.25
4 - 5	0	0	0	0	0
5 - 8	5	7	9	7	0.11
4 - 6	6	9	12	9	1
4 - 7	3	6	8	5.83	0.69
6 - 9	1	2	3	2	0.11
8 - 9	3	5	8	5.17	0.69
7 - 10	8	15	20	13	4
9 - 10	2	5	6	4.67	0.44

The critical path is 1 – 3 – 4 – 7 – 10. The project duration is, 33 days. That is, T_E = 33 days. The expected variance of the critical path is

$$= 0.11 + 2.25 + 0.69 + 4$$
$$= 7.05$$

So, the expected standard deviation

$$\sigma_c = \sqrt{7.05} = 2.65$$

$$\therefore \quad Z = \frac{T_S - T_E}{\sigma_c}$$

$$= \frac{35 - 33}{2.65} = 0.75$$

$$\therefore \quad P(T_S \leq 35) = P(Z \leq 0.75)$$

$$= 0.5 + \phi\ (0.75)$$

$$= 0.5 + 0.2734$$

$$= 0.7734 \approx 77.34\%$$

The various floats are calculated in Table 12.10.

table 12.10

Activity i-j (1)	Duration (2)	Earliest Start (3)	Earliest Finish (4)	Latest Start (5)	Latest Finish (6)	Total float (7)	Free float (8)	Independent float (9)
1 - 2	5	0	5	2.84	7.84	2.84	0	0
1 - 3	2	0	2	0	2	0	0	0
2 - 5	8.33	5	13.33	7.84	16.17	2.84	0.840	0
3 - 4	12.17	2	14.17	2	14.17	0	0	0
4 - 5	0	14.17	14.17	16.17	16.17	2	0	
5 - 8	7	14.17	21.17	16.17	23.17	2	0	0
4 - 6	9	14.17	23.17	17.34	26.34	3.17	0	0
4 - 7	5.83	14.47	20	14.17	20	0	0	0
6 - 9	2	23.17	25.17	26.34	28.34	3.17	1.17	0
8 - 9	5.17	21.17	26.34	23.17	28.34	2	0	0
7 - 10	13	20	33	20	33	0	0	0
9 - 10	4.67	26.34	31	28.33	33	2	2	0

Example 4

Three time estimates (in months) of all activities of a project are as given below:

Activity	:	1 – 2	2 – 3	2 – 4	3 – 4	4 – 5	5 – 6
a	:	0.8	3.7	6.2	2.1	0.8	0.9
m	:	1.0	5.6	6.6	2.7	3.4	3.4
b	:	1.2	9.9	15.4	6.1	3.6	2.7

(a) Find the expected duration and standard deviation of each activity.

(b) Construct the project network.

(c) Determine the critical path, expected project length and expected variance of project length.

(d) What is the probability that the project will be completed (i) two months earlier than expected? (ii) What due date has about 90% chance of being met?

Solution: The expected duration and expected standard deviation of each activity is calculated in Table 12.11.

table 12.11

Activity	a	m	b	te	σ_c^2	σ_c
1 - 2	0.8	1.0	1.2	1	0.0045	0.06
2 - 3	3.7	5.6	9.9	6	1.0609	1.03
2 - 4	6.2	6.6	15.4	8	2.3409	1.53
3 - 4	2.1	2.7	5.1	3	0.25	0.5
4 - 5	0.8	3.4	3.6	3	0.2209	0.47
5 - 6	0.9	1.0	1.1	1	0.0011	0.033

(b) The network is shown in Fig. 12.26.

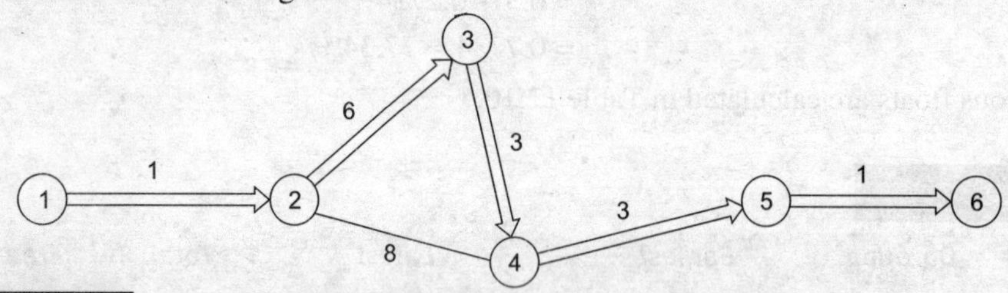

figure 12.26

(c) Critical path: 1 – 2 – 3 – 4 – 5 – 6

Expected project length = 14 months

$$\text{Expected variance} = 0.0045 + 1.0609 + 0.25 + 0.2209 + 0.0011$$
$$= 1.5374$$

Hence, the expected standard deviation is

$$\sigma_c = \sqrt{1.5374} = 1.2399$$

(d) (i) $T_S = 12,\ T_E = 14\ \sigma_c = 1.2399$

$$Z = \frac{T_S - T_E}{\sigma_c} = \frac{12 - 14}{1.2399} = -1.61$$

$$\therefore \quad P(T_S \le 12) = P(Z \le -1.61)$$
$$= 0.5\,0 - \phi(1.61)$$
$$= 0.5 - 0.4463 = 0.0537 \approx 5.3\%$$

Required probability = 5.4%

(*ii*) Z value for 90% area in the normal table = 1.28 (approximately)

Let T_S be the required time.

Then, $$Z = \frac{T_S - T_E}{\sigma_c}$$

i.e. $$1.28 = \frac{T_S - 14}{1.2399}$$

$$\therefore \quad T_S = 1.28 \times 1.2399 + 14$$
$$= 15.58 \text{ months (nearly).}$$

12.5 RESOURCE ANALYSIS IN NETWORK SCHEDULING

In the earlier discussions on PERT and CPM, it was assumed that the only constraint for an activity was the starting date and the completion date. Availability of resources and cost aspects were not considered. There are two kinds of costs which can influence project schedules—direct costs and indirect costs.

Direct costs are the costs directly associated with each activity such as machine costs, labour costs, and so on. Indirect costs are the costs due to management services, rentals, insurance including allocation of fixed expenses, cost of security, and so on.

While **direct costs** allocated to a project goes up with the increase in project duration in **direct costs** go high as the time for individual activity is reduced.

Crashing: The process of reducing an activity time by putting extra effort is called crashing the activity. It is the process of shortening a project and is usually achieved by adding extra resources to an activity.

It may be noted that for technical reasons, the duration of an activity cannot be reduced indefinitely. The crash time represents the fully expedited or the minimum activity duration time that is possible, and any attempts to further 'crash' would only raise the direct costs of the activity without reducing the time. The activity cost corresponding to the crash time is called the crash cost which equals the minimum direct cost required to achieve the crash performance time. These are in contrast to the normal time and the normal cost of the activity. The normal cost is equal to the absolute minimum of the direct cost required to perform an activity. The corresponding activity duration is known as the normal time.

The time-cost relationship can be represented by the time-cost trade-off curve shown in Fig.12.27.

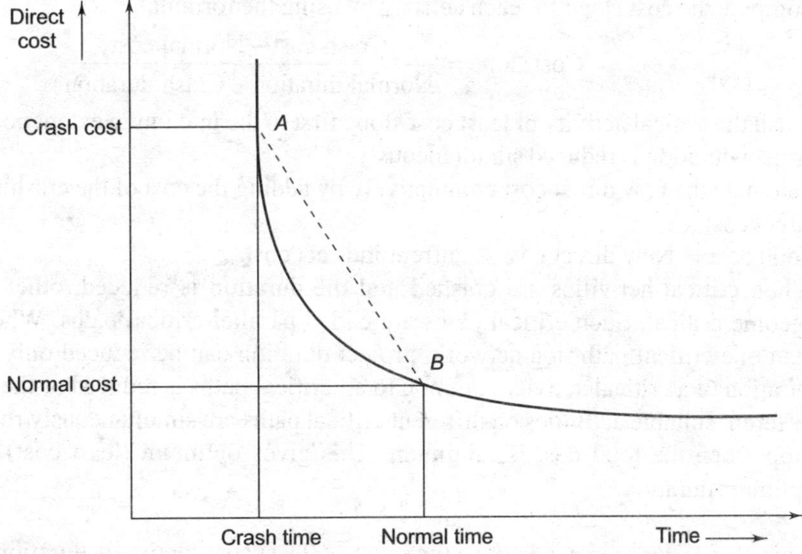

figure 12.27 **Time-Cost Trade-Off**

For simplicity, the relationship between normal-time cost and crash-time cost for an activity is assumed to be linear instead of being. Thus, the crash cost per unit of time can be estimated by computing the relative change in cost (cost slope) per unit change in time. From Fig. 12.27, it is clear that one must be interested in the central region or the curve contained between points *A* and *B*. To establish a trade-off between time and cost, the direct cost of completing an activity per unit time is computed as follows:

$$\text{Cost slope} = \frac{\text{Crash cost} - \text{Normal cost}}{\text{Normal time} - \text{Crash time}}$$

The cost slope represents the rate of increase in the cost of performing an activity per unit increase in time and is called cost/time trade-off. It varies from activity to activity.

There is a method for finding the optimum duration and the associated least cost called the least cost schedule. It should be observed that there is no need to crash all jobs to get a project done faster. Only critical activities need be expedited. Least cost schedule will indicate which critical activities are to be crashed and how much so as to get the optimum duration.

12.5.1 Time Cost Optimisation Algorithm (or) Time Cost Trade-off Algorithm (or) Least Cost Schedule Algorithm

Step 1: Draw the network.

Step 2: Determine the critical path and the normal duration. Also, identify the critical activities.

Step 3: Find the total normal cost and the normal duration of the project.

Step 4: Compute the cost slope for each activity by using the formula

$$\text{Cost slope} = \frac{\text{Crash cost} - \text{Normal cost}}{\text{Normal duration} - \text{Crash duration}}$$

Step 5: Crash the critical activity of least cost slope first to the maximum extent possible so that the project duration is reduced simultaneously.

Step 6: Calculate the new direct cost cumulatively by adding the cost of the crashing to the current direct cost.
Total cost = New direct cost + Current indirect cost.

Step 7: When critical activities are crashed and the duration is reduced, other paths may also become critical. Such critical paths are called parallel critical paths. When there is more than one critical path in a network, project duration can be reduced only when either the duration of a critical activity common to all critical paths is reduced or the durations of the different suitable activities on different critical paths are simultaneously reduced.

Step 8: Stop when the total cost is minimum. This gives optimum (least cost) schedule called optimum duration.

Remarks:

1. Minimum (least) duration need not be the same as (least cost) optimum duration.
2. When no indirect cost is given the least cost optimum duration is the same as the normal duration where no crashing is done but least cost duration may be found out.

Example 1

The following Table gives the activities in a construction project and other relevant information:

Activity	Immediate predecessor	Time (days)		Direct Cost (Rs)	
		Normal	Crash	Normal	Crash
A	–	4	3	60	90
B	–	6	4	150	250
C	–	2	1	38	60
D	A	5	3	150	250
E	C	2	2	100	100
F	A	7	5	115	175
G	D, B, E	4	2	100	240

Indirect costs vary as follows:

Days :	15	14	13	12	11	10	9	8	7	6
Cost (Rs) :	600	500	400	250	175	100	75	50	35	25

(i) Draw an arrow diagram of the project.

(ii) Determine the project duration which will result in minimum total cost.

Solution: With the help of normal time, network is constructed and is shown in Fig. 12.28.

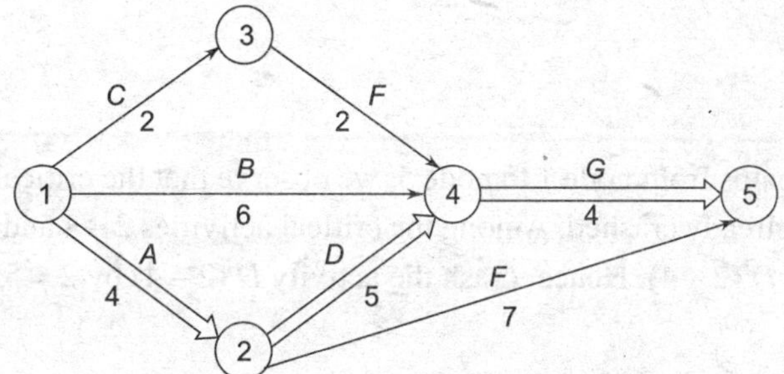

figure 12.28

To identify the critical path

Path	Duration (in days)
1 – 2 – 5	12
1 – 4 – 5	10
1 – 2 – 4 – 5	13
1 – 3 – 4 – 5	8

Hence, the critical path is 1 – 2 – 4 – 5, project duration is 13 days and the critical activities are 1 – 2, 2 – 4 and 4 –5.

Cost slope for each activity is calculated by using the formula

$$\text{Cost slope} = \frac{\text{Crash cost} - \text{Normal cost}}{\text{Normal time} - \text{Crash time}}$$

Activity :	A (1 – 2)	B (1 – 4)	C (1 – 3)	D (2 – 4)	E (3 – 4)	F (2 – 5)	G (4 – 5)
Cost slope :	30	50	22	50	–	30	70

Among the critical activities, activity with least cost slope is A (1 – 2). So, crash the activity A by 4 – 1 = 1 (Normal time – crash time) day. Thus, project duration reduces to 12 days and the total project cost = Total direct cost + Increased direct cost due to crashing.

Indirect cost for 12 days

$$= 713 \text{ (i.e. total of normal cost)} + 1 \times 30 + 250 = \text{Rs } 993.$$

The new network is shown in Fig. 12.29.

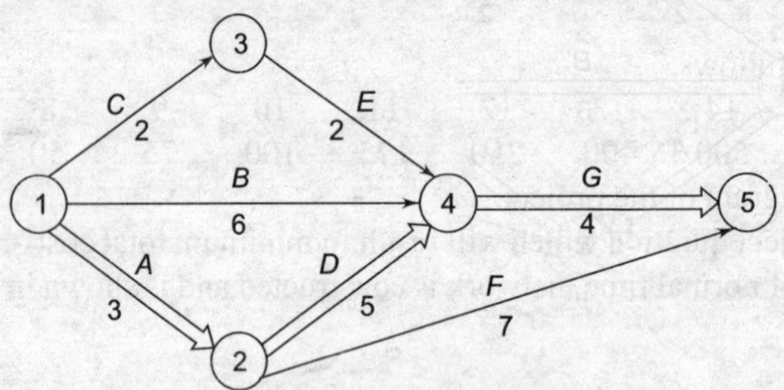

figure 12.29

By examining all the paths from node 1 to node 5, we observe that the critical path is again 1 – 2 – 4 – 5. The activity 1 – 2 is already crashed. Among the critical activities 2 – 4 and 4 – 5, the activity with the smallest cost slope is *D* (2 – 4). Hence, crash the activity *D* (2 – 4) by 2 = 5 – 3 days. The network is shown in Fig . 12.30.

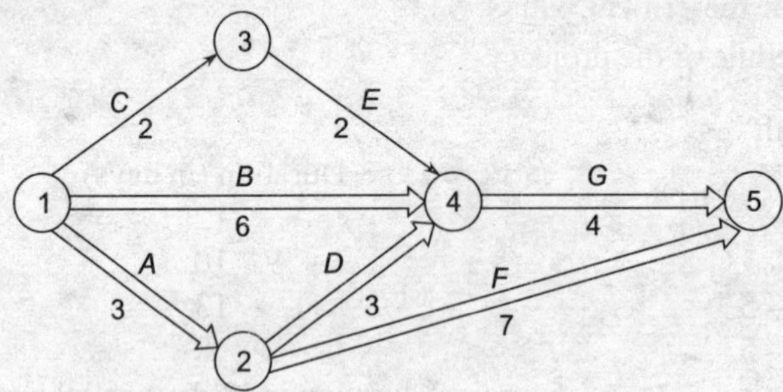

figure 12.30

Now, there are three critical paths

1 – 2 – 5

1 – 4 – 5

1 – 2 – 4 – 5

each with, duration 10 days.

Total cost = 743 + 2 × 50 + 100 = Rs 943.

Since there are more than one parallel critical path, crashing must be performed.

Since critical activities *A* and *D* cannot be crashed further, therefore, crash activities *F* and *G* by 2 days each to reduce the project duration by 2 days. The new network is shown in Fig. 12.31.

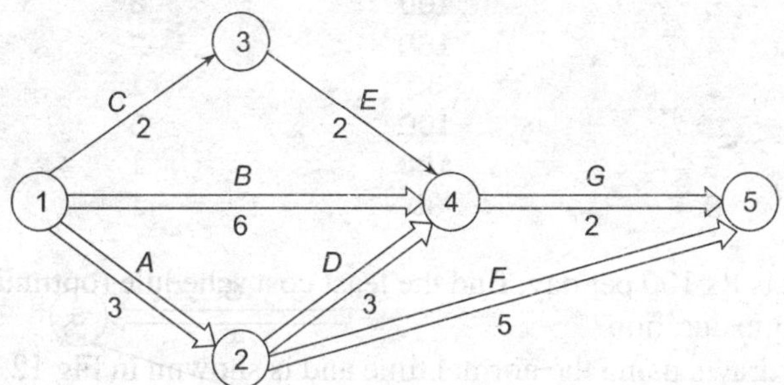

figure 12.31

The critical paths are

1 – 4 –5
1 – 2 – 5
1 – 2 – 4 – 5

each with duration 8 days.

Total cost = Total direct cost + Increased direct cost due to crashing + Indirect cost for 8 days
= 843 + (2 × 30 + 2 × 70) + 50 = Rs 1,093.

Since all the activities in the critical path 1 – 2 – 5 are crashed, further parallel crashing is not possible. The crashing-schedule of the project is as shown in Table 12.12.

table 12.12

Project duration (days)	Crashing activity and days	Direct Cost (Rs) Normal	Direct Cost (Rs) Crashing	Direct Cost (Rs) Total	Indirect cost (Rs)	Total cost (Rs)
13	–	713	-	713	400	1,113
12	1 - 2 (1)	713	30	743	250	993
10	2 - 4 (2)	713	30 + 2 × 50	843	100	943
8	2 - 5 (2) 4 - 5 (2)	713	130 + 2 × 30 + 2 × 70	1,043	50	1,093

From the Table we see that minimum total cost is Rs 943 and the optimum duration is 10 days.

Example 2

The following data is pertaining to a project with normal time and crash time.

Jobs	Normal		Crash	
	Time	Cost	Time	Cost
1 - 2	8	100	6	200
1 - 3	4	150	2	350
2 - 4	2	50	1	90
2 - 5	10	100	5	400
3 - 4	5	100	1	200
4 - 5	3	80	1	100

(a) If the indirect cost is Rs 100 per day, find the least cost schedule (optimum duration).
(b) What is the minimum duration?

Solution: The network is drawn using the normal time and is shownn in Fig 12.32.

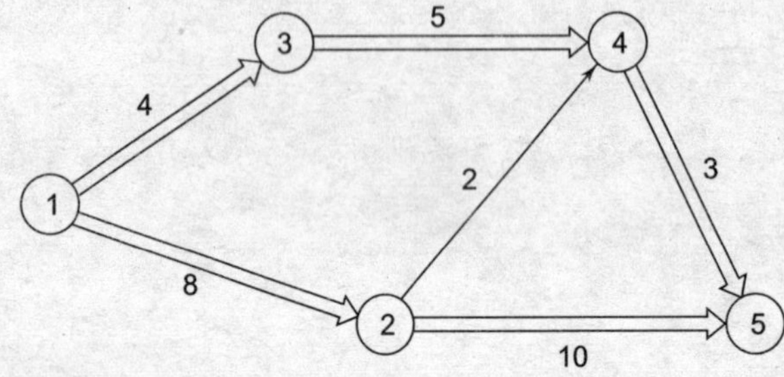

figure 12.32

We observe that the critical path is 1 – 2 – 5.

Normal duration = 18 days

Total cost = Direct cost + Indirect cost
= Rs 580 + Rs 1,800
= Rs 2,380

Cost slope for various activities are called as follows:

Activity :	1 – 2	1 – 3	2 – 4	2 – 5	3 – 4	4 – 5
Cost slope :	50	100	40	60	25	10

Since 1 –2 is the critical activity with least cost slope, crash the activity 1 – 2 by 2 (8 – 6) days. The new network is shown in Fig. 12.33.

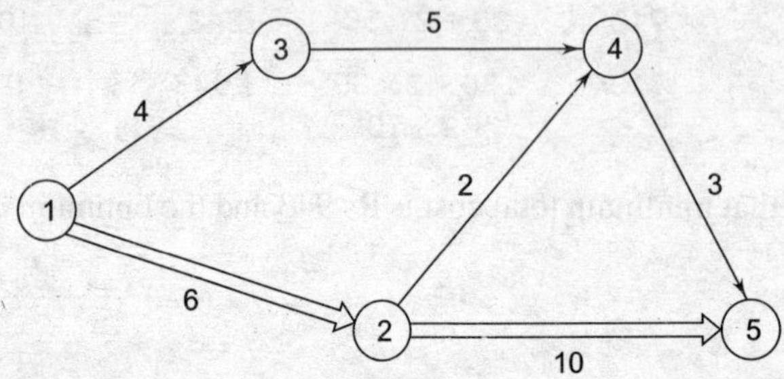

figure 12.33

Critical path is again 1 – 2 – 5.

Current duration is 16 days.

$$\text{Current total cost} = 580 + 2 \times 50 + 16 \times 100$$
$$= \text{Rs } 2{,}280$$

Since 1 – 2 cannot be crashed further, crash the critical activity 2 – 5 by 4 days. (Even though 5 days are available for crashing 2 – 5, we take 4 days, since if we crash 2 – 5 by 5 days then the path 1 – 2 – 5 will become non-critical.)

The new network is given in Fig. 12.34.

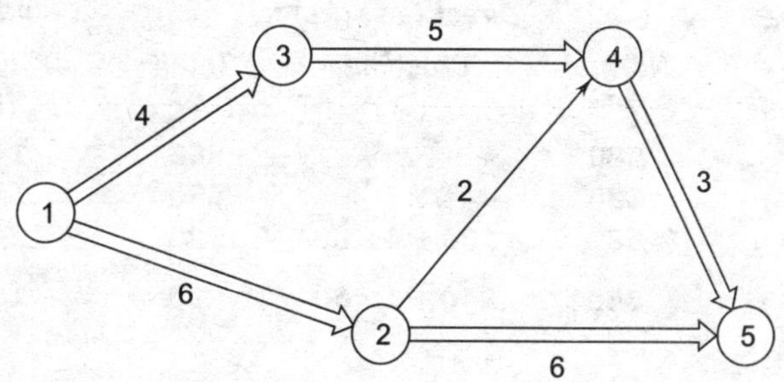

figure 12.34

Current critical path: (i) 1 – 2 – 5
(ii) 1 – 3 – 4 – 5

Current duration is 12 days.

$$\text{Current total cost} = \text{Rs } 680 + 4 \times 60 + 12 \times 100$$
$$= \text{Rs } 2{,}120.$$

Since there are more than one critical paths, only parallel crashing is allowed. Crash the critical activities 2 – 5 and 4 – 5 by 1 day each. The duration of the path 1 – 2 – 4 – 5 is 11 days and also the activity 2 – 5 can be crashed only by 1 day. The resulting network is shown in Fig. 12.35.

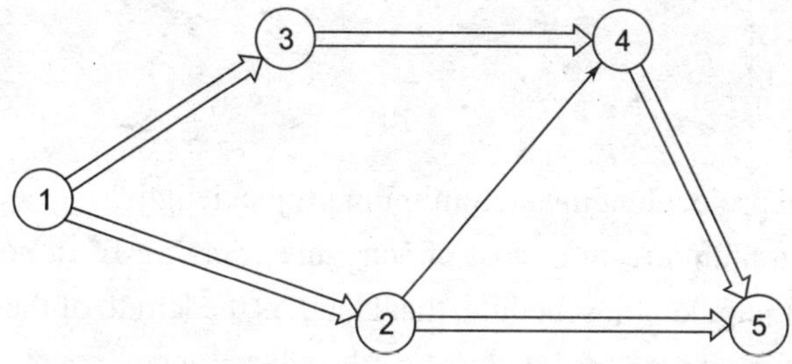

figure 12.35

Current critical paths: 1 – 2 – 5
1 – 3 – 4 – 5

Current duration: $12 - 1 = 11$ days

$$\text{Current total cost} = 920 + 1 \times 60 + 1 \times 10 + 11 \times 100 = \text{Rs } 2,090$$

No further crashing is possible since all the activities on the critical path 1 – 2 – 5 have been crashed to the maximum extent.

The crashing schedule of the project is as shown in Table 12.13.

table 12.13

Project duration (days)	Crashing activity and days	Direct Cost (Rs)			Indirect cost (Rs)	Total cost (Rs)
		Normal	Crashing	Total		
18	–	580	–	580	1,800	2,380
16	1 - 2 (4)	580	2×50	680	1,600	2,280
12	2 - 5 (4)	580	$100 + 4 \times 60$	920	1,200	2,120
11	2 - 5 (1)	580	$240 + 1 \times 60$	990	1,100	2,090
	4 - 5 (1)		$+ 1 \times 10$			

(b) From Table 12.13, we observe that the optimum duration is 11 days and the minimum duration is also 11 days.

Example 3

A small project consists of the jobs in the accompanying Table. With each job is listed its normal time and a minimum or crash time (in days). The cost (in Rs per day) of crashing each job is also given.

Job (i – j)	Normal duration (days)	Minimum (crash) duration (days)	Cost of crashing (Rs per day)
1 - 2	9	6	20
1 - 3	8	5	25
1 - 4	15	10	30
2 - 4	5	3	10
3 - 4	10	6	15
4 - 5	2	1	40

(a) What is the normal project length and minimum project length?

(b) Determine the minimum crashing cost of schedules ranging from normal length down to and including the minimum length schedule, that is, if L is the length of the normal schedule, find the cost of schedules which are L, $L - 1$, $L - 2$, and so days long.

Overhead costs total Rs 60 per day. What is the optimal length schedule duration on each job for the optimal solution?

Solution: Using the normal duration the network is drawn and is shown in Fig. 12.36.

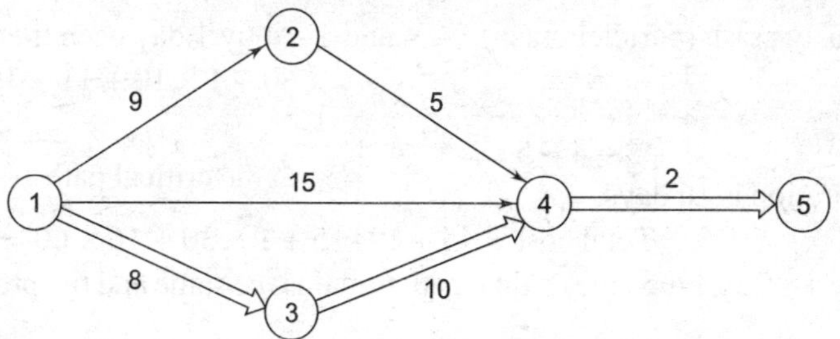

figure 12.36

The critical path is calculated as follows:

Path	*Duration*
1 – 2 – 4 – 5	16 days
1 – 4 – 5	17 days
1 – 3 – 4 – 5	20 days

The critical path is 1 – 3 – 4 – 5 and the critical activities are 1 – 3, 3 – 4 and 4 – 5.

$$\text{Total cost} = \text{Direct cost} + \text{Indirect cost}$$
$$= 0 + 20 \times 60 = \text{Rs } 1,200.$$

Among the critical activities, the activity with the least cost slope is 3 – 4. Hence, crash the critical activity 3 – 4 by 1 day. The new network is given in Fig. 12.37.

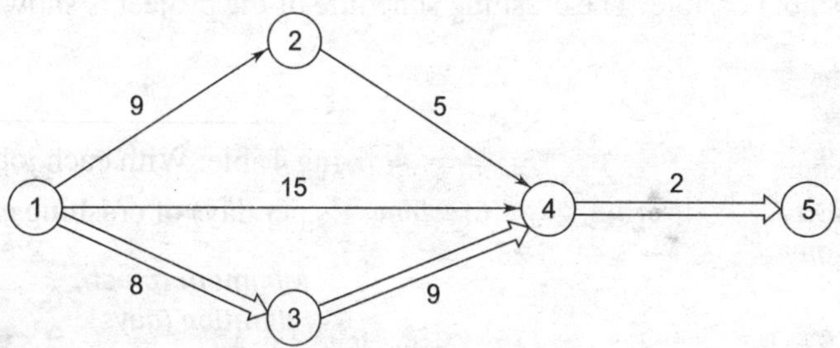

figure 12.37

Critical path: 1 – 3 – 4 – 5 and the project duration is 19 days.

$$\text{Total cost} = \text{Direct cost} + \text{Indirect cost}$$
$$= 1 \times 15 + 19 \times 60 = \text{Rs } 1,155$$

Now, if we crash the critical activity by 1 more day the critical path becomes 1 – 3 – 4 – 5 and the project duration 18 days.

$$\text{Total cost} = \text{Direct cost} + \text{Indirect cost}$$
$$= 15 + 1 \times 15 + 18 \times 60 = \text{Rs } 1,110$$

If we crash the critical activity by 1 more day, then the critical paths are

1 – 3 – 4 – 5

and, 1 – 4 – 5

The project duration is 17 days

$$\text{Total cost} = \text{Direct cost} + \text{Indirect cost}$$
$$= 30 + 1 \times 15 + 17 \times 60 = \text{Rs } 1,065.$$

If we simultaneously crash (parallel crash) 3 – 4 and 1 – 4 by 1 day each then the critical paths are
1 – 3 – 4 – 5
1 – 4 – 5
and, 1 – 2 – 4 – 5

The projection duration is 16 days.

$$\text{Total cost} = 45 + 1 \times 15 + 1 \times 30 + 16 \times 60 = \text{Rs } 1{,}050.$$

If we cash activity 4 – 5 by 1 day, the critical path remains the same and the project duration becomes 15 days.

$$\text{Total cost} = 90 + 1 \times 40 + 15 \times 60 = \text{Rs } 1{,}030.$$

If we perform parallel crash of activities 1 – 3, 1 – 4 and 2 – 4 by 1 day each, the project duration becomes 14 days.

$$\text{Total cost} = 130 + 1 \times 25 + 1 \times 30 + 1 \times 10 + 14 \times 60 = \text{Rs } 1{,}035.$$

If we perform parallel crash of activities 1 – 3, 1 – 4 and 2 – 4 by 1 day each, the critical path remains the same and the project duration becomes 13 days.

$$\text{Total cost} = 195 + 1 \times 25 + 1 \times 30 + 1 \times 10 + 13 \times 60 = \text{Rs } 1{,}040.$$

If we perform parallel crash of activities 1 – 3, 1 – 4 and 1 – 2 by 1 day each, then the critical path becomes the same and the project duration becomes 12 days.

$$\text{Total cost} = 260 + 1 \times 25 + 1 \times 30 + 1 \times 20 + 12 \times 60$$
$$= \text{Rs } 1{,}055.$$

Further crashing is not possible. The crashing schedule of the project is shown in the Table 12.14.

table 12.14

Project duration (days)	Crashing Activity and days	Direct Cost (Rs)			Indirect Cost (Rs)	Total Cost (Rs)
		Normal	Crashing	Total		
20	–	–	–	–	20 × 60	1,200
19	3 - 4 (1)	–	15	15	19 × 60	1,155
18	3 - 4 (1)	15	15	30	18 × 60	1,110
17	3 - 4 (1)	30	15	45	17 × 60	1,065
16	1 - 4 (1) 3 - 4 (1)	45	30 + 15	90	16 × 60	1,050
15	4 - 5 (1)	90	40	130	15 × 60	1,030
14	1 - 3 (1) 1 - 4 (1) 2 - 4(1)	130	25 + 30 + 10	195	14 × 60	1,035
13	1 - 3 (1) 1 - 4 (1) 2 - 4 (1)	195	25 + 30 + 10	260	13 × 60	1,040
12	1 - 3 (1) 1 - 4 (1) 1 - 2 (1)	260	25 + 30 + 20	335	12 × 60	1,055

From Table 12.14, we observe that the optimum duration is 15 days and the minimum duration is 12 days.

12.6 UPDATING A PROJECT

A project may not exactly follow the time schedule developed for it when it is actually executed. There are bound to be unexpected delays and difficulties in terms of delays in supply of material, non-availability of some machines and/or breakdown of machines, non-availability of skilled manpower, natural calamity, and so on. In such cases, it may be necessary to review the progress of network planning and scheduling. Such review helps in taking stock of the progress that has been made and making necessary change in the initial schedule in terms of time and resources required by incomplete activities in the project.

There is no rule about the time for updating a project network. The frequency of updating may be more when project duration is small because few slippages in detecting the progress will affect the project as a whole as the time for absorbing such slippages in less. But in case of large projects the frequency of updating may be less in the initial stages because a few initial slippages may be absorbed later in the project. However, to add dynamism to the nature and progress of work, updating should be carried out as frequently as economically possible.

The process of revising the network diagram of a project by incorporating in it the changes which have occurred due to replanning and rescheduling is called updating. When such changes are made, the original network diagram is no longer valid.

Updating can be done in two ways:

1. Use the revised estimates of incomplete activities and calculate from the initial event the earliest and latest completion of each event in the usual manner to know the project completion time.
2. Change the complete work to zero duration and represent all the activities already finished by an arrow called the elapsed time arrow. Events in the revised network diagram are renumbered. The completion time of remaining activities are taken the revised time.

Example

The network for a project is shown below. A review of the project after 15 days reveals that:

(i) Activities 1 – 2, 1 – 3, 2 – 3, 2 – 4 and 3 – 4 are complete.

(ii) Activities 3 – 5 and 4 – 6 are in progress and need 2 and 4 days more, respectively.

(iii) The revised estimate shows that activity 8 – 9 will take only 8 days but 7 – 9 will need 10 days.

Draw the new network after updating the project and determine the new critical path. The work completed may be shown by an elapsed time activity.

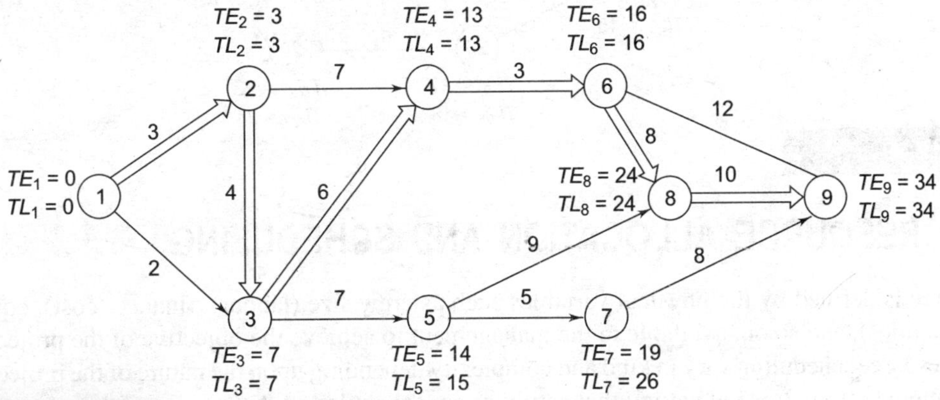

figure 12.38

Solution: The progress of the work noted at the end of 15th day may be summarised as shown in Table 12.15.

table 12.15

Activity	*Time required (days)*	*Situation*
1 - 2	0	Complete
1 - 3	0	Complete
2 - 3	0	Complete
2 - 4	0	Complete
3 - 4	0	Complete
3 - 5 (20 - 25)	2	In progress
4 - 6 (20 - 26)	4	In progress
5 - 8 (25 - 28)	9	Not started
5 - 7 (25 - 27)	5	Not started
6 - 8 (26 - 28)	8	Not started
6 - 9 (26 - 29)	12	Not started
7 - 9 (27 - 29)	10	Not started
8 - 9 (28 - 29)	8	Not started

In the new network diagram given in Fig. 12.38, activity 1 – 20 show the elapsed time of 15 days. Other activities are assigned the time required for their completion after review as given in Table 12.15. Based on these revised time estimates, the critical path: 1 – 20 – 26 – 28 – 29 is shown by double lines in Fig. 12.39. The total project duration has increased by 1 day.

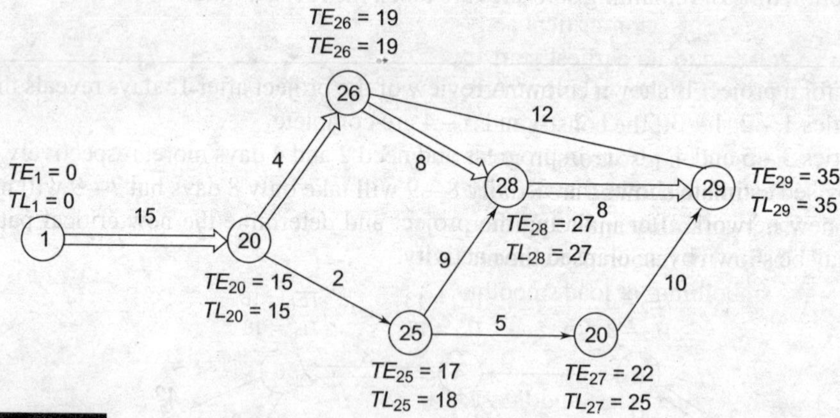

figure 12.39

12.7 RESOURCE ALLOCATION AND SCHEDULING

A resource is defined by the physical variables such as crew size (labour), finance (cost), equipment (shop facilities), and so on, available to the management to achieve the objective of the project. Problems of resource scheduling vary in kind and complexity depending upon the nature of the project and its organisational set up. It is but natural that activities are scheduled so that no two of them requiring the same facility occur at the same time, wherever possible. The problem of scheduling activities so that none of the precedence relations are violated is an extremely difficult task even for projects of modest size. The problem of scheduling project with the limited resources is usually large, combinatorially. Even the powerful techniques aided by the fastest, most sophisticated computer can solve only small projects having not more than about 100 activities. Analytical techniques are impractical for real world problems of this type. One turns to heuristic programmes for such cases.

Lacking time or inclination to pursue more through problem solving procedures, one employs a rule of thumb arising out of experience, expertise and commonsense. In some cases, the rule of thumb is insufficient. It must be combined with other rules to take into consideration additional factors or exceptional circumstances. A collection of such rules for solving a particular problem is called a heuristic programme. Such a programme may require a computer for solving complex problems.

Heuristic programmes for resource scheduling are

1. Resource levelling programmes
2. Resource allocation programmes

Resource Levelling Programmes

In the process of resource levelling, the activities are critically sequenced, subject to the constraint of availability of resources, and the minimum period of the project is redetermined accordingly.

Under this technique, the project manager may be able to lower the maximum resource requirements by shifting a non-critical activity between its earliest start time and latest finish time.

The two rules normally used in scheduling non-critical activities are as under:

1. If the total float of a non-critical activity is equal to its free float, then it can be scheduled anywhere between its earliest start and latest completion times.
2. If the total float of a non-critical activity is more than its free float, then its starting time can be delayed relative to its earliest start time by no more than the amount of its free float without affecting the scheduling of its immediately succeeding activities.

Resource smoothen: If the constraint is the total project duration, then the resource allocation only smoothens the demand on resources in order that the demand of any resource is as uniform as possible. The periods of maximum demand are located and the activities according to their float values are shifted for balancing the availability and requirement of resources. So, the intelligent utilisation of floats can smoothen the demand of resources to the maximum possible extent. Such type of resource allocation is called 'resource smoothing' or load smoothing'.

Procedure

1. Calculate the earliest start and latest finish times of each activity and then draw a time-scaled version (or squared) of the network. Critical path in the network is drawn along a straight line and non-critical activities are on either sides of this line. Resource requirement of each activity is given along the arrows.
2. Draw the resource histogram by taking the earliest start times or latest start times of activities on the *x*-axis and cumulative resources required on the *y*-axis.
3. Shift start times of non-critical activities having largest float first in order to smoothen the resources.

Example 1

Given below is the crew size (manpower) requirement for each activity in a project:

Activity	:	0 – 1	1 – 2	1 – 3	2 – 4	3 – 5	3 – 6	4 – 7	5 – 7	6 – 8	7 – 9	8 – 9
Normal time (days)	:	2	3	4	2	4	3	6	6	5	4	4
Crew Size (Manpower required)	:	4	3	3	5	3	4	3	6	2	2	9

(a) Draw the network of the project activities.

(b) Rearrange the activities suitably for reducing the exising total crew size.

(c) If only 9 men are available for the execution of the project, rearrange the activities suitably for levelling the manpower resource.

Solution: The network is given in Fig. 12.40.

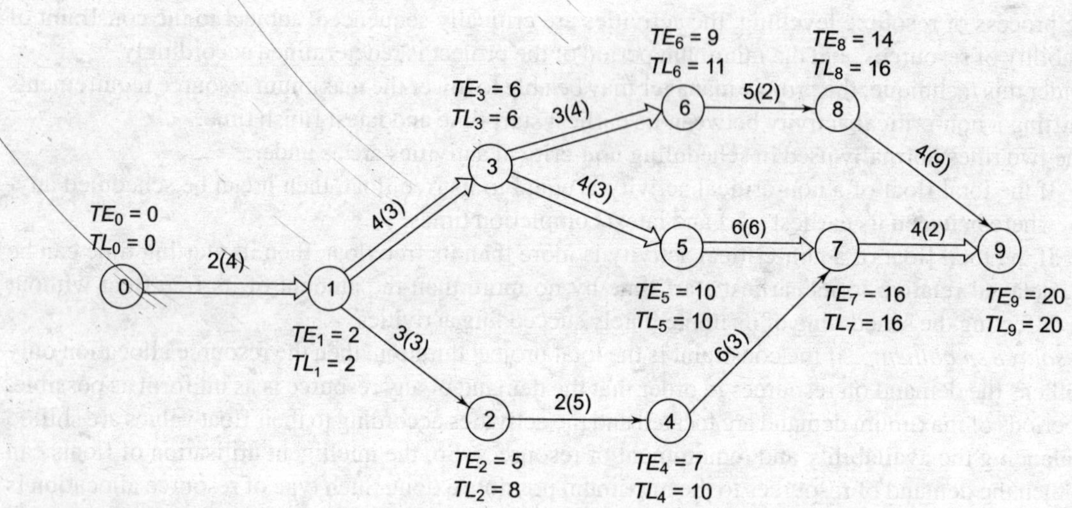

figure 12.40

Numbers in brackets with each activity arrow in the network represent the manpower requirement of the activity.

The earliest and latest completion times of each activity are indicated along the nodes of Fig. 12.40. The critical path is $0 \to 1 \to 3 \to 5 \to 7 \to 9$ and the project duration is 20 days.

In the time scaled version, the critical path is drawn along the straight line and non-critical are added in dotted lines, as shown in Fig. 12.41.

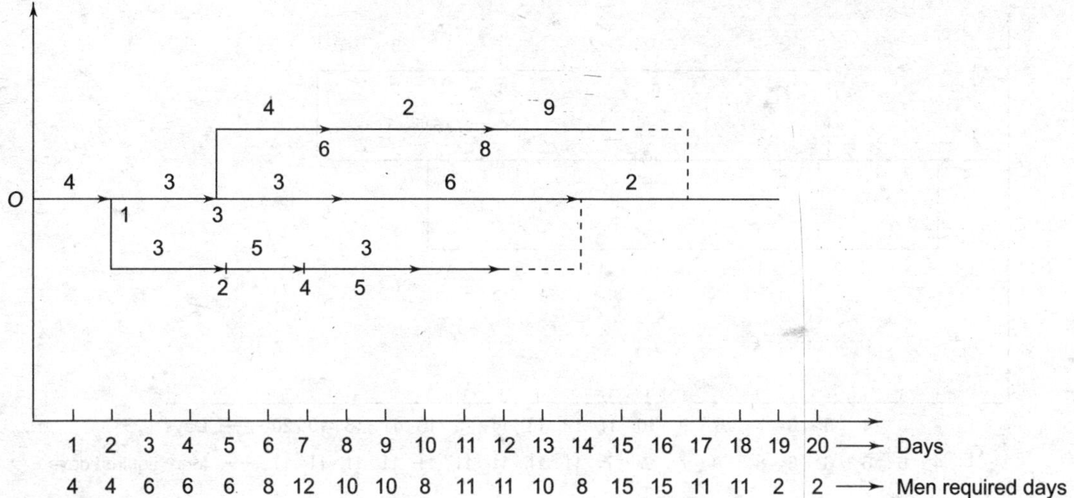

figure 12.41

(b) The numbers above the arrows indicate the crew size (the resource requirements), the numbers below the arrows indicate the days and the non-critical paths are added as shown in Fig. 12.41. The resource histogram can now be drawn by summing up the manpower requirements for each day. This is shown in Fig. 12.42.

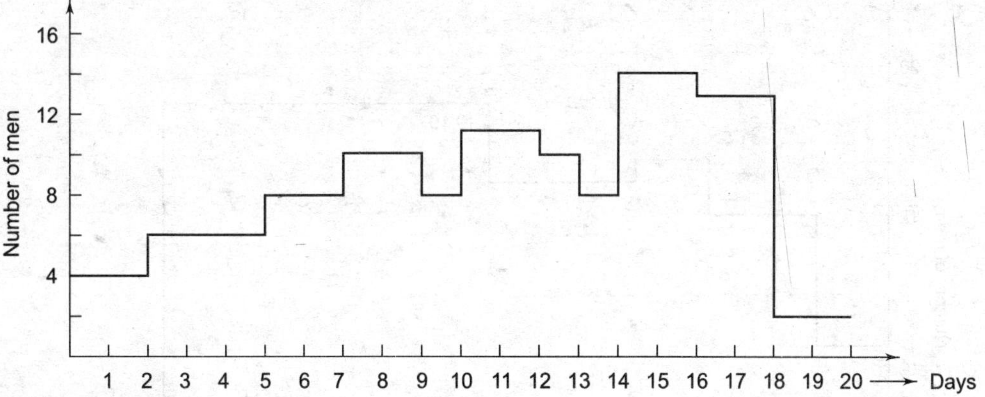

figure 12.42

From Fig. 12.43, we observe that the maximum demand of 15 men occurs on the 15th and 16th days. To smoothen load, the activities will have to be shifted depending upon the floats. Since the path 3 – 6 – 8 – 9 has a float of two days, the activities 6 – 8 and 8 – 9 are shifted to the right. Thus, the start of each of these activities is delayed by 2 days. Similarly, activity 4 – 7 can be shifted to the right so that it starts on the 10th day instead of starting on the 7th day. After making the necessary shifting, the network is drawn as shown in Fig. 12.43:

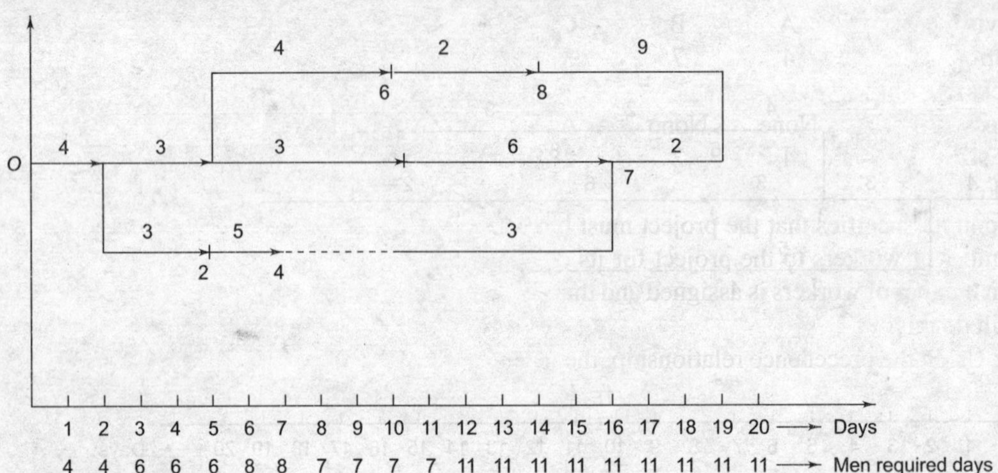

figure 12.43

Resource histogram for this squared network is given in Fig. 12.44. This indicates that the maximum manpower required is 11 men.

(c) Since only 9 men are available instead of 11, therefore, activities have to rearranged and expanded in term of their completion time. This would increase the total project duration from 20 days to 25 days.

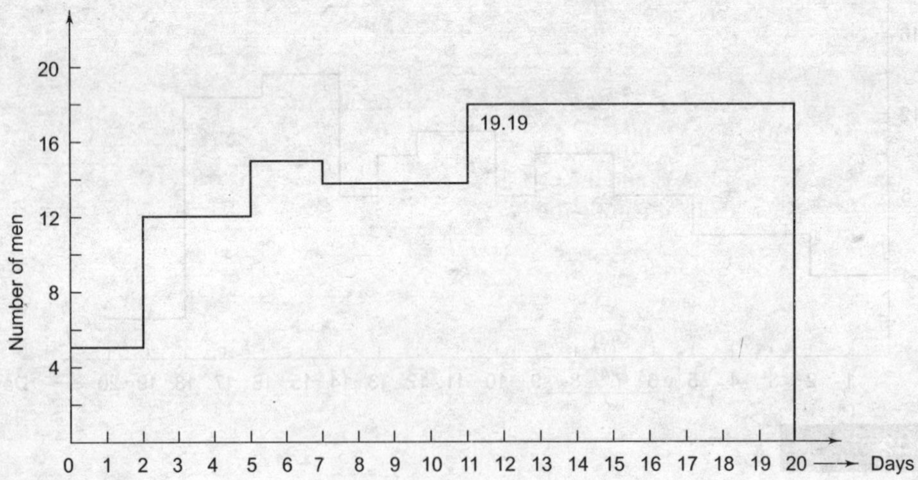

figure 12.44

Example 2

A Chennai construction company is bidding on a contract to instal a line of microwave towers. It has identified the following activities, along with their expected times, predecessor restrictions, and work requirements.

Activity	:	A	B	C	D	E	F	G	H
Duration (weeks)	:	4	7	3	3	2	2	2	3
Predecessors	:	None	None	A	A	B	B	D, E	F, G
Crew size (Workers)	:	4	2	2	4	6	3	3	4

The contract specifies that the project must be completed in 14 weeks. The company will assign a fixed number of workers to the project for its entire duration and so it would like to ensure that the minimum number of workers is assigned and that the project is completed in 14 weeks. Find a schedule which will do this.

Solution: Using the precedence relationship, the network diagram is given in Fig. 12.45.

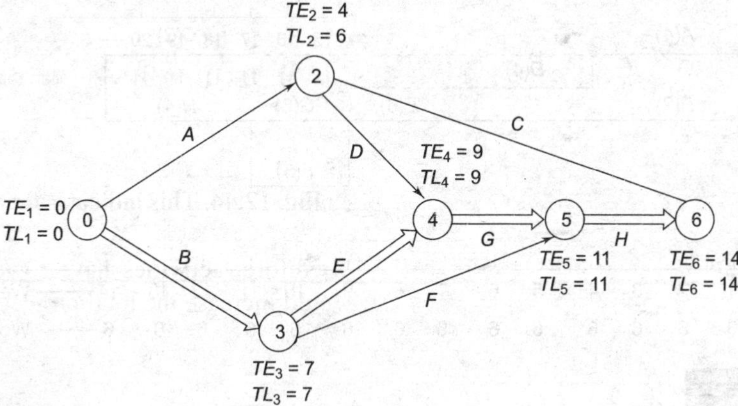

figure 12.45

The earliest and latest completion times of each activity are indicated along the nodes in Fig. 12.42. The critical path is B → E → G → H and the project duration is 14 weeks. The time-scaled variation of the network, showing the critical path along the straight line is shown in Fig. 12.46. Numbers above the arrows indicate the workers.

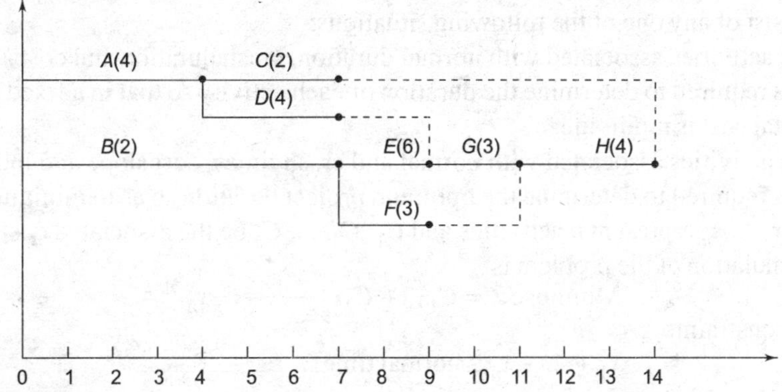

figure 12.46

To determine the minimum number of workers required to complete the project, the critical activities are scheduled first. Since non-critical activity A is scheduled at $t = 0$ followed by the non-critical activities C and D on the fourth day, workers required would rise to 8. The total float of activity A is more than its free float and so the earliest start time as well as free float is zero. Therefore, it has to be scheduled otherwise it would result in delay of the project.

Activity C has a float of 7 weeks and activity F has a float of 2 weeks. Requirement of workers in the 5th, 6th and 7th weeks is 8 and during the 8th and 9th weeks is 9 workers. This demand has to be smoothened. For this, we shift activity F to the 9th week. Thus, the total weekly requirement of workers after rearranging the activities comes to be of 6 during the entire project duration of 14 weeks as shown in Fig. 12.47.

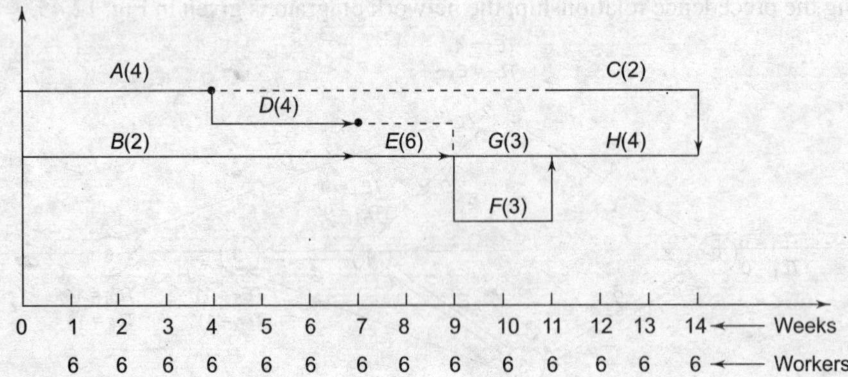

figure 12.47

12.8 SOME OTHER TECHNIQUES

12.8.1 Linear Programming Formulation

Usually a project may consist of a large number of activities resulting in a number of constraints. Therefore, it is sometimes desirable to formulate the problem in the linear programming format. The problem may consist of any one of the following situations:

(a) There are n activities associated with normal duration, crash duration and cost slope of a certain project. It is required to determine the duration of each activity so that in a fixed project duration time, the total cost is minimum.

(b) There are n activities associated with normal and crash times, cost slope and indirect cost of the project. It is required to determine the optimum project duration so as to minimise the total cost.

In general, let $x, ..., x_n$ represent n activities and $C_1, C_2, ..., C_n$ be the associated costs. Then, a linear programming formulation of the problem is

$$\text{Minimise } Z = C_1x_1 + C_2x_2 + ... + C_nx_n$$

Subject to the constraints

$$x_1 + ... + x_n \leq \text{normal time}$$
$$x_1 + ... + x_n \geq \text{crash time}$$
$$x ..., x_n \geq 0.$$

12.8.2 Management Operation System Technique (MOST)

Management Operation System Technique (MOST) is a graphical representation of PERT to a time-scale. The target date is taken as the zero date of the completion of the project. The expected date of start and completion of other activities based on the zero date of completion are shifted to the left side of the graph, indicating inter-dependencies of the activities by a line. The completion of various activities on a date of review are drawn on the same graph with solid horizontal lines so that target and completion, even in parts, may be shown on the same graph.

This enables the management to assess the present status of the project and the expected date of completion of various activities as on the date of review. Floats are also indicated by dotted lines. Critical path may be shown by a shaded line. This technique can also be applied for larger projects.

Example 1

Consider the network diagram in Fig. 12.48 involving 7 activities with the expected duration of each activity indicated above the arrow. Discuss MOST for this network diagram.

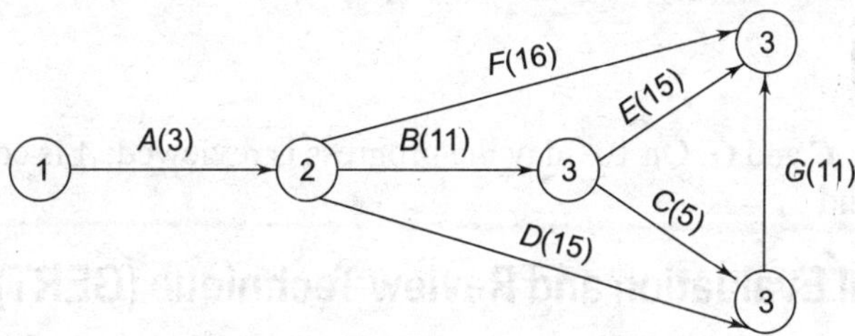

figure 12.48

Solution: The time-scale diagram of the above PERT network is shown in Fig. 12.49.

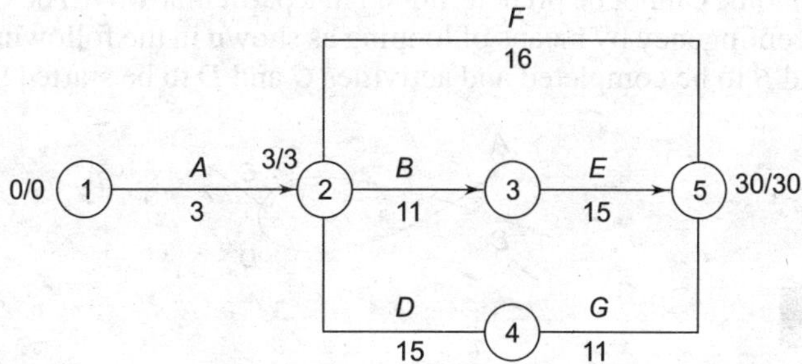

figure 12.49

Using MOST on Fig. 12.49, we get the required graphical representation of the given network. This is shown in Fig. 12.50.

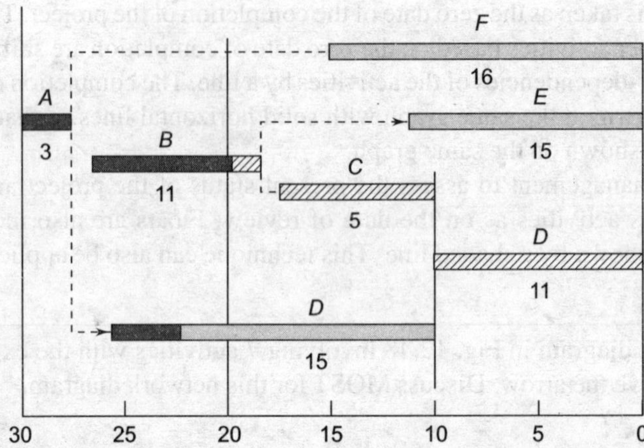

figure 12.50

Critical path is *A*, *B*, *C* and *G*. On 10th day the progress is reviewed; *A* is complete, *B* is ahead of the schedule and *B* is behind.

12.8.3 Graphical Evaluation and Review Technique (GERT)

Grpahical Evaluation and Review Technique (GERT) deals with probabilistic situations. Although PERT technique also deals with uncertainties in timing of the projects, its calculations are based on the assumption that each activity contemplated will materialise without any omission and there is no allowance for a probabilistic choice of a path between nodes. PERT cannot deal with a situation where the activities already dropped at a subsequent stage to occur again with another alternative, for example, projects where the outcome cannot be predetermined in a particular way. The GERT technique accommodates this kind of contingency by means of looping as shown in the following figure.

(a) Activities *A* and *B* to be completed and activities *C* and *D* to be started thereafter.

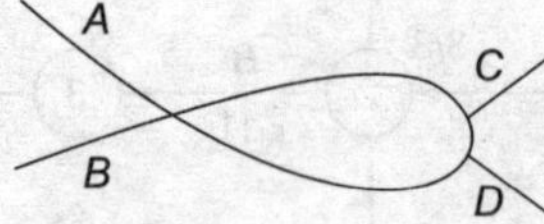

figure 12.51

(b) Activities *A*, *B*, *C* or *A*, *B*, *D* to be started but not *C* and *D* together.

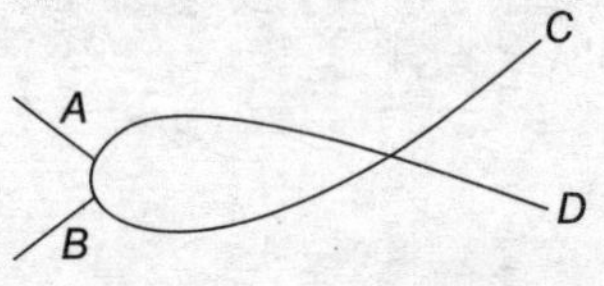

C
A
D
B

figure 12.52

(c) An applied research activity is conducted and table model experiment is executed. If the experiment is successful a profit plan is designed and erected. If the experiment is unsuccessful applied research activity is suitably modified and revised table model experiment is conducted. This can be indicated by GERT.

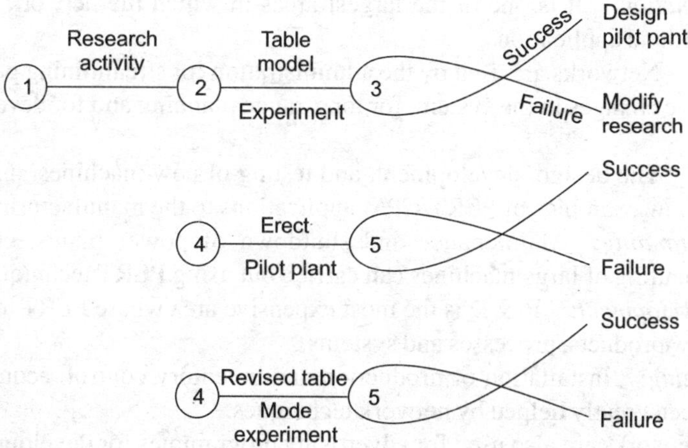

figure 12.53

Looping is done at nodes 3 and 5. GERT is prima facie a branching network where at a decision node different branches are evolved with probabilities estimates. It is, therefore, rightly said that GERT is in when PERT is out.

12.8.4 Line of Balance

Line of Balance (LOB) is a technique by which an attempt is made to maintain a balance between the availability and requirements of components/parts in a sub-assembly or assembly shop. The objective is to have better inventory control in the components/parts and reduce the total cost of the inventory of such items. The lead times required for bought out/manufactured items are ascertained. A time-scale PERT diagram is drawn showing the various activities. The time is shown as if the product is required on zero date. The network diagram will indicate the starting time for each activity. From the diagram the cumulative requirements of components/parts at periodical intervals of time are indicated in a bar chart. The physical position of inventory is also depicted in the same bar chart so that the management can identify what items are over/under stocked and take remedial action.

The procedure for line balancing may be summarised in the following six steps:

Step 1: Divide the total work (project) into sub-tasks called work elements (or element tasks).

Step 2: Establish precedence requirements between the element tasks and construct a precedence diagram.

Step 3: Select a feasible cycle time and determine the minimum number of workstations needed.

Step 4: Apply assignment heuristic and assignment elemental tasks to individual workstations.

Step 5: Check whether further improvements can be made.

Step 6: Calculate the effectiveness and efficiency of the proposed line.

12.9 APPLICATION AND DISADVANTAGES OF NETWORKS

12.9.1 Application Areas of PERT/CPM Techniques

PERT/CPM techniques are widely used in the following typical areas:

Building construction: It is one of the largest areas in which the network techniques of project management have found applications.

Administration: Networks are used by the administration for streamlining paper work, for making major administrative changes in the system, for long-range planning and for developing staffing plans, and so on.

Manufacturing: The design, development, and testing of new machines, installing machines and plant layouts are a few examples of PERT/CPM applications to the manufacturing function of a firm.

Maintenance planning: Maintenance and shutdown of power plants, chemical plants, steel furnaces and overhauling of large machines can carried out using PERT techniques.

Research & Development: R & D is the most expensive area where PERT techniques are used for development of new products, processes and systems.

Inventory planning: Installation of production and inventory control, acquisition of spare parts, and others, have been greatly helped by network techniques.

Marketing: Networks are also used for advertising programmes for development and launching of new products and for planning their distribution.

12.9.2 Uses of PERT/CPM for Management

1. PERT/CPM techniques help the management in properly planning the complicated projects, controlling the working plan and also keeping the plan up-to-date. These are also helpful in searching the potential spots and in taking corrective measures.
2. The network techniques provide a number of checks and safeguards against going astray in developing the plan for a project. Thus, there is little chance of oversight of certain activities and events.
3. These techniques help the management in reaching the goal in minimum time and at least cost and also in forecasting the probable project duration and the associated cost.
4. The networks clearly designate the responsibilities of different supervisors. The supervisor of an activity himself knows the time schedule and also the supervisors of other activities with whom he has to operate.
5. The flexibility of the network permits the management to make necessary alternations and improvements as and when needed. These allocations can be made during the deployment of resources or reviewing.
6. Application of network techniques has resulted in better managerial control, better utilisation of resources, improved communication and progress reporting, and better decision-making.
7. Application of PERT/CPM techniques have resulted in saving of time which directly results in the saving of cost. Also, saving in time or early completion of the project results in earlier return of revenue and introduction of the product or process ahead of the competitors, resulting in increased profits.

12.9.3 Disadvantages of Network Techniques

The following difficulties are faced by the management while using the network techniques:

1. The difficulty arises while securing realistic time estimates. In new and non-repetitive projects, the time estimates produced are often mere guesses.
2. It is sometimes troublesome to develop a clear logical network. This depends upon the data input and thus the plan can be better than the personnel who provides the data.
3. The natural tendency to oppose changes results in the difficulty of persuading the management to accept these techniques.
4. Determination of the level of network detail is another troublesome area. The level of detail varies from planner to planner and depends upon his judgment and experience.
5. The planning and implementation of networks require personnel trained in network methodology. Managements are reluctant to spare the existing staff to learn these techniques or to recruit trained personnel.

EXERCISES

Section 12.2

Construct the network for each of the projects whose activities and their precedence relationships are as given below:

1.

Activity	:	A	B	C	D	E	F	G	H	I	J	K
Immediate Predecessor	:	–	–	–	A	B	B	C	D	E	H, I	F, G

2. A < C, D, I; B < G, F; D < G, F; F < H, K, G, H < J, I, J, K < E
3. A, B, C can start simultaneously A < F, E; B < D, C, E, D < G
4.

Activity	:	A	B	C	D	E
Immediate Predecessor	:	–	A	A	A	B, C, D

5. A < C, D; B < C, D; C < E; D, E < F
6. A < C, D, B < E; C, E < F, G; D < H, G < I; H, I < J
7.

Activity	:	A	B	C	D	E	F	G	H	I
Immediate Predecessor	:	–	–	A, B	B	B	A, B	F, D	F, D	C, G

8. Draw the network for the project whose activities and their precedence relationships are given below:

Activity	:	P	Q	R	S	T	U
Predecessor	:	–	–	–	P, Q	P, R	Q, R

9. Draw the network for the project whose activities with their predecessor relationships are given below:
   A, C, D can start simultaneously; E > B, C; F, G, > D, H, I > E, F; J > I, G; K > H; B > A.
10. Construct the network for the project whose activities and their relationships are as given below:
    Activities: A, D, E can start simultaneously; B, C > A; G, F > D, C; H > E, F
11. Draw the network for the project whose activities and their precedence relationships are as given below:

Activities	:	A	B	C	D	E	F	G	H	I
Immediate Predecessor	:	–	A	A	–	D	B, C, E	F	E	H

12. Construct the network of the project whose precedence relationships are given below:
B < E, F; F < G; C < G, L; E, G < H, L, H < I; L < M; H, M < N, H < J; I, J < P; P < Q.
13. Construct the network diagram comprising activities B, C, ..., Q and N such that the following constraints are satisfied:
B > E, F; C < G, L; E, G < H; L, H < I, L < M; H < N; H < J; I, J < P; P < Q. The notation X < Y means that the activity X must be finished before Y can begin.
14. Listed in the Table are the activities and the sequence necessary for a maintenance job on heat exchangers in a refinery:

Activity	*Immediate Predecessor*	*Activity*	*Immediate Predecessor*
A	None	H	D, E, F
B	None	I	D
C	None	J	G
D	A	K	G
E	B, C	L	H, J
F	A	M	K
G	C	N	I, L

15. Construct the arrow diagram comprising activities *A*, *B*, ... and *L* such that the following relationships are satisfied:
    (i) *A*, *B* and *C*, the first activities of the project, can start simultaneously.
    (ii) *A* and *B* precede *D*.
    (iii) *B* precedes *E*, *F* and *H*.
    (iv) *F* and *C* precede *G*.
    (v) *E* and *H* precede *I* and *J*.
    (vi) *C*, *D*, *F* and *J* precede *K*.
    (vii) *K* precedes *L*.
    (viii) *I*, *G* and *L* are the terminal activities of the project.
16. A publisher has a contract with an author to publish a textbook. The (simplified) activities associated with the production of the textbook are given. Develop the associated network of the project.

	Activity	*Predecessor activity*	*Duration (weeks)*
A	Manuscript proof-reading by editor	–	5
B	Sample pages prepared by typesetter	–	3
C	Book cover design	–	5
D	Preparation of diagrams used in the book	–	4
E	Author's approval of edited manuscript and sample pages	A, B	3
F	Book typesetting	E	5
G	Author checks	F	3
H	Author checks artwork	D	2
I	Production of printing plates	G, H	3
J	Book printing and binding	C, I	7

17. An assembly is to be made from two parts X and Y. Both parts must be turned on a lathe and Y must be polished whereas X need not be polished. The sequence of activities together with their predecessors are given below:

Activity	*Description*	*Predecessor activity*
A	Open work order	–
B	Get material for X	A
C	Get material for Y	A
D	Turn X on lathe	B, C
E	Turn Y on lathe	E
F	Polish Y	D, F
G	Assemble X and Y	G
H	Pack	

Draw a network diagram for the project.

18. Draw a network diagram from the following list of activities:

Activity	*Immediate predecessor*	*Activity*	*Immediate predecessor*
A	–	L	K
B	A	M	K
C	B	N	K
D	C	O	D
E	B	P	O
F	E	Q	B
G	E	R	N
H	C	S	L, M
I	C, F	T	S
J	C, H, E	U	P, Q
K	J	V	U

19. A new type of water pump is to be designed for an automobile. Its major specifications are given in the Table.

Activity	*Description*	*Predecessor activity*
A	Drawing prepared and approved	–
B	Cost analysis	A
C	Tool feasibility (economics)	A
D	Tool manufactured	C
E	Favourable cost	B, C
F	Raw material procured	D, E
G	Sub-assemblies ordered	E
H	Sub-assemblies received	D, F
I	Parts manufactured	I, H
J	Final assembly	J
K	Testing and shipment	

Draw the arrow diagram of activities of the project.

20. Draw the network for jobbing production and indicate the critical path from the following:

Activity	*Description*	*Time (weeks)*	*Preceded by*
A	Market research	15	–
B	Make drawings	15	–
C	Decide production policy	3	A
D	Prepare sales programme	5	A
E	Prepare operation sheets	8	B, C
F	Buy material	12	B, C
G	Plan labour force	1	E
H	Make tools	14	E
I	Schedule production	3	D, G
J	Produce product	14	F, H, I

Section 12.3

21. Draw the network diagram for the following activities and find the critical path and total float.

Job	*Job time (days)*	*Immediate predecessors*
A	13	–
B	8	A
C	10	B
D	9	C
E	11	B
F	10	E
G	8	D, F
H	6	E
I	7	H
J	14	G, I
K	18	J

[*Answer*: Critical path is A → B → E → F → G → J – C and the project duration is 82 days. Total float:

A	B	C	D	E	F	G	H	I	J	K
0	0	0	0	5	0	0	0	2	0	0]

22. An architect has been awarded a contract to prepare plans for an urban renewal project. The job consists of the following activities and their estimated times:

Activity	*Description*	*Immediate predecessors*	*Time (Days)*
A	Prepare preliminary sketches	–	2
B	Outline specifications	–	1
C	Prepare drawings	A	3
D	Write specifications	A, B	2
E	Run off prints	C, D	1
F	Have specifications	B, D	3
G	Assemble bid packages	E, F	1

(a) Draw an arrow diagram for this project.
(b) Indicate the critical path, and calculate the total and free float for each activity.
[*Answer*: Critical path is A – D – F – G and project duration: 8 days.]

23. The following maintenance job is to be performed periodically on the heat exchangers in a refinery:

	Task	*Precedence*	*Duration in hours*
A	Dismantle pipe connections	–	14
B	Dismantle header, closure and floating head front	A	22
C	Remove tube bundle	B	10
D	Clean bolts	B	16
E	Clean header and floating heat front	B	12
F	Clean tube bundle	C	10
G	Clean shell	C	6
H	Replace tube bundle	F, G	8
I	Prepare shell pressure test	D, E, H	24
J	Prepare tube pressure test and make final reassembly	I	16

(a) Draw an arrow diagram for his project.
(b) Identify the critical path. What is its length?
(c) Find the total float and free float for each task.
[*Answer*: Critical path is A – B – C – F – H – I – J; project duration = 8 days.]

24. A project consists of a series of tasks labelled A, B, ..., H, I with the following relationships (W < X, Y means X and Y cannot start until W is complete; X, Y < W means W cannot start until both X and Y be complete). With this notation construct the network diagram having the following constraints

$$A < D, E;\ B, D < F;\ C < G;\ B, G < H,\ F, G < I$$

Also, find the minimum time of completion of the project, when the time (in days) of completion of each task is as follows:

Task :	A	B	C	D	E	F	G	H	I
Time :	23	8	20	16	24	18	19	4	10

[*Answer*: Project duration is 67 days.]

25. A small project consists of 7 activities for which the relevant data are given below:

Activity :	A	B	C	D	E	F	G
Preceding activities :	–	–	–	A, B	A, B	C, D, E	C, D, E
Activity duration (days) :	4	7	6	5	7	6	5

(i) Draw the network and find the project completion time.
(ii) Calculate total float for each of the activities.
(iii) Draw the time scaled diagram.
[*Answer*: Project duration is 20 days. Total float is as follows

Activity :	A	B	C	D	E	F	G
Total Float :	4	7	6	5	7	6	5.]

26. The following are the details of estimated time of activities of a certain project.

Activity	:	A	B	C	D	E	F
Immediate predecessors	:	–	A	A	B, C	–	E
Estimated time (weeks)	:	2	3	4	6	2	8

(a) Find the critical path and the expected time of the project.
(b) Calculate the earliest start and earliest finish time for each activity.
(c) Calculate the slack for each activity.

[*Answer*: Critical path : A → C → D → F

Activity	:	A	B	C	D	E	F
Slack	:	0	1	0	0	4	0.]

27. The following Table gives the activities in a construction project and the time duration:

Activity	:	1 – 2	1 – 3	2 – 3	2 – 4	3 – 4	4 – 5
Preceding activity	:	–	–	1 – 2	1– 2	1 – 3, 2 – 3	2 – 4, 3 – 4
Normal time (days)	:	20	25	10	12	5	10

(a) Draw the activity network of the project.
(b) Find the total float and free float for each activity.
(c) Determine the critical path and the project duration.

[*Answer*: Critical path : 1 – 2 – 3 – 4 – 5; project duration: 45 days

Activity	:	1 – 2	1 – 3	2 – 3	2 – 4	3 – 4	4 – 5
Total Float	:	0	5	0	3	0	0
Free Float	:	0	5	0	3	0	0.]

28. Determine early start (TE_i) and late start (TL_i) in respect of all node points and identify the critical path in respect of the following network.

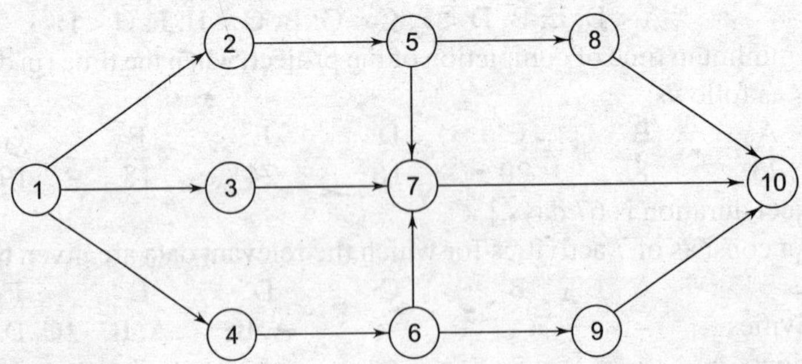

figure 12.54

[*Answer*: Critical paths are:
(i) 1 – 2 – 5 – 8– 10 (ii) 1 – 2 – 5 – 7 – 10.]

29. Find the critical path and calculate the slack time of each event for the following PERT diagram.

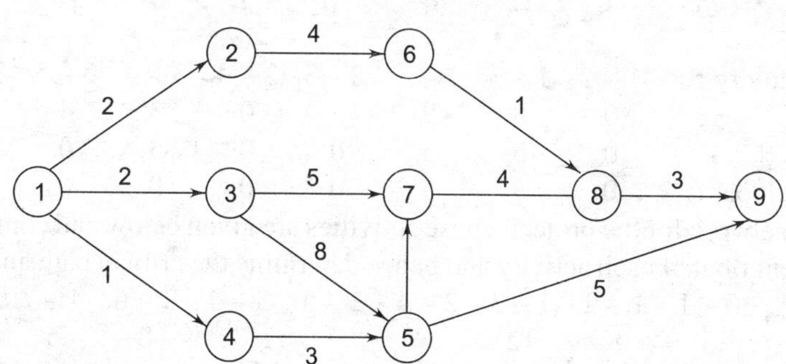

figure 12.55

[*Answer*: Critical path is 1 – 3 – 5 – 9; project duration is 15 days. Slack time is as follows:

Activity	:	1 – 2	1 – 3	1 – 4	2 – 6	3 – 7	3 – 5	4 – 5	5 – 9	6 – 8	7 – 8	8 – 3
Slack	:	5	0	6	5	1	0	6	0	5	1	1.]

30. Tasks A, B, C, …, H, I constitute a project. The notation X < Y means that the task X must be finished before Y can begin. With this notation A < D, A < E, B < F, D < F, C < G, C < H, F < I, G < I, draw a graph to represent the sequence of tasks and find the minimum time of completion of the project when the time (in days) of completion of each task is as follows:

Task	:	A	B	C	D	E	F	G	H	I
Time	:	8	10	8	10	16	17	18	14	9

[*Answer*: Critical path is A – D – F – I. The minimum time to complete the project is 44 days.]

31. Given the following information:

Activity	:	0 – 1	1 – 2	1 – 3	2 – 4	2 – 5	3 – 4	3 – 6	4 – 7	5 – 7	6 – 7
Duration (in days)	:	2	8	10	6	3	3	7	5	2	8

(i) Draw the arrow diagram.

(ii) Identify the critical path and find the total project duration.

(iii) Determine total, free and independent floats.

[*Answer*: Critical path is 0 – 1 – 3 – 6 – 7. Project duration = 27 days

Activity	:	0 – 1	1 – 2	1 – 3	2 – 4	2 – 5	3 – 4	3 – 6	4 – 7	5 – 7	6 – 7
Total Float	:	0	6	0	6	12	7	0	6	12	0
Free Float	:	0	0	0	0	0	1	0	6	12	0
Independent Float	:	0	0	0	0	0	1	0	0	0	0.]

32. Calculate the total float, free float and independent float for the project whose activities are given below.

Activity :	1 – 2	1 – 3	1 – 5	2 – 3	2 – 4	3 – 4	3 – 5	3 – 6	4 – 6	5 – 6
Duration (in weeks) :	8	7	12	4	10	3	5	10	7	4

[*Answer*:

Activity :	1 – 2	1 – 3	1 – 5	2 – 3	2 – 4	3 – 4	3 – 5	3 – 6	4 – 6	5 – 6
Total Float :	0	8	9	3	0	3	4	3	0	4
Free Float :	0	5	5	0	0	3	0	3	0	4
Independent Float :	0	5	5	0	0	0	0	0	0	0.]

33. Construct the network for the project whose activities are given below and compute the total, free and independent float of each activity and hence determine the critical path and project duration.

Activity :	0 – 1	1 – 2	1 – 3	2 – 4	2 – 5	3 – 4	3 – 6	4 – 7	5 – 7	6 – 7
Duration (in weeks) :	3	8	12	6	3	3	8	5	3	8

[*Ans*: Critical path is 0 – 1 – 3 – 6 – 7 and the project duration = 31 weeks.

Activity :	0 – 1	1 – 2	1 – 3	2 – 4	2 – 5	3 – 4	3 – 6	4 – 7	5 – 7	6 – 7
Total Float :	0	9	0	9	14	8	0	8	14	0
Free Float :	0	0	0	1	0	0	0	8	14	0
Independent Float :	0	0	0	0	0	0	0	0	0	0.]

34. A small maintenance project consists of the following 10 jobs. The precedence relationships are identified by their node numbers:

Job :	a	b	c	d	e	f	g	h	i	j
Node Numbers :	(1, 2)	(2, 3)	(2, 4)	(3, 5)	(3, 6)	(4, 6)	(4, 7)	(5, 8)	(6, 8)	(7, 8)
Estimated Duration :	2	3	5	4	1	6	2	8	7	4

(a) Draw an arrow diagram representing the project.
(b) Calculate early and late start and finish times for each job.
(c) Which jobs are critical?
(d) How much slack does job (3, 5) have?

[*Answer*: Critical jobs are (1, 2), (2, 4), (4, 6) and (6, 8) and the slack of (3, 5) is 3 days.]

35. Draw the network for the data given below and compute:
(i) Critical path
(ii) Early start and late start times for each activity
(iii) Total slack for each activity

Activity :	A	B	C	D	E	F	G	H	I
Predecessor :	–	–	–	A	B	C	D, E	B	H, F
Estimated Time: (weeks)	3	5	4	2	3	9	8	7	9

[*Answer*: Critical path C – E – I; project duration = 22 weeks

Activity :	A	B	C	D	E	F	G	H	I
Early Start :	0	0	0	3	5	4	8	5	13
Late Start :	9	1	0	12	11	4	14	6	13
Total Slack:	9	1	0	9	6	0	6	1	0.]

36. Draw the network for the following project and compute the earliest and latest times for each event and find the critical path.

Activity :	1 – 2	1 – 3	2 – 4	3 – 4	4 – 5	4 – 6	5 – 7	6 – 7	7 – 8
Immediate Predecessor :	–	–	1 – 2	1 – 3	2 – 4	2 – 4& 3 – 4	4 – 5	4 – 6	5 – 7& 6 – 7
Time (days):	5	4	6	2	1	7	8	4	3

[*Answer*: Critical path: 1 – 2 – 4 – 6 –7 – 8

Event :	1	2	3	4	5	6	7	8
TE_i :	0	5	4	11	12	18	22	25
EL_i :	0	5	9	11	14	18	22	25.]

37. A project schedule has the following characteristics

Activity :	1 – 2	1 – 3	2 – 4	3 – 4	3 – 5	4 – 9	5 – 6
Time (days) :	4	1	1	1	6	5	4
Activity :	5 – 7	6 – 8	7 – 8	8 – 10	9 – 10		
Time (days) :	8	1	2	5	7		

Construct PERT network and find the critical path.

[*Answer*: Critical path is 1 – 3 – 5 – 7 – 8 – 10 and the project duration = 22 days.]

38. Draw the network and determine the critical path for the given data:

Jobs :	1 – 2	1 – 3	2 – 4	3 – 4	3 – 5	4 – 5	4 – 6	5 – 6
Duration:	6	5	10	3	4	6	2	9

Find the total, free and independent floats of each activity.

[*Answer*: Critical path is 1 – 2 – 4 – 5 –6; project duration is 31 days.

Activity :	1 – 2	1 – 3	2 – 4	3 – 4	3 – 5	4 – 5	4 – 6	5 – 6
Total Float :	0	8	0	8	13	0	13	0
Free Float :	0	0	0	8	5	0	13	0
Independent Float :	0	0	0	0	0	0	13	0.]

39. What are the different types of floats associated with an activity in a CPM model? What are their uses?

40. The following refers to a project network:

Activity :	A	B	C	D	E	F	G	H	I	J
Predecessor :	–	A	A	A	A	E	D	G, F	C, H	B
Duration in Days :	1	4	2	3	2	3	2	1	3	2
Crew Required per Day :	7	1	5	4	3	6	2	9	10	8

(i) Draw the network.

(ii) Determine the critical path and project completion time

(iii) Draw the graph showing optimum requirements versus time.

[*Answer*: The critical paths are A – D – G – H – I and A – E – F – H – I and the project duration is 10 days.]

41. A maintenance activity consists of the following jobs. Draw the network for the project and calculate the total float and free float for each activity.

Jobs	:	1 –2	2 – 3	3 – 4	3 – 7	4 – 5	4 – 7	5 – 6	6 – 7
Duration (days)	:	3	4	4	4	2	2	3	2

[*Answer*:

Activity	:	1 – 2	2 – 3	3 – 4	3 – 7	4 – 5	4 – 7	5 – 6	6 – 7
Total Float	:	0	0	0	7	0	5	0	0
Free Float	:	0	0	0	7	0	5	0	0.]

42. A network consists of 9 cities A, B, C, ... to I, as nodes. The arc length data given below represents the amount of cable to be laid for establishing telephone links between the nodes. It is required to lay cable lines in such a manner that any node can connect to any other node (if necessary through another node) such that the total cable laid is minimum. Determine between what nodes the cable lines should be laid:

AB = 30,	BC = 15,	CD = 30,
AC = 45,	BE = 22,	CF = 23,
AD = 60,	BF = 22,	DF = 15,
DG = 38,	EF = 7,	EH = 15,
FH = 23,	EI = 60,	GI = 15
HI = 45,	GH = 15,	

[*Answer*: Cable lines must be laid by connecting the cities in the following order :
A – B – C – D – F – I – H;
A – C – D – F – I – H.]

43. Draw the network diagram for the following activities and find the critical path and total float of each activity.

Job	:	A	B	C	D	E	F	G	H	I	J	K
Immediate Predecessors	:	–	A	B	C	B	E	D, F	E	H	G, I	J
Job time (days)	:	13	8	10	9	11	10	8	6	7	14	18

[*Answer*: Critical path is 1 – 2 – 3 – 5 – 6 – 8 – 9 – 10

Activity	:	A	B	C	D	E	F	G	H	I	J	K
Total Float	:	0	0	0	0	5	0	0	0	2	0	0.]

44. Define an activity, event and dummy constraints in a PERT network. Using the following information plot a network, determine the critical path and compute slack for all events.

Activity	:	0 – 1	0 – 2	1 – 2	1 – 3	1 – 4	2 – 3	2 – 5
Duration (in weeks)	:	5	10	4	8	3	6	8
Activity	:	3 – 5	3 – 6	4 – 6	4 – 7	5 – 7	5 – 6	6 – 7
Duration (in weeks)	:	7	11	8	9	9	4	1

If the duration of activity 5 – 6 is increased to 6 and for activity 3 – 6 is reduced, what will be the new critical path?

Section 12.4

45. A project consists of the following activities and time estimates:

Activity	:	1 – 2	1 – 3	1 – 4	2 – 5	2 – 6	3 – 6	4 – 7	5 – 7	6 – 7
Least Time (days)	:	3	2	6	2	5	3	3	1	2
Greatest Time (days)	:	15	14	30	8	17	15	27	7	8
Most Likely Time (days)	:	6	5	12	5	11	6	9	4	5

(a) Draw the network.
(b) What is the probability that the project will be completed in 27 days.
[*Answer*: (a) Critical path: 1 – 4– 7; project duration is 25 days; (b) Required Probability = 0.6368.]

46. The following indicates [illegible] details of a project. The durations are in days; 'a' refers to optimistic time, 'm' refers to most [illegible] time and 'b' refers to pessimistic time duration.

Activity	:	1 – [illegible]	[illegible] – 3	1 – 4	2 – 4	2 – 5	3 – 4	4 – 5
a	:	2	[illegible]	4	8	6	2	2
m	:	4	4	5	9	8	3	5
b	:	5	6	6	11	12	4	7

(a) Draw the network.
(b) Find the critical path.
(c) Determine the expected standard deviation of the completion time.
[*Answer*: The critical path is 1 – 2 – 4 – 5. The expected standard deviation of the completion time = 1.09 (nearly).]

47. (a) One of the activities in a PERT project has an expected duration of 12 weeks with a standard deviation of 2 weeks. The most likely time estimate of this activity is 12 weeks. Calculate the optimistic and pessimistic time estimates for this capacity.
(b) Consider the following project.

Activity	:	A	B	C	D	E	F	G
t_o (week)	:	3	2	2	2	1	4	1
t_m (week)	:	6	5	4	3	3	6	5
t_p (week)	:	9	8	6	10	11	8	15
Predecessors	:	–	–	A	B	B	C, D	E

Find the path and a standard deviation. What is the probability that the project will be completed by 18 weeks?
[*Answer*: (a) $t_p = 18$ weeks, $t_o = 6$ weeks
(b) Critical path : A $\rightarrow$ C $\rightarrow$ F, $\sigma = 1.37$
Prob. (completing in 18 weeks) = 0.928.]

48. A civil engineering firm has to bid for the construction of a dam. The activities and time estimates are given below:

Activity	*Optimistic time*	*Most likely time*	*Pessimistic time*
1 – 2	14	17	21
2 – 3	14	18	21
2 – 4	13	15	18
2 – 8	16	19	28
3 – 4 (dummy)			
3 – 5	15	18	27
4 – 6	13	17	21
5 – 7 (dummy)			
5 – 9	14	18	20
6 – 7 (dummy)			
6 – 8 (dummy)			
7 – 9	16	20	41
8 – 9	14	16	22

The policy of the firm with respect to submitting bids is to bid the minimum amount that will provide a 95% probability of at best breaking even. The fixed costs for the project are Rs 8 lakhs and the variable costs are Rs 9,000 every day spent working on the project. The duration is in days.

What amount should the firm bid under this policy? (You may perform the calculations on duration, etc. up to two decimal places.}

[*Answer*: Critical path is 1 – 2 – 3 – 5 – 7 – 9 and the project duration = 77.49 days.

Amount of bid = Rs 8 lakh + Rs 9,000 × 86
= Rs 15,74,000.]

49. A construction company is preparing a PERT network for laying the foundation of a new art museum. Given the following set of activities, their predecessor requirements and three time estimates of completion time.

Activity	*Predecessors*	*Optimistic time (weeks)*	*Pessimistic time (weeks)*	*Most likely time (weeks)*
A	–	2	4	3
B	–	8	8	8
C	A	7	11	9
D	B	6	6	6
E	C	9	11	10
F	C	10	18	14
G	C, D	11	11	11
H	F, G	6	14	10
I	E	4	6	5
J	I	3	5	4
K	H	1	1	1

(i) Draw the PERT network.
(ii) Compute the slack for each activity and determine the critical path.

(iii) The contract specifies a Rs 5,000 per week penalty for each week the completion of the project extends beyond 37 weeks. What is the probability that this company will have to pay a maximum penalty of Rs 15,000?

[*Answer*:

(ii)

Activity :	A	B	C	D	E	F	G	H	I	J	K
Slack :	0	1	0	1	6	0	1	0	6	6	0

(iii) 93% probability of paying a maximum penalty of Rs 15,000.]

50. Assuming that the expected times are normally distributed, find the probability of meeting the schedule date as given for the network.

Activity $i-j$:	1 – 2	1 – 3	2 – 4	3 – 4	4 – 5	3 – 5
Optimistic Time (in days) :	2	9	5	2	6	8
Most Likely Time (in days) :	5	12	14	5	6	17
Pessimistic Time (in days) :	14	15	17	8	12	20

Scheduled project completion date is 30 days. Also, find the date on which the project manager can complete the project with a probability of 0.90.

[*Answer*: Required probability = 0.71 or 0.81.]

51. In the network shown in Fig. 12.56 the three time estimates for the activities are indicated. Number the events according to Fulkerson's rule and calculate the variance and expected time for each activity.

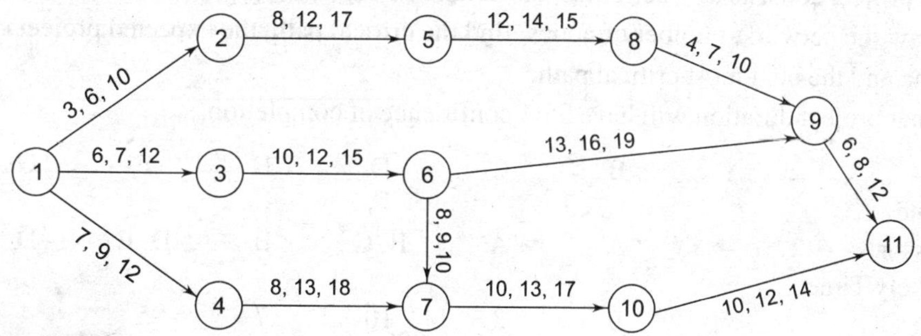

figure 12.56

52. A project has the following characteristics

Activity :	1 – 2	2 – 3	2 – 4	3 – 5	4 – 5	4 – 6	5 – 7	6 – 7	7 – 8	7 – 9	8 – 10	9–10
Most Optimistic Time (a) :	1	1	1	3	2	3	4	6	2	5	1	3
Most Pessimistic Time (b) :	5	3	5	5	4	7	6	8	6	8	3	7
Most Likely Time (m) :	1.5	2	3	4	3	5	5	7	4	6	2	5

Construct a PERT network. Find the critical path and variance for each event. Find the project duration at 95% probability.

[*Answer*: Critical path is 1 – 2 – 4 – 6 – 7 – 9 – 10.]

53. For the project

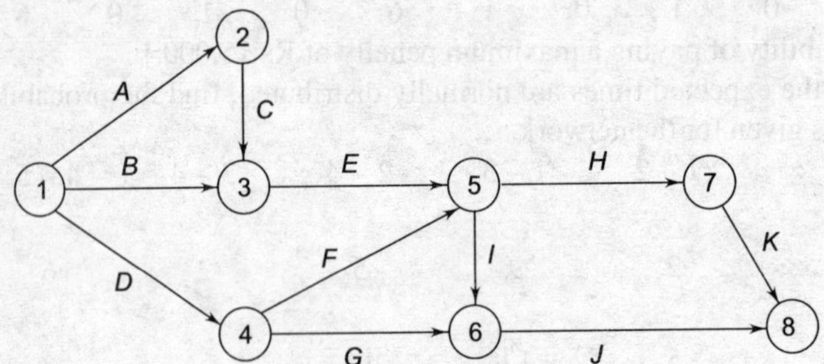

figure 12.57

Task	:	A	B	C	D	E	F	G	H	I	J	K
Least Time	:	4	5	8	2	4	6	8	5	3	5	6
Greatest Time	:	8	10	12	7	10	15	16	9	7	11	13
Most Likely Time	:	5	7	11	3	7	9	12	6	5	8	3

Find the earliest and latest expected times to each event and also the critical path of the network.

[*Answer*: Critical path is A → C → E → H → K.]

54. A small project consists of 7 activities, the details of which are given below.
    (i) Draw the network, number of nodes, find the critical path, the expected project completion time and the next most critical path.
    (ii) What project duration will have 95% confidence of completion.

Activity	:	A	B	C	D	E	F	G
Immediate Predecessor	:	–	A	A	B, C	B	D, E	D
Most Likely Time (in days)	:	3	6	3	10	7	5	4
Optimistic Time (in days)	:	1	2	3	4	3	2	4
Pessimistic Time (in days)	:	7	14	3	22	15	14	4

[*Answer*: Critical path A – B – D – F. Expected project duration is 27 days. Next critical path A – B – D – G with a duration of 25 days. Project duration which has 95% probability of completion if

$$Z = (1.64) = \frac{T_1 - 27}{\sqrt{18}} \text{ or } T_1 = 1.64 \times \sqrt{18} + 27 = 34 \text{ days.}]$$

55. A sociologist plans a questionnaire survey consisting of the following tasks:

Activity	*Description*	*Immediate predecessor*	*Duration in days*		
			Likely	*Minimum*	*Maximum*
A	Design of questionnaire	–	5	4	6
B	Sampling design	–	12	8	16
C	Testing of questionnaire and refinements	A	5	4	12
D	Recruiting interviewers	B	3	1	5
E	Training of interviewers	D, A	2	2	2
F	Allocation areas to interviewers	B	5	4	6
G	Conducting interviews	C, E, F	14	10	18
H	Evaluation of results	G	20	18	34

(a) For this PERT network find the expected task duration and the variance of task durations.
(b) Draw a network for this project and find the critical path. What is the expected length of the critical path? What is the variance of the length of the critical path?
(c) What is the probability that the length of the critical path does not exceed 60 days?

56. A project has the following activities and other characteristics:

Activity	*Preceding activity*	*Time estimates (in weeks)*		
		Optimistic	*Most likely*	*Pessimistic*
A	–	4	7	16
B	–	1	5	15
C	A	6	12	30
D	A	2	5	8
E	C	5	11	17
F	D	3	6	15
G	B	3	9	27
H	E, F	1	4	7
I	G	4	19	28

(i) Draw the PERT network diagram.
(ii) Identify the critical path.
(iii) Prepare the activity schedule for the project.
(iv) Determine the mean project completion time.
(v) Find the probability that the project is completed in 36 weeks.
(vi) If the project manager wishes to be 99 % sure that the project is completed on 30 May 2004, when should he start the project work?

57. The owner of a fastfood restaurant chain is considering a new computer system for accounting and inventory control. A computer company has sent the following information about computer system installation:

Activity	Activity Description	Immediate Predecessor	Time (Days) Optimistic	Most likely	Pessimistic
A	Select the computer model	–	4	6	8
B	Design I/O system	A	5	7	15
C	Design monitoring system	A	4	8	12
D	Assemble computer hardware	B	15	20	25
E	Develop main programmes	B	10	18	26
F	Develop input/output routines	C	8	9	16
G	Create data base	E	4	8	12
H	Instal system	D, F	1	2	3
I	Test and implement	G, J	6	7	8

(i) Construct an arrow diagram for this problem.
(ii) Determine the critical path and the expected completion time.
(iii) Determine the probability of completing the project in 55 days.

58. Consider a project having the following activities and time estimates:

Activity	Immediate predecessor	Expected time (in days) Most optimistic	Most likely	Most pessimistic
A	–	3	4	5
B	–	4	8	10
C	B	5	6	8
D	A, C	9	15	10
E	B	4	6	8
F	D, E	3	4	5
G	D, E	5	6	8
H	D, E	1	3	4
I	G	2	4	5
J	F, I	7	8	10
K	G	4	5	6
L	H	8	9	13
M	J, K, L	6	7	8

(a) Draw an arrow diagram for the project.
(b) Compute the expected project completion time.
(c) What should be the due date to have 0.90 probability of completion?
(d) Find the total float and free float (both early and late) for all non-critical activities.

59. The following Table lists the jobs of a network along with their time estimates.

Jobs	:	1 – 2	1 – 3	2 – 4	3 – 4	4 – 5	3 – 5
Optimistic time	:	2	9	5	2	6	8
Most likely time	:	5	12	14	5	6	17
Pessimistic time	:	14	15	17	8	12	20

(a) Draw the network.
(b) Calculate the length and variance of critical path.
(c) Find the probability that the project will be completed within 30 days.

60. The following Table lists the jobs of a network with three time estimates:

Job	*Duration*		
	Optimistic	*Most likely*	*Pessimistic*
i, j			
1 – 2	3	6	15
1 – 6	2	5	14
2 – 3	6	12	30
2 – 4	2	5	8
3 – 5	5	11	17
4 – 5	3	6	15
6 – 7	3	9	27
5 – 8	1	4	7
7 – 8	4	10	28

(i) Draw the project network.

(ii) What is the approximate probability that the jobs on the critical path will be completed by of 42 days?

61. The data for a small PERT project is as given below where *a* represents optimistic time, *m* most likely time and *b* pessimistic time estimates of the activities A, B, ..., J, K.

	A	B	C	D	E	F	G	H	I	J	K
a :	3	2	6	2	5	3	3	1	4	1	2
m :	6	5	12	5	11	6	9	4	19	2	4
b :	15	14	30	8	17	15	27	7	28	9	12

Precedence relationships: A, B, C can start immediately. A < D, I; B < G, F; D < G, F; C < E; E < H, K; F < H, K; G, H < J.

Draw the arrow network of the project. Calculate earliest and latest expected time to reach each node, and find the critical path. What is the probability that the project will be completed 2 days later than expected?

62. The three estimates for the activities of a project are given below:

Activity	*Estimated duration (days)*		
	a	*m*	*b*
1 – 2	5	6	7
1 – 3	1	1	7
1 – 4	2	4	12
2 – 5	3	6	15
3 – 5	1	1	1
4 – 6	2	2	8
5 – 6	1	4	7

Draw the project network. Calculate the total slack and free slack of each activity. Find out the critical path of the project and project duration. What is the probability that the project will be completes at least 5 days earlier than expected? What is the probability of completing the project on or before 22 days?

63. Consider a project consisting of 7 jobs A, B, ..., G with the following precedence relations and time estimates.

Job	*Predecessor*	*a*	*m*	*b*
A	–	2	5	8
B	A	6	9	12
C	A	5	14	17
D	A	5	8	11
E	C, D	3	6	9
F	–	3	12	21
G	E, F	1	4	7

Draw the project network and find the probability of completing the project in 30 days.

Section 12.5

64. The following Table gives the activities in a construction project and other relevant information.

Activity	*Description*	*Immediate Predecessor*	*Normal Time*	*Normal cost*	*Crash Time*	*Crash cost*	*Cost Slope*	*Cost Rank*
A	Dismantle and inspect	–	2	100	1	140	40	II
B	Sub-contract repairs	A	6	200	2	500	75	IV
C	Buy spares	A	3	100	1	140	20	I
D	Rebuild/repair	A	4	500	2	800	150	V
E	Assemble	B, C, D	2	200	1	250	50	III
F	Trial run	E	1	50	Not possible	250	–	–
	Total			1,150				

Fixed-cost data for the different project durations are given below:

Duration (in days) :	11	10	9	8	7	6	5	4
Cost (in Rs) :	1,100	1,000	900	800	700	600	500	400

(a) Draw a PERT network for the project.

(b) Determine the project duration which will result in minimum total project cost.

65. The following is a Table showing details of a project:

Activity	*Immediate Predecessor*	*Normal Time (weeks)*	*Normal Cost (Rs'000)*	*Crash Time (weeks)*	*Crash Cost (Rs '000)*
A	–	10	20	7	30
B	–	8	15	6	20
C	B	5	8	4	14
D	B	6	11	4	15
E	B	8	9	5	15
F	E	5	5	4	8
G	A, D, C	12	4	8	4

Indirect cost is Rs 400 per day. Find the optimum duration and the associated minimum project cost.

[*Answer*: The optimum duration is 19 weeks and the associated minimum cost is Rs 1,32,200.]

66. The activities of a project are tabulated below with immediate predecessors and normal crash time cost.

Activity	*Immediate predecessor*	*Normal*		*Crash*	
		Cost (Rs)	*Time (days)*	*Cost (Rs)*	*Time (days)*
A	–	200	3	400	2
B	–	250	8	700	5
C	–	320	5	380	4
D	A	410	0	800	4
E	C	600	2	670	1
F	B, E	400	6	950	1
H	B, E	550	12	1,000	6
G	D	300	11	400	9

(a) Draw the network corresponding to the normal time.

(b) Determine the critical path and the normal duration and cost of the project.

(c) Suitably crash the activities so that the normal duration may be reduced by 3 days at minimum cost. Also, find the project cost for this shortened duration if the indirect cost per day is Rs 25.

67. The time and cost estimates and precedence relationship of the difference activities constituting a project are given below:

Activity	*Predecessor activities*	*Time (in weeks)*		*Cost (in rupees)*	
		Normal	*Crash*	*Normal*	*Crash*
A	None	3	2	8,000	9,000
B	None	8	6	600	1,000
C	B	6	4	10,000	12,000
D	B	5	2	4,000	10,000
E	A	13	10	3,000	9,000
F	A	4	4	15,000	15,000
G	F	2	1	1,200	1,400
H	C, E, G	6	4	3,500	4,500
I	F	2	1	7,000	8,000

(a) Draw a project network diagram and find the critical path.

(b) If a dead line of 17 weeks is imposed for completion of the project, what activities will be crashed, what would be the additional cost and what would be critical activities of the crashed network after crashing?

68. The Table below shows jobs, their normal time and cost and crash time and cost for a project.

Job	*Normal Time (days)*	*Cost (Rs)*	*Crash Time (days)*	*Crash Cost (Rs)*
1 – 2	6	1,400	4	1,900
1 – 3	8	2,000	5	2,800
2 – 3	4	1,100	2	1,500
2 – 4	3	800	2	1,400
3 – 4	Dummy	–	–	–
3 – 5	6	900	3	1,600
4 – 6	10	2,500	6	3,500
5 – 6	3	500	2	800

Indirect cost for the project is Rs 300 per day.

(i) Draw the network for the project?
(ii) What is the normal duration cost of the project?
(iii) If all activities are crashed, what will be the project duration and corresponding cost?
(iv) Find the optimum duration and minimum project cost.

[*Answer*: The normal duration is 20 days. Minimum total cost is Rs 5,350 and the optimum duration of the project is 17 days.]

69. The following Table gives the activities in a construction project and other relevant information.

Activity i – j	*Preceding activity*	*Normal time (days)*	*Crash time (days)*	*Normal cost (Rs)*	*Crash cost (Rs)*
1 – 2	–	20	17	600	720
1 – 3	–	25	25	200	200
2 – 3	1 – 2	10	8	300	440
2 – 4	1 – 2	12	6	400	700
3 – 4	1 – 3, 2 – 3	5	2	300	420
4 – 5	2 – 4, 3 – 4	10	5	300	600

(i) Draw the activity network of the project.
(ii) Find the total float and free float of each activity.
(iii) Using the above information, crash or shorten the activity step by step until the shortest duration is reached.

[*Answer*: : Normal duration of the project = 45 days

Activity :	1 –2	1 – 3	2 – 3	2 – 4	3 – 4	4 – 5
Total Float :	40	0	70	50	40	60

Minimum duration of the project is 42 days with a cost of Rs 2,220.]

70. In the project network shown in Fig. 12.61 the nodes are denoted by numbers and the activities by letters. The normal and crash durations of the various activities along with the associated costs are shown below:

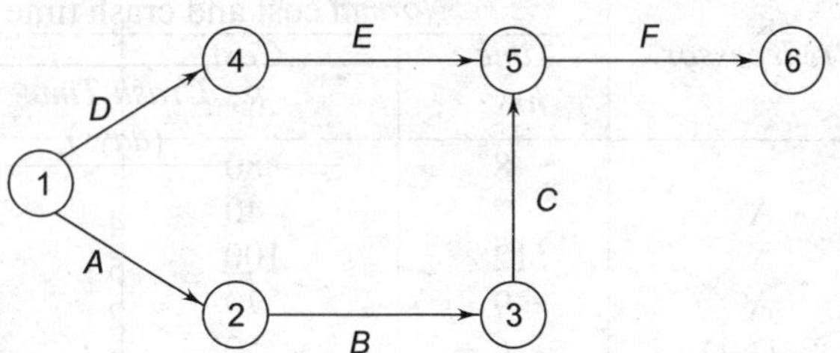

figure 12.58

Activity	Normal duration (days)	Normal cost (Rs)	Crash duration (days)	Crash cost (Rs)
A	8	1,800	6	2,200
B	16	1,500	11	2,200
C	14	1,800	9	2,400
D	12	2,400	9	3,000
E	25	800	14	2,000
F	10	2,000	8	4,000

Determine the least cost 36 days schedule.
[*Answer*: The least cost for 36 days is Rs 12,200.]

71. The following time-cost table (time in days, cost in rupees) applies to a project. Use it to arrive at the network associated with completing the project in minimum time at minimum cost.

Activity	Normal		Crash	
	Time	Cost	Time	Cost
1 – 2	2	800	1	1,400
1 – 3	5	1,000	2	2,000
1 – 4	5	1,000	3	1,800
2 – 4	1	500	1	500
2 – 5	5	1,500	3	2,100
3 – 4	4	2,000	3	3,000
3 – 5	6	1,200	4	1,600
4 – 5	3	900	2	1,600

[*Answer*: The minimum duration is 7 days with associated cost as Rs 11,800.]

72. A maintenance foreman has given the following estimate of time and cost of jobs in a maintenance project:

Job	Predecessor	Normal		Crash	
		Time hrs.	Cost Rs	Time hrs.	Cost Rs
A	–	8	80	6	100
B	A	7	40	4	94
C	A	12	100	5	184
D	A	9	70	5	102
E	B, C, D	6	50	6	50

Overhead cost is Rs 25 per hour. Find

(a) the normal duration of the project and associated cost.

(b) the minimum (optimum) duration of the project and associated cost.

(c) the least duration of the project and its cost.

(d) if all the activities are crashed what will be the project duration and the corresponding cost.

73. The following Table gives data on normal time-cost and crash time cost of a project.

Activity	Normal		Crash	
	Time (days)	Cost (Rs)	Time (days)	Cost (Rs)
1 – 2	6	600	4	1,000
1 – 3	4	600	2	2,000
2 – 4	5	500	3	1,500
2 – 5	3	450	1	650
3 – 4	6	900	4	2,000
4 – 6	8	800	4	3,000
5 – 6	4	400	2	1,000
6 – 7	3	450	2	800

The indirect cost per day is Rs 100.

(a) Draw the network and identify the critical path.

(b) What is the normal project duration and the associated cost?

(c) Crash the relevant activities systematically and determine the optimum project completion time and cost.

[*Answer*: Critical path is 1 – 2 – 4 – 6 – 7 and the normal project duration is 22 days with an associated cost of Rs 6,900. The optimum project duration is 13 days with an associated cost of Rs 11,750.]

74. A small project has 7 activities. The relevant data about these activities is given below:

Activity	Dependence	Normal duration (days)	Crash duration (days)	Normal cost (Rs)	Crash cost (Rs)
A	–	7	5	500	900
B	A	4	2	400	600
C	A	5	5	500	500
D	A	6	4	800	1,000
E	B, C	7	4	700	1,000
F	C, D	5	2	800	1,400
G	E, F	6	4	800	1,600

(a) Find out the normal duration and the minimum duration?

(b) What is the percentage increase in cost to complete the project in 21 days?

[*Answer*: The normal duration to complete the project is 25 days. The minimum duration is 21 days and the associated cost is Rs 5,100. The percentage increase in cost to complete the project in 21 days is 13.3%.]

75. The Table below provides cost and gives estimates of 7 activities of a project:

Activity	*Time estimates (weeks)*		*Direct cost estimates (Rs)*	
(i – j)	*Normal*	*Crash*	*Normal*	*Crash*
1 – 2	2	1	10	15
1 – 3	8	5	15	21
2 – 4	4	3	20	24
3 – 4	1	1	7	7
3 – 5	2	1	8	15
4 – 6	5	3	10	16
5 – 6	6	2	12	36

(a) Draw the project network corresponding to normal time.

(b) Determine the critical path and the normal durations and normal cost of the project.

(c) Crash the activities so that the project completion time reduces to 9 weeks, with minimum additional cost.

[*Answer*: Project duration = 16 weeks and the associated normal cost = Rs 82,000. After crashing the project duration reduces to quiet by incurring a total additional costs of Rs 36,000.]

76. The following Table gives data on normal and crashed time and cost for a project.

Activity	*Time (in weeks)*		*Cost (Rs)*	
	Normal	*Crash*	*Normal*	*Crash*
1 – 2	9	4	1,300	2,400
1 – 3	15	13	1,000	1,380
2 – 3	7	4	7,000	1,540
2 – 4	7	3	1,200	1,920
2 – 5	12	6	1,700	2,240
3 – 6	12	11	600	700
4 – 5	6	2	1,000	1,600
5 – 6	9	6	900	1,200

Find the optimum project time and corresponding minimum total project cost by crashing appropriate activities in proper order. Show the network on time-scale at each step. The indirect cost per day is Rs 400.

[*Answer*: Optimum project duration = 27 days, minimum total cost = Rs 19,750.]

77. Find the optimum schedule for the given project, with overhead cost of Rs 75 per day.

Activity	*Immediate predecessors*	*Duration (Days)* Normal time	*Crash time*	*Increase in cost (Rs) for crashing by one day*
A	–	3	2	150
B	–	4	3	100
C	A	5	4	200
D	A	7	5	300
E	B, C	3	3	0
F	B, C, D	6	2	75

(i) Draw the project network using normal duration.
(ii) Find the path and the project duration for case (i).
(iii) Find the optimal schedule and optimal project duration.
[*Answer*: Optimum project duration: 12 days with min. cost = Rs 1,760.]

78. The required data for a small project consisting of different activities are given below:

Activity	*Dependence*	*Normal duration (days)*	*Normal cost (Rs)*	*Crash duration (days)*	*Crash cost (Rs)*
A	–	6	300	5	300
B	–	8	400	6	600
C	A	7	400	5	600
D	B	12	1,000	4	1,400
E	C	8	800	8	800
F	D	7	400	6	500
G	D, F	5	1,000	3	1,400
H	F	8	500	5	700

(a) Draw the network and find out the normal project length and minimum project length.
(b) If the project is to be completed in 21 days with minimum crash cost, which activities should be crashed by how many days?

[*Answer*: Project duration = 26 days and the minimum project length = 21 days. Crash activities: A by 1 day, G by 2 days, H by 2 days, C by 2 days and D by 2 days; total additional cost for project duration of 21 days is Rs 1,233.33.]

79. Suggest optimal crashing schedule for the following project:

Activity	*Preceding activity*	*Time (in weeks)* Normal	Crash	*Cost (Rs)* Normal	Crash
A	–	7	4	2,100	3,000
B	–	5	3	1,400	1,800
C	–	8	5	2,700	3,900
D	A	2	1	900	1,000
E	B	3	1.5	1,200	1,500
F	C	3	1.5	1,000	1,300
G	D, E, F	4	2	1,200	1,700

Section 12.6

80. Consider the network shown below:

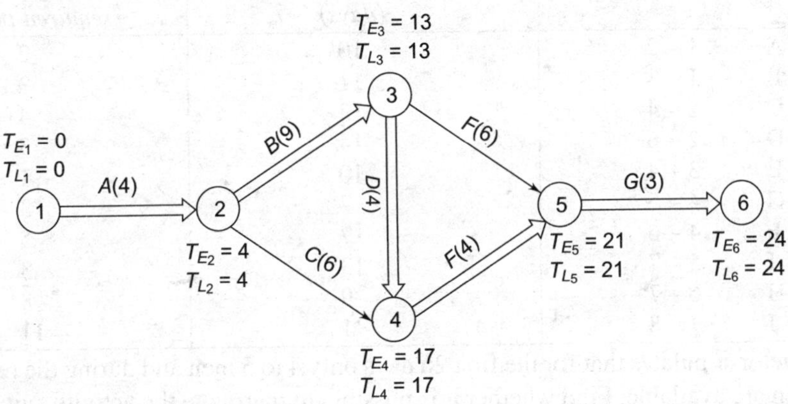

figure 12.59

The critical path is 1 – 2 – 3 – 4 – 5 – 6 and the project duration time is 24 days. At the end of 15th day, a review of the existing conditions is made and the observations are listed below.

(i) Activities 1 – 2, 2 – 3 and 2 – 4 are already complete.
(ii) Activity 3 – 5 is in progress since last 2 days and needs 6 more days for completion.
(iii) Activity 3 – 4 is in progress since last 2 days and needs 2 more days for completion.
(iv) Activity 4 –5 can be completed in 3 days as against originally planned 4 days.
(v) Activity 5 – 6 needs 4 days for completion.

Formulate a new project based on the review at the end of 15 days.

81. A project has the following activities and durations.

Activity :	12	13	14	24	25	34	36	47	57	67	68
Duration :	13	15	9	10	27	7	18	30	12	10	10

(a) Draw the network of the project and find its duration.
(b) At the end of 25 days it is observed that
(i) activities 12, 13, 14 have been completed.
(ii) activitity 24 is being done and will be completed in 5 more days.
(iii) activity 36 is in progress and will need 20 more days for completion.
(iv) activity 67 is presenting some problem and will take 15 days.

Draw the updated network and find out its revised duration. 12 denotes activity 1– 2 and same is true for other activities.

Section 12.7

82. The manpower required for each activity of a project is given in the following Table.

Activity		*Normal time (days)*	*Manpower required per day*
A	1 – 2	10	2
B	1 – 3	11	3
C	2 – 4	13	4
D	2 – 6	14	3
E	3 – 4	10	1
G	4 – 5	7	3
F	4 – 6	17	5
I	5 – 7	13	3
H	6 – 7	9	8
J	7 – 8	1	11

The contractor stipulates that for the first 26 days only 4 to 5 men and during the remaining days 8 to 11 men are available. Find whether it is possible to rearrange the activity suitably for levelling the manpower resources satisfying the above condition.

83. Following are the manpower requirements for each activity in a project.

Activity	*Normal time*	*Manpower required*
0 – 1	2	4
1 – 2	3	3
1 – 3	4	3
2 – 4	2	5
3 – 5	4	3
3 – 6	3	4
4 – 7	6	3
5 – 7	6	6
6 – 8	5	2
7 – 9	4	2
8 – 9	4	9

(a) Draw the network diagram of the project activities.

(b) Rearrange the activities suitably for reducing the existing total manpower requirement.

(c) If only 9 men are available for the execution of the project, then rearrange the activities suitably for levelling the manpower resource.

84. Following are the manpower requirements for each activity in a project.

Activity	*Normal time (days)*	*Manpower required per day*
1 – 2	10	2
1 – 3	11	3
2 – 4	13	4
2 – 6	14	3
3 – 4	10	1
4 – 5	7	3
4 – 6	17	3
5 – 7	13	5
6 – 7	9	8
7 – 8	1	11

(i) Draw the network and find out total float and free float for each activity.
(ii) The contractor stipulates that during the first 26 days only 4 to 5 men and during the remaining days 8 to 11 men only can be made available. Rearrange the activities suitably for levelling the manpower resources, satisfying the above condition.

85. The activities, activity durations and manpower requirements of a project are given below:

Activity	*Duration (days)*	*No. of men required*
1 – 2	2	5
1 – 3	2	4
1 – 4	0	0
2 – 5	2	2
2 – 6	5	3
3 – 7	4	6
4 – 8	5	2
5 – 9	6	8
6 – 9	3	7
7 – 8	4	4
8 – 9	6	3

There are eleven persons who can be employed for this project. Carry out appropriate manpower levelling so that the fluctuations of workforce requirement from day to day are as small as possible.

86. The activities comprising a certain project have been identified as follows:

Activity	*Preceding activity*	*Duration (weeks)*	*No. of men required*
A	–	4	1
B	–	7	1
C	–	8	2
D	A	5	3
E	C	4	1
F	B, E	4	2
G	C	11	2
H	G, F	4	1

(a) For the above project, draw the network. Determine the critical path and its duration.
(b) If there were only three men available at any one time, how long would the project take and how would you allocate the men to the activities?
(c) If there were no restrictions on the amount of labour available explain how you might schedule the activities.

87. A project consists of 10 activities, each of which requires either, or both, of the two types of resources R_1 and R_2 for its performance. The duration of the activities and their resource requirements are as follows:

Activity	*Duration (days)*	*Resource requirements*	
		R_1	R_2
1 – 2	3	3	2
1 – 3	2	6	–
1 – 4	6	4	–
2 – 6	4	–	4
3 – 5	2	2	2
4 – 5	1	4	–
4 – 8	4	4	–
5 – 7	3	3	2
6 – 7	2	1	3
7 – 8	4	4	5

Resource availability: 7 units of R_1 and 5 units of R_2. Determine the duration of the project under the given resource constraints. If the resources were not a problem, how long would the project take to complete in the normal course?

88. A project consists of 9 activities. The Table below shows the duration for each activity and the corresponding labour requirements. Draw a network and establish the critical path. Also, draw a square network assuming that all the activities begin at the earliest start time. Adjust the project work such that there will be smooth demand for the resources.

Activity	*Duration*	*No. of labourers*
1 – 2	3	5
2 – 4	2	3
2 – 3	3	7
3 – 4	0	0
3 – 5	3	2
4 – 5	7	2
3 – 6	2	1
5 – 6	6	6
4 – 6	3	5

89. The following data is given for a small project:

Activity	*Preceding activity*	*Duration (days)*	*Labour requirement time*
A	–	1	2
B	–	2	1
C	A	1	1
D	–	5	1
E	B	1	1
F	C	1	1

(a) Draw the PERT chart.

(b) Indicate the critical path without taking account of resource constraint.

(c) Give the bar diagram on time-scaled network.

(d) From the time-scaled chart given in (c) prepare a resource aggregation profile.

(e) Now, prepare a resource allocation profile with 2 men constraint and another diagram for 5 days constraint.

Section 12.8

90. A toy manufacturer intends to produce 10,000 pieces of a particular toy per year. He has identified 10 work elements with the following precedence restrictions and durations:

Elements	:	1	2	3	4	5	6	7	8	9	10
Immediate Predecessors	:	–	1	1	2, 3	4	4	6	5	7, 8	9
Duration (minutes)	:	5	8	3	2	5	5	8	2	3	6

(a) Draw a precedence diagram for the assembly of toys.
(b) Design the assembly line for a cycle time of 8 minutes.
(c) Calculate the line efficiency, balance delay and smoothness index for assembly line under (b).

13

Simulation

13.1 INTRODUCTION

There are many problems of real life which cannot be represented mathematically due to the stochastic nature of the problem, the complexity in problem formulation, or the conflicting ideas needed to properly describe the problem under study. In such situations simulations can be used.

Simulation analysis is a natural and logical extension of the analytical and mathematical techniques used for solving problems in Operations Research. Simulation, which can be called management laboratory, determines the effect of alternate policies without disturbing the real system. Recent advances in simulation methodologies, software availability, and technical developments have made this technique one of the most widely used and popularly accepted tools in Operations Research and System Analysis. Simulation helps us in deciding the best policy with the prior assurance that its implementation will certainly prove to be beneficial to the organisation.

The technique of simulation has long been used by designers and analysts in physical sciences and it promises to become an important tool for tackling the complicated problems of managerial decision making. Scale models of machines have been tested in wind tunnels to determine their aerodynamic characteristics. The first important application of simulation was made by Jon Von Neumann and Stannislaw Ulam for studying the tedious behaviour of neutrons in a nuclear shielding problem which was too complex for mathematical analysis. With the remarkable success of the technique on the neutron problem, simulation became popular and found many applications in business and industry. Development of the digital computer in the 1950s is further responsible for the rapid progress made by simulation techniques.

Simulation is one of the easiest tools of management science to use, but one of the hardest to apply properly and perhaps the most difficult from which to draw accurate conclusions. Due to widespread applications of digital computers, simulation becomes readily available to most engineers and managers engaged in Operations Research projects. Regardless of these drawbacks, simulation is a useful technique and one which is specially suitable for complicated Operations Research and System Analysis problems.

The purpose of this chapter is to examine the process of simulation and the necessary tools to perform such analysis. Special emphasis is given to the Monte-Carlo method of simulation. Some simple examples are discussed to explain the Monte-Carlo technique. To obtain reliable results, the use of computer is very essential in all simulation techniques; the numerical examples have been solved by hand computation only.

13.1.1 What is Simulation

Simulation is an imitation of reality.

Some examples of simulation:

1. *Children's cycling park* with various crossings and signals is a simulated model of the city traffic system.
2. Testing an aircraft model in a wind tunnel.
3. *Planetarium*: To determine the behaviour of a real system in true environments a number of experiments are performed on simulated models either in the laboratories or in the computer itself.
4. *Training of pilots*: A computer directs a student's handling of the controls in a simulated flight deck. The instruments are then operated by the computer to give the same readings as in a real flight. An instructor can intervene with 'catastrophes' like an engine failure or a storm and a television camera is moved over a model of some countryside to give the trainee visual feedback of how the aircraft is behaving.
5. *TV games*: The combination of computing and simulation has resulted in the production of TV games. Players interrupt the way a computer program moves various images around the screen with a keyboard or hand-held controller. The computer incorporates their responses into these movements in accordance with the rules of the particular game. Incidentally, such programs make extensive use of random numbers to find the deflection of tennis balls, the positioning of hostile space ships, and others.
6. *Chess*: Chess is a non-probabilistic simulation of a fight between black and white armies.
7. *Snakes and ladders game*: The game of snakes and ladders was initially proposed to simulate the moral progress of the players who moved up ladders when they were 'good' and fell down snakes, indicating temptation, when they were bad. Like in many other board games, dice are used as random generators.

In all these examples, reality is imitated to see what might happen under real operating conditions. This imitation of reality, which may be in the physical form or in the form of mathematical equations may be called simulation.

The simple examples cited above are of simulating the reality in physical form, and are referred to as analogue simulation.

13.1.2 Definitions of Simulation

Following are a few definitions of simulation.

1. Simulation is a representation of reality through the use of a model or other device which will react in the same manner as reality under a given set of conditions.
2. Simulation is the use of a system model that has the designed characteristics of reality in order to produce the essence of actual operation.
3. According to Donald G. Malcolm, a simulated model may be defined as one which depicts the working of a large-scale system of men, machines, materials and information operating over a period of time in a simulated environment of the actual real world conditions.
4. According to Naylor *et al.* (1966), simulation is a numerical technique for conducting experiments on a digital computer, which involves certain types of mathematical and logical relationships necessary to describe the behaviour and structure of a complex real world system over extended periods of time.

5. Churchman has defined simulation as follows:
   "X simulates Y" is true if and only if:
   (i) X and Y are formal systems;
   (ii) Y is taken to be the real system;
   (iii) X is taken to be an approximation to the real system; and
   (iv) The rules of validity in X are non-error-free, otherwise X will become the real system.

13.1.3 Types of Simulation

Simulation is mainly of two types.

1. *Analogue (environmental) Simulation*: The simple examples cited in 13.1.1 are of simulating the reality in physical form, which we refer as analogue (or environmental) simulation.
2. *Computer (or system) Simulation*: For complex and intricate problems of managerial decision making, analogue simulation may not be applicable, and actual experimentation with the system may be uneconomical also. Under these situations, the complex system is formulated into a mathematical model for which a computer program is developed and then the problem is solved by using high-speed electronic computer. Such type of simulation is called a computer simulation or system simulation.

Simulation models can be classified into following four categories:

Deterministic models: In these models input and output variables are not permitted to be random and models are described by exact functional relationship.

Stochastic models: In these models at least one of the variables or functional relationships is given by probability functions.

Static models: These models do not take variable time into consideration.

Dynamic models: These models deal with time varying interaction.

13.1.4 Why to Use Simulation

For solving various types of managerial decision-making problems in operations research, the following techniques are adopted:

(a) Scientific method,
(b) Analytical method, and
(c) Interactive method.

But each method has its own drawbacks and limitations.

1. *Drawbacks of Scientific Method*: The scientific method has the following limitations and difficulties:
   (i) It may be either impossible or extremely costly to observe certain processes in real-life situations.
   (ii) The observed system may be so complex that it may be impossible to describe it in terms of a set of mathematical equations.
   (iii) Event though a mathematical model can be formulated to describe the system under study, a straightforward analytical solution may not be available. Such situations may arise in complex queueing problems, job-shop problems, high order difference equations, complicated stochastic models, multi-integral problems, and so on.

(iv) It may be either impossible or very costly to perform validating experiments on mathematical models describing the system.

On account of these drawbacks, the scientific method cannot be used to solve complex managerial decision-making processes.

2. *Drawbacks of Analytical Method*: Analytical techniques used in dynamic programming, queueing theory, network models, and hence forth, are not sufficient to tackle problems requiring data analysis due to the following limitations:

(i) Dynamic programming models can be used to determine optimal strategies taking into account the uncertainties and can analyse multi-period planning problems. But, it has its own shortcomings. Dynamic programming models can only be used to tackle very simple situations involving very few variables. If the number of static variables becomes larger, the computation becomes complex and difficult.

(ii) Similar limitations also hold good for other mathematical models such as inventory and waiting lines. Only small-scale systems are amenable to these models. But by making a number of assumptions the systems are simplified to such an extent that in many cases the results thus obtained are only rough approximations.

3. *Drawbacks of Iterative method*

(i) In linear programming models, we assume that the data does not change over the entire planning horizon. This is a one-time decision process and assumes average values for the decision variables. If the planning horizon is long, say 15 years, the multi-period linear programming model may deal with the yearly averaged data, but will not take into account the variations over the months and weeks. Consequently, months to month and week to week operations are left implict.

(ii) Linear programming assumes that the data should be known with certainty. In many real-life situations, the uncertainties about the data are such that they cannot be ignored. In case the uncertainty relates to only a few variables, sensitivity analysis can be used to determine its effect on the decision. But in situations where uncertainty pervades the entire model, the sensitivity analysis may become too cumbersome and computationally difficult to determine the impact of uncertainty on the recommended plan.

From above mentioned drawbacks, we conclude that whenever characteristics like uncertainty, complexity, dynamic interaction between the decision and subsequent event, and the need to develop detailed procedures and finely divided time intervals are combined together in one situation, then the model becomes too complex to be solved by any of the techniques of mathematical programming and probabilistic models. Such complex models must be analysed by some other kind of quantitative technique which may give accurate and reliable results. Many new techniques have been tested but among all, the best available is simulation. In general, the simulation technique is a dependable tool in situations where mathematical analysis is either too complex or too expensive.

13.1.5 Limitations of Simulation

1. Quantification of the variables may be difficult.
2. Large number of variables makes simulation unwieldy and difficult.
3. Simulation may not yield optimum results.
4. It may not be economic.

5. It may not always be less time consuming.
6. One cannot rely too much on the results obtained from simulation models.

13.1.6 Advantages of Simulation

1. It is less complicated mathematically.
2. It is flexible.
3. It can be modified to suit the changing environments of the situation.
4. It can be used for training purposes.
5. It may be less expensive in several real-world situations.
6. It may be less time consuming.

13.1.7 Phases of Simulation Model

A simulation model consists of two basic phases:

Phase I: Data Generation Data generation involves the sample observation of variables and can be carried out with the help of any of the following methods:

(i) using random numbers,
(ii) resorting to mechanical devices, and
(iii) using electronic computers.

Phase II: Book Keeping The book-keeping phase of a simulation model deals with updating the system when new events occur, monitoring and recording the system states as and when they change, and keeping track of quantities of our interest (such as, idle time and waiting time) to compute the measures of effectiveness.

Steps in Simulation Process

Step 1: Identify the measure of effectiveness.

Step 2: Decide the variables which influence the measure of effectiveness, choose those variables which affect the measure of effectiveness significantly.

Step 3: Determine the probability distribution for each variable in step 2 and construct the cumulative probability distribution.

Step 4: Choose an appropriate set of random numbers.

Step 5: Consider each random number as a decimal value of the game cumulative probability distribution.

Step 6: Use the simulated values so generated into the formula derived from the measure of effectiveness.

Step 7: Repeat (5) and (6) until the sample is large enough to arrive at a satisfactory and reliable decision.

13.2 EVENT TYPE SIMULATION

The event-type simulation can be understood with the following two examples.

Example 1

Consider a situation where customers arrive at a one-man barber shop for cutting their hair. The problem is to analyse the system in order to evaluate the quality of service and the economic feasibility of offering the service. To measure the quality of service, assess the average waiting time per customer and the percentage of time the barber remains idle. Construct the simulation model.

Formulation of the model: To formulate the model of this system we observe that the changes involved in the analysis of the system can occur only when a customer arrives for service or departs after completion of service. When a customer reaches the barber's shop, he will have to wait if the server (barber) is busy. On the other hand, departure of a customer after the completion of his service (hair cut) indicates that the server is available to serve waiting customers, if present. From this, we conclude that only two types of events can occur, that is, an arrival event and a departure event. It shows that as the simulator progresses on the time scale, we must pay due attention to one system whenever an event occurs.

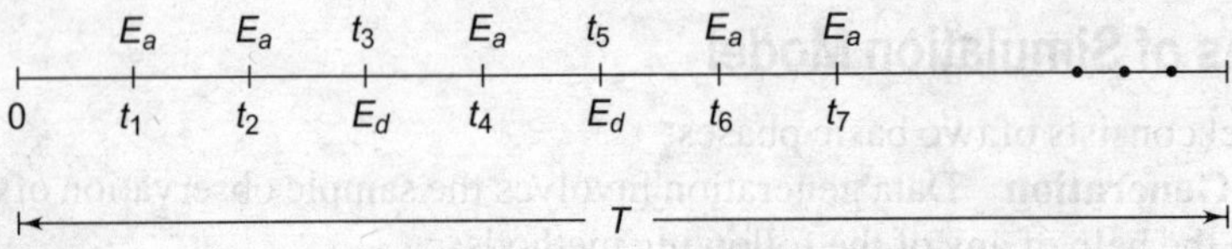

figure 13.1

Let us denote the arrival event by E_a, the departure event by E_d, and the simulated period (time span) by T. The simulator starts at time 0 and first progresses up to $t = t_1$ then $t = t_2$, and so on until the entire simulated period T is covered. Fig. 13.1 shows the occurrences of events E_a and E_d over the time period T, where the simulation starts by generating E_a to t_1. In the beginning, when the service facility is free, the service of the customer will begin immediately. Then, the following two new events must be generated:

(i) the next arrival may occur

(ii) the service of the customer may be completed.

The next arrival can be determined from the inter-arrival time. Thus, E_a is determined at time t_2. The departure time of the customer in service is obtained from service time and from this event E_d is generated at time t_3. Now, both the events E_a (at t_1) and E_d (at t_3) are listed in chronological order, so that the simulator may recognise that the event E_a occurs before E_d. The next event under consideration is E_a at t_2 and at this moment E_a at t is deleted from the stored list (because of past event). The event E_a at t_2 now generates E_a at t_4. Since the service facility is busy, the new arriving customer E_a (at t_2) joins a waiting line. Now, E_a at t_2 is deleted from the list and E_d at t_3 is considered next. At this time a customer is taken from the waiting line and departure event E_d at t_5 is generated. This process is repeated until the entire simulated period T is covered.

Example 2

Customers arrive at a milk booth for the required service. Assume that inter-arrival and service time are constants and given by 1.5 and 4 minutes, respectively. Simulate the system by hand computation for 14 minutes. (i) What is the waiting time per customer? (ii) What is the percentage idle time for the facility? (Assume that the system starts at $t = 0$).

Solution: First customer starts getting the service. So, its departure time becomes $t = 0 + 4 = 4$ minutes. Next event (arrival) occurs at $t = 0 + 15 = 15$ minutes, which is listed before d_1 at $t = 4$. The facility is still busy; the 2nd customer stands in the queue and the 1st one is to be considered in this queue. The 3rd arrival event a_4 (customer) at $t = 3.0 + 1.5 = 4.5$. This event succeeds d_1 at $t = 4$.

At this moment, the first customer departs leaving the servicing facility free. The 2nd customer who was the first to join the queue now gets service. The waiting time is the time period from the moment he joined the queue until his service is started.

This process is repeated until the simulated period is completed. The results of the simulation are given in Table 13.1.

table 13.1

Time	Event arrival/departure	Customer no.	Waiting time
0.0	a_1	1	
1.5	a_2	2	
3.0	a_3	3	
4.0	d_1	1	4 - 1.5 = 2.5 (customer 2)
4.5	d_4	4	
6.0	a_5	5	
7.5	a_6	6	
8.0	d_2	2	8 - 3 = 5 (customer 3)
9.0	d_7	7	
10.5	a_8	8	
12.0	a_9, d_3	9, 3	12 - 4.5 = 7.5 (customer 4)
13.0	a_{10}	10	
14.0	end	—	

For customers who are yet to get service after 14 minutes the waiting times are given in Table 13.2.

table 13.2

Customer no.	Waiting time in minutes
5	14 - 6 = 8
6	14 - 7.5 = 6.5
7	14 - 9.0 = 5.0
8	14 - 10.5 = 3.5
9	14 - 12.0 = 2.0
10	14 - 13.5 = 0.5

From this simulation table it is clear that

(i) Average waiting time for customer

$$= \frac{2.5 + 5 + 7.5 + 8 + 6.5 + 5.0 + 3.5 + 2 + 0.5}{10}$$

$$= \frac{40.5}{10} = 4.05 \text{ minutes.}$$

(ii) Average waiting time per customer for those who must wait $= \frac{40.5}{9} = 4.5$ minutes.

(iii) Percentage idle time of the facility = 0%.
(since service facility is always busy)

13.3 GENERATION OF RANDOM NUMBERS

Random numbers play a vital role in simulation models. In computer simulation, random numbers can be obtained by the following methods:

1. Random numbers may be drawn from the random number table stored in the memory of the computer. This process, however, is neither practicable nor economical. It is a very slow process and the numbers occupy a large portion of the computer memory. Even on methodological grounds it is objectionable to use the same set of random numbers again and again. Therefore, several methods of generating random numbers internally by the computer have been evolved. The necessary formulae occupy relatively little space.
2. An electronic device may be constructed as part of the digital computer to generate true random numbers. The method, however, is expensive.
3. In the mid-square method of generating pseudo-random numbers, a four-digit number is taken. By squaring it a high digit figure is obtained from which the middle four digits are picked. This yields the second random number and the process is repeated and a sequence of pseudo-random numbers is obtained. For instance, if

$$x_1 = 2421$$

then, $$x_1^2 = 05\underline{8612}41$$

$\therefore$ $$x_2 = 8612$$

$$x_2^2 = 74\underline{1665}44$$

$\therefore$ $$x_3 = 1665$$

$$x_3^2 = 02\underline{772225}$$

$\therefore$ $$x_4 = 7722 \text{ and so on.}$$

However, one may come across the following situations:

(i) The series may vanish because a random number obtained is 0000.

(ii) A random number reproduces itself. For example, $x_7 = 8625$, $x_8 = 8625$, $x_q = 8625$.

(iii) A loop occurs, for example, $x_{11} = 6100$, $x_{12} = 2100$, $x_{13} = 4100$, $x_{14} = 8100$, $x_5 = 6100$ and the process continues in a circle.

4. Pseudo-random numbers may also be generated by arithmetic operations. A frequently used method is the congruence method or the residue method, which is described by the expression

$$x_{i+1} = (ar_i + b) \text{ (modulo } m) \tag{13.3.1}$$

where, a, b and m are constants and r_i, r_{i+1} are the ith and $(i + 1)^{\text{th}}$ random numbers. The expression implies multiplying of a by r_i, addition of b and then dividing by m. Then, r_{i+1} is the remainder or residue. To begin the process of random number generation, in addition to a, b and m, the value of r_o is also required. It may be any random number and is called seed.

The congruence random number generator may be of the additive, multiplicative or mixed type. Equation (13.3.1) gives the mixed type.

If $a = 1$, the equation (13.3.1) reduces to the additive type

$$r_{i+1} = (r_i + b) \text{ (modulo } m) \tag{13.3.2}$$

If $b = 0$, the congruence method is called the multiplicative type

$$r_{i+1} = ar_i \text{ (modulo } m) \tag{13.3.3}$$

The multiplicative methods are considered better than the additive methods and as good as the mixed methods.

The selection of the values for the constants a, b and m is very important, because the length of the sequence of random numbers depends on it. It is not possible to generate a non-repeating sequence of numbers with these methods. However, a sufficiently long string of random numbers can be obtained by making a suitable selection of constants. Since the number r_{i+1} can be predicted from r_1 and the whole string of random numbers is reproducible, the numbers obtained are not truely random. They are called pseudo-random numbers and the method is termed as the pseudo-random number generator.

The validation of a pseudo-random number generator is very essential. A sequence of random numbers is considered to be adequately random if its uniformity is assured, and the successive numbers in the sequence are independent.

The Chi-square test of goodness of fit is employed to check if the sequence of numbers is generated from a (0, 1) uniform distribution. The randomness or independence test is used to check that the successive numbers are not correlated. One of the most effective methods for this purpose is the poker test. Most computer systems have a sub-routine available for generating random numbers.

In inventory models, the variables include customer's demand and delivery times, which may also be probabilistic. In this type of situations, for generating random numbers the *manual method* consisting of the following steps can be used.

(i) Collect the data related to the current problem.
(ii) Construct a frequency distribution with this data.
(iii) Construct the relative frequency distribution.
(iv) Assign a coding system that relates the identified events to generate random numbers.
(v) Select a suitable method for obtaining the required random numbers.
(vi) Match the random numbers to the assigned events and tabulate the results.
(vii) Repeat step (vi) until the desired number of simulation runs has been generated.

Example 1

MR Colour Lab uses an expensive grade of developing fluid when printing special colour portraits. Since the developing fluid cannot be stored for long periods, it is important to keep on hand only as much as is needed to fill anticipated demand. In the past few months, however, demand for the product has been fluctuating.

The owner has decided to simulate the demand for this service.

A study of MR Colour Lab appointment book resulted in the following frequency distribution:

Daily demand :	0	1	2	3	4	5
Number of days :	10	20	40	20	6	4

The data was taken for a 100-day period, during which no more than five special prints were requested on any given day. Using the data given above, generate a 10-day sequence of demand values.

Step 1: Collect the data relevant to the current problem.

Step 2: Using the data of step 1, construct the frequency distribution.

Step 3: Construct the corresponding relative frequency distribution.

Daily demand	*Relative frequency*	*Probability*
0	10/100	0.10
1	20/100	0.20
2	40/100	0.40
3	20/100	0.20
4	6/100	0.06
5	4/100	0.04

Step 4: Given the relative frequency distribution of step 3, assign a coding system that relates the identified events to generated random numbers. This coding system is the device whereby, given a random number, a particular event will be specified.

The most practical coding system is one which assigns random numbers in proportion to the probability value. In this case, 10% of the random numbers will be assigned to a daily demand of zero, 20% to a daily demand of one, and 40% to a daily demand of two. In a similar manner, random numbers will be assigned to the remaining daily demand figures.

table 13.3

Daily demand	Probability	Comulative Probability	Random number assignments	Numbers assigned
0	0.10	0.10	00 – 09	10
1	0.20	0.30	10 – 29	20
2	0.40	0.70	30 – 69	40
3	0.20	0.90	70 – 89	20
4	0.06	0.96	90 – 95	6
5	0.04	1.00	96 – 99	4

Step 5: *Select an appropriate method for generating the required random numbers*. For the limited purposes of this example it is feasible to use a manual method such as the spinning arrow. To use the approach, we divide the face of a clock into 100 equal parts and number the parts from 00 to 99, inclusive, centre an arrow on the clock in such a way that it can spin freely. At each spin of the arrow, record the number to which it points when it stops. This is the generated random number. Each spin thus corresponds to the simulation run.

Step 6: Using the method selected in step 5, generate the random numbers to be used in the simulation, match these to the assigned events, and summarise the results in an appropriate table. Since the owner wants a 10-day simulation, it is necessary to spin the arrow 10 times and record the random number generated each time. Once the random numbers have been generated, then with reference to the coded assignment, Table 13.4 gives the value of the generated demand. For example, if 35 is the first number generated, it falls in the range 30 – 69 and thus corresponds to the event of a daily demand for two portraits. Results for the full 10-day simulation are summarised in Table 13.4.

table 13.4

Day number	Generated random demand	Generated demand
1	35	2
2	92	4
3	68	2
4	03	0
5	51	2
6	05	0
7	72	3
8	84	3
9	98	5
10	34	2

Step 7: Repeat step 6 until the required number of simulation runs have been generated. Since the owner was only interested in one simulation run, this step is not required. However, if there were to be more than 10-day simulations, it would be necessary to repeat step 6 as needed.

Remark: The numbers generated by the computer are always predictable and reproducible and hence cannot be treated as random. For this reason they are called as pseudo-random numbers.

Example 2 Congruence Methods

1. Mixed congruence method

$$r_{i+1} = (ar_i + b) \text{ (modulo } m)$$

Take

$$r_0 = 1$$
$$r_1 = (16 \times 1 + 18) \text{ (modulo 23)} = 11$$
$$r_2 = (16 \times 11 + 18) \text{ (modulo 23)} = 10$$
$$r_3 = (16 \times 10 + 18) \text{ (modulo 23)} = 17$$
$$r_4 = (16 \times 17 + 18) \text{ (modulo 23)} = 14$$
$$r_5 = (16 \times 14 + 18) \text{ (modulo 23)} = 12$$
$$r_6 = (16 \times 12 + 18) \text{ (modulo 23)} = 3$$
$$r_7 = (16 \times 3 + 18) \text{ (modulo 23)} = 20$$
$$r_8 = (16 \times 20 + 18) \text{ (modulo 23)} = 16$$
$$r_9 = (16 \times 16 + 18) \text{ (modulo 23)} = 21$$
$$r_{10} = (16 \times 21 + 18) \text{ (modulo 23)} = 9$$
$$r_{11} = (16 \times 9 + 18) \text{ (modulo 23)} = 1$$

The string of random numbers obtained is 1, 11, 10, 17, 14, 12, 3, 20, 16, 21, 9 after which the sequence starts repeating.

2. Multiplicative congruential method

$$r_{i+1} = 16\, r_i \text{ (modulo 23)}$$

Put

$$r_0 = 1$$
$$r_1 = 16 \times 1 \text{ (modulo 23)} = 16$$

$$r_2 = 16 \times 16 \text{ (modulo 23)} = 3$$
$$r_3 = 16 \times 3 \text{ (modulo 23)} = 2$$
$$r_4 = 16 \times 2 \text{ (modulo 23)} = 9$$
$$r_5 = 16 \times 9 \text{ (modulo 23)} = 6$$
$$r_6 = 16 \times 6 \text{ (modulo 23)} = 4$$
$$r_7 = 16 \times 4 \text{ (modulo 23)} = 18$$
$$r_8 = 16 \times 18 \text{ (modulo 23)} = 12$$
$$r_9 = 16 \times 12 \text{ (modulo 23)} = 8$$
$$r_{10} = 16 \times 8 \text{ (modulo 23)} = 13$$
$$r_{11} = 16 \times 13 \text{ (modulo 23)} = 2$$

3. Additive congruential method

$$r_{i+1} = (r_i + 18) \text{ (modulo 23)}$$

put $r_0 = 1$

$$r_1 = (1 + 18) \text{ (modulo 23)} = 19$$
$$r_2 = (19 + 18) \text{ (modulo 23)} = 14$$
$$r_3 = (14 + 18) \text{ (modulo 23)} = 9$$
$$r_4 = (9 + 18) \text{ (modulo 23)} = 4$$
$$r_5 = (4 + 18) \text{ (modulo 23)} = 22$$
$$r_6 = (22 + 18) \text{ (modulo 23)} = 17$$
$$r_7 = (17 + 18) \text{ (modulo 23)} = 12$$
$$r_8 = (12 + 18) \text{ (modulo 23)} = 7$$
$$r_9 = (7 + 18) \text{ (modulo 23)} = 2$$
$$r_{10} = (2 + 18) \text{ (modulo 23)} = 10$$

13.4 MONTE-CARLO METHOD OF SIMULATION

The Monte-Carlo method is a simulation technique in which statistical distribution functions are created by using a series of random numbers. This approach has the ability to develop many months or years of data in a matter of a few minutes on a digital computer. The method is generally used to solve problems which cannot be adequately represented by mathematical models, or where the solution of the model is not possible by analytical method.

Monte-Carlo simulation yields a solution which should be very close to the optimal, but not necessarily the exact solution. However, it should be noted that this technique yields a solution that converges to the optimal or correct solution as the number of simulated trials lead to infinity.

Steps in Monte-Carlo Simulation:

Step 1: *Clearly define the problem*

(i) Identify the objectives of the problem

(ii) Identify the main factors which have the greatest effect on the objectives of the problem.

Step 2: *Construct an appropriate model*

(i) Specify the variables and parameters of the model.

(ii) Formulate the appropriate decision rules, that is, state the conditions under which the experiment is to be performed.

(iii) Identify the type of distribution that will be used—models use either theoretical distributions or empirical distributions to state the patterns and the occurrence associated with the variables.

(iv) Specify the manner in which time will change.

(v) Define the relationship between the variables and parameters.

Step 3: *Prepare the model for experimentation*

(i) Define the starting conditions for the simulation.

(ii) Specify the number of runs of simulation to be made.

Step 4: *Using steps 1 to 3, experiment with the model*

(i) Define a coding system that will correlate the factors defined in step 1 with the random numbers to be generated for the simulation.

(ii) Select a random number generator and create the random numbers to be used in the simulation.

(iii) Associate the generated random numbers with the factors identified in step 1 and coded in step 4 (i).

Step 5: *Summarise and examine the results obtained in step 4.*

Step 6: *Evaluate the results of the simulation.*

Select the best course of action.

Example 1

Using simulation, find the value of π.

Solution: Taking the origin O as the centre, draw an arc AB of unit random cutting the coordinate axes ox, oy at A and B and complete the square $OACB$ as shown in the Fig. 13.2.

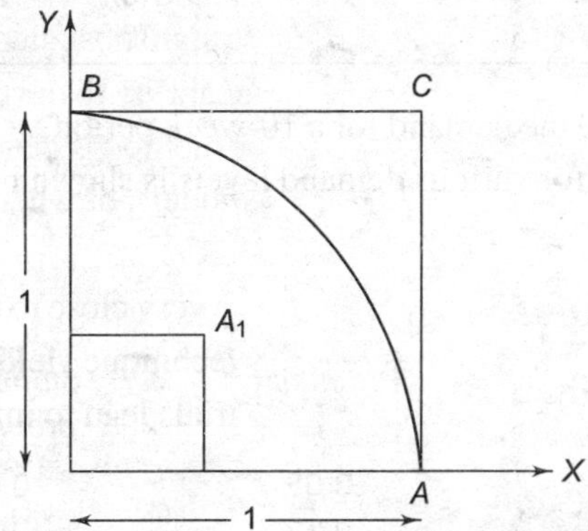

figure 13.2

Equation of arc AB is $x^2 + y^2 = 1, 0 \le x \le 1, 0 \le y \le 1$ and $0 \le x \le 1, 0 \le y \le 1$ is nothing but all the points are inside and on the square $OACB$.

Using the random number table, select any two random numbers, say 0.8262 and 0.9586 and let $x = 0.8262$ and $y = 0.9586$. The point (0.8262, 0.9586) will lie inside or on the arc AB if $x_2 + y_2 \leq 1$ or will be outside the arc and within the square if $x_2 + y_2 > 1$.

In this manner, a large number of pairs of random numbers are selected and whether the points representing the pairs lie in, on the arc or beyond the arc but inside the square is determined.

Let p be the total number of points considered (p pairs of random numbers are selected) and let q be the points which lie inside or on the arc.

Then, obviously,

$$\frac{q}{p} = \frac{\text{area enclosed by the arc}}{\text{area of the square}}$$

$$= \frac{\pi/4\,1^2}{1} = \frac{\pi}{4}$$

Hence, $$p = \frac{4q}{p}$$

Larger the value of p, closer will be the value obtained to p.

Example 2

A tourist car operator finds that during the past few months the car's use has varied so much that the cost of maintaining the car has varied considerably. During the past 200 days the demand for the car fluctuated as below:

Trips per week	*Frequency*
0	16
1	24
2	30
3	60
4	40
5	30

Using random numbers simulate the demand for a 10-week period.

Solution: The tag-numbers alloted for various demand levels is shown in the Table 13.5.

table 13.5

Trip/week (or) Demand/week	*Frequency*	*Probability*	*Cumulative probability*	*Tag numbers*
0	16	0.08	0.08	00 - 07
1	24	0.12	0.20	08 - 19
2	30	0.15	0.35	20 - 34
3	60	0.30	0.65	35 - 64
4	40	0.20	0.85	65 - 84
5	30	0.15	1.00	85 - 90

The simulated demand for the cars for the next 10 weeks is given in Table 13.6.

table 13.6

Week	Random number	Demand
1	82	4
2	95	5
3	18	1
4	96	5
5	20	2
6	84	4
7	56	3
8	11	1
9	52	3
10	03	0

Total demand = 28 cars

Average demand = $\frac{28}{10}$ = 2.8 cars per week.

Example 3

Suppose that the sales of a particular item per day is Poisson with mean 5. Generate 20 days of sales by the Monte-Carlo method.

Solution: The cumulative distribution of sales is calculated on the basis of the information that sales have Poisson distribution with mean $\lambda = 5$.

The probability sales is given by

$$p\,(x = r) = \frac{e^{-\lambda}\,\lambda^r}{r!} = \frac{e^{-5}\,5^r}{r!} \qquad (\because \lambda = 5)$$

table 13.7

r	Cumulative probability	Tag numbers
0	0.01	00 - 00
1	0.04	01 - 03
2	0.13	04 - 12
3	0 13	13 - 26
4	0.27	27 - 43
5	0.44	43 - 61
6	0.62	62 - 75
7	0.76	76 - 86
8	0.87	87 - 92
9	0.93	93 - 96
10	0.97	97 - 97
11	0.99	98 - 98
12	1.00	99 - 99

The simulated sales for the next 20 days is given in Table 13.10.

table 13.8

Day	Random number	Sales	Day	Random number	Sales
1	49	05	11	99	12
2	58	05	12	89	08
3	89	08	13	10	02
4	15	03	14	27	04
5	12	02	15	50	05
6	94	09	16	93	09
7	85	07	17	92	08
8	34	04	18	57	05
9	07	02	19	50	05
10	53	05	20	78	07

13.5 APPLICATIONS TO QUEUEING PROBLEMS

Queueing theory provides techniques for determining measures of effectiveness, such as queue length, average waiting time, and so on, when the distribution of inter-arrival times and service times are known. If costs be assigned to waiting time of customers and idle time of the service facility, the problem of establishing a proper balance between these costs can be determined.

However, many queueing problems cannot be solved explicitly by analytical methods. In such cases, the only possible method of solution is to simulate the experiment. We illustrate the use of simulation in the study of queues by the following example.

Example 1

Records of 100 truck-loads of finished jobs arriving in a department's check-out area show the following: checking out takes 5 minutes and checker takes care of only one truck at a time. The data is summarised in the following table.

Truck inter-arrival time (minute)	:	1	2	3	4	5	6	7	8	9	10
Frequency	:	1	4	7	17	31	23	7	5	3	2

As soon as the trucks are checked out, the truck drivers take them to the next department. Using Monte-Carlo simulation, determine:

(a) What is the average waiting time before service?

(b) What is likely to be the longest wait?

Solution: From the given distribution of truck arriving times, we construct a cumulative probability distribution, as shown in Table 13.11. The table enable us to select the range of random numbers for which we shall choose a prescribed value of time.

table 13.9

Period (length = 1min)	Frequency	Probability	Cumulative probability	Random numbers
1	1	0.01	0.01	0
2	4	0.04	0.05	1 - 4
3	7	0.07	0.12	5 - 11
4	17	0. 17	0.29	12 - 28
5	31	0.31	0.60	29 - 59
6	23	0.23	0.83	60 - 82
7	7	0.07	0.90	83 - 89
8	5	0.05	0.95	90 - 94
9	3	0.03	0.98	95 - 97
10	2	0.02	1.00	98 - 99

A simulation work-sheet can be prepared as in Table 13.12.

table 13.10

S. No.	Random number	Inter-arrival time (min)	Arrival time	Service begins	Service ends	Waiting time for	
						Checker	Truck
1	12	4	10.04	10.04	10.09	4	–
2	81	6	10.10	10.10	10.15	1	–
3	36	5	10.15	10.15	10.20	–	–
4	82	6	10.21	10.21	10.26	1	–
5	21	4	10.25	10.26	10.31	–	1
6	74	6	10.31	10.31	10.36	–	–
7	90	8	10.39	10.39	10.44	3	–
8	55	5	10.44	10.44	10.49	–	–
9	79	6	10.50	10.50	10.55	1	–
10	70	6	10.56	10.56	11.01	1	–
11	14	4	11.00	11.01	11.06	–	1
12	59	5	11.05	11.06	11.11	–	1
13	62	6	11.11	11.11	11.16	–	–
14	57	5	11.16	11.16	11.21	–	–
15	15	4	11.20	11.21	11.26	–	1
16	18	4	11.24	11.26	11.31	–	2
17	74	6	11.30	11.31	11.36	–	1
18	11	3	11.33	11.36	11.41	–	3
19	41	5	11.38	11.41	11.46	–	3
20	29	5	11.43	11.46	11.51	–	3

First, we select 20 two-digit random numbers from the random table. The first random number for the arrival time is 12. This number lies in the range (12 – 28). Therefore, this random number indicates the arrival time at 10.04 a.m., assuming that the checking starts at 10.00 a.m. Similarly, we can work out all the simulated arrivals and service times. Since the first truck arrives at 10.04 a.m, the checker waits for 4 minutes. This is indicated in the last column, as checker's waiting time in Table 13.12. The checker takes 5 minutes and thus the service for first truck will end at 10.09 a.m., which indicates that the checker waits for 1 minute. Whenever a truck has to wait because of the checker being busy in dealing with a previous truck, the waiting time is listed in the last column of Table 13.12. The similar procedure can be adopted to prepare the entire work sheet.

After completing the simulation, the following information can be obtained which describes the behaviour of a single server counter:

(i) Average waiting time of trucks before service

$$= \frac{\text{total waiting time}}{\text{total number of arrivals}} = \frac{16}{20} = 0.8 \text{ minutes}$$

(ii) Expected longest period of waiting = 3 minutes.

13.6 APPLICATIONS TO INVENTORY PROBLEMS

Many of the inventory problems, especially storage problems, cannot be solved analytically because of the complex nature of the distribution followed by demand or supply. It is, however, possible to get the solution by using simulation techniques. The basic approach would be to determine the probability distribution of the input and output functions from the past data, and run the inventory system artificially by generating the future observations on the assumption of the same distributions. Subsequently, the decision-making regarding the optimisation problems would be made by the trial and error method.

The artificial samples for future can be generated with the help of random numbers.

In inventory control, the recorder point is to be chosen with consideration for the demand during lead time to provide adequate service to customers. If both the lead time and demand of inventory per unit of time are random variables, then simulation technique can be used to investigate the effect of different inventory policies.

Example 1

A wholesaler dealing in stationery items wants to determine the order of size for desk calendars. The demand and lead time are probabilistic and their distribution are given below.

Demand/week (thousand)	*Probability*	*Lead time (weeks)*	*Probability*
0	0.2	2	0.3
1	0.4	3	0.4
2	0.3	4	0.3
3	0.1		

The cost of placing an order is Rs 50 per order and the holding cost of 1,000 calendars is Rs 2 per week. The shortage cost is Rs 10 per thousand. The inventory manager is considering a policy whereby whenever the inventory level is equal to or below 2,000, an order is placed equal to the difference between the current inventory balance and specified maximum replenishment level of 4,000.

Simulate the policy of the 20-week period assuming that (a) the beginning inventory is 3,000 units, (b) no back orders are permitted, (c) each order is placed at the beginning of the week following the drop in inventory level to (or below) the reorder point, and (d) the replenishment orders are received at the beginning of the week.

Solution: Using the weekly demand and lead time distributions, assign an appropriate set of random numbers to represent value or range of values of variables (Tables 13.13 and 13.14).

table 13.11

Demand/week (thousand)	Probability	Cumulative probability	Random numbers
0	0.2	0.2	00 – 19
1	0.4	0.6	20 – 59
2	0.3	0.9	60 – 89
3	0.1	1.0	90 – 99

table 13.12

Lead time (weeks)	Probability	Cumulative probability	Random numbers
2	0.3	0.3	00 – 29
3	0.4	0.7	30 – 69
4	0.3	1.0	70 – 99

The simulation for a period of 20 weeks concerning the inventory system and the related costs with a replenishment level of 4,000 units and reorder level of 2,000 units is given in Table 13.15. At the start of the a simulation, the first random number 31 generates a demand of 1,000 units, as determined from the cumulative probability values of calendar demand in Table 13.13, leaving 2,000 units on hand at the end of the first week. Since it is equal to the reorder level, an order for 4,000 – 2,000 = 2,000 units is placed. With a random number 29, the lead time is 2 weeks; refer to the lead time cumulative probability values in Table 13.14.

table 13.13

Week	Beginning inventory (in thousands)	Demand R.N.	Units	Ending inventory	Lead R.N.	Time (weeks)	Quantity ordered (in thousands)	Cost (Rs) Holding	Shortage
0				3	–	–	–	–	–
1	3	31	1	2	29	2	2	4	–
2	2	70	2	0	–	–	–	–	–
3	0	53	1	-1	–	–	–	–	10
4	2*	86	2	0	83	4	4	–	–
5	0	32	1	-1	–	–	–	–	10
6	0	78	2	-2	–	–	–	–	20
7	0	26	1	-1	–	–	–	–	10
8	0	64	2	-2	–	–	–	–	20
9	4*	45	1	3	–	–	–	–	–
10	3	12	0	3	–	–	–	6	–
11	3	99	3	0	58	3	4	6	–
12	0	52	1	-1	–	–	–	–	10
13	0	43	1	-1	–	–	–	–	10
14	0	84	2	-2	–	–	–	–	20
15	4*	38	1	3	–	–	–	6	–

Week	Beginning inventory (in thousands)	Demand R.N.	Units	Ending inventory	Lead R.N.	Time (weeks)	Quantity ordered (in thousands)	Cost (Rs) Holding	Shortage
16	3	40	1	2	41	3	2	4	–
17	2	19	0	2	–	–	–	4	–
18	2	87	2	0	–	–	–	–	–
19	0	83	2	–2	–	–	–	–	20
20	2	73	2	0	13	2	4	–	–

* Includes order quantity just received.

With 2,000 units to be held, the holding cost is Rs 4 and there is no shortage cost. In the next week, the random number 70 results in a demand of 2,000 units from Table 13.13, and 2000 units available in inventory at the beginning of the second week are reduced to zero units, by the end of the week. In the third week, demand is for 1,000 units but as the available inventory is zero, this results in the shortage cost of Rs 10. The 2,000 units ordered in the first week are received in the beginning of the fourth week. The demand in the fourth week is also for 2,000 units, and hence ending inventory is zero. The second shortage thus occurs in the fifth week and continues till the end of the eighth week. The units ordered at the end of fourth week are received only in the beginning of the nineth week. In this way, the simulation is continued for a period of 20 weeks.

To evaluate the performance of the policy simulated, we need to know the number of orders placed, the average inventory, and the number of units short. From Table 13.15, we note that orders were placed 5 times during the course of simulation.

The average inventory can be calculated by adding the weekly ending inventory balances (ignoring negative balances) and dividing by the number of weeks. Thus,

$$\text{Average inventory} = \frac{15{,}000}{20} = 750 \text{ units per week.}$$

The total average weekly cost can be calculated as follows:

Weekly average cost = Ordering cost + Inventory holding cost + Shortage cost

$$= \frac{(\text{Rs}\,50)(5)}{20} + \frac{(\text{Rs}\,2)(750)}{1{,}000} + \frac{(\text{Rs}\,10)(13)}{20}$$

$$= \frac{250}{20} + 1.50 + \frac{130}{20}$$

$$= \text{Rs } 12.50 + 1.50 + 6.50$$

$$= \text{Rs } 20.50.$$

In this case the average shortage cost is high as compared to the holding cost. This shortage cost can be reduced by increasing the reorder level.

The two decision variables, replenishment level and reorder level interact with each other and influence the three cost elements. Since replenishment level and reorder level depend on each other, experiments conducted with the simulation model should be so designed as to list the various combinations of both variables. In the present problem, the average lead time is 2.8 weeks and the average demand per week is 14,000 units. Hence, average demand during lead time is 3,920 units. On the other hand, maximum lead time is 4 weeks and maximum weekly demand is 3,000 units. Thus, maximum demand during lead time is 12,000 units. It follows that the best reorder point, irrespective of the replenishment level, should be somewhere in the range 3,920 and 12,000 units.

13.7 APPLICATIONS TO CAPITAL BUDGETING PROBLEM

An important decision of financial analysis is to select the optimum alternative among various capital investment policies and evaluation of risk involved with the specific decision. The main purpose of evaluating the risk is to determine the effect of various factors (e.g. selling price, market growth rate, market size, etc.) on financial parameters. The samples from the probability distributions of the factors involved can be drawn and analysed to determine the rate of return on investment.

Step 1: *To develop the probability distribution of uncertain factor*

The probability distributions are developed on the basis of assessment of the probable outcomes. The following factors are taken into consideration for evaluating an investment proposal:

(i) Market size
(ii) Selling price
(iii) Market growth rate
(iv) Share of market
(v) Investment required
(vi) Residual value of investment
(vii) Operating costs
(viii) Fixed costs
(ix) Useful life of facilities

Step 2: *To generate a set of random numbers*

A set of random numbers is generated and respective probability distribution is assigned to each factor in the system for calculating the expected values of these factors. Then, the values of all the factors are combined to determine the rate of return for the combination.

Step 3: *To simulate the process*

The above process is repeated several times to simulate a clear portrayal of investment risk.

Example 1

An investment corporation wants to study the investment projects based on three factors: market demand in units, price per unit minus cost per unit, and the investment required. These factors are felt to be independent of each other. In analysing a new consumer product, the corporation estimates the following probability distributions:

table 13.14

Annual demand		(Price – Cost) per unit		Investment required	
Units	Probability	Rs	Probability	Rs	Probability
20,000	0.05	3.00	0.10	17,50,000	0.25
25,000	0.10	5.00	0.20	20,00,000	0.50
30,000	0.20	7.00	0.40	25,00,000	0.25
35,000	0.30	9.00	0.20		
40,000	0.20	10.00	0.10		
45,000	0.10				
50,000	0.05				

Using simulation process, repeat the trial 10 times, compute the return on investment for each trial taking these three factors into accounts. What is the most likely return (appx.)?

Solution: The yearly return can be determined by the formula

$$\text{Return } (R) = \frac{(\text{Price} - \text{Cost}) \times \text{Number of units demanded}}{\text{Investment}}$$

To determine a cumulative probability distribution corresponding to each of the three factors, we assign an appropriate set of random numbers representing each of the three factors as shown in Tables 13.17, 13.18 and 13.19.

table 13.15

Annual demand	Probability	Cummulative probability	Random number
20,000	0.05	0.05	00 – 04
25,000	0.10	0.15	05 – 14
30,000	0.20	0.35	15 – 34
35,000	0.30	0.65	35 – 64
40,000	0.20	0.85	65 – 84
45,000	0.10	0.95	85 – 94
50,000	0.05	1.00	95 – 99

table 13.16

(Price – Cost) per unit	Probability	Cummulative probability	Random number
3.00	0.10	0.10	00 – 09
5.00	0.20	0.30	10 – 29
7.00	0.40	0.70	30 – 69
9.00	0.20	0.90	70 – 89
10.00	0.10	1.00	90 – 99

table 13.17

Investment required	Probability	Cumulative probability	Random number
17,50,000	0.25	0.25	00 - 04
20,00,000	0.50	0.75	25 - 74
25,00,000	0.25	1.00	75 - 99

The simulation worksheet for 10 trials is given in Table 13.20.

table 13.18

Trials	Random number of demand	Simulated demand (in thousands)	Random no. for profit (price – cost) per unit	Simulated profit	Random number for investment	Simulated investment (in thousands)	Simulated return (%) = Demand of profit per unit / investment
1	28	30	19	5.00	18	1,750	8.57
2	57	35	07	3.00	67	2,000	5.25
3	60	35	90	10.00	16	1,750	20.00
4	17	30	02	3.00	71	2,000	4.50
5	64	35	57	7.00	43	2,000	12.25
6	20	30	28	5.00	68	2,000	7.50
7	27	30	29	5.00	47	2,000	7.50
8	58	35	83	9.00	24	1,750	18.00
9	61	35	58	7.00	19	1,750	14.00
10	30	30	41	7.00	97	2,500	8.40

Table 13.19 shows that the highest likely return is 20%, which is corresponding to the annual demand of 35,000 units resulting in a profit of Rs 10 per unit. The required investment will be Rs 17,50,000.

13.8 APPLICATIONS TO PERT PROBLEMS

Simulation can be applied to PERT problems. This is illustrated with the following examples.

Example 1

A project consists of activities *A* to *H*. The completion time for each activity is a random variable. The data concerning probability distribution along with completion times for each activity is as follows:

Activity	Immediate Predecessor(s)	Time (days)/Probability								
		1	2	3	4	5	6	7	8	9
A	–	–	–	–	0.2	–	0.4	0.4	–	–
B	–	–	–	–	–	–	0.5	–	0.5	–
C	A	–	–	0.7	0.3	–	–	–	–	–
D	B, C	–	–	–	–	0.9	–	–	0.1	–
E	A	–	–	–	–	0.2	–	–	–	0.8
F	D, E	–	–	–	0.6	0.4	–	–	–	–
G	E	–	–	0.4	0.4	–	0.2	–	–	–
H	F	–	0.4	–	–	–	–	0.6	–	–

(a) Draw the network diagram and identify the critical path using the expected activity times.

(b) Simulate the project to determine the activity times. Determine the critical path and the expected project completion time.

(c) Repeat the simulation four times and state the estimated duration of the project in each of the trials.

Solution: (a) The network diagram based on the precedence relationships is shown in Fig. 13.3. The expected completion time of each activity is obtained by using the formula.

$$\text{Expected time} = \Sigma\,(\text{Activity} \times \text{Probability})$$

$$= 4 \times 0.2 + 6 \times 0.4 + 7 \times 0.4 = 6 \text{ days (Activity A)}$$

The critical path of the project is 1 – 2 – 3 – 4 – 5 – 6 – 7, with expected time of 23.6 days.
The random coding for each activity's expected time is shown in Table 13.21.

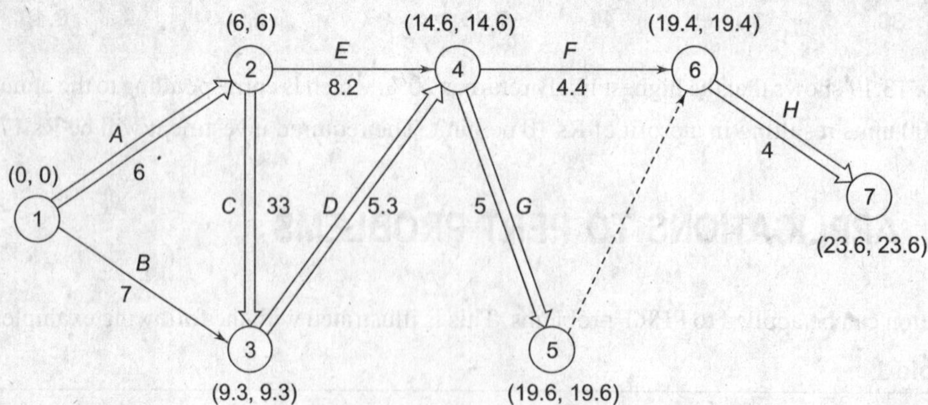

figure 13.3

table 13.19 **Simulation worksheet**

Breakdown number	*Random number for breakdown*	*Time between breakdown*	*Time of breakdown*	*Repair work begins at*	*Random number for repair time*	*Repair time required*	*Repair work ends*	*Total idle time (hours)*	*Waiting time (hours)*
(1)	*(2)*	*(3)*	*(4)*	*(5)*	*(6)*	*(7)*	*(8)*	*(9)*	*(10)*
1	61	2.5	02.30	02.30	87	3	05.30	3.00	–
2	85	3	05.30	05.30	39	2	07.30	2.00	–
3	16	1.5	07.00	07.30	28	2	09.30	2.30	0.30
4	46	2	09.00	09.30	97	3	12.30	3.30	0.30
5	88	3	12.00	12.30	69	2	14.30	2.30	0.30
6	08	1	13.00	14.30	87	3	17.30	4.30	1.30
7	82	3	16.00	17.30	52	2	19.30	3.30	1.30
8	56	2	18.00	19.30	52	2	21.30	3.30	1.30
9	22	1.5	19.30	21.30	15	1	22.30	3.00	2.00
10	49	2	21.30	22.30	85	3	01.30	4.00	1.00
11	44	2	23.30	01.30	41	2	03.30	4.00	2.00
12	33	2	01.30	03.30	82	3	06.30	5.00	2.00
13	77	2.5	04.00	06.30	98	3	09.30	5.30	2.30
14	87	3	07.00	09.30	99	3	12.30	5.30	2.30
15	54	2	09.00	12.30	23	2	14.30	5.30	3.30
				38.30		36.00		57.30	21.30

table 13.20 **Random number coding: Activity times**

Activity	*Time*	*Probability*	*Cumulative probability*	*Random number range*
A	4	0.20	0.20	00 - 19
	6	0.40	0.60	20 - 59
	7	0.40	1.00	60 - 99
B	6	0.50	0.50	00 - 49
	8	0.50	1.00	50 - 99
C	3	0.70	0.70	00 - 69
	4	0.30	1.00	70 - 99
D	5	0.90	0.90	00 - 89
	8	0.10	1.00	90 - 99
E	5	0.20	0.20	00 - 19
	9	0.80	1.00	20 - 99
F	4	0.60	0.60	00 - 59
	5	0.40	1.00	60 - 99
G	3	0.40	0.40	00 - 39
	4	0.40	0.80	40 - 79
	6	0.20	1.00	80 - 99
H	2	0.40	0.40	00 - 39
	7	0.60	1.00	40 - 99

The simulation worksheet for four simulation runs is shown in Table 13.23. For each run the project time is obtained as follows:

Total time = Larger of times for activities A, B and C+ Larger of times for activities D and F + Larger of times for activities F and G + Time for activity H.

Using the data given in Table 13.23, we have the simulation results shown in Table 13.24.

table 13.21 **Simulation Worksheet**

Run	A		B		C		D		E		F		G		H	
	R.No.	Time	R.No.	Time	R.No.	Time	R.No.	Time	R.No.	Time	R.No.	Time	R.No.	Time	R.No.	Time
1	22	6	17	6	68	3	65	5	84	9	68	5	95	6	23	2
2	92	7	35	6	61	3	09	5	43	9	95	5	06	3	87	7
3	02	4	22	6	57	3	51	5	58	9	24	4	82	6	03	2
4	47	6	19	6	36	3	27	5	59	9	46	4	13	3	79	7
5	93	7	37	6	66	3	85	5	52	9	05	4	30	3	62	7
Total		30		30		15		25		45		22		21		25
Average		6		6		3		5		9		4.4		4.2		5

table 13.22 **Simulation Results**

Simulation run	*Activity time*	*Project duration (days)*	*Longest (Critical) path*
1	6 + 9 + 6 + 2	23	1 - 2 - 3 - 4 - 5 - 6 - 7
			1 - 3 - 4 - 5 - 6 - 7
2	7 + 9 + 5 + 7	28	1 - 3 - 4 - 6 - 7
3	6 + 9 + 6 + 2	23	1 - 2 - 3 - 4 - 5 - 6 - 7
4	6 + 9 + 4 + 7	26	1 - 2 - 3 - 4 - 6 - 7
			1 - 3 - 4 - 6 - 7
5	7 + 9 + 4 + 7	27	1 - 3 - 4 - 6 - 7
		127	

Here, it may be noted that simulated mean project completion time 25.4 days is almost of two days longer than the 23.6 days completion time indicated using expected values alone.

13.9 HOSPITAL SIMULATION

Hospital simulation is an example of a manual simulation.

Example 1

Wednesday's schedule for operating room number 3 at a famous hospital is shown in Table 13.25. From looking at this schedule, the operating room head nurse concludes that it may not be possible to finish with the operating and clean-up schedule by 5 pm, the time at which this operating room must be available for emergency night service.

table 13.23

Time	Activity	Expected time
8:00 am	Appendectomy	40 min
8:40	Clean-up	20 min
9:00	Laminetormy	90 min
10:30	Clean-up.	20 min
10:50	Kidney removal	120 min
12:50 pm	Clean-up	20 min
1:10	Hysterectomy	60 min
2:10	Clean-up	20 min
2:30	Colostomy	100 min
4:10	Clean-up	20 min
4:30	Lesion removal	10 min
4:40	Clean-up	20 min

The hospital management analyst suggests that simulation might indicate whether the schedule for Wednesday is workable and if not, what change could be made. The analyst reviews the operating records for the past few months and finds that patients do not always arrive at the operating room at the scheduled time. They often have to wait for pre-op medication to be administered; sometimes operating room transportation personnel are late, and from time to time physicians forget to order the patient moved from the floor to the operating room. Investigation of the operating room log indicates that arrival expectations are as shown in Table 13.26. The management analyst finds thats operating times also vary according to surgical difficulties encountered, differences in surgical skills, and the effectiveness of the surgical team in general. An analysis of operations scheduled over the past few months produces the results shown in Table 13.27, which gives a good indication of this variation. He also recognises that any variation in the expected clean-up time will affect the schedule and checks the records once again. Here, he finds that about half the time, the clean-up crew finishes in 10 minutes. The rest of the time, it takes them 30 minutes.

table 13.24 **Arrival Expectation**

Patient arrives on time	*0.50 probability*
Patient arrives 5 minutes early	0.10 probability
Patient arrives 10 minutes early	0.05 probability
Patient arrives 5 minutes late	0.20 probability
Patient arrives 10 minutes late	0.15 probability

table 13.25 **Operation Time Expectations**

Operation is completed in the expected time	0.45 probability
Operation is completed in 90% of the expected time	0.15 probability
Operation is completed in 80% of the expected time	0.05 probability
Operation is completed in 110% of the expected time	0.25 probability
Operation is completed in 120% of the expected time	0.10 probability

Generating the variables in the system (Process Generators)

The analyst needs a way to generate arrival times, operating times, and clean-up times. The methods he uses to do this are called process generators. He decides to use a random number table. A random number table is the output we would expect to get from sampling a uniformly distributed random variable where all of its values (digits 0 through 9 in this case) are alike.

Generating arrival times

The analyst decides to use the first two digits of each 10-digit number in a random number table as his process generator for arrival times. Since there are 100 possible two-digit numbers from 00 through 99, he relates these two-digit numbers to arrival variation as shown in Table 13.28

table 13.26

Random numbers	*Arrivals*	
00 through 49	on time	(0.50 probability)
50 through 59	5 minutes early	(0.10 probability)
60 through 64	10 minutes early	(0.05 probability)
65 through 84	5 minutes late	(0.20 probability)
85 through 99	10 minutes late	(0.15 probability)

Generating operating times

The analyst now decides to use the last two digits of each 10-digit number in a random number table. He relates these two-digit numbers to operating times as shown in Table 13.29.

table 13.27

Random	*Numbers*	*Operating times*
00 - 44	On-time completion	(0.45 probability)
45 - 59	Completion in 90% of expected time	(0.15 probability)
60 - 64	Completion in 80% of expected time	(0.05 probability)
65 - 89	Completion in 110% of expected time	(0.25 probability)
90 - 99	Completion in 120% of expected time	(0.10 probability)

Generating clean-up times

Since there are only two values the random variable takes on here, the analyst decides to use the fourth digit of each 10-digit number in a random table as his process generator. If it's an odd number, he will let that represent a 10-minute clean-up an, even number would represent a 30-minute clean-up.

The simulation

The analyst now proceeds with the simulation. First, he generates an arrival-time deviation for the first patient; then he generates an operating time deviation for the first operation; finally, he generates a clean-up time for that operation. He continues with this process until the last operation has been performed and the operating room cleaned up for the final time that day. The results of the simulation are shown in Table 13.30.

table 13.28 **Results of Simulation of Activity in Operating Room No. 3**

Random number	First two digits	Last two digits	Fourth digit	Meaning	Outcome
15	x			On-time arrival of appendectomy patient	Appendectomy began at 8 am
96		x		Appendectomy completed in 120% of expected time (48 min)	Appendectomy completed at 8:48 am
1			x	Clean-up done in 10 min	Room ready for second operation at 8:58 am
9	x			On-time arrival of laminectomy patient (9 am)	Laminectomy began at 9 am
82		x		Laminectomy completed in 110% of expected time (99 min)	Laminectomy completed at 10:39 am
8			x	Clean-up done in 30 min	Room ready for third operation at 11:09 am
41	x			On-time arrival of kidney patient (10:50 am)	Kidney removal began at 11:09 am
56		x		Kidney removal completed in 90% of expected time (108 min)	Kidney removal completed at 12:51 pm
2			x	Clean-up done in 30 min	Room ready for fourth operation at 1:27 pm
75	x			Hysterectomy patient arrives 5 min late (1:15 pm)	Hysterectomy began at 1:27 pm
68		x		Hysterectomy completed in 110% of expected time (66 min)	Hysterectomy completed at 2:33 pm
7			x	Clean-up done in 10 min	Room ready for fifth operation at 2:43 pm

(Contd.)

Random number	First two digits	Last two digits	Fourth digit	Meaning	Outcome
00	x			On-time arrival of Colostomy patient (2:30 pm)	Colostomy began at 2:43 pm
58		x		Colostomy completed in 90% of expected time (90 min)	Colostomy completed at 4:13 pm
9			x	Clean-up done in 10 min	Room ready for sixth operation at 4:23 pm
72	x			Lesion patient arrivals 5 min late (4:35 pm)	Lesion operation begun at 4:35 pm
40		x		On-time completion of lesion operation (10 min)	Lesion operation completed at 4:45 pm
5			x	Clean-up done in 10 min	Operating room schedule for Wednesday completed at 4:55 pm

From the analyst's simulation, it appears that the scheduled operations can be completed and the room vacated by 5 pm. In fact, his simulation indicates that the day's schedule ends at 4:45 pm, a few minutes early.

Assumptions

The analyst simulated the day's operation only once, and it may be dangerous for us to draw conclusions from such a short simulation. If he had repeated the day's simulation several times using different random numbers, then we could feel better about generalising from his results. He also assumed that the variables in this simulation (arrival deviation, operating time deviation, and clean-up deviation) were independent of each other. If this is not the case, his simulation is not a valid one. Finally, he used discrete distributions of the three variables. In actual practice, were computation time not such a problem, continuously distributed random variables would be more appropriate.

13.10 COMPUTER SIMULATION

The role of computers in simulation is vital. They are used to generate random numbers, simulate the given problem with varying values of variables in a few minutes and help the decision-maker to prepare reports which enable making quick decision as well as drawing valid conclusions.

For the complex and intricate problems of managerial decision-making, the analogue simulation may not be practicable, and actual experimentation with the system may be uneconomical. Under such circumstances, the complex system is formulated into a mathematical model for which a computer program is developed, and the problem is solved by using high-speed electronic computer, and hence it is named as computer simulation or system simulation.

Simulation languages

The efficiency of programming and execution of a simulation project depends upon the programming language used. Computer languages available to help the simulation process can be divided into two categories:

1. General Purpose Programming Languages

The general purpose programming languages includes FORTRAN, BASIC, COBAL, PL/1, PASCAL, C, C++, Java, and so on. To use these languages for simulation process, an extensive programming experience is required as even a simple queueing problem involves many tedious details.

Being well known and commonly available on computer systems, FORTRAN is often used to write simulation programmes. It is efficient in computer time and storage requirements. However, programming in FORTRAN is more difficult and time consuming as compared to the special simulation languages. When the complexities of the simulation project increase, the book-keeping of the intricate details of simulation becomes difficult and this makes the programming in FORTRAN harder. Thus, for realistic situations, simulation programs should be written in specialised simulation languages which are designed to meet the following objectives:

(i) Conveniently describe the elements which commonly appear in simulation, such as the generation of random variates for most of the statistical distributions.

(ii) Flexibility of changing the design configuration of the system so as to consider alternate configurations.

(iii) Internal timing and control mechanism for book keeping of vital information during the simulation run.

(iv) Obtain conveniently the data and statistics about the behaviour of the system.

(v) Provide simple operational procedures, such as altering the initial state of the system and kind of output data to be generated, and so on.

2. Special Purpose Simulation Languages

Special purpose simulation languages have several advantages. (i) They reduce programme preparation time and cost with features specially designed for simulation models. Such features generally include a master sequencing routine to automatically maintain event sequence and to keep track of simulated time sub-routines to handle arrivals and departures in a queuing system, (ii) they have the capability to readly generate different types of random and certain types of statistical table, (iii) they require little or no prior programming knowledge for use. The main special purpose simulation languages are:

(a) *GPSS (General Purpose System Simulation)*: Usually, it does not require programme writing. The system model is constructed via block diagrams using block commands. The third version of this language, GPSS III consists of two parts. The first part is an assembly programme which converts the system descriptors into input for the second part that performs the simulation. This language was developed by IBM in early 1960s.

(b) *SIMSCRIPT*: This language neither depends on any predefined coding forms nor on any intermediate language such as FORTRAN for its implementation. This language was developed by RAND Corporation in the early 1960s.

(c) *DYNAMO*: It is a computer program which is capable of taking input in the form of a set of equations describing the system. These equations are evaluated continuously for each time interval to understand the behaviour of the system. This language was developed at MIT in 1959 and is best suited for econometric modelling of industrial complexes and urban, social and world systems planning.

The choice of a simulation package depends on the specific purpose, the availability of simulation languages on a particular computer, the training and experience in simulation modelling programming and the availability of experienced programmers.

13.11 SIMULATION OF JOB SEQUENCING

In this section, we shall illustrate the use of simulation in sequencing problems.

Example 1

A job has to be processed over two machines M_1 and M_2 in that order. The distribution of inter-arrival time of the jobs at the first machine is as follows:

Time (minutes)	*Probability*
1	0.2
2	0.2
3	0.2
4	0.4

The processing time at the two machines is as follows:

Machine M_1		*Machine M_2*	
Time (minutes)	*Probability*	*Time (minutes)*	*Probability*
1	0.1	4	0.2
2	0.2	5	0.3
3	0.3	6	0.4
4	0.3	7	0.1
5	0.1		

On the basis of 10 simulation runs find out the average queue length before machine M_1 and the average queue length before machine M_2.

Solution: The various steps for simulation are explained below:

1. Simulate the inter-arrival time of the *i*th job on machine M_1 with the help of random numbers ($i = 1, 2, ...$).
2. Arrival time of ith job
   = arrival time of $(i - 1)^{th}$ job ($i = 1, 2, ...$) + inter-arrival time of ith job.
   Arrival time of the first job = inter-arrival time of the first job.
3. Simulate the processing time on machine M_1 by random numbers.
4. Departure time on machine M_1
   = Arrival time for processing on Machine M_2
   = Max (Arrival time of ith job process completion time of $(i - 1)^{th}$ job on machine M_1) + processing time of ith job on Machine M_1.
5. If arrival time of *i*th < process completion time of $(i - 1)^{th}$, i^{th} arrival waits. If arrival time of ith < process completion time of $(i - 2)^{th}$, both i^{th} and $(i - 1)^{th}$ arrival wait.
6. Simulate the processing time of machine M_2 with the help of random numbers.
7. Process completion by machine M_2 of *i*th arrival
   = Max (Arrival time of *i*th, process completion time of $(i - 1)^{th}$
   + process time by machine M_2.

table 13.29 **Simulation Worksheet**

Simulation run	Random number	Inter-arrival time	Arrival time on M_1	Random number	Process time on M_1	Departure time on M_1	Random number	Process time on M_2	Completion time on M_2	Queue size before M_1	Queue size before M_2
1	8	4	4	0	1	5	3	5	10	–	–
2	7	4	8	6	4	12	0	4	16	–	–
3	8	4	12	9	5	17	0	4	21	–	–
4	9	4	16	3	3	20	0	4	25	1	1
5	4	3	19	7	4	24	0	4	29	1	1
6	4	3	22	0	1	25	2	5	34	1	1
7	0	1	23	0	1	26	4	5	39	2	2
8	6	4	27	5	3	30	5	6	45	–	2
9	8	4	31	6	4	35	1	4	49	–	2
10	7	4	35	4	3	38	2	5	54	–	3

8. If arrival time of ith < process completion time of $(i-1)^{th}$, i^{th} arrival waits. If arrival time of ith < process completion time of $(i-2)^{th}$, both i^{th} and $(i-1)^{th}$ arrivals wait and so on.

Using the procedure, the simulated processing data of 10 arrivals are displayed in the simulation worksheet in Table 13.31.

On the basis of simulation study, the average queue length before machine M_1 is 0.5, and the average queue length before machine M_2 is 1.2.

13.12 APPLICATION OF SIMULATION

There is a wide range of applications of computer-based simulation models because it is an approach rather than an application of specific techniques. The major use of computer-based Monte-Carlo simulation model has been in the solution of complex problems. Some of the major applications of simulation are:

1. Queuing problems
2. Inventory problems
3. Training programmes
4. Network problems (PERT)
5. Job sequencing
6. Capital budgeting and investment problems
7. Military studies of logistics, support planning and weapon system and effectiveness
8. Studies of individual and group behaviour
9. Financial studies involving risks investments
10. Testing of decision rules for hospital admission and operating policies
11. Carpet cutting application
12. Public school planning application.

EXERCISES

Section 13.1

1. What is simulation?
2. Explain simulation.
3. Explain the Monte-Carlo method.
4. What are the advantages of simulation?
5. What are the limitations of simulation?
6. Mention the uses of simulation.
7. When is simulation preferable?
8. What are the different steps involved in solving a problem by simulation?
9. Distinguish between solutions derived from simulation models and from analytical models.
10. "When it becomes difficult to use an optimisation technique for solving a problem, one has to resort to simulation technique." Discuss.
11. Can you in all situations use simulation exercise without a computer? Discuss.
12. Do you think that the application of simulation will increase in the next 10 years? Why?

13. What types of problems can be solved more easily by quantitative techniques other than simulation?
14. Explain the various types of simulation.
15. Why should one use simulation?
16. Explain the phases of a simulation model.

Section 13.2

17. With the help of a single server queueing model having inter-arrival and service times constant 1.4 and 3 minutes, respectively, explain the discrete simulation technique taking 10 minutes as the simulation period. Find from this the average waiting time of a customer (assume that initially the system is empty and the first customer arrives at time $t = 0$).
[*Answer*: Average time = 2.35 minutes per customer.]
18. At a telephone booth, customers arrive with an average time of 1.2 time units between two arrivals. Service times are assumed to be 2.8 time units. Simulate the system for 12 time units by assuming that the system starts at $t = 0$. What is the average waiting time per customer?
[*Answer*: Average time = 3.36 time units per customer.]

Section 13.3

19. Write short notes on generation of random numbers.
20. What do you mean by the mid-square method of generating pseudo-random numbers? What are the disadvantages of this method?
21. What do you mean by the congruence method of generating random numbers? Explain with an example.
22. What are random numbers? Why are random numbers useful in simulation models and solutions derived from analytical models?
23. Use the mixed congruential method to generate the following sequences of random numbers:
    (i) A sequence of five two-digit numbers, such that
    $r_{i+1} = (21r_i + 53)$ (modulo 100), take $r_0 = 46$.
    (ii) A sequence of five random numbers between 0 and 31, such that
    $r_{n+1} = (9r_n + 15)$ (modulo 32), take $r_0 = 12$.
24. Solve problem 24 when mixed congruential method is reduced to multiplicative method.

Section 13.4

25. Explain Monte-Carlo method of simulation with a suitable example.
26. (The chef example) The number of customers at a restaurant each evening is distributed as shown below:

Number of customers :	Lots	Average	Very few
Probability :	0.2	0.4	0.4

The chef refuses to work on an evening when there are very few customers and walks out. He will not return to work until at evening when there are lots of customers, although he always comes in on Friday because he gets paid then. We are interested in the fraction of evenings that the chef is at the restaurant.
[*Answer*: Chef is in for 49% of all days.]

27. A tourist car operator has 25 taxis in operation. He keeps three drivers as reserve to attend to calls, in case the scheduled drivers report sick. The probability distribution of sick drivers is as follows:

Number of sick drivers	:	0	1	2	3	4	5
Probability	:	0.20	0.25	0.20	0.15	0.12	0.08

Use the Monte-Carlo method to estimate the utilisation of reserve drivers and the probability that at least one taxi will be off the road due to non-availability of a driver. Compare with the correct answers.

Section 13.5

28. A company has a single service station which has the following characteristics: the mean arrival rate of customers and the mean service time are 6.2 minutes and 5.5 minutes, respectively. The time between an arrival and its service varies from 1 minute to 7 minutes. The arrival and service time distributions are given below:

Time (Minutes)	*Arrival (Probability)*	*Service (Probability)*
1 – 2	0.05	0.10
2 – 3	0.20	0.20
3 – 4	0.35	0.40
4 – 5	0.25	0.20
5 – 6	0.10	0.10
6 – 7	0.05	—

The queueing process starts at 11 am and closes at 12 pm. An arrival moves immediately into the service facility if it is empty. On the other hand, if the service station is busy, the arrival will wait in the queue. Customers are served on the first come, first served basis.

If the clerk charges Rs 6 per hour and the customer's waiting line costs Rs 5 per hour, would it be economical for the manager to engage a second clerk?

[*Answer*: Average queue length = 0.85.]

Average waiting time of customer before starting service = 2.80 minutes

Average service time = 2.70 minutes

Time a customer spends in the system = 5.5 minutes.]

29. Dr Anish is a dentist who schedules all his patients for 30-minute appointments. Some of the patients take more or less than 30 minutes depending on the type of dental work to be done. The following summary shows the various categories of work, their probabilities and the time actually taken to complete the work.

Category	*Time required (minutes)*	*Probability category*
Filling	45	0.40
Crown	60	0.15
Cleaning	15	0.15
Extraction	45	0.10
Check-up	15	0.20

Simulate the dentist's clinic for four hours and determine the average waiting time for the patients as well as the idle time of the doctor. Assume that all the patients show up at the clinic at exactly their scheduled arrival time starting at 8 am. Use the following random numbers handling the above. Problem: 40, 82, 11, 34, 25, 66, 17, 79.

[*Answer*: The average waiting time = 35.625 minutes.]

30. A company manufactures 2,000 mopeds. Depending upon the availability of raw materials and other conditions, the daily production has been varying from 196 mopeds to 204 mopeds, whose probability distribution is as given below:

Production per day :	196	197	198	199	200	201	202	203	204
Probability :	0.05	0.09	0.12	0.14	0.20	0.15	0.11	0.08	0.06

The finished mopeds are transported in a specially designed three-storeyed lorry that can accommodate only 200 mopeds. Using the given 15 random numbers, viz., 82, 89, 78, 24, 53, 61, 18, 45, 04, 23, 50, 77, 27, 54, 10 simulate the process to find out

(i) What will be the average number of mopeds waiting in the factory?

(ii) What will be the average number of empty spaces in the lorry?

[*Answer*: (i) 0.67 per day (ii) 0.93 per day.]

31. An owner of a petrol pump with single attendant wishes to perform a simulation of operations to see whether any improvement is possible. He studied the system and found that an average of 6 customers arrive for service with random arrival times and form a queue, and the attendant provides a service for exactly 9 minutes. For simulating the arrival times of customers, he has selected 10 random numbers with expected length of interval equal to one as

3.62, 1.78, 1.84, 1.31, 1.27 and

0.14, 1.71, 0.77, 0.97, 1.32

Find (i) The total idle time for the attendant,

(ii) Total waiting time for the customers, and

(iii) Maximum queue length during this period.

If the service time is reduced to 6 minutes what is the quality of service?

32. An automobile production line turns out about 100 cars a day, but deviations occur owing to many causes. The production is more accurately described by the probability distribution given below:

Production/day	*Probability*	*Production/day*	*Probability*
95	0.03	101	0.15
96	0.05	102	0.10
97	0.07	103	0.07
98	0.10	104	0.05
99	0.15		
100	0.20	105	0.03

Finished cars are transferred across the bay, at the end of each day, by ferry. If the ferry has space for only 101 cars, what will be the average number of cars waiting to be shipped, and what will be the average number of empty space on the boat?

33. An airline has 15 flights leaving a base per day, each with a hostess. The airline keeps three hostesses in reserve so that they may be called in case the scheduled hostess for a flight reports sick. The probability distribution for daily number of sick hostesses is as follows:

Number of sick	:	0	1	2	3	4	5
Probability	:	0.20	0.25	0.20	0.15	0.10	0.10

Use the Monte-Carlo method to estimate the utilisation of reserve hostesses and also the probability that at least one flight will be cancelled in a day because of non-availability of a hostess.

34. A firm has a single-channel service station with the following arrival and service time probability distributions:

Arrivals (min)	*Probability*	*Service time (min)*	*Probability*
1.0	0.35	1.0	0.20
2.0	0.25	1.5	0.35
3.0	0.20	2.0	0.25
4.0	0.12	2.5	0.15
5.0	0.08	3.0	0.05

The customer's arrival at the service station is a random phenomenon and the time between the arrival varies from one minute to five minutes. The service time varies from one minute to three minutes. The queueing process begins at 10:00 am and proceeds for nearly 2 hours. An arrival goes to the service facility immediately, if it is free. Otherwise, it will wait in a queue. The queue discipline is *FIFO*. If the attendant's wages are Rs 8 per hour and the customer's waiting time costs Rs 9 per hour, would it be an economical proposition to engage a second attendant? Answer on the basis of the Monte-Carlo simulation technique.

Section 13.6

35. Consider an inventory in a manufacturing concern. If the number of sales per day is Poisson with mean 5, then generate 30 days of sales by the Monte-Carlo method.

36. After studying the weekly receipts and payments over the past 200 weeks a retailer has developed the following information:

Weekly receipts (Rs)	Probability	Weekly Payment (Rs)	Probability
3,000	0.20	4,000	0.30
5,000	0.30	6,000	0.40
7,000	0.40	8,000	0.20
12,000	0.10	10,000	0.10

Using the following set of random numbers, simulate the weekly pattern of receipts and payments for the 12 weeks of the next quarter, assuming further that the beginning bank balance is Rs 8,000. What is the estimated balance at the end of the 12-week period? What is the highest weekly balance during the quarter? What is the average weekly balance for the quarter?

Random Numbers

For Receipts	:	03,	91,	38,	55,	17,	46,	32,	43,	69,	72,	24,	22
For Payments	:	61,	96,	30,	32,	03,	88,	48,	28,	88,	18,	71,	99

[*Answer*: The balance at the end of the 12-week period is a deficit of Rs 3,000. The highest weekly balance during the quarter is Rs 7,000 while the average balance works out to be Rs 3,750.]

37. Iyengar Bakery keeps stock of a popular brand of cake. Previous experience indicates the daily demand as given here:

Daily demand :	0	10	20	30	40	50
Probability :	0.01	0.20	0.15	0.50	0.12	0.02

Consider the following sequence of random numbers:

48, 78, 19, 51, 56, 77, 15, 14, 68, 09

Using this sequence, simulate the demand for the next 10 days. Find out the stock situation if the owner of the bakery decides to make 30 cakes every day. Also, estimate the daily average demand for this cake on the basis of simulated data.

[*Answer*: Stock situation: 0, 00, 20, 20, 20, 20, 40, 80, 80 and 80, respectively. Average demand = 22 units per day.]

38. A confectioner's past data of demand per week (in hundred kg) with frequency is given below:

Demand/week :	0	5	10	15	20	25
Frequency :	2	11	8	21	5	3

Using the following sequence of random numbers, generate the demand for the next 10 weeks. Also, find the average demand per week.

35, 52, 90, 13, 23, 73, 34, 57, 35, 83, 94, 56, 67, 66, 60

[*Answer*: Demand level: 10, 15, 20, 05, 05, 15, 10, 15, 10, 15, respectively. Average demand = 1,200 kg per week.]

39. A gas transport company controls pipe lines between several natural gas fields and out of state distributors. The company has a 1,00,000 unit storage capacity. Because of federal regulations, the company receives either 40,000 or 60,000 units per day. There is no equal probability of either quantity being stripped on a given day. The actual demand for natural gas is given by the following table of relative frequencies:

Daily demand	*Probability*
25,001 – 45,000	0.3
45,000 – 55,000	0.3
55,000 – 65,000	0.4

(a) What is the expected daily demand?

(b) Construct a model that can be used to simulate the company's daily receiving, storage and shipping activities.

40. The manager of a book store has to decide the number of copies of a particular tax law book to order. The book costs Rs 60 and is sold for Rs 80. Since some of the tax laws change year after year, any copies unsold while the edition is current must be sold for Rs 30. From past records, the distribution of demand for this book has been obtained as follows:

Demand (No. of copies)	:	15	16	17	18	19	20	21	22
Proportion	:	0.05	0.08	0.20	0.45	0.10	0.07	0.03	0.02

Using the following sequence of random numbers, generate data on demand for 20 time periods (years). Calculate the average profit obtainable under each of the courses of action open to the manager. What is the optimal policy?

14, 02, 93, 99, 18, 71, 37, 30, 12, 10

88, 13, 00, 57, 69, 32, 18, 08, 92, 73

[*Answer*: No. of Copies : 15 16 17 18 19 20 21 22

Expected Profit : 6,000 6,300 6,400 6,250 5,850 5,200 4,900 4,350

Optimal policy = 17 copies.]

41. A retail store distributes catalogues and takes orders on the telephone. Distribution of intervals between incoming calls and the length of time required to complete each call is given below. The store management has determined that they want a probability of no more than 5% that a caller will have to wait longer than 10 seconds for the telephone to be answered. Use simulation to determine how many sales representatives should be available to answer incoming calls.

Interval between calls (seconds)	*Probability*	*Length of call (seconds)*	*Probability*
10	0.08	60	0.07
12	0.11	65	0.12
14	0.14	70	0.18
16	0.16	75	0.16
18	0.14	80	0.15
20	0.12	85	0.12
22	0.08	90	0.08
24	0.07	95	0.06
26	0.04	100	0.06
28	0.04		
30	0.02		

42. ABC Company stocks certain products. The following data is available:
    (a) Weekly demand of the production has the following distribution:

Number of units	:	0	1	2	3
Probability	:	0.1	0.2	0.4	0.3

(b) The variation of the lead time has the following distribution:

Lead time (weeks)	:	1	2	3
Probability	:	0.30	0.40	0.30

The company wants to know (1) how much to order and (2) when to order.

Assume that the inventory on hand at the start of the experiment is 20 units, and 15 units are ordered as soon as the inventory level falls to 10 units. No back orders are allowed. Simulate the situation for 25 weeks.

[*Answer*: Average inventory = 17.92 units, total lost sales is zero units.]

43. A book store wishes to carry *Ramayana* in stock. Demand is probabilistic and replenishment of stock takes 2 days (i.e. if an order is placed on 1 March, it will be delivered at the end of the day on 3 March). The probabilities of demand are given below:

Demand (daily)	:	0	1	2	3	4
Probability	:	0.05	0.10	0.30	0.45	0.10

Each time an order is placed, the store incurs an ordering cost of Rs 10 per order. The store also incurs a carrying cost of Re 0.50 per book per day. The inventory carrying cost is calculated on the basis of stock at the end of each day.

The manager of the book store wishes to compare two options for his inventory decision:

(a) Order 5 books when the inventory at the beginning of the day plus orders outstanding is less than 8 books.

(b) Order 8 books when the inventory at the beginning of the day plus orders outstanding is less than 8.

Currently, (beginning 1st day) the store has a stock of 8 books plus 6 books ordered two days ago and expected to arrive next day.

Using Monte-Carlo simulation for 10 cycles, recommend which option the manager should choose. The two-digit random numbers are given below.

89, 34, 70, 63, 61, 81, 39, 16, 13, 73.

[*Answer*: Total cost of option A = Rs 59.50
Total cost of option B = Rs 52.50, option B should be chosen.]

Section 13.7

44. The director of finance of a farm cooperative is concerned about the yields per acre she can expect from this year's corn crop. The probability distribution of the yields for the current weather conditions is given below:

Yield in kg per acre	Probability
120	0.18
140	0.26
160	0.44
180	0.12

She would like to see a simulation of the yields she might expect over the next 10 years for weather conditions similar to those she is now experiencing.

(i) Simulate the average yield she might expect per acre using the following random numbers: 20, 72, 34, 54, 30, 22, 48, 74, 76, 02.

She is also interested in the effect of market price fluctuations on the cooperative's farm revenue. She makes this estimate of per kg price of corn.

Price per kg (Rs)	Probability
2.00	0.05
2.10	0.15
2.20	0.30
2.30	0.25

2.40	0.15
2.50	0.10

(ii) Simulate the price she might expect to observe over the next 10 years using the following random numbers:

82, 95, 18, 96, 20, 84, 56, 11, 52, 03

(iii) Assuming that prices are independent of yields, combine these two into revenue per acre and also find out the average revenue per acre she might except every year.

[*Answer*: Average revenue per acre is Rs 3,386/10 = Rs 338.60.]

45. (a) A businessman is considering taking over a new business. Based on past information and his own knowledge of the business, he works out the probability distributions of the monthly costs and sales revenues, as given here:

Cost (in Rs)	Probability	Sales Revenue (Rs)	Probability
17,000	0.10	19,000	0.10
18,000	0.10	20,000	0.10
19,000	0.40	21,000	0.20
20,000	0.20	22,000	0.40
21,000	0.20	23,000	0.15
		24,000	0.05

Use the following sequence of random numbers to be used for estimating costs and revenues. Obtain the probability distribution of the monthly net revenue.

Sequence 1 : 82, 84, 28, 82, 36, 92, 73, 91, 63, 29, 27
26, 92, 63, 83, 03, 10, 39, 10, 10

Sequence 2 : 39, 72, 38, 29, 71, 83, 19, 72, 92, 59, 49,
39, 72, 94, 01, 92, 72, 18, 09, 00.

(b) Repeat the analysis in (a) by using the following random number streams:

Sequence 1 : 20, 63, 46, 16, 45, 41, 44, 66, 87, 26, 78, 40,
29, 92, 21, 36, 57, 03, 28, 08

Sequence 2 : 23, 58, 99, 84, 51, 29, 41, 11, 66, 30, 41, 80, 62,
74, 64, 26, 41, 40, 97, 15

[*Answer*:

Net Revenue	:	– 2,000	–1,000	–1,000	2,000	3,000	4,000	5,000	6,000	
(a) Probability	:	0.05	0.00	0.15	0.30	0.15	0.25	0.05	0.00	0.05
(b) Probability	:	0.00	0.00	0.05	0.10	0.30	0.30	0.05	0.20	0.00.]

46. The investment required for introducing a new product is Rs 10,000. The probability estimates for variable cost, selling price and annual sales volume are given below:

Variable cost		*Selling price*		*Annual volume*	
Value (Rs)	*Probability*	*Value (Rs)*	*Probability*	*Value (Rs)*	*Probability*
1	0.20	3	0.20	4,000	0.20
2	0.50	4	0.60	6,000	0.50
3	0.30	5	0.20	8,000	0.30

Simulate the situation by generating random numbers to find out the selling price, variable cost, and annual sales volume. The random numbers for the three factors are:

00 – 19, 20 – 69, 70 – 99; 0 – 19, 20 – 79, 80 – 99; and
00 – 19, 20 – 69, 70 – 99, respectively.

Determine the average profit for 20 simulations (observations).
[*Answer*: Average profit = Rs 2,600.]

Section 13.8

47. Consider the network of a project given in the following figure:

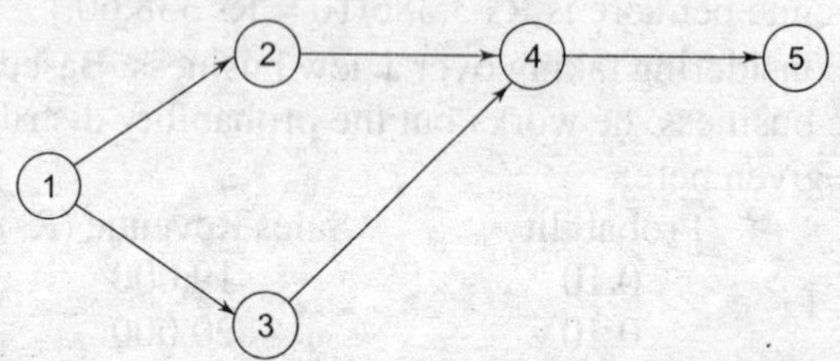

figure 13.4

Suppose, the duration of the activities is non-deterministic with the following probability distribution:

Activity	*Days*	*Probability*
1 – 2	1	.2
	4	.5
	8	.3
1 – 3	2	.3
	4	.5
	7	.2
2 – 4	2	.3
	4	.3
	6	.4
3 – 4	3	.3
	6	.4
	8	.3
4 – 5	2	.2
	3	.2
	4	.6

Simulate the duration of the project 10 times and estimate the chances of various paths to be critical. What is the average duration of the project?

48. For a project comprising activities *A*, *B*, ..., *H*, the following information is available:
Precedence relationships:
A and *B* are the first activities of the project. *C* succeeds *A* while *B* precedes *D*. Both *C* and *D* precede *E* and *F* activity follows activity *F* while *H* is the last activity of the project and succeeds *E* and *G*.
Time estimates and probabilities:

Activity	Time (days)					
	1	2	3	4	5	6
A		0.02	0.4	0.04		
B				0.3		0.7
C			0.3	0.3	0.4	
D		0.2	0.6		0.2	
E				0.6	0.4	
F		.08	0.2			
G	0.3		0.5	0.2		
H		0.1	0.2	0.3	0.4	

Simulate the project and determine, using random numbers, the activity times. Find the critical path and the project duration. Repeat five times. Does the critical path change? State the estimated duration of the project in each of the trials.

Section 13.10

49. Why do we need a computer to do simulation? Explain.
50. An assembly line has three work stations. The time required for each station to complete its operation is as follows:

Time	Probabilities		
(minutes)	Station 1	Station 2	Station 3
4	0.25	0.10	0.05
5	0.25	0.30	0.25
6	0.25	0.40	0.25
7	0.25	0.20	0.45

The times given are the only values the operation times take on. Simulate the flow of 20 items through the assembly line. What is the average time that an item takes to go through all the operations?

14

Non-Linear Programming

14.1 INTRODUCTION

One of the important assumptions of linear programming problem is the linear relationship among the decision variables. This means that each decision variable will have an independent effect on the objective and will contribute a constant and proportional amount to it and that each constraint will involve a linear relationship. In many real-world cases, decision variables do not always utilise a constant and independent amount. For instance, expansion programmes of a company forces work groups to specialise in particular tasks. This specialisation, in turn, creates productive efficiencies that can reduce rates of utilisation of resources, as the level of activity increases. For example, tripling production might require only a doubling rather than tripling of resource utilisation. Moreover, decision variables often have interrelated, instead of independent, effects on the system constraint relationships. For instance, consider an electrical service company in which the same personnel are engaged to serve both residential and commercial customers. That is, each customer group utilises joint rather than separate resources and facilities. For this reason, an increase in commercial calls could divert available service personnel from residential customer requests. If such a diversion delays service, the completion time per residential request would depend on the number of commercial calls. Thus, residential and commercial calls have an interrelated effect on the usage of service capacity.

In addition, each decision variable may not always make a constant and independent contribution to the objective function which we seek to optimise. A company, for example, can usually sell a larger quantity by lowering its price. Thus relevance per unit item will decrease as the sales quantity increases. In some other cases manufacturing cost per unit may decline as the output level increases. Thus a product's per unit contribution to an objective function will fluctuate rather than remain constant as the level of activity changes.

In some situations the decision variables frequently have interrelated, rather than, independent, effects on the objective criterion. For example, consider a company that sells razors and blades. Since these products complement each other in the market, one may expect that a growth in razor sales increases the demand for blades. As a consequence, the profit from blades may depend on razor sales, apart from the number of sales for blades. In other words, razor and blade sales make interrelated contributions to the objective criterion.

When any decision activity makes a variable unit contribution or interrelated contribution to the objective criterion the mathematical model. Objective function will not involve a linear relationship. In such a case a decision variable will have an exponent from one and/or more than one activity will appear in a single term of the function.

A variable utilisation rate or interaction among the decision activities creates a non-linear relationship in the corresponding constraint expression. As a result, the mathematical model consists of a non-linear objective function and/or at least one nonlinear constraint expression, which is offer referred to as a nonlinear program.

Thus, in a non-linear programming problem, either the objective function is non-linear or one or more constraints have non-linear relationship or both.

14.2 FORMULATION OF A NON-LINEAR PROGRAMMING PROBLEM (NLPP)

Some of the real-life situations can be formulated as a NLPP.

Example Production Allocation Problem

A manufacturing company produces two products: radios and TV sets. Sales-price relationships for these two products are given below:

table 14.1

Product	*Quantity demanded*	*Unit Price*
Radios	1,500 – 5 *p*	*p*
TV sets	3,800 – 10 *q*	*q*

The total cost functions for these two products are given by $200x + 0.1x^2$ and $300y + 0.1y^2$, respectively. The production takes place on two assembly lines. Radio sets are assembled on Assembly line I and TV sets are assembled on Assembly line II. Because of the limitations of the assembly line capacities, the daily production is limited to no more than 80 radio sets and 60 TV sets. The production of both types of products requires electronic components. The production of each of these sets requires five units and six units of electronic equipment components, respectively. The electronic components are supplied by another manufacturer, and the supply is limited to 600 units per day. The company has 160 employees, that is, the labour supply amounts to 160 man-days. The production of one unit of radio set requires 1 man-day of labour, whereas 2 man-days of labour are required for a TV set. How many units of radio and TV sets should the company produce in order to maximise the total profit? Formulate the problem as a non-linear programming problem.

Solution: Let us assume that whatever is produced is sold in the market. Let x and y stand for the quantities of radio sets and TV sets, respectively, produced by the firm. Then, we are given that

$$\begin{aligned} x &= 1{,}500 - 5p \\ y &= 3{,}800 - 10q \end{aligned} \quad \text{or} \quad \left\{ \begin{aligned} p &= 300 - 0.2x \\ q &= 380 - 0.1y \end{aligned} \right.$$

Further, if C_1 and C_2 stand for the total cost of production of these amounts of radio sets and TV sets, respectively, then we are also given that

$$C_1 = 200x + 0.1x^2 \text{ and } C_2 = 300y + 0.1y^2$$

Now, the revenue from radio sets is px and on TV sets is qy. Thus, the total revenue $R = px + qy$.

That is,

$$R = (300 - 0.2x)x + (380 - 0.1y)y$$
$$= 300x - 0.2x^2 + 380y - 0.1y^2$$

The total profit z is measured by the difference between the total revenue R and the total cost $C = C_1 + C_2$.

Thus, $z = R - C_1 - C_2 = 100x - 0.3x^2 + 80y - 0.2y^2$

The objective function thus obtained is a non-linear function.

In the present case, production is influenced by the available resources. The two assembly lines have limited capacity to produce radio and TV sets. Since no more than 80 radio sets can be assembled on assembly line I and 60 TV sets on assembly line II per day, we have the restrictions: $x \le 80$ and $y \le 60$.

There is another constraint in the daily requirement of the electronic components, so that $5x_1 + 6x_2 \le 600$. The number of available employees is limited to 160 man-days. Thus, $x_1 + 2x_2 \le 160$. Also obviously, since the manufacturer cannot produce negative number of units, we must have $x_1 \ge 0$ and $x_2 \ge 0$.

Hence, the required NLPP takes the form

$$\text{Maximise } Z = 100x - 0.3x^2 + 80y - 0.2y^2$$

Subject to the constraints

$$0 \le x \le 80,\ 0 \le y \le 60,\ 5x + 6y \le 600,$$
$$x + 2y \le 160 \text{ and } x, y \ge 0.$$

General Non-linear Programming Problem

Let z be a real value function of n variables defined by

$$z = f(x_1, ..., x_n)$$

Let $\{b_1, b_2, ..., b_m\}$ be a set of constraints such that

$$g_1(x_1, x_2, ..., x_n) \{\le, \ge \text{ or } =\} b_1$$
$$g_2(x_1, x_2, ..., x_n) \{\le, \ge \text{ or } =\} b_2$$
$$\vdots$$
$$g_m(x_1, x_2, ..., x_n) \{\le, \ge \text{ or } =\} b_m$$

where, $g_i s$ are real value functions of n variables $x_1, ..., x_n$. Finally, let $x_j \ge 0, j = 1, 2, ..., n$.

If either $f(x_1, ..., x_n)$ or some $g^i(x_1, ..., x_n), i = 1, 2, ..., m$, or both are non-linear, then the problem of determining the n-type $(x_1, ..., x_n)$ which makes z a maximum or minimum and satisfies the constraints is called as a general non-linear programming problem (GNLPP).

In matrix notations, the GNLPP may be written as:

Determine $X^T \in R^n$ so as to maximise or minimise the objective function $z = f(x)$, subject to the constraints

$$g^i(X) \{\le, \ge \text{ or } =\} b_i, X \ge 0 \quad i = 1, 2, ..., m$$

where, either $f(x)$ or some $g^i(X)$ or both are non-linear in X.

Sometimes it is convenient to write the constraints $g^i(x) \{\le, \ge \text{ or } =\} b_i$ as $h^i(X) \{\le, \ge \text{ or } =\}$ for $h_i(X) = g_i(X) - b_i$.

There is one simplex-like solution procedure for the solution of the general non-linear programming problem. However, numerous solution methods have been developed since the appearance of the fundamental theoretical paper by Khun and Tucker. A few primary types of available solution techniques will be discussed in the subsequent sections.

14.3 UNCONSTRAINED OPTIMISATION

14.3.1 Univariate Optimisation

In this section we deal with the optimisation of a single variable function $y = f(x)$. Figure 14.1, is a plot of a continuous function $y = f(x)$ which illustrates some of the concepts associated with the optimisation theory.

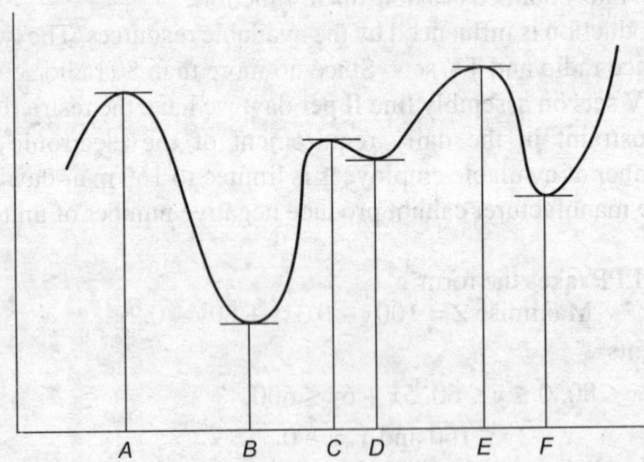

figure 14.1

We have shown some points *A*, *B*, *C*, *D*, *E* and *F* associated with peaks and troughs of the curve.

Point *A* is called the relative or local maximum of the curve.

Point *F* is called the relative or local minimum of the curve.

Point *E* is called absolute or global maximum of the curve.

Point *B* is called the absolute or global minimum of the curve.

Such maxima or minima are known collectively as extreme points. In the figure, there is a line drawn tangent to the curve $y = f(x)$ at *A*, *B*, *C*, *D*, *E* and *F*. The tangent lines are all flat, that is, they have a slope equal to zero. Such points are called stationary points.

Thus, extreme points are stationary points. But the converse is not true, since *C* is a stationary point but not an extreme point. The point *C* is an inflection point or saddle point.

Procedure for finding the saddle point:

Step 1: Find the first derivate of $f(x)$ denoted by $f'(x)$ or $\frac{d}{dx}(f(x))$.

Step 2: Solve $f'(x) = 0$ for x.

Step 3: Find the second derivative of $y = f(x)$ and determine its value at the stationary points. A negative (positive) value indicates that the stationary point is a maximum (minimum).

Example

Find the extreme points of $y = x^3 - 12x$.

Solution:

$$y = x^3 - 12x$$

$$\frac{dy}{dx} = 3x^3 - 12$$

$$\frac{dy}{dx} = 0 \quad \Rightarrow 3x^2 - 12 = 0$$

$$\Rightarrow x = \pm 2$$

Thus, $x = +2$ and $x = -2$ are the stationary points.

Now, $$\frac{d^2y}{dx^2} = 6x$$

$$\left.\frac{d^2y}{dx^2}\right/_{x=+2} = +12 > 0. \text{ Hence, } x = +2 \text{ is a point of minimum.}$$

$$\left.\frac{d^2y}{dx^2}\right/_{x=-2} = -12 < 0. \text{ Hence, } x = -2 \text{ is a point of maximum.}$$

14.3.2 Bivariate Optimisation

Let $f(x_1, x_2)$ be a function of two variables. Let f_{x_1} or (f_1) denote the partial derivative of f with respect to x_1 and let f_{x_2} or (f_2) denote the partial derivative of f with respect to x_2. $f_{x_1x_2}$ denotes derivative of y^f first with respect to x_2 and then with respect to x_1.

Procedure for obtaining extreme points

Step 1: Find the partial derivative f_{x_1}, f_{x_2} of the original function.

Step 2: Set the partial derivatives to zero and solve for x_1 and x_2. The solution yields stationary points denoted by $\bar{x}$.

Step 3: Evaluate the second derivative by substituting the stationary points and form a matrix of these values called the Hessian matrix at $\bar{x}$.

Step 4: The stationary point is a minimum if the leading minors of the Hessian matrix

$$f_{x_1x_1} > 0 \text{ and } \begin{vmatrix} f_{x_1x_1} & f_{x_1x_2} \\ f_{x_2x_1} & f_{x_2x_2} \end{vmatrix} > 0$$

but a maximum if the leading minors of the Hessian matrix

$$f_{x_1x_1} < 0 \text{ and } \begin{vmatrix} f_{x_1x_1} & f_{x_1x_2} \\ f_{x_2x_1} & f_{x_2x_2} \end{vmatrix} > 0$$

where | | denotes the determinant.

Example

Find the extreme points of the bivariate function $f(x_1, x_2) = 4x_1^2 + 8x_1x_2$.

Solution: We compute the first partial derivatives

$$f_{x_1} = 8x_1 + 8x_2$$
$$f_{x_2} = 8x_1$$

The second partial derivatives are

$$f_{x_1x_1} = 8 + 0 = 8, f_{x_1x_2} = f_{x_2x_1} = 8; f_{x_2x_2} = 0$$

The Hessian matrix is

$$H(x) = \begin{bmatrix} 8 & 8 \\ 8 & 0 \end{bmatrix}$$

Setting the first partial derivatives to zero

$$8x_1 + 8x_2 = 0$$

$$8x_1 = 0 \qquad \text{we get}$$

$$x_1 = 0,\ x_2 = 0.$$

Thus the stationary point is $\bar{x} = (0, 0)$. Since the Hessian matrix entries do not involve any variables, the evaluated Hessian matrix $\bar{x}$ is the above matrix itself.

Since $8 > 0$ and $\begin{vmatrix} 8 & 8 \\ 8 & 0 \end{vmatrix} < 0,$

we conclude that the stationary point (0, 0) is neither a minimum point nor a maximum point.

14.3.3 Multivariate Optimisation

Let us consider a function $f(x_1, x_2, x_3)$ of three variables x_1, x_2 and x_3. We shall indicate the natural extension to any number of variables. Here also, we evaluate the first partial derivatives, namely, fx_1, fx_2, fx_3 and the second partial derivatives, namely, $fx_1x_1, fx_1x_2, fx_1x_3, fx_2x_1, fx_2x_2, fx_2x_3, fx_3x_1, fx_3x_2$ and fx_3x_3. To find the optimum of $f(x_1, x_2, x_3)$ we follow the steps given below.

(i) Solve the equations

$$f_{x_1} = 0$$
$$f_{x_2} = 0$$
$$f_{x_3} = 0$$

to get the stationary point $\bar{x}$.

(ii) Evaluate the second partial derivaties at $\bar{x}$ and form the following Herssian matrix

$$H(\bar{x}) = \begin{bmatrix} f_{x_1x_1} & f_{x_1x_2} & f_{x_1x_3} \\ f_{x_2x_1} & f_{x_2x_2} & f_{x_2x_3} \\ f_{x_3x_1} & f_{x_3x_2} & f_{x_3x_3} \end{bmatrix}$$

(iii) Consider the leading minors of $H(\bar{x})$

$$f_{x_1x_2} \quad \begin{vmatrix} f_{x_1x_1} & f_{x_1x_2} \\ f_{x_2x_1} & f_{x_2x_2} \end{vmatrix} \quad \text{and} \quad \begin{vmatrix} f_{x_1x_1} & f_{x_1x_2} & f_{x_1x_3} \\ f_{x_2x_1} & f_{x_2x_2} & f_{x_2x_3} \\ f_{x_3x_1} & f_{x_3x_2} & f_{x_3x_3} \end{vmatrix}$$

(iv) If the sign of the above leading minors is positive, the stationary point $\bar{x}$ is a minimum point. On the other hand, if the signs are alternating, with minus sign for the first leading minor, then the stationary point is a maximum point.

Example

Examine the following function for extreme points $f(x_1, x_2, x_3) = x_1 + 2x_3 + x_2x_3 - x_1^2 - x_2^2 - x_3$.

Solution: We compute the first partial derivatives

$$fx_1 = 1 - 2x,$$
$$fx_2 = x_3 - 2x_2$$
$$fx_3 = 2 + x_2 - 2x_3$$

Setting these first derivatives equal to zero and solving them, we get the stationary point $\bar{x} = (1/2, 2/3, 4/3)$.

Next, we compute the second partial derivatives,

$$fx_1x_1 = 0 - 2 = -2$$
$$fx_1x_2 = 0, fx_1x_3 = 0$$
$$fx_2x_1 = 0, fx_2x_2 = 0 - 2 = -2$$
$$fx_2x_3 = 1 - 0 = 1, fx_3x_1 = 0$$
$$fx_3x_2 = 0 + 1 - 0 = 1, fx_3x_3 = 0 + 0 - 2 = -2$$

Here also, the second partial derivatives do not involve x_1, x_2, x_3 and evaluating these derivatives at $\bar{x}$ is not necessary.

Thus, the Hessian matrix is

$$\begin{bmatrix} -2 & 0 & 0 \\ 0 & -2 & 1 \\ 0 & 1 & -2 \end{bmatrix}$$

The leading minors whose sign we have to observe are

$$-2, \begin{vmatrix} -2 & 0 \\ 0 & -2 \end{vmatrix} \text{ and } \begin{vmatrix} -2 & 0 & 0 \\ 0 & -2 & 1 \\ 0 & 1 & -2 \end{vmatrix}$$

The exact values of the above quantifiers are – 2, + 4 and – 6

Since the signs are –, + and –, the stationary point $\bar{x}$ is a maximum point.

14.4 CONSTRAINED OPTIMISATION

14.4.1 Constrained Optimisation with Equality Constraints

If the non-linear programming problem is composed of some differentiable objective function and equality sign constraints, the optimisation may be achieved by the use of Lagrange multipliers (a systematic way of generating the necessary conditions for a stationary point) as illustrated below.

Consider the problem of maximising or minimising $z = f(x, x_2)$ subject to the constraints:

$$g(x_1, x_2) = c \text{ and } x_1, x_2 \geq 0, \text{ where } c \text{ is a constant.}$$

We assume that $f(x_1, x_2)$ and $g(x_1, x_2)$ are differentiable w.r.t. x_1 and x_2. Let us introduce a different function $h(x_1, x_2)$ differentiable w.r.t. x_1 and x_2 and defined by $h(x_1, x_2) = g(x_1, x_2) - c$. Then, the problem can be restated as follows:

Maximise $z = f(x_1, x_2)$ subject to the constraints

$$h(x_1, x_2) = 0 \text{ and } x_1, x_2 \geq 0.$$

To find the necessary conditions for a maximum (or minimum) value of Z, a new function is formed by introducing a Lagrange multiplier λ, as

$$L(x_1, x_2, \lambda) = f(x_1, x_2) - \lambda h(x_1, x_2).$$

The number λ is an unknown constant, and the function $L(x_1, x_2, \lambda)$ is called the Lagrangian function with Lagrange multiplier λ. The necessary conditions for a maximum or minimum (stationary value) of $f(x_1, x_2)$ subject to $h(x_1 x_2) = 0$ are thus given by

$$\frac{\partial L(x_1, x_2, \lambda)}{\partial x_1} = 0, \quad \frac{\partial L(x_1, x_2, \lambda)}{\partial x_2} = 0 \text{ and}$$

$$\frac{\partial L(x_1, x_2, \lambda)}{\partial \lambda} = 0.$$

Now, these partial derivatives are given by

$$\frac{\partial L}{\partial x_1} = \frac{\partial f}{\partial x_1} - \frac{\partial h}{\partial x_1},$$

$$\frac{\partial L}{\partial x_2} = \frac{\partial f}{\partial x_2} - \lambda \frac{\partial h}{\partial x_2}, \text{ and}$$

$$\frac{\partial L}{\partial \lambda} = -h,$$

where, L, f and h stand for the functions $L(x_1, x_2, \lambda), f(x_1, x_2)$ and $h(x_1, x_2)$, respectively, or simply by

$$L_1 = f_1 - \lambda h_1,\ L_2 = f_2 - \lambda h_2 \text{ and } L_\lambda = -h$$

The necessary conditions for maximum or minimum of $f(x_1, x_2)$ are thus given by

$$f_1 = \lambda h_1, f_2 = \lambda h_2 \text{ and } -h(x_1, x_2) = 0$$

Remark: The necessary conditions become sufficient conditions for a maximum (minimum) if the objective function is concave (convex) and the constraints are in the form of equalities. (We recall that, a function $f(x)$ on a non-empty subset of R^n is said to be convex if for any two vextors x_1 and x_2 is S,

$$f[\lambda x_1 + (1 - \lambda)x_2] \leq \lambda f(x_1), + (1 - \lambda)f(x_2), 0 \leq \lambda \leq 1.$$

A function $f(x)$ is said to be concave if $-f(x)$ is convex.

Example 1

Obtain the necessary and sufficient conditions for the optimum solution of the following NLPP:

$$\text{Minimise } Z = 3e^{2x_1 + 1} + 2e^{x_2 + 5}$$

subject to $$x_1 + x_2 = 7,\ x_1, x_2 \geq 0.$$

Solution: Let us introduce a new differentiable Lagrangian function $L(x_1, x_2, \lambda)$ defined by

$$L(x_1, x_2, \lambda) = f(x_1, x_2) - \lambda(x_1 + x_2 - 7)$$
$$= 3e^{2x_1+1} + 2e^{x_2+5} - \lambda(x_1 + x_2 - 7)$$

where, λ is the Lagrangian multiplier.

Since the objective function $Z = f(x_1, x_2)$ is convex and the side constraint an equality one, the necessary and sufficient conditions for the minimum of $f(x_1, x_2)$ are given by

$$\frac{\partial L}{\partial x_1} = 6e^{2x_1+1} - \lambda = 0 \quad \text{or} \quad \lambda = 6e^{2x_1+1}$$

$$\frac{\partial L}{\partial x_2} = 2e^{x_2+5} - \lambda = 0 \quad \text{or} \quad \lambda = 2e^{x_2+5}$$

$$\frac{\partial L}{\partial \lambda} = -(x_1 + x_2 - 7) = 0 \quad \text{or} \quad x_1 + x_2 = 7.$$

These imply, $\quad 6e^{2x_1+1} = 2e^{x_2+5} = 3e^{7-x_1+5}$

or, $\quad \log 3 + 2x_1 + 1 = 7 - x_1 + 5 \quad \text{or} \quad x_1 = 1/3\,(11 - \log 3)$

Thus, $\quad x_2 = 7 - 1/3(11 - \log 3).$

Example 2

Find the dimension of a rectangular parallelopiped with largest volume whose sides are parrallel to the coordinate planes to be inscribed in the ellipsoid

$$g(x, y, z) \equiv \frac{x^2}{a^2} + \frac{y^2}{b^2} + \frac{z^2}{c^2} - 1 = 0$$

Solution: Let the dimensions of a rectangular parallelopiped be x, y, and z. Its volume is then given by $v(x, y, z) = xyz$

Forming the Lagrangian function L, we have

$$L(x, y, z, \lambda) = v(x, y, z) - \lambda g(x, y, z)$$

Now, differentiating the above equation with respect to each variable and setting the results equal to zero, we obtain

$$\frac{\partial L}{\partial x} = yz - \frac{2\lambda x}{a^2} = 0$$

$$\frac{\partial L}{\partial y} = xz - \frac{2\lambda y}{b^2} = 0$$

$$\frac{\partial L}{\partial z} = xy - \frac{2\lambda z}{c^2} = 0$$

$$\frac{\partial \lambda}{\partial L} = -\left(\frac{x^2}{a^2} + \frac{y^2}{b_2} + \frac{z^2}{c^2} - 1\right) = 0$$

Multiplying first three equations by x, y, z, respectively, adding, and then making use of the last equation, we obtain,

$$3v(x, y, z) - 2\lambda = 0$$

Thus, $\lambda = \frac{3}{2} v(x, y, z)$

Now, with this value of λ substituted in first three equations, we have

$$x = a/\sqrt{3}, y = b/\sqrt{3}, z = c/\sqrt{3}$$

which is the required answer.

Generalised Lagrangian Method to *n*-Dimensional Case

Consider the general NLPP

Maximise (or minimise) $Z = f(x_1, x_2, ..., x_n)$ subject to the constraints

$$g_i(x_1, ..., x_n) = c_i \text{ and } x_i \geq 0, i = 1, 2, ... m \ (m < n)$$

The constraints can be written as

$$h_i(x_1, ..., x_n) = 0 \text{ for } i = 1, 2, ..., m$$

where, $h_i(x_1, ..., x_n) = g_i(x_1, ..., x_n) - c_i$

The problem can be written in the matrix form as

Maximise (or minimise) $Z = f(x)$, $x \in R^n$ subject to the constraints $h_i(x) = 0$, $x \geq 0$.

To find the necessary conditions for a maximum of minimum of $f(x)$, the Lagrangian function $L(x, \lambda)$ is formed by introducing m Lagrangian multipliers $\lambda = (\lambda_1, \lambda_2, ..., \lambda_m)$. This function is defined by

$$L(x, \lambda) = f(x) = \sum_{i=1}^{m} \lambda_i h_i(x).$$

Assuming that L, f and h_i are all partially differentiable w.r.t. $x_1, x_2, ..., x_n$ and $\lambda_1, \lambda_2, ..., \lambda_m$, the necessary conditions for a maximum (minimum) of $f(x)$ are

$$\frac{\partial L}{\partial x_j} = \frac{\partial f}{\partial x_j} - \sum_{i=1}^{m} \lambda_i \frac{\partial h_i(x)}{\partial x_j} = 0, \quad j = 1, 2, ..., n$$

$$\frac{\partial L}{\partial \lambda_j} = - h_i(x) = 0 \qquad i = 1, 2, ..., n$$

These $m + n$ necessary conditions can be represented in the following abbreviated form:

$$L_j = f_i - \sum_{i=1}^{m} \lambda_i h_{ij} = 0 \text{ or } f_j = \sum_{i=1}^{m} \lambda_i h_{ij}$$

$$j = 1, 2, ..., n$$

$$L\lambda_j = - h^i = 0 \text{ or } h^i = 0, \quad i = 1, 2, ..., m$$

where, $f_j = \frac{\partial f(x)}{\partial x_j}$, $h_i = h_i(x)$ and

$$h_{ij} = \frac{\partial h_i(x)}{\partial x_j}$$

Note: These necessary conditions also become sufficient for a maximum (minimum) of the objective function if the objective function is concave (convex) and the side constraints are equality ones.

Example

Obtain the set of necessary conditions for the non-linear programming problem

$$\text{Maximise } Z = x_1^2 + 3x_2^2 + 5x_3^2$$

subject to the constraints

$$x_1 + x_2 + 3x_3 = 2$$
$$5x_1 + 2x_2 + x_3 = 5$$
$$x_1, x_2, x_3 \geq 0$$

Solution: In this problem we are given that

$$x = (x_1, x_2, x_3), f(x) = x_1^2 + 3x_2^2 + 5x_3^2,$$
$$h_1(x) = x_1 + x_2 + 3x_2 - 2 = 0 \; h_2(x) = 5x_1 + 2x_2 + x_3 - 5 = 0$$

We construct the Lagrangian function for maximising $f(x)$

$$L(x, \lambda) = f(x) - \lambda_1 h_1(x) - \lambda_2 h_2(x).$$

This gives the following necessary conditions:

$$\frac{\partial L}{\partial x_1} = 2x_1 - \lambda - 5\lambda_2 = 0, \frac{\partial L}{\partial x_2} = 6x_2 - \lambda_1 - 2\lambda_2 = 0,$$

$$\frac{\partial L}{\partial x_3} = 10x_3 - 3\lambda_1 - \lambda_3 = 0, \frac{\partial L}{\partial \lambda_1} = -(x_1 + x_2 + 3x_3 - 2) = 0$$

$$\frac{\partial L}{\partial x_2} = -(5x_1 + 2x_2 + x_3 - 5) = 0$$

Sufficient Conditions for Maximum (Minimum) of Objective Function (with Single Equality Constraint)

Let the Lagrangian function for a general NLPP involving n variables and one constraint be:

$$L(x, \lambda) = f(x) - \lambda h(x).$$

The necessary conditions for a stationary point to be a maximum or minimum are

$$\frac{\partial L}{\partial x_i} = \frac{\partial f}{\partial x_j} - \lambda \frac{\partial h}{\partial x_j} = 0 \; (j = 1, 2, ..., n)$$

and,

$$\frac{\partial L}{\partial \lambda} = -h(x) = 0.$$

The value of λ is obtained by $\lambda = \dfrac{\partial f/\partial x_j}{\partial h/\partial x_j}$,

$$j = 1, 2, ..., n$$

The sufficient conditions for a maximum or minimum require the evaluation at each stationary point of $n - 1$ principal minors of the determinant given below:

$$\Delta_{n+1} = \begin{vmatrix} 0 & \frac{\partial h}{\partial x_1} & \frac{\partial h}{\partial x_2} & \cdots & \frac{\partial h}{\partial x_n} \\ \frac{\partial h}{\partial x_1} & \frac{\partial^2 f}{\partial x_1^2} - \lambda \frac{\partial^2 h}{\partial x_1^2} & \frac{\partial^2 f}{\partial x_1 \partial x_2} - \lambda \frac{\partial^2 h}{\partial x_1 \partial x_2} & \cdots & \frac{\partial^2 f}{\partial x_1 \partial x_n} - \lambda \frac{\partial^2 h}{\partial x_1 \partial x_n} \\ \frac{\partial h}{\partial x_2} & \frac{\partial^2 f}{\partial x_2 \partial x_1} - \lambda \frac{\partial^2 h}{\partial x_2 \partial x_1} & \frac{\partial^2 f}{\partial x_2^2} - \lambda \frac{\partial^2 h}{\partial x_2^2} & \cdots & \frac{\partial^2 f}{\partial x_2 \partial x_n} - \lambda \frac{\partial^2 h}{\partial x_2 \partial x_n} \\ \vdots & \vdots & \vdots & & \\ \frac{\partial h}{\partial x_n} & \frac{\partial^2 f}{\partial x_n \partial x_1} - \lambda \frac{\partial^2 h}{\partial x_n \partial x_2} & \frac{\partial^2 f}{\partial x_n \partial x_2} - \lambda \frac{\partial^2 h}{\partial x_n \partial x_2} & \cdots & \frac{\partial^2 f}{\partial x_n^2} - \lambda \frac{\partial^2 h}{\partial x_n^2} \end{vmatrix}$$

If $\Delta_3 > 0, \Delta_4 < 0, \Delta_5 > 0, \ldots$, the sign pattern being alternate, the stationary point is a local maximum.
If $\Delta_3 < 0, \Delta_4 < 0, \ldots \Delta_{n+1} < 0$, the sign being always negative, the stationary point is a local minimum.

Example

Solve the non-linear programming problem

Minimise $Z = 2x_1^2 - 24x_1 + 2x_2^2 - 8x_2 + 2x_3^2 - 12x_3 + 20$

subject to the constriants

$$x_1 + x_2 + x_3 = 1$$

$$x_1, x_2, x_3 \geq 0.$$

Solution: The Lagrangian function can be formulated as follows:

$$L(x_1, x_2, x_3, \lambda) = 2x_1^2 - 24x_1 + 2x_2^2 - 8x_2 + 2x_3^2 - 12x_3 + 200 - \lambda(x_1 + x_2 + x_3 - 11).$$

The necessary conditions for the stationary point are:

$$\frac{\partial L}{\partial x_1} = 4x_1 - 24 - \lambda \pm 0, \quad \frac{\partial L}{\partial x_2} = 4x_2 - 8 - \lambda = 0,$$

$$\frac{\partial L}{\partial x_3} = 4x_3 - 12 - \lambda = 0, \quad \frac{\partial L}{\partial \lambda} = -(x_1 + x_2 + x_3 - 11) = 0.$$

By solving these simultaneous equations, we get the stationary point $x = (x_1, x_2, x_3) = (6, 2, 3), \lambda = 0$.
The sufficient condition for the stationary point

$$x = (x_1, x_2, x_3) = (6, 2, 3); \lambda = 0.$$

The sufficient condition for the stationary point to be a minimum is that minors Δ_3 and Δ_4 both be negative. Now,

$$\Delta_3 = \begin{vmatrix} 0 & 1 & 1 \\ 1 & 4 & 0 \\ 1 & 0 & 4 \end{vmatrix} \quad \text{and} \quad \Delta_4 = \begin{vmatrix} 0 & 1 & 1 & 1 \\ 1 & 4 & 0 & 0 \\ 1 & 0 & 4 & 0 \\ 1 & 0 & 0 & 4 \end{vmatrix} = -48$$

Thus, $x = (6, 2, 3)$ is the solution to the given LPP.

Sufficient Conditions for Maximum (Minimum) of Objective Function (with More than One Equality Constraints)

Introducing the m Lagrangian multipliers $\lambda = (\lambda_1, \lambda_2, ..., \lambda_m)$, let the Lagrangian function for a general NLPP with more than one constraint be

$$L(x, \lambda) = f(x) = -\sum_{i=1}^{m} \lambda_i h_i(x) \qquad (m < n).$$

The reader may verify that the equations

$$\frac{\partial L}{\partial x_i} = 0 \text{ and } \frac{\partial L}{\partial \lambda_i} = 0 \; (i = 1, 2, m; j = 1, 2, ...n)$$

yield the necessary conditions for stationary points of $f(x)$. Thus, the optimisation of $f(x)$ subject to $h(x) = 0$ is equivalent to the optimisation of $L(x, \lambda)$.

We state here the sufficiency conditions for the Lagrange multiplier method of stationary point of $f(x)$ to be a maxima or minima without proof. For this, we assume that the function $L(x, \lambda)$, $f(x)$ and $h(x)$ all possess partial derivative of order one and two w.r.t. the decision variables.

Let $$V = \left(\frac{\partial^2 L(x,\lambda)}{\partial x_i \partial x_j}\right)_{n \times n}$$

be the matrix of the second order partial derivatives of $L(x, \lambda)$ w.r.t. O decision variables

$$U = [h_{ij}(x)]_{m \times n} \text{ where}$$

$$h_{ij}(x) = \frac{\partial h_i(x)}{\partial x_j}, \; i = 1, 2, ..., m; \; j = 1, 2, ..., n$$

Define the square matrix

$$H^B = \begin{bmatrix} O & \vline & U \\ \cdots & \vline & \cdots \\ U^T & \vline & V \end{bmatrix}_{(m+n) \times (m+n)}$$

where, O is a $m \times n$ null matrix. The matrix H^B is called the boardered Hessian matrix. Then, the sufficient conditions for maximum and minimum stationary point are given below.

Let (x_0, λ_0) for the function $L(x, \lambda)$ be its stationary point. Let H_0^B be the corresponding bordered Hessian matrix computed at this satisfactory point. Then, x_0 is a

(a) Maximum point, if starting with principal minor of order $(2m + 1)$, the last $(n - m)$ principal minors of H_0^B form an alternating sign pattern starting with $(-1)^{m+n}$; and

(b) Minimum point, if starting with principal minor of order $(2m + 1)$, the list $(n - m)$ principal minors of H_0^B have the sign of $(-1)^m$.

Note: It may be observed that the above conditions are only sufficient for identifying an extreme point, but not necessary condition. That is, a stationary point may be an extreme point without satisfying the above conditions.

Example

Solve the non-linear programming problem

Optimise $Z = 4x_1^2 + 2x_2^2 + x_3^3 - 4x_1x_2$ subject to the constraints $x_1 + x_2 + x_3 = 15$, $2x_1 - x_2 + 2x_3 = 20$

Solution:

$$\text{Now } f(x) = 4x_1^1 + 2x_2^2 + x_3^2 - 4x_1x_2$$

$$h_1(x) = x_1 + x_2 + x_3 - 15$$

$$h_2(x) = 2x_1 - x_2 + 2x_3 - 20$$

Construct the Lagrangian function

$$L(x, \lambda) = f(x) - \lambda_1 h_1(x) - \lambda_2 h_2(x)$$

$$= (4x_1^1 + 2x_2^2 + x_3^2 - 4x_1x_2) - \lambda_1(x_1 + x_2 + x_3 - 15) - \lambda_2(2x_1 - x_2 + 2x_3 - 20)$$

The stationary point (x_0, λ_0) has thus given the following necessary conditions:

$$\frac{\partial L}{\partial x_1} = 8x_1 - 4x_2 - \lambda_1 - 2\lambda_2 = 0$$

$$\frac{\partial L}{\partial x_2} = 4x_2 - 4x_1 - \lambda_1 + \lambda_2 = 0$$

$$\frac{\partial L}{\partial x_3} = 2x_3 - \lambda_1 - 2\lambda_2 = 0$$

$$\frac{\partial L}{\partial \lambda_1} = -[x_1 + x_2 + x_3 - 15] = 0$$

$$\frac{\partial L}{\partial \lambda_2} = -[2x_1 - x_2 + 2x_3 - 20] = 0$$

The solution to these simultaneous equations yields

$$x_0 = (x_1, x_2, x_3) = (33/9, 10/3, 8) \text{ and}$$

$$\lambda_0 = (\lambda_1, \lambda_2) = (40/9, 52/9).$$

The bordered Hessian matrix at this solution (x_0, λ_0) is given by

$$H_0^B = \begin{bmatrix} 0 & 0 & \vdots & 1 & 1 & 1 \\ 0 & 0 & \vdots & 2 & -1 & 2 \\ \cdots & \cdots & \vdots & \cdots & \cdots & \cdots \\ 1 & 2 & \vdots & 8 & -4 & 0 \\ 1 & -1 & \vdots & -4 & 4 & 0 \\ 1 & 2 & \vdots & 0 & 0 & 2 \end{bmatrix}$$

Hence, since $n = 3$ and $m = 2$ therefore $n - m = 1$, $(2m + 1 = 5)$. This means one needs to check the determinant H_0^B only; it must have the sign of $(-1)^2$.

Now, since det $H_0^B = 72 > 0$, x_0 is a minimum point.

14.4.2 Constrained Optimisation with Inequality Constraints

We now derive the Kuhn-Tucker conditions (necessary and sufficient) for the optimal solution of general NLPP. Consider the general NLPP:

Optimise $Z = f(x_1, x_2, ..., x_n)$, subject to the constraints.

$$g(x_1, .., x_n) \leq C \text{ and } x_1, x_2, ..., x_n \geq 0$$

where, C is a constant.

Introducing the function $h(x_1, ..., x_n) = g - C$, the constraint reduces to $h(x_1, ..., x_n) \leq 0$. The problem thus can be written as

$$\text{Optimise } Z = f(x) \text{ subject to } h(x) \leq 0 \text{ and } x \geq 0 \text{ where } x \in R^n.$$

We now slightly modify the problem by introducing 9 new variable S, defined by $S^2 = -h(x)$ or $h(x) + S^2 = 0$.

The new variable S is called slack variable and appears as its square in the constraint equation so as to ensure its being non-negative. This avoids an additional constraint $S \geq 0$. Now, the problem can be restated as

$$\text{Optimise } Z = f(x), x \in R^n \text{ subject to the constraints } h(x) + S^2 = 0 \text{ and } x \geq 0.$$

This is a problem of constrained optimisation in $n + 1$ variables and a single equality constraint and can thus be solved by the Lagrangian multiplier method.

To determine the stationary points, we consider the Lagrangian function defined by

$$L(x, s, \lambda) = f(x) - \lambda[h(x) + S^2],$$

where λ is the Lagrange multiplier. The necessary conditions for stationary points are

$$\frac{\partial L}{\partial x_j} \equiv \frac{\partial f}{\partial x_j} - \lambda \frac{\partial h}{\partial x_j} = 0 \text{ for } j = 1, 2, ..., n$$

$$\frac{\partial L}{\partial \lambda} \equiv -[h(x) + S^2] = 0$$

$$\frac{\partial L}{\partial S} \equiv -2S\lambda = 0$$

$\frac{\partial L}{\partial S} = 0$ requires either $\lambda = 0$ or $S = 0$. If $S = 0$, (2) implies that $h(x) = 0$. Thus, we have $\lambda h(x) = 0$.

The variable S was introduced merely to convert the inequality constraints into an equality one, and therefore may be discarded. Moreover, since $S^2 \geq 0$, (2) gives $h(x) \leq 0$ whenever $h(x) < 0$. We get $\lambda = 0$ and whenever $\lambda > 0$, $h(x) = 0$. However, λ is unrestricted in sign whenever $h(x) = 0$.

The necessary conditions for the point x to be a point of maximum are thus restated as (in the abbreviated form):

$f_j - \lambda h_j = 0$	$(j = 1, 2, ..., n)$
$\lambda h = 0$	Maximise f
$h \leq 0$	subject to:
$\lambda \geq 0$	$h \leq 0$

The set of such necessary conditions is called Kuhn-Tucker condition.

A similar argument holds for the minimisation non-linear programming problem:

Minimise $Z = f(x)$ subject to the constraints $g(x) \geq C$ and $x \geq 0$.

Introduction of $h(x) = g(x) - C$ reduces the first constraint to $h(x) \geq 0$. The new surplus variable so can be introduced in $h(x) \geq 0$ so that we may have the equality constraint $h(x) - S_0^2 = 0$. The appropriate Lagrangian function is

$$L(x, S_o, \lambda) = f(x) - \lambda[h(x) - S_0^2]$$

The following set of Kuhn-Tucker conditions is obtained:

$$f_j - \lambda h_j = 0 \qquad (j = 1, 2, \dots n)$$
$$\lambda h = 0 \qquad \text{Minimise} f$$
$$h \geq 0 \qquad \text{subject to}$$
$$\lambda \geq 0 \qquad h \geq 0.$$

Theorem (Sufficiency of Kuhn-Tucker Conditions)

The Kuhn-Tucker conditions for a maximisation NLPP of maximising $f(x)$ subject to the constraints $h(x) \leq 0$ and $x \geq 0$ are sufficient conditions for a maximum of $f(x)$, if $f(x)$ is concave and $h(x)$ is convex.

Proof: The result follows if we are able to show that the Lagrangian function

$$L(x, S, \lambda) = f(x) - [h(x) + S^2],$$

where, S is defined by $h(x) + S^2 = 0$, is concave in x under the given conditions.

In that case the stationary point obtained from the Kuhn-Tucker conditions must be the global maximum point.

Now, since $h(x) + S^2 = 0$, it follows from the necessary conditions that $\lambda S^2 = 0$. Since $h(x)$ is convex and $\lambda \geq 0$, it follows that $\lambda h(x)$ is also convex and $-\lambda h(x)$ is concave. Thus, we conclude that $f(x) - \lambda h(x)$ and hence

$$f(x) - \lambda[h(x) + S^2] = L(x, S, \lambda) \text{ is concave in } x.$$

Note: By a similar argument it can be shown that for the minimisation of NLPP, Kuhn-Tucker conditions are also the sufficient conditions for the minimum of the objective function, if the objective function $f(x)$ is convex and the function $h(x)$ is concave.

Example 1

Use the Kuhn-Tucker conditions to solve the NLPP

$$\text{Max. } Z = 8x_1 + 10x_2 - x_1^2 - 2x_2^2$$

subject to $\quad 3x_1 + 2x_2 \leq 6,\ x_1 \geq 0,\ x_2 \geq 0$

Solution: Here,

$$f(x) = 8x_1 + 10x_2 - x_1^2 - x_2^2$$
$$g(x) = 3x_1 + 2x_2,\ C = 6$$
$$h(x) = g(x) - C = 3x_1 + 2x_2 - 6$$

The Kuhn-Tucker conditions are

$$\frac{\partial f(x)}{\partial x_1} - \lambda \frac{\partial h(x)}{\partial x_1} = 0, \quad \frac{\partial f(x)}{\partial x_2} - \lambda \frac{\partial h(x)}{\partial x_2} = 0$$

$\lambda h(x) = 0$, $h(x) \leq 0$, $\lambda \geq 0$ where λ is the Lagrangian multiplier.

That is,

$$8 - 2x_1 = 3\lambda \tag{14.4.1}$$
$$10 - 2x_2 = 2\lambda \tag{14.4.2}$$
$$\lambda[3x_1 + 2x_2 - 6] = 0 \tag{14.4.3}$$

$$3x_1 + 2x_2 - 6 = 0 \quad \text{...(14.4.4)}$$

$$\lambda \geq 0 \quad \text{...(14.4.5)}$$

From equation (14.4.3) either $\lambda = 0$

or, $3x_1 + 2x_2 - 6 = 0$

Let $\lambda = 0$, then from (14.4.1) and (14.4.2), we get $x_1 = 4$ and $x_2 = 5$. With these values of x_1 and x_2, (14.4.4) cannot be satisfied. Thus, optimal solution cannot be obtained here for $\lambda = 0$. Let then $\lambda \neq 0$ and so $3x_1 + 2x_2 - 6 = 0$. This, together (14.4.1) and (14.4.2) yield the stationary value

$$x_o = (x_1, x_2) = \left(\frac{4}{13}, \frac{33}{13}\right)$$

Now, it is easy to observe that $h(x)$ is convex in x and $f(x)$ is concave in x. Thus, Kuhn-Tucker conditions are the sufficient conditions for the maximum. Hence, $x_o = \left(\frac{4}{13}, \frac{33}{13}\right)$ is the solution to the given LPP NLPP. The maximum value of Z (corresponding to x_o) is given by $Z_0 = 21.3$.

Kuhn-Tucker Conditions for General NLPP with $m(< n)$ constraints)

Introducing $S = (S_1, S_2, ..., S_m)$, let the Lagrangian function for a general NLPP with $m(< n)$ constraints be

$$L(x, s, \lambda) = f(x) - \sum_{i=1}^{m} \lambda_i[h_i(x) + s_i^2]$$

where, $\lambda = (\lambda_1, ..., \lambda_m)$ are the Lagrangian multipliers,

The necessary conditions for $f(x)$ to be a maximum are:

$$\frac{\partial L}{\partial x_i} = \frac{\partial f}{\partial x_j} - \sum_{i=1}^{m} \lambda_i \frac{\partial h_i}{\partial x_j} = 0 \text{ for } j = 1, 2, ..., n... \quad (14.4.6)$$

$$\frac{\partial L}{\partial \lambda_i} = h_i + s_i^2 = 0 \text{ for } i = 1, 2, ..., m \quad (14.4.7)$$

$$\frac{\partial L}{\partial S_i} = -2S_i\lambda_i = 0 \text{ for } i = 1, 2, ..., m \quad (14.4.8)$$

where, $L = L(x, S, \lambda), f = f(x)$ and $h_i = h_i(x)$

Equation (14.4.8) states that either $\lambda_i = 0$ or $S_i = 0$. By an argument parallel to that considered in the case of single constraint, the conditions (14.4.8) and (14.4.7) together are replaced by conditions (14.4.9), (14.4.10) and (14.4.11) below:

$$\lambda_i h_i = 0 \qquad \text{for } i = 1, 2, ..., m \quad (14.4.9)$$

$$h_i \leq 0 \qquad \text{for } i = 1, 2, ..., m \quad (14.4.10)$$

$$h_i \geq 0 \qquad \text{for } i = 1, 2, ..., m \quad (14.4.11)$$

The Kuhn-Tucker conditions for a maximum thus may be restated as

$$f_j = \sum_{i=1}^{m} \lambda_i h_{ij} \qquad (j = 1, 2, ..., n)$$

$$\lambda_i h_i = 0 \quad (i = 1, 2, ..., m)$$
$$h_i \geq 0 \quad (i = 1, 2, ..., m)$$
$$\lambda_i \geq 0$$

Maximise of subject $h_i \leq 0$

where, $$h_{ij} = \frac{\partial h_i}{\partial x_j} \quad (i = 1, 2, ..., m),$$

Theorem (Sufficiency of Kuhn-Tucker Conditions)

For the NLPP of maximising $f(x)$, $x \in R^n$ subject to the inequality constraints $h_i(x) \leq 0$ $(i = 1, 2, ..., m)$, the Kuhn-Tucker conditions are also the sufficient conditions for a maximum if $f(x)$ is concave and all $h_i(x)$ a convex function $f(x)$.

The Kuhn-Tucker conditions for a minimisation non-linear programming problem may be obtained in a similar manner. These conditions in that case come out to be

$$f_j = \sum_{i=0}^{m} \lambda_i h_{ij} \quad (j = 1, 2, ..., n)$$
$$\lambda_i h_i = 0$$
$$h_i \geq 0$$
$$\lambda_i \geq 0 \quad (i = 1, 2, ..., m)$$

Minimise f subject to $h_i \geq 0$

It can be shown that for this minimisation problem, Kuhn-Tucker conditions are also sufficient conditions for the minima if $f(x)$ is convex and all $h_i(x)$ are concave in x, that is, $-h_i(x)$ is also convex.

Note:

1. If $f(x)$ is strictly concave (convex), the Kuhn-Tucker conditions are sufficient conditions for an absolute maximum (minimum).
2. We may consider $x \geq 0$ or $-x \leq 0$ to have been included in the inequality constraint $h_i(x) \leq 0$.
3. In both maximisation and minimization NLPP, the Lagrange multipliers λ_i corresponding to the equality constraints $h_i(x) = 0$ must be unrestricted in sign.
4. A general NLPP may contain the constraints of the '≥' or '=' or '≤' type. In the case of maximisation NLPP, all constraints must be converted into those of '≤' type and in the case of minimisation NLPP, into those of '≥' type by suitable multiplying by – 1.

Example

Determine x_1, x_2, x_3 so as to

Maximise $Z = -x_1^2 - x_1^2 - x_3^2 + 4x_1 + 6x_2$

Subject to the constraints:

$$x_1 + x_2 \leq 2$$
$$2x_1 + 3x_2 \leq 12$$
$$x_1, x_2 \geq 0$$

Solution: Here,

$$f(x) = -x_1^2 - x_2^2 - x_3^2 + 4x_1 + 6x_2, x \in R^n$$
$$h_1(x) = x_1 + x_2 - 2$$
$$h_2(x) = 2x_1 + 3x_2 - 12.$$

First, we decide about the concavity-convexity of $f(x)$. For this, we compute the bordered Hessian matrix

$$H_B = \begin{bmatrix} -2 & 0 & 0 \\ 0 & -2 & 0 \\ 0 & 0 & -2 \end{bmatrix} \qquad \begin{array}{c} n = 3 \\ m = 2 \\ n - m = 1 \end{array}$$

$\therefore$

$$|H_B| = -8 < 0.$$

The objective function $f(x)$ is concave if the principal minors of matrix H_B alternate in sign, starting with the negative sign. If the principal minors are positive, the objective function is convex. So, in this case $f(x)$ is concave. Clearly, $h_1(x)$ and $h_2(x)$ are convex in x. Thus, the Kuhn-Tucker conditions will be the necessary and sufficient conditions for a maximum. These conditions are obtained by partial derivatives of Lagrangian function:

$$L(x, \lambda, s) = f(x) - \lambda_i[g_1(x) + s_1^2] - \lambda_2[g_2(x) + s_2^2]$$

where, $S = (s_1, s_2)$, $\lambda = (\lambda_1, \lambda_2)$, s_1 and s_2 being slack variables, and λ_1, λ_2 are Lagrangian multipliers.

The Kuhn-Tucker conditions are given by

1 (i) $-2x_1 + 4 = \lambda_1 + 2\lambda_2$ (ii) $-2x_2 + 6 = \lambda_1 + 3\lambda_2$ (iii) $-2x_2 = 0$
2 (i) $\lambda_1(x_1 + x_2 - 2) = 0$ (ii) $\lambda_2(2x_1 + 3x_2 - 12) = 0$
3 (i) $x_1 + x_2 - 2 \le 0$ (ii) $2x_1 + 3x_2 - 12 \le 0$
4 $\lambda_1 \ge 0, \lambda_2 \ge 0$

Now, four different cases may arise:

Case 1: $\lambda_1 = 0$ and $\lambda_2 = 0$ (i), (ii) and (iii) of 1 yield $x_1 = 2$, $x_2 = 3$, $x_3 = 0$. However, this solution violates both the inequalities of (3).

Case 2: $\lambda_1 = 0$ and $\lambda_2 \neq 0$. In this case (ii) of 2 will give $2x_1 + 3x_2 = 12$ and (i) and (ii) of 1 give $-2x_1 + 4 = 2\lambda_2$, $-2x_2 + 6 = 3\lambda_2$. The solution to these simultaneous equations gives $x_1 = 2/13$, $x_2 = 3/13$ $\lambda_2 = 24/13 > 0$, also (iii) of 1 gives $x_3 = 0$. However, this solution violates (i) of 3. So, this solution is discarded.

Case 3: $\lambda_1 \neq 0$, $\lambda_2 \neq 0$. In this case, (2) (i) and (ii) give $x_1 + x_2 = 2$ and $2x_1 + 3x_2 = 12$. These equations give $x_1 = -6$ and $x_2 = 8$. Thus, 1 (i), (ii) and (iii) yield $x_3 = 0$, $\lambda_1 = 68$, $\lambda_2 = -26$. Since $\lambda_2 = -26$ violates the condition (4), so this solution is also discarded.

Case 4: $\lambda_1 \neq 0$, $\lambda_2 = 0$. In this case 2 (i) gives $x_1 + x_2 = 0$. This together with 1 (i) and (ii) give $x_1 = 1/2$, $x_2 = 3/2$, $\lambda = 3 > 0$. Further, from 3 (iii) $x_3 = 0$. This solution does not violate any of the Kuhn-Tucker conditions.

Hence, the optimum (maximum) solution to the given problem is $x_1 = 1/2$, $x_2 = 3/2$, $x_3 = 0$ with $\lambda_1 = 3$, $\lambda_2 = 0$, the maximum value of the objective function is $Z^* = 17/2$.

14.5 Graphical Method of Solving a Non-Linear Programming Problem

In a linear programming problem the optimal solution was usually obtained at one of the corner (extreme) points of the convex region generated by the constraints and the objective function of the problem. But, it is not necessary to find the solution at a corner or edge of the feasible region of a non-linear programming problem. The graphical method of solving an NLPP involving only two variables is best illustrated by the following sample problems.

Example 1 Non-Linear Objective Function and Linear Constraints

Minimise the distance of the origin from the convex region bounded by the constraints. $x_1 + x_2 \geq 4$, $2x_1 + x_2 \geq 5$ and $x_1 \geq 0, x_2 \geq 0$. Verify that the Kuhn-Tucker necessary conditions hold at the point of minimum distance.

Solution: The problem of minimising the distance of the origin from the convex region is equivalent to minimising the radius of a circle with centre as origin, say $r^2 = x_1^2 + x_2^2$ such that it touches (passes through) the convex region bounded by the given constraints. Thus, the problem is formulated as

Minimise $Z (= r^2) = x_1^2 + x_2^2$ subject to the constraints $x_1 + x_2 \geq 4$, $2x_1 + x_2 \geq 5$, $x_1, x_2 \geq 0$.

Consider a set of rectangular cartesian axis $OX_1 X_2$ in the plane. Each point has the coordinates of the type (x_1, x_2) and conversely every ordered pair (x_1, x_2) of real numbers determine a point in the plane.

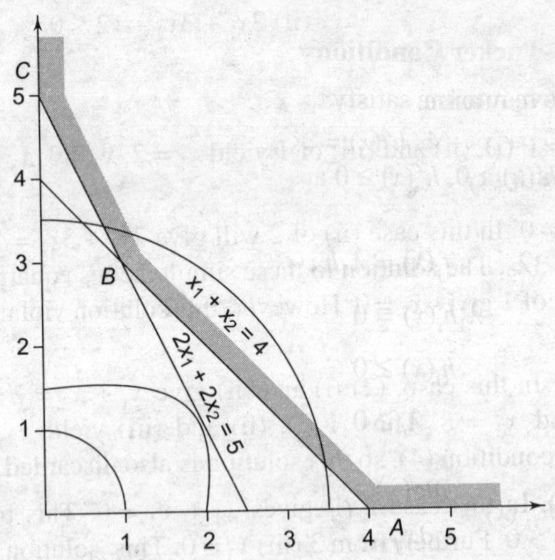

figure 14.2

Clearly, any point which satisfies the condition $x_1 \geq 0$ and $x_2 \geq 0$ lies in the first quadrant and conversely for any point (x_1, x_2) in the first quadrant, $x_1 \geq 0$ and $x_2 \geq 0$. Since $x_1, x_2 \geq 0$, consider the first quadrant only. Since $x_1 + x_2 \geq 4$ and $2x_1 + x_2 \geq 5$, the desired point must be somewhere in the unbounded convex region ABC shown in Fig. 14.1. Since our search for such a pair (x_1, x_2) which gives a minimum value of $x_1^2 + x_2^2$ and lies in the convex region the desired point will be that point of the region at which a side of the convex region is tangent to the circle.

Differenting $x_1^2 + x_2^2 = r^2$, we have $2x_1\, dx_1 + 2x_2 dx_2 = 0$, which gives

$$\frac{dx_2}{dx_1} = \frac{-x_1}{x_2} \quad .... \tag{14.5.1}$$

Considering the equalities $2x_1 + x_2 = 5$ and $x_1 + x_2 = 4$, we have on differentiation

$$2dx_1 + dx_2 = 0 \text{ and } dx_1 + dx_2 = 0$$

These yield $\frac{dx_2}{dx_1} = -2$ and $\frac{dx_2}{dx_1} = -1$ (14.5.2)

From (14.5.1) and (14.5.2) we obtain

$$\frac{-x_1}{x_2} = -1 \text{ or } x_1 = x_2 \text{ and } \frac{-x_1}{x_2} = -2 \text{ or } x_1 = 2x_2$$

This shows that the circle $r^2 = x_1^2 + x_2^2$ has a tangent to it

(i) the line $x_1 + x_2 = 4$ at the point (2, 2)

(ii) the line $2x_1 + x_2 = 5$ at the point (2, 1)

The point (2, 1) does not lie in the convex region and hence is to be discarded. Thus, the minimum distance from the origin to the convex region bounded by the constraints is $Z_0 = 2^2 + 2^2 = 8$ is given by the point (2, 2).

Verification of the Kuhn-Tucker Conditions

We now verify that the above minima satisfy the Kuhn-Tucker conditions also. Here we have,

$f(x) = x_1^2 + x_2^2$, $h'(x) = x_1 + x_2 - 4$, $h^2(x) = 2x_1 + x_2 - 5$ and the problem is that of maximising $f(x)$ subject to the constraints $h'(x) \geq 0$, $h^2(x) \geq 0$ and $x \geq 0$. The Kuhn-Tucker conditions for this minimisation NLPP are:

$$f_j(x) = \lambda_1 h_{ij}(x) + \lambda_2 h_{2j}(x) \qquad j = 1, 2$$

$$\lambda_1 h_i(x) = 0 \qquad i = 1, 2$$

$$h_i(x) \geq 0 \qquad i = 1, 2$$

$$\lambda_i \geq 0 \qquad i = 1, 2$$

where, $f_j(x) = \frac{\partial f(x)}{\partial x_j} = h_{ij}(x) = \frac{\partial h_i(x)}{\partial x_j}$, $(j = 1, 2)$ and λ_1, λ_2 are Lagrangian multipliers.

These conditions thus are as given below:

(a) $\begin{cases} 2x_1 = \lambda_1 + 2\lambda_2 \\ 2x_2 = \lambda_1 + \lambda_2 \end{cases}$

(b) $\begin{cases} \lambda_1 [x_1 + x_2 - 4] = 0 \\ \lambda_2 [2x_1 + x_2 - 5] = 0 \end{cases}$

(c) $\begin{cases}(x_1 + x_2 - 4) \geq 0 \\ (2x_1 + x_2 - 5) \geq 0\end{cases}$

(d) $\{\lambda_1 \geq 0, \lambda_2 \geq 0$

If the point (2, 2) satisfies these conditions, then we must have from (a), $\lambda_1 = 4$, $\lambda_2 = 0$. With these values, $(x_1, x_2) = (2, 2)$ and $(\lambda_1, \lambda_2) = (4,0)$, it is clear that the conditions (b), (c) and (d) are satisfied. Hence, the minima obtained by graphical method satisfies the Kuhn-Tucker conditions for a minima.

14.6 Quadratic Programming

Quadratic programming is concerned with the NLPP of maximising (or minimising) the quadratic objective function, subject to a set of linear inequality constraints. In solving LPPs, because of linearity we were able to develop a very efficient algorithm (simplex method). Unlike the linear programming case, no such general algorithms exist for solving all non-linear programming problems. However, for problems with certain suitable structures, efficient algorithms have been developed. Also, it is often possible to convert the given non-linear problem into one in which these structures become visible.

Unlike linear programming, the optimal solution to a NLPP can be found anywhere on the boundary of the feasible region and even at some interior point of it. In recent years, several solution methods of NLPP have been developed. But, an efficient-like technique for a GNLPP is still required to be developed. A few available techniques for some particular cases of GNLPP shall be discussed in the subsequent sections.

Kuhn–Tuker Conditions with Non-Negative Constraints

In the Section 14.4, we obtained the necessary conditions for a point $x^0 \in R^n$ to be a relative maximum of $f(x)$ subject to the constraints $h_i(x) \leq 0$, $i = 1, 2, ..., m$, $x \geq 0$. These conditions called Kuhn-Tucker conditions were found by converting each inequality constraint to an equation through the addition of a squared slack variable s_i^2, imposing the first-order conditions for maxima, on the first partial derivative of the Lagrangian function, and then simplifying the outcome. The following conditions resulted:

(a) $f_j - \sum_{i=1}^{m} \lambda_i h_{ij}$ $\quad j = 1, 2, ..., n$

(b) $-\lambda_i h_i(x) = 0$ $\quad i = 1, 2, ..., m$

(c) $h_i(x) \leq 0$ and $\quad i = 1, 2, ..., m$

(d) $\lambda_i \geq 0$ $\quad i = 1, 2, ..., m$

The reader may have observed that in obtaining these conditions, the non-negativity constraints $x \geq 0$ were completely ignored. However, we always had in mind to discard all such solutions of (a) to (d) that violate $x \geq 0$.

Now, we shall consider the non-negativity constraint $x \geq 0$ as one of the constraints, namely, $h(x) \geq 0$, where $h(x) = x$, and derive the Kuhn-Tucker conditions for the resulting problem.

We restate the problem as

Maximise $Z = f(x)$, $x \in R^n$ subject to the constraints $h_i(x) \geq 0$, $-x \leq 0$, $i = 1, 2, ..., m$. Clearly, there are $m + n$ inequality constraints, and thus we add the squares of $(m + n)$ slack variables $S_1, ..., S_m, S_{m+1}, ..., S_{m+n}$ in the inequalities so as to convert them into equations.

$$h_i(x) + S_i^2 = 0 \qquad \text{for } i = 1, 2, ..., m$$

$$-x_j + S_{m+j}^2 = 0 \qquad \text{for } j = 1, 2, ..., n$$

To find the necessary conditions for maximum of $f(x)$, we construct the associated Lagrangian function

$$L(x, S, \lambda) = f(x) - \sum_{i=1}^{m} \lambda_i[h_i(x) - S_i^2] - \sum_{i=1}^{n} \lambda_{m+1}[-x_j + S_{m+j}^2]$$

where, $S = (S_1, S_2, ..., S_{m+n})$, and $\lambda = (\lambda_1, ..., \lambda_{m+n})$
are the Lagrangian multipliers. The Kuhn-Tucker conditions are:

$$\frac{\partial L}{\partial x_j} = f_j - \sum_{i=1}^{n} \lambda_i h_{ij} + \lambda_{m+j} = 0 \quad \text{for } j = 1, 2, ..., n$$

$$\frac{\partial L}{\partial S_i} = -2\lambda_i S_i = 0 \qquad \text{for } i = 1, 2, ..., m$$

$$\frac{\partial L}{\partial S_{m+j}} = -2 S_{m+j} \lambda_{m+j} = 0 \qquad \text{for } j = 1, 2, ..., n$$

$$\frac{\partial L}{\partial \lambda_i} = -(h_i(x) + S_i^2) = 0 \qquad \text{for } i = 1, 2, ..., m$$

and,
$$\frac{\partial L}{\partial \lambda_{m+j}} = -[-x_j + S_{m+j}^2] = 0 \qquad \text{for } j = 1, 2, ..., n$$

The Kuhn-Tucker conditions, as obtained from these, upon simplification are

(a) $f_j = \sum_{i=1}^{m} \lambda_i h_{ij} - \lambda_{m+j}$ $\quad (i = 1, 2, ..., n)$

(b) $\lambda_i[h_i(x) = 0]$ $\quad (i = 1, 2, ..., m)$
(c) $-\lambda_{m+j} x_j = 0$ $\quad (j = 1, 2, ..., n)$
(d) $h_i(x) \leq 0$ $\quad (i = 1, 2, .., m)$
(e) $\lambda_i, \lambda_{m+j}, x_j \geq 0$ $\quad (i = 1, 2, ... m)$
$\quad (j = 1, 2, ..., n)$

Maximise of
subject to
$h_i(x) \geq 0$
$x \geq 0$

Note: As before these conditions are sufficient also if $f(x)$ is concave and all $h_i(x)$ are convex in x. Similarly, the Kuhn-Tucker conditions for GNLPP minimisation case are sufficient also if $f(x)$ is convex and all $h_i(x)$ are concave in x.

Definition (Quadratic Programming Problem)

Let x^T and $C \in R^n$. Let Q be a symmetric $m \times n$ real matrix. Then, the problem of maximising (determining x so as to maximise)

$$f(x) = Cx + \frac{1}{2} x^T Qx \text{ subject to the constraints.}$$

$$Ax \leq b \text{ and } x \geq 0.$$

where, $b^T \in R^n$ and A is a $m \times n$ real matrix is called a general quadratic programming problem.

Note: x^TQx represents a quadratic form. We recall that a quadratic form x^TQx is said to be positive-definite (negative definite) if $x^TQx > (<0)$ for $x \neq 0$ and positive semi-definite (negative semi-definite) if $x^TQx \geq 0$ (≤ 0) for all x such that there is one $x \neq 0$ satisfying $x^TQx = 0$.

It may be easily verified that

1. If x^TQx is positive semi-definite (negative semi-definite), then it is convex (concave) in x over all of R^n, and
2. If x^TQx is positive definite (negative definite) then it is strictly convex (strictly concave) in x over all of R^n.

These results help in determining whether the quadratic objective function $f(x)$ is concave (convex) and the implication of the same on the sufficiency of the Kuhn-Tucker conditions for constrained maxima (minima) of $f(x)$.

Now, let us consider a *QPP* written in the form

$$\text{Maximise} = f(x_1, .., x_n) = \sum_{j=1}^{n} c_j x_j + \frac{1}{2} \sum_{j=1}^{n} \sum_{k=1}^{n} x_j d_{jk} x_k$$

subject to the constraints

$$\sum_{j=1}^{n} a_{ij} x_j \leq b_i \qquad i = 1, 2, ..., m$$

$$x_j \geq 0 \qquad j = 1, 2, ..., n$$

where, $d_{jk} = d_{kj}$ for all j and k, and where $b_i \geq 0$. The Kuhn-Tucker conditions for maximum are:

(i) $f_j - \sum_{i=1}^{m} \lambda_i h_{ij} + \lambda_{m+j} = 0$ for $j = 1, 2, ..., n$

or $\left\{c_j + \frac{1}{2}\left[2 \sum_{k=1}^{n} d_{jk} x_k\right]\right\} - \sum_{i=1}^{m} \lambda_i a_{ij} + \lambda_{m+j} = 0$ $(j = 1, 2, ..., n)$

(ii) $\lambda_i h_i(x) = 0$ $(i = 1, 2, ..., m)$

or $\lambda_i \left[\sum_{i=1}^{n} a_{ij} x_j - b_i\right] = 0$ $(i = 1, 2, ..., m)$

(iii) $\lambda_{m+j}(-x_j) = 0$ $(j = 1, 2, ..., n)$

or $-\lambda_{m+j} x_j = 0$ $(j = 1, 2, ..., n)$

(iv) $h_i(x) \leq 0$ $(i = 1, 2, ..., m)$

or $\sum_{j=1}^{n} (a_{ij} x_j - b_i) \leq 0$ $(i = 1, 2, ..., m)$

and (v) $\lambda_i, \lambda_{m+j}, x_j \geq 0$ $(i = 1, 2, ..., m, j = 1, 2, ..., n)$

Thus, the Kuhn-Tucker necessary conditions for an optimal solution to be QPP are:

(a) $c_j + \sum_{k=1}^{n} d_{jk} x_k - \sum_{i=1}^{m} \lambda_i a_{ij} + \lambda_{m+j} = 0$ $(j = 1, 2, ..., n)$

(b) $\lambda_i \left[\sum_{j=1}^{n} a_{ij} x_j - b_i \right] = 0$ $\quad (i = 1, 2, ..., m)$

(c) $(-x_j)\, \lambda_{m+j} = 0$ $\quad (j = 1, 2, ..., n)$

(d) $\sum_{j=1}^{n} a_{ij}\, x_j \le b_i$ $\quad (i = 1, 2, ..., m)$

(e) $\lambda_i, \lambda_{m+j}, x_j \ge 0$ $\quad (i = 1, 2, ..., m\, j = 1, 2, ..., n)$

These expressions can be simplified by introducing some additional notation. Let $S_i^2 \ge 0$ be the slack variable introduced in the ith constraint (d) so that

$$(d)' = \sum_{j=1}^{n} a_{ij}\, x_j + S_i^2 = b_i \text{ for } i = 1, 2, ..., m$$

and, let $u_j = \lambda_{m+j}$ for $j = 1, 2, ..., n$. Then, (b) and (c) after simplification become

$$(b)'\ \lambda_j S_i^2 = 0 \text{ for } \quad i = 1, 2, ..., m$$

$$(c)'\ x_j u_j = 0 \text{ for } \quad j = 1, 2, ..., n$$

Further, since from (e) $S_i^2, \lambda_i, x_i, u_j$ must be all non-negative for $i = 1, 2, ..., m, j = 1, 2, ..., n$ $(b)'$ and $(c)'$ can be expressed by a single constraint:

$$(b)''\ \sum_{i=1}^{n} \lambda_i S_i^2\, \lambda_i S_i^2 + \sum_{j=1}^{n} x_j u_j = 0$$

(complementary slackness)

Using the newly defined variables u_j, we can rewrite (a) as

$$(a)'\ \sum_{k=1}^{n} d_{jk}\, x_k - \sum_{i=1}^{m} \lambda_i a_{ij} + u_j = -\, c_j, \text{ for } j = 1, 2, ..., n$$

If we assume that the quadratic form $\frac{1}{2} \sum \sum d_{jk}\, x_j\, x_k$ is negative semi-definite, then $f(x)$ is concave in x and hence the conditions (a) to (e) above become necessary and sufficient conditions for an optimal solution to *QPP*. Thus, under such assumption if we are able to find non-negative λ_i, s_i^2, x_j and u_j such that $(d)'$, $(b)''$ and $(a)'$ are satisfied, then such x_j determines an optimal solution to the given *QPP*.

14.6.1 Wolfe's Modified Simplex Method

The iterative procedure for the solution of a quadratic programming problem by Wolfe's modified simplex method can be summarised as follows:

Step 1: Convert the inequality constraints into equations by introducing the slack variable s_i^2 in the ith constraint $i = 1, 2, .., m$, and the slack variable s_{m+j}^2 in the jth non-negativity constraint $j = 1, 2, ..., n$.

Step 2: Construct the Lagrangian function

$$L(x, s, \lambda) = f(x) - \sum_{i=1}^{m} \lambda_i \left[\sum_{j=1}^{n} a_{ij}x_j - b_i + s_i^2 \right] - \sum_{j=1}^{n} \lambda_{m+j}(-x_j + s_{m+j}^2)$$

where $x = (x_1,..., x_n)$, $s = (s_1, ..., s_{m+n})$, $\lambda = (\lambda_1, ... \lambda_{m+1})$

Differentiate $L(x, s, \lambda)$ partially w.r.t the components of x, s and λ and equate the first order partial derivatives equal to zero. Derive the Kuhn-Tucker conditions from the resulting equations.

Step 3: Introduce the non-negative artificial variables A_j, $j = 1, 2, ..., n$ in the Kuhn-Tucker condition $c_j + \sum_{k=1}^{n} d_{jk}\, x_k - \sum_{i=1}^{m} \lambda_i\, a_{ij} + \lambda_{m+j} = 0$ for $j = 1, 2, ..., n$ and construct an objective function $Z = A_1 + A_2 + ... + A_n$.

Step 4: Obtain an initial basic feasible solution to the *LPP*:

Minimise $Z = A_1 + ... + A_n$ subject to

$$\sum_{k=1}^{n} d_{jk}\, x_k - \sum_{i=1}^{m} \lambda_i\, a_{ij} + \lambda_{m+j} + A_j = -c_j \qquad (j = 1, 2, ..., n)$$

$$\sum_{j=1}^{n} a_{ij}\, x_j + x_{n+i} = b_i \qquad (i = 1, 2, ..., m)$$

$$A_i, \lambda_i, \lambda_{m+j}, x_j \geq 0 \qquad (i = 1, 2, ..., m, j = 1, 2, ..., n)$$

where, $x_{n+1} = s_i^2$, $i = 1, 2, ..., m$, and satisfying the complementary slackness conditions:

$$\sum_{j=1}^{n} \lambda_{m+j}\, x_j + \sum_{i=1}^{m} x_{n+i}\, \lambda_i = 0$$

Step 5: Use two phase simplex method to obtain an optimum solution to the *LPP* of step 4, the solution satisfying the complementary slackness condition.

Step 6: The optimum solution obtained in step 5 is an optimum solution to the given *QPP* also.

Important Remarks on Wolfe's Method

1. If the quadratic programming problem is given in the minimisation form, then convert it into maximisation one by suitable modifications in $f(x)$ and the '$\geq$' constraints.
2. With the exceptional condition of complementary slackness, the problem constructed in step 4 is exactly a linear programming problem. So, we only need to modify the simplex algorithm to include the complementary slackness condition. Thus, while deciding to introduce $s_i(=q_i^2)$ we must first ensure that: (i) either λ_i is not in the solution (or) (ii) λ_i will be removed when s_i enters. This additional check is not difficult to perform within the simplex routine.
3. The solution to the above system is obtained by using phase I of the simplex method. Since our aim is to obtain a feasible solution, it does not require the consideration of phase II. The only necessary thing is to maintain the condition $\lambda_i s_i = 0 = \mu_j x_j$ all the time. This implies that if λ_i is in the basic solution with positive value, then s_i cannot be basic with positive value. Similarly, μ_j and x_j cannot be positive simultaneously.
4. It should be observed that phase I will end in the usual manner with the sum of all artificial variables equal to zero only if the feasible solution to the problem exists.

Example 1

Use Wolfe's method to solve the *QPP*

Maximise $Z = 2x_1 + 3x_2 - 2x_1^2$ subject $x_1 + 4x_2 \le 4$, $x_1 + x_2 \le 2$, $x_1, x_2 \ge 0$.

Solution: *Step 1*: First, convert the inequality constraints into equations by using slack variables s_1^2, s_2^2, s_3^2 and s_4^2, respectively. Thus, the problem becomes

$$\text{Max } z = 2x_1 + 3x_2 - 2x_1^2$$

subject to

$$x_1 + 4x_2 + s_1^2 = 4,\ x_1 + x_2 + s_2^2 = 2,\ -x_1 + s_3^2 = 0,\ -x_2 + s_4^2 = 0$$

Step 2: Construct the Lagrangian function

$$\begin{aligned} &L(x_1, x_2, s_1, s_2, s_3, s_4, \lambda_1, \lambda_2, \lambda_3, \lambda_4) \\ &= (2x_1 + 3x_2 - 2x_1^2) - \lambda_1(x_1 + 4x_2 + s_1^2 - 4) \\ &\quad - \lambda_2 (x_1 + x_2 + s_2^2 - 2) - \lambda_3 (-x_1 + s_3^2) \\ &\quad - \lambda_4(-x_2 + s_4^2) \end{aligned}$$

Step 3: As $-x_1^2$ represents a negative semi-definite quadratic form $Z = 2x_1 + 3x_2 - 2x_1^2$ is concave in x_1, x_2. Thus, maxima of L will be maxima of $Z = 2x_1 + 3x_2 - 2x_1^2$ and vice versa. To derive the necessary and sufficient conditions for maxima of L (and hence of z) we equate the first order partial derivatives of L with respect to the variables x_1, x_2, s_i's and λ_i's. Thus, we have

$$\frac{\partial L}{\partial x_i} = 2 - 4x_1 - \lambda - \lambda_2 + \lambda_3 = 0; \quad \frac{\partial L}{\partial x_2} = 3 - 4\lambda - \lambda_2 + \lambda_4 = 0$$

$$\frac{\partial L}{\partial s_1} = -2\lambda_1 s_1 = 0; \quad \frac{\partial L}{\partial \lambda_1} = x_1 + 4x_2 + s_1^2 - 4 = 0,$$

$$\frac{\partial L}{\partial s_2} = -2\lambda_2 s_2 = 0; \quad \frac{\partial L}{\partial \lambda_2} = x_1 + x_2 + s_2^2 - 2 = 0$$

$$\frac{\partial L}{\partial s_3} = -2\lambda_3 s_3 = 0; \quad \frac{\partial L}{\partial \lambda_3} = -x_1 + s_3^2 = 0$$

$$\frac{\partial L}{\partial s_4} = 2\lambda_4 s_4 = 0; \quad \frac{\partial L}{\partial \lambda_4} = -x_2 + s_4^2 = 0$$

Upon simplification and necessary manipulations these yield:

$$\left.\begin{aligned} 4x_1 + \lambda_1 + \lambda_2 - \lambda_3 = 2,\ 4\lambda_1 + \lambda_2 - \lambda_4 = 3 \\ x_1 + 4x_2 + s_1^2 = 4,\ x_1 + x_2 + s_2^2 = 2 \end{aligned}\right\} \quad (14.6.1)$$

$$\left.\begin{aligned} \lambda_1 s_1^2 + \lambda_2 s_2^2 + x_1\lambda_3 + x_2\lambda_4 = 0 \\ x_1, x_2, s_1^2, s_2^2, \lambda_1, \lambda_2, \lambda_3, \lambda_4 \ge 0 \end{aligned}\right\} \quad (14.6.2)$$

A solution x_j, $j = 1, 2$ to (14.6.1) and satisfying (14.6.2) shall necessarily be an optimal one for maximising L. To determine the solution to the above simultaneous equation (1), we introduce the artificial variables $A_1 \ge 0$ and $A_2 \ge 0$ in the first two constraints of (14.6.1).

Step 4: The modified LPP problem becomess

Maximise $Z = -A_1 - A_2$

subject to

$$\begin{aligned} 4x_1 + \lambda_1 + \lambda_2 - \lambda_3 + A_1 &= 2 \\ 4\lambda_1 + \lambda_2 - \lambda_4 + A_2 &= 3 \\ x_1 + 4x_2 + x_3 &= 4 \end{aligned}$$

$$x_1 + x_2 + x_4 = 2$$

$$x_1, x_2, x_3, x_4 \geq 0\ A_1, A_2, \lambda_1, \lambda_2, \lambda_3, \lambda_4 \geq 0$$

satisfying the complementary slackness condition $\sum \lambda_i x_i = 0$, where we replaced s_1^2 by x_3 and s_2^2 by x_4.

An initial basic feasible solution to the LPP is clearly given by

$$\begin{bmatrix} 1 & 0 & 0 & 0 \\ 0 & 1 & 0 & 0 \\ 0 & 0 & 1 & 0 \\ 0 & 0 & 0 & 1 \end{bmatrix} \begin{bmatrix} A_1 \\ A_2 \\ x_3 \\ x_4 \end{bmatrix} = \begin{bmatrix} 2 \\ 3 \\ 4 \\ 2 \end{bmatrix}$$

Step 5: *Initial Table of Phase I*

table 14.2

		C_i	0	0	0	0	0	0	0	0	-1	-1
C_B	Y_B	X_B	X_1 ↓	X_2	X_3	X_4	λ_1	λ_2	λ_3	λ_4	A_1	A_2
-1	A_1 ←	2	4	0	0	0	1	1	-1	0	1	0
-1	A_2	3	0	0	0	0	4	1	0	-1	0	1
0	X_3	4	1	4	1	0	0	0	0	0	0	0
0	X_4	2	1	1	0	1	0	0	0	0	0	0
$z_j -$	c_j	-5	-4	0	0	0	5	2	-1	-1	0	0

From the Table 14.2 we observe that x_1, λ_1 or λ_2 can enter the basis. But λ_1 and λ_2 will not enter the basis, because x_3 and x_4 are in the basis. This is in view of the complementary slackness conditions $\lambda_1 x_3 = 0$ and $\lambda_2 x_4 = 0$.

First Iteraction: Introduce x_1 and drop A_1.

table 14.3

		c_j	0	0	0	0	0	0	0	0	-1
C_B	Y_B	X_B	X_1	X_2 ↓	X_3	X_4	λ_1	λ_2	λ_3	λ_4	A_2
0	X_1	1/2	1	0	0	0	1/4	1/4	-1/4	0	0
-1	A_2	3	0	0	0	0	4	1	0	-1	1
0	X_3 ←	7/2	0	4	1	0	-1/4	-1/4	1/4	0	0
0	X_4	3/2	0	1	0	1	-1/4	-1/4	1/4	0	0
		-3	0	0	0	0	-4	-1	0	1	0

Here, we observe that either λ_1 or λ_2 can enter the basis. But x_3 and x_4 are still in the basis, therefore, these cannot enter the basis because of the complementary slackness conditions. However, since λ_4 is not in the basis, x_2 can enter the basis (because of $\lambda_4 x_2 = 0$).

Second Iteration: Introduce x_2 and drop x_3.

table 14.4

C_B	Y_B	X_B	x_1	x_2	x_3	x_4	λ_1 ↓	λ_2	λ_3	λ_4	A_2
0	x_1	1/2	1	0	0	0	1/4	1/4	- 1/4	0	0
- 1	A_2 ←	3	0	0	0	0	4	1	0	- 1	1
0	x_2	7/8	0	1	1/4	0	- 1/16	- 1/16	1/16	0	0
0	x_4	5/8	0	0	- 1/4	1	- 3/16	- 3/16	3/16	0	0
		- 3	0	0	0	0	- 4	- 1	0	+1	0

Since x_4 is in the basis, λ_2 cannot enter and hence λ_1 enter the basis.

Final Iteration:

table 14.5

C_B	Y_B	X_B	x_1	x_2	x_3	x_4	λ_1	λ_2	λ_3	λ_4
0	x_1	5/16	1	0	0	0	0	3/16	- 1/4	1/16
0	λ_1	3/4	0	0	0	0	1	1/4	0	- 1/4
0	x_2	59/64	0	1	1/4	0	0	- 3/64	1/16	- 1/64
0	x_4	49/64	0	0	- 1/4	1	0	- 9/64	3/16	- 3/64
		0	0	0	0	0	0	0	0	0

Hence, the optimum solution is

$$x_1 = 5/16,\ x_2 = 59/64 \text{ and}$$

$$\text{Maximum of } Z = 3.19.$$

Example 2

Apply Wolfe's method for solving the quadratic programming problem

Minimise $\quad f(x) = x_1^2 + x_2^2 - 2x_1 - 4x_2$

subject to $\quad x_1 + 4x_2 \leq 5$

$$2x_1 + 3x_2 \leq 6 \text{ and } x_1, x_2 \geq 0$$

Step 1: Converting the problem into maximisation and the inequalities into '≤', the *LPP* becomes

Maximise $\quad -f(x) = -x_1^2 - x_2^2 + 2x_1 + 4x_2$

subject to

$$x_1 4x_2 \leq 5;\ 2x_1 + 3x_2 \leq 6;\ -x_1 \leq 0 \text{ and } -x_2 \leq 0$$

Step 2: Introducing the slack variables $s_1^2, s_2^2, s_3^2, s_4^2$, our problem takes the form

Max. $\quad -f(x) = -x_1^2 - x_2^2 + 2x_1 + 4x_2$

subject to $\quad x_1 + 4x_2 + s_1^2 = 5$

$$2x_1 + 3x_2 + s_2^2 = 6$$

$$-x_1 + s_3^2 = 0$$
$$-x_1 + s_4^2 = 0$$

Step 3: Construct the Lagrangian function

$$L(x_1, x_2, s_1, s_2, s_3, s_4, \lambda_1, \lambda_2, \lambda_3, \lambda_4) = (-x_1^2 - x_2^2 + 2x_1 + 4x_2) - \lambda_1(x_1 + 4x_2 + s_1^2 - 5)$$
$$- \lambda_2(2x_1 + 3x_2 + s_2^2 - 6) - \lambda_3(-x_1 + s_3^2)$$
$$- \lambda_4(-x_2 + s_4^2)$$

The necessary and sufficient conditions are

$$\frac{\partial L}{\partial x_1} = -2x_1 + 2 - \lambda_1 - 2\lambda_2 + \lambda_3 = 0$$

$$\text{(or)} \quad 2x_1 + \lambda_1 + 2\lambda_2 - \lambda_3 = 2$$

$$\frac{\partial L}{\partial x_2} = -2x_2 - 4 - 4\lambda_1 - 3\lambda_2 + \lambda_4 = 0$$

$$\text{or} \quad 2x_2 + 4\lambda_1 + 3\lambda_2 - \lambda_4 = 4$$

$$\frac{\partial L}{\partial s_1} = -2\lambda_1 s_1 = 0; \quad \frac{\partial L}{\partial s_2} = -2\lambda_2 s_2 = 0$$

$$\frac{\partial L}{\partial s_3} = -2\lambda_3 s_3 = 0; \quad \frac{\partial L}{\partial s_4} = -2\lambda_4 s_4 = 0$$

$$\frac{\partial L}{\partial \lambda_1} = x_1 + 4x_2 + s_1^2 - 5 = 0$$

$$\frac{\partial L}{\partial \lambda_2} = 2x_1 + 3x_2 + s_2^2 - 6 = 0$$

Step 4: Introducing artificial variables $A_1, A_2 \geq 0$, the modified linear programming problem is

Max. $Z = -A_1 - A_2$

subject to
$$2x_1 + \lambda_1 + 2\lambda_2 - \lambda_3 + A_1 = 2$$
$$2x_2 + 4\lambda_1 + 3\lambda_2 - \lambda_4 + A_2 = 4$$
$$x_1 + 4x_2 + x_3 = 5$$
$$2x_1 + 3x_2 + x_4 = 6$$

where, $s_1^2 = x_3$ and $s_2^2 = x_4$

All the variables are non-negative and

$$\lambda_1 x_3 = 0,\ \lambda_2 x_4 = 0,\ \lambda_3 x_1 = 0 \text{ and } \lambda_4 x_2 = 0$$

Step 5: Initial table of Phase I

table 14.6

	c_j		0	0	0	0	0	0	0	0	-1	-1
C_B	Y_B	X_B	x_1	x_2	x_3	x_4	λ_1	λ_2	λ_3	λ_4	A_1	A_2
-1	A_1	2	2	0	0	0	1	2	-1	0	1	0
-1	A_2	4	0	2	0	0	4	3	0	-1	0	1
0	x_3	5	1	4	1	0	0	0	0	0	0	0
0	x_4	6	2	3	0	1	0	0	0	0	0	0
			-2	-2	0	0	-5	-5	1	1	0	0

Since x_3 and x_4 are in the basis we cannot introduce λ_1 and λ_2. Introduce x_1 and drop A_1.

Step 6: First Iteration—Introduce x_2 and drop A_1.

table 14.7

	c_j		0	0	0	0	0	0	0	0	-1
C_B	Y_B	X_B	x_1	x_2	x_3	x_4	λ_1	λ_2	λ_3	λ_4	A_2
0	x_1	1	1	0	0	0	1/2	1	-1/2	0	0
-1	A_2	4	0	2	0	0	4	3	0	-1	1
0	x_3	4	0	4	1	0	-1/2	-1	1/2	0	0
0	x_4	4	0	3	0	1	1	-2	1	0	0
			0	-2	0	0	-4	-3	0	1	0

Since x_3 and x_4 are in the basis, we cannot select λ_1 and λ_2. Introduce x_2 and drop A_2.

Step 7: Second Iteration

table 14.8

	c_j		0	0	0	0	0	0	0	0
C_B	Y_B	X_B	x_1	x_2	x_3	x_4	λ_1	λ_2	λ_3	λ_4
0	x_1	1	1	0	0	0	1/2	1	-1/2	0
0	x_2	2	0	1	0	0	2	3/2	0	-1/2
0	x_3	-4	0	0	1	0	7/2	-7	1/2	2
0	x_4	-2	0	0	0	1	-7	-13/2	1	3/2
			0	0	0	0	0	0	0	0

Since x_3 and x_4 occur in the solution with negative values, they must be eliminated. Hence, introduce λ_1 and drop x_3.

Step 8: Third Iteration

table 14.9

		c_j	0	0	0	0	0	0	0	0
C_B	Y_B	X_B	x_1	x_2	x_3	x_4	λ_1	λ_2	λ_3	λ_4
0	x_1	13/17	1	0		0	0			
0	x_2	18/17	0	1		0	0			
0	λ_1	8/17	0	0		0	1	14/17	- 1/17	- 4/17
0	x_4	22/17	0	0		1	0			
			0	0	0	0	0	0	0	0

Since all $z_j - c_j \geq 0$ and all $x_{B_i} \geq 0$ the current solution is optimum. The optimum solution is $x_1 = 13/17$, $x_2 = 18/17$

with, $$\text{Min. } f(x) = \left(\frac{13}{17}\right)^2 + \left(\frac{18}{17}\right)^2 - 2\left(\frac{13}{17}\right) - 4\left(\frac{18}{17}\right) = 69.$$

14.6.2 Beale's Method

Another approach to solving a quadratic programming problem has been suggested by Beale. Unlike Wolfe's method this approach does not use the Kuhn-Tucker conditions. Instead, this method involves the partitioning of variables into basic and non-basic variables and the results of classical calculus are used. At each iteration, the objective function is expressed in terms of non-basic variables only.

Beale's Algorithm for QPP

Let the *QPP* be given in the form

$$\text{Maximise } f(x) = C^T x + 1/2\, x^T Q x$$

subject to $$Ax\ \{\leq, \geq \text{ or } =\}\ b \text{ and}$$

$x \geq 0$, $x \in R^n$, A is $m \times n$, b is $m \times 1$, c is $n \times 1$ and Q is a $n \times n$ symmetric matrix.

Step 1: Convert the objective function to maximisation type (if it is in the minimisation type). Introduce slack and/or surplus variables in the inequality constraints, if any, and put the *QPP* in the standard form.

Step 2: Select arbitrary m variables as basic variables, so that the remaining $n - m$ variables become non-basic. Denote the basic and non-basic variables by

$X_B = (x_{B_1}, x_{B_2}, \ldots, x_{B_n})$ and $x_{N_B} = \left(x_{N_{B_1}}, x_{N_{B_2}}, x_{N_{B_3}}\right)$ respectively.

Step 3: Express each basic variable x_{B_i} entirely, in terms of the non-basic variables x_{N_B}'s (and u_i's, if any) using the given (as well as additional, if any) constraints.

Step 4: Examine the partial derivatives of $f(x)$ formulated above w.r.t the non-basic $x_{N_{B_1}}$'s (and u_i's, if any)

Step 5: Examine the partial derivatives of $f(x)$ formulated above with respect the non-basic variables at the point $x_{N_B} = 0$ (and $u = 0$)

(i) If $\left(\dfrac{\partial f(x)}{\partial x_{N_{B_k}}}\right)_{\substack{x_{N_B}=0 \\ u=0}} = 0$ for each $k = 1, 2, \ldots, n - m$

and $\left(\dfrac{\partial f(x)}{\partial u_j}\right)_{\substack{x_{N_B}=0\\ u=0}} = 0$ for each i the current basic solution is optimal.

Go to Step 8.

(ii) If $\left(\dfrac{\partial f(x)}{\partial x_{N_{B_k}}}\right)_{\substack{x_{N_B}=0\\ u=0}} > 0$ for at least one k,

choose the most positive one. The corresponding non-basic variable will enter the basis.

(iii) If $\left(\dfrac{\partial f(x)}{\partial x_{N_{B_k}}}\right)_{\substack{x_{N_B}=0\\ u=0}} < 0$ for each $k = 1, 2, \ldots, n - m$

but $\left(\dfrac{\partial f(x)}{\partial u_i}\right)_{\substack{x_{N_B}=0\\ u=0}} \neq 0$ for some $i = r$

Introduce a new non-basic variable u_j, defined by $u_j = \dfrac{1}{2}\dfrac{\partial f}{\partial u_r}$ and treat u_r as a basic variable (to be ignored later). Go to Step 3.

Step 6: Let $x_{N_{B_i}} = x_k$ be the entering variable identified in Step 5 (ii). Compute the minimum of the ratios

$$\min \cdot \left\{\frac{\alpha_{h0}}{|\alpha_{hk}|}, \frac{\gamma_{k0}}{|\gamma_{kk}|}\right\}$$

for all basic variables x_h, where x_{h0} is the constant term and α_{nk} is the coefficient of x_k in the expression of the basic variable x_h when expressed in terms of the non-basic ones; and n_{k0} is the constant term and v_k is the coefficient of x_k in $\dfrac{\partial f}{\partial x_k}$.

(i) If the minimum ratio occurs for some $\dfrac{\alpha_{h0}}{|\alpha_{h0}|}$, the corresponding basic variable x_h will leave the basis.

(ii) If the minimum ratio occurs for some $\dfrac{\gamma_{k0}}{|\gamma_{kk}|}$, the exit criterion correspond to a non-basic variable. In this case, introduce an additional non-basic variable, called a free-variable defined by

$$u_i = \frac{1}{2}\frac{\partial f}{\partial x_k} \quad (u_i \text{ is unrestricted})$$

which relation becomes an additional constraint equation.

Step 7: Go to step 3 and repeat the procedure until an optimal basic solution is reached.

Step 8: Determine the optimal value of x_B and $f(x)$ by setting $x_{N_B} = 0$ in the expressions obtained in steps 3 and 4.

Example 1

Use Beale's method to solve the following *NLPP*:

$$\text{Minimise } Z = 6 - 6x_1 + 2x_1^2 - 2x_1x_2 + 2x_2^2$$

subject to the constraints.

$$x_1 + x_2 \le 2$$

and,

$$x_1, x_2 \ge 0$$

Solution: Introducing a slack variable x_3, the constraint equation becomes $x_1 + x_2 + x_3 = 2$. Converting the minimisation objective function into a maximisation one, we have

$$f(x) = -6 + 6x_1 - 2x_1^2 + 2x_1x_2 - 2x_2^2$$

Arbitrarily, choosing x_3 as the basic variable, we have

$$x_B = (x_3),\ x_{N_B} = (x_1, x_2).$$

Expressing x_B and $f(x)$ in terms of x_{N_B}, we have

$$x_3 = 2 - x_1 - x_2$$

$$f = -6 + 6x_1^2 - 2x_1^2 + 2x_1x_2 - 2x_2^2$$

The partial derivatives are:

$$\left(\frac{\partial f}{\partial x_1}\right)_{x_{N_B}=0} = (6 - 4x_1 + 2x_2)_{\substack{x_1=0\\x_2=0}} = 6$$

$$\left(\frac{\partial f}{\partial x_2}\right)_{x_{N_B}=0} = (2x_1 - 4x_2)_{\substack{x_1=0\\x_2=0}} = 0$$

Clearly, x will enter the basis.

Now, $$\min.\left\{\frac{\alpha_{30}}{|\alpha_{31}|}, \frac{\gamma_{10}}{|\gamma_{11}|}\right\} = \min.\left\{\frac{2}{|-1|}, \frac{6}{|-4|}\right\} = 6/4$$

which corresponds to $\dfrac{\gamma_{10}}{|\gamma_{11}|}$. Thus, we cannot remove x_3 from the basis. We introduce a new non-basic free variable u_1, defined by

$$u_1 = \frac{1}{2}\frac{\partial f}{\partial x_1} = 3\ 2x_1 + x_2.$$

The current basis is $x_B = (x_3, x_1)$ and $x_{N_B} = (x_2, u_1)$.

Expressing the current x_B and $f(x)$ in terms of x_{N_B}

$$x_1 = \frac{3}{2} - \frac{1}{2}u_1 + \frac{1}{2}x_2,\ x_3 = \frac{1}{2} + \frac{1}{2}u_1 - \frac{3}{2}x_2$$

$$f = -6 + \left\{\frac{3}{2} - \frac{1}{2}u_1 + \frac{1}{2}x_2\right\}(6 - 3 + u_1 - x_2 + 2x_2) - 2x_2^2$$

$$= -6 + \left\{\frac{3}{2} - \frac{1}{2}u_1 + \frac{1}{2}x_2\right\}(3 + u_1 + x_2) - 2x_2^2$$

$$= -\frac{3}{2} - \frac{1}{2}u_1^2 - \frac{3}{2}x_2^2 + 3x_2$$

Evaluating the partial derivatives f with respect to x_{N_B}

$$\left(\frac{\partial f}{\partial x_2}\right)_{x_{N_B}=0} = (3 - 3x_2)_{x_2=0} = 3$$

$$\left(\frac{\partial f}{\partial u_1}\right)_{\substack{x_{N_B}=0\\ u_1=0}} = (-u_1)_{u_1=0} = 0$$

Clearly, x_2 will enter the basis.
Again, we compute the ratios,

$$\min\cdot\left\{\frac{\alpha_{10}}{|\alpha_{12}|}, \frac{\alpha_{30}}{|\alpha_{32}|}, \frac{\gamma_{20}}{|\gamma_{22}|}\right\}$$

$$= \min\cdot\left\{\frac{3/2}{1/2}, \frac{1/2}{|-3/2|}, \frac{3}{|-3|}\right\} = \frac{\alpha_{30}}{|\alpha_{32}|}$$

Thus, x_3 will leave the basis and the new variables are:

$$x_B = (x_1, x_2) \text{ and } x_{N_B} = (u_1, x_3)$$

Expressing the new basis x_B and $f(x)$ in terms of X_{N_B}

$$x_2 = \frac{1}{3} + \frac{1}{3}u_1 - \frac{2}{3}x_3,\quad x_1 = \frac{5}{3} - \frac{1}{3}u_1 - \frac{1}{3}x_3$$

$$f = -6 + \left(\frac{5}{3} - \frac{1}{3}u_1 - \frac{1}{3}x_3\right)\left(\frac{10}{3} + \frac{4}{3}u_1 - \frac{2}{3}x_3\right)$$

$$-2\left(\frac{1}{3} + \frac{1}{3}u_1 - \frac{2}{3}x_3\right)^2$$

$$= \frac{-2}{3} + \frac{2}{3}u_1 - \frac{4}{3}x_3 + \frac{2}{3}x_3u_1 - \frac{2}{3}u_1^2 + \frac{2}{3}x_3^2$$

The partial derivatives of f with respect to x_{N_B} are

$$\left(\frac{\partial f}{\partial x_3}\right)_{\substack{x_{N_B}=0\\u_1=0}} = \left(-\frac{4}{3}+\frac{2}{3}u_1-\frac{4}{3}x_3\right)_{\substack{x_3=0\\u_1=0}} = -\frac{4}{3}$$

$$\left(\frac{\partial f}{\partial u_1}\right)_{\substack{x_{N_B}=0\\u_1=0}} = \left(\frac{2}{3}+\frac{2}{3}x_3-\frac{4}{3}u_1\right)_{\substack{x_3=0\\u_1=0}} = \frac{2}{3}$$

Since $\left(\frac{\partial f}{\partial u_1}\right) \neq 0$, current solution can be further improved. However, the entry criterion does not qualify x_3 to enter the basis. Thus, we introduce another non-basic free variable u_2, defined by

$$u_2 = \frac{1}{2}\frac{\partial f}{\partial u_1} = \frac{1}{3}+\frac{1}{3}x_3-\frac{2}{3}u_1$$

Treating u_1 as a basic variable, and expressing the basic variables $f(x)$ in terms of non-basic variables we have,

$$u_1 = \frac{1}{2}-\frac{3}{2}u_2+\frac{1}{2}x_3, \quad x_2 = \frac{1}{2}-\frac{1}{2}u_2-\frac{1}{2}x_3,$$

$$x_1 = \frac{3}{2}+\frac{1}{2}u_2-\frac{1}{2}x_3$$

$$\begin{aligned} f &= -\frac{2}{3}+\frac{2}{3}u_1(1+x_3-u_1)-\frac{4}{3}x_3-\frac{2}{3}x_3^2 \\ &= -\frac{2}{3}+\frac{1}{6}(1-3u_2+x_3)(1+3u_2-x_3)-\frac{4}{3}x_3-\frac{2}{3}x_3^2 \\ &= \frac{-2}{3}+\frac{1}{6}\left(1-9u_2^2-5x_3^2+6u_2x_3-8x_1\right) \end{aligned}$$

Now, we see that

$$\left(\frac{\partial f}{\partial x_3}\right)_{\substack{x_{N_B}=0\\u_2=0}} = \frac{1}{6}(-10x_3-6u_2-8)_{\substack{x_3=0\\u_2=0}} = -\frac{8}{6}$$

$$\left(\frac{\partial f}{\partial u_1}\right)_{\substack{x_{N_B}=0\\u_2=0}} = \frac{1}{6}(-18u_2+6x_2)_{\substack{x_3=0\\u_2=0}} = 0$$

Thus, the current x_B provides an optimal solution.

Ignoring the free variable u_1, the optimal solution to the given quadratic programming problem is

$$x_1 = \frac{3}{2}, x_2 = \frac{1}{2}, z^* = -\left(-\frac{1}{2}\right) = \frac{1}{2}$$

14.7 SEPARABLE PROGRAMMING

Separable programming deals with non-linear programming problems in which the objective function as well as all the constraints are separable. In many decision-making situations, the non-linear profit or cost functions are related by relatively smooth curves, rather than sharp ones. Thus, breaking points are obtained on such curves. In such cases, the non-linear objective function with a smooth curve may be approximated by a series of piece-wise linear segments. This approximation will introduce some error which may be negligible in many cases. Such error can be reduced by increasing the number of linear segments. But on the other hand, excessive increase of linear segments will enlarge the size of the problem, thus consuming more computational time to obtain the optimal solution.

Separable functions: A function $f(x_1, ..., x_n)$ is said to be separable if it can be expressed as the sum of n single valued functions $f_1(x_1), f_2(x_2)..........f_n(x_n)$.

That is, $f(x_1, ..., x_n) = f_1(x_1) + f_2(x_2) + ... + f_n(x_n)$.

Example

The function $g(x_1, x_2, x_3, x_4) = x_1 + 5x_2 + 10x_3 + 2x_4$ is separable and the function $h(x_1, x_2, x_3) = x_1^3 + x_2^2 \cos(x_1 + x_3) + x_3 . 3x_2 + \log(x_1 + x_2)$ is not a separable function.

Definition (Separable Programming Problem): A *NLPP* in which the objective function can be expressed as a linear combination of several different single variable functions, of which some or all are non-linear, is called a separable programming problem.

Definition (Convex Programming): Non-linear programming which has the problem of minimising a convex objective function (or maximising a concave objective function) in the convex set of points is called convex programming.

Piece-wise Linear Approximations: In many decision-making situations, the non-linear profit or cost objective functions are related by relatively smooth curves, rather than sharp curves obtaining breaking points. In such cases, the non-linear objective function, with a smooth curve for it, may be approximated by a series of piece-wise linear segment. This approximation will introduce some errors that in many cases may be negligible. In general, care is to be taken while approximating a non-linear function by linear segments, to reduce such errors. One way of reducing such errors is to increase the numbers of linear segments. But excessive linear approximations will enlarge the size of the problem, thus increasing the computational time for an optimal solution.

Piece-wise linear approximations can be made for both convex and concave functions. However, as we have seen earlier, if all the linear segments of the objective function are concave (convex), a valid optimal solution requires the objective function to be maximised (minimised). Thus, a NLPP becomes a LPP with piece-wise linear approximations, and the simplex method can be applied to obtain an optimal solution.

We shall now study how piece-wise linear approximations reduce a given separable convex (concave) NLPP to an LPP and then solve it by using the simplex method.

Consider the separable programming problem

$$\text{Maximise } f(x) = \sum_{i=1}^{n} f_i(x_j) \text{ subject to the constraint}$$

$$\sum_{j=1}^{n} g_{ij}(x_j) \leq b_i, x_j \geq 0 \ (i = 1, 2, .., m \ j = 1, 2, ... n) \text{ where, some or all } g_{ij}(x_{ij}), f_j(x_j) \text{ are non-linear.}$$

To solve this NLPP by piece-wise linear approximation method, we shall reduce each non-linear $f_j(x_j)$ and $g_{ij}(x_j)$ to other linear approximations by fitting the linear segments to the original curve as shown in Fig. 14.4.

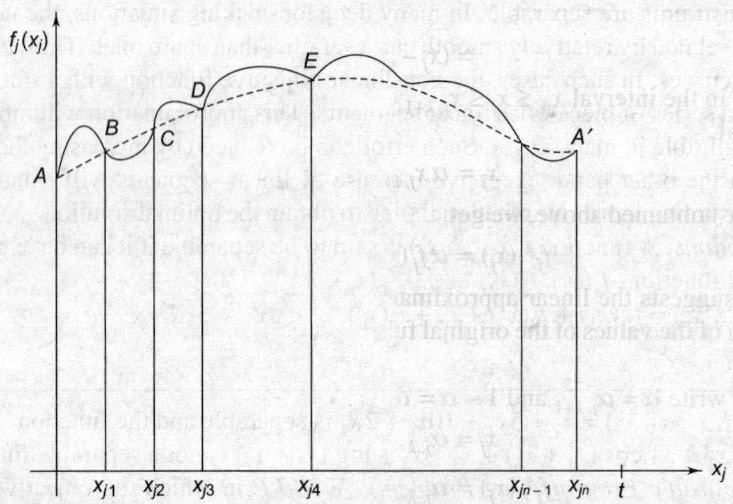

figure 14.3

This approximation can be carried out to any degree of accuracy by dividing the curve AA' into a suitable number of arcs, each are replaced by a linear segment. This is $\overset{\frown}{AB} \approx \overline{AB}$, $\overset{\frown}{BC} \approx \overline{BC}$, and so on.

Let the point x_{jk} in the interval $0 \le x_j \le t$ be so determined that

$$x_{j0} = 0,\ x_{jk} < x_{j1} < x_{j2} < \ldots < x_{jn} < t$$

For each point x_{jk}, we calculate the function values $f_j(x_{jk})$.

Connect the adjacent points $[x_{ji}, f_j(x_{j1})]$, $[x_{j2}, f_j(x_{j2})]$, ... $[x_{jn}, f_j(x_{jn})]$ .... by straight lines (shown dotted in Fig. 14.3) and denote this piece-wise linear approximation to $f_j(x_j)$ by $f_j^*(x_j)$. Naturally, $f_j^*(x_j)$ will be a better approximation to $f_j(x_j)$ if the distance between the adjacent x_{jk} is very small.

Assume that we have found an approximation of this kind for each $f_j(x_j)$ and $g_{ij}(x_j)$, $j = 1, 2, ..., n$.

Let these approximated new functions be f_j^* and $g_{ij}^*(x_j)$ for $j = 1, 2, .., n$, respectively. Thus, the given NLPP reduces to

Maximise $$f(x) = \sum_{j=1}^{n} f_j^*(x_j)$$

subject to $$\sum_{j=1}^{n} g_{ij}^*(x_j) \le b_i,\ x_j \ge 0\ (i = 1, 2, ..., m, j = 1, 2, ..., n)$$

where, $f_j^*(x_j)$ and $g_{ij}^*(x_j)$ are now linear in x_j.

Now, evidently, we have

$$\frac{f_j(x_{j,k+1}) - f_j(x_{jk})}{x_{j,k+1} - x_{jk}} = \frac{f_j^*(x_j) - f_j(x_{jk})}{x_j - x_{jk}}$$

for any x_j in the interval $(x_{jk}, x_{j,k+1})$

This yields $$f_j^*(x_j) = f_j(x_{jk}) + \frac{f_j(x_{j,k+1}) - f_j(x_{jk})}{x_{j,k+1} - x_{jk}}$$

$$= (x_j - x_{jk})$$

Also, for any x_j in the interval $x_{jk} \le x_j \le x_{j,\,k+1}$ can be expressed as a convex combination of the end point of the interval.

Thus, $$x_j = \alpha x_{j,\,k+1} + (1 - \alpha)x_{jk}.$$

Using it in $f_j^*(x_j)$ obtained above, we get

$$f_j^*(x_j) = \alpha f_j(x_{j,\,k+1}) + (1 - \alpha) f_i(x_{jk}).$$

The last result suggests the linear approximation $f_j^*(x_j)$ between the interval $x_{jk} \le x_j \le x_{j,\,k+1}$ as a linear combination of the values of the original function at the end points, namely, $f_j(x_{jk})$ at x_{jk} and $f_j(x_{j,\,k+1})$ at $x_{j,\,k+1}$.

If we choose to write $\alpha = \alpha_{j,\,k+1}$ and $1 - \alpha = \alpha_{jk}$, the above may be rewritten as

$$x_j = \alpha_{j,\,k+1}\, x_{j,\,k+1} + \alpha_{jk}\, x_{jk}\ (\alpha_{j,\,k+1} + \alpha_{jk} = 1)$$

$$f_j^*(x_j) = \alpha_{j,\,k+1} f_j(x_{j,\,k+1}) + \alpha_{jk} f_j(x_{jk}).$$

These two relations with the additional restrictions that $\alpha_{jk} + \alpha_{j,k+1} = 1$ and $\alpha_{jk}, \alpha_{j,\,k+1} \ge 0$ describe the approximating linear function $f_j^*(x_j)$ anywhere in the interval $x_{jk} \le x_j \le x_{j,\,k+1}$ in terms of the known values $f_j(x_{j,k+1})$ and $f_j(x_{jk})$.

Generalising this approximation concept for any x_j in the interval $0 \le x_j \le t$, we may write

$$x_j = \sum_{k=0}^{n} \alpha_{jk}\, x_{jk} \text{ and } f_j^*(x_j) = \sum_{k=0}^{m} \alpha_{jk} f_j(x_{jk})$$

where, $\sum_{k=0}^{m} \alpha_{jk} = 1$ and $\alpha_{jk} \ge 0$ $(k = 0, 1, 2, ..., m)$ provided that

(a) no more than two α_{jk} are positive, and

(b) if two α_{hk} are positive they must correspond to adjacent points.

Now, assume that for each x_j in the given problem, a finite maximum is either given or can be determined and call it t_j. Then, the interval $0 \le x_i \le t_j$ is sub-divided into mesh points x_{jk}, so that

$$x_{j0} = 0 \le x_{j1} < x_{j2} < \ldots < x_{jm} = t_j$$

This shows the division of each specific variable x_j. Thus, if the same mesh points (the same intervals) are chosen for f and all constraints g_{ij}, for any particular x_j, we shall have

(a) $x_j = \sum_{k=0}^{m_j} \alpha_{jk}\, x_{jk}$

(b) $f_j^*(x_j) = \sum_{k=0}^{m_j} \alpha_{jk} f_j(x_{jk})$ and $g_{ij}^*(x_j) = \sum_{k=0}^{m_j} \alpha_{jk}\, g_{ij}(x_{jk})$

(c) $\sum_{k=0}^{m_j} \alpha_{jk} = 1$ and, $\alpha_{jk} \ge 0$ $(j = 1, 2, .., n;\ k = 0, 1, ..., m)$

provided that (i) for each j, no more than two α_{jk} are positive and (ii) if two α_{jk} are positive, they must be adjacent. Thus, the piece-wise linear approximation of the given problem is

Maximise $$f^*(x) = \sum_{j=0}^{n} \sum_{k=0}^{m_j} f_j(x_{jk})\, \alpha_{jk}$$

subject to $$\sum_{j=0}^{n} \sum_{k=0}^{m_j} g_{ij}(x_{jk})\, \alpha_{jk} \le b_i \; (i = 1, 2, ..., m)$$

$$\sum_{k=0}^{m_j} \alpha_{jk} = 1,\; \alpha_{jk} \ge 0 \; (j = 1, 2, .., n.\; k = 1, 2, .., m)$$

with the additional restrictions

(i) for each j, no more than two α_{jk} are positive

(ii) if two α_{jk} are positive, they must correspond to adjacent $x_j^1 s$.

Separable Programming Algorithm

Let the NLPP be

Maximise $f(x) = \sum_{j=0}^{n} f_j(x_j)$ subject to the constraints

$$\sum_{j=0}^{n} g_{ij}(x_j) \le b_i,\; x_j \ge 0 \qquad (i = 1, 2, ..., m, j = 1, 2, ..., n)$$

The iterative procedure for solving such separate NLPP is given below in different steps:

Step 1: Convert the minimisation objective function into that of maximisation, if it is in the minimisation form.

Step 2: Check whether the function $f_j(x_j)$ and $g_{ij}(x_j)$ satisfy the concavity (convexity) conditions required for the maximisation NLPP. If the conditions are not satisfied the method fails, otherwise proceed on to the next step.

Step 3: Divide the interval $0 \le x_j \le t_j$ $(j = 1, 2, ..., n)$ into a number of mesh-points x_{jk}, $(k = 0, 1, ..., m_j)$ such that $x_{j0} = 0, x_{j0} < x_{j1} < ... < x_j m_j = t_j$.

Step 4: For each point x_{jk}, compute piece-wise linear approximation for each $f_j(x_j)$ and $g_{ij}(x_j)$ $(j = 1, 2, ..., n, i = 1, 2, ..., m)$

Step 5: Using the computations of step 4, write down the piece-wise linear approximation of the given NLPP.

Step 6: Solve the resulting LPP by two-phase simplex method. In this method treat α_{i0} $(i = 1, 2, ..., m)$ as the artificial variables and since the costs associated with them are not there, these are assumed to be zeros, and phase I of the method is automatically completed. Thus, the initial simplex table of phase I becomes optimum and represents the starting simplex table for phase II.

Step 7: Obtain the optimum solution $x°$ of the given NLPP by using the relations

$$x_{j0} = \sum_{k=1}^{m_j} \alpha_{jk}\, x_{jk} \qquad (j = 1, 2, .., n)$$

Example 1

Use separable convex programming to solve the NLPP:

Maximise $f(x) = 3x_1 + 2x_2$

subject to $g(x) = 4x_1^2 + x_2^2 \le 16$ and $x_1, x_2 \ge 0$

Solution: The constraints of the problem suggest that $x_1 \le 2$ and $x_2 \le 4$.

Let us suppose that $t_1 = 4$ and $t_2 = 4$ are the upper limits for the variables x_1 and x_2, respectively. Divide the closed interval [0, 4] into four equally sized sub-intervals for both x_1 and x_2. It is necessary that the number of sub-intervals for x_1 and x_2 be same. Also, it is not essential that any set of them be of equal size.

Letting $f_1(x_1) = 3x_1, f_2(x_2) = 2x_2, g_{11}(x_1) = 4x_1^2$ and $g_{12}(x_2) = 2x_2^2$, (we observe that these functions satisfy the concavity-convexity conditions required for the maximisation problem), the given problem reduced to

Maximise $\qquad f(x) = \Sigma f_j(x_j)\ (j = 1, 2)$

subject to $\qquad \sum_{j=1}^{n} g_{2j}(x_j) \le 16,\ x_1, x_2 \ge 0.$

To obtain the approximating LPP, for this NLPP let us divide the interval $0 \le x_j < 4$ into a number of mesh points x_{jk} such that $x_{j0} = 0, x_{j0} \le x_{j1} \le x_{jk} < x_{j4} = 4$ for each point x_{jk}. Compute the piece-wise linear approximation for each of $f_j(x_j)$ and $g_{1j}(x_j)$.

Thus, for $j = 1, 2$ and $k = 0, 1, 2, 3, 4$ we have

x_j	0	1	2	3	4
$f_1(x_1)$	0	3	6	9	12
$f_2(x_2)$	0	2	4	6	8
$g_{11}(x_2)$	0	4	16	36	64
$g_{12}(x_2)$	0	1	4	9	16
Mesh points	x_{j0}	x_{j1}	x_{j2}	x_{j3}	x_{j4}

Using the information, the approximating LPP of the given NLPP may be written as

Maximise $\qquad f^*(x) = (0 \cdot \alpha_{10} + 3\alpha_{11} + 6\alpha_{12} + 9\alpha_{13} + 12\alpha_{14})$

$$+ (0\ \alpha_{20} + 2\ \alpha_{21} + 4\alpha_{22} + 6\alpha_{23} + 8\alpha_{24})$$

subject to $\quad (0 \cdot \alpha_{10} + 4\alpha_{11} + 16\alpha_{12} + 36\alpha_{13} + 64\ \alpha_{14})$

$$+ (0.\ \alpha_{20} + \alpha_{21} + 4\alpha_{22} + 9\alpha_{23} + 16\alpha_{24}) \le 16$$

$$\alpha_{10} + \alpha_{11} + \alpha_{12} + \alpha_{13} + \alpha_{14} = 1$$

$$\alpha_{20} + \alpha_{21} + \alpha_{22} + \alpha_{23} + \alpha_{24} = 1$$

and $\alpha_{jk} \ge 0\ (j = 1, 2, k = 0, 1, ..., 4)$

with the additional restriction that

(i) for each j, no more than two α_{jk} are positive, and

(ii) if two α_{jk} are positive, they must correspond to adjacent points.

Introducing the slack variable $s_1 \ge 0$, the inequation is converted into an equation. The reduced LPP can now be solved by phase II of the two-phase simplex method by treating α_{10} and α_{20} as the artificial variables whose costs in the objective function are zero. Thus, the initial simplex table for phase II is as follows.

Phase II

Starting Table

	c_j		3	6	9	12	2	4	6	8	0	0	0
C_B	Y_B	X_B	α_{11}	α_{12}	α_{13}	α_{14}	α_{21}	α_{22}	α_{23}	α_{24}	s_1	α_{10}	α_{20}
0	s_1	16	4	16	36	64	1	4	9	16	1	0	0
0	α_{10}	1	1	1	1	1	0	0	0	0	0	1	0
0	α_{20}	1	0	0	0	0	1	1	1	1	0	0	1
			-3	-6	-9	-12	-2	-4	-6	-8	0	0	0

The table shows that α_{14} must enter the basis and s_1 must leave it. But this violates the additional restriction, so we examine the next best vector to enter the basis. The table indicates that α_{13} can enter the basis. But it again creates the same difficulty. The next best entering vector is α_{24} and this indicates that either s_1 or α_{20} can leaves the basis. We choose α_{20} to leave the basis.

First iteration: Introduce α_{24} and drop α_{20}.

			3	6	9	12	2	4	6	8	0	0
C_B	Y_B	X_B	α_{11}	α_{12}	α_{13}	α_{14}	α_{21}	α_{22}	α_{23}	α_{24}	s_1	α_{10}
0	s_1	0	4	16	36	64	-15	-12	-7	0	1	0
0	α_{10}	1	1	1	1	1	0	0	0	0	0	1
8	α_{24}	1	0	0	0	0	1	1	1	1	0	0
			-3	-6	-9	-12	6	4	2	0	0	0

The above table indicates that any of the vectors corresponding to $z_j - c_j, j = 1, 2, 3, 4$ can enter the basis. But in order to satisfy the additional restrictions, we find that α_{11} enters the basis and s_1 leaves it.

Second iteration: Introduce α_{11} and drop s_1.

			3	6	9	12	2	4	6	8	0	0
C_B	Y_B	X_B	α_{11}	α_{12}	α_{13}	α_{14}	α_{21}	α_{22}	α_{23}	α_{24}	s_1	α_{10}
3	α_{11}	0	1	4	9	16	-15/4	-3	-7/4	0	1/4	0
0	α_{10}	1	0	-3	-8	-15	15/4	3	7/4	0	-1/4	1
8	α_{24}	1	0	0	0	0	1	1	1	1	0	0
			0	6	8	36	-21/4	5	-13/4	0	3/4	0

α_{21} cannot enter into the basis. Introduce α_{23} and remove α_{10}.

Third iteration:

			3	6	9	12	2	4	6	8	0
C_B	Y_B	X_B	α_{11}	α_{12}	α_{13}	α_{14}	α_{21}	α_{22}	α_{23}	α_{24}	s_1
3	α_{11}	1	1	1	1	1	0	0	0	0	0
6	α_{23}	4/7	0	-12/7	-32/7	-60/7	15/7	12/7	1	0	-1/7
8	α_{24}	3/7	0	12/7	32/7	60/7	-8/7	-5/7	0	1	1/7
			0	3/7	22/7	67/7	12/7	4/7	0	0	9/14

Since all $\Delta_j \geq 0$, the optimal solution of the reduced LPP is: $\alpha_{11} = 1$, $\alpha_{23} = 4/7$ and $\alpha_{24} = 3/7$. Hence, the optimum solution to the given NLPP is given by

$$x_j^0 = \sum_{k=0}^{4} \alpha_{1k}\, x_{jk} \quad \text{for } j = 1, 2.$$

This gives

$$\begin{aligned} x_1^0 &= \alpha_{10}\, x_{10} + \alpha_{11}\, x_{11} + \alpha_{12}\, x_{12} + \alpha_{13}x_{13} + \alpha_{14}x_{14} \\ &= (0)(0) + (1)(1) + (0)(2) + (0)(3) + (0)(4) \\ &= 1 \end{aligned}$$

and,

$$\begin{aligned} x_2^0 &= \alpha_{20}\, x_{20} + \alpha_{21}\, x_2 + \alpha_{22}\, x_{22} + \alpha_{23}\, x_{23} + \alpha_{24}\, x_{24} \\ & (0)\,(0) + (0)\,(1) + (0)\,(2)\,(4/7)\,(3) + (3/7)(4) \\ &= 24/7 \end{aligned}$$

Hence, the optimum solution to the given NLPP is $x_1 = 1, x_2 = 24/7$ with max. $Z = 3 + 2(24/7) = 69/7$.

14.8 GEOMETRIC PROGRAMMING

In this section, we shall focus our attention on an interesting technique called geometric programming for solving a special type of non-linear problem. This technique is initially derived from inequalities rather than calculus and its extensions. It was given the name 'geometric programming' because the geometric–arithmetic mean inequality was the basis of its development. This programming was developed by R. Duffin and C. Zener (1964).

Geometric programming was developed in connection with problems of engineering design. The solution to engineering design problems requires a compromise among many alternatives. The engineer seeks an ideal compromise: one that optimises some specific characteristics. The characteristics of an engineering design depend on parameters both fixed and adjustable. The design engineer strives to set the adjustable parameters for optimum design. A typical optimum design is that for which the system performs a given function at minimum cost. An attempt to develop a rapid and systematic method of designing at minimum cost has been the stimulus for the development of geometric programming.

The approach requires that all engineering characteristics be expressed quantitatively as generalised positive polynomials with adjustable parameters. This restriction enables one to make effective use of the geometric mean and of geometrical concepts. It is for this reason that this approach is called geometric programming.

There are three advantages of using geometric programming:

(a) An overall picture of the relative importance of various design parameters emerges.
(b) This approach is more amenable to digital computers, and
(c) It imposes engineering discipline on its user.

Polynomial: In an engineering design the total costs u_j. Thus $Z = f(x) = u_1 + u_2 + .. + u_n$.
Generally, the component costs are expressed as $u_j = c_j x_1^{aij}\, x_2^{a2j} \ldots x_m^{amj}$ $(j = 1, 2, \ldots, n)$
where $c_j > 0$, x_i $(i = 1, 2, \ldots, m) > 0$ and
a_{ij} $(i = 1, 2, \ldots, m, j = 1, 2, \ldots, n)$ are unrestricted in sign. The function $f(x)$ is usually referred to as a polynomial.

Geometric–Arithmetic Mean Inequality

To obtain the geometric–arithmetic mean inequality, we proceed as follows:

Consider the optimal sub-division problem

Maximise $f(x) = x_1 x_2 \ldots x_n$ subject to

$$x_1 + \ldots + x_n = c \ (c > 0)$$

Using the method of recursive equations, we find that max. $f(x) = \left(\frac{c}{n}\right)^n$.

It then follows that

$$f(x) \le \left(\frac{c}{n}\right)^n \quad \text{or} \quad \prod_{j=1}^{n} x_j \le \left(\sum_{j=1}^{n} \frac{x_j}{n}\right)^n$$

Taking the nth root of each side, the inequality becomes

$$\left(\prod_{j=1}^{n} x_j\right)^{1/n} \le \frac{1}{n} \sum_{j=1}^{n} x_j$$

It is to be noted that the inequality becomes strict equation when $c_j = c/n$. The inequality is known as geometric–arithmetic mean inequality and is fundamental for the development of geometric programming.

A more general form of this inequality can be written as

$$\left[\prod_{j=1}^{n} (V_j)^{\alpha_j}\right]^{1/\alpha} \le \frac{1}{\lambda} \sum_{j=1}^{N} \alpha_j V_j \tag{14.8.1}$$

where, $\lambda = \sum_{j=1}^{n} \alpha_j$, V_j(j = 1, 2, ..., N) are N positive numbers and α_j (j = 1, 2, ..., N) are N positive rational numbers. The numbers x_j are called the weight functions.

Now, if we put $\alpha_j = \lambda t_j$, the inequality becomes

$$\prod_{j=1}^{N} (V_j)^{t_j} \le \sum_{j=1}^{n} t_j V_j \sum_{j=1}^{n} t_j = 1 \tag{14.8.2}$$

The numbers t_j (j = 1, 2, ..., N) are called the normalised weight functions.

Further, by putting $V_j t_j = u_j$, we get

$$\prod_{j=1}^{N} \left(\frac{u_j}{t_j}\right)^{t_j} \le \sum_{j=1}^{N} u_j \sum_{j=1}^{N} t_j = 1 \tag{14.8.3}$$

In terms of non-normalised weights α_j, the inequality (14.8.1) can be rewritten as

$$\prod_{j=1}^{N} \left(\frac{\lambda u_j}{\alpha_j}\right)^{\alpha_i/\lambda} \le \sum_{j=1}^{n} u_j$$

or,

$$\prod_{j=1}^{N} \left(\frac{u_j}{\alpha_j}\right)^{\alpha_j} (\lambda)^{\lambda} \le \left(\sum_{j=1}^{n} u_j\right)^{\lambda} \tag{14.8.4}$$

Note: It is important to note that (14.8.3) becomes an equation if all u_i/t_j are equal. Similarly, (14.8.4) is an equation if all u_j/α_j are equal.

14.8.1 Unconstrained Geometric Programming Problem

The unconstrained geometric programming problem is defined as follows:

$$\text{Minimise } f(x) = \sum_{j=1}^{n} c_j \prod_{i=1}^{m} x_i^{a_{ij}}$$

where, $c_j(j = 1, 2, ..., n) > 0$, $x_i(i = 1, 2, ..., m) > 0$ and
$a_{ij}(i = 1, 2, ..., m; j = 1, ..., n)$ are unrestricted in sign.
Now, using the geometric–arithmetic inequality (14.8.3), the objective function defined can be re-written as $u_1 + u_2 + ... + u_n \geq \left(\frac{u_1}{t_2}\right)^{t_1}\left(\frac{u_2}{t_2}\right)^{t_2} \cdots \left(\frac{u_n}{t_n}\right)^{t_n}$ (14.8.5)

where, $u_j = u_j(x)$ for $j = 1, 2, ...,$ 1 the weights $t_j(j = 1, 2, ..., n)$ satisfy the relation $t_1 + ... + t_n = 0$.
The left-hand side of the inequali 14.8.5) is known as the primal function whereas the right-hand side is called the pre-dual function.

Since $u_j = c_j \prod_{j=1}^{m} x_i^{a_{ij}}$, the pre-dual function can be written as

$$\prod_{j=1}^{n}\left(\frac{u_j}{t_j}\right)^{t_j} = \prod_{j=1}^{n}\left[\frac{C_j \prod_{i=1}^{m} x_i^{a_{ij}}}{t_j}\right]^{t_j}$$

$$= \prod_{j=1}^{n}\left(\frac{c_j}{t_j}\right)^{t_j} \prod_{j=1}^{n}\left(\prod_{i=1}^{m} x_i^{a_{ij}}\right)^{t_j}$$

$$= \prod_{j=1}^{n}\left(\frac{c_j}{t_j}\right)^{t_j}\left\{\left(\prod_{i=1}^{m} x_i^{a_{ij}\,t_1}\right)\left(\prod_{i=1}^{m} x_i^{a_{ij}\,t_2}\right)\cdots\left(\prod_{i=1}^{m} x_i^{a_{in}\,t_n}\right)\right\}$$

$$= \prod_{j=1}^{n}\left(\frac{c_j}{t_j}\right)^{t_j}\left\{\left(\sum_{x_1^{j=0}}^{m} a_{ij}\,t_j\right)\left(\sum_{x_2^{j=0}}^{m} a_{ij}\,t_j\right)\cdots\left(\sum_{x_n^{j=0}}^{m} a_{mj}\,t_j\right)\right.$$

$$= \prod_{j=1}^{n}\left(\frac{c_j}{t_j}\right)^{t_j} \prod_{i=1}^{m} \sum_{x_1^{j=1}}^{n} a_{ij}\,t_j$$

Now, making use of the conditions $\sum_{j=1}^{n} a_{ij}\, t_j = 0, (i = 1, 2, ..., m)$ together with the normality condition $\sum_{j=1}^{n} t_j = 1$, the above expression reduces to

$$\prod_{j=1}^{n}\left(\frac{u_j}{t_j}\right)^{t_j} = \prod_{j=1}^{n}\left(\frac{c_j}{t_j}\right)^{t_j} \tag{14.8.6}$$

using (14.8.6), the inequality (14.8.5) becomes

$$u_1 + \cdots + u_n \geq \left(\frac{c_1}{t_j}\right)^{t_j}\left(\frac{c_2}{t_j}\right)^{t_j}\cdots\left(\frac{c_n}{t_n}\right)^{t_n}$$

or, $f(x) \geq \phi(t)$, where t is a vector with components $t_1, t_2, ..., t_n$. Since equality is possible, we have from above

$$\text{Minimum } f(x) = \text{Maximum } \phi(t).$$

In this way, we have converted the problem of minimising $f(x)$ into a new problem of maximisation $\phi(t)$. The original problem of minimisation is referred to as the primal one and the related problem of maximation and its dual.

In the present discussion $\phi(t)$ is just a constant, and therefore,
minimum $f(x) = \phi(t)$
which can be easily evaluated for given numerical values of c_j.

Example 1

Solve the following problem:

$$\text{Minimise } f(x) = c_1x_1x_2x_3 + c_2x_1x_2^{-1} + c_3x_2x_3^{-2} + c_4x_1^{-3}\, x_2$$

where $c_j > 0$, $x_j > 0$; $j = 1, 2, 3, 4$ and $i = 1, 2, 3$.

Solution: Let $u_1 = c_1x_1x_2x_3$, $u_2 = c_2x_1x_2^{-1}$

$u_3 = c_3x_2x^{-2}_3$ and $u_4 = c_4x_1^{-3}x_3$

Then, the objective function is $\sum_{j=1}^{4} u_j$

Using the geometric mean inequality, we can write

$$\prod_{j=1}^{n}\left(\frac{c_j}{t_j}\right)^{t_j} \leq \sum_{j=1}^{4} u_j = f(x)$$

Substituting the value of u_j ($j = 1, 2, 3, 4$), this inequality gives

$$\prod_{j=1}^{4}\left(\frac{c_j}{t_j}\right)^{t_j} \cdot x_1^{(t_1+t_2-3t_4)}\, x_2^{(t_1-t_2+t_3)}\, x_3^{(t_1-2t_2+t_4)} < f(x)$$

Here, the values t_j are arbitrary and can be chosen according to convenience subject to $t_j > 0$ and $\sum_{j=1}^{4} t_j = 1$.

Let us set the powers of x_i for each i equal to zero. Then, the right-hand side of the above inequality will become independent of x_i. For this, we require $t_1 + t_2 - 3t_4 = 0$, $t_1 - t_2 + t_3 = 0$ and $t_1 = 2t_3 + t_4 = 0$.

These are called necessary conditions for orthogonality. Now, the orthogonality conditions together with the normality condition $\sum_{j=1}^{4} t_j = 1$ are sufficient to give the values of t_j.

Now, the orthogonality–normality conditions yield

$$t_1 = 1/5,\ t_2 = 2/5,\ t_3 = 1/5 \text{ and } t_4 = 1/5.$$

Using these values of t_j ($j = 1, 2, 3, 4$), the geometric–arithmetic mean inequality reduces to

$$\left(\frac{c_1}{1/5}\right)^{1/5}\left(\frac{c_2}{2/5}\right)^{2/5}\left(\frac{c_3}{1/5}\right)^{1/5}\left(\frac{c_4}{1/5}\right)^{1/5} < f(x)$$

Since equality is possible, we have from the above

$$\text{Minimum} f(x) = \text{Maximum}\ \phi(t)$$

$$= \left(\frac{c_1}{1/5}\right)^{1/5}\left(\frac{c_2}{2/5}\right)^{2/5}\left(\frac{c_3}{1/5}\right)^{1/4}\left(\frac{c_4}{1/5}\right)^{1/5}$$

It is to be noted that the minimum value of the object function has been determined without having determined the optimal value of the variable x_1, x_2 and x_3. To obtain the optimal value of $x_i (i = 1, 2, 3)$ we proceed as follows:

For equality in the geometric–arithmetic mean inequality (14.8.3) all the u_j/t_j must be equal. Therefore, if we let $(u_j/t_j) = N$, $j = 1, 2, 3, 4$ then we shall have

$$\sum_{j=1}^{4} U_{ij} = N \sum_{j=1}^{4} t_j = N$$

Hence, $\qquad \text{Min. } f(x) = f(x_0) = N$

Thus, $\qquad u_j = t_j f(x_0)$

To evaluate $x_i (i = 1, 2, 3)$, we shall define the new variable y_i by substituting $y_i = \log x_i \quad i = 1, 2, 3$. We have,

$$y_1 + y_2 + y_3 = \log (u_1/c_1)$$
$$y_1 - y_2 = \log (u_2/c_2)$$
$$y_2 - 2y_3 = \log (u_3/c_3)$$
$$3y_1 + y_3 = \log (u_4/c_4)$$

These are four equations in three unknowns and are not linearly independent, the values of y_i and hence of x_i are uniquely determined. Thus we get the optimal solution.

Now, substituting $c_1 = 16$, $c_2 = 4$, $c_3 = 2$ and $x_4 = 8$, the optimal solution of the above problem is obtained as follows:

Minimum $f(x) = f(x_0)$

$$= \left(\frac{16}{1/5}\right)^{1/5}\left(\frac{4}{2/5}\right)^{2/5}\left(\frac{2}{1/5}\right)^{1/5}\left(\frac{8}{1/5}\right)^{1/5}$$

$$16\, x_1x_2x_3 = u_1 = \frac{1}{5}(20), \qquad 4x_1x_2^{-1} = u_2 = \frac{2}{5}\,(20)$$

$$2x_2\, x_3^{-2} = u_3 = \frac{1}{5}(20), \qquad 8x_1^{-3}x_3 = u_4 = \frac{1}{5}\,(20)$$

The solution of these equations gives us

$$x_1 = 1,\ x_2 = 0.5 \text{ and } x_3 = 0.5$$

14.8.2 Constrained Geometric Programming Problem

The constrained geometric programming problem, in general, is stated as follows:

$$\text{Minimise } f(x) = \sum_{j=1}^{n_0} c_{0j} \prod_{i=1}^{m} x_i^{a_{0ij}} \text{ subject to} \tag{14.8.7}$$

The constraints:

$$g_k(x) \equiv \sum_{j=1}^{n_k} c_{kj} \prod_{i=1}^{m} x_i^{a_{kij}} \{\le, =, \ge\}$$

where, x_i $(i = 1, 2, ..., m) > 0$, the constraints c_{0i} $(j = 1, 2, ..., n_0)$ and c_{kj} $(k = 1, 2, ..., N; j = 1, 2, ..., n_k)$ are strictly positive, the exponents a_{oij} $(i = 1, 2, .., m, j = 1, 2, .., n_0)$ and a_{kij} $(k = 1, 2, ..., N, i = 1, 2, ..., m, j = 1, 2, ..., n_k)$ are real numbers.

The symbol N indicates the total number of constraints; n_0 represents the number of terms in the objective function and n_k denotes the number of terms in the kth constraints as follows:

For the solution of the above-mentioned constrained minimisation problem, we rewrite the constraints:

$$f_k = \sigma_k(1 - g_k(x)] \ge 0,\ k = 1, 2, ..., N \tag{14.8.8}$$

where, σ_k is called the signum function and is introduced for the kth constraints so that it takes on the value + 1 or –1 depending on whether $g_k(x)$ is ≤ 1 or $\ge$.

The problem now reduces to minimising the objective function (14.8.7), subject to the constraints given in (14.8.8). This problem is called the primal problem and can be replaced by an equivalent problem (the dual problem) with linear constraints.

The method of solving constrained geometric programming problem is best illustrated through the following sample problems:

Example 1

Solve the GP problem

$$\text{Minimise } f(x) = 40x_1^{-1}x_2^{-1/2}x_3^{-1} + 20x_1x_3 + 20x_1x_2x_3$$

subject to the constraints

$$x_1^{-2}x_2^{-2} + 4x_1^{-1/2}x_3^{-1} \le 3,\ x_1,\ x_2\, x_3 > 0.$$

Solution: Let $u_1 = 40x_1^{-1}x_2^{-1/2}x_3^{-1}$, $u_2 = 20x_1x_3$, $u_3 = 20x_1x_2x_3$, $u_4 = 1/3x_1^{-2}x_2^{-2}$ and $u_5 = 4/3\,x_2^{1/2}x_3^{-2}$

Then, $$f(x) = u_1 + u_2 + u_3,\ g(x) = u_4 + u_5$$

Using the arithmetic–geometric inequality, we have

$$\left(\frac{u_1}{t_1}\right)^{t_1}\left(\frac{u_2}{t_2}\right)^{t_2}\left(\frac{u_3}{t_3}\right)^{t_3} \le u_1 + u_2 + u_3 = f(x)$$

and, $$\left(\frac{u_4}{\alpha_1}\right)^{\alpha_1}\left(\frac{u_5}{\alpha_2}\right)^{\alpha_2}\lambda^{\lambda} \le (u_4 + u_5)^{\lambda} = [g(x)]^{\lambda}$$

where, $t_1 + t_2 + t_3 = 1$ and $\alpha_1 + \alpha_2 = \lambda$.

Now, since the constraint is $g(x) \le 1$, we write $1 \ge [g(x)]^{\lambda}$ where $t_1 + t_2 + t_3 = 1$ and $\alpha_1 + \alpha_2 = \lambda$.

Now, since the constraint is $g(x) \le 1$, we write $1 \ge [g(x)]^{\lambda}$

Multiplying it on both sides by $f(x)$, we get

$$f(x) \ge f(x)\ [g(x)]^{\lambda}$$

or, $$f(x) \ge \left(\frac{u_1}{t_1}\right)^{t_1}\left(\frac{u_2}{t_2}\right)^{t_2}\left(\frac{u_3}{t_3}\right)^{t_3}\left(\frac{u_4}{\alpha_1}\right)^{\alpha_1}\left(\frac{u_5}{\alpha_2}\right)^{\alpha_2}\cdot\lambda^{\lambda}$$

Substituting the values of u_j ($j = 1, 2, 3, 4$), the inequality reduces to

$$\left(\frac{40}{t_1}\right)^{t_1}\left(\frac{20}{t_2}\right)^{t_2}\left(\frac{20}{t_3}\right)^{t_3}\left(\frac{1}{3\alpha_1}\right)^{\alpha_1}\left(\frac{4}{3\alpha_2}\right)^{\alpha_2} x_1{}^{d}{}_1 x_2{}^{d}{}_2 x_3{}^{d}{}_3 \lambda^{\lambda} \le f(x)$$

where, d_1, d_2, d_3 are functions of t_j ($j = 1, 2, 3$) and α_i($i = 1, 2$) are to be chosen such that

$$d_1 \equiv t_1 + t_2 + t_3 - 2\alpha_1 = 0,\ d_2 = -\frac{1}{2}t_1 + t_3 - 2\alpha_1 + \frac{1}{2}\alpha_2 = 0,$$

$$d_3 \equiv -t_1 + t_2 + t_3 - \alpha_1 = 0.$$

These three equations (orthogonality conditions) together with the normality condition $t_1 + t_2 + t_3 = 1$, yield

$$t_1 = 1 - (\alpha_{2+1})/2,\ t_2 = (1 + \alpha_2)/2,\ t_3 = (1 + \alpha_2)/4 \text{ and } \alpha_1 = \alpha_2/2.$$

Substituting these values of t_j ($j = 1, 2, 3$) and α_i($j = 1,2$), the abbove inequality further reduces to

$$\left(\frac{80}{1-c_2}\right)^{\frac{1-\alpha_2}{2}}\left(\frac{80}{1+\alpha_2}\right)^{\frac{1+\alpha_2}{2}}\left(\frac{80}{1+\alpha_2}\right)^{\frac{1+\alpha_2}{4}}\left(\frac{2}{3\alpha_2}\right)^{\frac{\alpha_2}{2}}\left(\frac{4}{\alpha_2}\right)^{\alpha_2}\lambda^{\lambda} \le f(x)$$

After simplification, it becomes

$$80(1 - \alpha_2^2)^{-1/2}\,2^{\alpha_2}(1 - \alpha_2)^{\alpha_2/2}\,(1 + \alpha_2)^{-\alpha_2/2} \le f(x)$$

or, $$\phi(\alpha) \le f(x)$$

Since equality is possible, we have minimum of $f(x)$ = minimum of $\phi(\alpha)$.

As $\phi(\alpha)$ becomes a function of a single variable α_2, the search for the maximum of $\phi(\alpha)$ will be a single variable search.

To determine the optimal values of the primal variables, we utilise the fact that $f(x)$ attains its minimum when $\phi(\alpha) = f(x)$. Also, it implies that $g(x) = 1$. Thus, for an optimal solution we must have

$$u_1 + u_2 + u_3 = f(x) = \phi(\alpha)$$

and,
$$u_4 + u_5 = g(x) = 1.$$

To determine the values of x_j, we take algorithms of $u_i(i = 1, 2, 3, 4, 5)$ and put $\log x_j = z_j(j = 1, 2, 3)$, thereby getting five linear equations in z_j. Out of these only three are linearly independent which give us a unique solution.

14.8.3 Complementary Geometric Programming

The method of geometric programming was extended by Avriel and Williams to include the terms of rational function of posynomials and was called complementary geometric programming. The case in which some of the terms may be negative will then become a special case of complement geometric programming. While geometric programming problems have the remarkable property that every constrained local minimum is a global minimum, no such claim can generally be made for complementary geometric programming. However, in many practical situations, it is sufficient to find a local minimum.

The algorithm that solves any complementary programming problem (CGPP) consists of successive approximating rational functions of posynomial terms by posynomials. Thus, solving a CGP problem by this algorithm involves the solution of a sequence of ordinary geometric programming problems. The algorithm produces a sequence whose limit is a local minimum of the CGPP.

The complementary geometric programming problem, in general, is defined as follows:

Minimise $z = R_0(x)$ subject to the constraints:

$$R_k(x) \le 1, x \ge 0, k = 1, 2, ..., n$$

where,
$$R_k(x) = \frac{A_k(x) - B_K(x)}{C_k(x) - D_k(x)} \quad k = 1, 2, ..., n$$

where, the terms $A_k(x)$, $B_k(x)$, $C_k(x)$ and $D_k(x)$ are posynomials in x and possibly some of them may be absent. However, we assume that $R_0(x) > 0$ for all feasible x. This assumption can always be satisfied by adding, if necessary, a sufficiently large constraint to $R_0(x)$.

To solve the above-mentioned problem we first introduce a new variable $x_0 > 0$, constrained to satisfy the relation $x_0 \ge R_0(x)$, i.e. $\dfrac{R_0(x)}{x_0} \le 1$. The problem is, therefore, restated as

Minimise x_0 subject to the constraints

$$\frac{A_k(x) - B_k(x)}{C_k(x) - D_k(x)} \le 1, \qquad k = 1, 2, ..., n,$$

where, $A_0(x) = R_0(x)$, $B_0(x) = 0$ $C_0(x) = x_0$ and $D_0(x) = 0$.

It is to be noted that the constraints having meaning only if $C_k(x) - D_{kj}(x)$ has a constant sign throughout the feasible region. Therefore, if $C_k(x) - D_k(x)$ is positive for some feasible x, it must be positive for all feasible x. Accordingly, as $C_k(x) - D_k(x)$ is positive or negative, we rewrite the constraints of the problem as

$$\frac{A_k(x)+B_k(x)}{C_k(x)+C_k(x)} \le 1 \text{ or } \frac{B_k(x)+C_k(x)}{A_k(x)+D_k(x)} \le 1$$

Thus, any complementary geometric programming problem can be expressed in its standard form as follows:

Minimise x_0 subject to the constraints

$$\frac{P_k(x)}{Q_k(x)} \le 1,\ k = 1, 2, \ldots, n \qquad \text{....(14.8.9)}$$

and, $x = (x_0, x_1, \ldots, x_m) > 0$

where, $P_k(x)$ and $Q_k(x)$ are posynomials of the form

$$P_k(x) = \sum_j p_{jk}(x) = \sum_{j=1}^{n} c_{jk} \prod_{j=1}^{m} x_i^{a_{ijk}} \qquad \text{... (14.8.10)}$$

and,

$$Q_k(x) = \sum_j q_{jk}(x) = \sum_{j=1}^{n} d_{jk} \prod_{i=1}^{n} x_i^{b_{ijk}} \qquad \text{... (14.8.11)}$$

Complementary G.P. Algorithm

The algorithm for solving CGP problem is based on the assumption that with constraints of the type $P_k(x)/Q_k(x) \le 1$ $(k = 0, 2, \ldots, n)$ if we could approximate each of the $Q_k(x)$ by a posynomial term, we would then have constraints that require a posynomial to be less than or equal to 1. This follows because a posynomial term divided by a posynomial term is again a posynomial. Thus, with this approximate the problem reduces to an ordinary geometric programming problem.

The iterative procedure for solving any CGP problem is described in the following two steps:

Step 1: Approximate each of the polynomial $Q(x)$, omitting the subscript k, by a single term polynomial. Choose any $\bar{x} > 0$ and let

$$u_j = q_i(x),\ t_j = \bar{q}_j(x)/Q(\bar{x})$$

where, q_j denotes the jth term of the polynomial $Q(x)$.

Therefore, by using the arithmetic geometric inequality (14.8.3), we write

$$Q(x) \equiv \sum_{j=1}^{N} q_j(x) \ge \prod_{j=1}^{n} \left[\frac{q_j(x)}{q_j(\bar{x})} Q(\bar{x}) \right]^{t_j} \qquad \text{...(14.8.12)}$$

Using (14.8.11) and (14.8.12), we get

$$Q(x) \ge \bar{Q}\ (x, \bar{x}) = Q(\bar{x}) \prod_{i=0}^{m} (x_i/\bar{x}_i)^{\lambda}, \text{ where}$$

$$\lambda = \sum_{j=1}^{n} \{b_{ij}\, q_j(\bar{x})/Q(\bar{x})\}$$

where, the equality sign holds true if $x_i = \bar{x}_i$. The function $Q(x, \bar{x})$ is the approximation for $Q(x)$ of $\bar{x}$.

Step 2: At any feasible point $x^{(1)}$, replace $Q_k(x)$ in (14.8.9) with their approximations $\overline{Q}_k(x, x^{(1)})$, and solve the resulting ordinary geometric programming problem to obtain the next point $x^{(2)}$. Continuing in this way, we generate a sequence $\{x^{(\alpha)}\}$ where $x^{(\alpha+1)}$ is an optimal solution for the ordinary geometric programming problem

Minimise x_0 subject to the constraints:

$$\frac{p_k(x)}{Q_k\left(x_1 x^{(\alpha)}\right)} \leq 1, \quad k \geq 0, 1, 2 \ldots, n$$

$$x = (x_0, x_1, \ldots, x_m) > 0$$

Avriel and Williams have shown that under certain mild restrictions, the sequence of points $\{x^{(\alpha)}\}$ converges to a local minimum of the complement geometric programming problem. The computational procedure of the complementary geometric programming is illustrated through the following sample problem.

Example

Use the method of complementary geometric programming to solve the following problems:

Minimise x_1 subject to

$$-4x_1^2 + 4x_2 \leq 1, x_1 + x_2 \geq 1, x_1, x_2 > 0.$$

Solution: In terms of the complementary geometric programming, the problem is restated as follows:

Minimise x_1 subject to the constraints:

$$4x_2/(1 + 4x_1^2) \leq 1$$

$$x_1^{-1}/(1 + x_1^{-1} x_2) \leq 1$$

$$x_1, x_2 > 0$$

and let $Q_1(x)$ and $Q_2(x)$ denote the denominators of the constraints, such that

$$Q_1(x) = 1 + 4x_1^2 \text{ and } Q_2(x) = 1 + x_1^{-1} x_2$$

Both these expressions can be approximated by a single term posynomial with the help of (14.8.12) as

$$Q_1(x, \bar{x}) = (1 + 4\bar{x}_1^2)(x_1/\bar{x}_1)x^2/(1 + 4x_1)$$

and,

$$Q_2(x, \bar{x}) = (1 + \bar{x}_2/\bar{x}_1)(x_1/\bar{x}_1)^{-\alpha}(x_2/\bar{x}_2)^{\alpha}$$

$$\alpha = \bar{x}_2/(\bar{x}_1, \bar{x}_2)$$

We now start the iterative procedure from the point $x^1 = (1, 1)$. Clearly, this point is feasible. By taking $x = x^1$, we have

$\overline{Q}_1(x, x^{(1)}) = 5x_1^{8/5}$ and $Q_2(x_1, x_2^{(1)}) = x_1^{-1/2} x_2 1/2$ using (14.8.13), we formulate the first ordinary geometric programming problem can be obtained by solving a square system of linear equations, namely,

$$t_1 = 1$$

$$t_1 - 8/5\, t_2 - 1/2\, t_3 = 0$$

$$t_2 - 1/2\, t_3 = 0$$

Solving these linear equations, we obtain

$$t_1 = 1, t_2 = 5/13 \text{ and } t_3 = 10/13.$$

Substituting these values of $t_j (j = 1, 2, 3)$ into the dual objective function, we get

$$v(t) = (4/5)^{5/13} (1/2)^{10/13} \cong 0.5385$$

Making use of the duality relations, therefore, we shall have $x_1^0 \cong 0.5385$, and $x_2^0 = 5/4\ x_1^{8/15}$ $\cong 9.4683$.

Next, we choose $x^{(2)}$ to be the optimal solution of (OGP_2), i.e. $x^{(2)} = (0.5383, 0.4683)$ and approximate Q_1 and Q_2 about this point solve (OGP_2), and so on. The sequence of optimal solution of (OGP_α) as generated by the iterative procedure is shown below:

Iteration	x_1	x_2
0	1.0	1.0
1	0.5385	0.4643
2	0.5019	0.5007
4	0.5000	0.5000

The optimal value of the variables for the *CGP* are $x_1^* = 0.5$ and $x_2 = 0.5$. It can be seen that in three iterations, the solution of the approximate geometric programming problems (OGP_α) is correct to four significant figures.

EXERCISES

Section 14.2

1. A company manufactures two products *A* and *B*. It takes 30 minutes to process one unit of product *A* and 15 minutes for each unit of *B* and the maximum machine time available is 35 hours per week. Products *A* and *B* require 2 kg and 3 kg of raw material per unit, respectively. The available quantity of raw material is considered to be 180 kg per week. The products *A* and *B* which have unlimited market potential sell for Rs 200 and Rs 500 per unit, respectively. If the manufacturing for products *A* and *B* are $2x^2$ and $3y^2$, respectively, find how much of each product should be produced per week, where

   x = quantity of product *A* to be produced

   y = quantity of product *B* to be produced

   [*Answer*: Max. $Z = (200 - 2x^2) + (500 - 2y^2)$; subject to $0.5x + 0.25y \leq 35$, $2x + 3y \leq 180$ and $x \geq 0$, $y \geq 0$.]

2. *One-potato, two-potato problem*: A frozen-foods company processes potatoes into packages of french fries, hash browns and flakes (for mashed potatoes). At the beginning of the manufacturing process, the raw potatoes are sorted by length and quality and then allocated to separate product lines.

   The company can purchase potatoes from two sources which differ in their yields of various sizes and quality. Each source yields different fractions of the products french fries, hash browns and flakes. Suppose that it is possible, at different costs, to alter these yields somewhat. Let f_1, f_2 and f_3 be the fractional yield per unit of weight of source 1 potatoes made into three products, Similarly, let g_1, g_2, g_3 be the yields of source 2. Suppose that each f_i and g_i can vary within ± 10% of the yields shown below.

Product	*Source 1*	*Source 2*	*Purchase limitation*
French fires	0.2	0.3	1.8
Hash browns	0.2	0.1	1.2
Flakes	0.3	0.3	2.4
Relative Profit	5	6	

Let $C_i(f_1, f_2, f_3)$ and $C_2(g_1, g_2, g_3)$ be the expense associated with obtaining these yields. The problem is to determine how many potatoes should the company purchase from each source? Formulate the problem as NLPP.

[*Answer*: Minimise $Z = 100 - 1.2x_1 - 1.5x_2 + 0.3x_1^2 + 0.5x_2^2$ subject to $6x_1 + x_2 \geq 35$, $x_1 \geq 0$ and $x_2 \geq 0$.]

3. A manufacturing concern operates its two available machines to polish metal products. The two machines are equally efficient, although their maintenance costs are different. The daily maintenance and operation cost of the machines is given in rupees as the non-linear function:

$$f(x, y) = 100 - 1.2x - 1.5y + 0.3x^2 + 0.4y^2$$

where, x and y are the number of hours of operation of machine I and machine II, respectively. The past records of the firm indicate that the combined operating hours of two machines should be at least 35 hours a day in order to perform a satisfactory job. However, the production manager wishes to operate machine I at least 6 hours more than machine II because of the higher repair cost of the latter. Find the optimal hours of operating the two machines and the minimum daily cost. Formulate the problem as a NLPP.

4. The ABC Investment Company's manager is trying to create a portfolio of two stocks that promise annual returns of 16 and 20%, respectively. Additionally, he estimates that the risk of each investment will vary directly with twice the square of the size of investment. Each lot (100 shares) of the first stock costs Rs 3,000 and exact lot (100 shares) of the second stock cost Rs 2,000. The total amount of cash available for investment is Rs 6,000. Formulate and solve this problem as an NLPP to maximise return and minimise risk.

5. The operations research team of the ABC Company has come up with the mathematical data (daily basis) needed for two products which the firm manufactures. It has also determined that this is a non-linear programming problem having linear constraints and objective function which is the sum of a linear and a quadratic form. The pertinent data, gathered by the *OR* team, is

$$\text{Maximise } Z = 12x + 21y + 2xy - 2x^2 - 2y^2$$

subject to the constraints

$$8 - y \geq 0$$
$$10 - x - y \geq 0$$
$$x, y \geq 0$$

Find the maximum contribution and number of units that can be expected from these two products which are a part of the firm's total output, where x and y represent the number of units of the two products.

Section 14.3

6. Find the extreme points of the function

$$f(x_1, x_2) = 900x_1 - 10x_1^2 + 900x_2 - 10x_2^2 - 10x_1x_2$$

7. Find the maximum and minimum of the function

$$f(x_1, x_2, x_3) = x_1^2 - 12x_1 + x_2^2 - 8x_2 + x_3^2 - 4x_3$$

8. Find the maximum and minimum of the function

$$f(x_1, x_2, x_3) = 2x_1 + x_1x_2 + x_3 - x_1^2 - x_2^2 - x_3^2$$

Section 14.4

9. (a) Examine $z = 6x_1 x_2$ for maxima and minima under the requirement $2x_1 + x_2 = 10$.
   (b) What happens if the problem becomes that of maxima $Z = 6x_1 x_2 - 10x_3$ under the constraint equation $3x_1 + x_2 + 3x_3 = 10$?
10. Verify that the function

$$f(x_1, x_2, x_3) = 2x_1 x_2x_3 - 4x_1x_3 - 2x_2x_3 + x_1^2 + x_2^2 + x_3^2 - 2x_1 - 4x_2 + 4x_3$$

has the stationary points (0, 3, 1), (2, 1, 1), (2, 3, – 1), (0, 1, – 1) and (1, 2, 0) use the sufficiency condition to check for the extreme points.

Solve the following non-linear programming problems using the method of Lagrangian multipliers:

11. Minimise $z = 6x_1^2 + 5x_2^2$ subject to

$$x_1 + x_2 = 3,\ x_1, x_2 \geq 0$$

[*Answer*: $x_1 = 3/31$, $x_2 = 18/31$; Minimum $z = 54/31$.]

12. Minimise $z = 2x_1^2 + x_2^2 + 3x_3^2 + 10x_1 + 8x_2 + 6x_3 - 100$ subject to

$$x_1 + x_2 + x_3 = 20,\ x_1, x_2, x_3 \geq 0$$

[*Answer*: $x_1 = 5$, $x_2 = 11$ and $x_3 = 4$; minimum of $z = 281$.]

13. Minimise $z = x_1^2 + x_2^2 + x_3^2$ subject to

$$x_1 + x_2 + 3x_3 = 2,\ 5x_1 + 2x_2 + x_3 = 5,\ x_1, x_2, x_3 \geq 0$$

[*Answer*: $x_1 = 0.81$, $x_2 = 0.35$, $x_3 = 0.28$; minimum of $z = 0.857$.]

14. Write the Kuhn-Tucker necessary conditions for the following problems:

   (i) Max. $f(x) = x_1^3 - x_2^2 + x_1x_3^2$
   subject to

$$x_1 + x_2^2 + x_3 = 5,$$
$$5x_1^2 - x_2^2 - x_3 \geq 2$$
$$x_1, x_2, x_3 \geq 0$$

   (ii) Min. $f(x) = x_1{}^4 + x_2{}^2 + 5x_1 x_2 x_3$
   subject to

$$x_1^2 - x_2^2 - x_3^2 \leq 0,$$
$$x_1^2 + x_2^2 + 4x_3^2 \geq 20.$$

15. Use the Kuhn-Tucker conditions to solve the following non-linear programming problems:

(i) Max. $z = 2x_1^2 + 12x_1 x_2 - 7x_2^2$
subject to $2x_1 + 5x_2 \le 98$
$x_1, x_2 \ge 0$
[*Answer*: $x_1 = 44$, $x_2 = 2$; max $Z = 4{,}900$.]

(ii) Max. $z = 2x_1 - x_1^2 + x_2$
subject to
$2x_1 + 3x_2 \le 6$
$2x_1 + x_2 \le 4$, $x_1, x_2 \ge 0$
[*Answer*: $x_1 = 2/3$, $x_2 = 14/9$; max. $Z = 22/9$.]

(iii) Min. $z = -\log x_1 - \log x_2$
subject to $x_1 + x_2 \le 2$
$x_1 \ge 0$
[*Answer*: $x_1 = 1$, $x_2 = 1$; min. $Z = 0$.]

(iv) Max. $z = 8x_1^2 + 2x_2$
subject to
$x_1^2 + x_2^2 \le 9$
$x_1 \le 2$
$x_1, x_2 \ge 0$
[*Answer*: $x_1 = 2$, $x_2 = \sqrt{5}$; max $z = 42$.]

(v) Maximise $z = 2x_1 - x_1^2 + x_2$ subject to
$2x_1 + 3x_2 \le 6$, $2x_1 + x_2 \le 4$ and $x_1, x_2 \ge 0$
[*Answer*: $x_1 = 2/3$, $x_2 = 14/9$, maximum of $z = 22/9$.]

(vi) Max $z = 3x_1 + x_2$ subject to
$x_1^2 + x_2^2 \le 5$, $x_1 - x_2 \le 1$ and $x_1, x_2 \ge 0$
[*Answer*: $x_1 = 1.43$, $x_2 = 0.48$ maximum of $z = 0.77$.]

(vii) Minimise $f(x, x_2) = (x_1 - 1)^2 + (x_2 - 5)^2$
subject to
$-x_1^2 + x_2 \le 4$ and $-(x_1 - 2)^2 + x_2^1 \le 3$
[*Answer*: $x_1 = 1$, $x_2 = 5$; minimum of $f(x_1, x_2) = 0$.]

Solve the following non-linear programming problem graphically:

16. Maximise $z = 8x_1 - x_1^2 + 8x_2 - x_2^2$ subject to
$x_1 + x_2 \le 12$, $x_1 - x_2 \ge 4$ and $x_1, x_2 \ge 0$
[*Answer*: $x_1 = 6$, $x_2 = 2$; maximum of $z = 24$.]

17. Maximise $z = -(2x_1 - 5)^2 - (2x_2 - 1)^2$ subject to
$x_1 + 2x_2 \le 1$, $x_1 \ge 0$ and $x_2 \ge 0$
[*Answer*: $x_1 = 0$, $x_2 = 0.5$; maximum of $z = -25$.]

18. Maximise $z = x_1$, subject to
$(1 - x_1)^3 - x_2 \ge 0$, $x_1, x_2 \ge 0$.

Also show that the Kuhn-Tucker necessary conditions for the maxima do not hold. What do you conclude?

[*Answer*: $x_1 = 1, x_2 = 0$; maximum of $z = 1$. Constraint qualification is not satisfied.]

19. Maximise $z = 8x_1^2 + 2x_2^2$
subject to
$$x_1 x_2 \geq 8,\ x_1^2 + x_2^2 \leq 9,\ x_1 \leq 2,\ x_1, x_2 \geq 0$$

20. Minimise $z = (x_1 - 1)^2 + (x_2^1 - 2)^2$ subject to
$$0 \leq x_1 \leq 2,\ 0 \leq x_2 \leq 1$$
[*Answer*: $x_1 = 0, x_2 = 1$; min. $z = 2$.]

Section 14.6

21. Apply Wolfe's method to solve the quadratic programming problem
Maximise $z = 2x_1 + x_2 - x_1^2$
subject to
$$2x_1 + 3x_2 \leq 6$$
$$2x_1 + x_2 \leq 4$$
$$x_1, x_2 \geq 0$$
[*Answer*: $x_1 = 2/3, x_2 = 14/9$ with max $z = 22/9$.]

22. Use Wolfe's method to solve the following problems:
(i) Minimise $z = x_1^2 + x_2^2 + x_3^2$
subject to
$$x_1 + x_2 + 3x_3 = 2$$
$$5x_1 + 2x_2 + x_3 = 5$$
$$x_1, x_2, x_3 \geq 0$$
[*Answer*: $x_1 = 0.81, x_2 = 0.35, x_3 = 0.35$; min. $Z = 0.857$.]
(ii) Min. $z = -x_1 - x_2 - x_3 + 1/2\,(x_1^2 + x_2^2 + x_3^2)$
subject to
$$x_1 + x_2 - x_3 \leq 0$$
$$4x_1 + 2x_2 - 7/2 \leq 0$$
$$x_1, x_2, x_3 \geq 0$$
[*Answer*: $x_1 = x_2 = x_3 = 1/3$; min. $z = -15/18$.]
(iii) Max. $z = 8x_1 + 10x_2 - x_1^2 - x_2^2$
subject to
$$3x_1 + 2x_2 \leq 6$$
$$x_1\ x_2 \geq 0$$
[*Answer*: $x_1 = 4/13, x_2 = 33/13$; max $z = 267/13$.]
(iv) Maximise $z = 6x_1 + 3x_2 - 4x_1x_2 - 2x_1^2 - 3x_2^2$
subject to $x_1 + x_2 \leq 1,\ 2x_1 + 3x_2 \leq 4,\ x_1, x_2 \geq 0$
[*Answer*: $x_1 = 1, x_2 = 0$; maximum $z = 4$.]

(v) Maximise $z = -2x_1^2 + 2x_1 + 3x_2$, subject to

$$x_1 + 48x_2 \le 8$$
$$x_1 + x_2 \le 4$$
$$x_1, x_2 \ge 0$$

[*Answer*: $x_1 = 0, x_2 = 1$ with maximum $z = 3$.]

23. Solve the following quadratic programming problems by using Beale method

(i) Maximise $z = 4x_1 + 6x_2 - 2x_1^2 - 2x_1x_2 - 2x_2^2$

subject to $x_1 + 2x_2 \le 2$

$$x_1, x_2 \ge 0$$

[*Answer*: $x_1 = 1/3, x_2 = 5/6, x_3 = 0$; max. $Z = 25/6$.]

(ii) Max. $z = 10x_1 + 25x_2 - 10x_1^2 - x_2^2 - 4x_1x_2$

subject to $x_1 + 2x_2 + x_3 = 10$

$$x_1 + x_2 + x_4 = 9$$
$$x_1, x_2, x_3, x_4 \ge 0$$

[*Answer*: $x_1 = 0, x_2 = 5$, maximum of $z = 100$.]

(iii) Min. $z = 183 - 44x_1 - 42x_2 + 8x_1^2 - 12x_1x_2 + 17x_2^2$

subject to $2x_1 + x_2 \le 10, x_1 x_2 \ge 0$

[*Answer*: $x_1 = 3.8, x_2 = 2.4$, min $z = 19$.]

(iv) Min. $z = 4t_0^2 + 4t_0t_1 + 6t_1^2 + 4t_0 - 56t_1 + 180$

subject to $t_0 \ge -3$ and $t_1 \le 6$

[*Answer*: $t_0 = -3, t_1 = 17/3$, minimum $z = 34/3$.]

(v) Maximise $z = 2x_1 + 3x_2 - x_1^2$

subject $x_1 + 2x_2 \le 4$ and $x_1, x_2 \ge 0$

[*Answer*: $x_1 = 1/7, x_2 = 15/8$, maximum $z = 97/16$.]

(vi) Maximise $f(x) = 2x_1 + 3x_2 - 2x_2^2$

subject to $x_1 + 4x_2 \le 4, x_1 + x_2 \le 2, x_1, x_2 \ge 0$

[*Answer*: $x_1 = 0, x_2 = 3/4$, maximum $Z = 9/8$.]

(vii) Maximise $f = 4x_1 + 6x_2 - x_1^2 - 3x_2^2$

subject to

$$x_1 + x_2 \le 4 \text{ and } x_1, x_2 \ge 0$$

[*Answer*: $x_1 = 2, x_2 = 1, x_3 = 0$, maximum $f = 7$.]

(viii) Minimise $z = -4x_1 + x_1^2 - 2x_1x_2 + 2x_2^2$

subject to $2x_1 + x_2 \ge 6, x_1 - 4x_2 \ge 0$ and

$$x_1, x_2 \le 0$$

[*Answer*: $x_1 = 38/31, x_2 = 2/13$, minimum of $Z = 116/23$.]

Section 14.7

24. Use the method of separable convex programming to solve the following non-linear programming problems:

(i) Max. $Z = (x_1 - 2)^2 + (x_2 - 2)^2$
subject to

$$x_1 + 2x_2 \leq 4$$
$$x_1, x_2 \geq 0$$

[*Answer*: $x_1 = 1.6, x_2 = 1.2$, max. $Z = 0.8$.]

(ii) Minimise $f(x) = (x_1 - 2)^2 + (x_2 - 1)^2$ subject to

$$-x_1^2 + x_2 \geq 0, -x_1 - x_2 + 2 \geq 0\ x_1, x_2 \geq 0$$

[*Answer*: $x_1 = 1, x_2 = 1$, minimum of $f(x) = 1$.]

(iii) Minimise $f(x) = (x_1 - 2)^2 + (x_2 - 1)^2$
subject to

$$x_1 - 2x_2 + 1 = 0$$

$$-1/4\ x_1^2 - x_2^2 + 1 \geq 0, x_1, x_2 \geq 0$$

[*Answer*: $x_1 = 0.823, x_2 = 0.94$, minimum $f(x) = 1.393$.]

(iv) Maximise $f(x) = x_1 + x_2^2$
subject to

$$3x_1 + 2x_2^2 \leq 9; x_1 \geq 0 \text{ and } x_2 \geq 0$$

[$x_1 = 0, x_2 = 2.1$, maximum $f(x) = 22.5$.]

(v) Maximize $f(x) = 16 - 2(x_1 - 3)^2 - (x_2 - 7)^2$
subject to the constraints

$$x_1^2 + x_2 \leq 16, x_1, x_2 \geq 0$$

[*Hint*: $8 - 2(x_1 - 3)^2$ and $f_2(x_2) = 8 - (x_2 - 7)^2$
[*Answer*: $x_1 = 3, x_2 = 7$, maximum $f(x) = 16$.]

Section 14.8

25. Using geometric programming, sove the following problems:

(i) Minimise $f(x) = c_1x_1^{-1}x_2^{-1} + c_2x_2x_3 + c_3x_1x_3 + c_4x_1x_2$
where $c_1 > 0, x_j > 0, i = 1, 2, 3, 4$ and $j = 1, 2, 3$
[*Answer*: Minimum of $z = 5(c_1/2)^{2/5}(5c_2)^{1/5}(5c_3)^{1/5}(5c_4)^{1/5}$.]

(ii) Minimise $f(x) = 5x_1\,x_2^{-1}\,x_3^2 + x_1^{-2}x_3^{-1} + 10x_2^3 + 2x_1^{-1}x_2\,x_3^{-3}$.
[*Answer*: Minimum of $f(x) = 10.28, x_1 = 1.26, x_2 = 0.41, x_3 = 0.59$.]

(iii) Minimise $z = 7x_1\,x_2^{-1} + 3x_2\,x_3^{-2} + 5x_1^{-3}\,x_2x_3 + x_1x_2x_3, x_1, x_2, x_3 > 0$

26. In the economic lot size problems minimise

$$f(q) = 1/2c_1q + c_3R/q,$$

to find the optimum inventory level.

$$\left[\textit{Answer}: q^* = \sqrt{\frac{2C_3R}{C_1}}.\right]$$

27. Using geometric programming, solve the following problems:
 (i) Minimise $f(x) = 5x_1^{-1} x_2^{-1} x_3^{-1} + 5x_2x_3$
 subject to the constraints
$$2x_1 x_3 + x_1 x_2 = 4,\ x_1, x_2, x_3 > 0$$
 [*Answer*: Minimum of $f(x) = 15\sqrt[3]{2}$.]
 (ii) Minimise $f(x) = 2x_1 x_2^{-2} + 4x_1^{-1} x_2^{-2} + 10.7x_1x_2$
 subject to $10x_1^{-1} x_2^{2} = 1,\ x_1, x_2 > 0$
 [*Answer*: Minimum of $f(x) = 49.2$.]
 (iii) Minimise $f(x) = 20x_1x_3 + 40x_2x_3 + 80x_1x_2$
 subject to $8x_1^{-1} x_2^{-1} x_3^{-1} \le 1,\ x_1, x_2, x_3 > 0$
 [*Answer*: Minimum of $f(x) = 480$, $x_1 = 2$, $x_2 = 1$, $x_3 = 4$.]

APPENDIX 1

Random Numbers Table

2181922396	2068577984	8262130892	8374856049	14837657422
1128105582	7295088579	9586111652	7055508767	5172382962
7112077556	3440672486	1882412963	0684012006	0633147925
6557477468	5435810788	9670852913	1291265730	2290031331
4199520858	3090908872	2039593181	5973470495	8076135523
3545174840	2275698645	8416549348	4676463101	5629367908
1749420382	4832630032	5670984959	5432114610	0666095693
9103161011	7413686599	1198757695	0414294470	9240121544
0764238934	7666127259	5263097712	5133648980	5111966912
3493969525	0272759769	0385998136	9999089966	1344056826
1292054466	0700014629	5169439659	8408705169	6574373193
4397426117	6488888550	4031652526	8123543276	6027534501
3807950579	9564268448	3457416988	1531027886	5116633717
4984768758	2389278610	3859431781	3643768456	5041314549
1340145867	9120831830	7228567652	1267173884	1320651658
0590453442	4800088084	1165628554	5407921254	9468932498
9566554338	5585265145	5089052204	9780623691	5795448061
7615116284	9172824179	5544814334	0016943666	2628538741
8508771938	4035554324	0840126299	4942059208	7875623913
6970024586	9324732696	1186263397	4425453189	3316653259
5799997185	0135968939	7678931194	1351031403	6002561840
6364375912	8383232768	1892850701	2323673751	3188881718
4165492027	6349104233	3382569662	4579426926	1513082455
0354683246	4765104877	8149224168	5468631609	6474393896
9130555058	5255147182	3519287786	2481675649	8907598697
5826984369	4725370390	9641916289	5049082870	7463807244
6285048453	3646121751	8436077768	2928794356	9956043516

(*Contd.*)

7527791048	5765558107	8762562043	6185670830	6363845920
8976470693	0441608934	8749472723	2202271078	5897002653
2327991661	7936797054	9527542791	4711871173	8300978148
1182095589	5535798279	4764439855	6279247618	4446895088
3659397698	1056981450	8416606706	8234013222	6426813469
5924779358	1333750468	9434074212	5273692238	5902177065
3941092295	5726289716	3420847871	1820481234	0318831723
1955104281	0903099163	6827824899	6383872737	5901682626
2117595534	1634107293	8521057472	1471300754	3044151557
7471564123	7344613447	1128117244	3208461091	1699403490
8674262892	2809456764	5806554509	8224980942	5738031833
9061122871	0746980892	9285305274	6331989649	8764467686
6438538678	3049068967	6955157269	5482964330	2161984904
1834182305	6203476893	5937802079	3445280195	3694915658
1884227732	2923727501	8044389132	4611203081	6072112445
6791857341	6696243386	2219599137	3193884246	8224729918

APPENDIX 2

Proportion of Total Area under the Normal Curve from ∞ to t where $t = x - \mu/\sigma$

t	$\psi(t)$	t	$\psi(t)$	t	$\psi(t)$	t	$\psi(t)$
0.00	0.5000	0.65	0.7422	1.30	0.9032	1.95	0.9744
0.01	0.5040	0.66	0.7454	1.31	0.9049	1.96	0.9750
0.02	0.5080	0.67	0.7486	1.32	0.9066	1.97	0.9756
0.03	0.5120	0.68	0.7517	1.33	0.9082	1.98	0.9761
0.04	0.5160	0.69	0.7549	1.34	0.9099	1.99	0.9767
0.05	0.5199	0.70	0.7580	1.35	0.9115	2.00	0.9772
0.06	0.5239	0.71	0.7611	1.36	0.9131	2.02	0.9783
0.07	0.5279	0.72	0.7642	1.37	0.9147	2.04	0.9793
0.08	0.5319	0.73	0.7673	1.38	0.9162	2.06	0.9803
0.09	0.5359	0.74	0.7703	1.39	0.9177	2.08	0.9812
0.10	0.5398	0.75	0.7734	1.40	0.9192	2.10	0.9821
0.11	0.5438	0.76	0.7764	1.41	0.9207	2.12	0.9830
0.12	0.5478	0.77	0.7794	1.42	0.9222	2.14	0.9838
0.13	0.5517	0.78	0.7823	1.43	0.9236	2.16	0.9846
0.14	0.5557	0.79	0.7852	1.44	0.9251	2.18	0.9854
0.15	0.5596	0.80	0.7881	1.45	0.9265	2.20	0.9861
0.16	0.5636	0.81	0.7910	1.46	0.9279	2.22	0.9868
0.17	0.5675	0.82	0.7937	1.47	0.9292	2.24	9.9875
0.18	0.5714	0.83	0.7967	1.48	0.9306	2.26	0.9881
0.19	0.5753	0.84	0.7995	1.49	0.9319	2.28	0.9887
0.20	0.5793	0.85	0.8023	1.50	0.9332	2.30	0.9893
0.21	0.5832	0.86	0.8051	1.51	0.9345	2.32	0.9898
0.22	0.5871	0.87	0.8078	1.52	0.9357	2.34	0.9904
0.23	0.5910	0.88	0.8106	1.53	0.9370	2.36	0.9909
0.24	0.5948	0.89	0.8133	1.54	0.9382	2.38	0.9913
0.25	0.5987	0.90	0.8159	1.55	0.9394	2.40	0.9918
0.26	0.6026	0.91	0.8186	1.56	0.9406	2.42	0.9922

(Contd.)

t	$\psi(t)$	t	$\psi(t)$	t	$\psi(t)$	t	$\psi(t)$
0.27	0.6064	0.92	0.8212	1.57	0.9418	2.44	0.9927
0.28	0.6103	0.93	0.8238	1.58	0.9429	2.46	0.9931
0.29	0.6141	0.94	0.8264	1.59	0.9441	2.48	0.9934
0.30	0.6179	0.95	0.8289	1.60	0.9452	2.50	0.9938
0.31	0.6217	0.96	0.8315	1.61	0.9463	2.52	0.9941
0.32	0.6255	0.97	0.8340	1.62	0.9474	2.54	0.9945
0.33	0.6293	0.98	0.8365	1.63	0.9484	2.56	0.9948
0.34	0.6331	0.99	0.8389	1.64	0.9495	2.58	0.9951
0.35	0.6368	1.00	0.8413	1.65	0.9505	2.60	0.9953
0.36	0.6406	1.01	0.8438	1.66	0.9515	2.62	0.9956
0.37	0.6443	1.02	0.8461	1.57	0.9525	2.64	0.9959
0.38	0.6480	1.03	0.8485	1.68	0.9535	2.66	0.9961
0.39	0.6517	1.04	0.8508	1.69	0.9545	2.68	0.9963
0.40	0.6554	1.05	0.8531	1.70	0.9554	2.70	0.9965
0.41	0.6591	1.06	0.8554	1.71	0.9564	2.72	0.9967
0.42	0.6628	1.07	0.8577	1.72	0.9573	2.74	0.9969
0.43	0.6664	1.08	0.8599	1.73	0.9582	2.76	0.9971
0.44	0.6700	1.09	0.8621	1.74	0.9591	2.78	0.9973
0.45	0.6736	1.10	0.8643	1.75	0.9599	2.80	0.9974
0.46	0.6772	1.11	0.8665	1.76	0.9608	2.82	0.9976
0.47	0.6808	1.12	0.8686	1.77	0.9616	2.84	0.9977
0.48	0.6844	1.13	0.8708	1.78	0.9625	2.86	0.9979
0.49	0.6879	1.14	0.8729	1.79	0.9633	2.88	0.9980
0.50	0.6915	1.15	0.8749	1.80	0.9641	2.90	0.9981
0.51	0.6950	1.16	0.8770	1.81	0.9649	2.92	0.9982
0.52	0.6985	1.17	0.8790	1.82	0.9656	2.94	0.9984
0.53	0.7019	1.18	0.8810	1.83	0.9664	2.96	0.9985
0.54	0.7054	1.19	0.8830	1.84	0.9671	2.98	0.9986
0.55	0.7088	1.20	0.8849	1.85	0.9678	3.00	0.99865
0.56	0.7123	1.21	0.8869	1.86	0.9686	3.20	0.99931
0.57	0.7157	1.22	0.8888	1.87	0.9693	3.40	0.99966
0.58	0.7190	1.23	0.8907	1.88	0.9699	3.60	0.999841
0.59	0.7224	1.24	0.8925	1.89	0.9706	3.80	0.999928
0.60	0.7257	1.25	0.8944	1.90	0.9713	4.00	0.999968
0.61	0.7291	1.26	0.8962	1.91	0.9719	4.50	0.999997
0.62	0.7324	1.27	0.8980	1.92	0.9726	5.00	0.999997
0.63	0.7357	1.28	0.8997	1.93	0.9732		
0.64	0.7389	1.29	0.9015	1.94	0.9738		

Index

A

B

C

I

L

M

N

O

P

Q

R

S

T

U

V

W